Bridge Design

Bridge Design

Concepts and Analysis

António J. Reis
IST – University of Lisbon and Technical Director GRID Consulting Engineers
Lisbon
Portugal

José J. Oliveira Pedro
IST – University of Lisbon and GRID Consulting Engineers
Lisbon
Portugal

This edition first published 2019

Registered Offices
John Wiley & Sons, Inc., 111 River Street, Hoboken, NJ 07030, USA
John Wiley & Sons Ltd, The Atrium, Southern Gate, Chichester, West Sussex, PO19 8SQ, UK

Editorial Office
The Atrium, Southern Gate, Chichester, West Sussex, PO19 8SQ, UK

For details of our global editorial offices, customer services, and more information about Wiley products visit us at www.wiley.com.

Wiley also publishes its books in a variety of electronic formats and by print-on-demand. Some content that appears in standard print versions of this book may not be available in other formats.

Library of Congress Cataloging-in-Publication Data

Names: Reis, António J., 1949– author. | Oliveira Pedro, José J., 1968– author.
Title: Bridge design : concepts and analysis / António J. Reis, IST – University of Lisbon and Technical Director GRID Consulting Engineers, Lisbon, José J. Oliveira Pedro, IST – University of Lisbon and GRID Consulting Engineers, Lisbon.
Description: First edition. | Hoboken, NJ : John Wiley & Sons, Ltd, 2019. | Identifiers: LCCN 2018041508 (print) | LCCN 2018042493 (ebook) | ISBN 9781118927656 (Adobe PDF) | ISBN 9781118927649 (ePub) | ISBN 9780470843635 (hardback)
Subjects: LCSH: Bridges–Design and construction.
Classification: LCC TG300 (ebook) | LCC TG300 .R45 2019 (print) | DDC 624.2/5–dc23
LC record available at https://lccn.loc.gov/2018041508

Cover Design: Wiley
Cover Image: © Ana Isabel Silva

Set in 10/12pt Warnock by SPi Global, Pondicherry, India

Printed in the UK by Bell & Bain Ltd, Glasgow

10 9 8 7 6 5 4 3 2 1

Contents

About the Authors

António J. Reis became a Civil Engineer at IST – University of Lisbon in 1972 and obtained his Ph.D at the University of Waterloo in Canada in 1977. He was Science Research Fellow at the University of Surrey, UK, and Professor of Bridges and Structural Engineering at the University of Lisbon for more than 35 years. Reis was also Visiting Professor at EPFL Lausanne Switzerland in 2013 and 2015. In 1980, he established his own design office GRID where he is currently Technical Director and was responsible for the design of more than 200 bridges. The academic and design experience were always combined in developing and supervising research studies and innovative design aspects in the field of steel and concrete bridges, cable stayed bridges, long span roofs and stability of steel structures. A. Reis has design studies and projects in more than 20 countries, namely in Europe, Middle East and Africa and presented more than 150 publications. He received several awards at international level from IABSE, ECCS, ICE and Royal Academy of Sciences of Belgium.

José J. Oliveira Pedro became a Civil Engineer at IST – University of Lisbon in 1991, concluding his Master's degree in 1995 and Ph.D in 2007, with the thesis "Structural analysis of composite steel-concrete cable-stayed bridges". He joined the Civil Engineering Department of IST in 1990, as a Student Lecturer, and is currently Assistant Professor of Bridges, Design of Structures and Special Structures. In 1999, he was Researcher at Liège University / Bureau d'Etudes Greisch and, in 2015, Visiting Professor at EPFL Lausanne. In 1991, he joined design office GRID Consulting Engineers, and since then is very much involved in the structural design of bridges and viaducts, stadiums, long span halls and other large structures. He is the author/co-author of over seventy publications in scientific journals and conference proceedings. In 2013, he received the Baker medal, and in 2017 the John Henry Garrood King Medal, from the Institute of Civil Engineers, for the best papers published in *Bridge Engineering* journal.

Preface

About 15 years ago, the first author, A. J. Reis, was invited by Wiley to write a book on *Bridge Design* that could be adopted as textbook for bridge courses and as a guideline for bridge engineers. The author's bridge course notes from the University of Lisbon, updated over almost 30 years, were the basis for this book. For different reasons, the completion of the book was successively postponed until a final joint effort with the second author, J. J. Oliveira Pedro, made this long project a reality. The book mainly reflects the long design experience of the authors and their academic lecturing and research activities.

Bridge design is a multidisciplinary activity. It requires a good knowledge and understanding of a variety of aspects well beyond structural engineering. Road and railway design, geotechnical and hydraulic engineering, urban planning or environmental impact and landscape integration are key aspects. Architectural, aesthetic and environmental aspects are nowadays recognized as main engineering issues for bridge designers. However, these subjects cannot be studied independently of structural and construction aspects, such as the bridge erection method. On the other hand, what differentiates bridge design from building design, for example, is generally the role of the bridge engineer as a leader of the design process. Hence, the first aim of this book is to present an overview on all these aspects, discussing from the first bridge concepts to analysis in a unified approach to bridge design.

The choice of structural materials and the options for a specific bridge type are part of the design process. Therefore, the second aim of the book is to discuss concepts and principles of bridge design for the most common cases – steel, concrete or composite bridges. Good bridge concepts should be based on simple models, reflecting the structural behaviour and justifying design options. Sophisticated modelling nowadays adopts available software, most useful at advanced stages of the design process. However, it should be borne in mind that complex modelling does not make necessarily a good bridge concept.

The methodology to select the appropriate bridge typology and structural material is discussed in the first four chapters of this book. Examples, mainly from the authors' design experiences, are included. General aspects and bridge design data are presented in Chapter 2. Actions on bridges are included in Chapter 3 with reference to the Eurocodes. Structural safety concepts for bridge structures and limit state design criteria are also outlined in this chapter. Chapter 4 includes the conceptual design of bridge super- and substructures. Basic concepts for prestressed concrete, steel or steel concrete composite bridges, with slab, slab-girder and box girder decks are dealt with. These topics are discussed in relation to superstructures and execution methods such as classical falsework, formwork launching girders, incremental launching and balanced cantilevering. Bridge substructures are referred to in Chapter 4 as well, namely for the basic typologies of bridge piers, abutments and foundations.

Architectural, environmental, and aesthetic aspects that could be adopted as primary guidelines when developing a bridge concept are addressed in Chapter 5. Principles are explained on the basis of design cases from the authors' design practices. Of course, this could have been done on the basis of many other bridges. However, it is sometimes difficult to comment on bridge aesthetics while not being aware of design, cost or execution constraints faced by other designers.

Specific aspects of structural analysis and design are dealt with in Chapters 6 and 7. Particular reference is made in Chapter 6 to simplified approaches to the preliminary superstructure design. These approaches can also be adopted to check results from sophisticated numerical models at the detailed design stages. The influence of the erection method on structural analysis and design of prestressed concrete, steel and composite bridge superstructures is considered in Chapter 6. Particular reference is made to safety during construction stages and redistribution of internal forces due to time dependent effects. Chapter 6 ends with some design concepts and analysis for bowstring arch bridges and cable-stayed bridges. Of course, due to the scope of the book, the aspects dealt with for these specific bridge types are introductory in nature.

The substructure structural analysis and design is presented in Chapter 7. The distribution of horizontal forces between piers and abutments due to thermal, wind and earthquake actions is discussed. Stability of bridge piers and reinforced concrete design aspects are dealt with. Bridge bearing typologies and specifications are introduced. Particular reference is made to bridge seismic isolation and different types of seismic isolation devices are presented.

The book ends with Chapter 8, which presents a simple design case with two different superstructure solutions – a prestressed concrete deck and a steel-concrete composite deck. The application of design principles presented throughout the book is outlined.

The authors expect readers may find this book useful and in some way it will contribute to bridges reflecting the 'art of structural engineering'.

António J. Reis and José J. Oliveira Pedro
Lisbon, May 2018

Acknowledgements

This book is the result of the authors' activities at IST–University of Lisbon and at GRID Consulting Engineers. The support of both institutions is a pleasure to acknowledge and special thanks are due to Professor Francisco Virtuoso from IST and to our colleagues from GRID.

During 45 years of professional life as designer, the first author, A. Reis, had the privilege of meeting a few outstanding bridge engineers. Particular reference is made to Jean Marie Cremer, from Bureau d'Études Greisch, with whom A. Reis had the pleasure of working with on a few bridge projects but, most important, developing a friendship with.

Part of this book was written by the first author, A. Reis, during his stays in 2013 and 2015 as Visiting Professor at EPFL École Polytechnique Féderale de Lausanne, Switzerland. The second author, J. Pedro, had a similar opportunity in 2015. Thanks are due to EPFL and, in particular, to Professor Alain Nussbaumer for these opportunities.

The authors are also grateful to all sources and organizations allowing the reproduction of some figures and pictures with due credit referenced in the text.

Last, but not least, thanks are due to our families for the time this book has taken from being with them.

António J. Reis and José J. Oliveira Pedro
Lisbon, May 2018

1 Introduction

1.1 Generalities

Bridges are one of the most attractive structures in the field of Civil Engineering, creating aesthetical judgements from society and deserving, in many cases, the Latin designation in the French language of *Ouvrages d'Art*.

Firstly, a set of definitions and appropriated terminology related to bridge structures is established before discussing bridge design concepts. A short historical view of the topic is included in this chapter to introduce the reader to the bridge field, going from basic concepts and design methods to construction technology.

A bridge cannot be designed without an appropriated knowledge of general concepts that go well beyond the field of structural analysis and design. The concept for a bridge requires from the designer a general knowledge of other aspects, such as environmental and aesthetic concepts, urban planning, landscape integration, hydraulic and geotechnical engineering.

The designer very often has to discuss specific problems for a bridge design concept with specialists in other fields, such as the ones previously mentioned, as well as from aspects of more closely related fields like highway or railway engineering.

Introducing the reader to the relationships between all the fields related to bridge design, from the development of the bridge concept to more specific aspects of bridge construction methods, is one of the aims of this book.

Most of the bridge examples are based on design projects developed at the author's design office. Some of these design cases have been summarized in the chapters in order to illustrate the basic concepts developed throughout the book.

1.2 Definitions and Terminology

A bridge may be defined as a structure to traverse an obstacle, namely a river, a valley, a roadway or a railway. The general term *bridge* is very often left for the first case, that is, a structure over a river leaving the more specific term of *viaduct* for bridges over valleys or over other obstacles. So, the relevance of the structure very often related to its length or main span has nothing to do with the use of the terms bridge or viaduct. One may have bridges of only 20 m length and viaducts 3 or 4 km long. In highway bridge terminology, it is usual to differentiate between viaducts passing over or under a main

Bridge Design: Concepts and Analysis, First Edition. António J. Reis and José J. Oliveira Pedro.

road by designating them as *overpasses* or *underpasses*. So, one shall adopt the term 'bridge' to designate bridges in particular, or viaducts. Figure 1.1 shows a bridge over the river Douro that is 703 m long, 36 m width for eight traffic lanes and has a main span of 150 m, and also a viaduct in Madeira Island, 600 m long for four traffic lanes and with a typical span of 45 m. The decks of these structures are made of two parallel box girders supported by independent piers.

(a)

(b)

Figure 1.1 (a) The Freixo Bridge over the river Douro in Oporto, 1993, and (b) a viaduct in Madeira Island, Portugal, 1997 (*Source:* Courtesy GRID, SA).

In Europe, important bridges have been built over the sea in recent years, such as fixed links across large stretches of water, for example, the Öresund Bridge (7.8 km long) between Sweden and Denmark, and the Vasco da Gama Bridge (12 km long) in the Tagus river estuary, Lisbon, shown in Figure 1.2.

Many of these structures include main spans as part of cable-stayed or suspension bridges and many typical spans repeated along offshore or inland areas. If that occurs over the riversides, it is usual to designate that part of the bridge the *approach viaduct*.

(a)

(b)

Figure 1.2 (a) The Öresund link between Sweden and Denmark, 2000 (*Source:* Soerfm, http://commons.wikimedia.org/wiki/File:Öresund_Bridge_-_Öresund_crop.jpg#mediaviewer/File:Öresund_Bridge_-_Öresund_crop.jpg. CC BY-SA 3.0.), and (b) the Vasco da Gama Bridge, in Lisbon, Portugal, 1998 (*Source:* Photograph by José Araujo).

A bridge integrates two main parts:

- the *superstructure*; the part traversing the obstacle and
- the *substructure*; the part supporting the superstructure and transferring its loads to the ground through the foundations.

The superstructure is basically made of a deck transferring the loads to the piers by bearings or by a rigid connection between the deck and the pier; the substructure includes the piers, foundations and the abutments, as shown in Figure 1.3. The piers transfer the loads from the superstructure due to permanent and variable actions, namely dead weight, traffic loads, thermal, wind and earthquake action, to the foundations. The abutments establish the transition between the superstructure and the earthfill of the highway or the railway and retain the filling material. The abutments transfer the loads induced by the superstructure, generally transmitted by the bearings, and supporting the soil impulses generated by the embankments.

The deck is, in general, supported by a set of bearings, some located at the abutments, as previously referred to, and others located at the top of the piers as shown in Figure 1.3. Nowadays, these bearings are generally made of elastomeric materials (natural rubber or synthetic rubber – chloroprene) and steel.

The foundations of the bridge piers and abutments may be by footings, as in Figure 1.3 (shallow foundations) or by piles (deep foundations). A different type of foundation include caissons made by lowering precasted segmental elements in a previous excavated soil, a method adopted sometimes for deep bridge piers foundations in rivers.

1.3 Bridge Classification

Bridges may be classified according several criteria namely:

- *the bridge function*, dependent on the type of use of the bridge, giving rise to designations of highway or railway bridges, canal bridges for the transportation of water, quay bridges in ports, runway or taxiway bridges in airports, pedestrian bridges or pipeline bridges. The function of the bridge may be twofold as for example in the case of the Oresund Bridge, for railway and highway traffic (Figure 1.2).
- *the bridge structural material*, like masonry bridges, as used in the old days since the Romans, timber bridges, metal bridges in steel or aluminium or in iron as adopted in the nineteenth century, concrete bridges either in reinforced concrete or prestressed concrete (more precisely, *partially prestressed concrete* as preferred nowadays) and, more recently, composite steel-concrete bridges.
- *the bridge structural system*, which may be distinguished by:
  - the longitudinal structural system;
  - the transversal structural system.

The former, the longitudinal system, gives rise to beam bridges, frame bridges, arch bridges and cable supported bridges; namely, cable-stayed bridges and suspension bridges. The last, the transversal structural system, is characterized by the type adopted for the cross section of the superstructure, namely slab, girder or box girder bridges. A preliminary discussion on bridge structural systems is presented in next section.

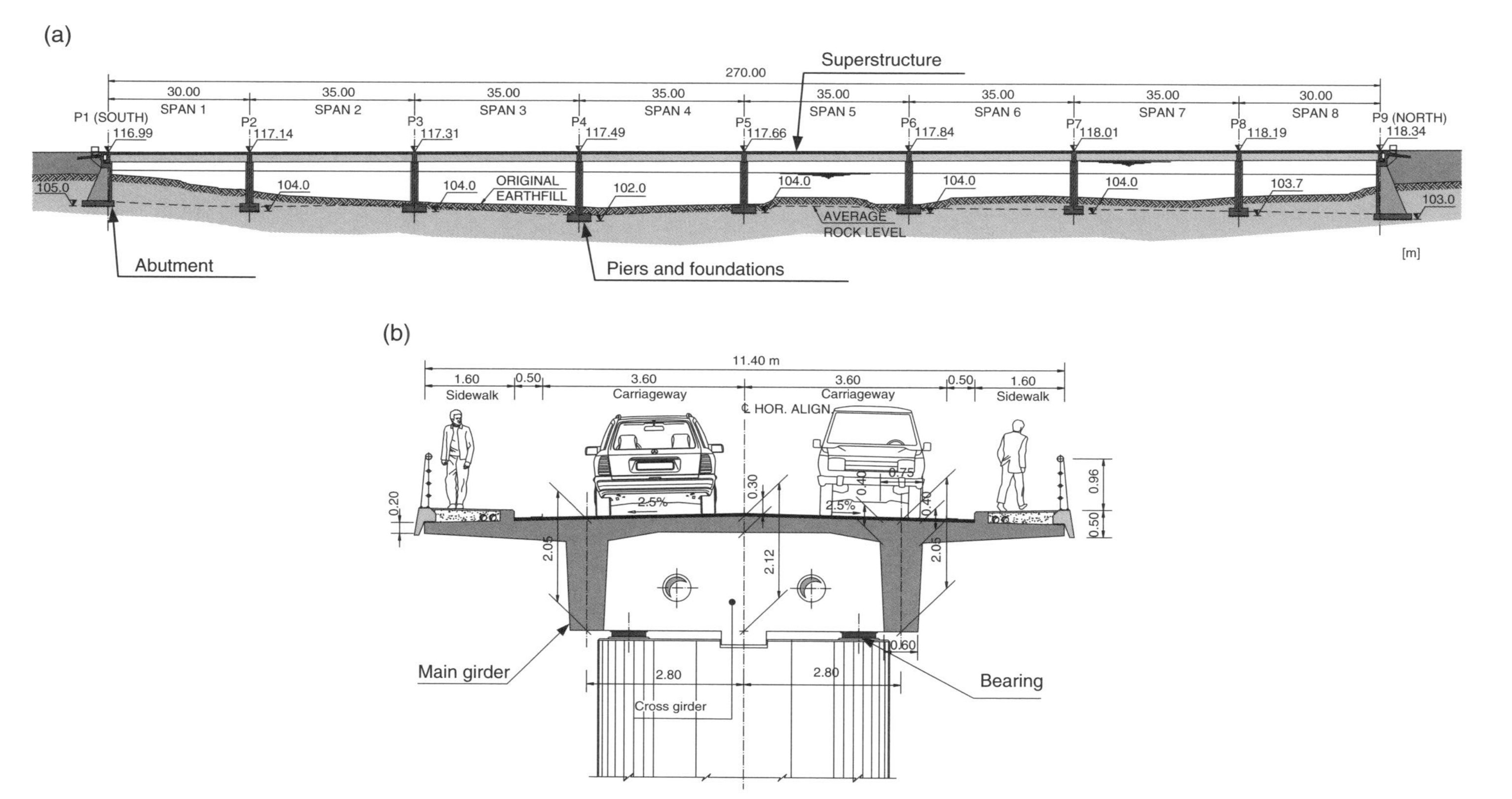

Figure 1.3 Section elevation and typical cross section of a bridge – The Lugela bridge in Mozambique, 2008. Superstructure (deck) and Substructure (piers, abutments and foundations).

- Another type of classification is often adopted, according to:
  - the predicted lifetime of the bridge, namely temporary (made in general in wood or steel) or definitive bridges.
  - the fixity of the bridge, namely fixed or movable bridges, like lift bridges if the deck may be raised vertically rolling bridges if the deck rolls longitudinally, or swing bridges if the deck rotates around a vertical axis.
  - the in-plan geometry of the bridge, like straight, skew or curved bridges.

1.4 Bridge Typology

Different bridge typologies, namely concerning the longitudinal structural system or the deck cross section, may be adopted with different structural materials. The concept design of a bridge is developed mainly in Chapters 4 and 5, but a brief description of the variety of bridge options is presented here in order to introduce the topics of Chapters 2 and 3 concerning the basic data and conditions for design.

Nowadays, a beam bridge is the most usual type where the deck is a simple slab, a beam and slab (Figure 1.3) or a box girder deck. Beam bridges may be adopted in reinforced concrete for small spans (l), generally up to 20 m, or in prestressed concrete or in steel-concrete composite decks (Figure 1.4) for spans up to 200 m or even more. The superstructure may have a single span, simply supported at the abutments, or multiple continuous spans (Figure 1.5). Between these two cases, some other bridge solutions are possible like multiple span decks, in which most of the spans are continuous, but some spans have internal hinges like in the so called 'Gerber' type beam bridges, shown in Figure 1.6. However, the general trend nowadays is to adopt, as far as possible, fully continuous superstructures, to reduce maintenance of the expansion joints and to improve the earthquake resistance of the bridge if located in a seismic region. Continuous decks more than 2000 m long have been adopted for beam bridges, either for road or rail bridges. Yet, in long continuous bridges, the distance between expansion joints is generally restricted to 300–600 m to reduce displacements at the expansion joints. In a beam bridge, the connection between the superstructure and the piers is made by bearings, as in Figure 1.3, which allow the relative rotations between the deck and the piers; the relative longitudinal displacements between the deck and the piers may or may not be restricted, depending on the flexibility and slenderness of the piers, as discussed in Chapters 4 and 7.

If the deck is rigidly connected to the piers, one has a frame bridge (Figure 1.7). The superstructure may be rigidly connected to some piers and standing in some bearings, allowing rotations, or rotations and displacements, between the deck and some of other piers.

In frame bridges, the piers are in most cases vertical. However, frame bridges with slant legs, exemplified in Figure 1.8a, are a possible option. For a frame bridge with slant legs or arch bridges, the main condition for adopting these typologies is the load bearing capacity of the slopes of the valley to accommodate, with very small displacements, the horizontal component H (the thrust of the arch) of the force reactions induced by the structure, as shown in Figure 1.8b.

An arch is likely to be a very efficient type of structure, an aesthetically pleasant solution for long spans in deep valleys, provided the geological conditions are appropriate. The ideal shape of the arch, if the load transferred from the deck is considered as a

Figure 1.4 Steel-concrete composite plate girder decks: Approach viaducts three-dimensional model of the Sado River railway crossing, in Portugal (Figure 1.12).

uniformly distributed load q (valid for closed posts), is a second degree parabola because the arch for the permanent load is free from bending moments. In this case, the arch is only subjected to axial forces; that is, the arch follows the 'pressure line'. It is easy to show using simple static equilibrium (bending moment condition equal to zero at the crown) that the thrust is given by $H = ql^2 / (8f)$.

Arch bridges may have different typologies and be made of different structural materials. In the old days, masonry arches made of stones were very often adopted for small to medium span bridges. More recently, iron, steel and reinforced concrete bridges replaced these solutions with spans going up to several hundred metres. One of the

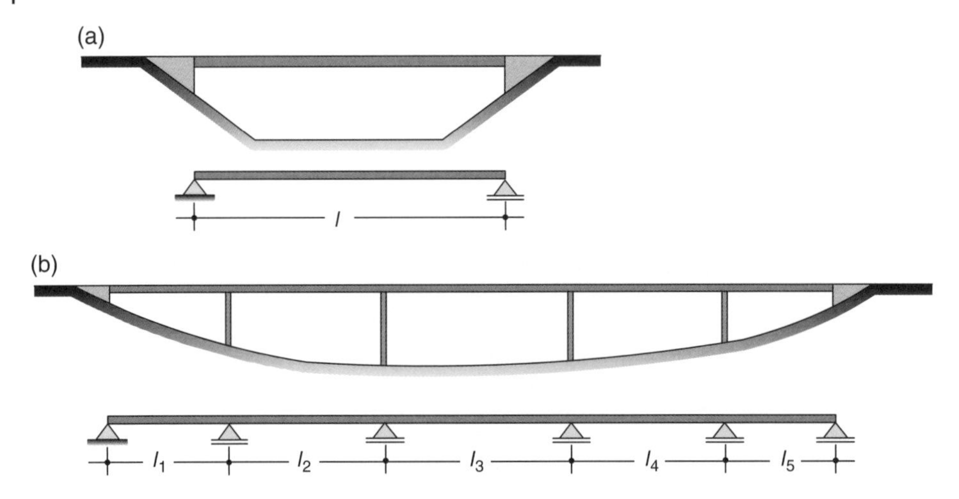

Figure 1.5 Beam bridges – Elevation and longitudinal model: (a) single span and (b) multiple spans.

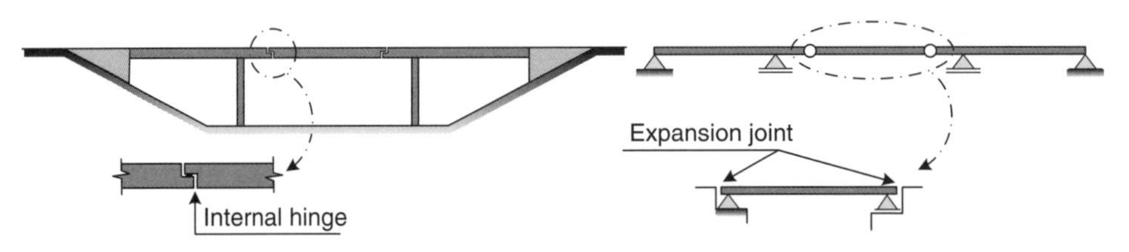

Figure 1.6 Beam bridge – Gerber type.

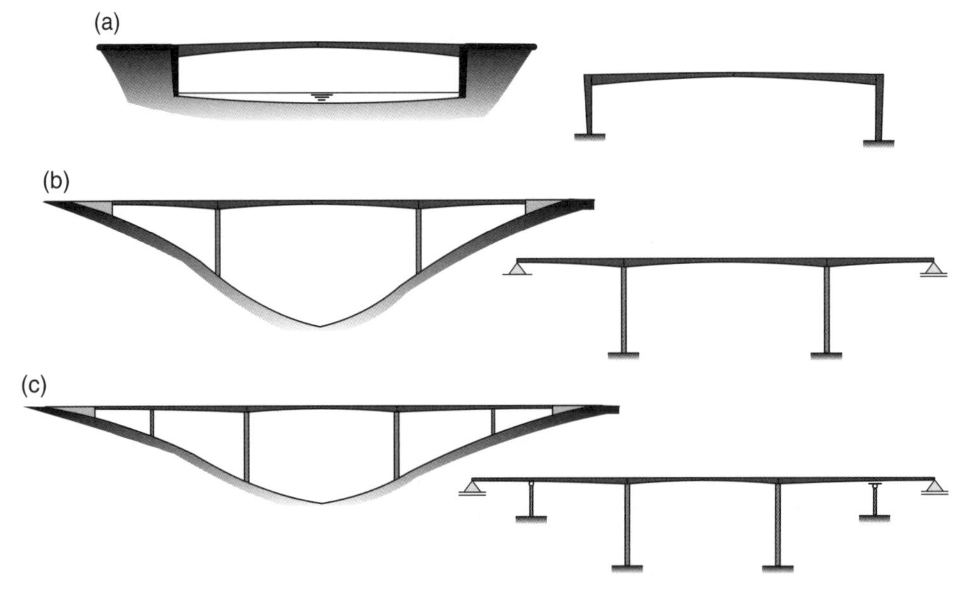

Figure 1.7 Beam bridges – elevation view and longitudinal structural model: (a) single span, (b) and (c) multiple spans.

(a)

(b)

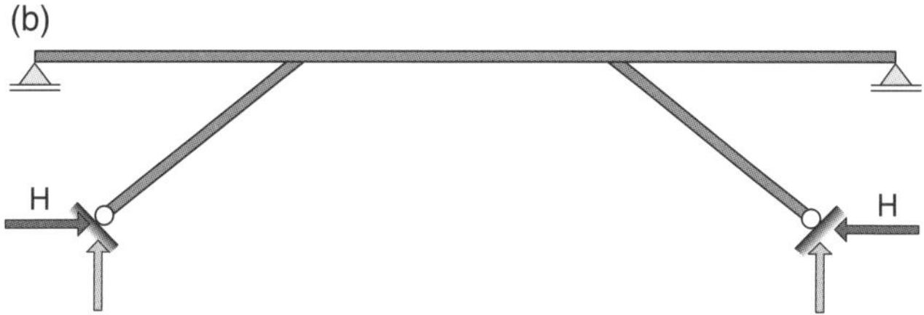

Figure 1.8 A frame bridge with inclined (slant) legs: (a) Reis Magos Bridge and (b) longitudinal structural model.

most beautiful arch bridges is Arrábida Bridge, in Oporto (Figure 1.9), designed at the end of the 1950s, the beginning of the 1960s and opened to traffic in 1963. The bridge, at the time the longest reinforced concrete arch bridge in the world, has a span of 270 m and a rise of 54 m ($f/l = 1/5$).

The arch bridge may have the deck working from above or from below, as shown in Figures 1.10 and 1.11. This last solution is adopted for traversing rivers at low levels above the water, with particular restrictions for the vertical clearance h for navigation channels. The horizontal component of the reaction at the base of the arch, at the connection between the arch and the deck, is taken by the deck. A *bowstring arch bridge* is the designation for this bridge type, in which the deck has a tie effect, together with its beam behaviour. Figure 1.12 shows a multiple bowstring arch bridge, with a continuous deck composed of a single steel box section. The deck, with spans of 160 m, is a steel-concrete composite box girder to allow the required torsion resistance under eccentric traffic loading. However, the classical solution for bowstring arches is made of a beam and slab deck suspended from above by two vertical or inclined arches, as presented in Chapter 6.

The main restriction nowadays for the construction of arches is the difficulty of the execution method, when compared to a long span frame bridge with vertical piers, built by the balanced cantilever method referred to in Chapter 4.

For spans above 150 m and up to 1000 m, cable-stayed bridges, as previously shown in Figure 1.2, are nowadays generally preferred to beam or frame bridges, for which the

Figure 1.9 The Arrabida Bridge in Oporto, Portugal, 1963 (*Source:* Photograph by Joseolgon / https://commons.wikimedia.org / Public Domain).

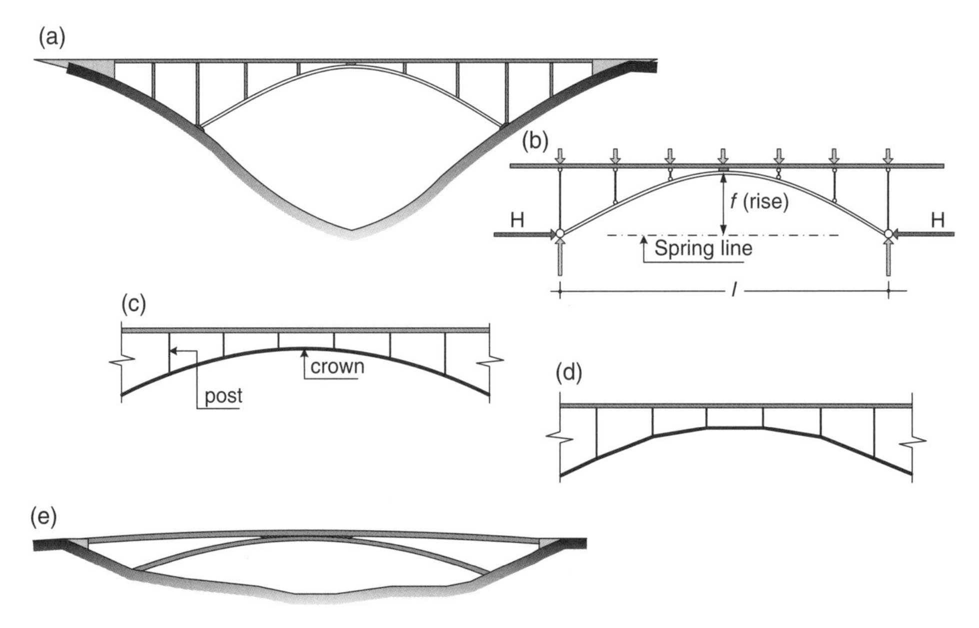

Figure 1.10 Arch bridges: (a) the classical parabolic two hinges arch bridge; (b) structural longitudinal model; (c) independent arch and deck at the crown; (d) segmental arch and (e) low rise arch for a pedestrian bridge without posts.

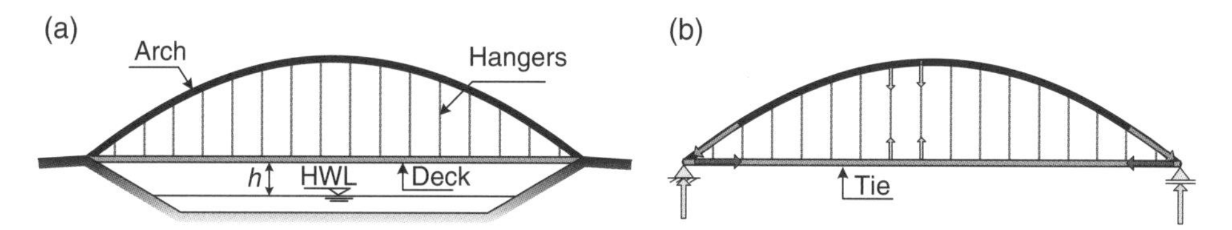

Figure 1.11 A bowstring arch bridge: elevation and longitudinal structural model.

(a)

(b)

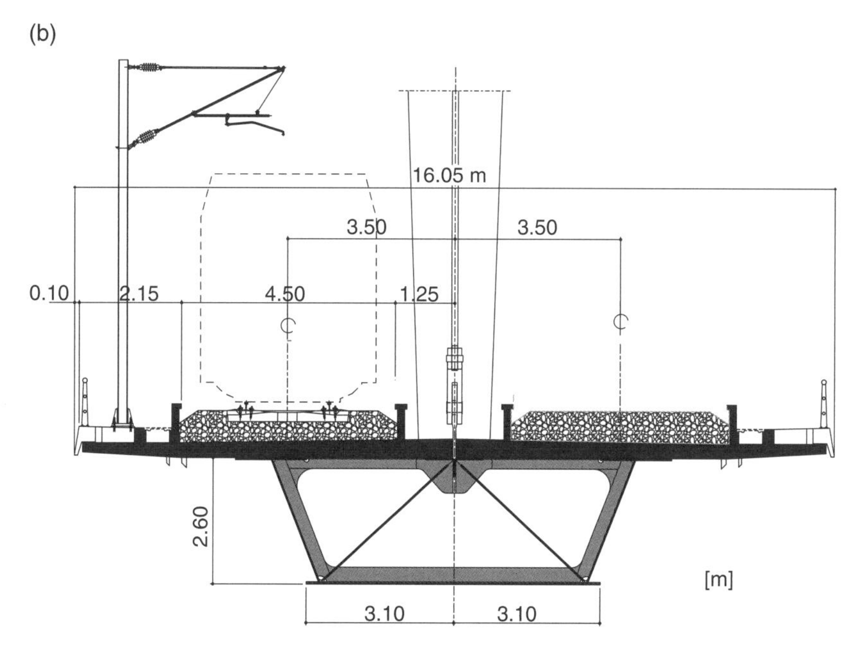

Figure 1.12 A bowstring arch for a railway bridge: (a) the crossing of the Sado River in Alcacer do Sal, Portugal, 2010 and (b) deck cross section – a steel concrete composite box girder (*Source:* Courtesy GRID, SA).

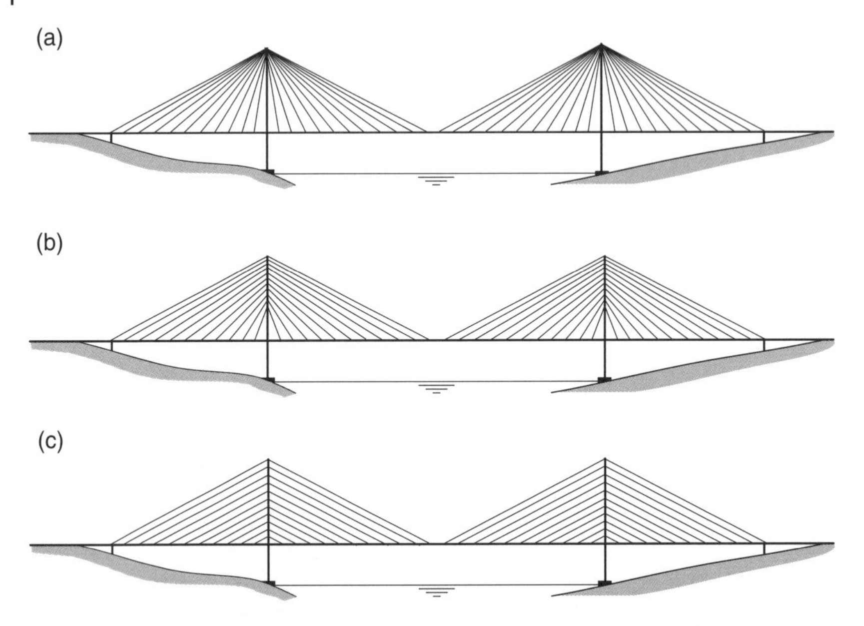

Figure 1.13 Stay arrangements for cable-stayed bridges: (a) fan, (b) semi-fan and (c) harp systems.

longest span is about 300 m. Even for spans bellow 100 m, cable-stayed bridges have been adopted as structurally efficient and aesthetically pleasant solutions; for example, in urban spaces where very slender decks are required. The basic schemes for the stay arrangement in cable-stayed bridges are shown in Figure 1.13 – the fan, semi-fan or harp arrangement. The semi-fan arrangement is the most adopted one for economy of stay cable quantity. For aesthetics, usually the harp arrangement is the preferred one, since it reduces the visual impact of crossing cables for skew views of the bridge, as is apparent from Figure 1.2

In cable-stayed bridges with spans up to 500 m the deck may be made of concrete, but above this span length, steel or steel-concrete composite decks are preferred, to reduce the dead weight of the superstructure. The cable-stayed bridge deck is subjected to large compressive forces induced by the stay cables, as shown in Figure 1.14. In the first generation of cable-stayed bridges, the decks where in steel and the stay-cable anchorages were kept at a considerable distance at the deck level. In these bridges, the beam load effect in the deck was relevant. In the last few decades, a new generation of cable-stayed bridges has been developed with multiple stay cables anchored at small distances at the deck level, very often between 5 and 15 m, allowing a considerable reduction of the beam internal forces in the deck. This is the case for cable-stayed bridges with very slender concrete decks with a span/height relationship for the deck that can reach very high values up to 400.

In cable-stayed bridges, the deck is supported by a single plan or two plans of stay cables. The former suspension scheme is designated *axial* or *central suspension* while the last one is called *lateral suspension*. The axial suspension scheme requires a deck

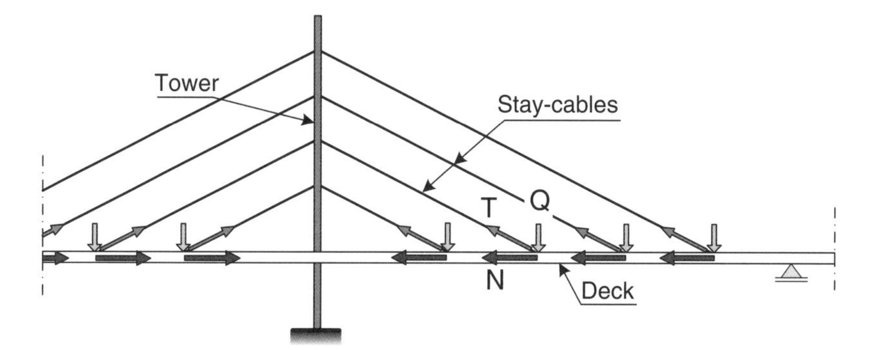

Figure 1.14 Cable-stayed bridges: static equilibrium at deck level with axial forces induced by the stays.

with considerable torsion rigidity, generally a box girder deck, to support the asymmetric traffic loadings as well as to improve its aerodynamic stability. Figure 1.15 shows one of the most remarkable bridges built in 1977, at the time a world record for cable-stayed bridges with a prestressed concrete deck. On the contrary, in lateral suspension cable-stayed bridges, the deck may be reduced to a simple slab supported laterally by stay cables, or to a slab and girder section with an open configuration, since torsion resistance is assured by the staying scheme, as is the case for the Vasco da Gama Bridge (Figure 1.2b).

The other type of cable supported bridge is, as previously referred to, the suspension bridge. A suspension bridge includes a stiffening girder, the main cables, towers, hangers and anchoring blocks, as shown in Figure 1.16.

The main cables are usually externally anchored (earth anchored cables), but in some bridges with smaller spans it is possible to adopt a *self-anchored suspension bridge* by anchoring the cables at the deck, as shown in Figure 1.16. In the former, the tension forces are transferred directly to the ground while, in the latter, the cable forces are transferred (as large compression forces) to the deck.

The stiffening girder of a suspension bridge may be made of two parallel trusses, as the more classical suspension bridges developed in the North America during the twentieth century. A second generation of suspension bridges was introduced in Europe in the second part of the twentieth century, by replacing the truss-stiffening girder by a streamline steel box girder deck and diagonal hanger ropes greatly improving the aerodynamic stability of the deck. In Figure 1.17, two of these bridge stiffening girder typologies are shown – the Akashi Kaikyō Bridge completed in 1998 in Japan, presently the world's longest span at 1991 m, and the Great Belt East Bridge, also completed in 1998 but in Denmark, with a central span of 1624 m.

In a suspension bridge, the permanent loads of the deck are taken by the cable system. The main cables adopt a parabolic configuration under the uniform dead load transmitted from the deck through the hangers. The live loads are basically taken by combined local bending action of the deck, between the hangers, and the main cable.

(b)

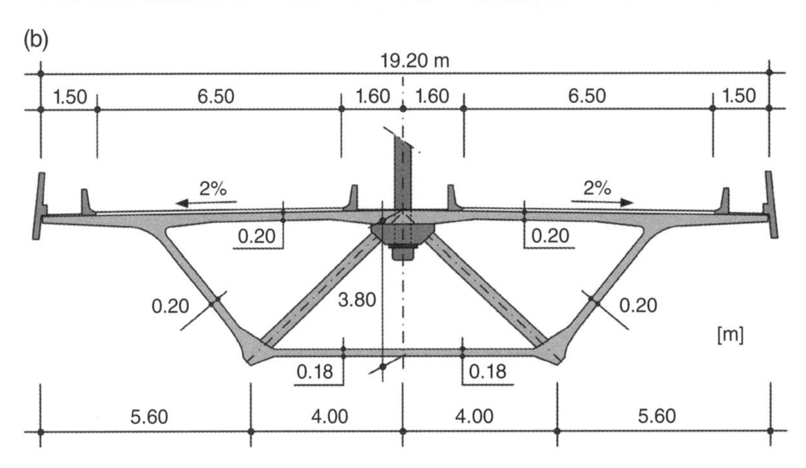

Figure 1.15 An example of a cable-stayed bridge with axial suspension: (a) the Brotonne Bridge, France, 1977, with a main span of 320 m (*Source:* Photograph by Francis Cormon) with (b) a concrete box girder deck cross section.

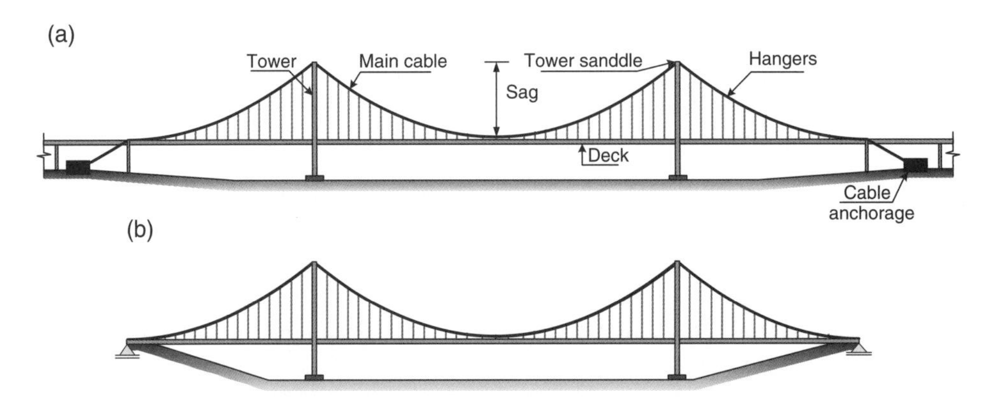

Figure 1.16 (a) Basic layout and notation of suspension bridges. (b) Externally anchored and self-anchored suspension bridges.

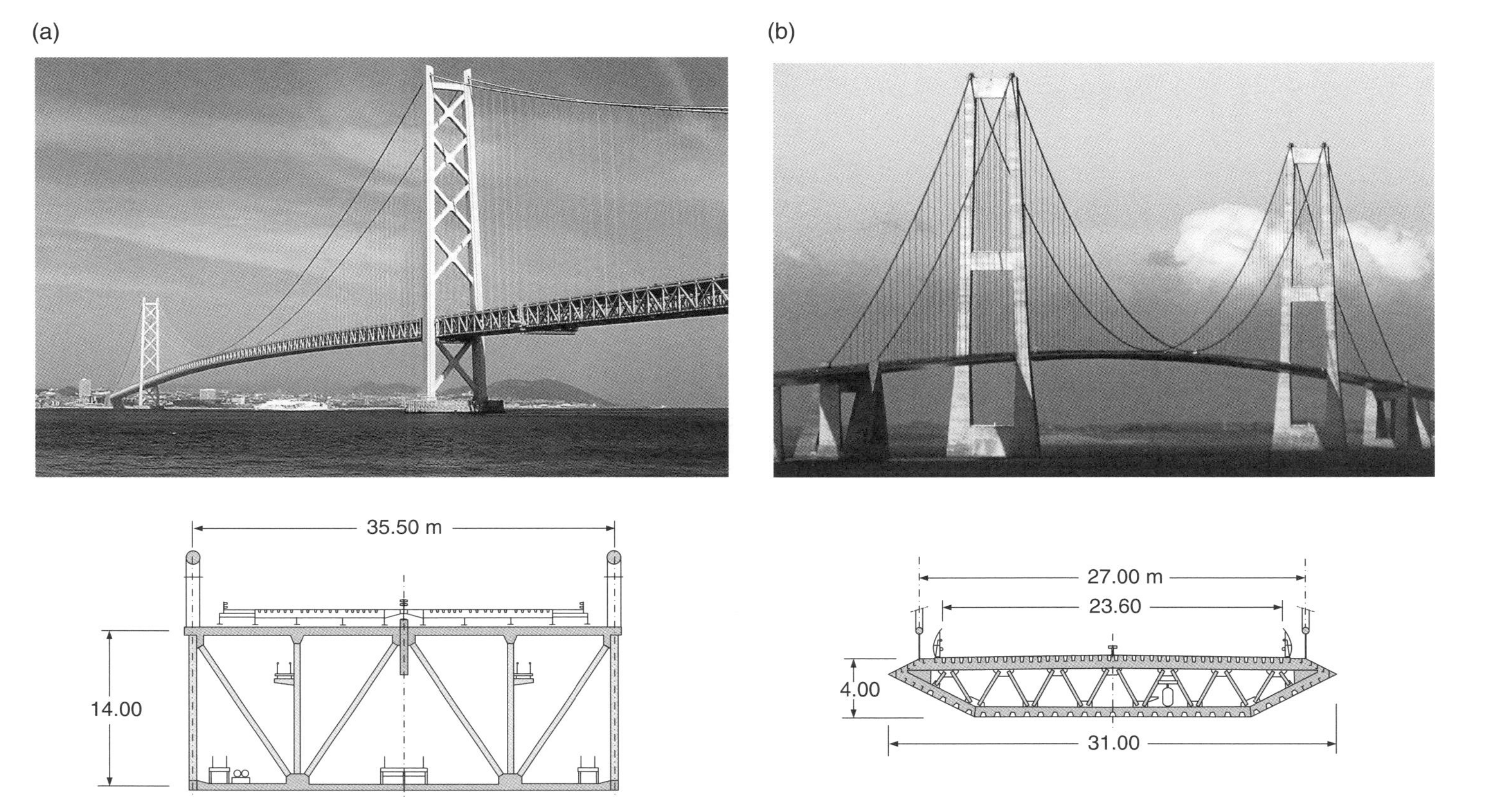

Figure 1.17 (a) Akashi Kaikyō Bridge, Japan, 1998, with a truss-stiffening girder deck (*Source:* Photograph by Pinqui/ https://commons.wikimedia.org) and (b) Great Belt Bridge, Denmark, 1998, with a streamline box girder deck (*Source:* Photograph by Tone V. V. Rosbach Jensen/ https://commons.wikimedia.org).

1.5 Some Historical References

The history of bridges is well-documented in a variety of excellent references [1–4]. To keep this section within the scope envisaged for this book, a short historical review will present the roles of some architects, engineers and bridge builders.

Materials and structural shapes have always been the key elements to understand how bridges were conceived and built throughout the centuries. For bridge structural materials, one may considerer different stages grouped as follows:

- Stone/masonry and wood bridges;
- Metal–iron and steel bridges;
- Concrete reinforced concrete and prestressed concrete bridges.

The first group, stone/masonry and wood bridges, includes most of the bridge history from the Romans to the eighteenth century, while the second and third groups (metal and concrete bridges) may be included in bridge history from the eighteenth century to the present.

Structural shapes and static schemes were also key factors in bridge historical development that, in short, may be summarized as:

- Arch bridges;
- Beam, frame and truss bridges;
- Cable supported bridges – suspension and cable-stayed bridges.

Contrary to what happens with bridge structural materials, it is not possible to include these structural typologies in different ages along bridge history. Even for cable supported bridges, one may find references to the use of ropes made with natural fibres to achieve a resistant structure in the early ages, even before the Roman Bridges. Nowadays, arches, introduced by the Romans, are made with different materials (concrete or steel), still reflecting the art of structural engineering.

1.5.1 Masonry Bridges

Apart from the earliest records of bridges, the first appears to be 600 BCE, engineering bridge history may be considered to have been initiated by the Romans. The Romans were the introducers of science in arch construction providing many examples of magnificent masonry bridges and aqueducts such as the 'Pont du Gard' in France and 'Puente de Alcantara' (Figure 1.18) over the river Tagus, Spain, from the second century [4]. The former is an aqueduct with a total length of 275 m, a maximum depth of 49 m and spans 22 m. The last is a masonry bridge as well, but with spans reaching 28 m and piles reaching 47 m high, appointed the most significant achievement of Roman engineering. Roman bridges are based on the concept of semi-circular arches, transferring the thrust to piers, and large piers require about one-third of the spans in multiple arch bridges. The Romans also introduced lime mortar and pozollanic cement for realizing voussoir arches, contributing to increasing span lengths and durability.

Up until the end of the eighteenth century, bridges were built with masonry or wood. Some beautiful examples built along the centuries may still be appreciated, such as Santa Trinita Bridge in Florence (Figure 1.19) built with masonry in the sixteenth century and

Figure 1.18 The Roman Alcantara Bridge in Spain, second century (*Source:* Photography by Dantla/ https://commons.wikimedia.org).

Figure 1.19 Santa Trinita Bridge in Florence, sixteenth century (*Source:* Photography by Bruno Barral/ https://commons.wikimedia.org).

Figure 1.20 Chapel Bridge, Lucerne, Switzerland, 1333 (rebuilt after fire in 1993) (*Source:* Photograph by José O. Pedro).

designed by the architect Bartolomeo Ammanati. While the classical Roman Bridges were made with circular arches, the elegance of Santa Trinita Bridge is due to three flattened elliptic arches, which are the oldest in the world. At the beginning of the twentieth century, in 1905 to be precise, the span record for masonry bridges was reached with the construction of the Plauen Bridge [1, 3], in Germany, which has a main span of 90 m.

1.5.2 Timber Bridges

In the eighteenth century, some remarkable wood bridges were built. In Switzerland, due to a long tradition of wood construction builders, remarkable structures were built, like the Shaffhouse Bridge, designed in 1775 by Grubleman, with two continuous spans of 52 and 59 m. In the USA, one of the most famous wood bridges was built in 1812, over the river Schuylkill with a main span of 104 m, unfortunately destroyed by a fire some years after its construction. Most old timber bridges have disappeared along history; an exception, presently the oldest wooden bridge to our knowledge, is the Kappellbruke (Chapel Bridge) erected originally in 1333 over Lake Lucerne, Switzerland (Figure 1.20). This footbridge was also destroyed by a fire in 1993 and completely rebuilt, standing now as a world heritage site.

1.5.3 Metal Bridges

With the Industrial Revolution of the nineteenth century, metal structures became more and more competitive and, finally, iron and steel started to be used as bridge materials. From this time on, metal truss bridges and suspension bridges were adopted with increasing span lengths.

The first metallic bridge built in the world may be considered to be the Coalbrookdale Bridge [4] with a span of 30 m, designed by Abraham Darby III, a cast iron structure built in Great Britain in 1779. But development had to wait until the end of the nineteenth century, when the prices for steel production has dropped down, to realize the first main steel bridges. One of the most decisive contributions for the development of metal bridges was from Thomas Telford in the nineteenth century. Telford was an English engineer with a large number of relevant projects, always taking aesthetics as a key issue for his work, in parallel with the development of new structural and construction schemes. Craigelachie Bridge, in Scotland, shown in Figure 1.21, is an excellent example of the aesthetical relevance of Telford's design work. In 1826, a famous eye bar wrought iron chain suspension bridge (Figure 1.22), with a main span of 176 m, also from Thomas Telford, was completed in the UK over the Menai Strait, achieving the world record for span length. The iron chain was replaced in 1938 by pin steel bars, allowing the bridge to remain in service up until now. The construction of long span beam bridges is considered to have been initiated with the first box girder bridge in wrought iron, the Britannia Bridge in Wales, UK, built in 1850 by Robert Stephenson (Figure 1.23). The Britannia Bridge, with two main spans of 146 m, had a rectangular cross section with a railroad inside. After a fire in 1970, this famous bridge was modified in 1971 by inserting a truss arch underneath and

Figure 1.21 Craigellachie Bridge, 45 m span, over the Spey River, Scotland, 1815 (*Source:* Photograph by Craig Williams/ https://commons.wikimedia.org).

Figure 1.22 Menai Suspension Bridge, 1826, Wales (*Source:* Photography by Bencherlite/ https://commons.wikimedia.org).

Figure 1.23 The Britannia Bridge, 1850, North Wales (*Source:* Courtesy of the Los Angeles County Museum of Art).

carrying road and rail traffic, but seriously affecting the architecture of the bridge. The type of box girder solution adopted for the Britannia Bridge was not retained due to the development of truss arch bridges at the end of the nineteenth century.

Reductions in steel prices allowed the increased development of steel arch bridges, or more precisely, wrought iron, with particular reference to long span bridges in the USA. The most famous one was the San Luis Bridge, over the Mississippi, concluded in 1874 with three arches of 152 + 157 + 152 m, and in Europe with several famous arch bridges from the Gustave Eiffel Society. The first relevant Eiffel's bridges were the Maria Pia Bridge in Oporto, Portugal [5], in 1877, and the Viaduc du Gabarit, in France, in 1884. The former (Figure 1.24) 563 m long and with a main span of 160 m, has a truss arch for the railway deck of only 6 m. Particular reference should be made to the erection scheme of the arch using cantilever construction. The Viaduct du Gabarit has a main span arch of 165 m, 52 m rise and a total length of 564 m. This arch was also erected by cantilever as per the Maria Pia Bridge. In 1885, the Belgium engineer Théophile Seyrig, who had worked with Eiffel, designed the Luiz I Bridge [6], also in Oporto and very close to the Maria Pia Bridge. The Luiz I Bridge, an arch bridge, was the largest span in the world at its time – 174.5 m (Figure 1.25a). This bridge with two road decks has a two hinged arch at the base, in spite of its appearance, due to the need of inserting the lower deck. The bridge is located in a classified UNESCO World Heritage Site. A great deal of strengthening of the bridge was done in 2005 (Figure 1.25b), including its arch, truss girder and piers, with an entire replacement of the upper deck with a new steel deck, adapting it for the Metro of Oporto trains (Figure 1.26). The reduction in dead weight of the existing deck, which included concrete and brick elements inserted many years after the original deck, was achieved. The reductions in deck dead load allowed a significant reduction in the strengthening of the arch required for the new functional conditions and required

Figure 1.24 Maria Pia Bridge, 1877, Oporto, Portugal (*Source:* Photograph by José O. Pedro).

(a)

(b)

Figure 1.25 (a) Luiz I Bridge, 1885, Oporto, Portugal: (a) before and (b) during the upgrading works (*Source:* Courtesy GRID, SA).

strengthening of the bridge. The design of the upgrading of the bridge was made without affecting its appearance as required by the design and build conditions of the bid.

With the development of industrial production of steel, after the introduction by Henry Bessemer in 1856 of its patent, the Bessemer Converter, transforming cast iron in to the much better steel material, in terms of tensile resistance and ductility, allowed the building of long span trusses in the United States and Europe. The first long span European bridge was the Firth of Forth Bridge (Figure 1.27) built between 1881 and 1890, with two main spans of 521 m and two side spans of 207 m. The two main spans were made with two cantilevers of 207 m supporting a central part made of truss beam 107 m in length. The development of these types of bridges was not unfortunately made

Figure 1.26 Luiz I Bridge after being upgraded (*Source:* Photograph by José Araujo).

Figure 1.27 Firth of Forth Bridge, 1890, Scotland (*Source:* Photograph by Andrew Shiva/Godot13, https://commons.wikimedia.org/wiki/File:Scotland-2016-Aerial-Edinburgh-Forth_Bridge.jpg. CC BY-SA 4.0).

without the occurrence of some historical accidents, such as the one occurring in the railway bridge over the river Tay, Scotland, in 1879 and the accident during the erection of the Quebec Bridge, Canada, in 1907. The Quebec Bridge with spans of 549 m was concluded only after a second accident during construction in 1917.

1.5.4 Reinforced and Prestressed Concrete Bridges

The lack of sufficient tensile resistance makes concrete inadequate as a bridge structural material. However, concrete offers the advantage of being a *plastic* material in the sense that it is poured like fluid in a mould before it hardens, allowing a variety of shapes for beams and piers that could not be achieved in a simple way with steel materials and steel profiles. Some arch bridges with relevant spans, using mass concrete, were made at beginning of the twentieth century, such as the Willeneuve – Sur-Lot Bridge, in 1919, with an arch span of 97 m, by Eugéne Freyssinet. François Hennebique was one of the first to understand the role of steel reinforcement in concrete to create a suitable bridge material. In 1911, he 'made' (Hennebique was more a builder than a designer) the Risorgimento Bridge in Rome, achieving a 100 m span, but the development of reinforced concrete bridges occurred with the work of the great Swiss engineer Robert Maillart who was the first to explore the potentialities of reinforced concrete to build magnificent arch bridges. He explored the concept of relative stiffness between the arch and the deck structure, introducing the concept of Maillart's arch, an *arch without bending stiffness*, achieved by reducing the stiffness in such a way that the arch follows the pressure line working only under normal forces. However, its thickness is enough to resist to arch buckling. The arch works in this case as the opposite of a cable in a suspension bridge. Among the bridges based on Maillart's arch concept, the longest arch span, 43 m, was the one of the Tshiel Bridge built in 1925. Robert Maillart explored a variety of arch structures from built-in arches to three-hinged arches, such as those he adopted for his classical bridge, the Salginatobel Bridge, in Switzerland (Figure 1.28), built in 1930 with a 90 m span. In 1901, Maillart was also the first to adopt a concrete box section for arches, in the Zuoz Bridge with a 30 m span structure. At the beginning, concrete arch bridges were always associated with large wood scaffolding structures, some of them initially developed by R. Coray, built by cantilever construction from each side until the closure at the mid-section was reached. One may even say the most relevant wood bridges in the world were the scaffolding of concrete arch bridges. A detailed discussion on the historical development of scaffolding structures is presented in the excellent book by Troyano [4].

The concept of multiple arch bridges in reinforced concrete was adopted by Eugéne Freyssinet in 1930, in Plougastel Bridge. This bridge, over the river Elorn in France, is made with three arches of a 186 m span each, following the concepts of previously built multiple arch metal bridges. Many years before, in 1910, Freyssinet designed a beautiful multiple three arch bridge with very flat arches and a rise-span ratio of only 1/14.7. The arches, each with a 77.5 m span, were initially built as three hinged arches but a few months after completion Freyssinet detected alarming increased deflections at the mid-span sections due to creep effects in the concrete. He decided to recover the initial shape of the bridge by inserting jacks at the mid span hinge sections and to eliminate these hinges. The beauty of the bridge was the result of the flatness of the arches but also due to the triangulated system adopted to transfer the loads from the deck to the arches. Unfortunately, this bridge was destroyed during the Second World War.

The span lengths of concrete arch bridges were increasing, and one of the largest spans is still the 390 m long arch of the Krk Bridge (Figure 1.29), completed in 1979 in the former Yugoslavia.

Figure 1.28 The Salginatobel Bridge, 1930, Switzerland (*Source:* Photograph by Rama, https://commons.wikimedia.org/wiki/File:Salginatobel_Bridge_mg_4077.jpg. CC BY-SA 2.0 FR).

Figure 1.29 The Krk reinforced concrete arch bridge, 1979, former Yugoslavia (*Source:* Photograph by Zoran Knez/ https://commons.wikimedia.org).

Since the end of nineteenth century, the idea of prestressing was always present as a potential scheme to induce initial compressive stresses in concrete and therefore to improve its lack of resistance under tensile stresses. However, it was only in 1928 that Freyssinet decided to concentrate all his activities to explore the potentialities of prestressing for concrete structures. The difficult experience he had with time dependent deformations of concrete in the Veudre Bridge has shown the influence in concrete. Freyssinet understood the need to adopt high strength steel for prestressing and he develop specific systems to anchor prestressing wires or steel bars.

The first long span bridge made in prestressed concrete was the Luzancy Bridge (Figure 1.30) in France, over the river Marne, with a span of 55 m and concluded in 1946. The bridge has been entirely precast in prestressed segmental construction and erected by mechanical devices without any scaffolding. The development of prestressed concrete bridges was very fast in the second half of the twentieth century. Apart from the use of prestressing for prescasting beams, developed in France after the pioneering works of Freyssinet, a special reference should be made to the German contributions in the reconstruction of many bridges after the Second World War. Most of the progress in this field was due to the development of new construction schemes based on the potentialities brought by the prestressing technique in concrete bridges. To achieve large spans avoiding any scaffolding, the idea of cantilever construction of concrete bridges was developed in Germany by Finsterwalder. In 1950, he designed the Balduintein Bridge over the Lahn River with a span of 62 m. It should be noted the cantilever system, a

Figure 1.30 The Luzancy Bridge, 1946, France (*Source:* Photograph by MOSSOT, https://commons.wikimedia.org/wiki/File:Pont_de_Luzancy_-1.JPG. CC BY-SA 3.0).

construction scheme adopted since 1874 for steel bridges, was adopted by the first time in concrete bridges for the construction of the bridge over the river Peixe, completed in 1931 in Brazil, with a main span of 68.5 m. The balanced cantilever scheme, in which the bridge is built by cast in-place segmental construction for each side of each pier allowing for large spans to be reached using simple and small steel equipment, opened a new field in bridge construction technology. The spans quickly increased passing from more than 100 m to reach the main spans of 250 m in several road bridges nowadays. The longest span in full prestressed concrete built by cantilever construction is 301 m in the Stolmasundet Bridge [7], concluded in Norway in 1998. This bridge adopted lightweight concrete in the middle section of the main span to reduce dead weight. Other relevant long span road bridges are the Gateway Bridge in Australia, built in 1985, and the Varodd Bridge in Norway, both with 260 m main spans. In railway bridges, the longest span is the São João Bridge [8], in Oporto, Portugal, a design due to Edgar Cardoso, with a span of 250 m, completed in 1991 (Figure 1.31).

Precasted segmental construction was also a relevant contribution to the development of prestressed concrete bridges, either made by balanced cantilever or with moving scaffoldings to erect the segments. The first bridge with this construction technology was made in 1966 for the Oleron Bridge in France, a long bridge (2900 m) connecting the continent to the island of Oleron, with typical spans of 79 m.

Figure 1.31 The São João Railway Bridge, 1990, Oporto, Portugal (*Source:* Photograph by Joseolgon/ https://commons.wikimedia.org / Public Domain).

1.5.5 Cable Supported Bridges

Suspension bridges were adopted in the early days of bridge history, namely in pedestrian bridges with natural fibres for the cables. The suspension system was made by a chain of eye bars in cast and wrought iron, already previously referred to with respect to the Menai suspension bridge. But it was the early wire suspension system, adopting wrought iron and strand spinning technology due to John Roebling [9], which gave rise to a new generation of suspension bridges. In 1883, the Brooklyn Bridge, designed by Roebling with a span of 487 m and built in New York, was the first steel wire suspension bridge in the world. At the beginning of the twentieth century, a variety of long span suspension bridges were built in USA. In 1931, George Washington Bridge, designed by the Swiss O. Ammann over the Hudson River in New York, used parallel wire strands rather than rope strand cables and reached a span longer than 1000 m for the first time. This bridge, with a 1067 m main span, doubled the longest existing span but retained this record for only six years when the famous Golden Gate Bridge [9], San Francisco, designed by Joseph Strauss, concluded with a main span of 1280 m.

Contrary to the robustness appearance of the existing suspension bridges at that time, the Tacoma Narrows Bridge was built in 1940, in the state of Washington, with a much slender deck than the typical ones adopted at the time. The main difference was the slender plate girder deck adopted for its superstructure. After a few months in service, the bridge collapsed because of aerodynamic instability under wind action with a much lower wind speed than the one adopted in its design. Aerodynamic phenomena were known at the time by aeronautical engineers but not by bridge engineers. The bridge was rebuilt with a deep stiffening truss deck in 1950. Many of the existing suspension bridges at the time had been strengthened and all new ones built since the Tacoma Bridge accident have been subjected to detailed aerodynamic studies.

While the US suspension bridge concepts have kept the rigid truss superstructure, the European evolution, mainly due to English engineers, tends towards the use of streamline bridge decks for aerodynamic stability. The Severn Bridge, concluded in 1966, with a main span of 988 m, was the first achievement, eliminating the need to adopt the deep stiffening truss decks. The Tagus River Bridge was opened to traffic in 1966 using a typical rigid truss girder designed by Steinman, New York. The design considered, according to the bid specifications, the possibility of a railroad addition at the lower chord level. With a 1013 m main span, it was, at the time, the longest span in Europe. Thirty years later, the railroad addition was implemented with a completely different concept for this stage than initially foreseen. The trains crossing the bridge nowadays are about 2.5 times heavier than the ones initially foreseen. In the original design, the additional loads due to the railroad were foreseen to be supported by a set stay cables; when the project was implemented, a second cable was added and one new hanger between each of two existing ones was inserted.

In 1981, the Humber Bridge was concluded in the UK. With the longest 1410 m main span at the time, a streamline box girder deck was used as was for the Severn Bridge, but adopting a triangular suspension system for the hangers. It kept the span world record for years until the Great Belt Bridge, 1624 m, and the Akashi Kaikyō, 1991 m, were built in 1998; referred to in Section 1.5.4.

In the second half of the twentieth century, a new generation of cable supported bridges was developed: cable-stayed bridges. The Strömsund Bridge, concluded in Sweden in 1955, was one of the first cable-stayed bridges, followed some years later by the three bridges in Dusseldorf, Germany; the North Bridge (260 m span) with two towers and the Knie and Oberkassel Bridges each with a single tower and spans of 320 m and 258 m, respectively [1]. All three Dusseldorf bridges, over the Rhine River, adopted steel decks suspended by stay cables with a harp arrangement and spaced about 30 m at the deck level. Following these bridges, a variety of longitudinal configurations and deck cross sections have been implemented for this bridge typology. However, the most important innovation was the introduction of closed spaced stay cables at the deck level (5–10 m in concrete bridge decks and about 12–20 m in steel decks), allowing the deck to work as a beam on elastic foundations with small bending moments.

At the end of the 1980s in the twentieth century, maximum spans increased, in particular with Annacis Bridge in Vancouver, Canada [10], with a main span of 465 m with a steel-concrete composite deck. A large step was achieved with the construction of the Normandy Bridge, in France in 1995, following a concept design from Michel Virlogeux [11]. With a main span of 856 m, this bridge adopted a steel box girder deck in the central part (624 m) of the main span continued to the towers and lateral spans with a prestressed concrete box girder deck. The 1000 m span length for cable-stayed bridges was first reached by two bridges. The first, the Stonecutters' Bridge concluded in 2009 in Hong Kong, with a 1018 m steel main span, and short 299 m lateral concrete spans [12]. This bridge introduced the option of splitting the deck cross section in to two separate boxes for aerodynamic stability.

The longest span length for cable-stayed bridges is presently the second, the Russky Bridge, built across the Eastern Bosphorus strait in Vladivostok in 2012, adopting a single box girder deck with a main span of 1104 m and lateral stay cable suspension, as in all other very long cable-stayed bridges [13].

The increased development of cable supported bridges, either suspension bridges or cable-stayed bridges, will continue in the coming years. Suspended spans above 3000 m, like the one considered in the design of the Messina Bridge [14], may be achieved in the near future.

Even if span length is often taken as key parameter to evaluate historical bridge evolution, one should take into consideration that many other aspects and solutions had decisive contributions to the development of bridge design throughout history. Even considering only cable supported bridges, a variety of bridge innovative solutions may be referred to as:

- The three-dimensional arrangement of stay cables – the Lérez River Bridge, Spain [4];
- The multiple cable-stayed bridges – Millau Viaduct, France [15], and Rion-Antirion Bridge, Greece [16];
- The combination of suspended and cable-stayed solutions – the Third Bosphorus Bridge, Turkey [17].[1]

---

1 In fact, it is not a really new structural solution since it has been applied before with success in the Brooklyn Bridge, E.U.A. 1883.

References

1 Leohnardt, F. (1982). *Bridges – Aesthetics and Design*, 308. London: The Architectural Press Ltd.
2 Wittfoht, H. (1984). *Building Bridges – History, Technology, Construction*, 327. Beton Verlag GmbH.
3 Menn, C. (1990). *Prestressed Concrete Bridges*, 536. Springer.
4 Troyano, L.F. (2003). *Bridge Engineering: A Global Perspective*, 775. Thomas Telford.
5 Ordem dos Engenheiros – Região Norte (2005). Ponte Maria Pia – A obra-prima de Seyrig 150 pp.
6 Reis, A., Lopes, N., and Ribeiro, D. (2007). The new metro of Oporto: a variety of bridge projects. *IABSE Symposium Report* 93 (13): 18–25.
7 Tang, M.-C. (2007). Evolution of bridge technology. *IABSE Symposium Report* 93 (31): 38–48.
8 Lousada Soares, L. (2003). *Edgar Cardoso: Engenheiro Civil*, 370. FEUP.
9 Gimsing, N.J. and Georgakis, C. (2012). *Cable Supported Bridges – Concept and Design*, 3e, 590. Wiley.
10 Taylor, P. and Torrejon, J. (1987). Annacis Bridge. *Concrete International* 9 (7): 13–22.
11 Virlogeux, M. (1994). The Normandie Bridge, France: a new record for cable-stayed bridges. *Structural Engineering International* 4 (4): 208–213.
12 Falbe-Hansen, K., Hauge, L., and Kite, S. (2004). Stonecutters Bridge – detailed design. *IABSE Symposium Report* 88 (6): 19–24.
13 Svensson, H. (2012). *Cable-Stayed Bridges – 40 Years of Experience*, 430. Wilhelm Ernst & Sohn, Germany.
14 Brown, D.J. (1993). *Bridges – Three Thousand Years of Defying Nature*, 176. Mitchell Beazley.
15 Virlogeux, M., Servant, C., Cremer, J.M. et al. (2005). Millau Viaduct, France. *Structural Engineering International* 15 (1): 4–7.
16 Teyssandier, J-P. (1997). The Rion-Antirion Bridge. Proceedings of the FIP International Conference, Vol. 2, 1163–1170.
17 Klein, J.F. (2017). Third Bosphorus Bridge – a masterpiece of sculptural engineering. *Stahlbau* 86 (2): 160–166.

2

Bridge Design

Site Data and Basic Conditions

2.1 Design Phases and Methodology

The design of a bridge should be developed in a sequence of phases that are relevant for selection between several design options. Joint decisions by designer and owner should be taken along all design phases.

Input data for design is needed. Hence, a data acquisition process is required, namely, for topographic surveys, geological, hydraulic data and environmental issues. The acquisition process requires preliminary concepts about bridge design options. This is the case for geotechnical investigations at the location of bridge piers, requiring, for example, the development of geological surveys at two design phases – one at the preliminary design phase and a second one after a decision about the retained option.

Design practice varies from country to country but, in short, the following design phases may be considered: preliminary design, base case design and the final/execution design.

The designer should start by fixing the conditions to develop the design options, namely the possible locations for the bridge (if it is not fixed in the contract documents), vertical and horizontal clearances, typical width for the deck (from the number of lanes and width of the walkways to be accommodated) and basic design criteria (design codes) to be adopted.

The *preliminary design* is the key phase for the success of a bridge design. The most important design options are investigated and proposed to the owner. Different options should be compared from the technical and cost point of view, including all relevant technical aspects like functionality, structural performance and durability, execution methods and associated construction issues, aesthetics and environmental integration. A certain number of options (two to three) should be proposed, but this number should be limited (usually to no more than five) to allow a decision by the owner within the time schedule.

The *Base Case Design* consists of the development of the solution, or at most two of the options, selected by the owner from the proposed preliminary design solutions. The base case design may be adopted as a basic document for the construction bid and should integrate all the results from the studies developed up to the end of that phase; namely, geological, hydraulic and environmental studies. This is the usual practice in some countries where the contractor is responsible for the final detailed design.

Bridge Design: Concepts and Analysis, First Edition. António J. Reis and José J. Oliveira Pedro.

If that is not the case, the owner may decide to skip the base case design and instruct the designer to go directly to final design after a solution from the preliminary study has been chosen. The *final design* includes all the design notes and drawings to detail all aspects of the proposed bridge solution. The former include the technical report (describing the solution to be executed), the technical specifications, the bill of quantities and the cost estimation; the latter are the execution drawings, including general and detailed drawings. The results of all the complementary studies developed at the design phases, like geological, geotechnical and environmental studies, should be added as specific reports.

2.2 Basic Site Data

2.2.1 Generalities

For the development of a bridge design, basic data should be collected that is specifically related to topographic, geological and geotechnical conditions. Hydraulic requirements, if relevant to the design case, possible seismic data, environmental and any other conditions related to the bridge site like, for example, local availability of the materials, may also be needed.

Apart from these data, it is necessary to establish the design requirements related to geometrical conditions for the longitudinal and transverse alignments, generally fixed by the highway or railway design, vertical and horizontal clearances when the bridge or viaduct crosses other roads, navigation channels in rivers or in bridge links and any other requirements relevant to the bridge case. Finally, elements that should be integrated in the bridge deck, like the surfacing and the waterproofing, the walkways, the parapets (guardrails) and handrails, the fascia beams, the drainage system, the expansion joints (EJs), the lighting system and any other equipment for the bridge, should be considered. From the beginning of the design process, due account should be taken on all these elements, since they affect the bridge deck geometry. These subjects are considered in this chapter, although the reader is referred to the literature for detail aspects, like some bridge accessories [1, 2].

2.2.2 Topographic Data

The development of a bridge design requires a detailed topographic survey on the site nowadays converted from aerial photographs in a series of digitalized files, from which it is possible to obtain all the necessary plans at different scales as well as the longitudinal alignments along the bridge axis and transverse sections at the pier or abutment locations.

Three dimensional (3D) models of the bridge site, obtained from the topographic digitalized data by using specific software, may be quite useful to study the integration of the bridge with the landscape, as shown in Figure 2.1.

The topographic survey shall lead to a detailed definition of the bridge site, but scales shall be conveniently selected at each design phase taking into consideration the type of element to be studied. For example, for a long bridge, one shall never, at a preliminary design phase, start with small scales, like 1/100 or 1/200, because one loses the overview

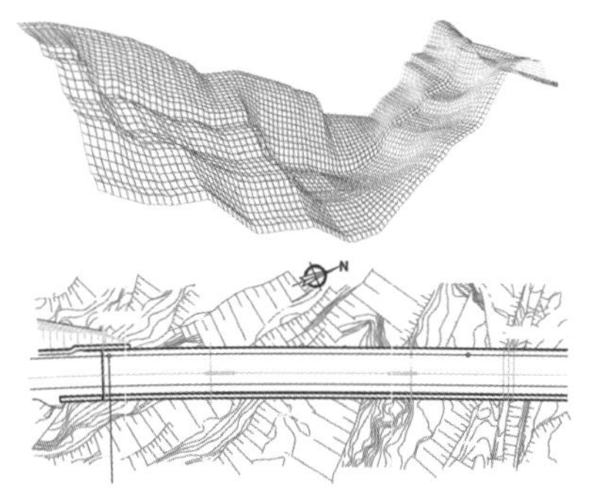

Figure 2.1 Model for landscape integration of a bridge.

of the site. In such a case scales like 1/1000 or 1/500 may be the most convenient ones to begin with. The general plans for a long bridge, shall include at least an extension of about 1 km for each side of the possible location the bridge abutments, with the definition of the horizontal highway or railway alignments.

The horizontal and vertical alignments of the highway or railway design are generally drawn on a non-deformed scale – 1/1000, 1/500 or 1/200, as shown in Figure 2.2. It is quite important to work on a single drawing with the plan and vertical alignments along the bridge axis, as shown in Figure 2.2. In the zone where the bridge is likely to be located, it is necessary to have plans on more detailed scales, like 1/100 for the zones at the abutments or at possible pier locations. Transverse sections to the bridge axis at the

(a)

Figure 2.2 Vertical and plan alignments of a bridge: Santa Cruz Bridge, Madeira Island – design by GRID. Longitudinal cross section with geotechnical profile and plan view of the deck level.

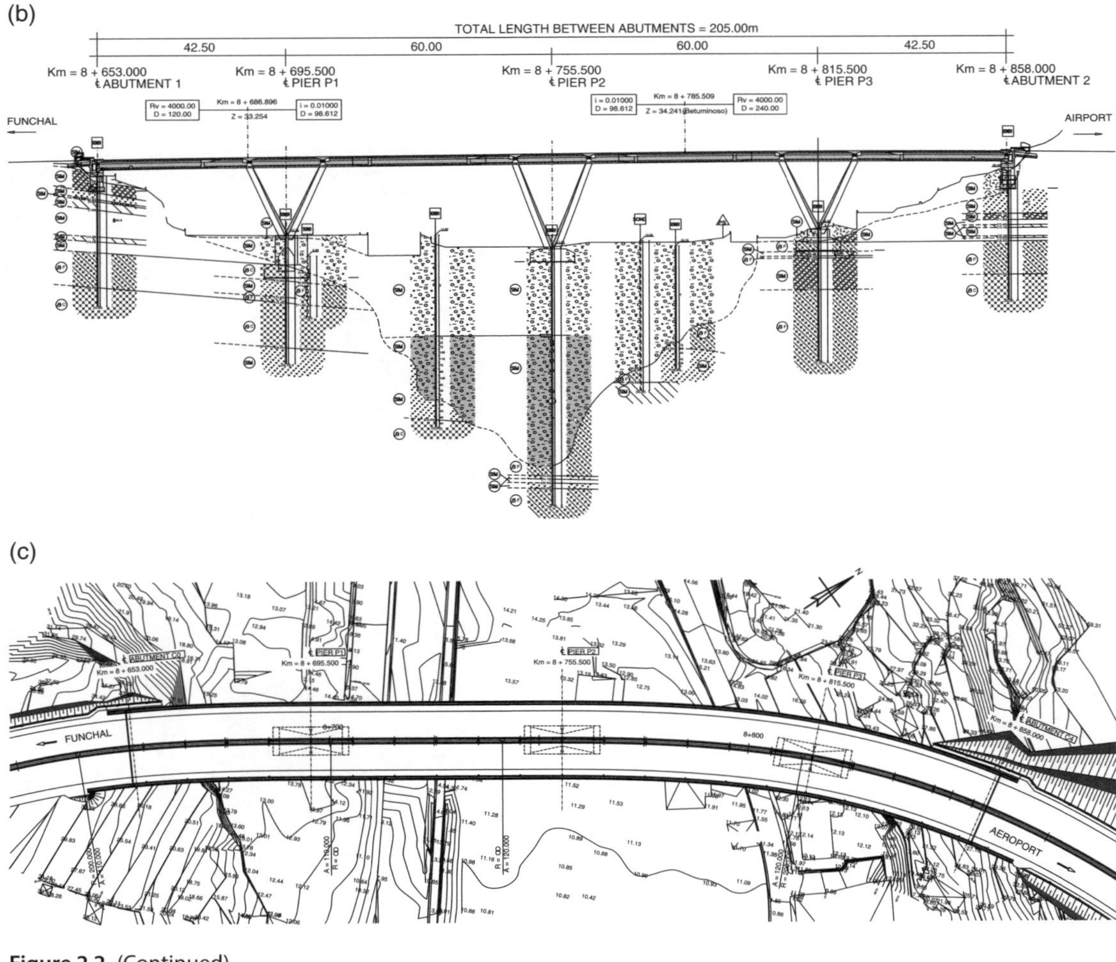

Figure 2.2 (Continued)

abutment locations are generally necessary to define the geometry of the substructure, as well as the exact length of the bridge, as shown in Figure 2.2.

The location of piers near the river bed should always be made on the basis of a detailed plan at this zone in order to take into consideration possible problems of skew between the river and the bridge axis. Even if no piers are likely to be located in the river, it is necessary to investigate the topography of the river bed, that is, the *bathymetry*, because some decisions may be taken for the execution scheme involving the use of the river bed during construction.

This is the case for possible location of temporary piers for construction. In this case, or if definitive piers are located in or near by the river bed, one should have a longitudinal section along the river bed, at least 500 m each side of the bridge axis. Transverse cross sections of the river – for example, at 100 m distance apart – are usually required in order to proceed with the hydraulic studies.

To locate the bridge, one shall have, from the highway or railway design, the coordinates M, P and Z (or X, Y and Z) of reference points of the alignments. For bridge execution, plans should define all the coordinates at reference points of the foundations and at deck level. These coordinates at the foundation level should be calculated from the coordinates at the deck level, taking into consideration possible transverse slopes of the superstructure if an in-plan curvature exists.

2.2.3 Geological and Geotechnical Data

The geological and geotechnical conditions to be considered have a very strong influence on the selection of the bridge option, in the layout of the spans, in the design of the abutments, in the choice of the foundation system for the piers, as well as in the decision about the construction or erection scheme to be adopted for the deck. Even the selection of the structural material for the bridge deck – concrete, steel or steel-concrete composite deck –may be influenced by the type of geotechnical conditions.

Two stages of soil investigation are generally considered in a bridge design: the preliminary and the detailed. The former is done at the preliminary design of the bridge, without knowing the exact location of the elements of the substructure, and the latter is made after the exact location of the piers and abutments is fixed.

The geological and geotechnical reports should include the results of the survey, with a general description of the soils and bed-rock levels, the location of the bore holes and equipment adopted in the soil investigation, depth levels reached, nature and thickness of the layers encountered, existence and evolution of water levels. The results of the in situ or laboratory tests, to define the physical and mechanical properties of the soils and rock layers encountered at the geological investigation, should also be included.

For the bridge design, the results of the soil investigation should be converted to a geological profile along the bridge axis, as shown in Figure 2.2. Some geological cross sections at the abutment locations are most useful to define the geometry of these elements. At pier locations, possible variations of the geotechnical conditions, transverse to the bridge axis, should be defined on the basis of different bore holes, two or three at each pier location. Different depths for pile foundations or an adjusted geometry to the bed-rock conditions may result from this variation of geological conditions in the transverse direction.

2.2.4 Hydraulic Data

In a bridge design, if the alignment is such that the crossing of a river has to be considered, one has to take into consideration the following problems:

- The influence of the bridge on the hydraulic flow;
- The hydrodynamic forces induced by the flow on piers located in the river bed;
- The stream bed erosion at the bridge site, generally designated by *scour*.

As a reference for the type of elements for the hydraulic study, one refers to:

- Bathymetry (topography of the river bed) at the bridge site, defined by a number of cross sections upstream and downstream, as well as a longitudinal alignment at the axis of the river bed;
- Available information about high water level dates and seasonal floods;
- Definition of the nature of the river bed, as well as of the river banks;
- Information on the stability of the river banks – erosion effects at the bridge site;
- Location upstream of particular conditions influencing the natural flow, like dams and reservoirs.

Apart from these data, it is important to collect data about other bridges previously built near the bridge site, namely flood-frequency curves, types of foundations adopted and possible problems of scours observed.

It may be relevant to obtain information about possible removal of materials from the river banks, like sand, near the bridge site, which may induce problems like *contraction scour* [2–4]. This phenomenon is induced by an acceleration of the water flow approaching the bridge site, removing the material from the river bed.

The hydraulic studies should be performed for two phases – the execution phase and the final stage at the bridge site, with possible permanent piers or abutments affecting the stream flow conditions under the bridge. Nowadays, for small rivers, the best solution is generally to increase the spans at the river crossing, in order to avoid any permanent pier in the river. However, for the execution phase, some temporary piers may be envisaged in the river bed. In wide rivers the construction of the piers may affect temporarily the flow conditions by reducing the stream cross section with some temporary 'islands' or 'peninsulas' made by embankments, as shown in Figure 2.3. The hydraulic conditions are usually critical for these phases, even considering a reduced *flood return period* for the execution phase with respect to the *design food return period* (T = 100–1000 years).

As a result of the hydraulic studies, the following data for the bridge design should be obtained:

- Maximum high water levels for return periods of at least 100 years flood;
- Maximum high water levels under normal flows during the life of the structure;
- Flow velocities under the bridge, under maximum flood conditions;
- Study of the flow conditions during the execution phase of the bridge.

Local scour consists of dragging of the river bed material by flow, around the piers and abutment zones, due to increased acceleration of river flow and induced vortices as shown in the scheme of Figure 2.4. The local scour tends to stabilize at a certain depth of the erosion cavity as the intensity of the vortices decreases with the reduced capacity of the flow for dragging material of the affected zone. *Generalized scour* is a dragging

Figure 2.3 Temporary embankments for the Quintanilha Bridge pier foundations.

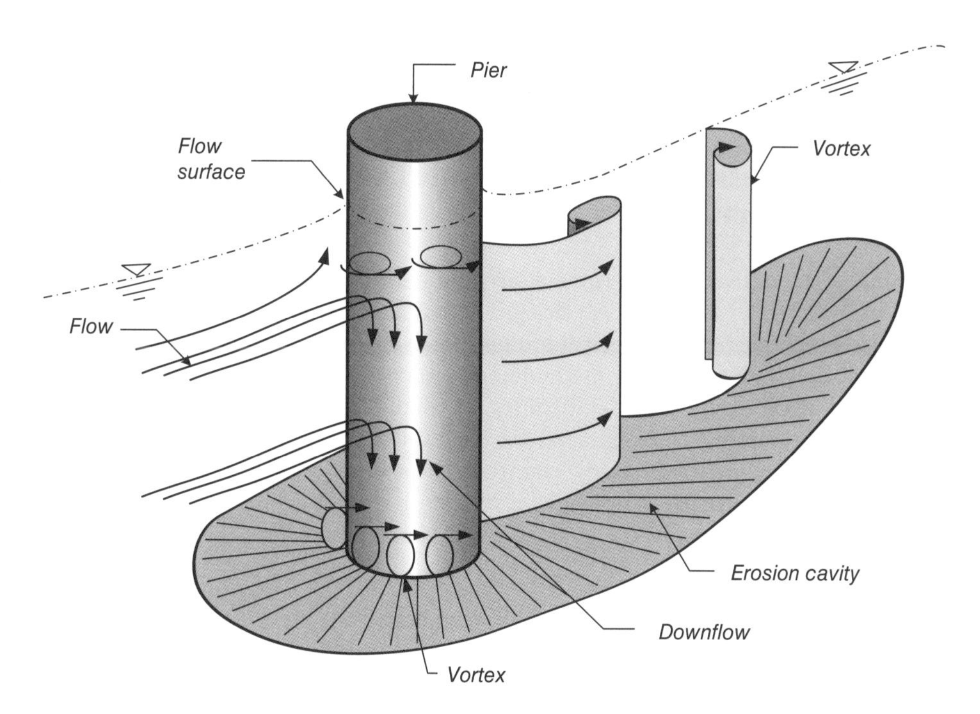

Figure 2.4 Local scour (*Source:* Adapted from Ref. [4]).

phenomenon affecting all the river beds. Local and generalized scours during floods are one of the main causes of bridge failures.

The reader is referred to Chapter 4 for the evaluation of the depth of scour, the protective systems and specific problems related to foundations in rivers. Chapter 3 presents an assessment of hydrodynamic pier forces.

2.2.5 Other Data

Depending on the bridge site and design case, many other conditions may require collecting complementary data, related to:

- *Local conditions* – access to the bridge site, availability of construction materials and execution technologies from local contractors, water disposal on the site, requirements for public services like pipelines, electrical cables and so on.
- *Existence of aggressive elements* – attack of the water to the construction materials; such as, for example, to the submersed reinforcement previous to concreting operations, attack by molluscs to temporary wood supports in some rivers and splashing and spraying of salt water in the concrete bridge structures located in marine environments.
- *Seismic conditions* – bridges located in seismic zones with earthquake actions affecting the concept design of the substructure or bridges located in zones with geological faults or in zones with risks of *soil liquefaction* affecting the design of the foundations.
- *Environmental conditions* – conditions for the integration of bridges with natural environments affecting the relationship between aesthetics and structural typology, as well as specific issues in the urban spaces relating aesthetics, noise impact and construction methods.

2.3 Bridge Location. Alignment, Bridge Length and Hydraulic Conditions

A first step in the design process is the study of the location of the bridge. Very often that is done at the highway or railway design without any particular interference of the bridge designer; in particular, for small bridges. If so, the bridge designer has very little choice to locate the bridge. This methodology is not desirable but occurs in design practice. This may yield increased difficulties for small bridges, with very complex and undesirable locations due to constraints for location of the piers, abutments or even because the spans cannot be properly balanced. The skew alignment of a small viaduct over a certain road or railway may require a very large span to keep the necessary horizontal clearances. In short, for the study of the location of a bridge, one may face two extreme situations (Figure 2.5):

- The bridge is constrained to a fixed alignment of a highway or a railway (Figure 2.5a);
- The location of the bridge is a constraint for the highway or the railway (Figure 2.5b).

Between these extreme situations, intermediate cases exist that can only be discussed for the specific conditions associated to the design case. The actual tendency is of

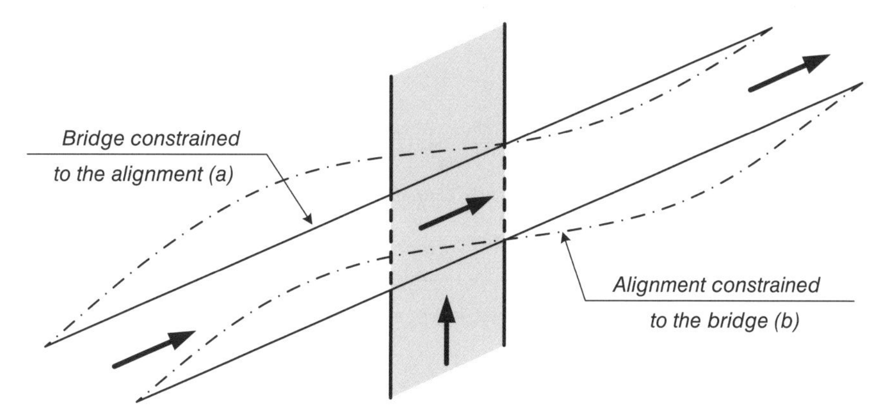

Figure 2.5 Bridge constrained to the alignment (a) or alignment constrained to the bridge (b).

Figure 2.6 The International Bridge of Quintanilha, at Portugal and Spain frontier; bridge-river skew crossing with piers located outside the river bed.

course to adapt the location and the alignment of the bridge to the highway or railway alignments. The best solution results in general from the compatibility of the best option for the bridge location and from the conditions established by the road or railway design. This is particularly true for the case of long bridges that have a considerable cost for the overall project.

For a bridge crossing a river, one shall locate the bridge in such a way so as to cause the least possible impact on water flow (Figure 2.6). In the case of a river that under normal conditions has a permanent sinuous river bed, one of the solutions illustrated in Figure 2.7 may be adopted. The in-plan alignment of the bridge should be, if possible, perpendicular to the major river bed in order to reduce the bridge length and to reduce the impact on the flow under flood conditions.

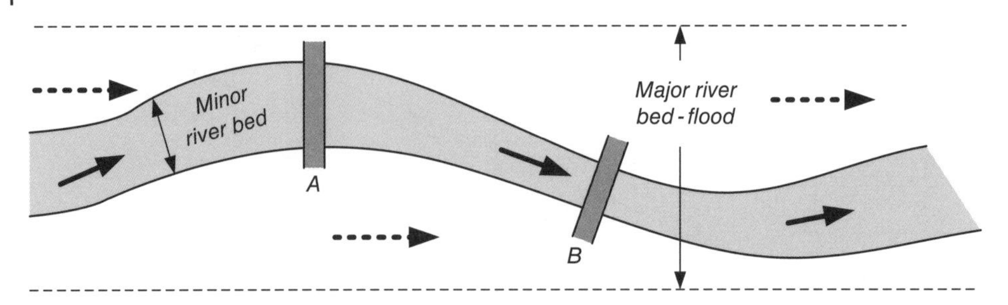

Figure 2.7 River bridge crossing. Possible locations A and B.

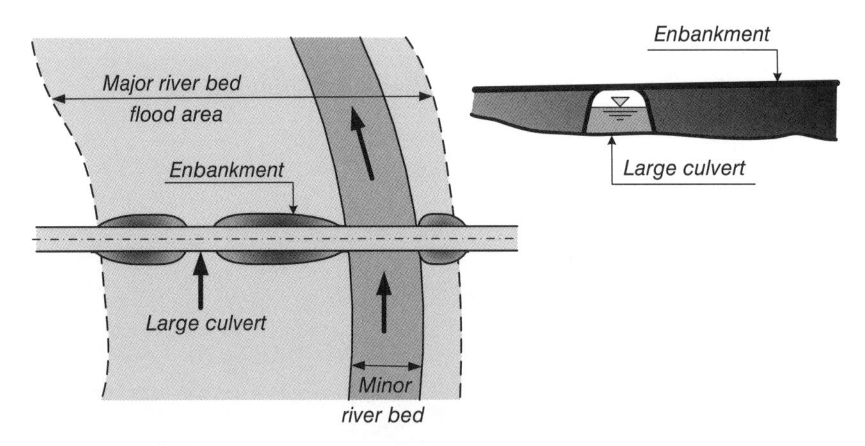

Figure 2.8 Embankment to reduce a bridge length.

If the river is usually located at the permanent river bed, or if the major river bed is only reached in particular flood conditions, one may reduce the bridge length by embankments (Figure 2.8). Usually, the cost of the earthfills is much lower than the cost of the bridge works. Of course, some large culverts should be constructed along the alignment in order to fulfil the discharge requirements. In this case, the bridge piers should be oriented, if possible, along the flow direction in the permanent river bed, to reduce scour problems. The embankments should be protected against flow erosion and be stable under flood conditions.

The number of piers located in the river bed should be as small as possible to minimize the impact on the flow. The following parameters may be defined:

- Coefficient of constriction – the ratio between the area of the section occupied by the bridge and the area of the section for flow without the bridge;
- Wet surface (A_M) – the area of the cross section defined by the free surface of the flow and the contour of the channel;
- Wet contour (P_M) – contour of the wet surface of the channel;
- Hydraulic radius ($R_h = A_M/P_M$) – the ratio between the wet surface and the wet contour.

The *design discharge* is the flow rate expressed as volume per unit of time, and is given by

$$Q = A_M \times V_m \tag{2.1}$$

Figure 2.9 Vertical velocity profile.

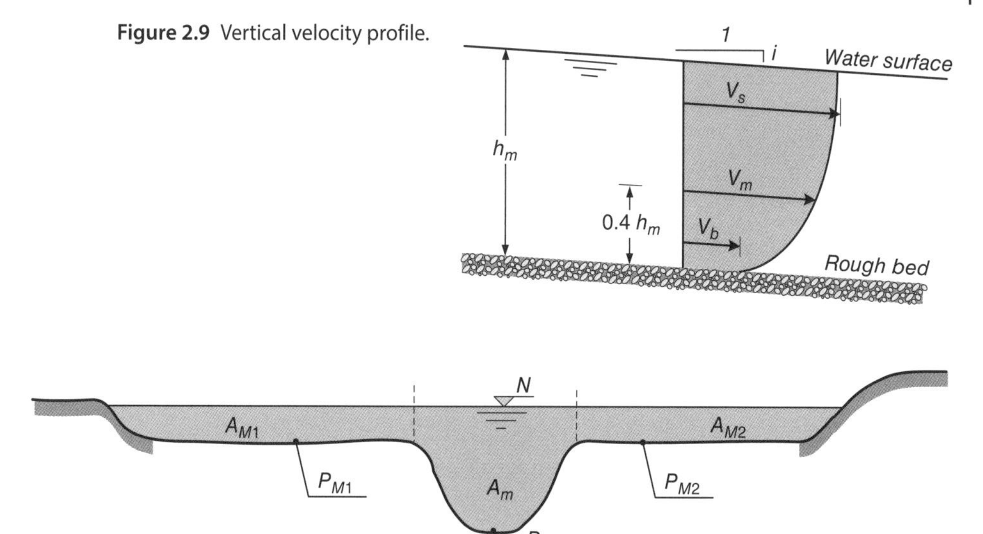

Figure 2.10 Design discharge in a river bridge crossing. Generalization of the Manning–Strickler formula.

where V_m is the average velocity of the flow at a free channel section. This average speed, V_m, should not be confused with the water surface speed, V_s, or the speed at the rough bed, V_b, as shown in Figure 2.9 at the vertical velocity profile.

The basic Manning–Strickler formula [2–4] allows the calculation of the average flow velocity for a straight free channel under uniform flow conditions,

$$V_m = K\, i^{1/2}\, R_{\mathrm{h}}^{2/3} \tag{2.2}$$

where i is the 'slope of the flow' and K is the global roughness Strickler coefficient, taking into consideration the contour and roughness conditions of the walls and the bottom of the channel. Under uniform roughness conditions at the wet contour, one has:

$$K = 21\left(\frac{d_m}{h}\right)^{1/24} / \, d_m^{1/6} \tag{2.3}$$

where d_m is the average dimension (in metres) of the particles under which 50% of the granulometry is encountered and h_m is the flow depth.

The Manning–Strickler formula may be generalized [4] for the specific case of a bridge crossing where there is a need to distinguish between the minor and major river beds (wet areas A_m and A_M, hydraulic radius R_m and R_M and roughness coefficients n_m and n_M) as shown in Figure 2.10:

$$\frac{Q}{i^{1/2}} = \frac{1}{n_m} K\, A_m R_m^{2/3} + \frac{1}{n_M}\sqrt{A_M^2 + A_M A_m \left(1 - K^2\right)}\, R_M^{2/3} \tag{2.4}$$

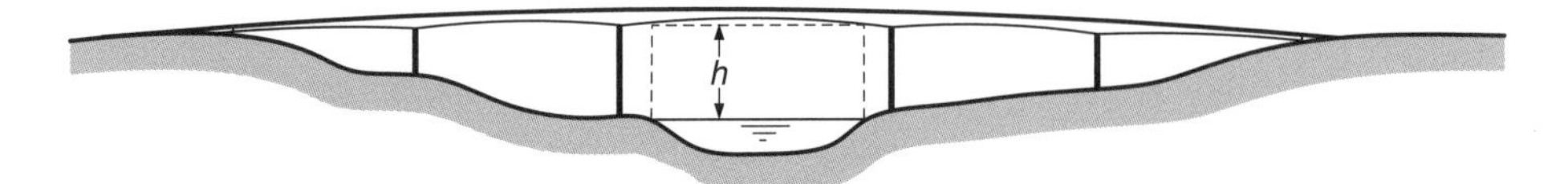

Figure 2.11 Vertical clearance at a river bridge crossing.

where the coefficient K is defined as a function of $r = R_M/R_m$ by

$$K = K_o = 0.9\left(\frac{n_m}{n_M}\right)^{\frac{1}{6}} \quad \text{for } r \geq 0.3 \tag{2.5}$$

$$K = \frac{1-K_o}{2}\cos\frac{\pi r}{0.3} + \frac{1+K_o}{2} \quad \text{for } r < 0.3 \tag{2.6}$$

The reader is referred to the literature [3, 4] for the values of the roughness coefficients n_m and n_M under different specific conditions of river beds. As an order of magnitude, one may say the coefficients n for minor river beds (n_m) up to 30 m width under flood conditions vary for normal conditions of roughness between 0.03 and 0.05.

If the flow conditions are changed by the presence of the piers or abutments or by temporary embankments during construction, the water level is increased. This phenomenon is known in hydraulics as the *back water effect*. In modern bridge construction, back water effects are generally more important during construction, due to temporary embankments (Figure 2.3) since at the final stage only a few piers, if any, exist at the river bed. In old bridges, where small spans were adopted, back water effects could be relevant under flood conditions. The *vertical clearance* is defined as the distance h (Figure 2.11) between the free water surface and the lowest point of the deck at the river bed. Under flood conditions, a minimum clearance of at least 1.0–2.0 m should be respected when fixing the bridge vertical alignment, in order to allow materials transported by the river to pass freely under the deck. Extreme events, like the one shown on the Zambezi River Bridge in Figure 2.12, may cause the bridge deck to be completely surrounded by water.

2.3.1 The Horizontal and Vertical Alignments

The definition of the horizontal and vertical alignments of the bridge axis is generally a result of the highway or railway design, as previously referred to. However, the bridge designer should participate in the study of the bridge alignments. If so, difficulties on the development of the bridge design, namely, from structural, aesthetical and execution points of view, are minimized. Some examples may be used to illustrate this point.

Let's take first the case of horizontal alignment. An in-plan curved alignment may favour aesthetics and view of the bridge by the users. The example of the S alignment of the Vasco da Gama Bridge in Lisbon, Figure 2.13, is a very good one.

From the execution point of view, horizontal alignments with constant radius of curvature are required if the bridge is to be erected by the *incremental launching method* to be discussed in Chapter 4. Variable radius of curvature, such as when transition curves are inserted in the bridge, are possible for a concrete bridge built by the *balanced*

Figure 2.12 Zambezi River Bridge, in Mozambique, under extreme flood conditions during construction (*Source:* GRID, SA archive).

Figure 2.13 The S alignment of the Vasco da Gama Bridge, in Lisbon (*Source:* Photograph by Till Niermann, https://commons.wikimedia.org).

cantilever method or even if the bridge is executed span by span using a *formwork launching girder* as is seen in Chapter 4.

The vertical alignment is not so critical with respect to the relationships with the execution method, but has a strong influence on the aesthetics of the bridge. Of course, one has to take into consideration all the requirements for highway or railway design, namely the maximum slopes allowed and the minimum radius and developments at concave or convex alignments, which are a function of the design speed. When the levels of the abutments are similar, one should adopt minimum longitudinal slopes

(a)

(b)

Figure 2.14 The Macau-Taipa Bridge (Governador Nobre de Carvalho Bridge, design by E. Cardoso): (a) Vertical alignment (*Source:* Photograph by lidxplus https://commons.wikimedia.org) and (b) aerial view (*Source:* Photograph by A. Pereira).

obtained, for example, from the adoption of a convex alignment (Figure 2.11) in order to guarantee good drainage conditions as well as an improved appearance from front views.

If the levels of the abutments are very low, one may adopt strong slopes from both sides by inserting the bridge in a convex alignment to match, for example, a certain vertical clearance on a river. The maximum slopes are of course constrained to the requirements for the highway or railway design. The Macau-Taipa Bridge, Figure 2.14, a highway bridge designed in the 1960s, is an extreme example of this design philosophy. The bridge has been designed to achieve a sufficient vertical clearance for the navigation channel at the 74 m main span. A very low level on each side of the main span and very pronounced slopes from there on to comply with the navigation channel. The bridge has a much smoother appearance if it is not seen from a longitudinal view but from front or aerial views (Figure 2.14).

The maximum allowable slopes in highway bridges are generally between 4 and 9% depending on the design speeds. For railway bridges, these slopes do not go above 2.5% having a strong influence on designing a bridge in a plan region to cross a river with vertical clearance defined by hydraulic requirements or navigation channels. If the soil conditions are difficult, for example, when a bridge crosses an alluvium region, the maximum levels of the earthfills at the abutment zones are fixed in order to control maximum allowable settlements at the soil to superstructure transitions. The length required for the bridge may be a result of the maximum depth allowed for the earthfill of the approaches at the abutments and/or the vertical alignment required to fulfil vertical clearances. If the bridge is inserted in a concave alignment it is better for aesthetics to adopt a constant depth deck rather than a variable depth deck.

(a)

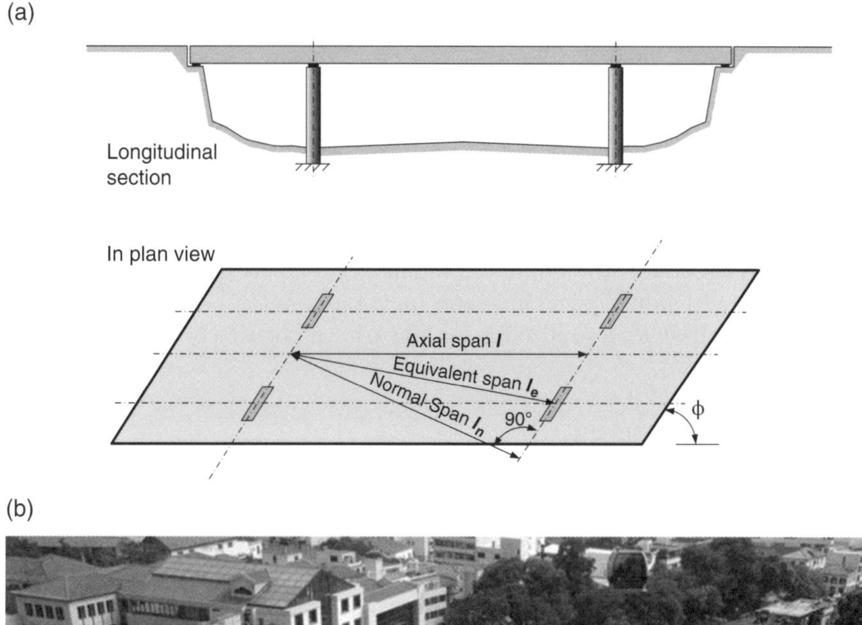

(b)

Figure 2.15 Skew bridge (a) curved Viaduct in Funchal, Madeira Island and (b) design by GRID.

Still with respect to the horizontal alignment, two special cases should be considered: (i) skew bridges and (ii) curved bridges. These two cases, resulting from requirements of the in-plan alignment or from existing site constraints at the bridge site, are illustrated in Figure 2.15.

For straight bridges the skew angle ϕ is 90° while it may be as low as 30° for skewed bridges. The span length ($\boldsymbol{l}$) may be measured along the axis of the bridge between the transverse pier alignments or in the normal ($\boldsymbol{l}_n$) to the transverse pier alignment; for structural analysis, one may define from the angles $\phi/2$ an intermediate equivalent span, $\boldsymbol{l}_e$, according to Figure 2.15a. For uniform loads on the deck as the dead load, the longitudinal bending moments may be obtained approximately adopting these equivalent spans, $\boldsymbol{l_e}$, rather than the span lengths, $\boldsymbol{l}$.

Curved bridges are inserted in a single constant curvature alignment or in an alignment composed by straight and curved stretches, usually requiring transition curves with clothoids between constant radius and strait segments (from R = const. to $R = 0$). For the urban viaduct shown in Figure 2.15b, the deck is a continuous two-span prestressed concrete single rib voided slab, with a single pier between two roadways at different levels; the deck is built at the left end side abutment to reduce torsion effects due to the in-plan curvature. The minimum radius R_{min} of a curved bridge is defined from the design speed V (km h^{-1}) and increases quite rapidly with V. For example, for a road bridge, one may have $R_{min} = 120$ m for $V = 60$ km h^{-1} increasing to $R_{min} = 700$ m for $V = 120$ km h^{-1}.

2.3.2 The Transverse Alignment

The transverse alignment over the bridge deck should satisfy the requirements from road or railway design, namely the width for lanes and the shoulders. Generally, one starts, for example, from the current section of the roadway, shown in Figure 2.16a, and defines the necessary transverse alignment on the bridge by assuming or not the existence of walkways (Figure 2.16b and c). In addition, kerbs, parapets, handrails, fascia beams and other accessories to be installed on the bridge deck should be considered. The

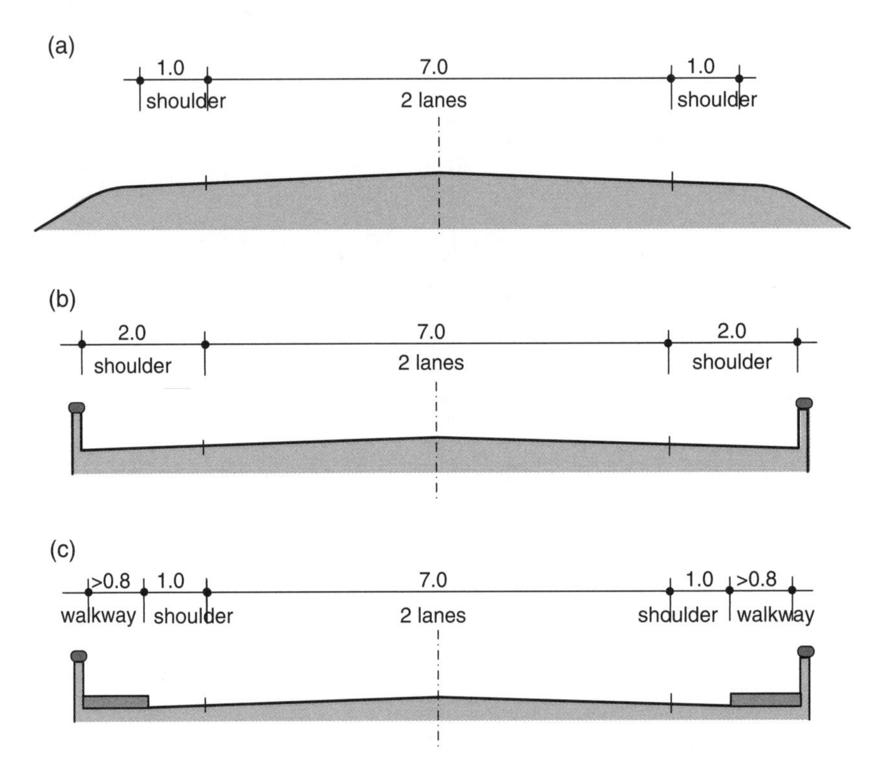

Figure 2.16 Definition of the transverse alignment on the bridge deck (b) without walkway; (c) with walkways; from the road transverse alignment (a).

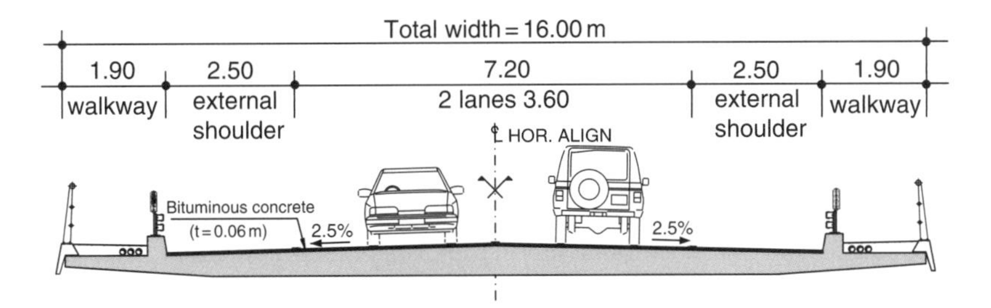

Figure 2.17 Transverse alignment on the bridge deck showing the impact on deck width of kerbs, guardrails, handrails and fascia beams.

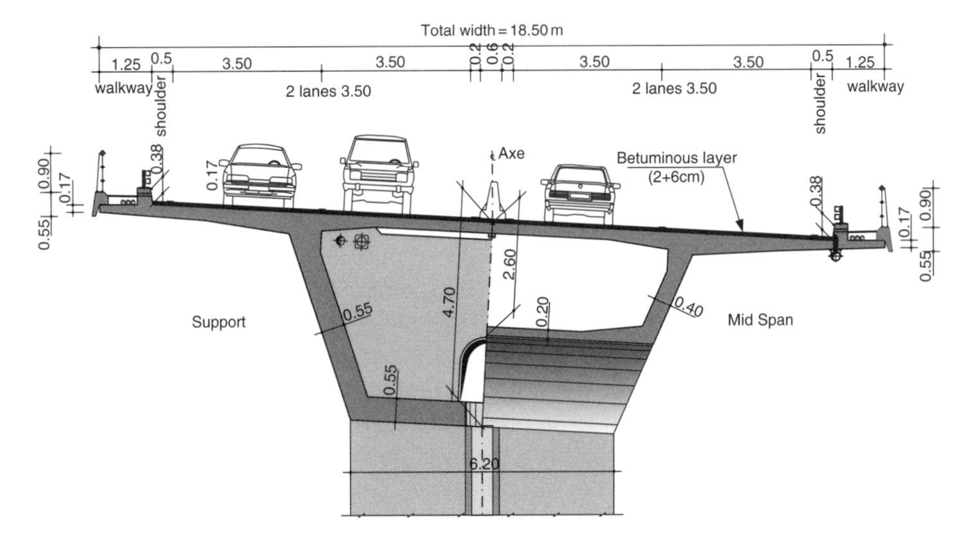

Figure 2.18 Deck cross section of the bridge of Figure 2.6 with two separated carriageways of 7 m.

detailing and characteristics of these elements are discussed in Section 2.4. For now, one should concentrate on its influence on the required width for the bridge deck. The example shown in Figure 2.17 illustrates a bridge with 2 × 3.6 m lanes, 2 × 2.5 m shoulders, and two 1.9 m walkways. The total width for the deck is 16.0 m.

Another example is shown in Figure 2.18 for a highway, for 2 × 2 road traffic lanes requiring a total bridge deck width of 18.5 m.

In bridges with pedestrian traffic, the minimum width of the footpaths is usually 0.8 m but it may reach 3.0 m or even more on urban bridges. If pedestrian traffic is not required, inspection walkways, 0.6 m width minimum, may be inserted at each side of the bridge deck.

Some general rules for establishing the transverse alignment on the bridge deck are:

- The total width between kerbs on highway bridges should exceed in 2.0 m for each side, the total width of the deck.
- The width between kerbs adjacent to the walkways should exceed the width of the carriageway with a minimum width of 7.0 m; only for secondary roads could a minimum of 6.0 m possibly be accepted.

In straight bridges, for drainage purposes, the transverse slopes are usually 2.0–2.5% for each side of the deck axis. In curved bridges, the transverse alignment over the deck can only be established taking into consideration two additional geometrical requirements related to the transverse slope and to the deck width. These are defined from the parameters, S_e (%), the required transverse slope for centrifugal force effects on the vehicles, as a function of the radius, R, design speed, V, and a possible extra widening, S_L, of the deck required for safe circulation of the vehicles on the bridge deck. Depending on road design specifications, one may have, for $V > 80\,\text{km h}^{-1}$ ($R = 350\,\text{m}$; $S_e = 6\%$) and for ($R > 1500\,\text{m}$; $S_e = 2\%$). Concerning S_L, one may have, for example, $S_L = 1\,\text{m}$ if $R = 40\,\text{m}$ and $S_L = 0.25\,\text{m}$ if $R = 350\,\text{m}$. For a very large radius, $S_L = 0$. In the example shown in Figure 2.18, the curved in-plan alignment of the bridge with $R = 700\,\text{m}$ required a variable transverse slope of the deck along the bridge length but no extra widening was required ($S_L = 0$).

The transverse alignment under a viaduct crossing a highway should take into consideration the following requirements:

- The dimensions of the carriageway cannot be reduced under the bridge;
- A minimum distance of 1.0 m should be adopted between the carriageway and the abutments or piers;
- If a pier is located in the inter-lane strip of a highway, between two or two sets of road traffic lanes, the minimum distance between the lanes and the pier shaft should be at least 2.0 m.

In overpasses, the vertical clearance under the bridge (to be measured as exemplified in Figure 2.19) should be at least 5.0 m. The horizontal clearances between pier shafts and the external lanes is usually 3.5 m, but in some cases, namely in urban bridges, this distance may be reduced when protections of pier shafts against vehicle impact are adopted.

In the case of bridges crossing railways, the minimum vertical clearances to accommodate the catenary is generally fixed at 7.25 m measured with respect to the Track Base Level (TBL). The minimum horizontal clearances between the tracks and piers or abutments of a viaduct is in the order of 3.0 m (see the design example in Chapter 8) but this distance depends on safety requirements related to risks of train accidental impact loads and is determined according to railway norms [5].

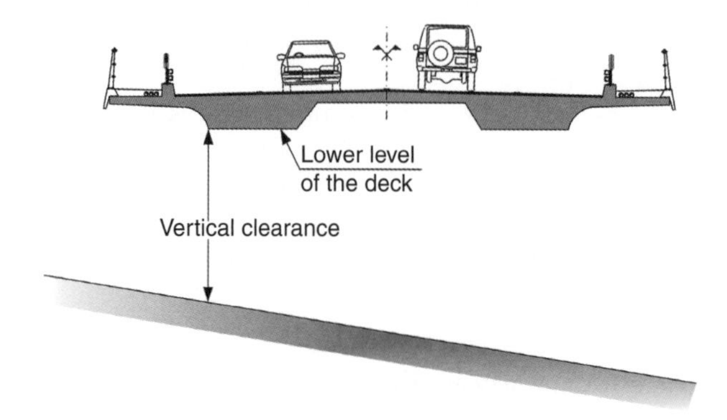

Figure 2.19 Vertical clearance of a bridge deck over a roadway with transverse slope.

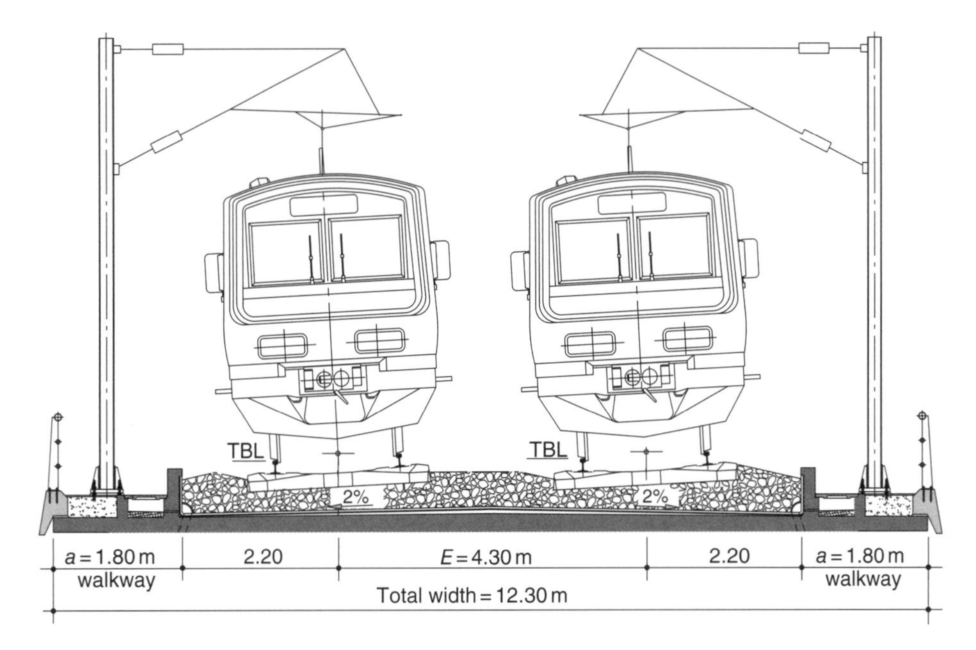

Figure 2.20 Example of Horizontal clearances in a double track railway deck.

In the case of railway bridges, the rules for establishing the deck cross width are generally fixed by the railway administration. In Figure 2.20, a deck with a double track with transverse slope due to the in-plan curved alignment is presented to exemplify the influence of the railway horizontal clearances on the bridge deck width. The minimum width, a, of the walkways is dependent on the design speed varying from 1.5 m for $V < 160\,\mathrm{km\,h^{-1}}$ to 2.0 m for $V > 160\,\mathrm{km\,h^{-1}}$. The minimum distance between the axis of any track and any rigid obstacle on the bridge should not be less than 2.15 m, in general. For a single track deck, a minimum width of approximately 8.0 m is usually required. In double track decks, the distance E between the track axes is 4.3 m, in general. A minimum deck width of 12.3 m is usually needed for a double track deck. The reader is addressed to UIC Norms in Ref. [5] for additional information on minimum horizontal and vertical clearances to be respected in rail bridges. Some additional information is given on Section 2.4.2.

2.4 Elements Integrated in Bridge Decks

2.4.1 Road Bridges

The elements to be integrated in the bridge deck (*bridge accessories*) should be established at preliminary design stages, since deck width and superimposed dead loads are influenced by those elements.

Bridge accessories for road bridge decks include:

- the surfacing and the deck waterproofing;
- the walkways, parapets and handrails;
- the fascia beams;

- the drainage system;
- the lighting system;
- the Expansion joints.

2.4.1.1 Surfacing and Deck Waterproofing

The surfacing of highway bridges with a concrete deck slab is generally made of a layer of bituminous concrete with a thickness between 50 and 80 mm. Nowadays, at least in some countries, it is usual to adopt a surfacing made of two layers – 30 + 50 mm or 40 + 40 mm, being the usual option – the wearing course, a layer with drainage characteristics in order to avoid 'aquaplaning' effects of the vehicles. This surfacing may be installed directly on the concrete surface if the environment is not very severe for durability of the concrete, as very often adopted in some countries in southern of Europe. However, in countries were icing salts are used for maintenance or for bridges located near a sea environment (often with risks of splashing of salt water) specific waterproofing membrane is mandatory.

A variety of deck waterproofing systems exists, but in short we may divide these into *waterproofing membranes* and *liquid systems* [1, 2, 6] The former, made of bituminous materials, polymer or elastomeric based membranes, are bonded to the surface in overlapping strips and require a bituminous protection layer; a total thickness – membrane + protective layer – of about 30–35 mm. The waterproofing membrane thickness is about 4–8 mm while the protective layer may range between 22 and 31 mm. As an example, one may specify:

- 20 mm for the waterproofing system;
- 20 mm for the bituminous protection of the waterproofing;
- 80 mm (40 + 40 mm) for the wearing and base course.

The so-called *liquid systems*, which may be identified as thin layers, are generally more costly and more complex to execute. Liquid systems are mainly acrylic, epoxy, polyurethane or bitumen based. The total thickness of the waterproofing system itself is no more than 1.5–3 mm.

2.4.1.2 Walkways, Parapets and Handrails

For the walkways, parapets, kerbs and handrails, a typical scheme for a concrete highway bridge is shown in Figure 2.18. The variety of details is quite large, changing from country to country and very often among highway administration departments within a country. Surveys of these bridge accessories are given in [1, 2, 7].

The effective width of walkways, measured from the kerbs to the internal surface of the handrail, should be at least 0.8 m, and is better at 1.0 m. In urban bridges with intense pedestrian traffic, the width of the walkway may reach 3 m or more. For highway bridges, where pedestrian traffic is limited to inspection and maintenance of the bridge, an inspection walkway is adopted with an effective minimum width of 0.6 m.

If the walkway is elevated with respect to the level of the wearing surface of the bridge deck, as shown in Figures 2.18 and 2.21, the walkway may be used to install services like electric and telecommunication cables inserted in 100 mm PVC ducts; precasted concrete slabs may be adopted for the wearing surface of the walkway. The other solution, generally better for durability, is to fill the walkway with lightweight concrete or sand stabilized by 6% of cement. In this case, the wearing surface of the walkway may be made of a simple mortar layer inclined at 2% to the road, as shown in Figure 2.21.

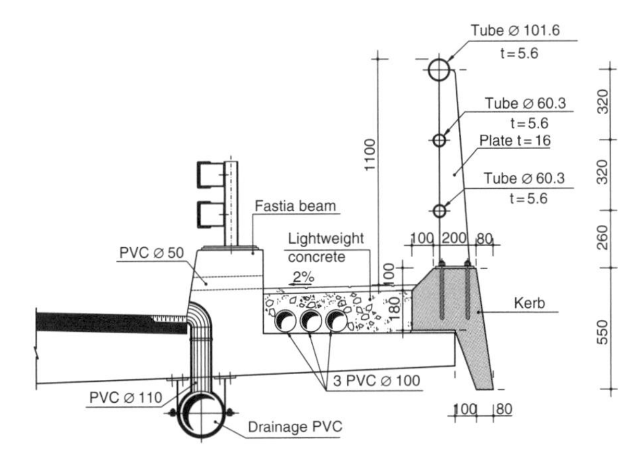

Figure 2.21 Walkways, guardrails, handrails, fascia beam and drainage system details, for a highway bridge.

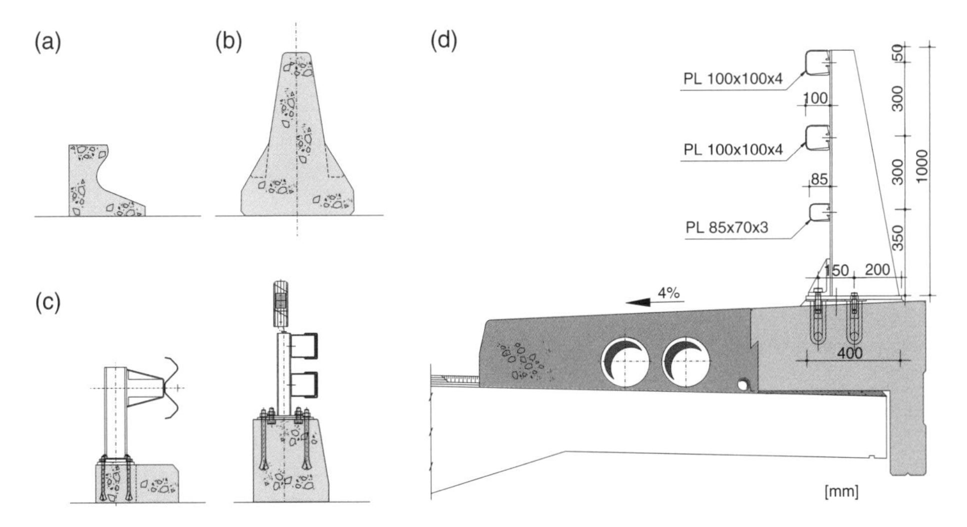

Figure 2.22 Guard rails for roadway bridges: (a) and (b) rigid; (c) flexible and (d) Guardrail type BN4.

For steel decks, the solution described for walkways should be modified by adopting, for example, prefabricated thin concrete slabs for the wearing surface of the walkway.

The solutions adopted for barriers for bridges, also designated as bridge parapets, are adopted for the protection of users. They may be divided in rigid and flexible, as illustrated in Figure 2.22. With a limited depth the rigid barriers, the *Trief* type (Figure 2.22a) for example, can only be adopted for limited design speeds like 60 km h^{-1} in urban bridges. As a central barrier, several bridges make use of *New Jersey* safety barriers (Figure 2.22b). These rigid road safety barriers, made of concrete, have an additional disadvantage due to their high weight, between 2.5 and 6.0 kN m^{-1}, compared to flexible barriers, which are made of steel weighing less than 0.5 kN m^{-1} (Figure 2.22c). Apart from the weight, flexible barriers are more efficient for energy dissipating by plastic deformation under vehicle impact loads. The distance between

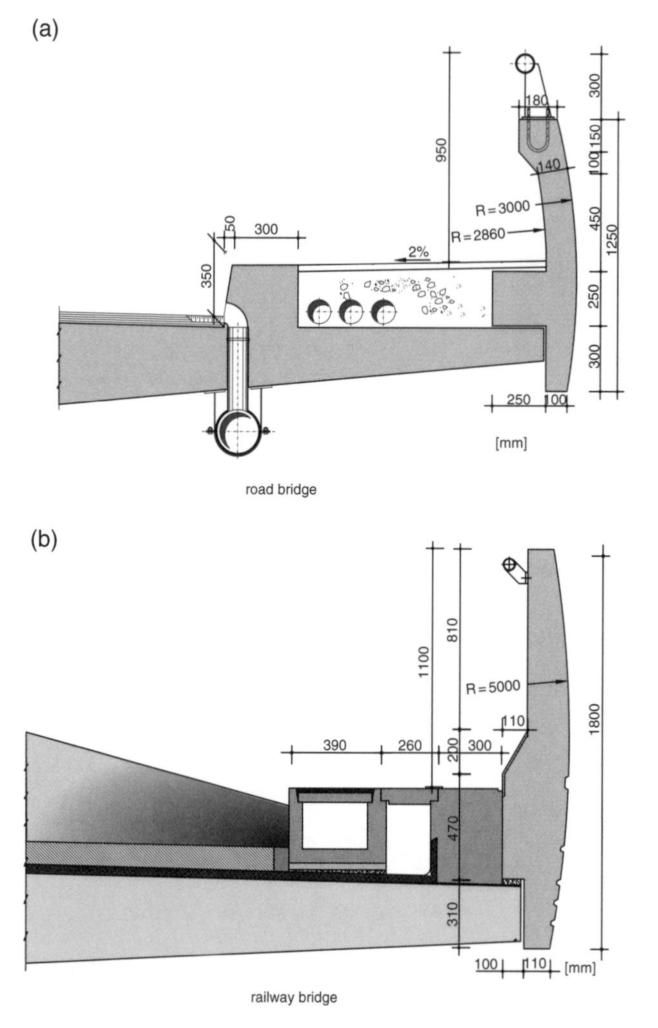

Figure 2.23 Walkway with a rigid vehicle and pedestrian parapet for a road bridge (a) and a railway bridge (b).

the vertical posts of the flexible road safety barriers on the bridge deck is usually 1.5 m. Figure 2.22d presents a typical parapet often adopted in some European countries and in Africa, designated BN4 [2]. Its function is twofold, since it works simultaneously as a barrier and hand rail. For geometrical details and fixation systems of guard rails, the reader is addressed to specific references on the subject as, for example, fabricators' catalogues.

In some bridges, namely in urban and rail bridges, the parapet function may be replaced by a concrete or steel barrier. It is made of a resistant concrete wall with a steel element on the top to complement its function of handrail as shown in Figures 2.23a and b. It was used frequently in USA and still is in some European countries, in spite of its large weight of approximately 6 kN m^{-1}, compared to modern steel barriers, weighing approximately 0.6–0.7 kN m^{-1}.

One may be in favour of concrete barriers due to their double function of providing safety and noise protective barriers. But noise emission in bridges is due to *structural noise* induced by the vibration of the superstructure. This occurs in particular in steel structures yielding *aerial noise* directly transmitted by the rolling stock on the wearing course. Only aerial noise may be controlled by the side barrier. At some bridge locations, noise environmental impact may require specific opaque noise barriers, negatively affecting the appearance of the bridge. If it is made of glass or any other transparent material, the appearance is much better. However, for noise absorption a thick opaque barrier may be required.

Handrails are made usually of steel elements or, more rarely, aluminium. When significant pedestrian traffic exists on the bridge, the maximum distance between vertical bars in handrails cannot be larger than 150–180 mm for child safety. The height of the handrail should always be higher than 1.0 m above the level of the walkway. For fixing the vertical posts of the handrail to the fascia beam, several typical details may be adopted as shown in Figure 2.23 (handrail/guardrail) or in Refs. [2] and [7].

2.4.1.3 Fascia Beams

The fascia beam, usually in precasted reinforced concrete, has a large impact on the aesthetics of the bridge. It is placed on the top of a mortar layer (Figures 2.24 and 2.25) with a thickness adjusted to give a continuous appearance to bridge longitudinal

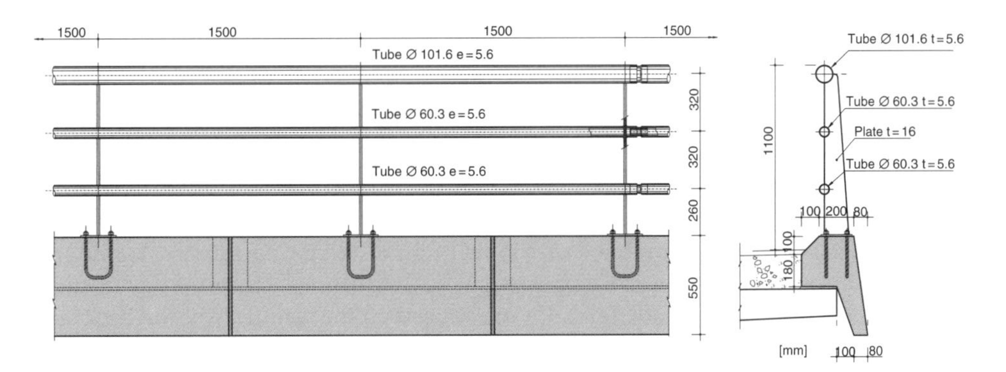

Figure 2.24 Handrail for an inspection footway.

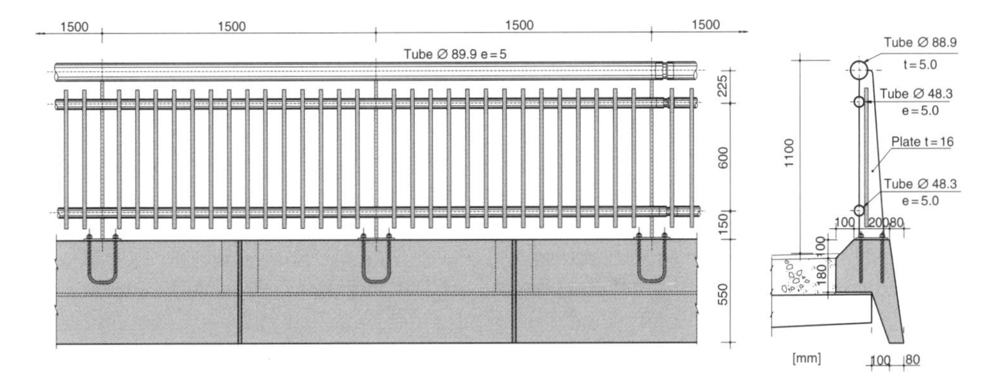

Figure 2.25 Handrail for a footway in a bridge with pedestrian traffic (urban bridge).

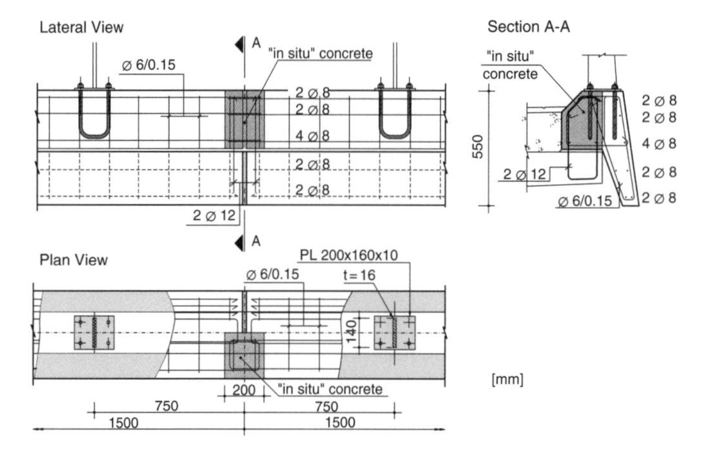

Figure 2.26 Example of the reinforcement and fixation of a fascia beam on a concrete bridge deck.

alignment, reducing the unavoidable vertical deviations of the theoretical alignment of the edge of the cast 'in situ' concrete deck slab. The shape of the fascia beam is mainly an architectural detail, its height being sufficient to cover the deviations of the concrete slab. The fascia beam is usually erected in 1.5–2.0 m pieces and its weight may vary between 1.5 and 3.0 kN m^{-1}.

The standard solution consists of adopting fascia beams in reinforced concrete, but nowadays there is a tendency to replace the steel reinforcement with fibres. A possible reinforcement detailing and fixation of the fascia beam to a concrete deck slab is shown in Figure 2.26. Sometimes, steel fascia beams may be adopted, as in the case of the viaduct in Figure 2.27.

2.4.1.4 Drainage System

The drainage system on the bridge deck should be predicted for the safety of traffic and for durability of the structure. As said before, the surface of the deck should have a minimum transverse slope of 2%, better 2.5%, introduced on the geometry of the deck slab and not by varying the thickness of the wearing and base course. The water is then conducted to the edges of the bridge deck and then by kerb drainage unities to the outside, if acceptable from an environment point of view, or alternatively may be collected into a pipe that is located under the deck overhang (a very inconvenient solution for aesthetics), under the deck but between girders in beam and slab bridges or at the inside of the superstructure in the case of a box girder bridge. These pipes conduct the water to the piers or to the abutments where it is collected into a general drainage system.

The decision regarding the distance between drainage unities at the deck level, as well as the diameter of the pipes for collecting the water, is one of the aims of the drainage study, and it depends on the deck geometry, namely its surface and in the rainfall at the bridge site. As a preliminary indication, the kerb drainage unities are located about 15–30 m apart if the longitudinal slope is at least 1%. The diameter of these pipes is at least 100 mm and for the longitudinal pipe the diameter may vary along the bridge axis, but it may be between 200 and 600 mm depending on the quantity of water to be collected. As an indication, the total section of the pipes collecting the water at the deck

Figure 2.27 A steel fascia beam for a steel-concrete composite deck – Viaduct over IC19 in Lisbon – design by GRID.

level may be in the order of 1/10 000 of the total deck surface. Some design examples are shown in Figures 2.18 and 2.21. Special attention should be paid to the detail of kerb drainage unities concerning their integration with the waterproofing system.

2.4.1.5 Lighting System

The lighting system consists of a series of lamp posts, usually located at the edges of the deck at the walkway or at the central strip. The electric wires are usually inserted in PVC tubes under the walkway. In cable-stayed bridges, lighting unities may be attached to the stay cables avoiding, for aesthetic reasons, visual crossing of posts and stay cables.

2.4.1.6 Expansion Joints

The EJs are located at the end sections of the bridge deck at the transition with the abutments, as well as at specific internal sections between parts of the bridge deck. This last case (EJ at internal sections) occurs if the deck is not fully continuous between abutments, as in the case of very long bridges or simply supported spans. EJs at internal intermediate sections should be avoided because they are one of the main critical issues for durability and maintenance.

In long highway bridges, it is usual to adopt maximum distances between EJs of 600–700 m. However, in some cases it may be interesting to go well beyond this distance, as in the example of (Figure 2.28) where the deck is fully continuous along 1260 m. The reduction of the number of intermediate EJs is usually beneficial for the reduction of maintenance problems in these elements, as well as for seismic behaviour, although one may expect an increasing cost for EJs required by long continuous decks. In fact, the

Figure 2.28 A bridge over the Guadiana River in Portugal, with a continuous prestressed concrete box girder deck 1260 m long (typical spans of 115 m built by cast in place balanced cantilever segments) – design by GRID.

main parameter affecting the cost of an EJ is the required displacement amplitude along the bridge axis. This amplitude is a function of the so-called *dilatation length*, L_d, defined as the distance between the fixed point and the section where the EJ is located. The fixed point is located at one of the abutments, at one of the piers or at a specific intermediate point to be determined in Chapter 7 as the *centre of stiffness of piers*. The thermal actions, with added shrinkage and creep effects for concrete bridge decks, are represented by a total equivalent temperature variation ΔT_{eq}, discussed in Chapter 3. The EJ should accommodate displacement amplitudes of

$$\Delta u = \alpha L_d \, \Delta T_{eq} \tag{2.7}$$

where α is the thermal expansion coefficient of deck ($\alpha = 10^{-5}$ per °C for concrete and composite steel-concrete decks, and 1.2×10^{-5} per °C for steel decks).

If the EJ is not perpendicular to the direction of the displacement, as in the case of skew decks, a displacement component in the direction of the EJ exists. The type of EJ should be able to accommodate this transverse displacement.

The selection of an EJ should take into consideration several criteria apart from movement capacity; namely, the bearing capacity under traffic loading, fatigue resistance, maintenance requirements and noise emission.

A variety of criteria are adopted to classify EJs in terms of expected movement amplitudes. Small movements, according to [8], are usually identified as up to 25 mm and large movements above 80 mm; those between 25 and 80 mm are considered medium movements. However, these limits are extended according to other criteria; large movements being considered as movements above 130 mm. The aim of these criteria is to define, for each range of movement, the most convenient type of EJ. With the present tendency for increasing the continuity of bridge decks, the requirements for EJs for medium to large amplitudes have been increased.

EJs for small movements can be reduced to a simple compression seal elastomeric joint that is held in place against smooth concrete faces or steel elements by friction. So, it is necessary to evaluate from the joint uncompressed width, w, and the expansion gap width at installation, w_{inst}, the values of the maximum and minimum gap widths, w_{max} and w_{min}, to check if these values are within the allowable values specify by the fabricator. As indicative values, one may take $w_{max} = 0.85\ w$ and $w_{min} = 0.4\ w$. Specifications for installation of the joints are provided by the fabricator [1, 2, 8], but at 20°C it is generally assumed the width of the seal is about 60% of the uncompressed width. Shear displacements in this type of joints occurring in skew bridges are generally limited to values of about 20% of the uncompressed width.

For medium displacements, a variety of types of EJs can be adopted [8] but one of the most common type is shown in Figure 2.29. The so-called *bolt down panel joints* or *elastomeric cushion joint* are reinforced neoprene plates where the movement's amplitudes are accommodated at gap widths at the neoprene panels. One of the disadvantages of this type of joint is the possibility of loosening and breaking the bolts and nuts holding the joint against the bridge deck under high speed traffic.

In the *elastomeric strip seal EJ* (Figure 2.30) the movements are accommodated by the folding of the elastomeric V-shaped sealing, which is fixed to steel elements anchored to the concrete deck. The disadvantage of this type of EJ is of course the possible accumulation of debris in the elastomeric elements. The use of a steel cover plate on the top of the sealing element to avoid this is not, in general, a satisfactory solution due to the risk of noise emission, corrosion and hindering accessibility to the seals.

Finally, for medium displacement amplitudes one can adopt *steel finger joints*, also called toothed joints, as shown in Figure 2.31. The main disadvantages are the limited capacity for accommodating displacements in the cross-wise direction, as well as

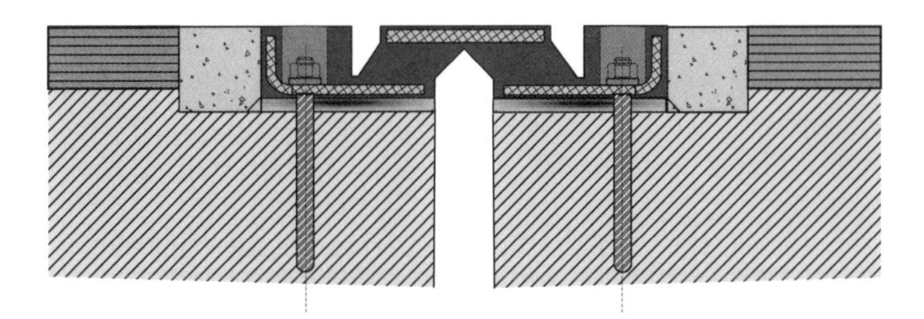

Figure 2.29 Elastomeric expansion joint.

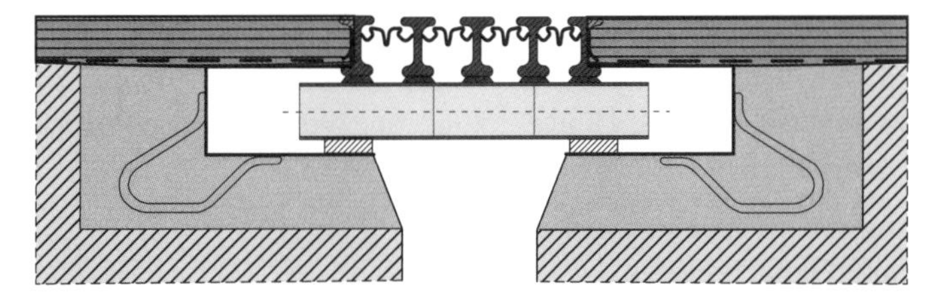

Figure 2.30 Elastomeric strip seal expansion joint.

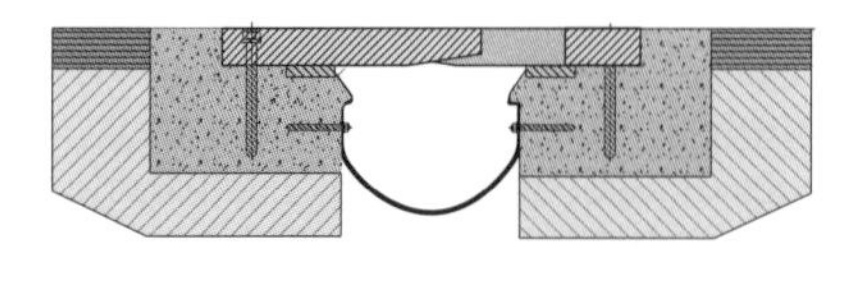

Typical section

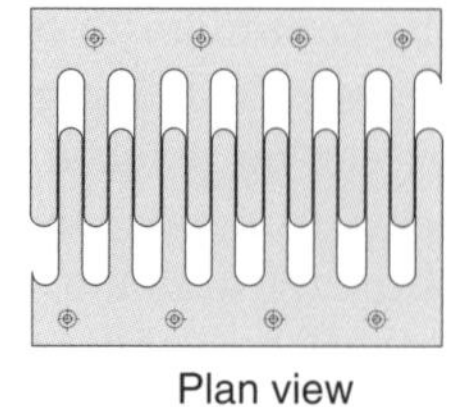

Plan view

Figure 2.31 Steel finger type expansion joint.

Figure 2.32 Railway Bridge over São Martinho River, in Portugal, with a prestressed girder deck 852 m long (typical spans of 28.4 m) – design by GRID.

possible issues for traffic due to small differential vertical deflections and rotations across the joint.

Large movement EJs include the previously referred to elastomeric EJs, or even the steel finger joints, to the so-called *multiple seal EJs*, which can accommodate displacements up to 400 mm or even more. The main concept for this type of EJ is the coupling of sealing elastomeric elements with steel rail elements. Additional springs or linkage elements to guarantee equal gap widths between sealing elements avoiding any overextending effects may be adopted at least for large movement joints. Details of this type of EJ may be found in [8].

For very large displacements such as the ones generally occurring in long span bridges, like cable-stayed or suspension bridges, very special EJs are required. Figure 5.22 shows the Tagus suspension bridge in Lisbon where the EJs are able to accommodate displacements of up to 1500 mm, including seismic displacements.

2.4.2 Railway Decks

The elements integrated in railway bridge decks have some differences compared to those discussed in the case of road bridges. Hence, taking the typical railway bridge deck shown in Figure 2.32 one has:

- the track system, including the rails, ballast and sleepers;
- the power traction system (catenary system);
- the footways, parapets/handrails, drainage and lighting systems;
- signalling and telecommunication cables, pipes and other equipment.

2.4.2.1 Track System

The standard rails are UIC 60 rails weighing 60 kg m^{-1} but other rail types like UIC 54 or UIC 70 may also be adopted. Nowadays, rails are *Continuous Welded Rails* (CWR) made by welding rail segments and avoiding any joints along the track, except if the *track-structure interaction effects* require some *Rail Expansion Devices* at the bridge abutments or at some internal sections. The reader is referred to [9, 10] for specific track-structure interaction effects in rail bridges requiring rail EJs.

The cross section of a rail and its direct fixation to the sleepers is shown in Figure 2.33. The track is inclined at 1/20 to the inside. The standard *track gauge*, that is, the distance between the inner faces of the rails, is 1435 mm for most European railways. As the top surface of the rail is approximately 70 mm, the distance between axes of the rails is 1505 mm. In double track railways the minimum distance between the centrelines of the track axes is 3.4 m, although a larger distance is usually adopted, as shown in Figure 2.20. The minimum distances between the rails and any other element integrated in the deck, like parapets, catenaries posts or retention ballast walls, have to be established at the preliminary design stages of the bridge, because the total deck width is a function of minimum required clearances. The boundary including all the necessary clearances should be established at the commencement of the design works and defines the so-called *structure gauge*. It depends on the railway's safety specifications as a function of the track geometry and design speed. If the bridge is inserted in a curved alignment, the railway track is usually super elevated to compensate for centrifugal forces, as already referred to for road bridges. This results in an increased level of the outer tracks with respect to inner tracks; that value being known as *the cant* (Figure 2.20).

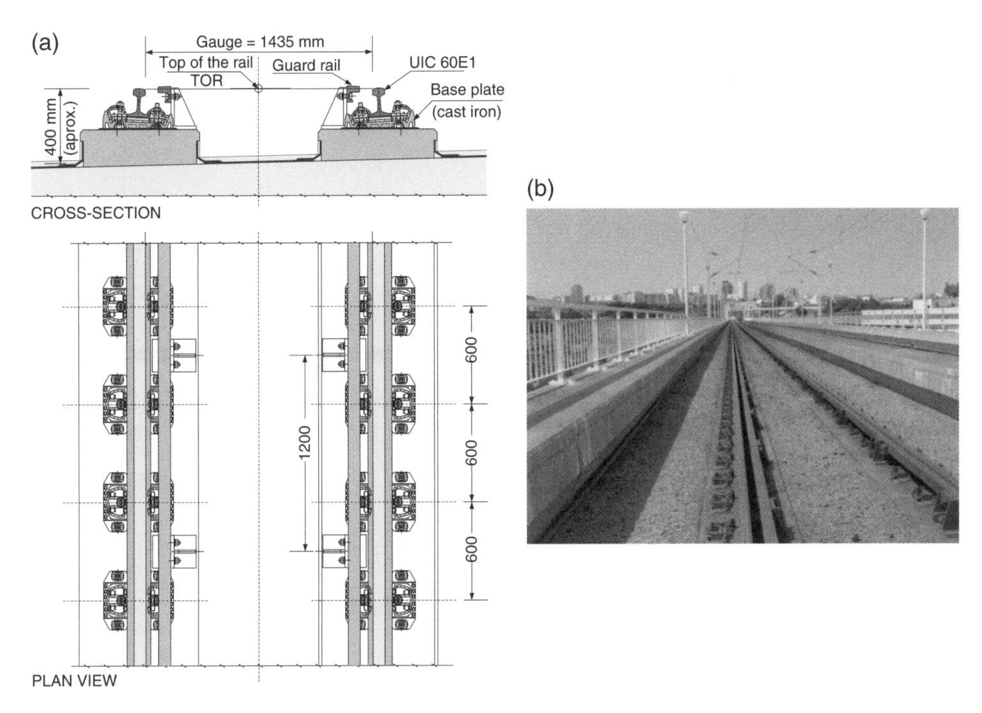

Figure 2.33 (a) Cross section and plan view of a rail with direct fixation to the sleepers and (b) view of the rail track at São João railway Bridge in Oporto.

The ballast provides a resilient bed for the sleepers and is made of crushed stones in 50–65 mm pieces well compacted under the rails. It weighs approximately 20 kN m^{-3} in the case of granite or gneiss stones; its thickness underneath the track is at least 300 and usually 400 mm for High Speed Railways. The average thickness of ballast in a railway deck is approximately 0.5 m. The ballast lies on a waterproofing membrane over the deck slab.

For sleepers, which are placed at 60 cm apart along the rail, the most common solution nowadays is using precasted prestressed concrete sleepers of a mono- or duo-block type. The dead weight for prestressed concrete sleepers with track fastenings is approximated at 4.8 kN m^{-1}. For a double track railway bridge, the dead weight of four rails UIC 60 and two mono-block sleepers is approximately 12 kN m^{-1}.

Wooden sleepers, that are much lighter than concrete sleepers, are made of hard and heavy types of wood like azobe (1.1 kN m^{-3}), these have been adopted for use for a long time and are still used frequently in North America. Of course, the drawbacks of wooden sleepers for bridges compared to concrete sleepers are the durability and high cost. The simplest alternative option is to replace the wooden sleepers by other composite materials that have recently become more accepted, such as *Fibre-reinforced Foamed Urethane* (FFU) presenting mechanical characteristics comparable to wood but lighter than of dense timbers, such as azobe.

The additional dead load on a bridge deck due to a ballasted track is quite large compared to the structural dead weight. For long span bridges, instead of ballasted tracks, one of the following systems may be envisaged:

- Direct fixation on concrete stringers;
- Direct fixation on concrete slab systems;
- Embedded rails.

In Figure 2.34, some examples of these systems are shown. For further details on ballastless systems for bridges, the reader is referred to specific literature [12]. For safety reasons, *guard rails* are very often adopted on bridge decks. The guard rail is usually fixed to the sleeper, concrete stringer or concrete plate in a direct fixation system.

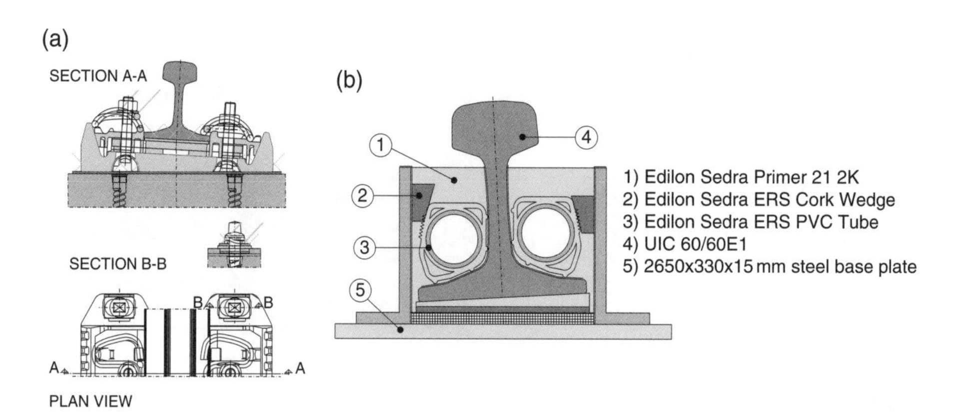

Figure 2.34 Ballastless systems for railway bridges: (a) direct fixation on concrete stringers and (b) embedded rail. (*Source:* UIC, adapted from Ref. [11]).

2.4.2.2 Power Traction System (Catenary System)

The most common type of power traction system is by *overhead line electrification* (Figure 2.20). The electric cables stand as catenaries between vertical posts fixed at the deck level and usually spaced at 20–25 m and require extra width to be inserted in the deck slab, which this should be accounted for when establishing the required width for the bridge deck. The example shown in Figure 2.20 illustrates the required extra width. The loads coming from these posts, namely under-wind loading, should be considered for design of the overhangs of the bridge deck slab.

Electrical bonded devices are required, in particular for steel bridges, to take into consideration traction current return rails or earth wires. Even in concrete bridges, special consideration should be given in railway bridges to the electrical bonding of individual structural components.

2.4.2.3 Footways, Parapets/Handrails, Drainage and Lighting Systems

Although all the aspects related to footpaths, kerbs, handrails, fascia beams, drainage devices and lighting systems may have a strong impact when establishing the cross width required for the bridge, it should be noted these issues are not so different from the ones discussed for road bridges. The reader is addressed to some of the requirements in railway norms such as the UIC recommendations in Ref. [11] for additional information on these aspects.

References

1 Parke, G. and Hewson, N. (2008). *ICE Manual of Bridge Engineering*, 2e. Thomas Telford Lim.

2 Calgaro, J. A. (2000). Projet et construction des ponts – Généralités, Fondations, Appuis et Ouvrages courants. Presse des Ponts et Chaussées, Paris.

3 Neil, C.R. ed. (1973). *Guide to Bridge Hydraulics*. Roads and Transportation Association of Canada, University of Toronto Press.

4 Matias Ramos, C. (2005). *Drenagem em infra-estruturas de transportes e hidráulica de pontes – Ed*, 262 pp. LNEC.

5 UIC Code 777-1 (2002). *Measures to Protect Railway Bridges Against Impact from Road Vehicles, and to Protect Rail Traffic from Road Vehicles Fouling the Track*, 2e, 16 pp. International Union of Railways.

6 Fascicule N°67: Titre I (1985). L'étanchéité des ponts routes support en béton de ciment. Ed. CCTG – Cahier des clauses techniques générales, 104 pp.

7 SETRA (1997). Garde-corps – Collection du Guide Technique GC. Ed. Service d'Études Techniques des Routes et Autoroutes, 118 pp.

8 Ramberger, G. (2002). *Structural Engineering Documents SED 6: Structural Bearings and Expansion Joints for Bridges*. Zurich, Switzerland: IABSE, International Association for Bridge and Structural Engineering.

9 UIC Code 774-3-R (2001). *Track/Bridge Interaction – Recommendations for Calculation*, 2e, 70 pp. International Union of Railways.

10 EN 1991-2 (2005). *Eurocode 1: Actions on Structures – Part 2: Traffic Loads on Bridges*. Brussels: CEN.

11 UIC (2008). *Recommendations for Design and Calculation of Ballastless Track – Ed*, 40 pp. ETF.
12 Ryjáček, P., Howlader, Md M. and Vokáč, M. (2015). The Behaviour of the Embedded Rail in Interaction with Bridges. IOP Conference Series: Materials Science and Engineering. Vol. 96. No. 1. IOP Publishing. Doi: 10.1088/1757-899X/96/1/012052.

3

Actions and Structural Safety

3.1 Types of Actions and Limit State Design

In bridge design, a variety of design actions should be considered at the construction stage or at the final stage (final static system). Construction loads are associated with equipment like *formwork launching girders* in span by span construction or *moving scaffoldings* in cantilever construction, as explained in Chapter 4. Under service conditions, the bridge is subjected to permanent actions (dead loads (DL), creep and shrinkage effects) and variable actions, like traffic loads, thermal, wind and earthquake actions, friction at the bearings or settlements of the foundations. Accidental actions may be considered at the final stage of the bridge or at the construction stages.

Nowadays, bridge design verifications are usually based on a semi-probabilistic approach yielding a *limit state design format*. This is the approach adopted by most design codes, namely the Eurocodes and specifically Parts 2 of Eurocode 2 (denoted by EC2-2 [1]) for concrete bridges, Eurocode 3 (denoted by EC3-2 [2]) for steel bridges, and Eurocode 4 (denoted by EC4-2 [3]) for steel-concrete composite bridges. Loads are quantified by *characteristic values*, except for accidental actions usually defined on the basis of *nominal values*. A characteristic value is a value with a certain probability of not being exceeded; say 5%, as illustrated in Figure 3.1. Actions are classified as *permanent actions*, *variable actions* and *accidental actions*.

At any stage, in service or during construction and according to a limit state design format, the safety of the structure should be verified at *Ultimate Limit States* (ULS) and at *Serviceability Limit States* (SLS).

At ULS, safety should be checked for resistance, fatigue and stability under *design values* of the permanent and variable loads.

At SLS, design verifications are related to limits on the stresses and deflections, on vertical accelerations under traffic loading and crack widths in partial prestressed or reinforced concrete structures. In Europe, as in many other parts of the world, bridge safety is verified according to the Eurocodes [1–5]. The limit state design format at ULS is defined for fundamental load combinations, as

$$\sum_{j\geq 1}\gamma_{g,j}\,G_{k,j}+\gamma_q P+\gamma_{q,1}\,Q_{k,1}+\sum_{i>1}\gamma_{q,i}\psi_{0,i}\,Q_{k,i} \tag{3.1}$$

where $\gamma_{g,j}$ and γ_p are the partial safety factors of each of the permanent actions and prestressing, defined by its characteristic values $G_{k,j}$ and P; $\gamma_{q,i}$ are the partial safety factors

Bridge Design: Concepts and Analysis, First Edition. António J. Reis and José J. Oliveira Pedro.

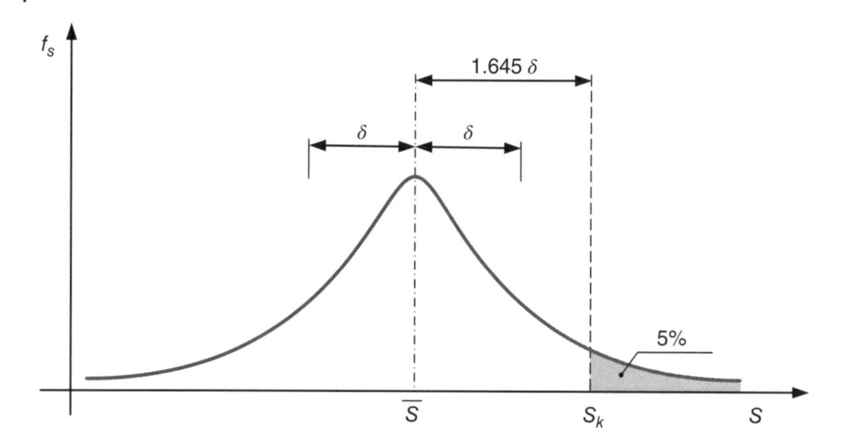

Figure 3.1 Concept of characteristic value $\boldsymbol{S_k}$ of an action $\boldsymbol{S}$. Example of an action with a normal distribution (density $\boldsymbol{f_s}$), average value $\boldsymbol{\overline{S}}$ and a standard deviation $\boldsymbol{\delta}$.

of the variable loads, involving the characteristic value $Q_{k,1}$ of the basic variable action and the combination of the other $(n-1)$ variable actions defined by its reduced values, $\psi_{0,i}\,Q_{k,i}$. In the following sections, characteristic load values are defined and examples of $\psi_{0,i}$ are given for road and railway bridges. A complete set of values for design practice may be taken from Annex A2 of Eurocode 0 (denoted as EC0 [4]).

Load combinations at serviceability limit states, accordingly to EC0, are defined for

$$\text{Characteristic (rare) combination} \quad \sum_{j\geq 1} G_{k,j} + P + Q_{k,1} + \sum_{i>1} \psi_{0,i}\,Q_{k,i} \tag{3.2}$$

$$\text{Frequent combination} \quad \sum_{j\geq 1} G_{k,j} + P + \psi_{1,1} Q_{k,1} + \sum_{i>1} \psi_{2,i} Q_{k,i} \tag{3.3}$$

$$\text{Quasi permanent combination} \quad \sum_{j\geq 1} G_{k,j} + P + \sum_{i\geq 1} \psi_{2,i} Q_{k,i} \tag{3.4}$$

where the values of the permanent G, prestressing P and variable actions Q are quantified adopting the reduction coefficients ψ_0, ψ_1 and ψ_2 for each variable action (i). A characteristic combination is expected to be reached (or exceeded) on a structure with a very low probability, representing in terms of duration, with respect to the design life of the structure, no more than a few hours. A frequent combination may represent a total duration, Σt_{1i}, of 5% of the design life of the structure, while the number of times reached by a quasi-permanent load combination, may represent a total duration, Σt_{2i}, of 50% of the design life of the structure. In Figure 3.2 this general concept is illustrated.

One important concept for bridges is the *return period*, T, of a variable action. The return period may be related to annual probability, p, of the occurrence of a variable action. For example, the wind loading is usually defined with an annual probability

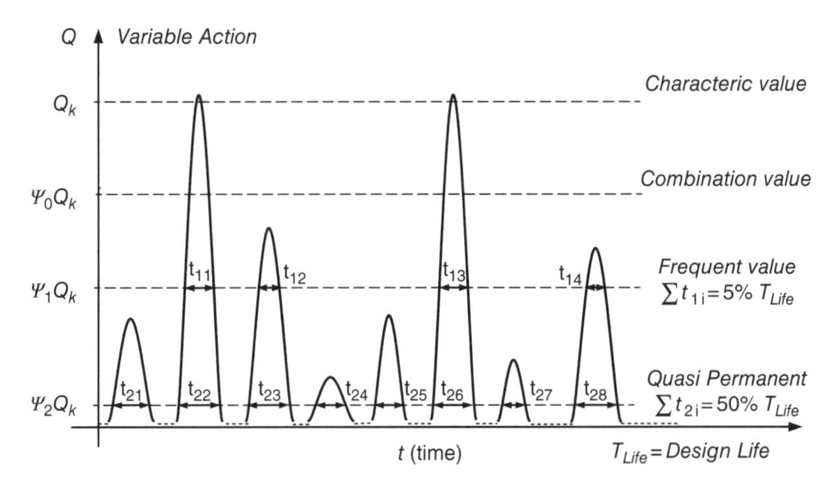

Figure 3.2 Concept of characteristic, combination, frequent and quasi-permanent value of a variable action Q.

of 2% of being exceeded. A variable action is reached during a certain period R (1 year, 10 years etc.) with a certain probability, p. Approximately, one may take $T = R/p$, which refers to the characteristic values of wind loading or thermal loading that are usually defined with a probability $p = 2\%$ being reached in 1 year, one has $T = 50$ years. Just for comparison, the return period of the characteristic value of the so called *Load Model 1* (LM1) in a road bridge, according to EC1-Part 2, corresponds to a probability of 10% of being reached in 100 years, which means a return period $T = 100/0.1 = 1000$ years.

Finally, one refers to the accidental load combinations, like the impact on bridge piers, defined from the *nominal values* (not defined on a probabilistic basis) of the accidental loads A_d as

$$\text{Accidental combination} \quad \sum_{j\geq 1} G_{k,j} + P + A_d + \left(\psi_{1,1} \text{ or } \psi_{2,1}\right) Q_{k,1} + \sum_{i>1} \psi_{2,i} Q_{k,i} \tag{3.5}$$

3.2 Permanent Actions

These actions include:

- Dead loads – DL, due to the dead weight of structural materials;
- Superimposed dead loads – SDL, from all non-structural materials, like road surfacing, waterproofing, footways, kerbs, parapets, services, lighting posts and any other materials and equipment included in the bridge;
- Permanent action effects due to prestressing or due to materials like imposed deformations due to creep and shrinkage of concrete or even permanent imposed deformations due to differential settlements.

Road Bridge: Example

Let us consider the standard concrete highway box girder deck from Figures 2.6 and 2.18. The DL and SDL, for structural analysis of the longitudinal model under DL, are:

1) *Deck DL:*

$$\text{At support section}\quad A = 13.66\,\text{m}^2\ \ DL = A\gamma_c = 341.5\,\text{kNm}^{-1}$$

$$\text{At mid span section}\quad A = 9.51\,\text{m}^2\ \ DL = A\gamma_c = 237.8\,\text{kNm}^{-1}$$

where A and γ_c = 25 kNm^{-3} are, respectively, the cross-sectional area and the unit weight of normal concrete.

2) *SDL:*

Bituminous layer (total 8 cm)	45.00 kN m^{-1}
Sidewalks (light concrete)	11.25 kN m^{-1}
Kerbs	6.80 kN m^{-1}
Handrails	1.50 kN m^{-1}
Safety steel railings	1.00 kN m^{-1}
Fascia beams	5.00 kN m^{-1}
Central 'New Jersey' barrier	7.50 kN m^{-1}
Drainage pipes	0.45 kN m^{-1}
Total SDL =	78.50 kN m^{-1}

3) *Permanent actions due to other effects*, such as creep and shrinkage or differential settlements, are detailed in specific sections of this chapter.

Rail Bridges: Example

Permanent loads in rail bridges are usually much higher than for road bridges. Most of the increase comes from the need to have a stiffer bridge deck and from the increase in SDL associated to track equipment, like the ballast. The following examples, a prestressed concrete viaduct and a steel concrete composite viaduct (both have double tracks as shown in Figures 3.3 and 3.4), illustrate the main SDL for rail bridges.

Due to different requirements for deck clearances and equipment for normal traffic rail bridges and High Speed Rail (HSR) bridges, associated SDL may differ due to width and thickness of ballast resulting from a variety of requirements, namely minimum distances to tracks, inspection walkways and location of catenary posts.

The unit weights and DL for track equipment evaluated from Annex A of Part 1-1 of Eurocode 1 (referred as EC1-1-1 [6]) are as follows for the cases of single track or double track railway decks:

1) *Ballast, track and sleepers*
   Ballast (Annex A of EC1-1-1) $\gamma = 20$ kN m^{-3} and average thickness of 0.50 m
   Width of ballast container: 9.30 m double track for HSR, that is, $9.3 \times 0.50 \times 20 = 93$ kN m^{-1}
   4.30 m for single track (normal traffic), that is, $4.3 \times 0.50 \times 20 = 43$ kN m^{-1}
   'Monoblock' sleepers + 2 rails UIC (*Union Internationale de Chemins de Fer*) 60 = 6.50 kN m^{-1} per track
2) *Water proofing layer with bituminous protection (30 mm)*
   Average thickness = 60 mm
   Unit weight = 1.10 kN m^{-2}

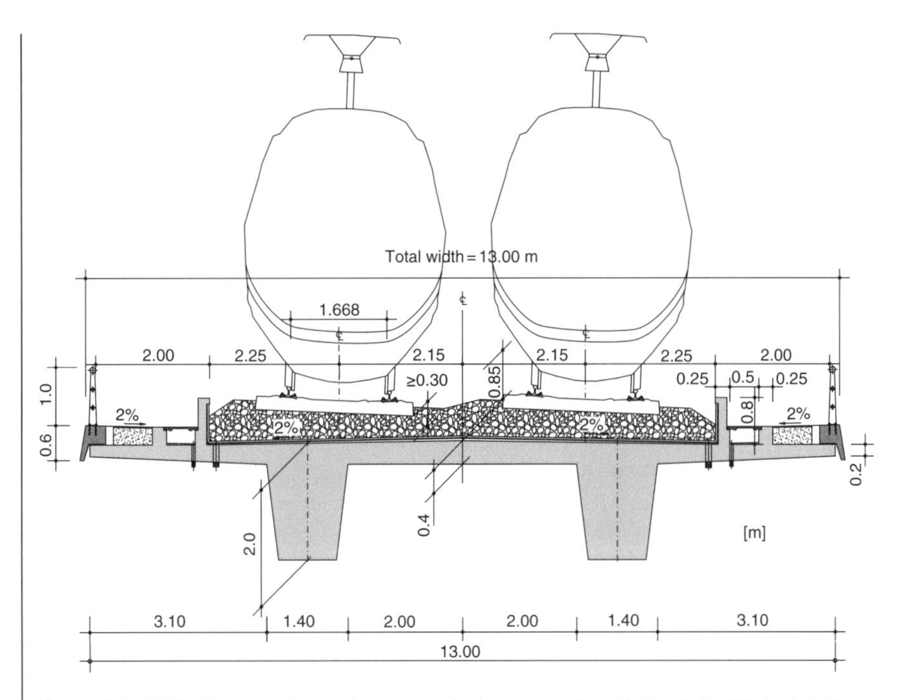

Figure 3.3 Rail bridge, prestressed concrete deck cross section: S. Martinho Viaduct, in the southern line, Portugal – design by GRID.

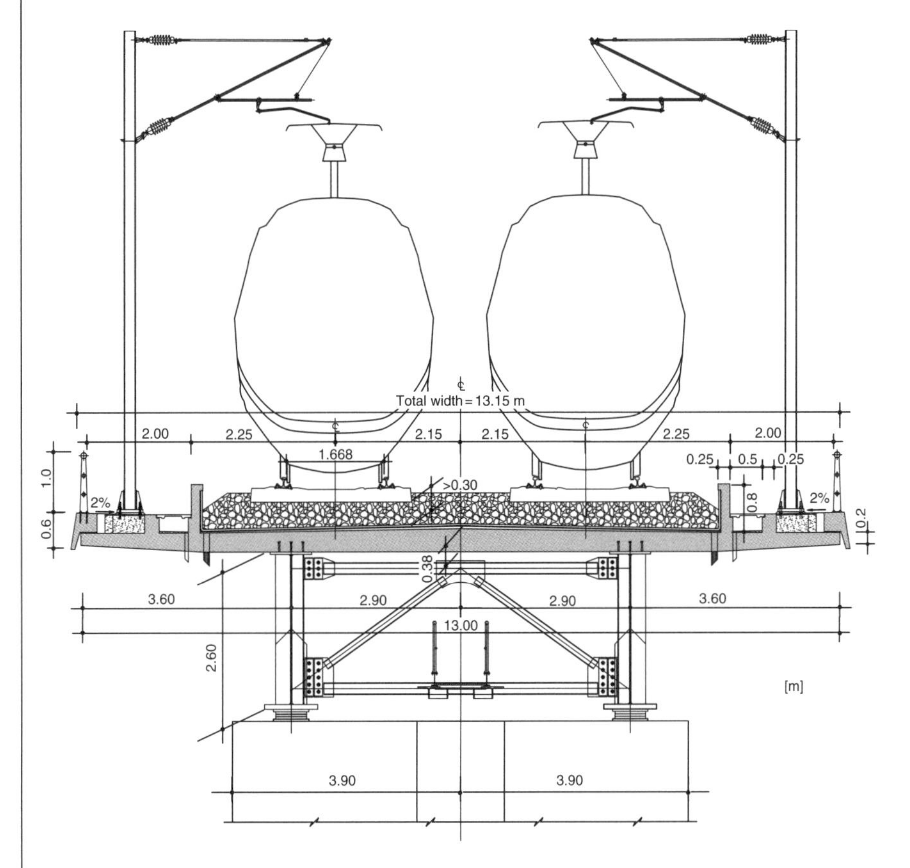

Figure 3.4 Rail bridge, composite deck cross section: bridge over the River Sado, Portugal. Approach viaducts – design by GRID.

3) *SDL of water proofing layer*:
   HSR double track railway decks: 10.3 kN m^{-1}
   Normal traffic single track railway decks: 4.8 kN m^{-1}
4) *Catenary, posts and signs*
   Post and catenary: 10.2 kN per post
   Concrete blocks for posts: 0.85 kN per post

   SDL (for 30 m distance between posts): 0.90 kN m^{-1} for HSR double track railway decks 0.45 kN m^{-1} for normal traffic single track railway decks.

For comparison, for a double track and considering the DL of walkways, concrete walls for the ballast container, fascia beams and hand rails, one may easily reach values for the total SDL in the order of 150–200 kN m^{-1}; in a road bridge with four traffic lanes one may have values of 40–60 kN m^{-1}; that is, approximately only 30% of the SDL of a double track rail bridge.

3.3 Highway Traffic Loading – Vertical Forces

These actions include the loads due to traffic, giving rise to static vertical and horizontal forces, including its dynamic load effects. Just a few years ago, considerable differences between national codes existed and different load models were adopted, even among most European countries. That was the case, for example, between the UK British Standard, the French instructions, the Spanish instruction or the German DIN (Deutsches Institut für Normung e.V.) code. Most of these codes specify traffic loading quite differently from the American AASHTO (The American Association of State Highway and Transportation Officials) – the standard specification for highway bridges. However, most of the codes were based on loading models specifying standard lane and truck loadings. In the last few years, Europe has made an attempt to harmonize traffic loading on bridges. The result was Eurocode 1- Part 2 (EC1-2 [5]), which will be considered here.

Today, highway loading is defined in the Eurocodes based on probabilistic models. The basic definition of traffic loading takes into consideration the categories of the vehicles using the bridges, namely the number, location and average load per axle, the level of loading at each vehicle, namely full or partially loaded, as well as the distance between vehicles. All these variables are characterized by statistical data resulting from measurements of traffic.

One relevant aspect is the influence of the length of the bridge on the design traffic loading. It is understood, for example, if the traffic loading is modelled by a load per square metre, for a suspension bridge with 1000 m of main span, that may not be reasonable modelling traffic loading for a bridge with a typical span of 40 m. The probability of having, for example, the full span and all the lanes loaded with a specified uniformly distributed load (UDL) is quite different between the two cases.

Road traffic load models do not attend to describe actual loads but load effects, such as internal forces and displacements and induced actual traffic loadings. These load models may include directly the dynamic load effects of the traffic, as in EC1-2 [5].

In the EC1-2, the load models are defined on the basis of the so-called *notional lanes*, defined as lanes of 3 m width, as shown in Figure 3.5. The number of notional lanes, n_l, in the full width w of the carriageway defined, for $w > 6$ m, is the greatest possible integer obtained as Int $(w/3)$. The width of the remaining carriageway is defined as $w - 3$ Int $(w/3)$ for $w \geq 5.4$ m. For $w < 5.4$ m, $n_l = 1$, and for $5.4\,\text{m} \leq w < 6\,\text{m}$, $n_l = 2$.

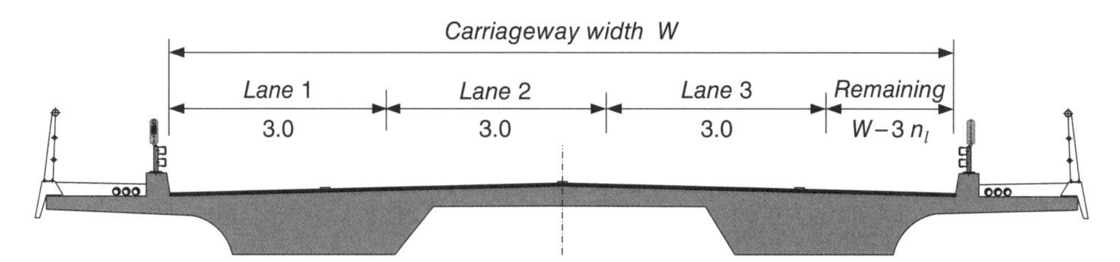

Figure 3.5 Definition of notional lanes, according to EC1-Part2 [5].

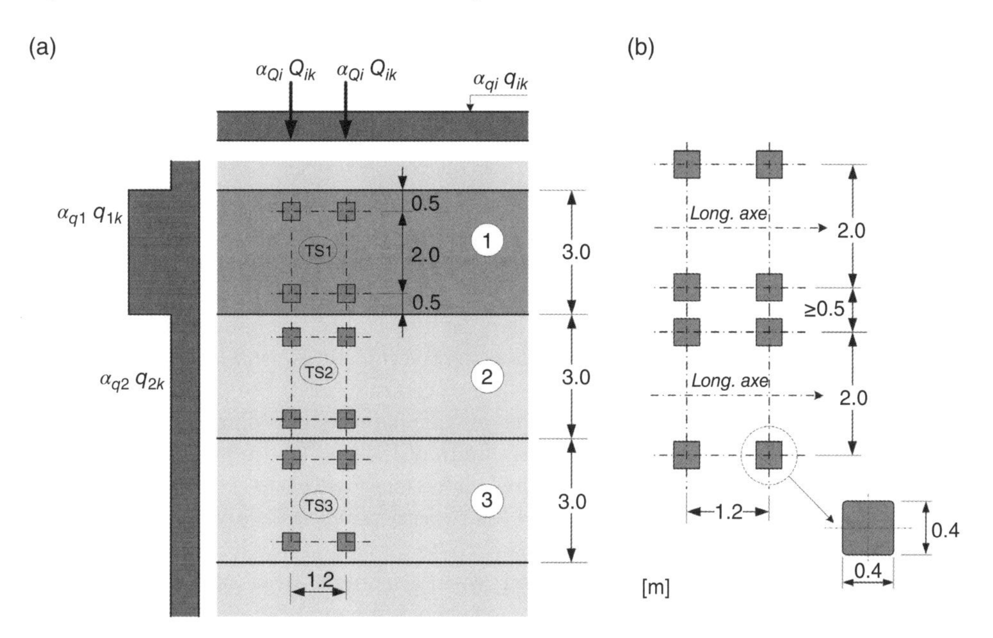

Figure 3.6 Load Model 1, LM1, in EC1-Part 2 (*Source:* Adapted from Ref. [5]): (a) Tandem System TS and Uniform Distributed Loads UDL; (b) geometrical definition of the TS for the verification of short structural members.

At a bridge deck with two parts, separated by a fixed central barrier, each part including all hard shoulders or strips is divided into sequentially numbered notional lanes, but only one numbering should be used for the whole carriageway. However, the notional lanes are not necessarily adjacent and the number of loads, their location on the carriageway and their numbering should be chosen to produce the most adverse effects from the load models.

The following load models, with loads specified as characteristic load values and including the dynamic amplification factors, are considered in EC1-2 [5]:

LM1 – covering most of the effects of the traffic of lorries and cars and intended for general and local verifications. It is defined by a UDL of $9\,\text{kN}\,\text{m}^{-2}$ on the Notional Lane 1 and by a double axel concentrated loads (tandem system) of $2 \times 300\,\text{kN}$, centred at the first notional lane, as shown in Figure 3.6. In the second and third notional lanes, the UDL is reduced to $2.5\,\text{kN}\,\text{m}^{-2}$ and the tandem systems are reduced to 200 and 100 kN per axle, as shown in Figure 3.6 and in Table 3.1. These values should be multiplied by α_{Qi}, α_{qi} adjustment factors for different lanes, defined by national authorities but not smaller that 0.8.

Table 3.1 Load model 1: characteristic values according with EC1-2 [5].

	Tandem system TS	Uniform distributed load UDL
Location	**Loads per axe Q_{ik} (kN)**	**q_{ik} (or q_{rk}) (kN m^{-2})**
Notional lane 1	300	9.0
Notional lane 2	200	2.5
Notional lane 3	100	2.5
Other lanes	0	2.5
Remaining area (q_{rk})	0	2.5

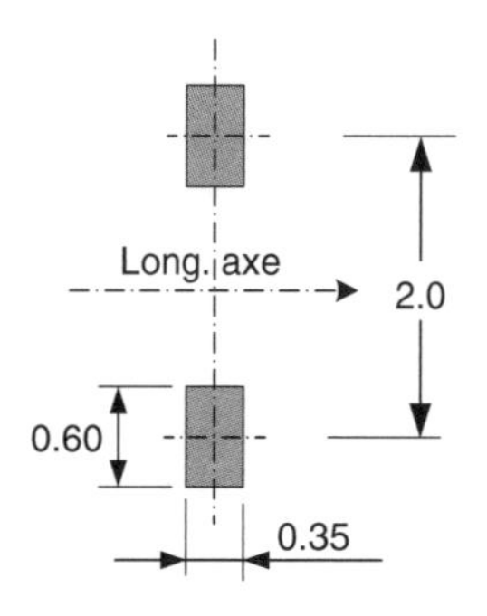

Figure 3.7 Load Model 2, LM2, in EC1-Part 2 (*Source:* Adapted from Ref. [5]).

The contact area of each wheel (Figure 3.6b) is $0.4 \times 0.4\,\text{m}^2$. In all the remaining area, a UDL = $2.5\,\text{kN}\,\text{m}^{-2}$ is applied; in all other possible existing lanes this UDL may be neglected, as well as the tandem system (TS) in the third notional lane, under the responsibility of the competent authority. LM1 was defined and calibrated in order to be usable for both general and local verifications. For general verifications, as mentioned earlier, the tandem systems travel centrally along the lanes, but for local effects, two tandems belonging to two different lanes can be closer with a minimum distance of 0.50 m between the axis of two neighbouring wheels (see Figure 3.6b).

Load model 2 (LM2) – with a single axle load, for local effects, equal to 400 kN to be applied at any location on the carriageway and with a contact area of each wheel of $0.35 \times 0.60\,\text{m}^2$, as shown in Figure 3.7.

Load model 3 (LM3) – to be taken into account when one or more of the standardized models is specified by the relevant authority, defined as conventional classes of special vehicles, corresponding to usual abnormal loads and with characteristic loads taken as nominal values associated with transient design situations. These special vehicles are defined in the Annex A of EC1-2 [5] ranging from vehicles with total weights of 600–3600 kN and with axle lines and location as shown in EN1991-2. Each notional lane and the remaining area of the bridge deck are loaded by the main loading system, under the conditions specified in EC1 and taken with its *frequent values.*

Load model 4 (LM4) – representing a crowd loading associated solely with transient design situations, if relevant, of $5\,\text{kN}\,\text{m}^{-2}$ to be placed on relevant parts of the length and width of the road bridge deck, the central reservation being included where relevant, as well as on the footways or across pedestrian and cycle tracks.

Numerical Example

Consider the case shown in Figure 3.8. The aim is to estimate the maximum bending moment occurring in one of the main girders of this steel-concrete composite deck due to LM1. The total width of the carriageway is 11 m, so, the number of notional lanes is $n_l = \text{Int}(11/3) = 3$ and the width of the remaining area is $11 - 3 \times 3 = 2\,\text{m}$.

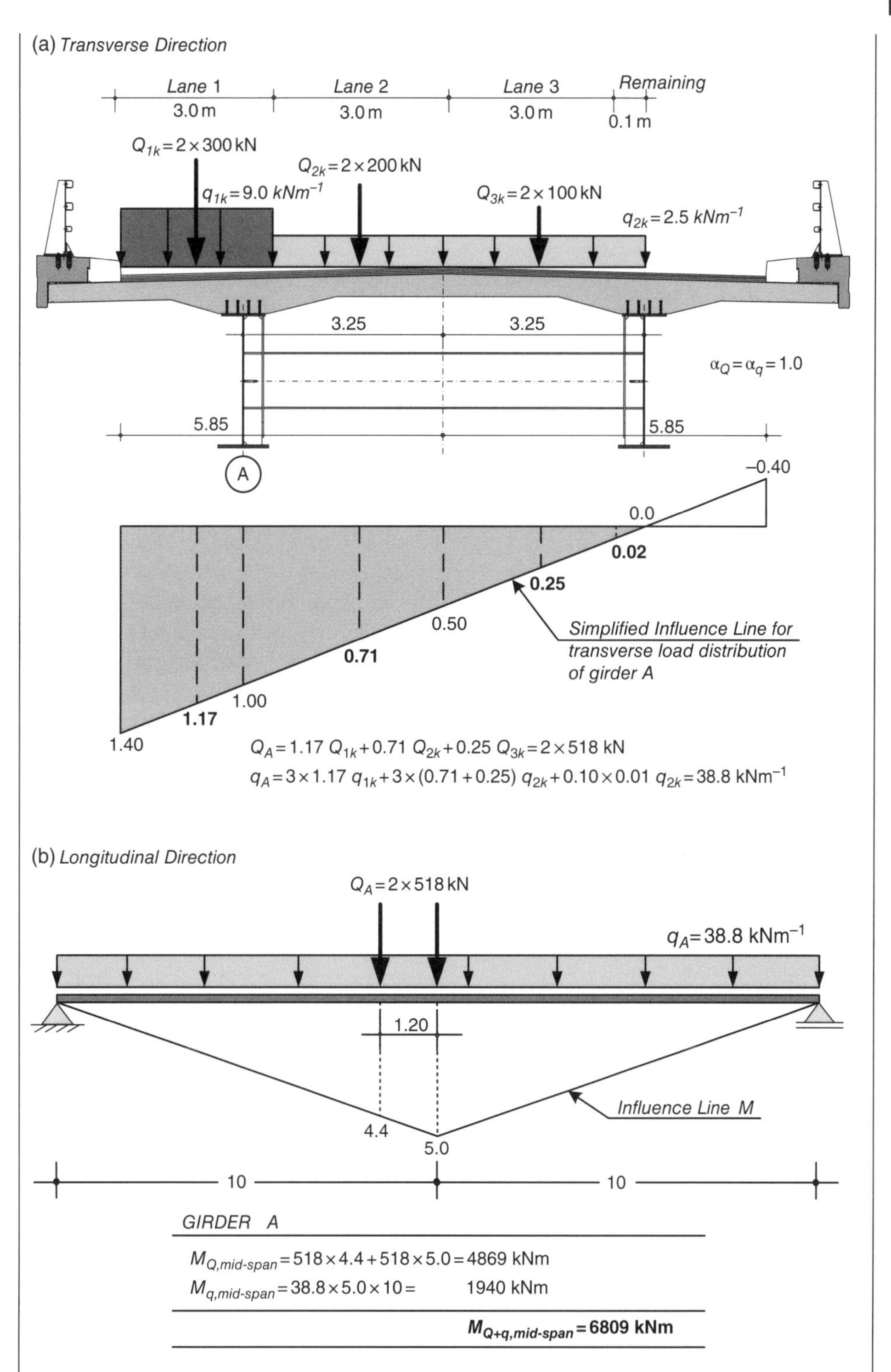

Figure 3.8 Example of a simply supported road deck: (a) Transverse LM1 action at girder A and (b) longitudinal bending moment at mid span cross section of girder A due to LM1.

One shall now locate the traffic loading in the cross section to produce the maximum bending moment at, for example, the left-hand side main girder. A more accurate analysis for the transverse load distribution, between girders, when a truck load is applied at any point on the deck is addressed in Chapter 6. However, the basic concept may be introduced in this example. For that, let the effect of a unit concentrated load $P = 1$ in the slab deck be considered as shown in Figure 3.8a. When this load is moving transversely at a certain section of the deck, the maximum bending moment at the left-hand side girder varies according to a *transverse influence line* as shown in Figure 3.8a. Longitudinally, one may think of this unit load as being distributed between the two girders according to some proportion that depends on the transverse stiffness of the slab as well as on the geometrical and material characteristics of the deck. Of course, no matter the transverse distribution considered, the overall bending moment at the mid span section is always obtained by $M = PL/4$. The bending moment M is then taken by the two main girders according to $M_A = k_A M = k_A\,PL/4$ and $M_B = k_B M = k_B\,PL/4$, where k_A and k_B (with $k_A + k_B = 1$) are the transverse load distribution coefficients of the load P between the two girders. The simplest approach to find out these influence lines corresponds to a statically determined distribution of the load between the two girders. This approach, to be discussed in Chapter 6, yields a straight influence line, while a more accurate solution yields a curved influence line.

Longitudinally, the loads here obtained Q_A and q_A, for girder A, should be located to produce the maximum bending moment in the girder that occurs at the mid-span section, as shown in Figure 3.8b. Hence, the well-known concept of influence line for M_{max} at the mid-span section of a simply supported girder is now adopted. The two concentrated loads of 518 kN each and the uniform load of 38.8 kN m^{-1} are located to induce the maximum bending moment of 6809 kNm at the mid-span section.

3.4 Braking, Acceleration and Centrifugal Forces in Highway Bridges

The acceleration and braking actions are associated, respectively, to positive and negative accelerations of vehicles and are quantified by means of horizontal longitudinal forces applied at the level of the pavement.

Let the case of braking be considered. When the brakes are activated, a friction force between the wheels and the surfacing is developed, which reduces the acceleration of the mass of the vehicle (Figure 3.9) according to the basic equation of dynamics

$$F = ma = Qa/g \tag{3.6}$$

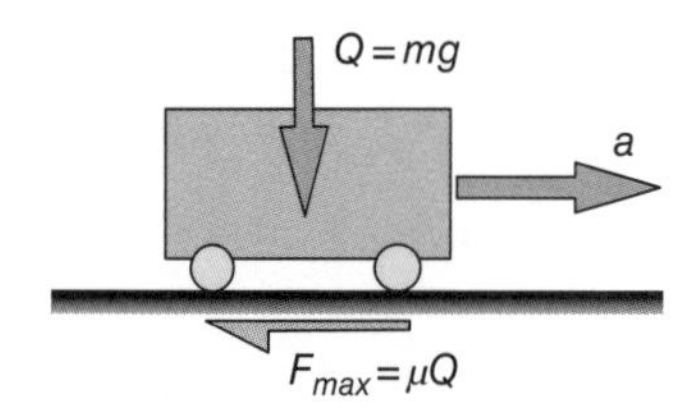

Figure 3.9 Concept for determining braking and acceleration forces.

where Q and a are, respectively, the weight and the acceleration of the vehicle and g is the acceleration of gravity.

When a maximum value of the force F is reached (F_{max}), consistent with the sliding friction coefficient, the vehicle slides on the surfacing keeping $F = F_{max}$. The friction coefficient μ increases with the reduction of the speed of the vehicle. In the case of a positive acceleration, the contact between the wheels and the surfacing induces horizontal forces – acceleration forces – which may be quantified under the same principle of braking forces, although with a smaller friction coefficient.

In short, the acceleration and braking forces may be quantified as a fraction of the characteristic values of the traffic loading LM1. Hence, in EC1-2, these actions are quantified by a force Q_l taken as a longitudinal force acting at the finished carriageway level and uniformly distributed over the loaded length, L, according to the following expression:

$$Q_l = 0.6\alpha_{Q1}(2Q_{1k}) + 0.10\alpha_{q1}q_{1k}\, w_1\, L \tag{3.7}$$

Such that

$$180\alpha_{Q1}(kN) \le Q_l \le 900(kN) \tag{3.8}$$

where L is the length of the deck or the part of it under consideration and w_1 the width of lane No. 1 (3 m in normal cases). Note that the values and limits of the force Q_l of Eqs. (3.7) and (3.8) are dependent on the adjustment factors α_{Q1} and α_{q1} adopted by the National Annexes. The upper limit 900 kN, adopted as the maximum force induced by military vehicles, may also be adjusted in the National Annex. A diagram for Q_l as a function of L, obtained from Eqs. (3.7) and (3.8), is shown in Figure 3.10 using adjustment factors equal to 1.

In plan-curved bridges, the centrifugal force associated with traffic actions should be considered. Let a vehicle of mass M and design speed v describe a curve of radius R, as shown in Figure 3.11, the centrifugal force Q_t is given by

$$Q_t = Mv^2/R = Qv^2/(gR) = cQ \tag{3.9}$$

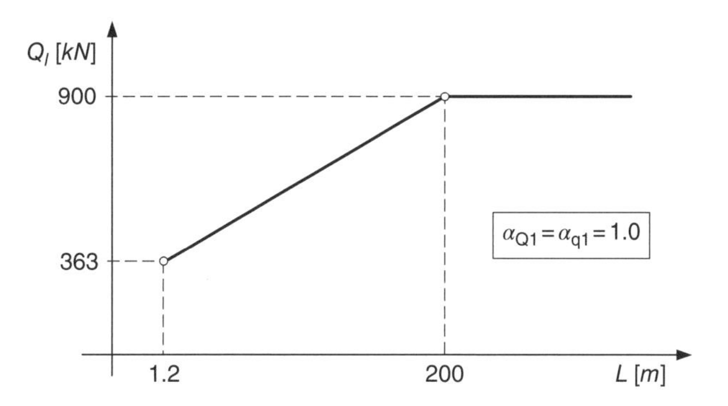

Figure 3.10 Braking and acceleration force Q_l in road bridges as a function of the loaded length L according to EC1-Part 2 [5].

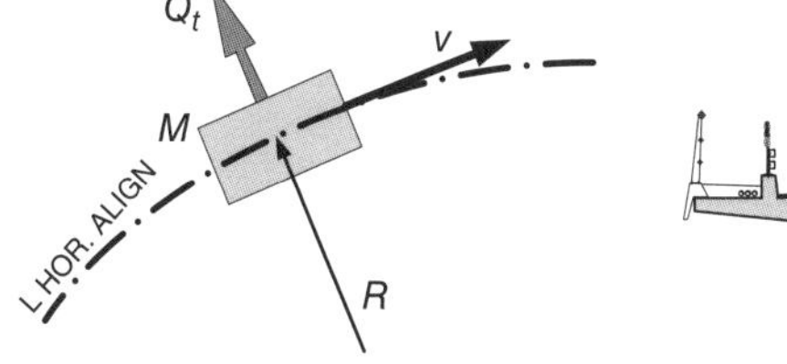

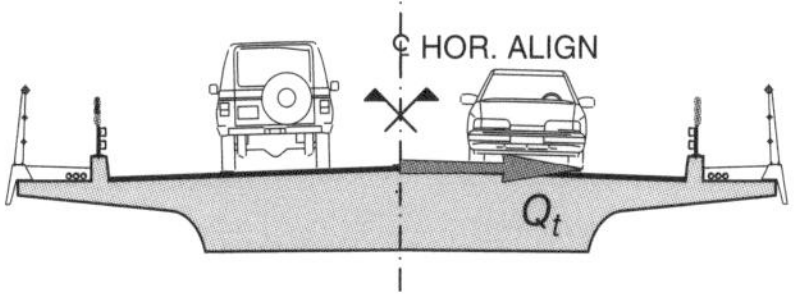

Figure 3.11 Centrifugal forces in road bridges.

where Q is the weight of the vehicle and g is the acceleration of gravity. If v is expressed in km h^{-1} and taking $g = 9.81$ m s^{-2}, the coefficient c in Eq. (3.9) is given by

$$c = (1000 \times 3600)^2 \, v^2 / (9.81 R) = v^2 / (127 R) \tag{3.10}$$

In EC1-2, the characteristic values of the centrifugal forces are taken as

$$\begin{aligned} Q_t &= 0.2Q(kN) \; if \; R < 200m \\ Q_t &= 40Q/R(kN) \; if \; 200 \le R \le 1500m \\ Q_t &= 0 \; if \; R > 1500m \end{aligned} \tag{3.11}$$

where Q is the total maximum weight of vertical concentrated loads of the tandem systems of LM1, that is, 1200 kN for $\alpha_{Q1} = 1$. Taking into consideration Eq. (3.10), one may conclude that the expression given in EC1-2 for 200 m ≤ R ≤ 1500 m, corresponds to a centrifugal force for a vehicle with a load equal to the tandem system and a design speed of about 70 km h^{-1}. For $R < 200$ m, a centrifugal force equal to 20% of the weight of the vehicle is taken and for $R > 1500$ m the centrifugal force is neglected.

The centrifugal force should be taken as shown in Figure 3.11, that is, as a transverse force acting at the carriageway level and radial to its axis.

3.5 Actions on Footways or Cycle Tracks and Parapets, of Highway Bridges

These actions are summarized in Figure 3.12, on the basis of the actions defined in detail in EC1-2 [5], for footways, cycle tracks and footbridges. It consists of a uniform distributed load of 5 kN m^{-2} specified to cover the static effects of a continuous dense

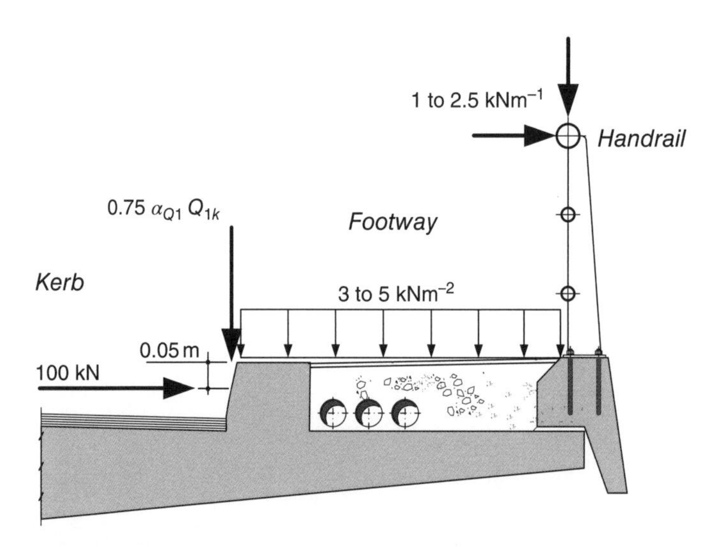

Figure 3.12 Actions on footways or cycle tracks, parapets and kerbs of road bridges according to EC1-Part 2 [5].

crowd, which should be applied in the relevant parts of the bridge to induce the maximum load effects for the element under study. A Uniformly distributed load of 3.0 kNm^{-2} is recommended as a combination value together with the LM1 on road bridges. For local effects only, a concentrated load of 10 kN should be considered.

For the design of pedestrian parapets, a pedestrian barrier, EC1-2 recommends a minimum value of 1.0 kN m^{-1} (no pedestrian traffic, inspection only) and a maximum of 2.5 kN m^{-1}, depending on the Class of the Parapet (see CEN/TR 1317-6 [7]) specifically on its width. These forces may be supposed to act horizontally or vertically, as shown in Figure 3.12. In some codes a fixed value is specified, for any kind of bridge parapets, as a minimum horizontal force of 1.5 kN m^{-1}.

3.6 Actions for Abutments and Walls Adjacent to Highway Bridges

The vertical forces applied on the carriageway located at the abutment zone, induces horizontal actions on the walls of the abutments, namely on side and front walls, which should be estimated from the appropriated load model adopted for the design of the bridge. The vertical loads induce the *impulse on rest* at the walls of the abutments. The simplest load model consists of simulating the effect of highway traffic loading as a UDL at the carriageway and to evaluate the earth pressure at the vertical walls of the abutments through the appropriated earth pressure coefficients. The uniformly vertical distributed load at the carriageway is taken in some codes as 10 kN m^{-2}.

In EC1-2 the appropriated load models are left to the National Annexes, but LM1 is recommended, although simplifying the tandem system loads by an equivalent UDL, q_{eq}, spread over a rectangular surface 3 m wide and 2.20 m long if, for a properly consolidated backfill, the dispersal angle from the vertical is taken equal to 30°. It should be noted the characteristic values of LM1 include a dynamic amplification that is not usually relevant for roads. Therefore, the characteristic values of LM1 may be multiplied by a reduction factor of 0.7. For example, considering the lane N°1, and using $\alpha_Q = \alpha_q = 1.0$, $q_{eq} = 0.7 \times 600/(2.2 \times 3) = 63.6\,\text{kN m}^{-2}$, and outside this area $q_{eq} = 0.7 \times 9 = 6.3\,\text{kN m}^{-2}$. A longitudinal braking force should also be taken for the design of the abutment stand walls (Figure 3.13) considering that a lorry may brake when arriving on the bridge, which in EC1-2 is $0.6\,\alpha_{Q1}Q_{1k}$, with the backfill assumed to be not loaded simultaneously.

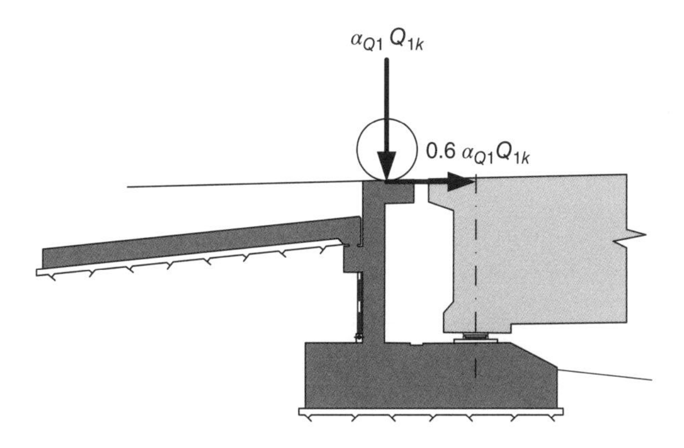

Figure 3.13 Longitudinal braking force for the design of abutments up stand walls in EC1-Part2. (*Source:* Adapted from Ref. [5]).

3.7 Traffic Loads for Railway Bridges

3.7.1 General

Railway bridges have many specific aspects to be taken into account in design, some of them are out of the scope of this book. However, some important aspects related to actions, to be referred to here, are:

- load models for railway traffic
- dynamic load effects
- traction, braking, centrifugal forces and nosing forces
- action effects due to track–structure interaction

3.7.2 Load Models

Most European countries adopt the UIC Recommendations, considered also in EC1-2 [5], where the following load models are specified:

- Load Model 71 and load model SW/0 for normal rail traffic on main railways,
- Load model SW/2 to represent heavy loads,
- Load model HSLM for high speed passenger trains, considered to be trains with speeds exceeding 200 km h^{-1},
- Load model 'Unloaded Train', to represent the effect of an unloaded train.

It is important to note that the load models tend to induce the effects of railway traffic, but they do not attempt to represent by themselves the actual loads from trains. The UIC 71 train [5], here denoted as the load model 71, or simply LM 71, is the typical train load model for design of most railway bridges. The reader may refer to EC1-2 [5] for the consideration of the remaining load models.

Load model 71 is represented in Figure 3.14. It consists of four axle loads of 250 kN, with the distribution shown in the figure and a UDL of 80 kN m^{-1} under an unlimited length. These values are characteristic values that, according to EC1-2, may be multiplied by a factor $0.75 \leq \alpha \leq 1.46$ on lines carrying rail traffic that is lighter or heavier than the normal rail traffic. This load model should be located at the most unfavourable position for the element under study, taking into consideration that the UDL, or even the concentrated loads, may not act at a certain loaded length if these are the most severe loading conditions for the element under consideration. An example is given in Figure 3.15 for a loading pattern case for the positive bending moment of a continuous beam bridge deck.

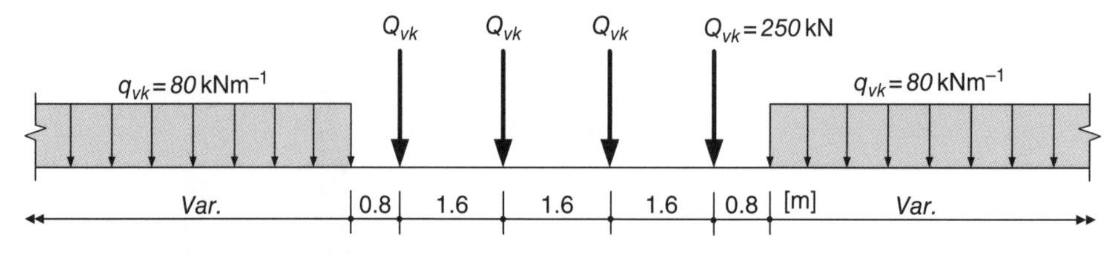

Figure 3.14 Load Model 71 for railway bridges.

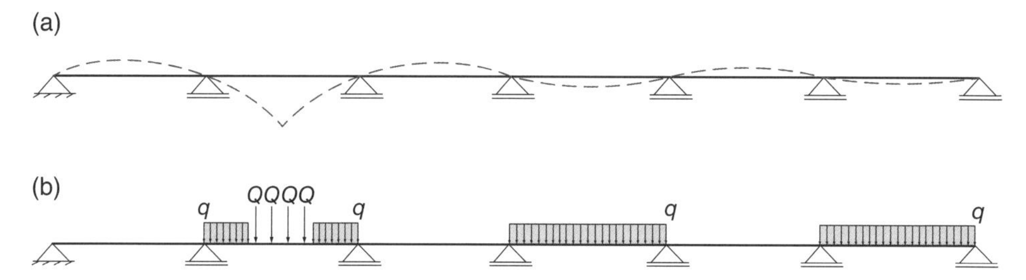

Figure 3.15 Continuous railway deck: (a) Influence line for M^+ at the second mid span and (b) application of LM 71 to obtain the maximum positive bending moment at the second mid span.

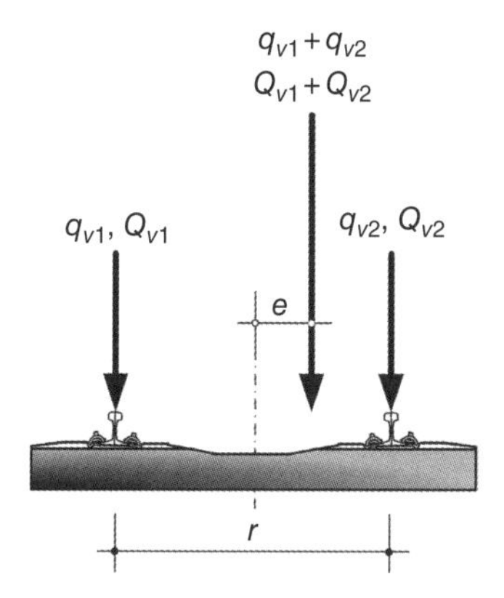

Figure 3.16 Eccentricity of vertical railway loads.

The loads of Load Model 71 should be multiplied by the appropriated dynamic load factor, Φ, with values between 1.0 and 2.0, which takes into account the dynamic amplification of load effects due to the overall vibration of the bridge or the local vibration modes of specific structural elements under consideration. This aspect of dynamic factors is considered in more detail in the next section.

For bridges carrying two tracks, Load Model 71 should be applied to one or two tracks in order to induce the most severe effects to the element under study. For bridges with three or more tracks, LM 71 should be applied to any one track, any two tracks or 0.75 times the load model to three or more tracks.

The loads of LM 71 shall be applied with a certain eccentricity to take into account possible lateral displacements of the vertical wheel loads on the tracks. In EC1-2, this is considered by adopting a ratio of wheel loads on all axles up to 1.25/1.00 on any one track. This applies both to the concentrated load Q_{vk} as well as to the distributed load q_{vk} of LM 71 (Figure 3.16). For maximum eccentricity, one obtains $e_{max} = 0.055\ r = r/18$. Hence, for a standard European gauge track of $r = 1450$ mm, one has an eccentricity of 81 mm, inducing a torsional moment on the deck under traffic loading.

3.8 Braking, Acceleration and Centrifugal Forces in Railway Bridges: Nosing Forces

The traction and braking forces in railway bridges should be considered to be longitudinal forces along the track, acting at the top of the rails, as UDLs along the associated influence length for the element under study. According to EC1-2 and for the LM 71, the values of these forces are, respectively, 33 and 20 kN m^{-1} for traction and braking forces. These are characteristic values, which according to EC1-2 may be multiplied by the factor α referred to in Section 3.7.2. For tracks restricted to high speed passenger traffic, the traction and braking forces may be taken as 25% of the sum of the axle loads acting on the influence length for the element under consideration.

The total values of traction and braking forces, are, in any case, limited to 1000 kN and 6000 kN, independently of the influence length of the element under study.

Concerning the centrifugal forces, these forces are determined from the general expression already given for road bridges – Eqs. (3.9) and (3.10), being the vertical load of the vehicle Q_v identified here with the axle loads Q_{vk} and q_{vk} of the LM 71. For train speeds higher than 120 km h^{-1}, a reduction factor is allowed subject to a minimum value of 0.35. This factor, denoted by f in EC1-2, is dependent on the influence length, L_f, of the loaded part of curved track on the bridge for the element under consideration, and is given by:

$$f = \left[1 - \frac{V-120}{1000}\left(\frac{814}{V} + 1.75\right)\left(1 - \sqrt{\frac{2.88}{L_f}}\right)\right] \geq 0.35 \quad (3.12)$$

where V is the maximum line speed at the site in km h^{-1}, and L_f in m. For design speeds V less or equal to 120 km h^{-1} there is no reduction ($f = 1.0$) and increasing the design speed above $V \geq 300$ km h^{-1}, f is kept at 0.35 for $L_f \geq 80$ m.

The centrifugal forces in railway bridges are applied as a horizontal force at 1.8 m above the track level, normal to the axle of the track and in correspondence with the vertical traffic loads of the LM 71 without any dynamic coefficient.

To take into account the impact lateral forces of the wheels on the tracks, namely due to gaps between the wheel and the rail, *nosing forces* are considered. These forces are applied at the top level of the track, as a concentrated force acting horizontally and perpendicularly to the track. Its characteristic value in most codes, namely in UIC recommendations and EC1-2, is Q_{sk}=100 kN to be applied without any dynamic factor. According to EC1-2, this value should be multiplied by the factor α referred to in Section 3.7.2 and always combined with the vertical traffic load.

3.9 Actions on Maintenance Walkways and Earth Pressure Effects for Railway Bridges

Pedestrian, cycle and general maintenance loads should be represented by a UDL with a characteristic value $q_{fk} = 5$ kN m^{-2}. For the design of local elements, a concentrated load $Q_{fk} = 2.0$ kN acting alone should be taken into account and applied on a square surface with a 200 mm side.

Concerning the loads to be considered in the evaluation of global earth pressure effects, EC1-2 specifies that they may be taken as the vertical loads of LM 71, without any dynamic coefficient, uniformly distributed over a width of 3 m at a level 0.7 m below the running surface of the track. For the case of one LM 71, and using $\alpha = 1$, the concentrated loads induce $q_{eq} = 1000/(6.4 \times 3) = 52.1$ kN m^{-2} and outside this area $q_{eq} = 80/3 = 26.67$ kN m^{-2}. An uniformly vertical distributed load of 30 kN m^{-2} is taken in some codes.

3.10 Dynamic Load Effects

3.10.1 Basic Concepts

To introduce the problem of dynamic load effects in roadway and railway bridges, a single degree of freedom system is adopted, with mass M, elastic stiffness K and viscous damping (i.e. damping proportional to the velocity) defined by the coefficient C. When a constant force P is suddenly applied, the basic equation for dynamic equilibrium [8, 9] is:

$$M\ddot{u} + C\dot{u} + Ku = P \tag{3.13}$$

where $\ddot{u}$ and $\dot{u}$ denote, respectively, the second and first derivatives of the displacement with respect to time, that is, the acceleration and the velocity. Assuming the system is at rest at $t = 0$, that is, initial conditions $\dot{u} = 0$, $u = 0$, the dynamic response for the undamped system is:

$$u_{dyn} = (P/K)(1 - cos\omega t) \tag{3.14}$$

where $\omega = 2\pi f = \sqrt{\frac{K}{M}}$ is the circular frequency of the undamped system. The term P/K represents the static deflection u_{sta} and, hence, the dynamic amplification factor Φ, defined as the ratio between the maximum dynamic deflection u_{dyn} and the static one $u_{sta} = P/K$ is given for the undamped system by:

$$\Phi = u_{dyn}/u_{sta} = \max\{1 - cos\,\omega t\} = 2 \tag{3.15}$$

Let the viscous damping be introduced through the damping ratio ξ, where $C_{cr} = 2M\omega$ is the *critical damping coefficient*. It represents the smallest amount of damping for which the free vibration response changes from an oscillatory to exponential decay response. The system now vibrates with a response as shown in Figure 3.17, including a negative exponential term yielding a decaying vibration, defined by:

$$u_{dyn} = \left(\frac{P}{K}\right)\left[1 - e^{-\xi\omega t}\left(\cos\,\omega_d t + \frac{\xi}{\sqrt{1-\xi^2}} sin\,\omega_d t\right)\right] \tag{3.16}$$

Here, $\omega_d = \omega\sqrt{1-\xi^2}$ is the damped vibration frequency of the system, differing very little from the undamped vibration frequency, since for most bridges $\xi << 5\%$.

The dynamic coefficient Φ is therefore reduced with ξ and is still obtained from Eq. (3.15), yielding for the damped system at instant $t = \pi/\omega_d$:

$$\Phi = u_{dyn}/u_{sta} = \left(1 + e^{-\pi\cdot\xi\cdot\omega/\omega_d}\right) \approx \left(1 + e^{-\pi\cdot\xi}\right) \tag{3.17}$$

as shown in Figure 3.17. The internal elastic force $R = K\,u$ is reduced as well in the same proportion. Now consider a moving load $P = M\,g$ on a simply supported bridge, with velocity v as shown in Figure 3.18. The response of the bridge, modelled here as a simple supported beam with bending stiffness EI and mass per unit length constant μ, is defined by the vertical displacement $w = w(x, t)$ obtained by solving the partial differential equation:

$$EI\frac{\partial^4 w}{\partial x^4} + \mu\frac{\partial^2 w}{\partial t^2} + 2\mu\omega_d\frac{\partial w}{\partial t} = \delta(x - v\,t)P \tag{3.18}$$

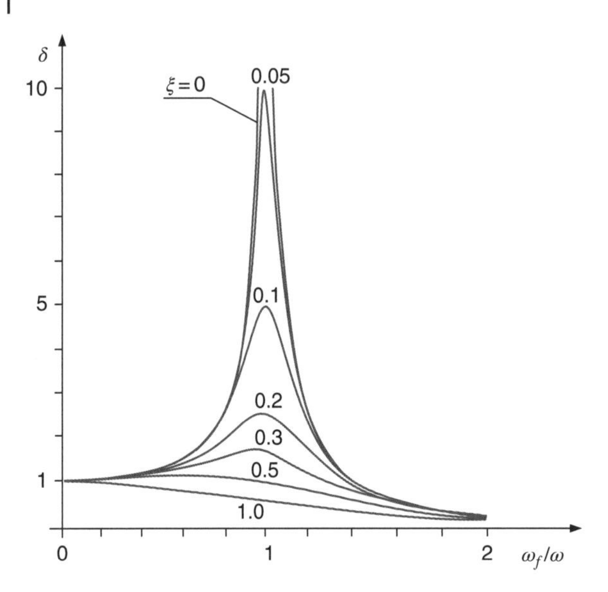

Figure 3.17 Variation of the dynamic coefficient with the viscous damping ratio and the relationship between the applied force frequency, ω_f, and the natural frequency, ω.

where $\delta(x - v\,t)$ represents the Dirac delta function [8, 10], that is, $\delta = \infty$ for $x = v\,t$ (at the point of application of the concentrate load) and $\delta = 0$ elsewhere, such that $\int_0^{\infty} \delta(x - v\,t) P dx = P$. Besides, in this equation, ω_d represents the damped system circular frequency depending on ξ.

The undamped beam has natural sinusoidal vibration modes (Figure 3.18) and natural circular frequencies ω_n, obtained from Eq. (3.18) for $\xi = 0$, $\omega_d = \omega$ and $P = 0$, by:

$$\omega_n = n^2 \pi^2 \sqrt{\frac{EI}{\mu L^4}} \qquad \text{for } n = 1,2,3\ldots \tag{3.19}$$

Equation (3.18) may be solved by analytical methods and classical solutions that are available in the literature [10]. The solutions involve the parameter:

$$\alpha = \frac{v}{v_{cr}} \qquad \text{with} \qquad v_{cr} = \frac{\pi}{L}\sqrt{\frac{EI}{\mu}} \tag{3.20}$$

where v_{cr} represents the critical speed of the moving load introducing an unbounded vertical deflection in the undamped system, that is, the resonance condition as it happens in a single degree of freedom (SDF) system when the frequency of the force tends to be equal to the natural frequency of the system. Taking into consideration the value of the first natural frequency of the beam $f_1 = \omega_1/(2\pi)$, one obtains $v_{cr} = 2f_1L$. Hence, the maximum displacement yields a dynamic amplification factor Φ dependent on the natural frequency of the bridge, as well as on the speed, v, of the mass running on the bridge. Of course, Φ is dependent on the bridge damping ratio and tends to increase with the natural frequency for v to approach v_{cr}. For velocities sufficiently far away from v_{cr}, the dynamic amplification factor for mid-span deflection, is approximately given by $\Phi = 1/(1 - \alpha)$ for an undamped bridge.

Figure 3.19 shows some numerical results obtained for a simply supported span under a moving load at different speeds with respect to the critical speed and for a damping

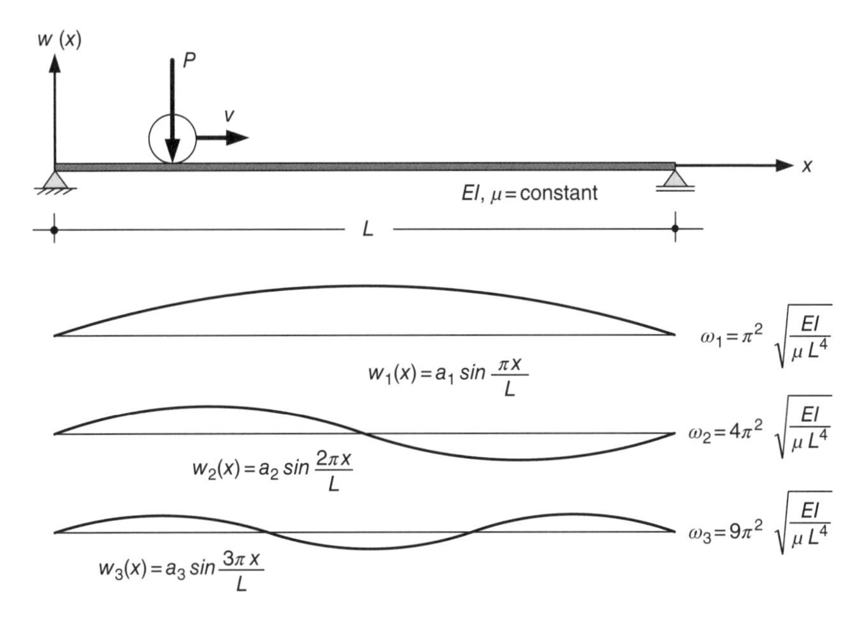

Figure 3.18 Simple supported deck vibration analysis: basic properties and first three natural vibration modes.

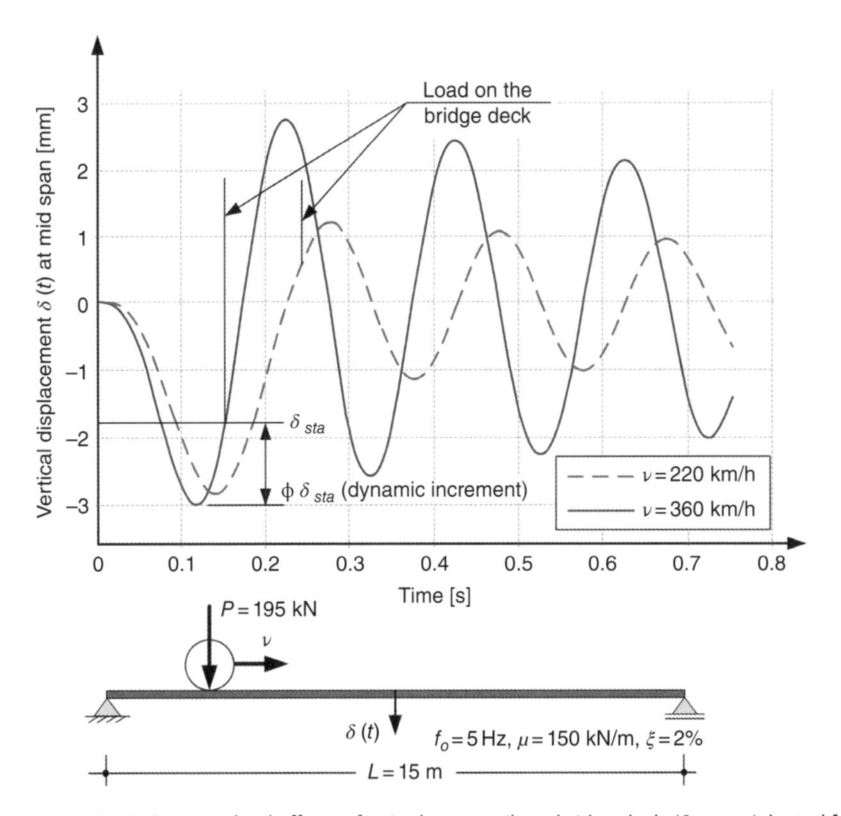

Figure 3.19 Dynamic load effects of a single span railway bridge deck. (*Source:* Adapted from Ref. [11]).

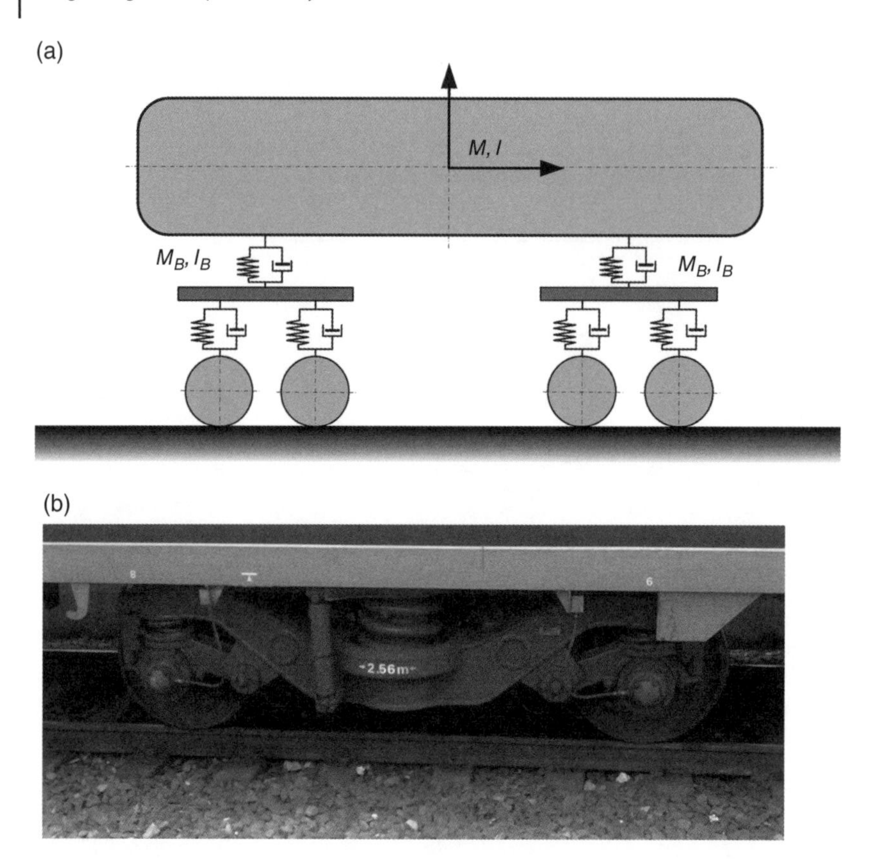

Figure 3.20 Model for bridge dynamic analysis with vehicle – structure interaction and real train boggy.

coefficient of 2%. If the irregularities of the carriageway of a highway bridge, or of the rail of a railway bridge are considered, the dynamic amplification factor is dependent on these irregularities as well. Besides, if the mass of the vehicle is not negligible with respect to the bridge mass, the effect of bridge-vehicle interaction may be relevant with due account for the elastic and damping suspension system of the vehicle. A model is shown in Figure 3.20 for this case. All these effects can only be considered through a dynamic numerical model that considers the structure-vehicle interaction and the irregularities of the carriageway of the track as a random process. This subject is out of the scope of this book and the reader is referred to specific literature [10–13].

The load models defined for road bridges already include the dynamic coefficient in the value specified for the characteristic traffic loads. Hence, specific consideration of dynamic load effects is relevant for railway bridges only.

3.10.2 Dynamic Effects for Railway Bridges

For railway bridges, the dynamic behaviour of the bridge should be considered, either on the basis of a dynamic analysis or based on a simplified procedure in which a static analysis is adopted but with the traffic loads (Section 3.6) multiplied by a dynamic coefficient.

For maximum design speeds not exceeding $200\,\text{km}\,\text{h}^{-1}$, the requirements to allow a static analysis are established in EC1-2 [5], and are related to the natural frequency of the bridge f_1 to be between a lower limit governed by dynamic impact criteria, defined by (Figure 3.21):

$$\begin{aligned} f_{1,lower} &= 80 / L_\Phi && \text{for } 4 \leq L_\Phi \leq 20\,\text{m} \\ f_{1,lower} &= 23.58 L_\Phi^{-0.592} && \text{for } 20 < L_\Phi \leq 100\,\text{m} \end{aligned} \tag{3.21}$$

and an upper limit, governed by dynamic increments due to track irregularities, defined as

$$f_{1,upper} = 94.76\, L_\Phi^{-0.748} \tag{3.22}$$

The natural frequency of the bridge should take into account the mass due to permanent actions. In Eqs. (3.21) and (3.22), L_Φ(m) is the span length for simple supported bridges or an equivalent span for other bridge types. Some simplified expressions to evaluate the natural frequencies of typical bridge cases are given in [14].

A detailed table is given in EC1-2 to define L_Φ. It is important to note that L_Φ is related to the vibration mode of the element under consideration. So, for example, if one takes the requirements to allow an overall static analysis of a continuous girder or slab bridge deck, or a continuous portal frame, over n spans, L_Φ is the average span length. However, if one considers the cross girders of a steel grid made of a ballastless open deck, L_Φ is twice the length of the cross girder. For continuous bridge decks (slabs, slab girders or box girder type) with span lengths L_i, $i = 1, 2, 3, \ldots, n$, one may take for $L_\Phi = k\, L_m$, where $k = 1.2$ for $n = 2$ spans, $k = 1.3$ for $n = 3$, $k = 1.4$ for $n = 4$ and $k = 1.5$ for $n \geq 5$, and L_m the average span. In any case, one should never take L_Φ to be smaller than the maximum span length L_i.

For cases where static analysis is allowed, the dynamic factors by which the loads of LM 71 should be multiplied are defined as Φ_2 or Φ_3, dependent, respectively, on whether

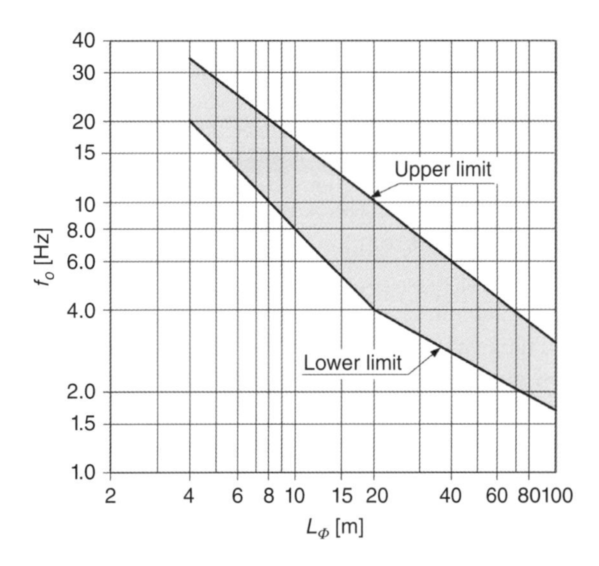

Figure 3.21 Lower and upper limits to allow a static analysis in a railway bridge for spans up to 100 m (*Source:* Adapted from Ref. [5]).

this is the case of a carefully maintained track or a track under standard maintenance. The expressions as proposed by EC1-2 and UIC specification are as follows:

$$\Phi_2 = \frac{1.44}{\sqrt{L_\Phi} - 0.2} + 0.82 \quad \text{and} \quad \Phi_3 = \frac{2.16}{\sqrt{L_\Phi} - 0.2} + 0.73 \tag{3.23}$$

with $1.00 \leq \Phi_2 \leq 1.67$ and $1.00 \leq \Phi_3 \leq 2.0$, where L_Φ is the 'determinant' length referred to earlier. These dynamic coefficients have been determined for simply supported girders; the concept of determinant length L_Φ allows generalizing the expressions for Φ_2 or Φ_3 to other bridge design typologies. In some way, L_Φ is the span of the equivalent simply supported beam, defined from the influence line of the deflections in the actual bridge deck.

3.11 Wind Actions and Aerodynamic Stability of Bridges

In most cases of bridge design, wind actions are considered static as equivalent forces, provided the slenderness of the superstructure is 'moderate'. Cable-stayed bridges and suspension bridges are typical examples where wind action must be considered on the basis of a dynamic analysis.

Cases of aerodynamic instability, such as the well-known historical collapse of the Tacoma Narrows suspension Bridge in 1940 (Figure 3.22a) under wind speeds about three times lower than the static design wind action, have shown the need for an in-depth study of aeroelastic problems in bridge engineering. At the time of this accident, aerodynamic instabilities were considered in the field of aeronautics, but not for civil engineering structures. The open deck cross section adopted in the Tacoma Narrow Bridge is compared in Figure 3.22b with a streamline box girder deck adopted in modern suspension bridges.

Nowadays, bridge designers need to know the type of phenomena involved even when dealing with ordinary structures, as well as the limits of application of design rules, to evaluate wind action on bridge decks. Therefore, some basic concepts on this topic will be developed here.

3.11.1 Design Wind Velocities and Peak Velocities Pressures

The wind is specified at a certain point $X = \{x_i\}$ acting on a structure by its instantaneous velocity $v = v\{x_i, t\}$, which may be considered to be composed by an average velocity, v_m, and a turbulent component, v_t:

$$v\{x_i,t\} = v_m\{x_i\} + v_t\{x_i,t\} \tag{3.24}$$

The average value v_m referred is to a certain time interval, Γ, during which wind speed is measured. Generally, $\Gamma = 10$ minutes and after a certain number of measurements one has a statistic distribution of the wind speed measured during a certain number of intervals. The *basic wind velocity*, v_b, is a reference velocity for a certain probability of being exceeded (usually 2%, i.e. with a return period T = 50 years) defined in the codes for the bridge site as a function of wind direction and time of the year, at 10 m above the ground and for a certain type of terrain (open field). From v_b,

(a)

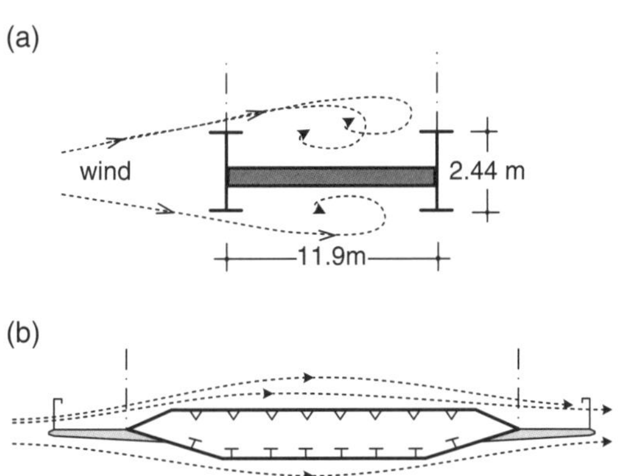

Figure 3.22 The aerodynamic instability that lead to the Tacoma Narrow Bridge collapse, in 1940: (a) Cross section of the deck of Tacoma bridge and (b) compared to the deck cross section of a modern streamline suspension bridge.

the so called *mean wind velocity* refers to a time interval of 10 minutes, with a probability, p, of being exceeded (usually 2%). A different matter is the *design wind speed*, v, defined in the codes by its characteristic value, v_k, which is associated with the *gust wind speed* on the site and which may be evaluated for a certain return period related to the design situation under consideration. For example, in the case of erection phases, the design wind speeds are generally lower. For checking the structure during these phases, one adopts a shorter return period, T, say $T = 10$ years for a construction time no greater than 1 year. The associated characteristic wind speed for $T = 10$ years and $T = 50$ years, may be related by [15]:

$$v_{k,T} = v_{k,50}\left(1 - 0.071 \ln \frac{50}{T}\right) \tag{3.25}$$

For $T = 10$ years, as for the erection phases, $v_{k,10} = 0.88\ v_{k,50}$, that is, a 12% reduction. In some cases, and for very relevant bridges, one may increase the return period from

50 to more than 100 or 120 years, generally the *reference lifetime* of the structure. For $T = 120$ years, one obtains $v_{k,120} = 1.06\ v_{k,50}$.

For design, and if no aerodynamic problems are relevant, only the peak wind velocities pressures are needed. However, aerodynamic instabilities should be investigated on the basis of mean wind velocities.

The *design mean wind velocity*, at a certain level z above the ground, may be determined by:

$$v_m(z) = c_r(z)c_0(z)v_b \tag{3.26}$$

where $c_0(z)$ is the orography factor, taken as 1.0 unless otherwise specified and $c_r(z)$ is the roughness factor defined, for example, in Part 1-4 of EC1 (denoted by EC1-1-4, [15]) by:

$$c_r(z) = k_r \ln\frac{z}{z_0} = 0.19\left(\frac{z_0}{z_{0,II}}\right)^{0.07} \ln\frac{z}{z_0} \tag{3.27}$$

where k_r is the *terrain factor*, which is a function of the *roughness length*, z_0. In EC1-1-4 there are five terrain categories, being categories 0 and I associated to a bridge site on the sea or in a coastal area. Category IV is for a bridge at an urban site. Intermediate categories II and III correspond, respectively, to open country with low vegetation and isolated obstacles, or areas with regular cover of vegetation or buildings (such as villages, suburban terrain and permanent forests).

For example, one has $z_0 = 0.003$ m for roughness length in terrain category 0, $z_0 = 0.05$ m for category II and 1.0 m for category IV. The expression in Eq. (3.27) holds for $z > z_{min}$; that is, respectively, 1.0 m for terrain category 0 and 10 m for category IV. Otherwise, one takes $c_r(z) = c_r(z_{min})$. Using $c_0(z) = 1.0$ and $c_r(z)$, according to the expressions given in EC1-1-4 [15], yields the curves given in Figure 3.23 for terrain roughness categories between 0 and IV. The design wind speed may be defined by multiplying the mean wind velocity by a factor to take turbulence into consideration:

$$v = \sqrt{1 + 2gI_v(z)}\ v_m \tag{3.28}$$

where g is the 'peak factor' (that may be taken as 3.5 as in EC1-1-4) and $I_v(z)$ the turbulence intensity at level z defined as:

$$I_v(z) = \frac{\sigma_v}{v_m} = \frac{k_l}{c_0 \ln\left(\frac{z}{z_0}\right)} \tag{3.29}$$

where σ_v is the standard deviation of the turbulence assumed constant with z and k_l is a turbulence coefficient ($k_l = 1.0$ in [15]) and c_0 is the orography factor already defined. The expression is only valid above a minimum level z_{min} (between 1.0 and 15 m, respectively, for terrain roughness of categories 0 and IV) above the ground and for $z \leq 200\ m$. Below z_{min}, one assumes a constant minimum turbulence evaluated from Eq. (3.29) for $z = z_{min}$. From these considerations, one may define the peak wind speed, including the turbulence effects, as:

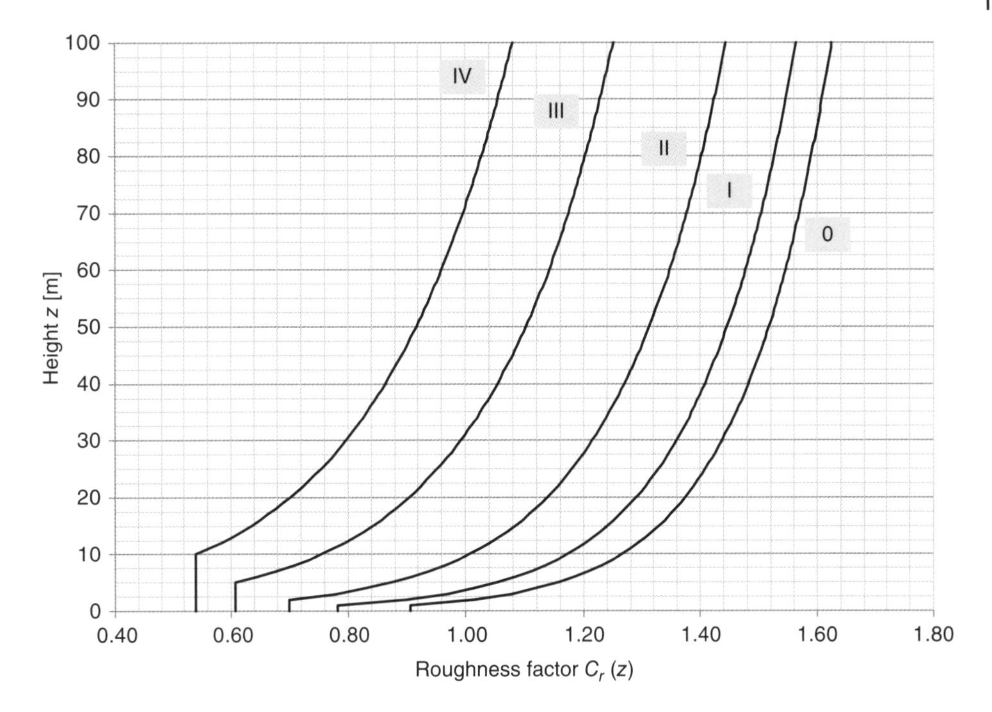

Figure 3.23 Roughness factor $c_r(z)$ as a function of the height z and for $c_o(z) = 1.0$.

$$v = \sqrt{1+7I_v(z)}\, v_m \tag{3.30}$$

From the design wind speed, v, the peak velocity pressure on an element of the structure is evaluated by:

$$q = \tfrac{1}{2}\rho v^2 \tag{3.31}$$

where ρ is the unit mass of the air taken, under normal conditions, as $\rho = 1.25\,\mathrm{kg\,m^{-3}}$. From these expressions, the peak pressure can be written in terms of the wind pressure, q_b, associated to the basic wind velocity, v_b, by:

$$q_p(z) = c_e(z) q_b \text{ where } q_b = \tfrac{1}{2}\rho v_b^2 \tag{3.32}$$

where $c_e(z)$ is the *exposure factor* given by:

$$c_e(z) = \left[1+7I_v(z)\right] c_r^2(z) c_0^2(z) \tag{3.33}$$

Figure 3.24 presents the values for the exposure factor as given in EC1-1-4 [15], allowing a direct calculation of the peak velocity pressure from the basic wind velocity, v_b, defined in the code for the bridge site.

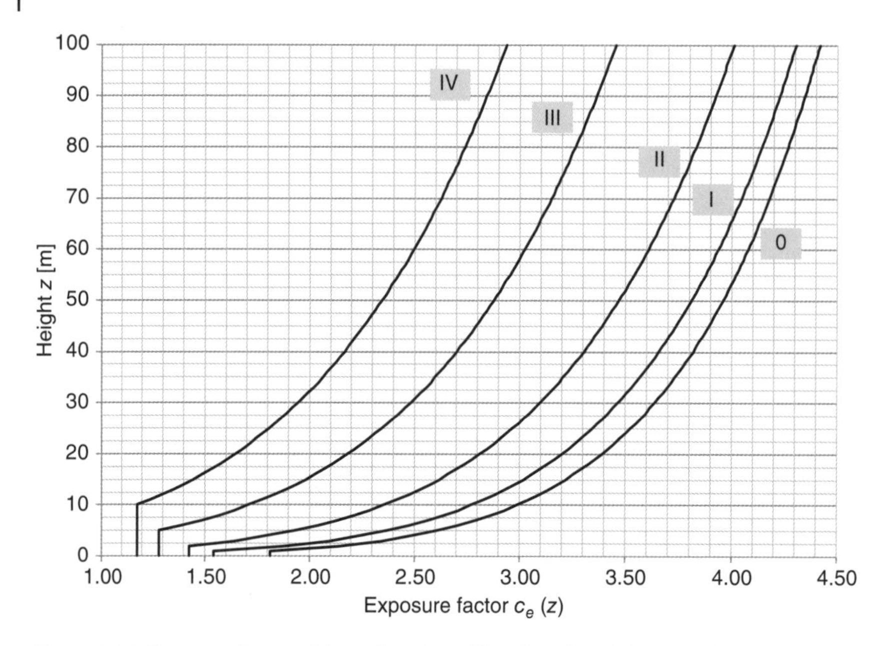

Figure 3.24 Exposure factor $c_e(z)$ as a function of height z, for $c_o(z) = 1.0$ and $k_I = 1.0$.

Numerical Example

Let a bridge deck be considered that is 50 m above the ground on a site near by the sea coast, with a basic wind speed $v_b = 30\,\text{m s}^{-1}$. The basic pressure is $q_b = \frac{1}{2} \times 1.25 \times 30^2 = 0.56$ kN m^{-2}. From Figure 3.24, one directly obtains $c_e(z = 50) = 3.95$ for a terrain of category 0. Hence, the peak velocity pressure is $q_p(z) = 2.22$ kN m^{-2}. This pressure includes the gust wind effects. Let the associated mean wind velocity be calculated. One has for the roughness factor from Eq. (3.27):

$$c_r(z = 50) = 0.19\left(\frac{0.003}{0.05}\right)^{0.07} \ln\frac{50}{0.003} = 1.517$$

Assuming $c_0 = 1.0$, as usually, $v_m = 1.517 \times 1.0 \times 30 = 45.5$ m s^{-1} = 164 km h^{-1}. The turbulence intensity and the design gust wind velocity come from Eqs. (3.29) and (3.30):

$$I_v = \frac{1.0}{1.0 \ln\left(\dfrac{50}{0.003}\right)} = 0.103$$

$$v = \sqrt{1 + 7 \times 0.103} \times 45.5 \text{m s}^{-1} = 1.312 \times 45.5 = 59.7 \text{m s}^{-1} \cong 215 \text{km h}^{-1}$$

Hence, the design wind velocity, including turbulence effects is 31% higher than the mean wind speed, which means the turbulence component represents 14.2 m s^{-1}. The peak wind pressure for design is therefore obtained by Eq. (3.32):

$$q_p(z = 50) = \tfrac{1}{2} \times 1.25 \times 59.7^2 = 2.22 \text{kN m}^{-2}$$

the same result previously obtained using the exposure factor $c_e(z = 50)$.

3.11.2 Wind as a Static Action on Bridge Decks and Piers

Let a bridge deck (Figure 3.25) be considered under the action of a uniform wind flow of speed v_α acting at a certain angle of attack, α. The deck is subjected per unit length, to the following forces as denoted in Figure 3.25:

$$F_D = C_D q_b d \qquad F_L = C_L q_b d \qquad M = C_M q_b d^2 \tag{3.34}$$

where $q_b = \frac{1}{2}\rho v_\alpha^2$ is the wind pressure associated to v_α, and C_D, C_L and C_M are nondimensional coefficients designated as *shape aerodynamic coefficients* dependent on the geometry of the cross section of the deck and on the angle of attach α of the wind flow. The dimension d is a characteristic dimension of the cross section, as its depth, to which the shape coefficients are referred to. However, the shape coefficients can also be referred to the deck width b or, for example,C_M may be defined by expressing the moment as $M = C'_M \frac{1}{2}\rho v_\alpha^2 b d$ where b is the width of the cross section and $C'_M = C_M \left(\frac{b}{d}\right)$.

For current bridge design cases, the drag force induced in a bridge component by the wind action is the most relevant one, and it is evaluated by (Figure 3.26):

$$F_D = C_D q_b A_{ref,x} \tag{3.35}$$

where F_D is the horizontal wind force and $A_{ref,x}$ is the reference area of the element exposed to the wind action and projected on a plan normal to the wind direction. For the wind actions on the bridge deck it is usual to define the forces F_D per unit length,

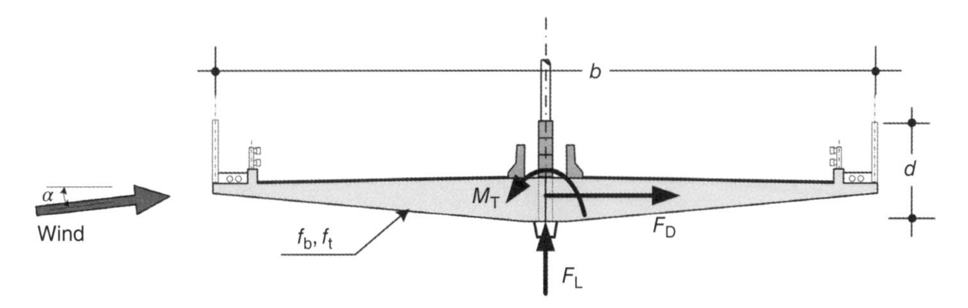

Figure 3.25 Wind forces on a bridge deck: drag F_D, lift F_L and moment M_T.

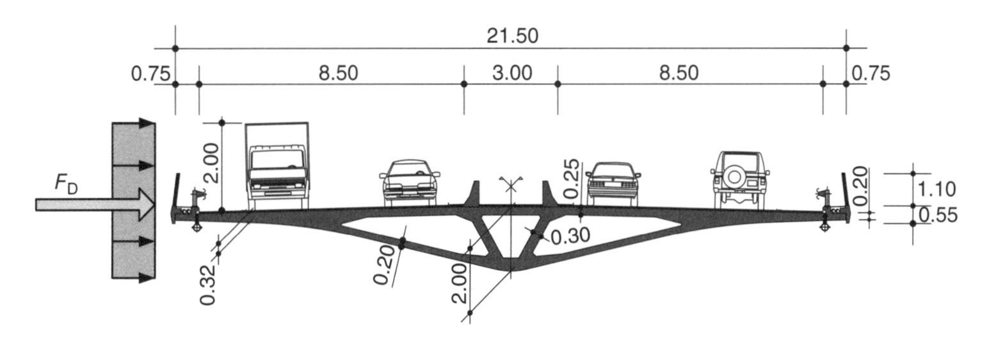

Figure 3.26 Drag wind force, F_D, on the envelope of a cable-stayed bridge deck, Funchal. Design by GRID.

where the deck is simulated by a rectangular envelope of width b and depth d. The drag coefficient, C_D is given as a function of the ratio b/d, and for a reference area $1 \times d$ one may adopt the values given in Figure 3.27 taken from the EC1-1-4 [15] for open or box sections.

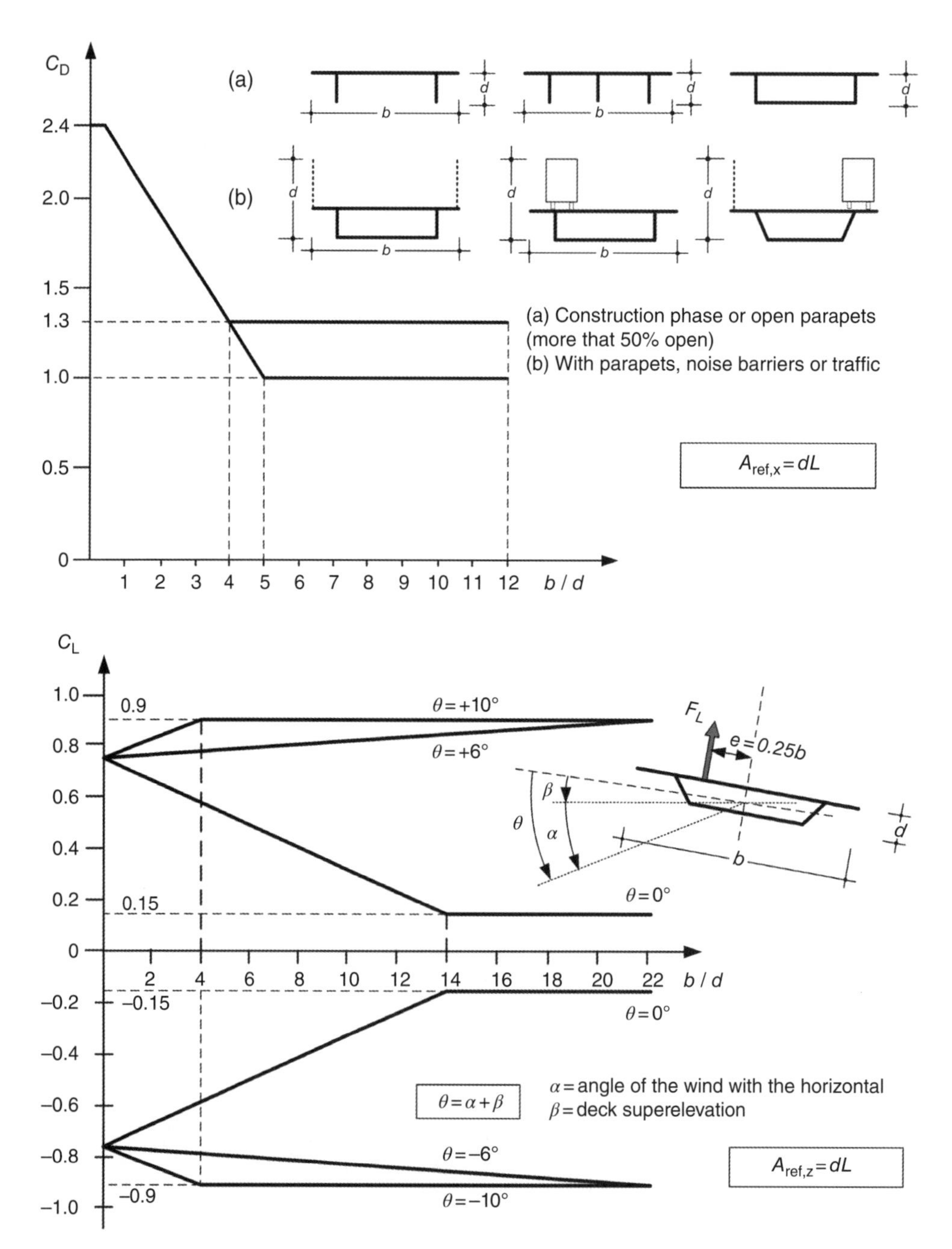

Figure 3.27 Drag coefficients, C_D and C_L, or unloaded or bridge decks under traffic (*Source:* Adapted from Ref [15], EN1991-4).

In most cases, b/d is above 4 and C_D lies between 1.0 and 1.3. For box sections with inclined webs, C_D may be reduced at 0.5% per degree of web inclination with the vertical, with a maximum reduction of 30% in any case.

When the bridge is under traffic (Figure 3.27) the depth of the vehicles must be considered when evaluating the exposed area, $A_{ref,x}$. For road bridges, this is considered by taking an additional depth of 2.0 m above the deck, while for railway bridges this is done by taking an additional depth of 4.0 m from the top of the rails on top of the total length of the bridge. For bridges under traffic, the dynamic wind pressure is reduced by taking a *reduced basic wind velocity* (v_b^* is limited to 23 m s^{-1} for road bridges and 25 m s^{-1} for railway bridges) to take into consideration the reduced probability of having the bridge under the characteristic value of the live loads together with winds blowing at very high speeds. For specific design cases, the maximum *peak wind velocity* may be limited (e.g. to 100 km h^{-1}) as a safety condition for the vehicles, which means the bridge should be closed to traffic if higher peak wind velocities are forecast.

For lift forces, unless some wind tunnel tests have been performed for the deck section or a specific C_L coefficient is defined in EC1-1-4 [15]; from Figure 3.27 one may take $C_L = +0{,}9$ or $C_L = -0{,}9$, and the vertical action as

$$F_L = C_L \, q \, A_{ref,z} \tag{3.36}$$

where $A_{ref,z} = b\,L$ is the in-plan area of the deck length L.

The moment due to wind actions on the deck may be evaluated by assuming the drag force, F_D, acting at 60% of the equivalent depth (d_{eq}), measured from below and the lift force, F_L, at 25% of the width (b) measured from windward (Figure 3.27).

For the piers, one may adopt the force (drag) coefficients given for rectangular or polygonal sections in EC1-1-4 [15] or, in a simplified approach, take the equivalent rectangular envelope shape as given in Figure 3.28. The force coefficients are in general between 1.0 and 1.8. In tall piers, these forces are evaluated with a variable wind pressure, q, with respect to the height above the ground.

3.11.3 Aerodynamic Response: Basic Concepts

Wind-tunnel tests, done on section models such as the one shown in Figure 3.29, are most useful to interpret and determine the dynamic response of the deck under wind flow-aerodynamic response. A section model for a wind-tunnel test consists of a scale model of a certain length of the deck, L, which is determined by the available width of the tunnel in order to avoid any blockage effects under the test wind flow due to the proximity of the walls of the tunnel. This sectional model is then attached at its extremities to springs simulating the elastic stiffness of the remaining structure. These springs are modelled with scale stiffness in order to reproduce the scale of the main frequencies of the deck, which for these types of problem are, basically, the vertical and torsion frequencies, f_b and f_t, respectively. Figure 3.29 presents the variation of the shape coefficients C_D, C_L and C_M with the angle of attack, α. These values may be adopted for the evaluation of the equivalent static wind forces on the deck referred to in the previous section. For example, for $\alpha = 0$, one has $C_D = 1.17$, $C_L = -0.494$, $C_M = -0.006$.

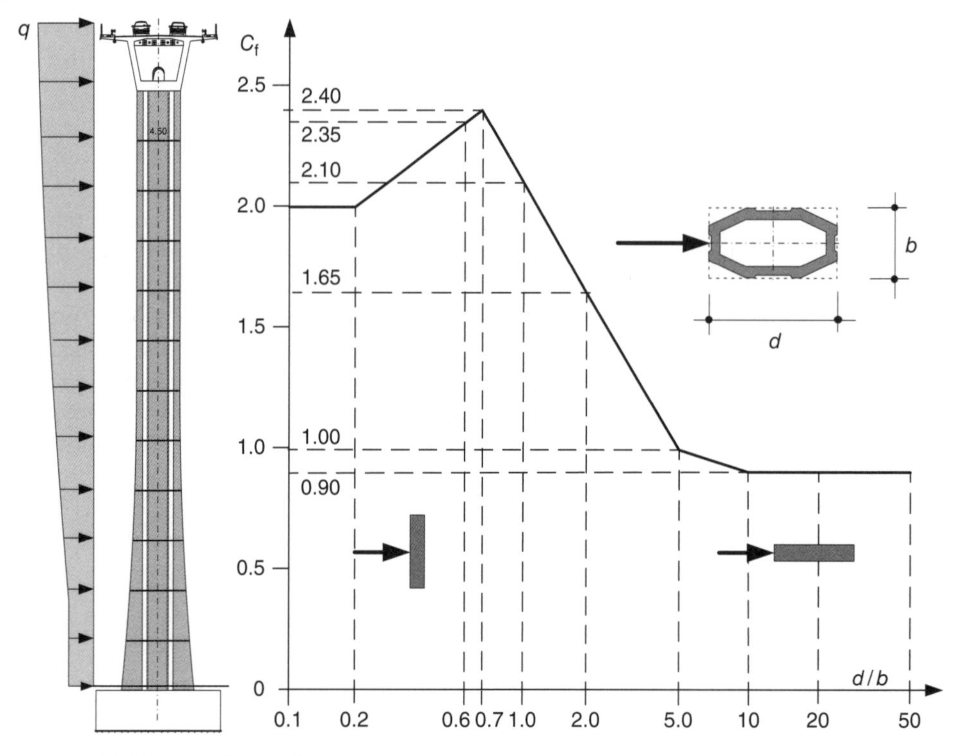

Figure 3.28 Force coefficients for bridge piers.

Under certain wind flow conditions, the deck (or the sectional model in the wind tunnel) may experience several types of aerodynamic excitation (Figure 3.30), which are characterized by a motion of limited or divergent amplitudes of oscillation. In the first case, *limited amplitude response*, which may induce unacceptable stresses or fatigue, different cases may occur:

- *Vortex induced oscillations* associated with the so-called vortex shedding alternately from the upper and lower surface of the bridge deck (Figure 3.30). The frequency of the vortices, dependent on the magnitude of the wind velocity, v, being close to the bending frequency of the deck, induce the risk of resonance if structural damping is not enough to reduce the amplitude of the oscillations.
- *Turbulence response*, due to the turbulent nature of the forces induced by wind gusts, the deck may oscillate if sufficient energy is induced at certain frequency bands close to the natural frequencies of the structure. This phenomenon may usually be neglected if both bending and torsion natural frequencies, f_b and f_t, are greater than 1 Hz. Otherwise, the dynamic effects of turbulence response should be considered on the basis of a dynamic analysis.

The case of *divergent amplitudes of oscillation* gives rise to aerodynamic instability of the deck, namely by *galloping* or *flutter*, as referred in the sequel. For details on basic references on bridge aerodynamic stability, the reader is referred to Refs. [17–19].

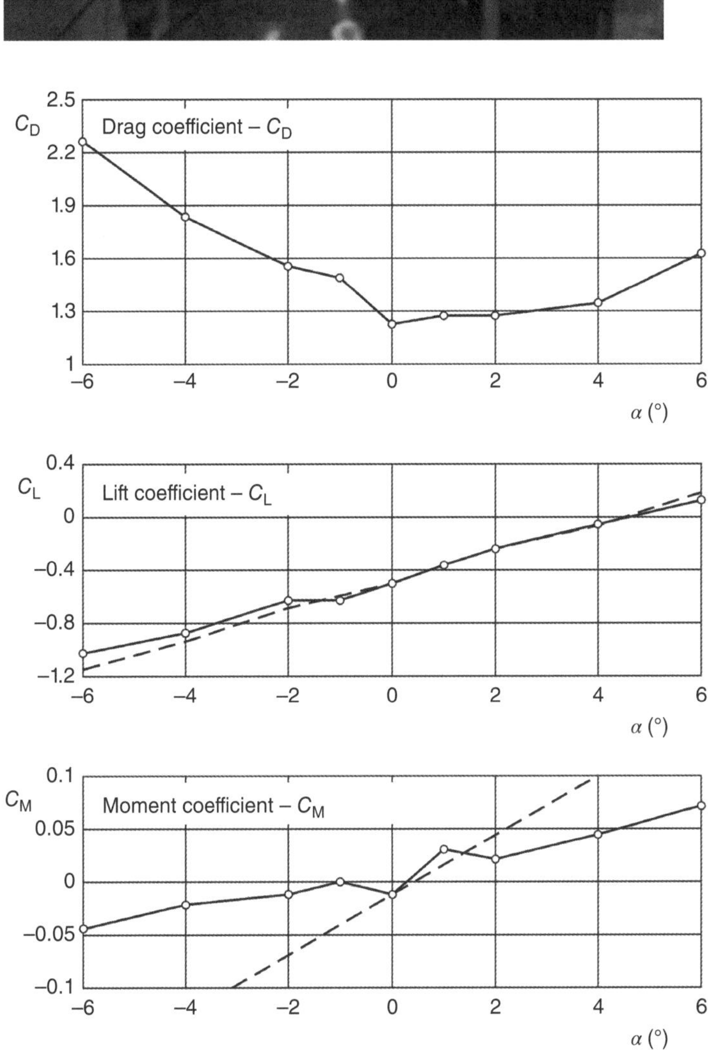

Figure 3.29 Wind tunnel test on a sectional model of a bridge deck of Figure 3.26: Drag C_D, lift C_L and moment C_M coefficients, as a function of the angle of attack α of the wind [16]. Reproduced with permission from GRID, SA.

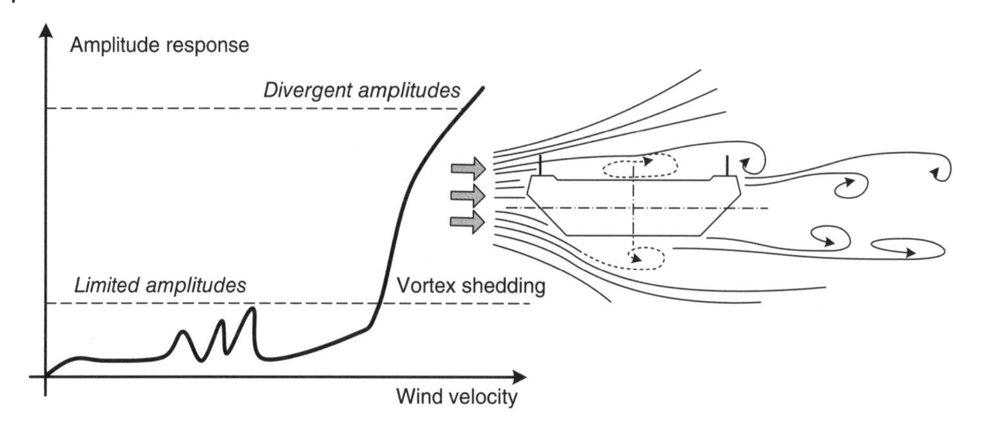

Figure 3.30 Aerodynamic instabilities.

3.11.3.1 Vortex Shedding

The wind flow around a bridge deck, a pier or even a stay cable, as shown in Figure 3.30, may induce vortex shedding. The pressure difference between the two sides of the element under wind flow, due to vortex shedding, produces transverse forces normal to the direction of the wind and so possible bending vibrations. These vortices are known as von Karman vortices after studies of this phenomenon by von Karman at the beginning of twentieth century. If the frequency of the vortices is close to one of the natural frequencies of the structure, only damping can limit large displacements induced by a resonance phenomenon.

In *vortex shedding*, the intensity and frequency of the vortices are dependent on the wind velocity v and on the geometry (shape) of the deck, namely its depth/width ratio d/b . The critical wind speed for a bridge deck, to induce vortex shedding, is given by:

$$v_{cr,VS} = \frac{f_o d}{S_t} \tag{3.37}$$

where f_o is the natural frequency of the fundamental mode, d is the depth of the cross section and S_t is the *Strouhal* number [15, 19], equal to 0.2 for a cylinder; for a rectangular section (that may be taken as bridge deck envelope) this only varies between 0.1 and 0.09, when $0.5 < d/b < 10$. If the critical wind speed, $v_{cr,VS}$, given by Eq. (3.37) is low, transverse oscillations may occur under normal wind conditions.

However, the induced transverse forces due to deck oscillations are small for low critical wind velocities and if the structural damping is enough to reduce the amplitude of the oscillations the consequences of the phenomena of vortex shedding are reduced. The relevance of vortex shedding may increase when the deck critical wind speeds are high, yielding high transverse forces in the structure under wind flow oscillations.

In practice, safety precautions against resonant critical conditions are satisfied if the *critical wind velocity for vortex shedding* is at least 1.25 times greater than the *mean wind velocity* as defined in Section 3.11.1.

3.11.3.2 Divergent Amplitudes: Aerodynamic Instability

The cases of *divergent amplitudes*, inducing oscillations that may lead to collapse by aerodynamic instability of the structure, are associated with the following basic phenomena:

- Instability in pure torsion – *torsional divergence*
- Instability in pure bending – *galloping*
- Instability by flutter in a single mode – *torsional flutter*
- Interactive instabilities, in torsion and bending – *classic flutter*

The first mode corresponds to an induced torsional rotation of the deck that cannot be 'reduced' by its elastic torsion resistance; is like a static instability mode and is not critical for a bridge deck because it is associated with very high wind velocities.

The second mode, *galloping*, is a self-induced crosswind bending mode vibration occurring when the apparent variable incidence of the wind, due to the motion of the deck, induces vertical forces amplifying the bending movement. Some cases of practical examples of galloping are referred in Ref. [20] for decks and in particular for pylons of cable stayed bridges with a single plane of cables. The example of the wind tunnel testing undertaken for the Second Panamá Bridge (Figure 3.31) is presented. Galloping may be appraised on the basis of the lift coefficient and to its variation with the angle of attack, α, of the wind flow. A *necessary condition* for gallop may be taken from sectional wind-tunnel tests. For that, observing at $\alpha = 0$ the value of the drag coefficient and the variation with α of the lift coefficient, the necessary condition for gallop, the *den Hartog criterion*, reads:

Figure 3.31 Centennial Bridge – second crossing of the Panama Canal (*Source:* Photograph by Wsvan/https://commons.wikimedia.org).

$$C_D + \left(\partial C_L / \partial \alpha \right) < 0 \tag{3.38}$$

As $C_D > 0$, gallop cannot occur if the derivative of the lift coefficient at $\alpha = 0$ is positive. For example, this condition is satisfied by the results for C_L presented in Figures 3.26 and 3.29. For the wind-tunnel studies on the second Panama Bridge mast [20], for wind angles of attack between 5 and 6°, $\partial C_L / \partial \alpha = -5{,}34$. However, further stability studies have shown the resulting damping remains positive for a 100-year wind return period, assuring aerodynamic stability [20].

If wind-tunnel tests are not available, the critical wind speed for galloping may be estimated for bridges with main spans up to 200 m by,

$$v_{cr,G} = \frac{f_b C_g m \delta_s}{\rho d} \tag{3.39}$$

where f_b is the natural bending frequency, C_g is a galloping coefficient, m is the mass per unit length of the deck, δ_s is the logarithmic decrement of structural damping, ρ the density of the air = 1.25 kg m^{-3} and d is the depth of the cross section of the deck. Galloping coefficients C_g may be taken from specific literature [15, 21]. In Ref. [21], values equal to 1.0 or 2.0 are proposed, respectively, for bridges with side overhangs less than or greater than 0.7 d, where d is the depth of the bridge deck. For aerodynamic stability problems, the logarithmic decrement δ_s may be taken to be 3% for steel bridges, 4% for composite bridges and 5% for concrete bridges. It should be noted that the damping coefficient is $\xi \cong \delta_s/(2\pi)$ [17], results in $\xi = 0.5\%$ for steel bridges and 0.8% for concrete bridges.

Flutter in a single mode, *stall flutter*, is similar to gallop. It is a self-induced vibration in a single periodic torsion mode of the deck; it is induced by a periodic torsion oscillation originating from a periodic variation of the flow around the deck. The stall flutter critical wind speed may be evaluated from the torsion natural frequency f_t (Hz) on the basis of the following simplified formula [19, 21]:

$$v_{cr,SF} = \tau f_t b \tag{3.40}$$

where τ is a coefficient depending only on the type of the cross section, such that $\tau = 5$ in most cases. But, for cases of slab or box girder sections with $b < 4d$, then $\tau = 12$.

Classical flutter is mainly an aerodynamic instability due to the interaction of the bending and torsion forced vibration modes of the deck under wind action. In slender bridge decks, namely with large values of the length/width ratio (L/b) as is often the case in suspension bridges, and with a low frequency ratio f_t/f_b between the torsion and bending natural frequencies, there is risk of interaction between the basic oscillation modes of the deck. The structure, for a certain critical wind speed, may experience large amplitude coupled oscillations and in these cases, structural damping may not be enough to avoid a collapse by aerodynamic instability by classical flutter.

The theoretical background of this phenomenon goes to the interaction of two oscillation modes in dynamic stability of structures, as shown in a schematic form in Figure 3.22 for a simple two degrees of freedom system [22]. The excitation load parameter λ is, for example, the dynamic wind pressure, inducing a forced oscillation in the

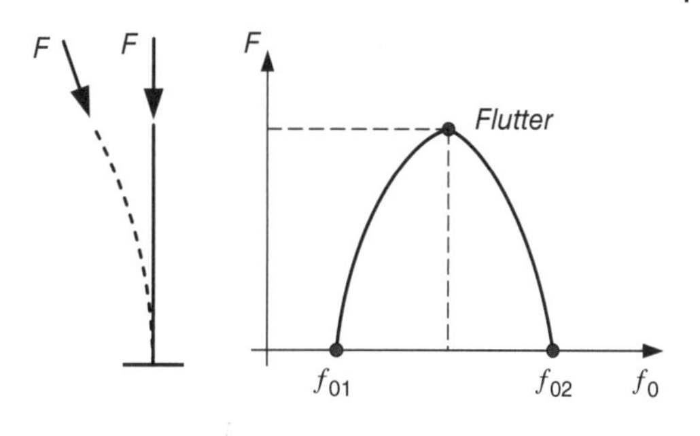

Figure 3.32 Flutter in a 2 DOF nonconservative system (F – follower force; f_{0i} – natural frequencies).

dynamic equilibrium of the structure. For $\lambda = 0$, the system has two natural vibration frequencies, f_{01} and f_{02}; under increasing load λ, these frequencies tend to coalesce, but above a certain load level dynamic equilibrium is no longer possible, inducing unlimited amplitude oscillations. The critical load for flutter instability is then associated to the load at which the two forced vibration frequencies tend to coalesce, as shown in Figure 3.32. Classical flutter of bridge decks may be investigated by a variety of approaches, namely simple analytical approaches [15, 17, 19, 21] – Selberg formula, Kloopel diagrams, CECM (Convention Européenne de la Construction Métallique) tables, Scanlan Derivatives, Numerical Approaches or Wind tunnel testing [23].

The simplest approach, for the critical wind speed of flutter instability of a bridge deck, is to adopt the basic case of an ideal thin plate aerofoil for which the *Selberg* formula [21] holds,

$$v_{cr,thin\ plate} = 3{,}7 b f_t \sqrt{\frac{mr}{\rho b^3}\left[1-\left(\frac{f_b}{f_t}\right)^2\right]} \tag{3.41}$$

where m is the mass per unit length of the deck, r is the polar radius of gyration of the bridge cross section, that is, $\sqrt{I_o/A}$, with $I_o = \rho(I_y + I_z)$ the polar mass moment of inertia of the cross section and A its area, ρ is the air unit mass = 1.25 kg m^{-3}, b is the width of the cross section and f_b and f_t are the natural bending and torsion frequencies of the deck. The *Selberg* formula can also be adapted to evaluate the flutter critical wind speed of bridges with main spans less than 200 m through a correlation factor η [19] (Figure 3.33),

$$v_{cr,F} = \eta\, v_{cr,thin\ plate} \tag{3.42}$$

The *Selberg* formula gives good approximations for the critical wind speed, provided that the ratio $f_b/f_t > 1.5$ and δ_s is about 0.05. From Eq. (3.41) the influence of the following parameters is apparent:

- the ratio f_b/f_t, yielding a low critical wind speed when the frequencies are close to each other; in practice, it is convenient to have $f_b/f_t > 2.5$ as a starting point to warranty aerodynamic stability; the η values may vary between 0.2 for a slab girder type deck to 0.6 for a standard box girder deck; for a streamline box girder deck the values may approach 0.9. Some values are given in Figure 3.33 and more detailed values may be taken from [19];

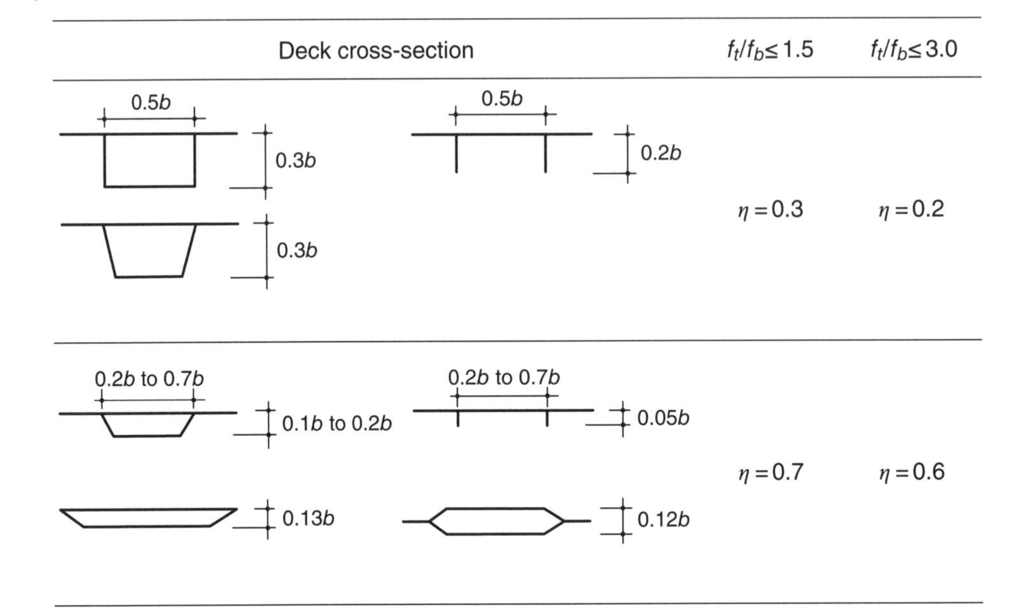

Figure 3.33 Values of the reduction factor η to determine the critical classical flutter velocity with respect to the critical velocity of a thin plate ($\eta = 1.0$) by the *Selberg* formula.

- the parameter $m\ r/\rho\ b^3 = [m/(\rho\ b^2)]\ (r/b)$, from which the influence of the ratios $m/(\rho\ b^2)$ and (r/b) are apparent; the former $m/(\rho\ b^2)$ is dependent on the cross section of the deck but it varies approximately inversely with the width b; the last (r/b) is about 0.3 for a practical range of bridge decks varying only between 0.27 for single box girders with large deck overhangs to 0.32 for decks supported by main girders or cables at the outer edges, as referred to in [21].

It is clear from the *Selberg* formula that aerodynamic stability is very much improved by the torsion stiffness of the deck, which increases the torsion frequency, f_t. This is the case for modern suspension bridges with 'streamline' box deck sections as shown in Figure 3.22b.

For all cases of aerodynamic phenomena (gallop, stall flutter or classical flutter), safety should be checked by assuring a *safety factor for the critical wind speed* of at least 1.5 with respect to the mean wind velocity measured at the deck level.

3.12 Hydrodynamic Actions

Bridge piers and foundations located in river beds are subjected to hydrostatic and hydrodynamic actions. The hydrostatic action on an immersed element is an upward vertical force equal to the weight of the volume of water occupied by the element. Hydrodynamic actions are determined from the water pressure against the element. If v is the water flow velocity in (m s^{-1}), ρ the unit weight of the water and g the acceleration of gravity, one has:

$$p = \frac{1}{2} c \rho v^2 = \frac{1}{2} c \gamma v^2 / g \qquad (3.43)$$

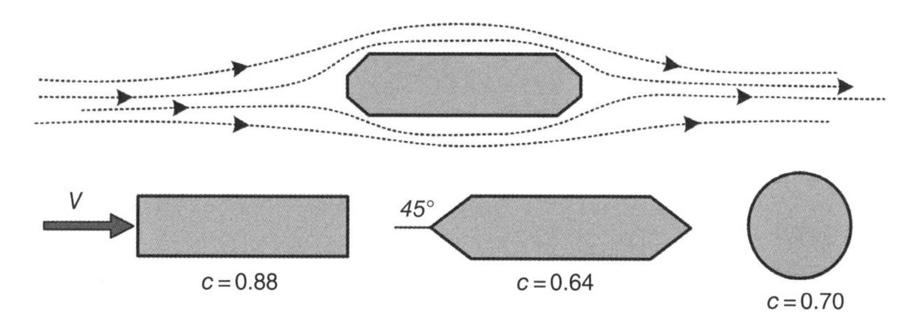

Figure 3.34 Hydrodynamic actions on immersed piers. Shape factors.

where c is a shape factor dependent on the geometry of the immersed element. Taking $g = 9.81\,\text{m s}^{-2}$ and $\gamma = 10\,\text{kN m}^{-3}$, one has from Eq. (3.43): $p = 0.51\ c\ v^2$ (kN m^{-2}).

Figure 3.34 presents values for the coefficient c for typical pier shapes. In flood conditions, it is quite important to take into consideration the effect of the solid material being transported by the water, yielding an equivalent unit weight (density) with values between 10 and 12 kN m^{-3}. Another relevant effect of hydrodynamic actions, as referred to in Chapter 2, is the *scour effect* around bridge pier foundations.

3.13 Thermal Actions and Thermal Effects

3.13.1 Basic Concepts

Temperature actions in bridges are due to temperature distributions within the structural elements, namely induced by solar radiation, as shown in Figure 3.35a. The heat of hydration of cement in concrete bridges is another relevant action, in particular up to three to four days after concreting the element. In general, one has to deal with a nonlinear distribution of temperature within the element (Figure 3.35b), which is dependent on several parameters, namely related to local climatic conditions, the orientation of the element, geometrical and material properties of the element as well as on specific characteristics of the surfacing of the bridge deck. The effects of thermal actions may be evaluated by splitting it into the following components:

- the *uniform temperature variation*, measured from a reference temperature, T_0, being the temperature with respect to which the action effect is considered; for example, the erection (or completion) temperature or more generally the temperature at the relevant stage of the structural element at its restraint.
- the *differential temperature variation*, which yields a *thermal gradient* within a bridge component, namely in a bridge deck due to solar radiation or due to the heat of hydration of cement in a concrete bridge. This is illustrated in Figure 3.36, where a nonlinear distribution of the temperature in a bridge deck results from a certain period of heating and cooling of the upper surface of the bridge deck under the solar radiation effect, while at the bottom surface the temperature is controlled by the shade temperature [24].

The uniform temperature variations are associated with seasonal changes, while the thermal gradients within the deck, or within any other bridge element such as a pier shaft, are associated with daily thermal variations.

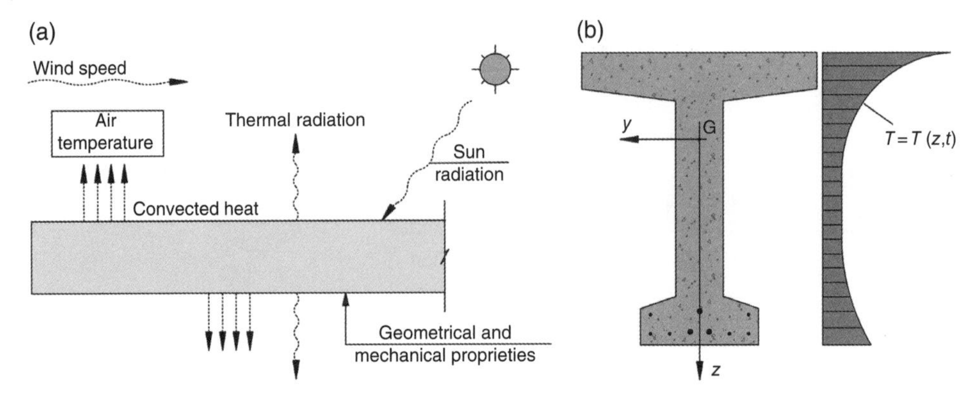

Figure 3.35 Effect of sun radiation on a bridge deck: (a) Parameters influencing the temperature distribution and (b) nonlinear temperature distribution in a prestressed concrete bridge girder.

The uniform and differential temperature actions in bridges are specified in codes for actions. These temperature distributions are dependent on a set of parameters, namely, solar radiation, air temperature, the degree of nebulosity and wind velocity, bridge location and its orientation, the shape of the cross section, thermal properties of the material and the type and colour of the surfacing of the bridge deck.

Formally, the temperature distribution, T, in a bridge deck is determined from the Fourier heat equation [25]:

$$k\Delta T + Q = \rho c \frac{\partial T}{\partial t} \tag{3.44}$$

Where:

- $\Delta T = \partial^2 T/\partial x^2 + \partial^2 T/\partial y^2$ is the Laplacian of the temperature $T\,(x,y,t)$ at each instant, t, if assumed only dependent on the cross section coordinates x and y; that is, the same for all cross sections,
- k is the thermal conductivity, representing the rate of heat flow per unit area and per unit temperature gradient, k (W m^{-1} °C),
- Q is the amount of the heat generated inside the deck per unit volume and per unit time, Q (W m^{-3}), as for example the heat of hydration of cement in a concrete bridge deck,
- ρ (kg m^{-3}) is the material density (unit weight), and finally,
- c is the specific heat of the material, that is, the quantity of heat required to increase the temperature of the unit mass by one degree, c (J kg^{-1} °C^{-1}).

Equation (3.44) is a partial differential equation to be solved taking into consideration the temperature and the heat flow conditions at the boundary of the deck surface, which may be expressed as

$$k\frac{\partial T}{\partial n} + q = 0 \tag{3.45}$$

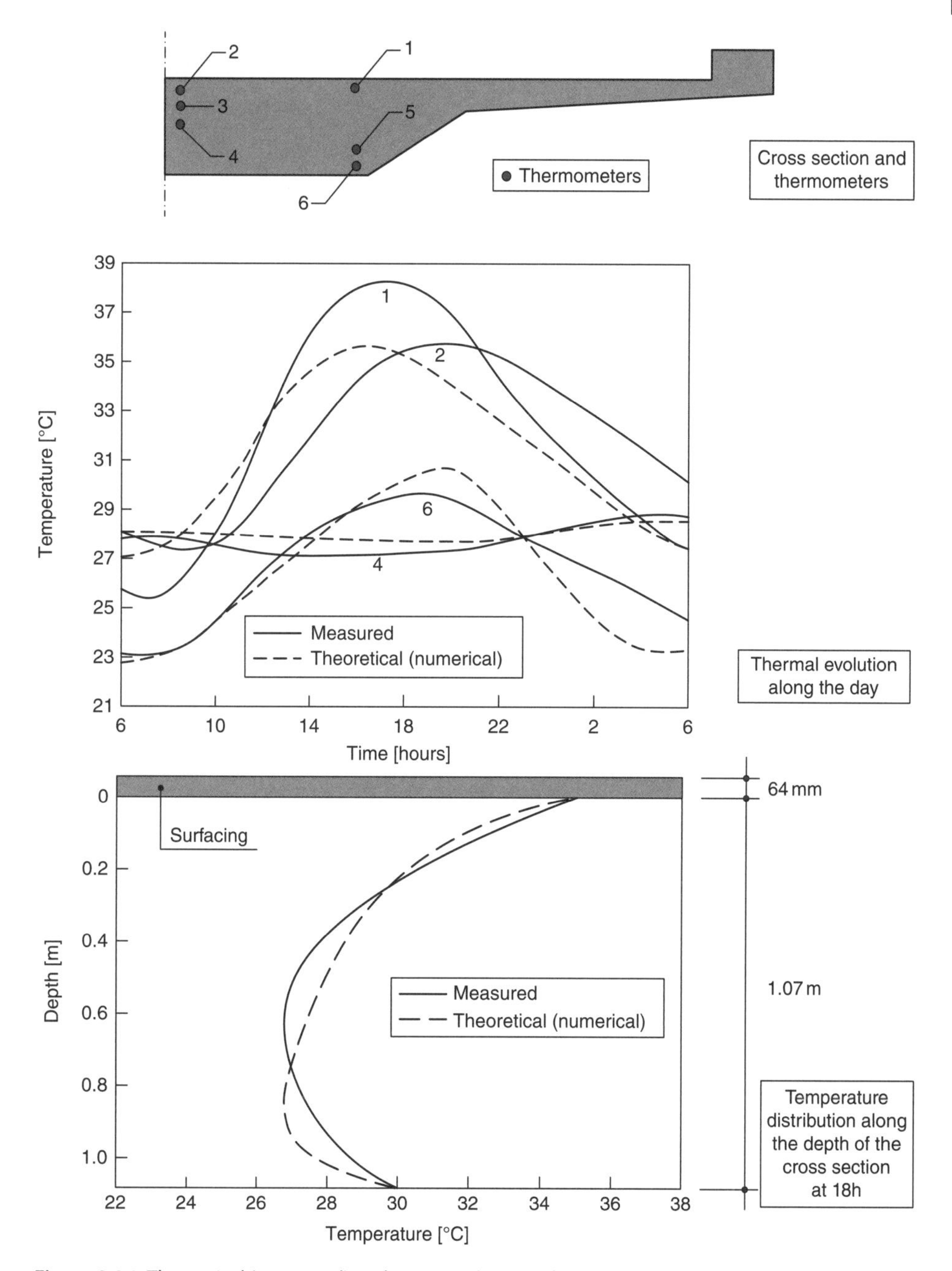

Figure 3.36 Theoretical (numerical) and measured results for the temperature distribution in a concrete bridge deck cross section [24].

where $\frac{\partial T}{\partial n}$ is the derivative of temperature in the direction normal to the boundary defined by the unit vector n, that is,

$$\frac{\partial T}{\partial n} = \boldsymbol{grad\,T.n} = \frac{\partial T}{\partial x}n_x + \frac{\partial T}{\partial x}n_y \tag{3.46}$$

where n_x and n_y are the direction cosines of n and ***grad T*** denotes the temperature gradient at the boundary. In Eq. (3.45), q denotes the amount of heat transfer per unit time and per unit area of the boundary, with the units $\mathrm{W\,m^{-2}}$.

To solve Eq. (3.46), the quantity of heat q in Eq. (3.44) has to be defined as the sum of three basic quantities – solar radiation, q_s, which is dependent on the total heat from solar rays reaching the surface per unit area and per unit time, the amount of heat transfer by convection, due to the temperature difference between the bridge surface and the air, which is dependent on the wind speed, the type of material and on the configuration and orientation of the surface, and finally on the amount of radiation from the surface to the surrounding air. The reader is referred to Ref. [26] for a detailed discussion on thermal and meteorological properties involved in the use of Eq. (3.44). This equation may be solved, for example, on the basis of a finite element approach, to find the temperature at the bridge deck at each instant and at each point. Standard computer programs [27] may be adopted for specific bridge problems and for parametric studies to define the thermal field in typical bridge decks. Figure 3.36 shows a typical example for a bridge deck. The temperatures at several hours of the day were evaluated from a finite element program along the depth of the cross section and measured in a deck model. The basic geometrical and thermal characteristics adopted for the numerical model are referred to in the figure.

3.13.2 Thermal Effects

For design practice, the temperature distributions in bridge decks are usually defined from values defined in the codes. Only in very particular cases numerical models based on the approach referred in the previous section to evaluate the thermal field have to be developed. Let the effect of a nonlinear temperature distribution $T(z,t)$ be assumed in the case of a statically determinate structure as shown in Figure 3.37 – a simply supported bridge deck with a constant cross section. One assumes the temperature distribution T at each time instant t, as being the same for all the cross sections and dependent only on the z coordinate of the sections. For the case under consideration, the nonlinear temperature distribution yields a self-equilibrated stress field (eigenstresses) in the bridge deck, but no internal forces since the structure is statically determinate. To show this, one may consider the evaluation of these stresses, by using the principle of superposition of elastic effects, as shown in Figure 3.37. Firstly, one introduces a longitudinal restraint at the end sections by introducing applied stresses defined by:

$$\sigma_0 = -E\alpha T(z,t) \tag{3.47}$$

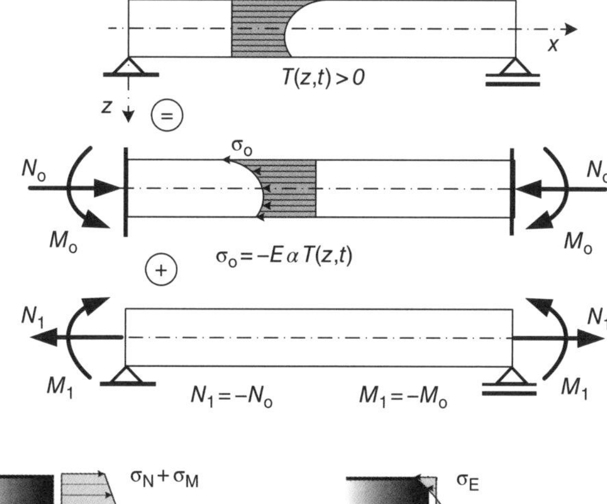

Figure 3.37 Nonlinear temperature distribution in a simple supported bridge deck – Analysis scheme to determine the self-equilibrated normal stresses.

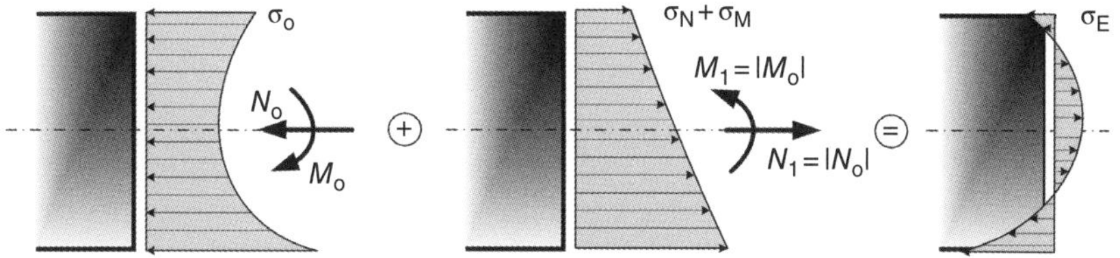

Figure 3.38 Self-equilibrated thermal stresses in a simple supported deck.

where E is the Young's modulus of the material and α is the coefficient of thermal expansion. The thermal strains $\varepsilon = \alpha\ T(z,\ t)$ are restrained by σ_0. At the end sections with area A of the cross section, the stress resultants are given by:

$$N_0 = \int A\sigma_0(z,t)dA \quad M_0 = \int A\sigma_0(z,t)z\,dA \tag{3.48}$$

By applying a force $N_1 = -N_0$ and a moment $M_1 = -M_0$ at the end sections and superposing the two stress fields, σ_0 and σ_1 (due to N_1 and M_1), the stress field as indicated in Figure 3.38 is obtained by simple beam theory:

$$\sigma_E(z,t) = -E\alpha T(z,t) + \frac{N_1}{A} + \frac{zM_1}{I_y} \tag{3.49}$$

where I_y is the moment of inertia with respect to the neutral y-axis of the cross-section. By introducing the definitions of N_1 and M_1 and taking into consideration Eq. (3.48), one obtains the stress distribution

$$\sigma_E(z,t) = E\alpha\left[-T(z,t) + \frac{\int T(z,t)dA}{A} + \frac{z\int T(z,t)z\,dA}{I_y}\right] \tag{3.50}$$

So, the resulting stresses may be defined as

$$\sigma_E(z,t) = -E\alpha\left[T(z,t) - \Delta T_U - \Delta T_M\right] = -E\alpha\,\Delta T_E \tag{3.51}$$

where ΔT_U and ΔT_M are the average (uniform) and linear temperature components defined as:

$$\Delta T_U = \frac{\int T(z,t)dA}{A} \qquad \Delta T_M = \frac{z\int T(z,t)z\,dA}{I_y} \tag{3.52}$$

and ΔT_E is a nonlinear temperature component defined as

$$\Delta T_E = T(z,t) - \Delta T_U - \Delta T_M \tag{3.53}$$

In conclusion, the temperature distribution within a specific bridge element $T(z,t)$, may be split into an uniform temperature component, ΔT_U, a linear varying temperature component, ΔT_M, and a non-linear temperature difference component, ΔT_E, the last one giving rise to a self equilibrated stress field σ_E – eigenstresses. On a more general form, as represented in Figure 3.39, the linear varying temperature component, ΔT_M, is a temperature gradient along the vertical z-axis, ΔT_{Mz}, but also a temperature gradient along the y-axis, ΔT_{My}.

For statically determined structures, ΔT_U and ΔT_M do not induce any internal forces or stresses in the structure, but only strains and deflections. The uniform temperature only affects the design of the bearings and expansion joints through the imposed displacements.

Effectively, ΔT_U and ΔT_M induce an elongation and a constant curvature deflection in the beam, according to the Bernoulli hypothesis of plane sections in a remaining plane. Since the structure is statistically determined, these deformations, as shown in Figure 3.40, do not induce any internal forces. The bridge deck elongates with a total displacement u at the free end section and takes a constant curvature, $1/R_M$, given by

$$u = \alpha L \Delta T_U \quad 1/R_M = \frac{\varepsilon_1 - \varepsilon_2}{h} = \alpha\frac{T_1 - T_2}{h} = \alpha\frac{\Delta T_M}{h} \tag{3.54}$$

where ε represents the strain at a distance z of the neutral axis and ΔT_M is the difference of temperature between the top and bottom surfaces of the bridge deck with depth h, as shown in Figure 3.40. Only the nonlinear temperature component gives rise to a nonlinear stress field σ_E, which has the same shape of ΔT_E distribution as results from Eq. (3.51).

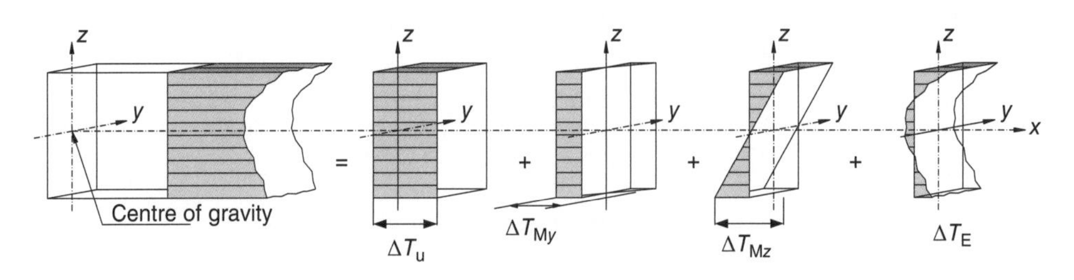

Figure 3.39 Decomposition of a nonlinear temperature distribution in a uniform, linear and self-equilibrated component.

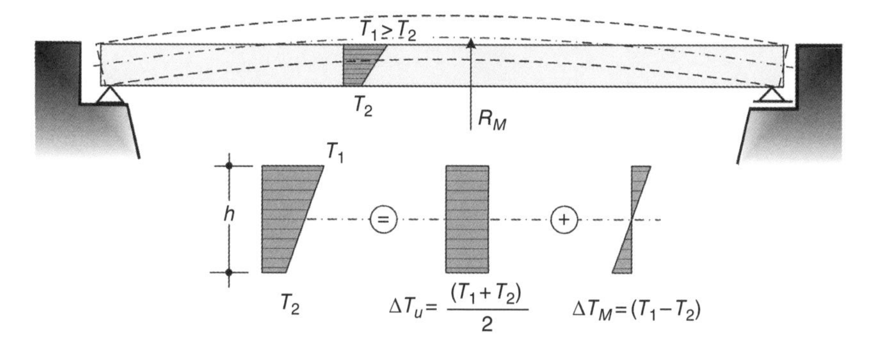

Figure 3.40 Effect of a uniform and a linear temperature distribution in a simple support bridge deck with a rectangular cross section. Analysis scheme to determine the self-equilibrated normal stresses.

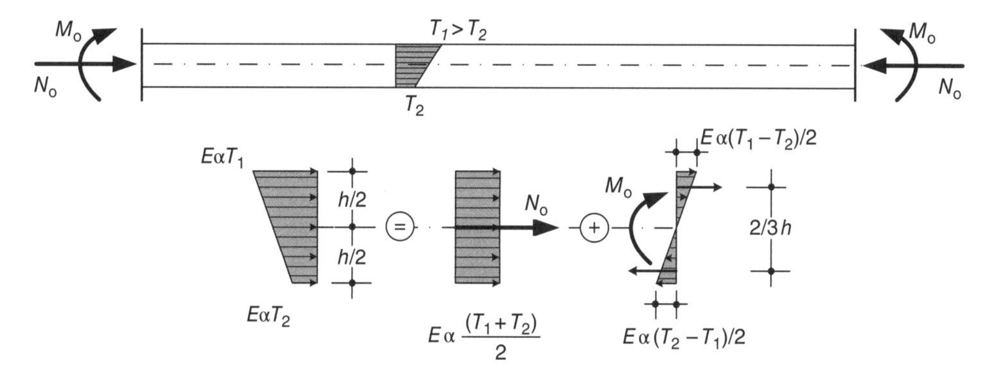

Figure 3.41 Thermal actions and actions effects of a linear temperature distribution on a built-in deck with a rectangular cross section.

Let the case of statistically undetermined structures be considered. If the same bridge deck is now assumed to be longitudinally built in at the end sections (Figure 3.41), which, however, are assumed to be free for the displacements in the plane of the cross section, the uniform and linear thermal components do not induce any deformations (the longitudinal displacements are totally restrained) but internal forces are generated that are equal to the stress resultants, N_0 and M_0, due to the stresses, σ_0, restricting the longitudinal displacement and curvature in the simply supported case:

$$N_0 = EAu = EA\alpha\, L\Delta T_U \quad M_0 = EI\left(1/R_M\right) = EI\left(\alpha\,\Delta T_M/h\right) \tag{3.55}$$

Now let the case of a continuous bridge deck be considered, as shown in Figure 3.42. The uniform temperature component induces a total displacement at the expansion joint, u, which is still given by Eq. (3.54), now with $2.5L$ being the length of the bridge deck. The nonlinear temperature distribution, ΔT_E, induces the same stresses as in the simply supported case shown in Figure 3.37.

The linear temperature component, ΔT_M, induces a linear bending moment diagram in the bridge deck, as shown in Figure 3.43, due to the induced reactions R that may be easily evaluated (Figure 3.44) by the force method of linear structural analysis. Using the symmetry and the principle of virtual work for evaluating displacements for the

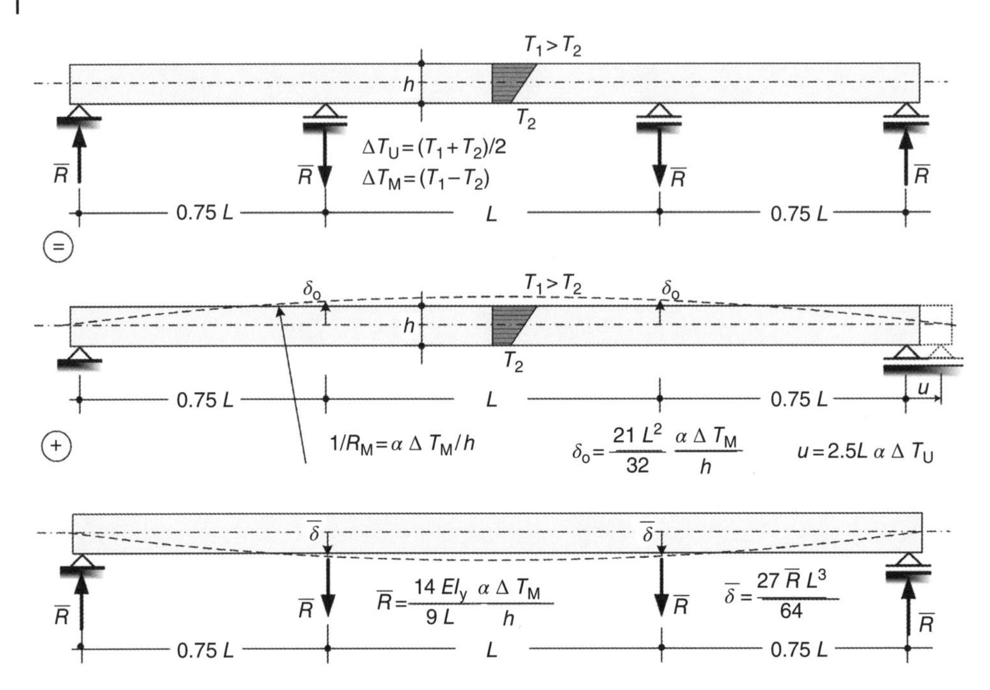

Figure 3.42 Linear temperature distribution in a continuous bridge deck.

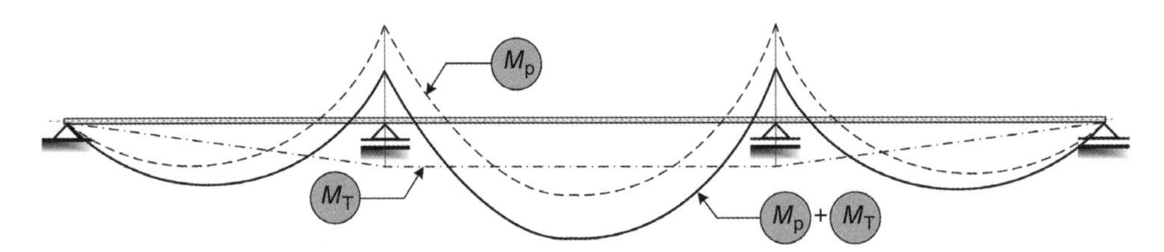

Figure 3.43 Bending moment diagrams, due to permanent loads (M_p) and a linear temperature distribution (M_T) in a three span continuous bridge deck.

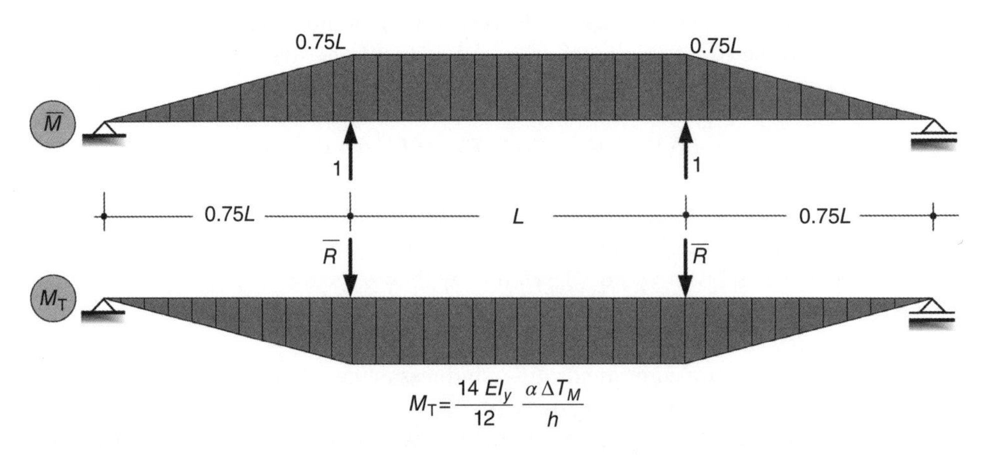

Figure 3.44 Evaluation of the bending moment diagram by the Force Method in a three span continuous bridge deck.

deflections at the internally supported sections of the released structure, one assumes a simply supported beam and, as $1/R_M$ given by Eq. (3.54),

$$\delta_0 = \frac{1}{R_M} \int \bar{M} dx = \frac{21L^2}{32} \frac{\alpha \Delta T_M}{h} \tag{3.56}$$

The displacements $\bar{\delta}$ in the base system due to the forces $\bar{R}$ are

$$\bar{\delta} = \frac{\bar{R}}{EI_y} \int \bar{M}\,\bar{M} dx = \frac{27\bar{R}L^2}{64EI_y} \tag{3.57}$$

From compatibility $\bar{\delta} = \delta_0$, one obtains

$$\bar{R} = \frac{14EI_y}{9L} \frac{\alpha \Delta T_M}{h} \tag{3.58}$$

Figure 3.43 shows the final elastic bending moments due to the dead load p and the linear temperature variation ΔT_M. Then, for a positive ΔT_M, the bending moments increase at the span sections due to ΔT_M and are reduced over the supports.

3.13.3 Design Values

The uniform temperature component depends on the maximum and minimum temperature variation experienced by the bridge, which is dependent on the maximum and minimum shade air temperature, T_{max} and T_{min}, and on the initial bridge temperature, T_0. So, the maximum expansion and contraction ranges of uniform bridge temperatures are defined as:

$$\Delta T_{U,exp} = T_{e,max} - T_0 \quad \Delta T_{U,cont} = T_0 - T_{e,min} \tag{3.59}$$

where the values of $T_{e,max}$ and $T_{e,min}$ are the uniform bridge temperature components that are dependent on T_{max} and T_{min}, as well as on the type of structural material. Figure 3.45 shows the correlation between the characteristic values of the shade air temperature and the uniform bridge temperature component, for steel, composite and concrete bridge decks. Characteristic values of shade air temperatures shall be obtained at the site location with reference to national maps of isotherms usually given in each country national annex of Part 1-5 of EC1 (denoted as EC1-1-5 [28]).

As an example, for the case of a bridge site where T_{max} and T_{min} are, respectively, 40 and −10°C, if one assumes T_0 = 10°C, one obtains for a steel bridge deck, using Eq. (3.59) and Figure 3.45, $\Delta T_{U,exp}$ = 55–10 = 45°C and $\Delta T_{U,cont}$ = 10 – (−13) = 23°C. For the same site conditions but for a concrete bridge deck, one obtains $\Delta T_{U,exp}$ = 42 – 10 = 32°C and $\Delta T_{U,cont}$ = 10 – (−2) = 12°C.

The vertical temperature component with nonlinear effect is defined as well in EC1-1-5 for steel, composite and concrete bridge decks on the basis of typical temperature diagrams along the depth of the deck. An example of these diagrams for a steel box girder deck is represented on Figure 3.46. It is important to note that these diagrams include the linear temperature gradient ΔT_M and the nonlinear temperature gradient ΔT_E, together with a small part of component ΔT_U, which is included in the uniform

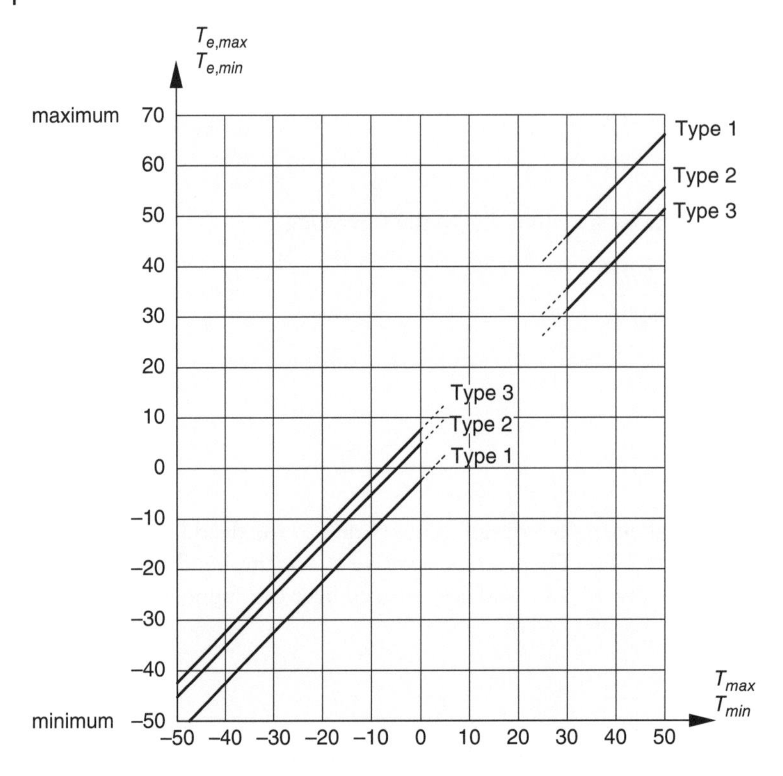

Figure 3.45 Correlation between T_{max} and T_{min} site temperatures and bridge deck uniform temperature on steel (type 1), composite (type 2) and concrete (type 3) bridge decks.

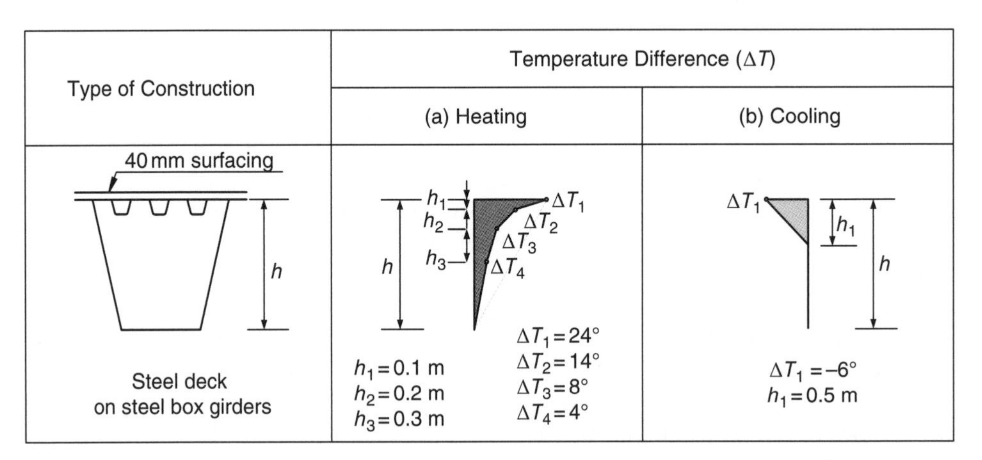

Figure 3.46 Nonlinear temperature distribution in a box girder steel deck according to EN1991-1-5.

bridge component. So, from Eqs. (3.52) and (3.53), these diagrams may be used to evaluate ΔT_M and ΔT_E.

In practice, the most important component of the non-uniform temperature variation is the linear component ΔT_M. Reference values of the linear temperature difference component for different types of bridge decks – steel, composite and concrete

Table 3.2 Recommended values of linear temperature difference component for different types of bridge decks for road, foot and railway bridges, in EC1-1-5.

Type of deck	Top warmer than bottom	Bottom warmer than top
	$\Delta T_{M,heat}$ (°C)	$\Delta T_{M,cool}$ (°C)
Type 1: Steel deck	18	13
Type 2: Composite deck	15	18
Type 3: Concrete deck:		
Concrete box girder	10	5
Concrete beam	15	8
Concrete slab	15	8

decks – are given in the Table 3.2, based on upper bound values of the linear temperatures for representative samples of bridge geometries. In this table, ΔT_M represents the characteristic value of temperature difference between the top and bottom surface of the deck.

These values are based on depth of surfacing of 50 mm for road and railway bridges. For other depths of surfacing, these values should be multiplied by the factor k_{surf} as proposed in EC1-1-5 [28]. For example, for the case of a concrete box girder roadway bridge with a surface thickness of 75 mm the linear temperature difference is 8.5°C (= 0.85 × 10°C) for the top warmer than the bottom, and 5°C (= 1.0 × 5°C) for the bottom warmer than the top.

It is sometimes necessary to consider horizontal temperature gradients, for example, taking into consideration possible action effects due to different temperatures between the webs of a box girder deck or between opposite outer faces of concrete piers. If no other information is available, a linear temperature difference of 5°C between the outer edges of the bridge independently of the width of the bridge, or between the outer faces of concrete piers, is recommended.

Differences in the uniform temperature components between different main structural elements may have to be considered, for example, between the stay cables and the deck of a cable-stayed bridge with values in the order of ±15°C.

3.14 Shrinkage, Creep and Relaxation in Concrete Bridges

Shrinkage and creep effects may be considered in the design of concrete and composite bridges, namely according to EC2-1 [29], CEB-FIP Model Code 90 [30] or *fib* Model Code 2010 [31]. Using this last code, the total shrinkage strain of concrete between the age t, in days, and the age t_s at the beginning of drying, is given by:

$$\varepsilon_{cs}\left(t-ts\right)=\varepsilon_{cbs0}\,\beta_{bs}\left(t\right)+\varepsilon_{cds0}\,\beta_{ds}\left(t-t_s\right) \tag{3.60}$$

where ε_{cbs0} and ε_{cds0} represent, respectively, the *basic* and *drying shrinkage* occurring in the long term. The basic shrinkage takes place at concrete young ages as the time function $\beta_{bs}(t)$ converges to 1 in the first year, that is, usually during bridge construction.

Therefore, it is the drying shrinkage that really influences the bridge behaviour, and this depends on the mean concrete compressive strength, the type of cement and the ambient relative humidity. The function $\beta_{ds}(t - t_s)$ represents the time development of drying shrinkage ranging between 0 and 1, and is a function of $(t - t_s)$ and of $h_0 = 2A_c/u$ – the *notional size of the member* in mm (dependent on the area A_c of the cross section and the perimeter u of the member in contact with the atmosphere). All these functions and coefficients are defined by mathematical expressions in the codes. The shrinkage strain at long term (ε_{cds0}) ranges usually between -10×10^{-5} and -40×10^{-5} (as shown in Figure 3.47 for a typical twin girder concrete bridge deck), that is equivalent to a temperature reduction of −10 to −40°C.

However, in practice, a considerable difference is frequently observed between measured and calculated values; in particular, concerning the time development of shrinkage. In fact, shrinkage development may be two to three times quicker, even if ε_{cds0} in the long term is not far from that predicted. This is likely to be due to the particular type of additives adopted in prestressed concrete bridge construction to allow rapid concrete hardening for stressing prestressing cables. For segmental bridges built by the cast in place cantilever method, the prestressing cables are very often stressed after 36–48 hours of concreting the segment.

The effect of shrinkage in concrete bridges is twofold – at the level of the structure and at the level of the cross sections. The former is illustrated in Figure 3.48 by the bending moments induced in the piers by the shrinkage of the deck in a concrete viaduct, as well as by the axial forces in the deck, which is subjected to bending moments as well if the eccentricity of the tensile forces introduced by the bearings is considered.

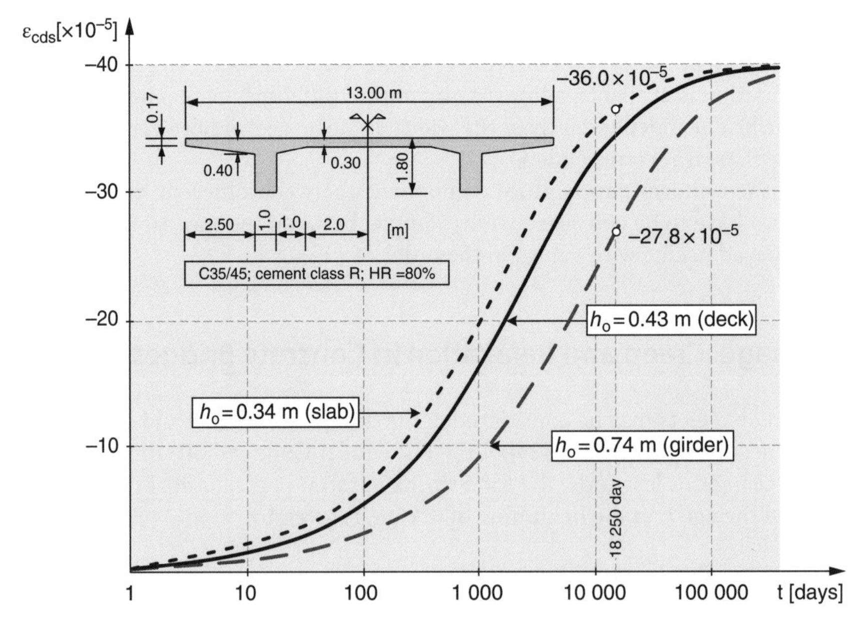

Figure 3.47 Evolution of free shrinkage in a concrete deck for the slab and for the girder independently, as a function of time *t* after casting, and the equivalent shrinkage at the centre of the gravity of the composed section.

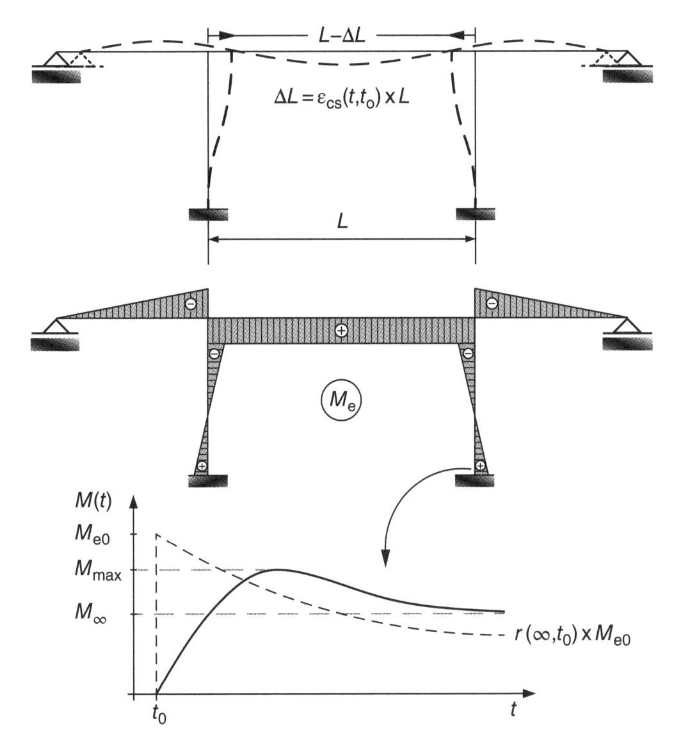

Figure 3.48 Self equilibrated elastic stresses due to differential shrinkage effects between the slab and the girders in the deck of Figure 3.47.

The shrinkage effects at cross sections is exemplified by the case of a beam-slab bridge deck of Figure 3.47, where the differential shrinkage between the slab and the girders – the uniform shrinkage is higher for the slab than for the girders if considered as independent –due to h_0 or if the slab is not cast in a single stage with the girders. Normal stresses at the cross sections are induced by *differential shrinkage effects*. These stresses may be evaluated by the same method – the principle of superposition, explained in Section 3.11.2 for the effect of non-linear temperature distribution. Hence, first one restraints the uniform shrinkage at the level of the slab and at the girder – the stresses at the restraints end sections are shown in Figure 3.49a – then, the resultant force and moment at the level of the cross section are evaluated. These stress resultants are then eliminated by applying $N_1 = -N_0$ and $M_1 = -M_0$ at the end sections, and the two stress states are added (Figure 3.49b) to obtain the self-equilibrated stresses due to differential shrinkage between the slab and the girders (Figure 3.49c).

A similar example of differential shrinkage may be given for a composite bridge deck where the free uniform shrinkage is restricted to the slab, which, however, induces (bending + axial) stresses in the girders as shown in Figure 3.50. In bridge decks where precasted girders are adopted, the effect of differential shrinkage between the slab and the girders also induces a state of normal stresses that should be considered at the SLS.

In reinforced concrete sections with a symmetric distribution of steel reinforcement, the effect of shrinkage in the concrete induces compressive stresses in the steel and tensile stresses within it. If the steel reinforcement is not symmetrically

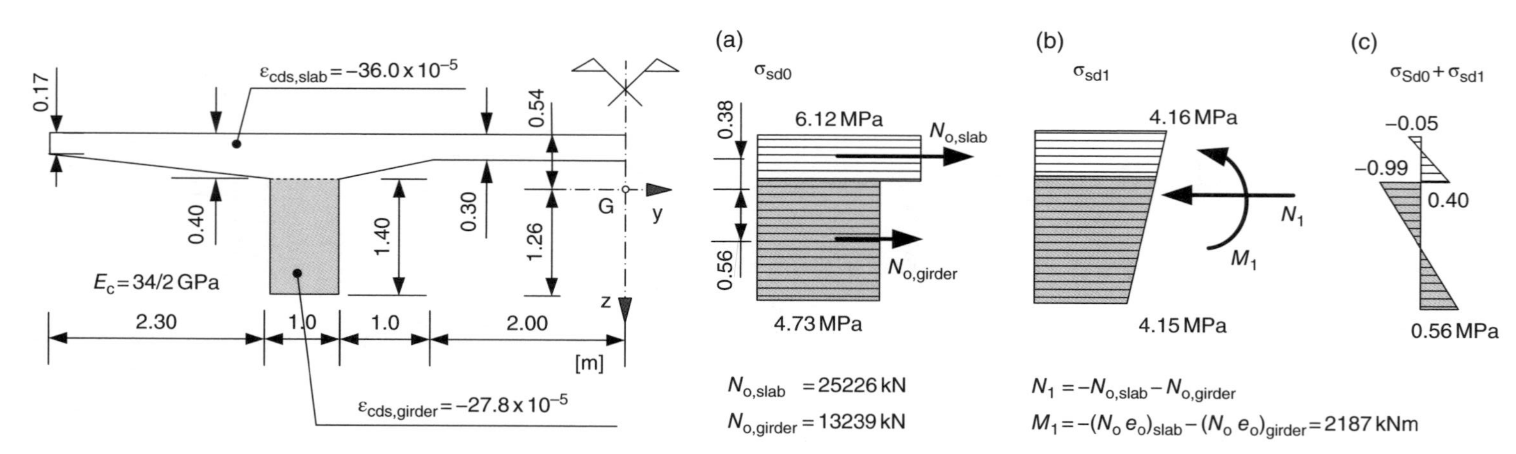

Figure 3.49 Induced bending moments in piers and deck in a concrete bridge frame structure due to uniform shrinkage of the deck and effect of relaxation of concrete stresses in the piers.

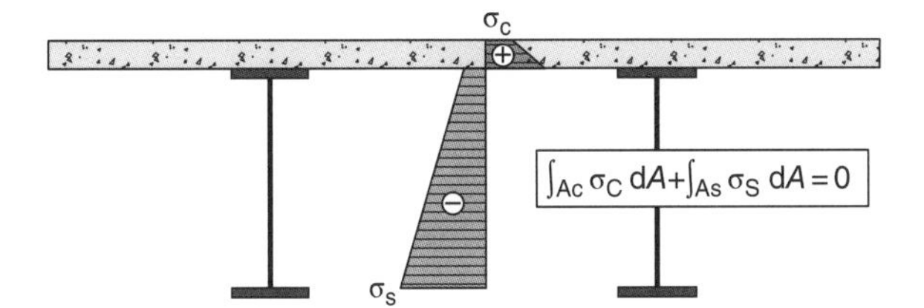

Figure 3.50 Normal elastic stresses induced in a composite bridge deck due to differential shrinkage between the concrete slab and the steel girders.

distributed, the concrete shrinkage induces a curvature at the cross section that may be expressed as [26]

$$1/R = k_s\,\varepsilon_{cs}/d \tag{3.61}$$

where d is the effective depth of the concrete section and k_s is a nondimensional coefficient depending on the geometrical characteristics of the cross section of the areas of concrete and steel reinforcement and on the ratio n between the modulus of elasticity of steel reinforcement and concrete $n = E_s/E_c$. For cracked sections, the extension of Eq. (3.61) and diagrams for the evaluation of curvature due to shrinkage are available [26].

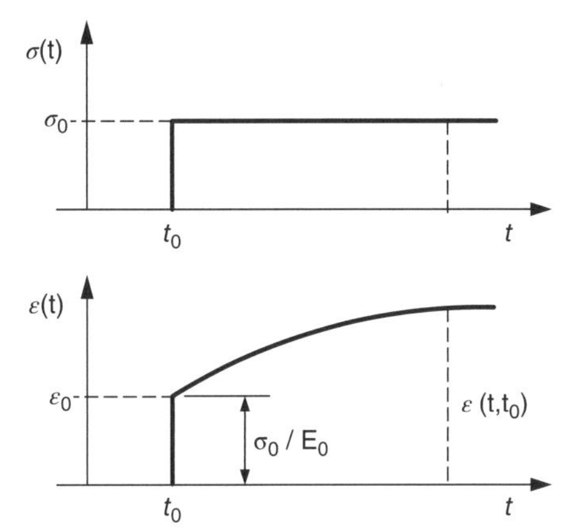

Figure 3.51 Creep deformations and definition of creep function $\Phi(t,t_0)$.

The structural overall effects or the cross sections effects of shrinkage of concrete cannot be separated from the *creep effect*. Basically, *creep* represents the increase in strains under sustained constant applied stress as illustrated in Figure 3.51. The *creep function* $\phi(t,t_0)$ is defined from the strains $\varepsilon(t,t_0)$ at time t_0 due to a constant applied stress σ_0 at time t_0

$$\varepsilon(t,t_0) = \sigma_0\,\phi(t,t_0) \tag{3.62}$$

The creep function is usually defined from the *creep coefficient* $\varphi(t,t_0)$ by

$$\phi(t,t_0) = 1/E_{c0} + \varphi(t,t_0)/E_{c28} \tag{3.63}$$

where E_{c0} and E_{c28} represent the modulus of elasticity of concrete at the age of loading t_0 and at 28 days. The creep coefficient $\varphi(t,t_0)$, defining creep between time instants t_0 and t, is evaluated from

$$\varphi(t,t_0) = \varphi_0(t,t_0)\beta_c(t - t_s) \tag{3.64}$$

where φ_0 is the *notional creep coefficient* and $\beta_c(t - t_s)$ is a function describing the development of creep with time after loading. These coefficients depend on the class of concrete and type of cement, the relative humidity, on the notional size of member h_0, and other parameters as defined in [26, 29, 30]. Figure 3.52 shows the evolution of

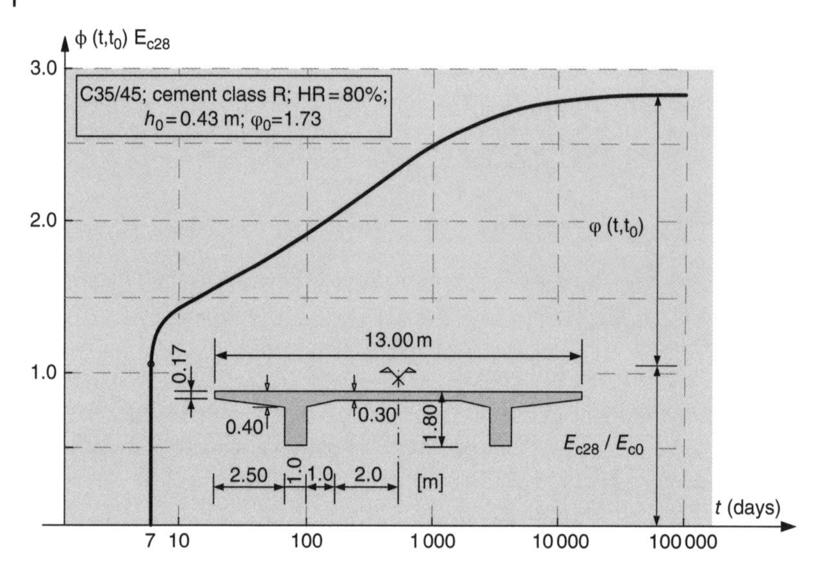

Figure 3.52 Creep function for the concrete deck of Figure 3.47.

the creep function with time for the deck cross section of Figure 3.47 after casting based on the creep parameters as per [29].

Let the case of imposed deformations, such as thermal strains or foundation settlements, be considered. These may be interpreted from an imposed strain at a certain time instant or along a certain time interval. Hence, let a strain ε_0 be applied at time t_0 and kept constant along time t, as shown in Figure 3.53. The induced stress σ_0 at time t_0, which is dependent on the instantaneous modulus of elasticity E_0 at time t_0, is reduced with time due to the *relaxation effect*; in some way an inverse of the creep effect. The relaxation function is defined as

$$\sigma(t,t_0) = r(t,t_0)\varepsilon_0 \tag{3.65}$$

where $r(t,t_0)$ is the relaxation function of concrete, representing the variation of stress due to an applied 'unit' strain at time t_0. This function may be determined from the creep function as is shown in the sequel.

Up to now, we have been considering the cases of creep or relaxation under constant stress or strain, respectively. In practice, one has to consider the cases of variable applied stress or variable imposed strain. The *principle of superposition of viscoelastic effects*, by Boltzmann, is used as illustrated in Figure 3.54 for creep effects. The stress at time t due to a series of increments of stresses $\Delta\sigma_i$, applied from t_0 to t, is given by

$$\varepsilon(t) = \Delta\sigma_0\varphi(t,t_0) + \Sigma\varphi_i(t,\tau_i)\Delta\sigma_i \tag{3.66}$$

Mathematically, the stress 'history' may be defined as a continuous function $\sigma(\tau)$ where τ denotes the time variable. So, to pass from the discrete case to the continuous one, one replaces $\Sigma\varphi_i(t,\tau_i)$ in (3.66) by an integral $\left(\int_{t_0}^{t}\varphi_i(t,\tau_i)d\tau\right)$ and $\Delta\sigma_i$ by $d\sigma = (\partial\sigma/\partial\tau)\,d\tau$, yielding

$$\varepsilon(t) = \sigma(t_0)\Sigma\varphi_i(t,t_0) + \int_{t_0}^{t}\varphi(t,\tau)(\partial\sigma/\partial\tau)d\tau \tag{3.67}$$

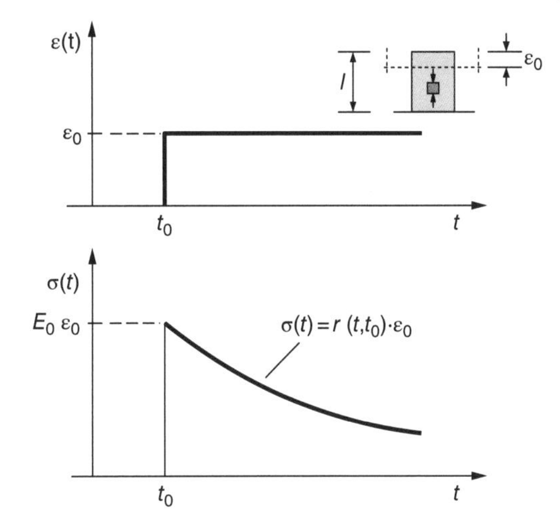

Figure 3.53 Stress relaxation in concrete and definition of the relaxation function $r(t,t_0)$.

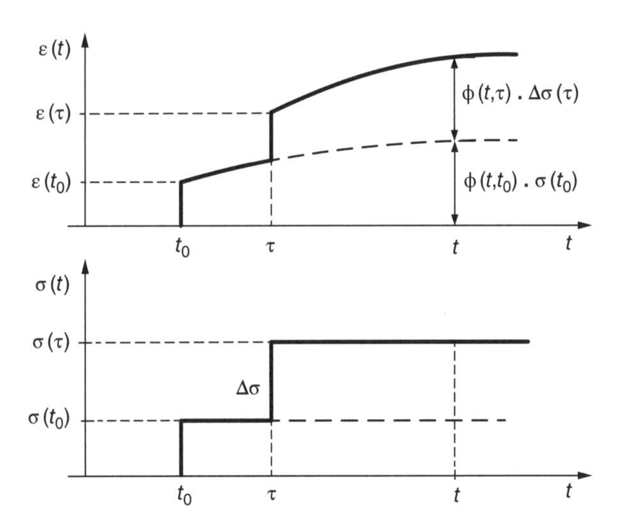

Figure 3.54 Superposition of viscoelastic effects – time dependent deformations, due to stress increments $\Delta\sigma$ applied at different time instants τ.

A similar equation may be written for the case of relaxation, as illustrated in Figure 3.55,

$$\sigma(t) = \varepsilon(t_0) r(t,t_0) + \int_{t_0}^{t} r(t,\tau)(\partial\varepsilon / \partial\tau)\mathrm{d}\tau \tag{3.68}$$

The relaxation function, taking into consideration the fact it represents the stress developed at time t due to a unit strain applied at time t_0, may be determined, knowing the creep function, by solving the integral Eq. (3.68) for $\sigma(t)$ after setting $\varepsilon(t)$ in (3.67) as a unit function.

A method of viscoelastic analysis, from Trost and Bazant [26], consists of rewriting Eq. (3.67) in the form

$$\varepsilon(t,t_0) = \sigma(t_0)/E_c^{*} + \Delta\sigma/E_c^{**} \tag{3.69}$$

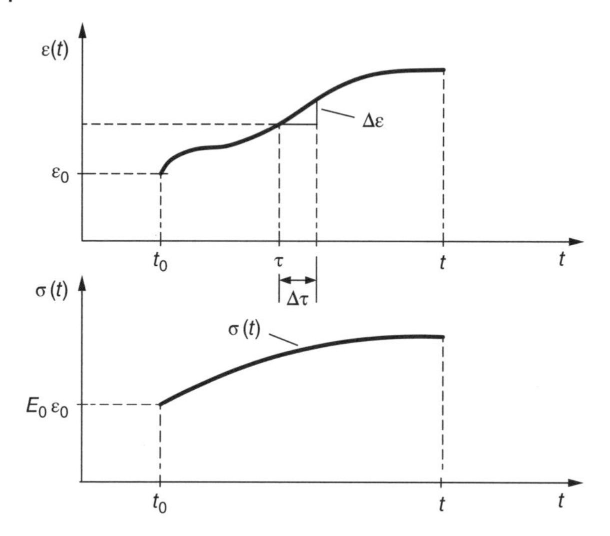

Figure 3.55 Superposition of relaxation effects.

where $\Delta\sigma = \sigma(t) - \sigma(t_0)$ is the total variation of stress between t_0 and t, and E_c^* and E_c^{**} are, respectively, the so called *effective modulus* and *adjusted effective modulus* given by

$$E_c^* = E_{c0}/\left[\left(1+E_{c0}/E_{c28}\right)\varphi\right] \tag{3.70}$$

$$E_c^{**} = E_{c0}/\left(1+\chi\varphi\right) \tag{3.71}$$

where χ *is the ageing coefficient* defined next. It may be shown, under certain simplified assumptions [26], that the ageing coefficient is independent on the history of stress and may be written in terms of the relaxation function $r(t,t_0)$ by

$$\chi(t,t_0) = \left(E_0/E_{28}\right)E_0/\left[E_0 - r(t,t_0)\right] - \left[1/\varphi(t,t_0)\right] \tag{3.72}$$

The functions $\varphi(t, t_0)$ and $\chi(t, t_0)$ for practical applications are calculated and plotted graphically [26] for different values of the creep coefficient and of the notional size of the section h_0.

Equation (3.69) is written in Ref. [26], including a term $\varepsilon_n(t)$ independent of the stress and denoting an imposed deformation like the shrinkage or a thermal strain, as

$$\varepsilon(t,t_0) = \varepsilon_n(t) + \sigma(t_0)\phi(t,t_0) + \left[\sigma(t) - \sigma(t_0)\right]\left[\left(1/E_{c0}\right) + \chi\varphi(t,t_0)/E_{c28}\right] \tag{3.73}$$

In current cases the ageing coefficient χ may be taken as 0.8, at least for cases of pure relaxation and in cases where only long-term effects are considered. Besides, if the stresses in concrete only vary slightly one obtains from Eq. (3.73), neglecting the second term correspondent to the stress variation, and introducing the concept of *effective modulus of elasticity* ($E_{c,\text{eff}}$),

$$\varepsilon(\text{t}) = \sigma(t_0)/E_{c,eff} \quad \text{where} \quad E_{c,eff} = E_{c0}/\left[1+\varphi(t,t_0)\right] \tag{3.74}$$

Equation (3.69) may be adopted to understand the effect of creep on stresses developed due to shrinkage effects. For that case $\sigma(t_0) = 0$ and therefore the stress developed at the instant, t, $\sigma(t) = \Delta\sigma$ is given by,

$$\sigma(t) = E_{c0}\varepsilon_{cs}(t - t_s)/[1 + \chi\varphi(t,t_0)] \tag{3.75}$$

Equation (3.61) may be shown [26] to include the effect of creep on curvatures developed due to shrinkage. In this case, the coefficient k_c is dependent on $\chi\ \varphi$. The reader is referred to Refs. [26, 29, 30] for additional information on creep and shrinkage effects in concrete structures.

3.15 Actions Due to Imposed Deformations. Differential Settlements

Let's take the case of the prestressed concrete bridge represented in Figure 3.48. Due to the shortening of the deck induced by shrinkage, elastic and creep deformations due to prestressing, also due to negative temperature variations, the top of the piers experience imposed horizontal displacements, dependent on time t, that is,$u = u(t)$, and bending moments develop along the pier shaft. If one assumes a pier of constant cross section, with a bending stiffness $EI << (EI)_{deck}$, the maximum induced elastic bending moment at the base section is given by

$$M_e = (6EI/L^2)u \tag{3.76}$$

If the displacement had been imposed with its final value u and kept constant along the time, the stresses σ_0 at time t_0 at the base section, are (as per Section 3.13) reduced proportionally to the relaxation function $r(t, t_0)$. Since the bending moments are

$$\begin{aligned} M(t) &= \int_A \sigma(t,t_0) z\, dA \\ &= \int_A \varepsilon_0 r(t,t_0) z\, dA = (1/E) r(t,t_0) \int_A \sigma_0 z\, dA = (1/E) r(t,t_0) M_e \end{aligned} \tag{3.77}$$

Hence, the elastic bending moments are then reduced proportionally to the relaxation function. The same conclusion holds for the bending moments induced in the deck due to differential settlements of the pier foundations, as illustrated in Figure 3.56. However, in both problems the displacement u or the settlement Δ does not occur immediately with its final value but the tendency is for the final values to increase continuously with time t from 0 at age t_0, to its final value Δ at age t_1. The superposition principle of viscoelastic effects and the concept of ageing coefficient previously introduced allow a solution for this problem as presented in [26]. The values at t_1 and at subsequent time t_2 of the action effects (reaction, bending moment etc.) due to support movements may be determined from the following equations

$$F(t_1) = F_{\text{sudden}}[1/(1 + \chi\varphi(t,t_0))] \tag{3.78}$$

$$F(t) = F(t_1)\{1 - [E(t_1)/E[t_e]][\varphi(t_2,t_e) - \varphi(t_1,t_e)]/(1 + \chi\varphi(t_2,t_1)\} \tag{3.79}$$

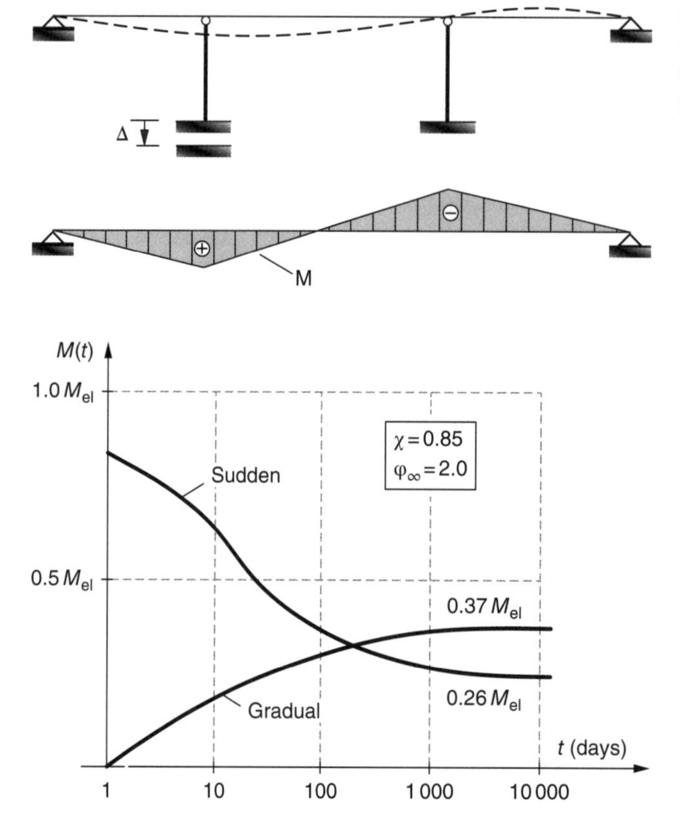

Figure 3.56 Bending moments induced in a bridge deck due to settlements of a pier foundation.

Figure 3.57 Effects of sudden or gradually pier foundation settlements in bending moments in a bridge deck (*Source:* Adapted from Ref. [32], with permission from IABSE).

where t_e is a time instant determined by solving the following equation

$$\left[1/E(t_e)\right]\left[1+\varphi(t_1,t_e)\right]=\left[1/E(t_0)\right]\left[1+\chi\varphi(t_1,t_0)\right] \tag{3.80}$$

The effective time t_e represents the instant such that, if a stress increment is introduced at that instant and kept constant up to t_1, it produces the same deformation of a gradually introduced stress increment introduced from zero at t_0 to the final value at t_1.

Figure 3.57 presents an example of the variation of the bending moment at support in a continuous bridge deck due to a sudden or a gradually introduced foundation settlement [32]. As may be concluded from the figure, the maximum bending moment if the settlement is introduced gradually is not more than 50% of the correspondent value if introduced suddenly with the final value. Of course this percentage tends to decrease with time corresponding to period of settlement. Another relevant aspect is the effect of cracking in reducing the elastic effects introduced in concrete structures due to imposed deformations.

A simplified approach to the effect of imposed gradually deformations in concrete-bridge thermal strains or foundation settlements, consists of considering the values obtained from an elastic analysis replacing the elastic modulus of concrete by the *reduced modulus of elasticity*. Imposed deformation effects may be considered, according to some codes and recommendations, with its elastic values divided by 2, that is, with $E_{c,eff} = E_c/2$.

3.16 Actions Due to Friction in Bridge Bearings

The loads transferred from the superstructure to piers and abutments by bearings, induce friction forces dependent on the type of bearing namely, a roller bearing or a sliding bearing; see Figure 3.58. The former are usually metal bearings while the latter are elastomeric bearings nowadays ('neoprene bearings') with a sliding thin layer of a low friction material like polytetrafluorethylene (PTFE). These bearings are designated in practice as 'neoprene-teflon bearings' as will be discussed in Chapter 7.

The friction forces developed at the bearings should be quantified from the vertical force reactions V, by the well-known friction law $H = \mu V$, where μ is the friction coefficient. For PTFE sliding bearings, a design value of 5% (i.e. $\mu = 0.05$) may be adopted.

3.17 Seismic Actions

3.17.1 Basis of Design

The design of bridges in seismic zones, such as in southern European countries, requires earthquake actions to be taken into consideration at concept, analysis and details of a design. At the concept stage, one aims to achieve a structure with a reliable response under seismic actions. During analysis, by modelling the seismic actions by static equivalent forces (acceptable for simple and small bridges) or by a dynamic analysis based on *response spectrums, power spectrums* or by a *time history analysis* [8]. In the detail design stage, by considering structural elements and connections with sufficient resistance and ductility under seismic actions.

It is understood bridges generally behave well for earthquakes of moderate intensity, say up to a magnitude in the order of five to six. For earthquakes with a higher magnitude, the following types of damage have occurred in several cases:

- collapse of the foundations by soil liquefaction under seismic actions
- damage in abutments

Figure 3.58 Friction in bridge bearings.

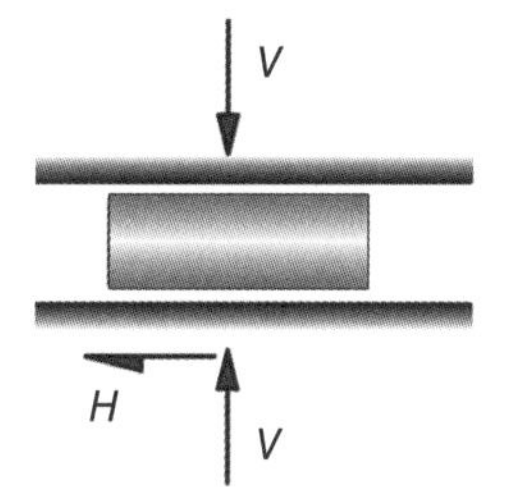

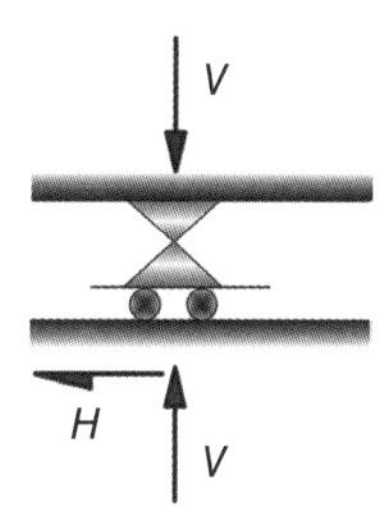

- falling of superstructure elements
- structural damage in piers or, less frequently, in the deck

Damages in abutments are generally caused by dynamic overpressures of the adjacent soil or due to forces induced by the deck. Falling of deck elements is usually due to characteristics of bridge bearings not being compatible with the order of magnitude of the horizontal displacements induced by the earthquake. Finally, damages in bridge piers are usually due to lack of ductility of these structural elements.

For current bridges, it is sufficient to consider the seismic action in horizontal directions; the vertical component should be considered in a few structures. That is the case for structures with vertical vibration modes associated to natural frequencies less than 1 Hz. In these cases, if seismic action is considered by response spectrums, the ordinates of the response spectrum correspondent to the vertical component may be considered to be two-thirds of the ordinates of horizontal components. The simplified methods of seismic analysis, based on *static equivalent forces,* are currently limited to bridges with vertical piers, bridges with a straight or almost straight in plan alignment or slightly skew bridges.

Besides, the span lengths should not be significantly different and should be approximately symmetrical with respect to a vertical plan perpendicular to the axis of the bridge. Static equivalent seismic analyses are not applicable to curved bridges, frame bridges with inclined piers, arch bridges, cable-stayed bridges and suspension bridges.

In the static equivalent seismic force method, actions may be considered separately in two horizontal directions, one along the longitudinal axis and the other perpendicular to the bridge axis. The equivalent static forces are obtained by multiplying the permanent loads of the bridge by a seismic coefficient β, currently with values between 0.02 and 0.3. This coefficient depends on the seismicity of the bridge site, namely on the peak rock acceleration, the type of soils at the foundations, the natural frequencies of the bridge (in particular, the fundamental natural frequency) and the *behaviour coefficient* to be adopted.

For simple bridges, the longitudinal seismic action in the deck may be simulated by a horizontal force applied at the deck with a value βG, where G represents the permanent load of the deck. This may be done because the longitudinal mode of vibration of the bridge may be modelled by a single degree of freedom system, represented by the deck translation.

However, for the transverse direction, one must take into consideration the transverse mode of vibration and a discrete model is adopted with concentrated masses (m_i) localized at the most flexible parts of the structure.

The static equivalent force distribution, which is close from the inertia force distribution associated to the fundamental mode of vibration, may be obtained by evaluating the deformed configuration of the structure under the action of forces $\beta\, G_i$ proportional to the permanent loads and acting along the direction under consideration.

The discrete model of the deck is as shown in Figure 3.59 and the effective acceleration of a mass m for a SDF system is $a_{ef} = (2\pi f)^2\, u$, where $u(t)$ is the time dependent displacement and f the fundamental frequency. The associated inertia force is $F = m\, a_{ef} = (2\pi f)^2\, m\, u$.

For a MDOF system with masses m_i, the characteristic value of the static equivalent forces are

$$F_{ki} = (2\pi f)^2 (\beta G_i / g) d_i \tag{3.81}$$

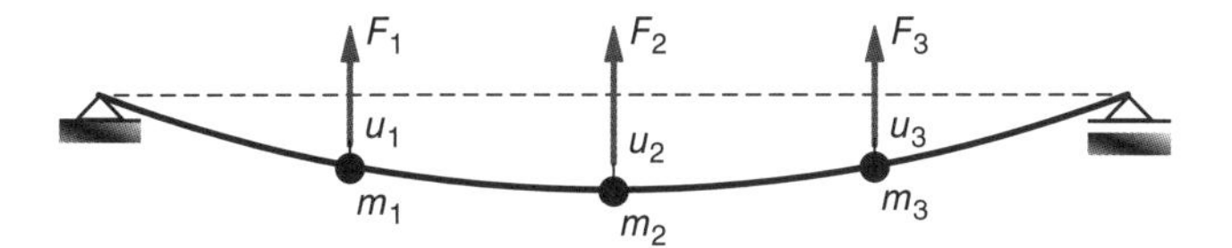

Figure 3.59 Transverse horizontal forces due to seismic actions induced in a bridge deck as a function of masses in a discrete deck model.

where d_i are the displacements due to the loads G_i acting simultaneously in the direction where the seismic action is being considered and g the acceleration of gravity. In loads G_i, one includes the permanent loads as well as the quasi-permanent values of the variable loads correspondent to the masses m_i.

Longitudinal and transverse behaviour should be analysed, due to different influences of the support conditions of the superstructure. Another aspect of seismic actions refers to the hydrodynamic forces induced by water surrounding piers. This is a complex solid–fluid interaction problem, but the simplest model consists of considering a mass of a cylinder of water, which vibrates with the pier.

When *response spectrums* or *power spectrums* need to be considered for the seismic analysis of bridges, a modal analysis is considered. The reader is referred to the specific literature [8, 33, 34] for these methods of seismic analysis. Design codes such as the Part 2 of Eurocode 8 (denoted as EC8-2 [35]) establish the basis for seismic analysis of bridges.

3.17.2 Response Spectrums for Bridge Seismic Analysis

The most adopted method for seismic analysis of bridges is the Response Spectrum Method. The basis of the method is outlined here. A SDF system may be considered, for example, for a bridge with a three span continuous deck supported by two flexible piers and with longitudinal sliding bearings at the abutments. The deck under seismic longitudinal actions is assumed to be rigid and hence the structure can be modelled by a SDOF system (Figure 3.60). Under a seismic action defined by the ground motion record in terms of accelerations, $\ddot{u}_g = \ddot{u}_g(t)$ -acting accelerograms, and the relative displacements of the mass M with respect to the foundation level is u. Hence the dynamic equation of motion (3.13) for $P = 0$ but with the inertia term $M\ddot{u}$ replaced by $M(\ddot{u}_g + \ddot{u})$ is written as

$$M\ddot{u} + C\dot{u} + ku = -M\ddot{u}_g \tag{3.82}$$

Taking into consideration, $C = 2M\omega\xi$ and $k = M\omega^2$ where $\omega = 2\pi f$, (3.82) is rewritten as

$$\ddot{u} + 2\xi\omega\dot{u} + \omega^2 u = -\ddot{u}_g \tag{3.83}$$

This differential equation may be solved for a certain ground motion record $\ddot{u}_g = \ddot{u}_g(t)$ to find the displacements $u = u(t)$ of the mass M. The value of peak response (maximum displacement) of the SDOF defined by its frequency f and damping coefficient ξ defines the *response spectrum* of displacements S_d (f, ξ). Solving for different frequencies f, a diagram ($S_d = S_d(f)$ or in terms of the period $T = 1/f$, ($S_d = S_d(T)$) can be plotted for the

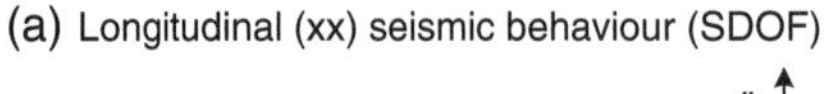

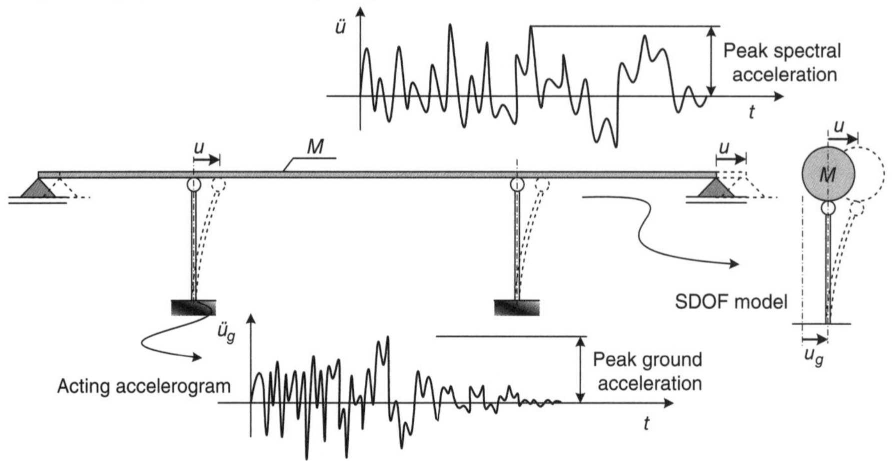

(b) Horizontal tranverse (yy) seismic action (MDOF)

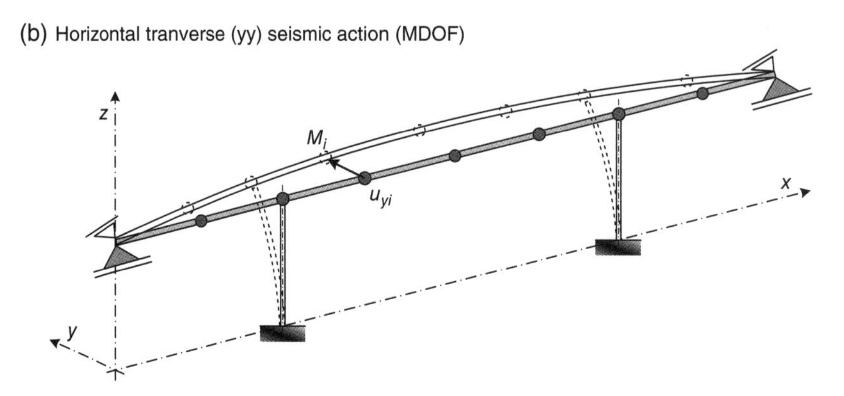

Figure 3.60 Seismic actions induced in a bridge: (a) longitudinal modelling by SDOF and (b) transverse modelling by MDOF.

response spectrum for different damping ratios ξ. The response spectrums are defined in codes (EC8-1, e.g. Figure 3.61) for different geotechnical conditions and zones usually associated to *peak ground accelerations* $a_g = \max \ddot{u}_g$ at the bridge site.

If the maximum displacement of the mass M is designated by S_d (f, ξ), the SDOF system is subjected to a maximum force

$$F_{max} = k S_d(T,\xi) = M\frac{4\pi^2}{T^2} S_d(T,\xi) = M S_a(T,\xi) \tag{3.84}$$

The term S_a $(T, \xi) = (4\pi^2/T^2)$ S_d is the acceleration spectrum of the mass; for $T = 0$, that is, a system with infinite frequency (totally rigid system) like a rigid mass supported directly on the rock, S_a is the peak ground acceleration of the mass M. Figure 3.61 shows different response spectra in terms of damping ratios.

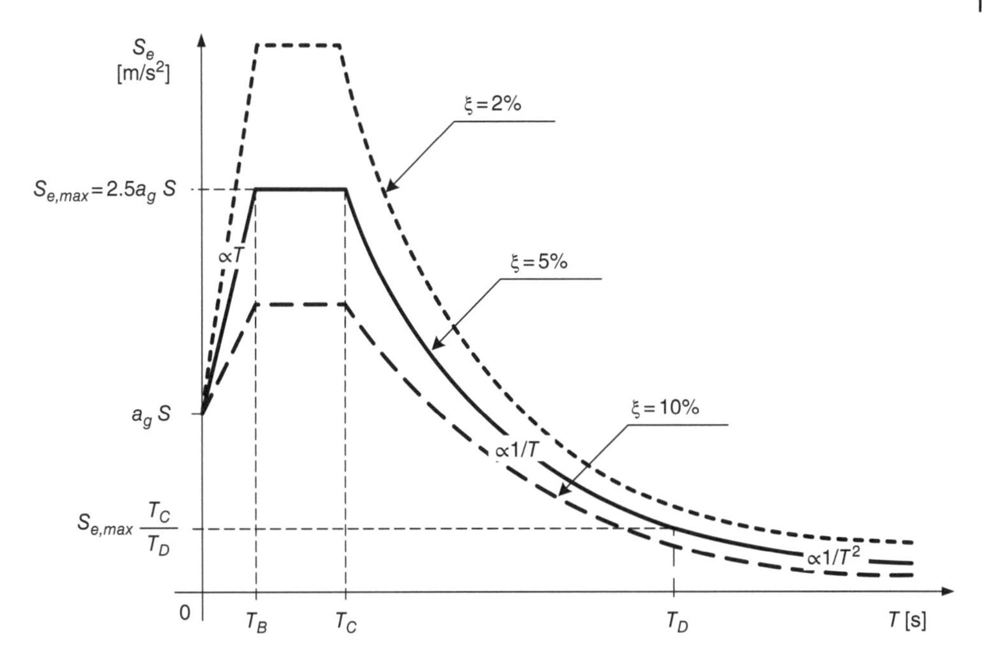

Figure 3.61 Typical Elastic Response Spectrum S_e in EC8 – T_B,T_C,T_D in (s) depending on geotechnical ground conditions; a_g is the peak ground acceleration at the bridge site and η is the damping correction factor ($\eta = 1.0$ for $\xi = 5\%$ viscous damping ratio).

For a Multi Degree of Freedom System (MDFS), a modal analysis [8, 35] has to be carried out and Eq. (3.83) is generalized for each normalized modal coordinate $U_{\mathrm{i}} = 1$, with $i = 1$ to n, as

$$\ddot{U}_i + 2\xi\omega_i\,\dot{U}_i + \omega_i^{\,2}\,U_i = -\frac{L_i}{M_i}\ddot{u}_g \tag{3.85}$$

where (L_i/M_i) is the *participation factor* [8, 35] of mode i in the response, for a specific direction of ground motion, depending on the mode shape and mass distribution. Denoting the modal displacement j in the mode i by φ_{ij}, one has

$$\frac{L_i}{M_i} = \frac{\sum_{i=1}^{n}\varphi_{ij}M_j}{\sum_{i=1}^{n}\varphi_{ij}^2 M_j} \tag{3.86}$$

The *effective masses* in mode i are defined as

$$M_{eff,i} = \frac{\left(\sum_{i=1}^{n}\varphi_{ij}M_j\right)^2}{\sum_{i=1}^{n}\varphi_{ij}^2 M_j} \tag{3.87}$$

A *multi degree spectrum analysis* is carried out for the complete seismic analysis of the structure. By defining the spectral acceleration for each mode i as S_a (T_i, ξ_i), the maximum elastic forces in mode i may be calculated by using the participation factors. The simplest approach to obtaining a 2D bridge structure modelled as n lumped masses (Figure 3.59 or the bridge structure in Figure 3.60 under a horizontal seismic action transverse to the plane of the structure), the maximum *base shear* due to mode i is obtained from the response spectrum and generalized masses, as $F_{max,\,i} = M_{eff,\,i}S_a\,(T, \xi)$. However, total response cannot be obtained by adding maximum force values at each mode, because these maximum values do not occur at the same time. The simplest and most adopted combination rule to obtain maximum values is the *square root of the sum of the squares* of the modal responses. The reader is referred to [8, 35] for developments on the subject.

A final remark is made concerning the use of response spectrums as defined. These are *elastic response spectrums* that should be modified by ductility/behaviour factors to take into consideration the nonlinear behaviours of the bridge structure during an earthquake. The nonlinear behaviours are associated to cracking or yielding reducing the stiffness and modifying the effective vibration frequencies of the bridge structure. The spectrum accelerations are then modified and this nonlinear effect is usually considered by reducing the elastic forces calculated on the basis of the elastic spectrums or reducing directly the spectrum by the behaviour factor. In bridge structures these behaviour factors q depend on piers and abutments ductility, and are in the order of 1.5–2.5, yielding a considerable reduction in the force values obtained from the elastic spectrums. Indicative values can be found in EC 8-2 [35]. However, inaccessible parts of the bridge, like foundations, and bearing design in general, total elastic forces are usually required by the codes. The reader is referred to [34–37] for further developments.

3.18 Accidental Actions

Accidental actions were already referred to in Section 3.1. The impact on bridge piers due to vehicles collision is the most current case. Ship impact, for bridges with piers near navigation channels, is another example of an accidental action. Mitigation measures for ship impact are currently adopted by designing ship impact defences. A variety of accidental actions are quantified by nominal values in Part 1-7 of Eurocode 1 [38].

3.19 Actions During Construction

During bridge construction the evolution of the static system should be considered in conjunction with specific actions due to construction equipment, like formwork launching girders or moving scaffoldings. These actions are referred in next chapter, for execution and construction methods for concrete and steel bridge decks. Construction live loads and accidental actions should also be considered when checking the bridge safety during erection stages. The reader is referred to specific literature on the subject [39, 40].

3.20 Basic Criteria for Bridge Design

A bridge must be designed to attain its functional objectives; resisting permanent, variable and accidental actions during its *life span* with minimum maintenance costs. Three types of *design situation* should be considered for checking the safety of the structure; namely,

- *persistent*, correspondent to normal functional conditions during its life span;
- *transitory*, correspondent to the execution period or time for repairing or strengthening the structure, that may be taken as having a time duration of one year;
- *accidental*, which may be considered to be instantaneous situations.

As referred to in Section 3.1, bridges should be designed to warranty safety at the ULS and an acceptable behaviour at SLS. The former (ULS) are associated to loss of static equilibrium, rupture and fatigue, while the last (SLS) are associated to excessive deformations, vibrations and crack width limitations in concrete and composite bridges. In prestressed concrete bridges, a SLS is defined when attaining a zero compressive stress in the concrete for certain duration of the SLS, as referred in Section 3.1 for the *characteristic (rare)*, *frequent*, or *quasi-permanent* load combinations. In steel and composite bridges, one should consider at SLS the localized plastic deformations, which may induce damage and irreversible deformations, as well as the SLS, associated with the slip resistance of prestressed bolted connections. The basic expressions for load combinations for ULS and SLS were presented in Section 3.1.

Design criteria for prestressed concrete decks, steel or composite decks will be developed in Chapter 6, after introducing the concept design and execution methods in Chapter 4.

References

1 EN 1992-2 (2005). *Eurocode 2 – Design of Concrete Structures – Part 2: Concrete Bridges – Design and Detailing Rules*. Brussels: CEN.
2 EN 1993-2 (2006). *Eurocode 3 – Design of Steel Structures – Part 2: Steel Bridges*. Brussels: CEN.
3 EN 1994-2 (2005). *Eurocode 4 – Design of Composite Steel and Concrete Structures – Part 2: General Rules and Rules for Bridges*. Brussels: CEN.
4 EN 1990 (2002). *Eurocode – Basis of Structural Design, and Amendment A1 (2005) Annex A2 – Application for Bridges*. Brussels: CEN.
5 EN 1991-2 (2005). *Eurocode 1: Actions on Structures – Part 2: Traffic Loads on Bridges*. Brussels: CEN.
6 EN 1991-1-1 (2002). *Eurocode 1 – Actions on Structures – Part 1–1: General Actions Densities, Self-Weight, Imposed Loads for Buildings*. Brussels: CEN.
7 TR 1317-6 (2012). *Technical Report – Road Restraint Systems – Part 6: Pedestrian Restraint System – Pedestrian Parapets*. Brussels: CEN.
8 Clough, R.W. and Penzien, J. (1993). *Dynamics of Structures*. New York: McGraw Hill.
9 Irvine, H.M. ed. (1986). *Structural Dynamics for the Practicing Engineer*. London and New York: E&FN Spon.

10 Frýba, L. (1996). *Dynamics of Railway Bridges*. London: Thomas Telford.
11 Goicolea, J. M. (2007). Dynamic behaviour of high speed railway bridges; Research at Escuela de Caminos, UPM. Seminar at Instituto Superior Técnico. Lisbon, 6 November 2007.
12 Frýba, L. (1999). *Vibration of Solids and Structures under Moving Loads*. London: Thomas Telford.
13 Frýba, L. (2001). A rough assessment of railway bridges for high speed trains. *Engineering Structures* 23: 548–556.
14 Menn, C. (1990). *Prestressed Concrete Bridges*, 536. Springer.
15 EN 1991-1-4 (2005). *Eurocode 1: Actions on Structures – Part 1–4: General Actions – Wind Actions*. Brussels: CEN.
16 Branco, F., Mendes, P. and Ferreira, J. (1997). Ensaios em Túnel de Vento do Tabuleiro do Viaduto sobre o Caminho do Comboio – Report IC-IST, EP N°10/97 (in Portuguese).
17 Dyrbye, C. and Hansen, S.O. (1997). *Wind Loads on Structures*. New York: Wiley.
18 Jurado, J.A., Hernández, S., Nieto, F., and Mosquera, A. (2011). *Bridge Aeroelasticity. Sensitivity Analysis and Optimal Design*. WIT Press.
19 CECM (1987) Recommandation pour le calcul des effets du vent sur les construction. CECM (Convention Européenne de la Construction Métallique), N°52, 2nd edn.
20 Svensson, H. (2012). *Cable-Stayed Bridges – 40 Years of Experience*, 430. Germany: Wilhelm Ernst & Sohn.
21 ICE (1981). *Bridge Aerodynamics*. London: Thomas Telford Limited.
22 Ziegler, H. (2013). *Principles of Structural Stability*, vol. 35. Birkhäuser.
23 Cremona, C. and Foucriat, J.C. (2002). *Comportement au vent des Ponts*. Presses des Ponts et Chaussées, AGFC.
24 Mendes, P., Reis, A. J. and Branco, F. (1988). Acções Térmicas Diferenciais em Tabuleiros de Pontes em Laje: Caracterização da Situação Portuguesa. Revista Portuguesa de Engenharia de Estruturas n. 28 – 1ª Série, pp. 23–30 (in Portuguese).
25 Scanlan, R.H. and Tomko, J.J. (1971). Airfoil and bridge deck flutter derivatives. *Journal of the Engineering Mechanics Division* 97 (EM6): 1717–1737.
26 Ghali, A., Favre, R., and Elbadry, M. (2002). *Concrete Structures: Stresses and Deformations*, 3e. London and New York: E&FN Spon.
27 SAP2000® v18 (2016). Integrated software for structural analysis and design. Available online at https://www.csiamerica.com/products/sap2000 (accessed July 2018).
28 EN 1991-1-5 (2003). *Eurocode 1: Actions on Structures – Part 1-5: General Actions – Thermal Actions*. Brussels: CEN.
29 EN 1992-1-1 (2004). *Eurocode 2 – Design of Concrete Structures – Part 1-1: General Rules and Rules for Buildings*. Brussels: CEN.
30 CEB-FIP (1991). Model Code 1990, Design Code. Comité Euro-International du Béton, Paris.
31 fib (2013). *Model Code for Concrete Structures 2010*. Wilhelm Ernst & Sohn.
32 Schlaich, J. and Scheef, H. (1982). Structural Engineering Documents SED 1: Concrete box-girder bridges. IABSE, International Association for Bridge and Structural Engineering, Zurich, Switzerland.
33 SETRA (2012). Ponts en zone sismique – Conception et dimensionnement selon l'Eurocode 8 – Guide méthodologique.
34 SETRA (2000). Ponts Courants en Zone Sismique. Guide de Conception.

35 EN 1998-2 (2005). *Eurocode 8 – Design of Structures for Earthquake Resistance – Part 2: Bridges*. Brussels: CEN.
36 Maguire, J.R. and Wyatt, T.A. (1999). *ICE Design and Practice Guide – Dynamics: An Introduction for Civil and Structural Engineers*. Thomas Telford.
37 EN 1998-1 (2004). *Eurocode 8 – Design of Structures for Earthquake Resistance – Part 1: General Rules, Seismic Actions and Rules for Buildings*. Brussels: CEN.
38 EN 1991-1-7 (2006). *Eurocode 1: Actions on Structures – Part 1–7: General Actions – Accidental Actions*. Brussels: CEN.
39 EN 1991-1-6 (2005). *Eurocode 1: Actions on Structures – Part 1–6: General Actions – Actions during Execution*. Brussels: CEN.
40 Calgaro, J. A. (2000). Projet et construction des ponts – Généralités, Fondations, Appuis et Ouvrages courants. Presse des Ponts et Chaussées, Paris.

4

Conceptual Design and Execution Methods

4.1 Concept Design: Introduction

Concept bridge design is an iterative process requiring a comparative analysis of a variety of possible options satisfying a set of data (constraints), namely those related to topographic, geometrical (road or rail alignments), hydraulics, geotechnical, environmental or any other constraints as referred to in Chapter 2. This comparative analysis is usually done at the preliminary design phase. Concept design should take into consideration the objectives of *Functionality, Safety and Durability, Economy, Aesthetics* and *Environmental Integration.*

In this chapter, a synthesis of multiple aspects of bridge concept design is presented. Options for the longitudinal and transverse structural systems for the bridge structure are discussed, as well as the choice of structural materials and options for superstructure and substructure. Finally, execution methods and their influence on the concept design is discussed.

The aspect of bridge location, longitudinal and transverse alignments and the elements that should be integrated in the bridge deck were already discussed in Chapter 2. The next step in the design process should be the definition of the main concepts to be taken into consideration in the study of bridge options.

Concepts – This aspect is discussed based taking the bridge case of Figure 4.1. The bridge is located in a deep valley. Piers should be avoided in the central part of the valley due to increased difficulties in execution, height of the piers and aesthetic and environmental integration of the bridge in the landscape. The last aspect is discussed in Chapter 5.

The relationship between the height of the piers and the span lengths is a key parameter for bridge options. Taking into consideration the symmetry of the valley, it is preferable to adopt span lengths decreasing from the centre of the valley towards to the abutments. This is a basic concept for integration in the landscape. If one adopts a main span length in the order of 115 m, the adjacent spans could be approximately 77.5 m and the end spans 40 m, as shown in Figure 4.1). With respect to this solution, designated *Option A*, the following issues may be defined:

Bridge Design: Concepts and Analysis, First Edition. António J. Reis and José J. Oliveira Pedro.

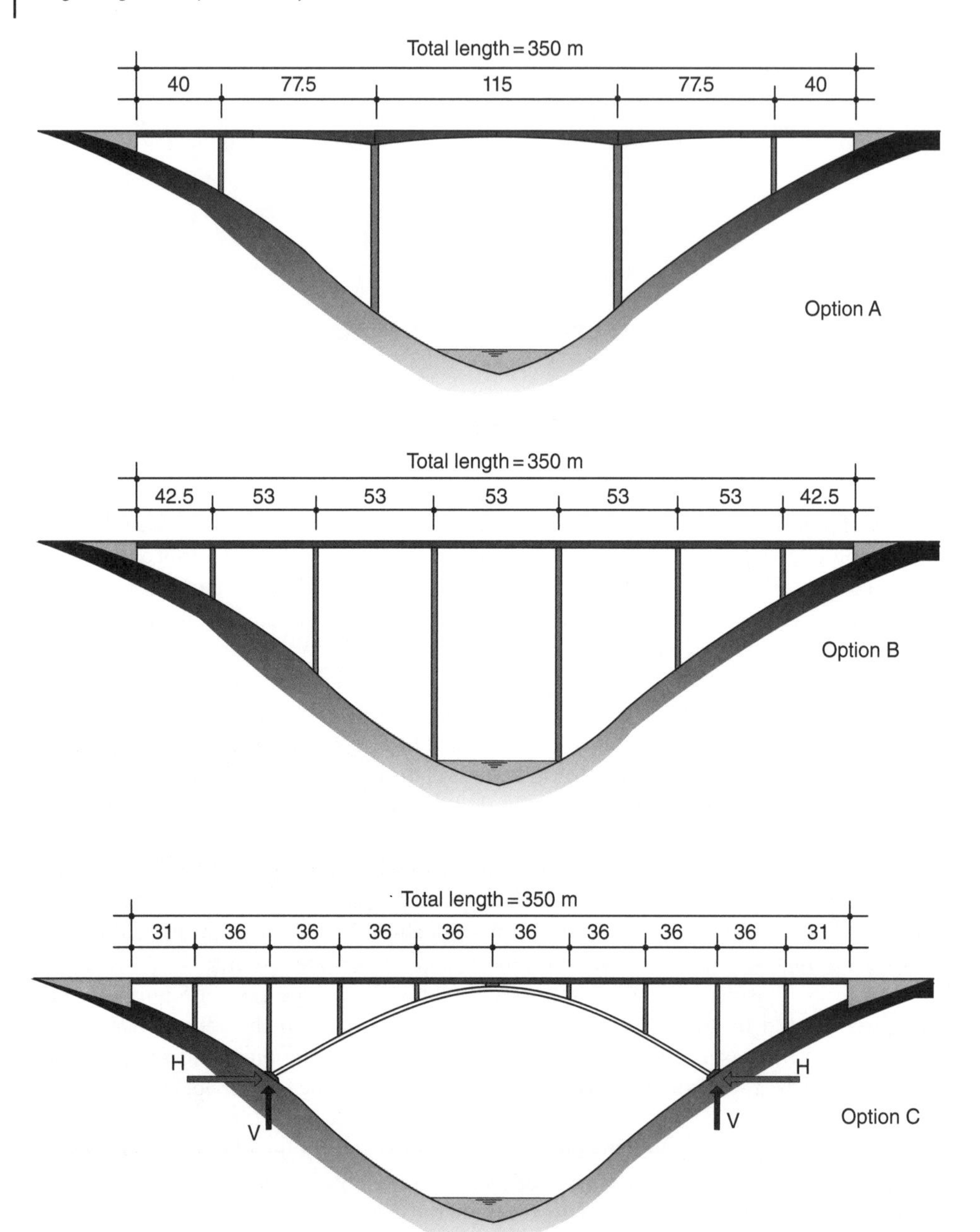

Figure 4.1 Concept design for bridge located in a deep valley: three different options.

- What should be the execution method for the deck considering its height to the ground and the structural material selected?
- What should be the cross section for the bridge deck – an open section (slab-girder deck) or a closed one (box girder deck)?
- What type of connection should be adopted between the deck and the piers – monolithic or articulated through bridge bearings?

As an alternative to Option A, a bridge deck with continuous multiple spans in the order of 60 m and end spans in the order of 40 m, may be considered (Figure 4.1). This option, *Option B*, with possible advantages of a reduced cost for the deck should be executed through a method different from option A. Of course, option B has some disadvantages with respect to option A concerning aesthetics and landscape integration. With respect to option B, the following issues may be stated:

- Based on the span lengths should a bridge deck with prefabricated girders in steel or in prestressed concrete, that could be launched from the abutments, or a cast in place prestressed concrete (PC) deck, that could be executed by a moving scaffolding supported by the piers,
- For the case of launched steel girders, what should be the number of steel girders at the cross sections?
- For the case of precasted PC girders, what should be the number of girders at the cross sections in order to be compatible with the launching capacity of the equipment? Should the span length be reduced for this case in order to reduce the dead weight of the girders?
- In both options – steel and concrete, should the slab deck be in RC or in PC, casted using the girders as scaffolding?

In principle, for this option B if a concrete deck is adopted it would be preferable to reduce the span length to a value between 40 and 45 m for the benefit of the cost of the launching or moving scaffolding equipment that are very much conditioned by the deck self-weight.

As an alternative to options A and B, a different *Option C* consisting of an arch bridge (Figure 4.1), or a frame bridge with inclined legs (Figure 1.8) may be considered. Of course, from a landscape integration point of view, these types of solutions would be preferable. However, these bridge typologies may raise the following issues:

- Are the geotechnical conditions at the slopes sufficiently good enough to absorb high horizontal component reactions (impulse H in Figure 4.1) from the arch?
- How can the arch be executed from the slopes assuming scaffolding supported from the ground is not feasible due to its high cost?

These issues show the interaction between selection of a bridge option and execution methods. No concept design should be made without going through the method to be adopted for the execution.

4.2 Span Distribution and Deck Continuity

4.2.1 Span Layout

As previously discussed, the most current longitudinal structural systems are a continuous girder bridge or a multiple span frame bridge. Span distribution is a basic issue that cannot be defined independently of the execution method.

Starting from the approximate location of the abutments that should be made in order to reduce its height, the total length of the superstructure is defined. Consideration must be made for a minimum clearance in the order of 2 m should retained between the underneath of the deck and the ground in order to protect the bearings at the abutments from any vandalism acts, particularly on urban bridges, and accumulation of litter as well. In bridges located in flat areas without any particular requirement for a main span, such as crossing a river, a navigation channel or a motorway, approximately equal inner spans should be the most adequate solution for a continuous girder bridge. The end spans should be in the order of 0.6–0.8 of the typical span lengths (Figure 4.1). This span arrangement contributes to a better distribution of permanent bending moments at the end spans and to avoid any uplift at the bearings located at the abutments. Of course, increasing span lengths would increase the cost of the superstructure and reduce the cost of the substructure. However, the former may represent 60–80% of the total bridge cost. The cost of piers and foundations may represent 20–40% of the total bridge cost, depending on the geotechnical conditions. The piers themselves usually do not represent more than approximately 10% of the total cost. The cost of deep foundations, by piles, may significantly increase the cost of the substructure. The same happens when piers are located in water or alluvium zones were the cost of execution of the pile caps is quite significant. Cofferdams are usually required for piers located in rivers and this condition may justify increasing the span length of the superstructure. The balance between the cost of superstructure and substructure should be a main issue for the conceptual design of the bridge.

If the bridge is located in a deep valley, a span modelling decreasing from the centre towards the abutments is the best solution as referred to in Section 4.1. However, assuming, for example, a cast in place girder deck, the variation of spans introduces an increased difficulty in construction. The depth of the superstructure should vary as shown in Figure 4.1 for Option A. The prestressing layout and the formwork should be different at each span. In Option B of Figure 4.1 the typical prestressing layout may be the same for most of the internal spans.

4.2.2 Deck Continuity and Expansion Joints

Decks should be continuous as far as possible to avoid internal expansion joints that are costly to maintain and very often need to be replaced more than once during the life of a structure. If possible, one should adopt only one expansion joint at each end span at the abutment-deck connections. The limits for continuity of the decks are of course the displacements to be accommodated at the expansion joints. For example, for a PC bridge with a total continuous deck in the order of 700 m, as in the bridge case of Figure 4.2 (Freixo Bridge, in Oporto), the deck is monolithic with the adjacent piers to the main span (150 m long) and supported by bearings at all the other piers. However, pier P4 has fixed bearings, while all the others have sliding unidirectional bearings. At the abutments, unidirectional bearings are introduced, as well as *seismic dampers* (a subject discussed in Section 7.4) to allow slow movements due to temperature, shrinkage and creep effects; high frequency movements due to earthquakes are restricted by the seismic dampers. The *centre of stiffness of the piers* (Section 7.2.2) is approximately located at the centre of the main span. The maximum distance from the centre of stiffness to the south abutment (Gaia) is approximately 400 m. Shrinkage, creep and thermal variations, induce amplitudes of movements at the south expansion joint equivalent to a thermal variation in the order of +15°/−60°C resulting in a movement in the order of 300 mm. With a safety coefficient of at least 1.2–1.3, this means the expansion joint

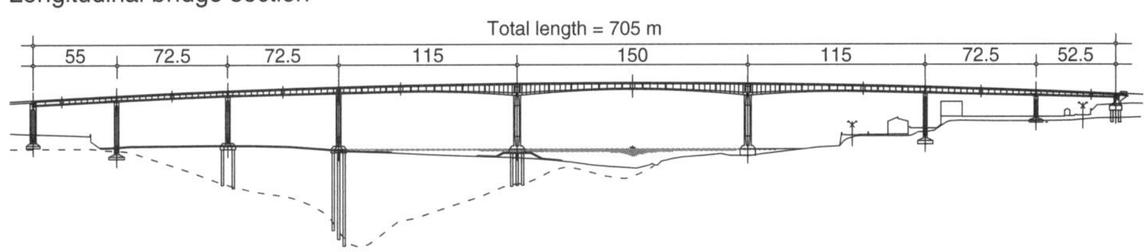

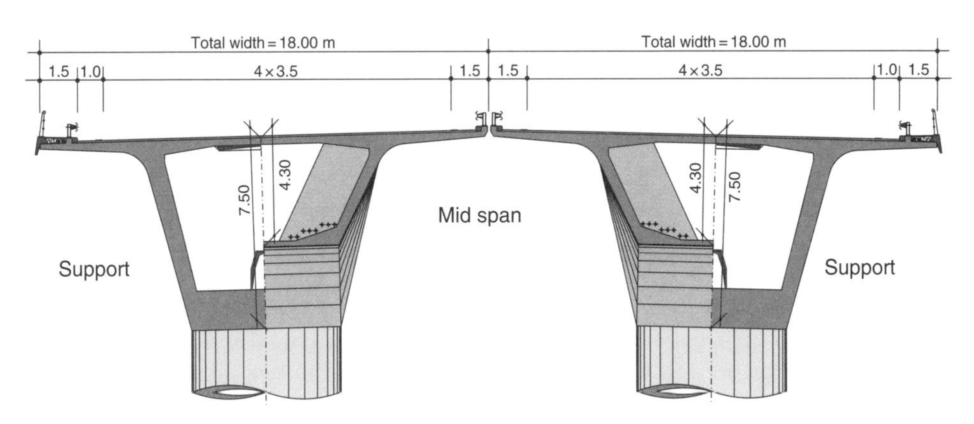

Figure 4.2 The Freixo Bridge in Oporto (150 m main span) (*Source:* Courtesy GRID, SA).

should be designed for approximately 400 mm of amplitude. A large part of this movement is due to shrinkage and creep effects inducing a shortening of the deck. When the expansion joint is installed, a *preset* should be specified taking into consideration the structure temperature at time of joint installation.

When the continuity of the deck is too large, the expansion joints need to accommodate very large amplitude displacements and are too costly. As an order of magnitude, continuous bridge lengths of 600–700 m nowadays (although in some cases the continuity may reach 1000 m or even more, as in Millau road Viaduct in France: see Figure 4.33 later) have a multi-cable-stayed steel deck of 2460 m, or in the HSR (High Speed Railway) Viaduct de las Piedras in Spain, having a steel-concrete composite deck of 1209 m or in the Bridge over Guadiana in Alqueva (Figure 5.8), a PC box girder deck with a span of 1230 m. That depends, as well, where the centre of stiffness of the top of the piers is located. It should also be noted that when an internal expansion joint is adopted at specific section on a pier, one has movements coming from both parts of the deck adjacent to the expansion joint. This may double the movement to be accommodated at the expansion joint. In this case, the continuity of the deck may need to be reduced. The referred value for continuity of 600–700 m may be reduced to 300–350 m unless the expansion joint is designed for larger movements. Another aspect of deck continuity is the displacements that should be accommodated at the sliding bearings on the piers. The top sliding plates at the bearings (see Section 7.3) should be large enough to accommodate the movements of the superstructure.

One may designate the distance from the centre of stiffness to the expansion joint location as the *dilatation length*. In railway bridges, an additional problem exists related to the *rail-structure interaction*. The continuity of *continuous welded rails* on ballast integrated on the slab deck requires long dilatation lengths, the adoption of *rail expansion joints*. Otherwise, the stresses induced in the rail due to differential movements of the superstructure with respect to the rail may be too large. The reader is referred to specific literature [1] on this subject.

4.3 The Influence of the Execution Method

To illustrate the influence of the construction method on the design of a bridge structure, two examples are considered:

- A concrete bridge built by the balanced cantilever method;
- A composite bridge with the slab deck cast in place on the previously erected steel girders.

4.3.1 A Prestressed Concrete Box Girder Deck

For the first example, a three span bridge, the evolution of the static system is shown in Figure 4.3. In Figure 4.4 a design case illustrates this construction scheme. The bridge deck is executed in symmetrical segments from the two piers with segment lengths between 2.5 and 5.0 m in length. For that there is a pair of *moving scaffoldings* from each pier in a balanced cantilever scheme. Reaching nearby the mid-span, the end sections of two cantilevers are connected through a closure segment with a length in the order of 2–3 m generally. At the end spans, the deck length to connect the end section of the cantilever with the support section at the abutments is cast on scaffolding supported from the ground. The internal forces at the execution phases are determined based on a simple cantilever. The negative bending moments during the execution phases of the deck increase towards the support section at the pier where the deck is built in.

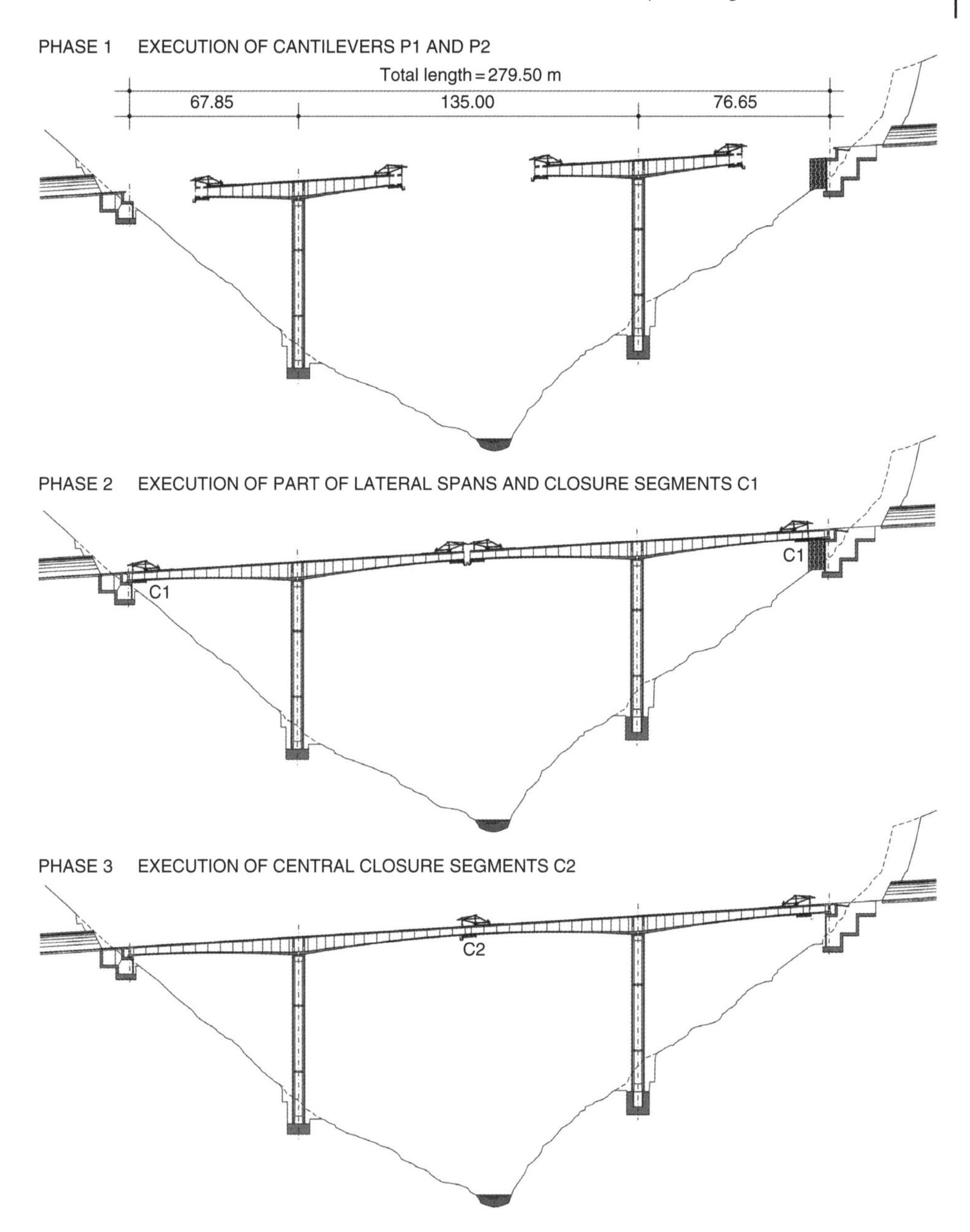

Figure 4.3 Evolution of the static system in a bridge built by balanced cantilever.

The shape of the bending moment diagram requires an increased depth, with a parabolic variation, towards the support section. These negative bending moments control the required depth of the section at the supports, and the overall geometry of the deck cross section. Thus, a thick lower flange is required to accommodate high compressive stresses that are developed during the execution phases. The ideal cross section is a box with increased thickness of the lower flange towards the support section. In a second phase, after completing the balanced cantilever scheme, the

Figure 4.4 The bridge over the Ribeira Funda, Madeira Island, built by balanced cantilever (135 m main span) – Approaching the execution of the closure segment at mid-span (*Source:* Courtesy GRID, SA).

deck lengths at the end spans are cast on formwork supported from the ground as shown in Figure 4.3. In a phase 3, the closure segment at mid-span is cast. If behaviour was elastic, the bending moments at the closure mid-span section would remain zero. However, due to creep effects of the concrete, the bending moments in the deck tend to approach the bending moment diagram of a continuous girder deck with three spans as if it was cast in a single phase on formwork supported from the ground and the formwork was also removed in a single phase. Therefore, at the mid-span, a positive bending moment exists after the closure segment is cast. The bending moment diagram is time dependent but it also tends in the long term towards the bending moment diagram of a continuous deck.

4.3.2 A Steel-Concrete Composite Steel Deck

In the second example, a steel-concrete composite deck is shown in Figure 4.5. After the erection of the steel girders, erected, for example, by cranes from the ground, the slab deck is cast using the girders as scaffolding. The stresses on the steel girders due to the dead weight of the concrete are evaluated as if the structure was a steel-only structure. However, after hardening of the concrete, the composite steel-concrete action is developed and for all the remaining loads, like superimposed dead loads and live loads (traffic loads), the stresses are evaluated for a steel-concrete composite section. The cross section dimensions of the steel girders should be established taking due account of the execution phases.

4.3.3 Concept Design and Execution: Preliminary Conclusions

From these examples, some conclusions can be drawn:

- The evolution of the static system during construction should be taken into account for the design of the bridge structure;
- The geometry of the final structure and the longitudinal and transverse models can only be established taking into consideration the execution method.

In next sections, the most common typologies for the superstructure of concrete, steel and composite bridges are discussed and typical pre-design rules are referred to.

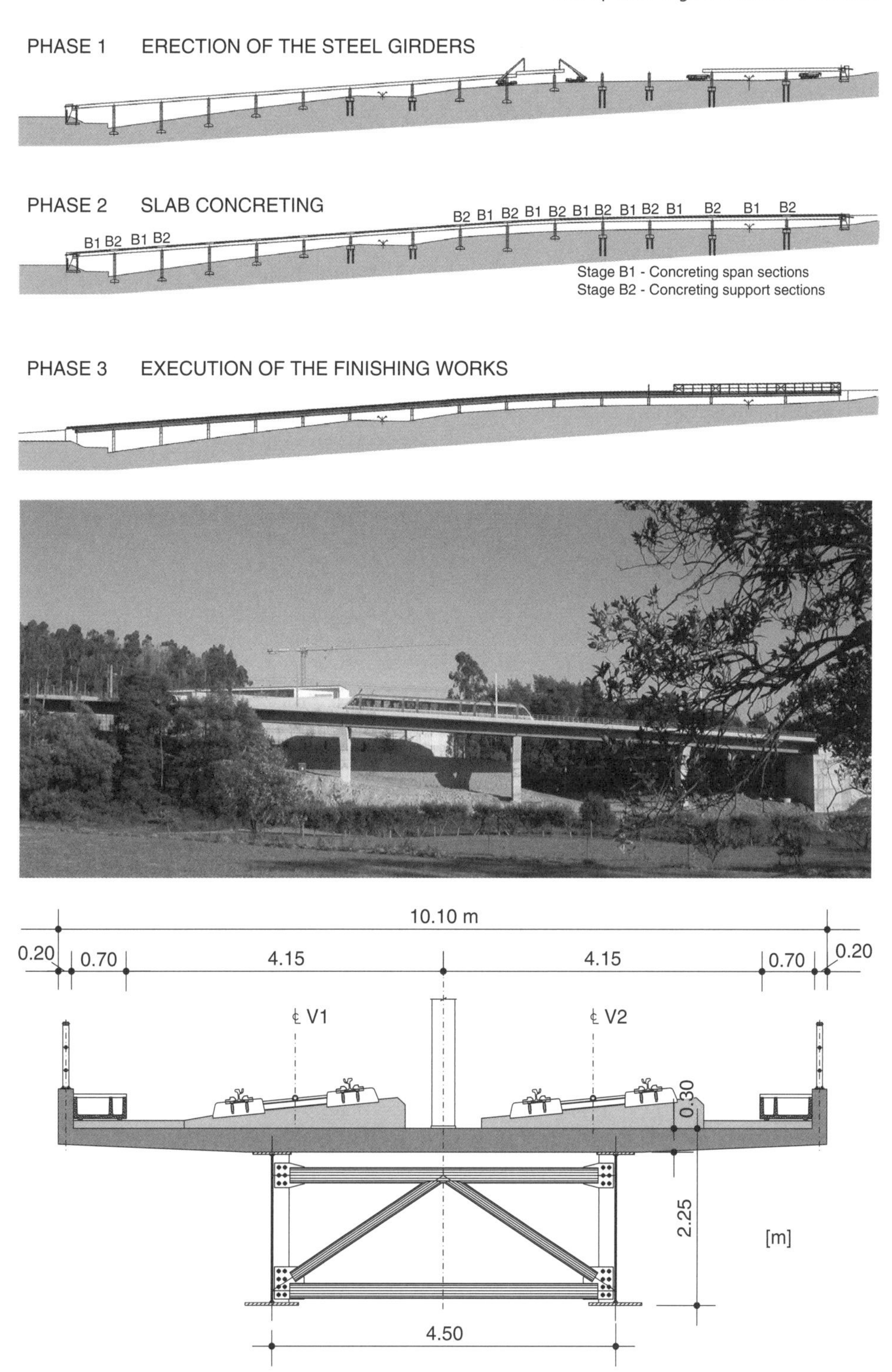

Figure 4.5 A composite steel-concrete deck for an Oporto Metro viaduct (total length 498.2 m with a main span of 48.5 m) (*Source:* Courtesy GRID, SA).

4.4 Superstructure: Concrete Bridges

4.4.1 Options for the Bridge Deck

For the conceptual design of the superstructure, options should be taken with respect to: (i) the structural materials, (ii) the longitudinal structural system, (iii) the transverse cross section of the deck, (iv) the connection between the deck and the piers and abutments and (v) the execution method. Several of these options were considered in previous sections and chapters. In this section, a detailed discussion is presented for the design of deck cross sections, considering concrete as the structural material for the deck. The option for the transverse deck cross section depends on:

- The width of the deck (b);
- The span length (l);
- The slenderness (l/h) envisaged for the bridge deck with a depth (h);
- The execution method and construction equipment to be adopted.

The following options may usually be considered for a concrete deck, in RC or PC:

- Slab and voided slab decks
- Ribbed slab and slab-girder decks
- Box girder decks

These options (Figure 4.6) may be adopted for the cross section of bridges with different longitudinal structural systems, namely simply supported or continuous beam bridges, frame bridges, arches or even for cable-stayed bridges.

One way of comparing different cross sections options, as shown in Figure 4.6, is to compare the so called *equivalent depth of the deck*, given by:

$$h_{eq} = A/b \tag{4.1}$$

where A is the area of the deck cross section of width b. Hence, h_{eq} is the depth of a rectangular solid slab with the same weight. The upper and lower stresses at the extreme fibres of the cross section deck (at distances v_i and v_s to the centroid) under bending, depend on the section module I/v_i and I/v_s where I is the relevant moment of inertia. The ideal most efficient cross section of area A, to resist to a bending moment is made of two separate thin layers (flanges) each of area $A/2$, at a distance h. The moment of inertia of this ideal section is approximately $I = A\ i^2$ where $i = h/2$ is the *radius of gyration* with respect to the neutral axis. The efficiency of a cross section depends how much v_i and v_s differ from the i value of the actual cross section. Taking the ratios i/v_i and i/v_s an *efficiency parameter*, ρ, may be adopted for comparing different cross section options:

$$\rho = \frac{i}{v_i}\frac{i}{v_s} = \frac{I}{A v_i v_s} \tag{4.2}$$

The ideal cross section has $\rho = 1$; for a real deck cross section $\rho < 1$. Different deck cross sections are discussed and compared in the next sections, with respect to efficiency. Of course, the efficiency of a deck cross section cannot be measured only by its h_{eq} and ρ values. Labour costs come into play, namely concreting rate, formwork and

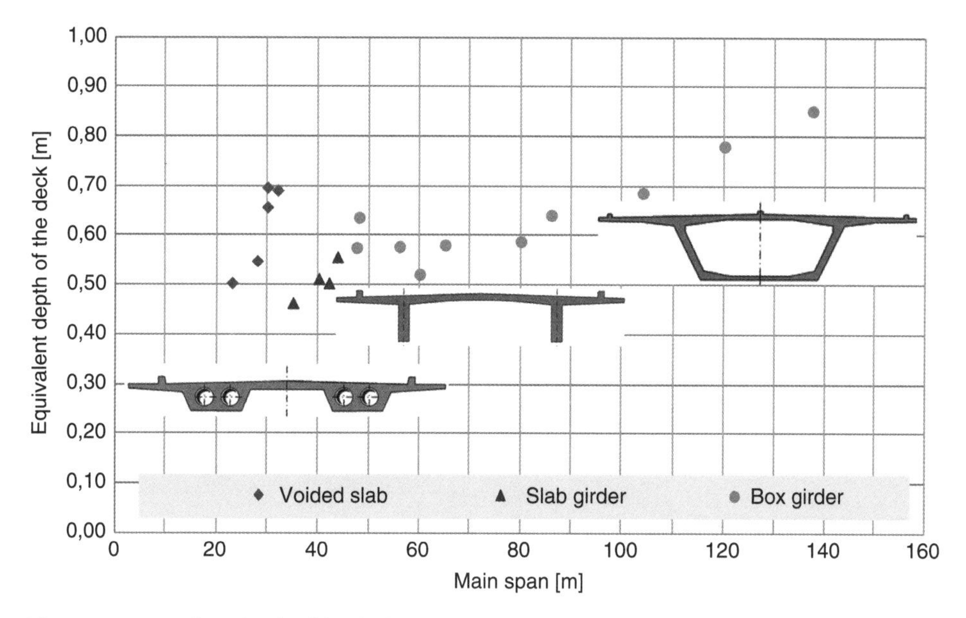

Figure 4.6 Equivalent depth of the deck (h_{eq}).

placing of the ordinary and active (prestressing) reinforcement. First, the concrete material should be considered.

4.4.2 The Concrete Material – Main Proprieties

4.4.2.1 Concrete

The compressive strength of concrete is denoted by concrete strength classes that relate to the characteristic (5%) cylinder strength f_{ck} at 28 days (or the cube strength $f_{ck,cube}$) in accordance with EN 206-1 [2]. The strength classes for concretes frequently used in bridges are presented in Table 4.1.

The mean tensile strength f_{ctm} refers to the highest stress reached under concentric tensile loading. The flexural tensile strength can be increased for members with depths h up to 0.6 m, by using the relationship: $f_{ctm,fl} = \max\{(1.6 - h)\, f_{ctm}; f_{ctm}\}$.

The elastic deformations of concrete largely depend on its composition (especially the aggregates). Therefore, values for the secant modulus of elasticity E_{cm}, between $\sigma_c = 0$ and $0.4 f_{cm}$, given in Table 4.1 for concretes with quartzite aggregates, should be regarded as indicative and specifically assessed if the structure is likely to be sensitive to deformations. For limestone and sandstone aggregates, the values should be reduced by 10 and 30%, respectively. For basalt aggregates, the values should be increased by 20%. Poisson's ratio may be taken equal to 0.2 for uncracked concrete and 0 for cracked concrete. The linear coefficient of thermal expansion is usually taken equal to $10^{-5}\, C^{-1}$. The value of the design compressive is defined from these values as $f_{cd} = f_{ck}/\gamma_c$, where γ_c is the partial safety factor for concrete, usually 1.50.

Finally, for executing the structural analysing and computing the resistance of concrete cross sections under compression the stress–strain diagrams of Figure 4.7a and b may be respectively adopted, using the properties from Table 4.1.

Table 4.1 Main proprieties for concrete frequently used in bridges.

Strength class for concrete	C25/30	C30/37	C35/45	C40/50	C45/55	C50/60
Characteristic compressive strength f_{ck} (MPa)	25	30	35	40	45	50
Mean compressive strength $f_{cm} = f_{ck} + 8$ (MPa)	33	38	43	48	53	58
Mean tensile strength f_{ctm} (MPa)	2.6	2.9	3.2	3.5	3.8	4.1
Secant modulus of elasticity E_{cm} (GPa)	31	33	34	35	36	37
Strain at peak stress ε_{c1} (‰)	2.1	2.2	2.25	2.3	2.4	2.45

4.4.2.2 Reinforcing Steel

Reinforcing steel can take the form of bars, de-coiled rods or welded fabric. Figure 4.7c presents the typical stress–strain diagram for reinforcing hot rolled steel (dashed) with the characteristic yield strength f_{yk} and tensile strength f_{tk}, and the correspondent strain, ε_{su}. The same figure also includes the stress–strain diagram usually used for design (for tension and compression), with a yield strength $f_{yd} = f_{yk}/\gamma_s$ and $\gamma_s = 1.15$, the partial safety factor for reinforcing steel at ULS (Ultimate Limit State).

The design value of the modulus of elasticity E_s is assumed to be 200 GPa. Typical values of strains ε_{s1} and ε_{ud} are, respectively, 2.174‰ and 10‰ for reinforcing steel B500 (in accordance with the EN 10080), the most frequently used in bridges.

4.4.2.3 Prestressing Steel

Prestressing steel usually appears in concrete decks as wires, bars or strands, with low level of relaxation (class 2) and susceptibility to corrosion. Additionally, prestressing steel should be adequately and permanently protected against corrosion in sheaths or ducts. These ducts are grouted with cement grout (both bonded and external tendons), with petroleum wax or mineral-oil-based grease (for external unbounded tendons only).

Figure 4.7d presents the typical stress–strain diagram for prestressing steel (dashed) with the characteristic 0.1% proof stress $f_{p0.1k}$ and tensile strength f_{pk} and the correspondent strain, ε_{uk}. The properties of prestressing steels are given for example in EN 10138. The same figure includes also the stress–strain diagram usually used for design, with a strength $f_{pd} = f_{p0.1k}/\gamma_s$, with $\gamma_s = 1.15$ at ULS.

The most commonly prestressing steel used in concrete bridges consists of a bundle of seven-wire strands, 15.3 or 15.7 mm nominal diameter (designated T15 normal or T15 super), with nominal tensile strengths of 1770 or 1860 MPa, respectively.

The design value for the modulus of elasticity E_p may be assumed equal to 195 GPa for strands, although the actual value can range from 185 to 205 GPa, depending on the manufacturing process. Typical values of strains $\varepsilon_{s1} = 0.75$‰, $\varepsilon_{ud} = 10$‰ and proof stress $f_{p0.1k} = 0.9\,f_{pk} = 1680$ MPa for the prestressing steel Y1860 (in accordance with the EN 10138-3 [3]).

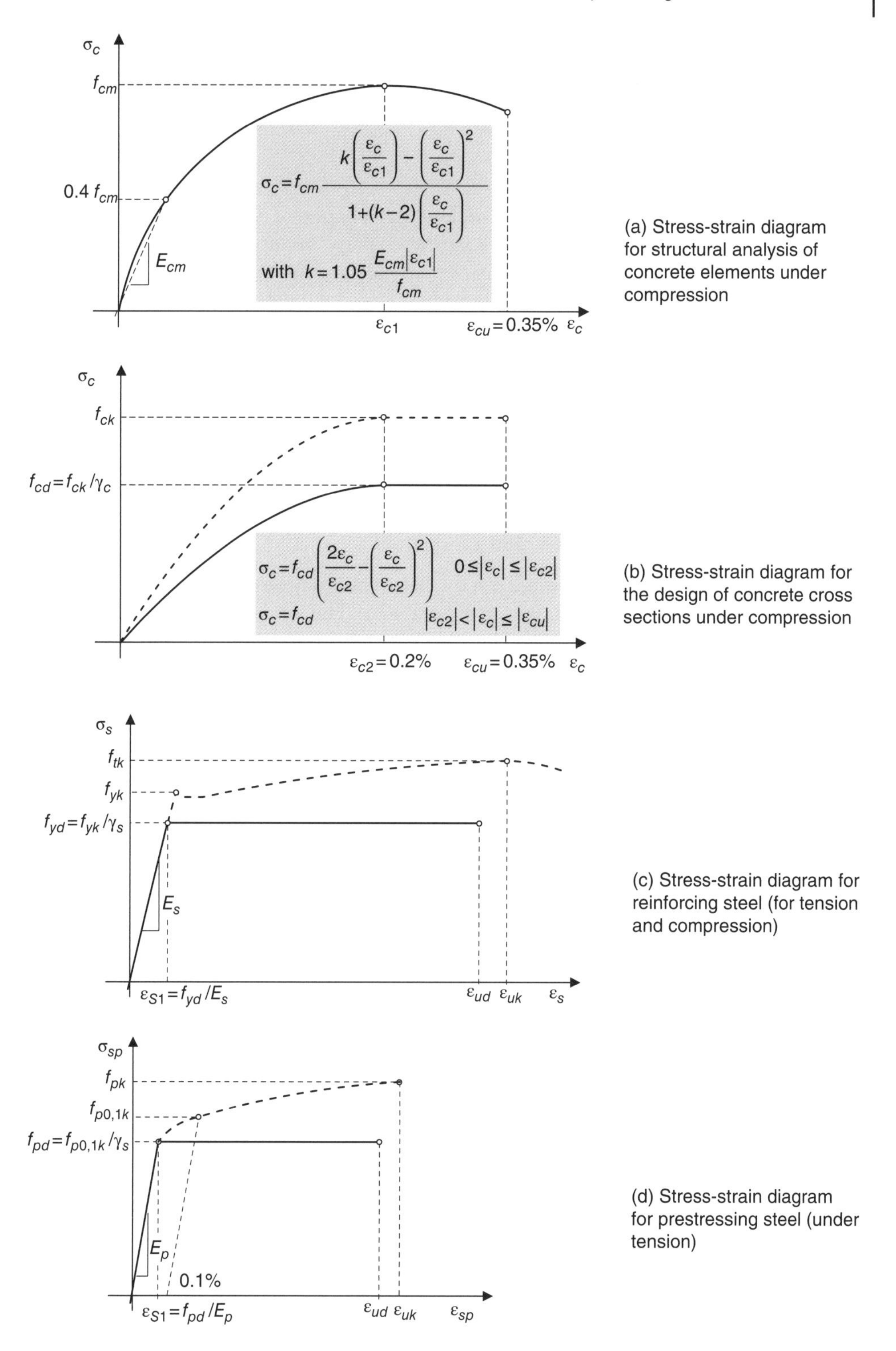

Figure 4.7 Stress–strain diagrams for concrete, reinforcing and prestressing steel.

The maximum forces in each strand should be obtained with the stressing limit at tension stage $f_{po} = \min\ \{0.8\ f_{pk};\ 0.9\ f_{p0.1k}\}$. Bars have usually lower stress limits due to higher susceptibility to fatigue and brittle failure.

4.4.3 Slab and Voided Slab Decks

Slab bridge decks (Figure 4.8a) with rectangular shapes or with overhangs may be adopted for spans up to 30 m. For small isostatic spans, say up to 20 m, RC-reinforced concrete may be adopted. Increasing spans generally require the use of PC-prestressed concrete. If the plan geometry is complex, namely in skew slab or curved bridge decks, the solid rectangular slabs have the advantage of simplicity of execution – formwork, reinforcement, concreting and installation of prestressing cables. The main disadvantages are the large dead weight and small eccentricities for prestressing cables, generally requiring large quantities of concrete, steel reinforcement and prestressing. Typical values for pre-design areas are as follows:

- Slab depth h – 0.25–0.80 m;
- Spans l – up to 15–20 m in RC and 20–30 m in PC;
- Slenderness – $l/h = 15–25$ in RC and 15–35 in PC.

Some of the disadvantages pointed out for slab decks may be overcomed by the adoption of a *voided slab deck* as shown in Figure 4.8a. The voids are usually circular or rectangular tubes in polyvinyl chloride (PVC), expanded polystyrene, thin spiral steel tubes or even in wood formwork. Some minimum distances, as indicated in Figure 4.9, should be respected in locating the tubes. Rectangular tubes may yield a reduced dead weight for the cross section, but have the disadvantage of difficulty for concreting underneath the tubes. In this case, the maximum width of the tubes should not exceed 1.0 m. When prestressing cables are adopted, the minimum distance of voids should be at least three times the diameter of the ducts. Special care should be taken to avoid tubes floating during concreting due to upward forces induced by concrete before hardening. For a void with a diameter Φ the upward force is $(\Phi\gamma_c)$ where γ_c is the unit weight of concrete. For example, a void of 0.7 m diameter has an upward force of approximately $10\,\text{kN}\,\text{m}^{-1}$. Because of that it is convenient to fix the tubes by specific reinforcement bars to the underneath layer of the main reinforcement. Care should also be taken to prevent water entering the voids during concreting. The voids may be interrupted at some transverse cross sections, in particular when large voids are adopted with respect to the total cross section. These cross beams (Figure 4.9) avoid any distortional effects at the cross sections. Over the piers, the voids are necessarily interrupted in order to transfer the bearing reaction, as in Figure 4.10.

The main advantages of the voided slab decks, with respect to a solid slab, are the reduced self-weight (reduced h_{eq}) and the increased depth h, allowing an increased eccentricity for prestressing cables and even for the ordinary reinforcement. The ρ factor of a voided slab is larger than in a solid slab of the same self-weight, that is, with the same h_{eq}. The main disadvantages of a voided slab with respect to a solid slab deck are an increased difficulty for the execution in particular for curved plane geometries.

In Figure 4.11 a comparison is made of the efficiency parameter ρ (Eq. 4.2) for different deck cross sections. Solid and voided slabs have a ρ in the order of 0.3–0.4 that may be compared to a box girder deck where ρ values are usually between 0.5 and 0.7.

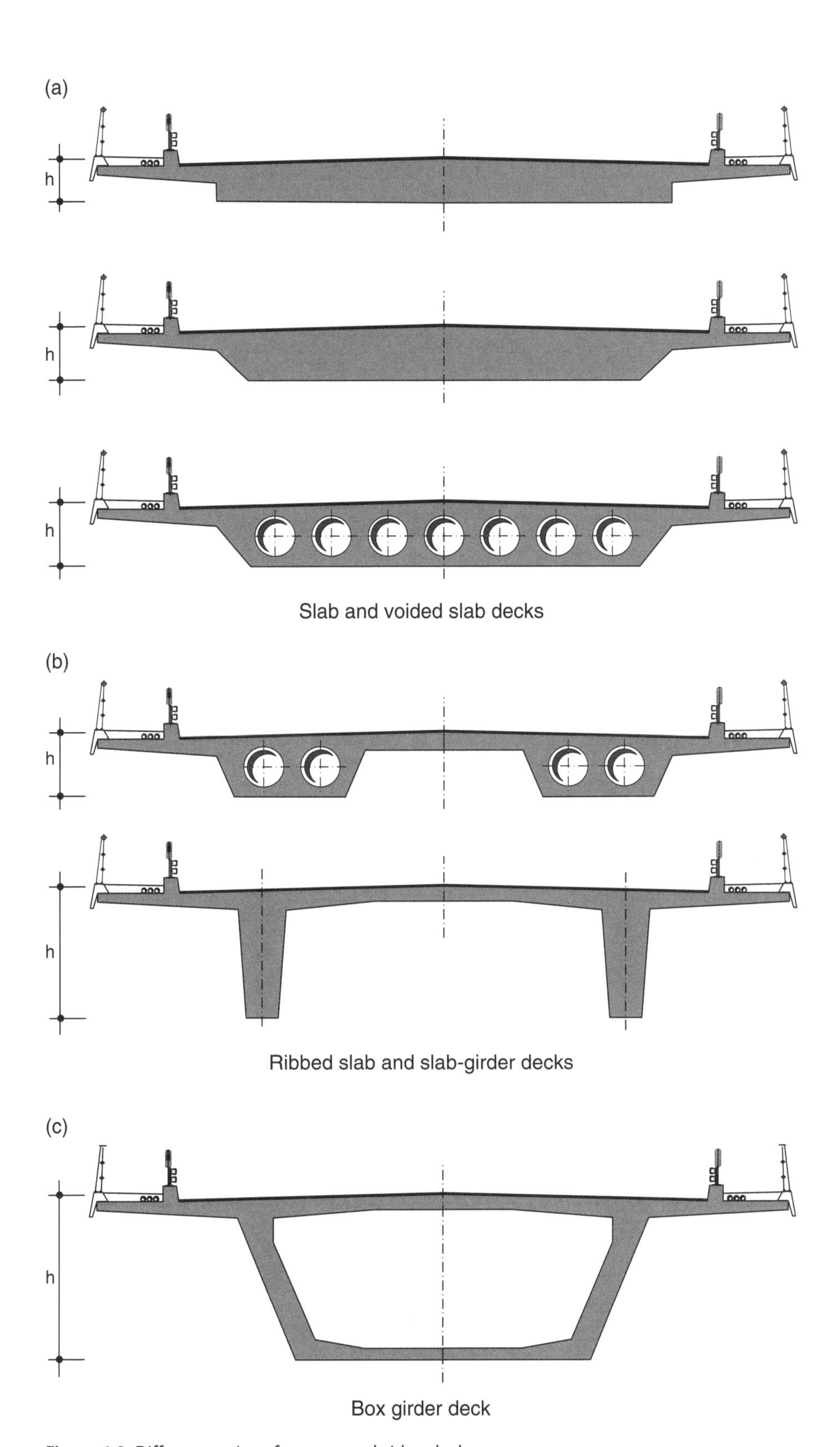

Figure 4.8 Different options for concrete bridge decks.

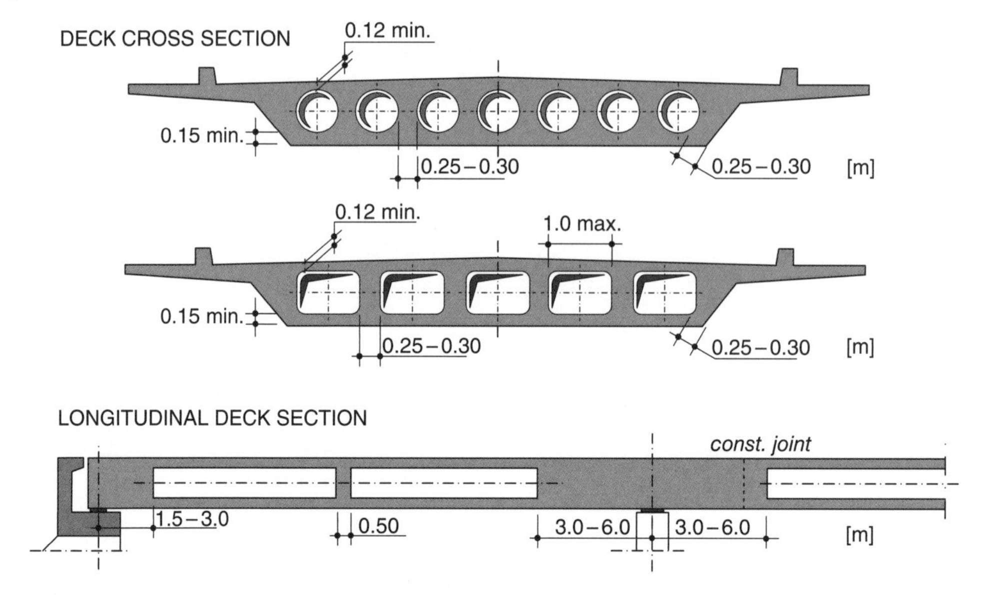

Figure 4.9 Voided slab decks: minimum distances for location and void dimensions.

4.4.4 Ribbed Slab and Slab-Girder Decks

To increase the efficiency of a slab deck under increasing spans, say above 30 m, it is necessary to reduce the self-weight of the cross section ($A\ \gamma_c$, kN m^{-2}) and to increase the eccentricity of the prestressing cables. This can be achieved by transforming solid slab or voided slab sections in a *ribbed slab section* or in a *slab girder deck* (Figure 4.8b). The difference between these two options is the width of ribs. When the ribs are narrow, depth may be increased keeping the same h_{eq}, resulting in a slab girder deck. In a ribbed slab deck, the number of ribs depends on the width, b, of the deck. To keep the slab deck in RC in the transverse direction, this requires limiting the transverse free span between ribs to about 5–6 m. With two ribs of 2–3 m width (at the slab deck level) and *overhangs* (transverse cantilevers of the slab deck) in the order of 2 m, a total deck width of 14–15 m may be reached. The number of ribs may be increased to three or even four in the limit if the deck width is very large. This type of deck option, in PC, may reach span lengths up to 35 m with constant deck height. Spans up to 40 m may be reached with a variable height depth, for example, in a three-span continuous deck. The ribbed slab option in PC is a more efficient solution than a voided slab deck for the same span length. The ρ values may reach 0.35–0.40.

Typical values for pre-design of a *ribbed slab deck* are as follows:

- Spans l – up to 35 m in PC
- Slab deck thickness (between ribs) – 0.22–0.30 m
- Width of the ribs at the slab deck level – 1.5–3.0 m
- Slenderness – l/h = 20–35 in PC.

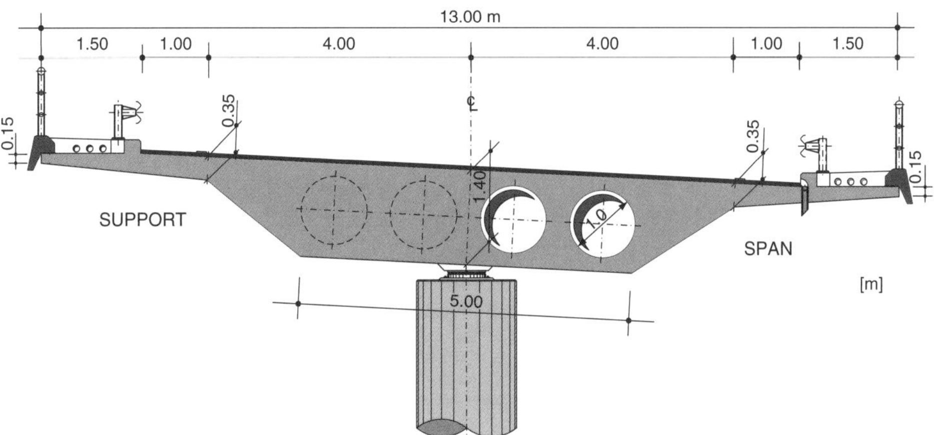

Figure 4.10 A design example of a continuous voided slab deck with single piers: overpass at the Funchal – Airport Highway, Madeira Island, Portugal (*Source:* Courtesy GRID, SA).

The lower slenderness values are usually adopted for simply supported spans. Ribbed slab decks are a quite convenient option for overpasses on highways where continuous three-span decks may require a central span up to 35 m, avoiding any pier in the central strip of the highway.

A combination of two schemes, rib and voided slabs, is an option in particular for increasing span lengths and width of the deck, as shown in the example of Figure 4.12.

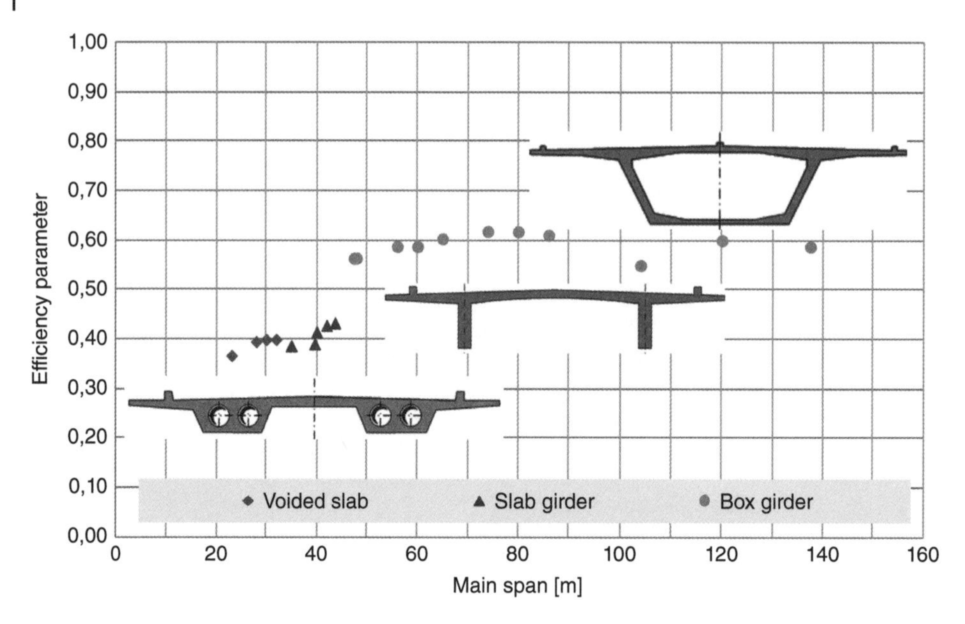

Figure 4.11 Efficiency parameter for different concrete deck cross sections.

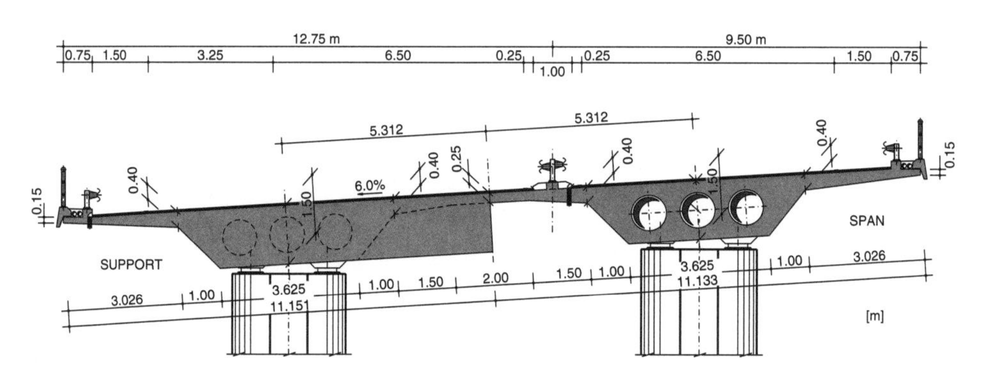

Figure 4.12 A two rib voided slab deck for a continuous prestressed concrete curved viaduct in a highway in Funchal, with typical spans of 30 m.

For medium spans, say 40–70 m, constant depth or variable depth slab girder bridges in PC are needed, unless box girder sections are adopted (Figure 4.8c). The slab girder deck may be adopted with two girders only and a RC slab deck (Figure 4.13). That is, in general, the more economic option. The main girders are located at 4–8 m in general if the slab deck is in RC. Multiple girder bridge decks, with three or four main girders, are only justified for large deck widths or very limited vertical clearances. If the depth h of the deck should be limited due to clearances or aesthetical reasons, yielding a slenderness l/h above 17 (constant depth cross sections) box girder options are usually preferred.

In slab-girder decks, the slab deck may be prestressed in the transverse direction to keep the deck with two girders only. The free transverse span length between girders,

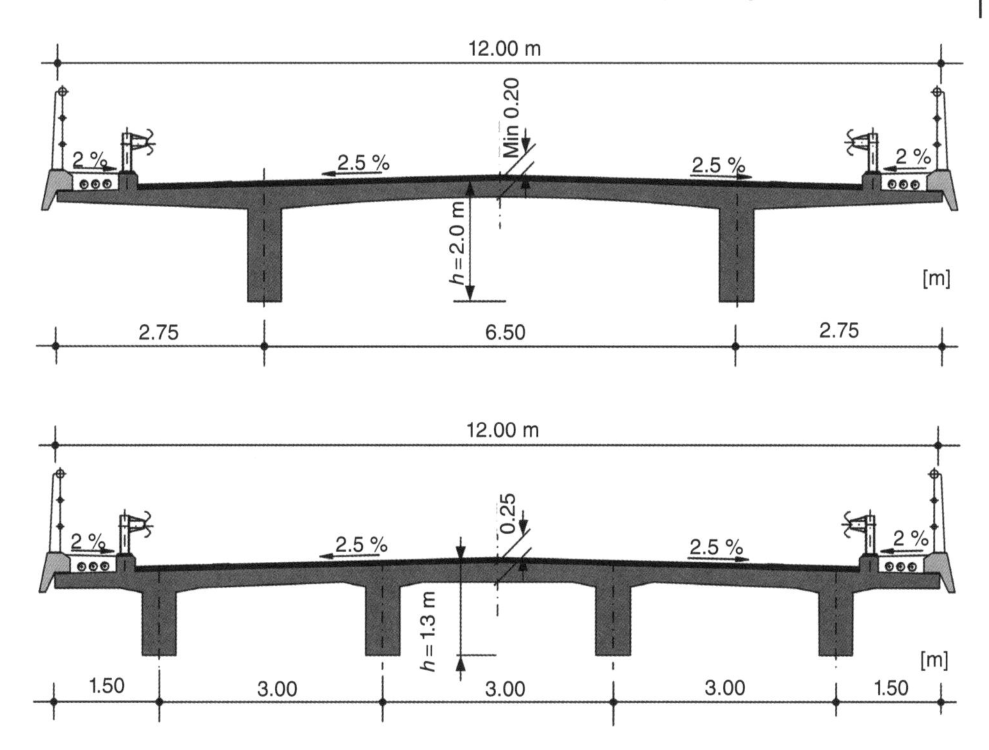

Figure 4.13 Slab-girder deck cross sections – with two or four girders.

may reach values as high as 15 m, or even more, for very wide highway decks up to 30 m (Figure 4.14). Large span overhangs and a variable thick slab deck are required, in these cases, to reduce transverse prestressing. The slenderness of the main girders (l/h) may also be reduced to values as low as 12–15 in constant depth decks, with only two girders and very wide decks.

Cross beams between the main girders (Figure 4.15) have been a solution very much adopted in the past. Nowadays, to simplify execution, namely when *formwork launching girders* are adopted (see Section 4.6). It is possible to achieve the *transverse load distribution* (Chapters 3 and 6) between main girders by the slab only. Even in very wide decks, like the one in Figure 4.14, the transverse load distribution may be made by the slab only for easiness of the execution scheme. If span cross beams are adopted, for example, when the deck is concreted on a formwork supported from the ground, its number should be limited to one at each mid-span or two cross beams at one- and two-thirds of the intermediate spans and one cross beam at 0.4 of the side spans (Figure 4.15). The cross and main girders work as a grid structural system. It should be noted that at the end sections of the side spans (i.e. at the abutment sections) a cross beam is usually adopted to support the slab deck. Otherwise, the slab deck has a free edge and the transverse bending moments due to local traffic loads are very high. These end cross girders are the support for the expansion joints as well. At intermediate support sections, it is also convenient to adopt cross girders, transferring pier reactions to the superstructure and helping to distribute horizontal transverse loads when two pier

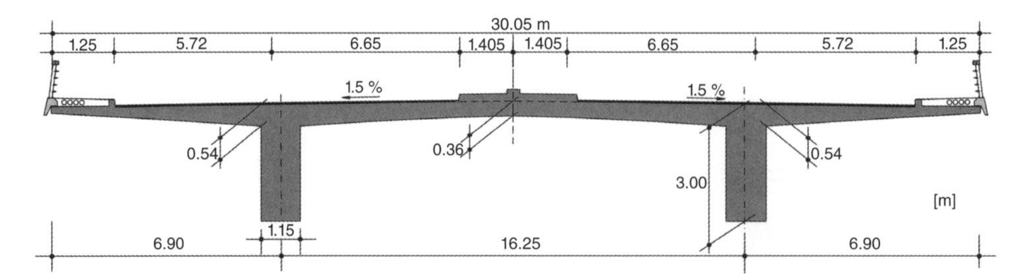

Figure 4.14 Deck cross section with only two girders for a very wide deck (Vorland Bridge, Germany, 1968; main span 39 m, deck slenderness of 13) (*Source:* Adapted from Ref. [4]).

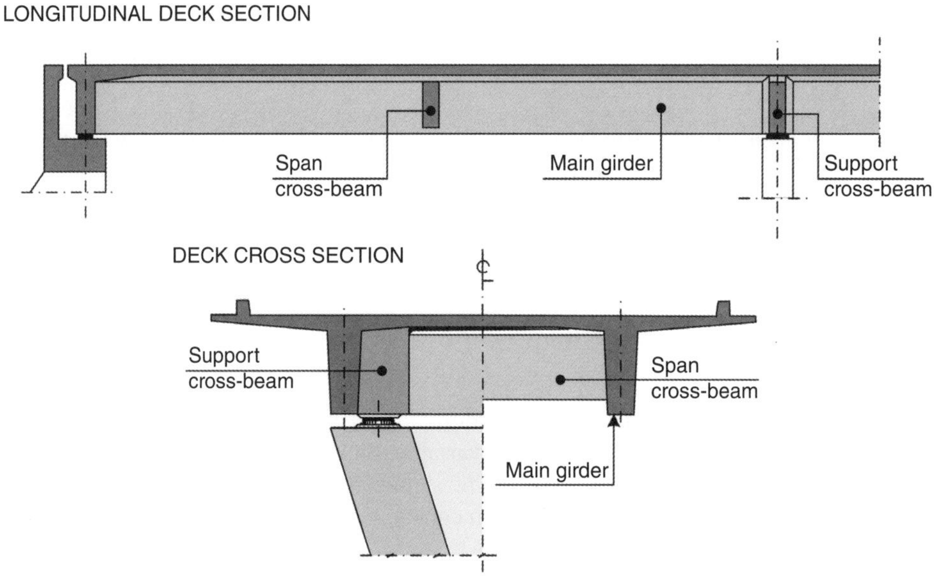

LONGITUDINAL DECK SECTION

Span cross-beam

Main girder

Support cross-beam

DECK CROSS SECTION

Support cross-beam

Span cross-beam

Main girder

Figure 4.15 Longitudinal and transverse cross sections of a slab-girder deck with cross beams.

shafts are adopted at support sections. The intermediate cross girders may also have the function of accommodating bridge bearings in the case of a single pier shaft with a reduced width and no pier cap. The main girders work as T beams in positive bending moment regions – sagging moments. In negative (hogging) bending moment regions, the slab deck is considered to be cracked at least at ULS and the section works as a rectangular cross section.

In most cases, at intermediate piers of slab-girder decks, the deck is supported by bearings. The cross girders may also work in frame action with two pier shafts, that are monolithically connected to each girder. However, this is not so common due to greater execution difficulties and induced longitudinal bending moments in the piers.

When the span increases, the negative bending moments at pier sections are not easy to accommodate because the negative bending compressive stresses in concrete tend to be too high. These negative bending moments of permanent loads are reduced by the prestressing effects (positive primary bending moments) but negative hyperstatic (secondary) bending

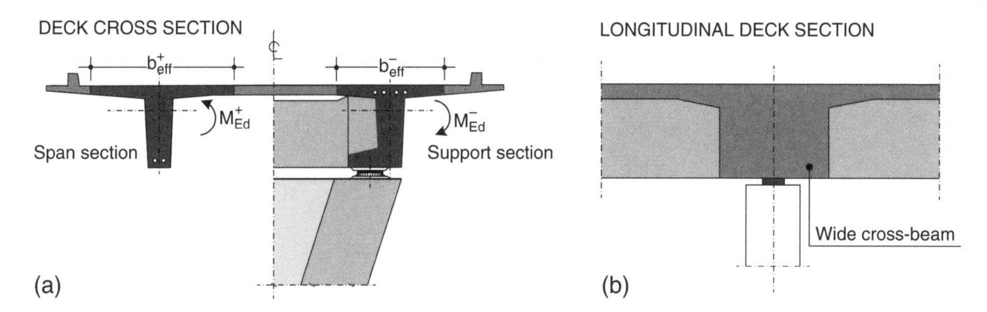

Figure 4.16 Different options for reducing compressive stresses at the bottom fibres of a slab-girder deck.

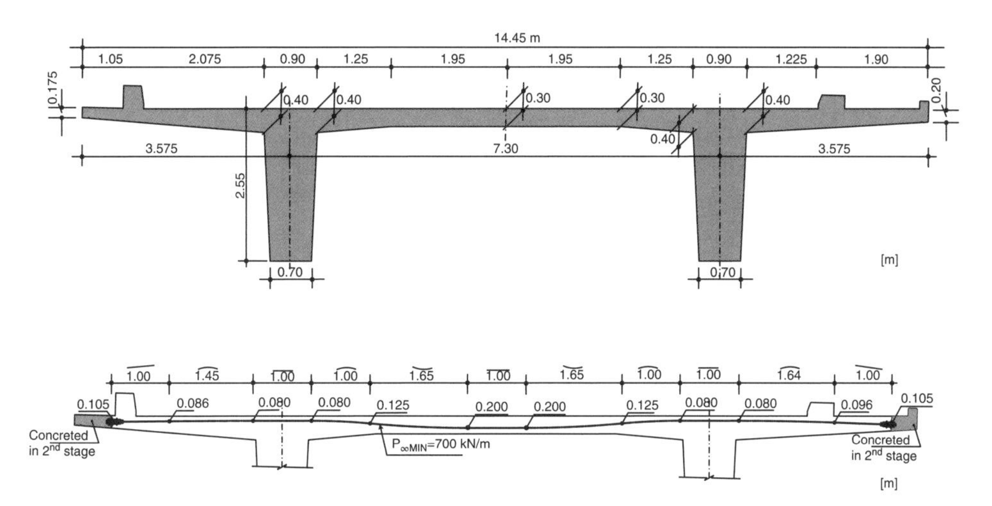

Figure 4.17 Transverse prestressing in the slab-girder decks of Figure 4.20.

moments are developed in continuous girder decks reducing the beneficial effect of the prestressing. Traffic loads and negative thermal gradients also induce compressive stresses at the bottom of the main girders. To reduce the compressive stresses at the bottom of the main girders, there are the following options:

- To increase the depth of the main girders
- To increase the width of the main girders
- To adopt a compressive flange at support sections
- To adopt a compressive reinforcement at the bottom of the girder (Figure 4.16a)
- To introduce a wide cross beam between main girders at support sections (Figure 4.16b)

The first option is the most efficient, although a compressive flange is a structurally very good solution, but difficult to execute with risks of not reaching at the bottom flange a well compacted concrete. If this option is adopted it is recommended to leave some openings at the formwork of the bottom flange to have an easy access for vibration of concrete.

The slab deck may be prestressed transversally, at least for decks with two girders only and widths above 10–12 m (Figure 4.17). This allows adopting slender slab decks, threby reducing self-weight. The transverse prestressing should be made by cables with a

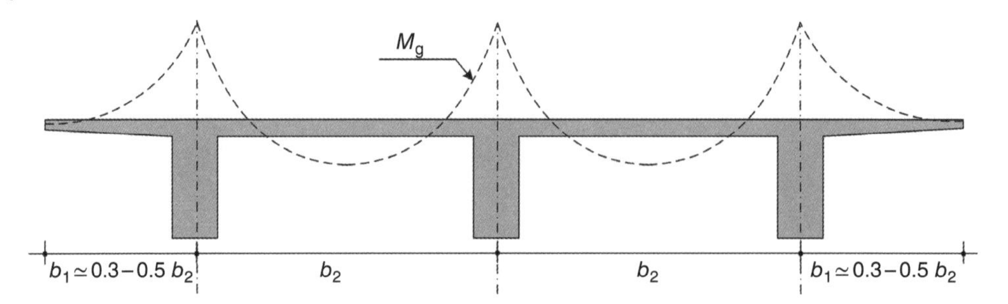

Figure 4.18 Location of the main girders in slab-girder decks to avoid permanent torsion in the girders.

reduced number of strands (usually three or four strands per cable) and close spacing (0.5–1.0 m). It improves the behaviour at serviceability limit states (SLS), namely cracking effects that may be avoided under frequent or even characteristic load combinations in the case of severely aggressive environments.

A main aspect for the concept design of slab-girder decks is the location of the main girders at the cross section. The main rule is to avoid torsion effects in the main girders under permanent actions (dead loads – self-weight, walkway, fascia beam, handrail and transverse prestressing). The transverse bending moment coming from the slab cantilever should be balanced by the transverse bending moment from the slab between girders (Figure 4.18). In a slab-girder deck, the cantilever span length should be between 0.3 and 0.5 of the distance between main girders. If the deck has two girders only, the resultant force from the dead loads (transverse prestressing included) of the half deck (Figure 4.17), should pass through the centre of gravity of each girder. The condition of zero torsion moment at each girder under permanent action is of course a target and very often one needs to accept some moderate permanent torsion in the main girders. When the transverse bending moments, to the 'left and right' of the main girders, are fully balanced, the slab does not rotate at the slab-girder connection, which means the slab between girders may be considered for dead loads, as built in along both longitudinal edges.

The transverse cantilevers should be designed with variable thickness with a minimum of 0.15–0.20 m at the free edge. It is quite important to take into account the plate anchorage dimensions for transverse prestressing if it exists. A design example is shown in Figure 4.19 for the deck of a cable-stayed bridge where the transverse cantilevers of the slab deck (overhangs) reach 5.5 m.

The slenderness of the girders, defined for continuous spans as l/h, should be reduced for simply supported spans l_i/h to values in the order of 70%, that is, $l_i/h = 0.7\ l/h$. In the following, some indicative values for pre-design are given for RC and PC decks.

Due to execution constraints, to be discussed in Section 4.6, most slab-girder decks have a constant depth, particularly if they are built adopting a *formwork launching girder*. However, if they are cast on fixed scaffolding supported from the ground (*classical scaffolding or stationary falsework*) a variable depth girder may be a good option from both the aesthetics and structural points of view. Table 4.2 present indicative values of slenderness l/h, the lowest values may be adopted for support sections while the highest values are usually adopted for mid-span sections. A parabolic variation of the girder depth or a linearly variable depth from the support section to one-fifth or a quarter of the span may be also adopted, keeping a constant depth in the central parts of the spans. These

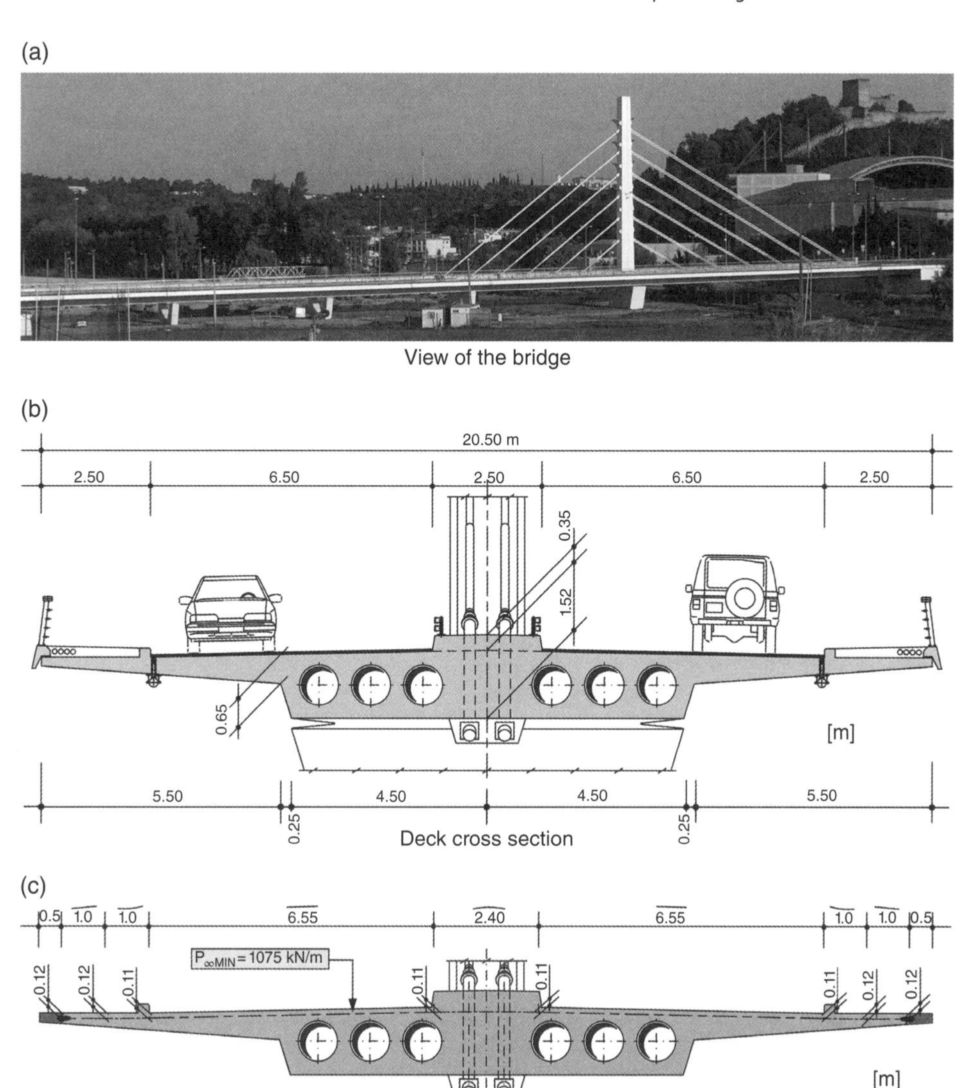

Figure 4.19 A voided slab deck in the approach spans of a cable stayed bridge (over river Lis, in Leiria, Portugal) with an axial scheme for the stay cables. Transverse prestressing of the slab deck with long (5.50 m length) overhangs. (*Source:* Courtesy GRID, SA).

variations in girder depth have a positive impact in reducing the quantity of longitudinal prestressing.

Another relevant aspect is the required width of the girders to accommodate the prestressing cables, particularly at sections where continuity anchorages are located. This may be concluded from minimum free clearances for the anchor plates if they are located side by side. Even for small 12-strand cables, a minimum width of 0.7 m width should be considered. If the continuity anchorages are not side by side, the width of the

Table 4.2 Slenderness for reinforced and prestressed concrete continuous girder decks.

Deck cross section	Slenderness	RC deck	PC deck
	l/h	12–17	14–25
	l/h	14–20	17–34

girder may be reduced to a minimum of 0.4–0.5 m. A minimum value of 0.6 m is recommended for installation of prestressing cables and to increase the concreting rate and good vibration conditions of the concrete. It is also convenient, for easiness of removing the formwork, to design the girders cross section slightly tapered to the bottom as shown for the design case in Figure 4.20.

Typical values for pre-design of a *slab-girder deck* are as follows:

- Spans l – up to 70 m in PC
- Slab deck thickness (between girders) – 0.22–0.30 m to 0.40–0.50 m at the connection with the girders
- Width of the girders at the slab deck level – 0.5–0.9 m
- Slenderness – $l/h = 15–20$ in PC.

4.4.5 Precasted Slab-Girder Decks

Main girders of a slab-girder deck may be precasted as shown in Figure 4.21. The number of girders to be adopted at cross sections should be studied with respect to the function of the self-weight of each girder and capacity of erection equipment, namely cranes or *launching girders*, as discussed in Section 4.6. The distance between axes of precasted girders usually varies between 2.5 and 5.0 m. The slenderness of the girders may vary between 1/15 and 1/20 of the span lengths. The slab deck, usually made in RC, is cast on top of a classical formwork, suspended from the girders or on top of a precasted panel. These panels may be fibre reinforced cement for small distance between girders or precasted RC panels in composite action (participating precasted panels) with the concrete cast in situ of the slab deck for large spacing of the precasted girders. The precasted girders may be pre- or post-tensioned. The geometry of precasted girders is currently an I shape with a wide top flange and a narrow lower flange, as shown in Figure 4.22. The lower flange accommodates the pretension wires or the post-tensioned prestressing cables. The web should have sufficient width for shear forces but particularly to allow for the prestressing cables to "go up" in the webs of continuous decks. If a prestressing cable needs to 'go up' in the web, the minimum web width is three times the diameter of the duct. For example, for a 12-strand 15 mm prestressing cable, this would require a minimum width of about 0.27 m for the web. The next solution to obtain lighter

(a)

View of the bridge

(b)

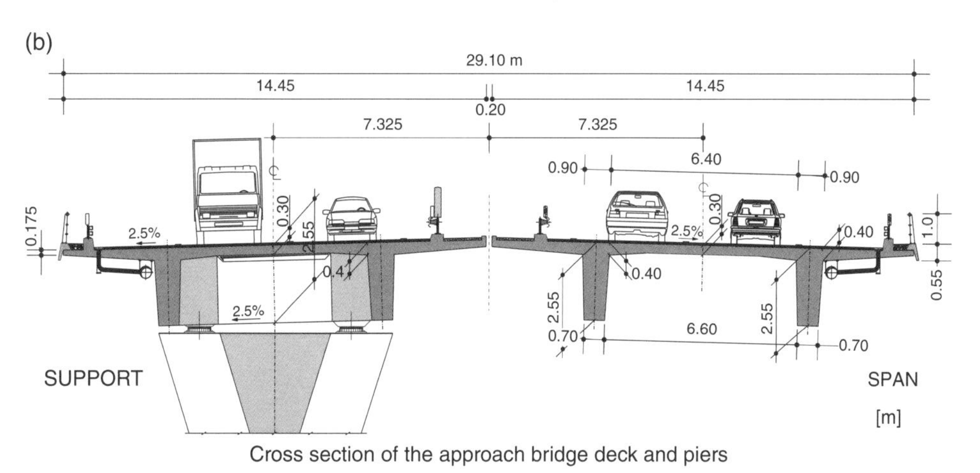

Cross section of the approach bridge deck and piers

Figure 4.20 Design example of a slab-girder deck. Approach spans of the bridge over Sorraia River in the highway A13 in Portugal (typical spans of 43.7 m) (*Source:* Courtesy GRID, SA).

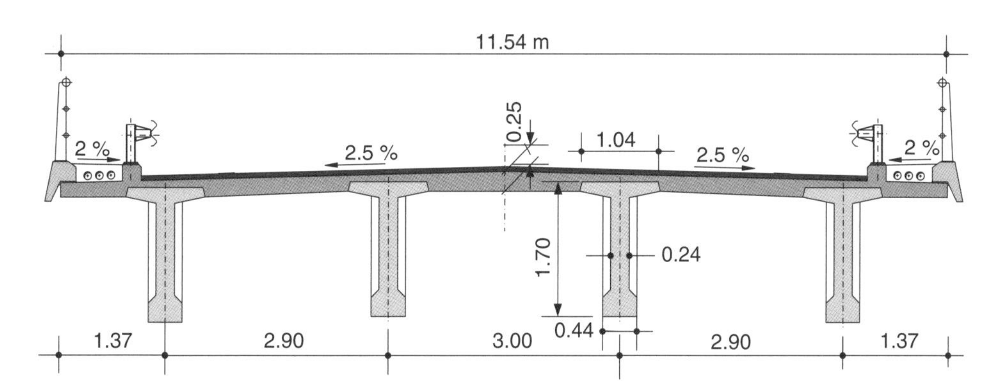

Figure 4.21 Slab-girder deck with prestressed precasted girders.

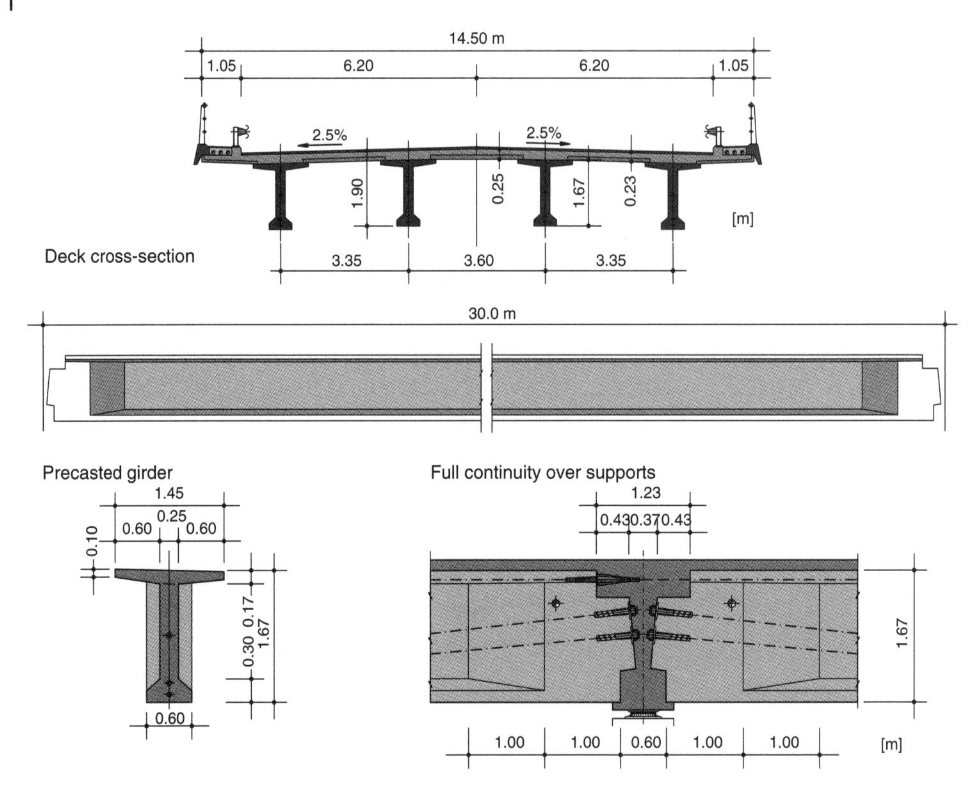

Figure 4.22 Typical geometry of a precasted girder for a bridge deck, with the slab concreted on precasted concrete panels.

prestressed girders is to increase the width of the web when approaching the end sections. In this case, the prestressing cable should go up only at the wider web sections. In any case, minimum widths of 0.2 m are recommended for easiness of installing the ordinary reinforcement and keeping minimum covers for durability. Compacting the concrete is very often made by adopting vibrators attached to the formwork. The lower flange of precasted girders is subjected to high compressive stresses, and its geometry and class of concrete should be studied in conformity.

The reader is referred to specific literature on precasted girder decks [5], but a main subject should also be discussed here: the option of simply supported multiple span decks or continuous spans decks, as shown in the schemes of Figure 4.23. The former are easy to erect but have a multiplicity of disadvantages, namely requiring twice the number of bearings and expansion joints at each support section usually requiring frequent maintenance. Besides, the simply supported independent girders are usually less slender and if they are supported by cross beams or pier caps with large disadvantages for the appearance. The best solution, both aesthetically and structurally is to adopt a continuous deck. The girders and the slab are continuous at the support sections, achieved by concreting a transverse cross beam at each support section. A second phase of prestressing is required for reaching full continuity of the deck. This solution proved to yield the most aesthetically pleasant solutions, increasing the life span of the

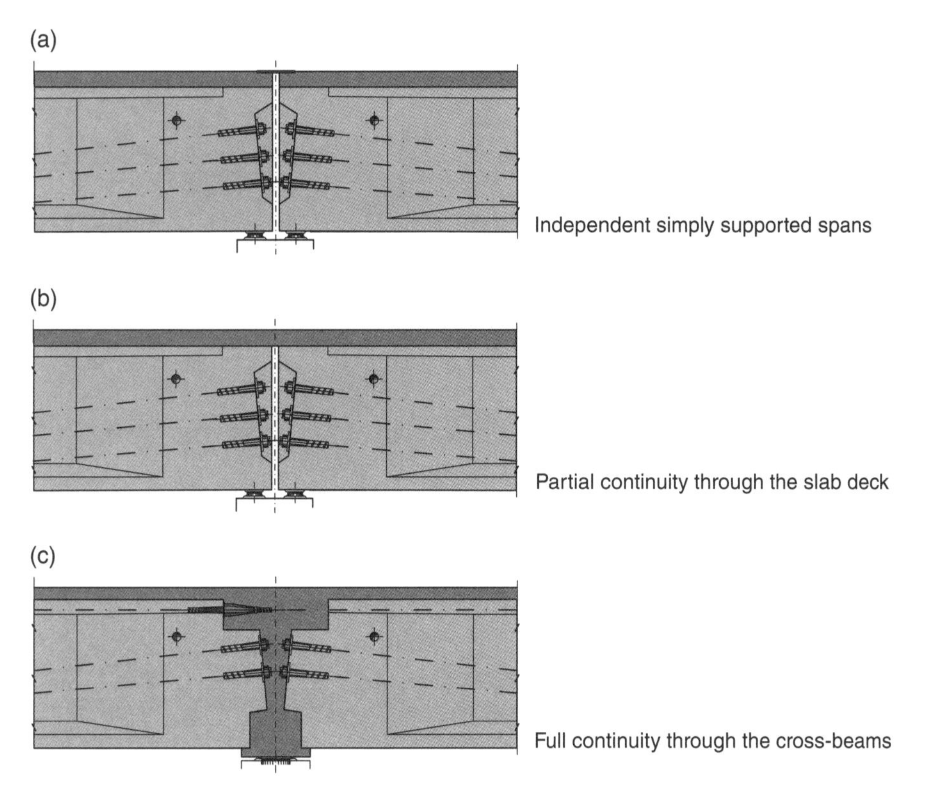

Figure 4.23 Different schemes for slab-precasted girder decks at support sections: (a) independent simply supported spans; (b) partial continuity through the slab deck and (c) full continuity through the cross beams.

bridge and yielding improved comfort for the users. Of course, the continuous solution is more difficult to execute and more expansive. An intermediate solution is to make continuity by the slab only. The girders tend to work as simply supported and the slab needs to have sufficient ordinary reinforcement to accommodate tensile stresses and to control cracking effects, induced by the traffic, differential thermal gradients and time-dependent effects of the concrete. Being an attractive solution from an execution point of view, the partial continuity by slab only has the same aesthetical inconvenience as the simply supported independent span option.

4.4.6 Box Girder Decks

For long spans, the compressive stresses at the lower flange of slab-girder decks tend to reach very high values. In a box section, a complete lower flange to accommodate these stresses exists (Figure 4.8c). Besides, in curved bridges where torsion effects are usually relevant, box sections have a large torsion rigidity and high capacity to resist to shear torsion stresses. Other cases, where box girder decks are the most convenient option, are bridge cases where a reduced depth is required for vertical clearances or aesthetical reasons, such as in urban viaducts or bridges over navigation channels.

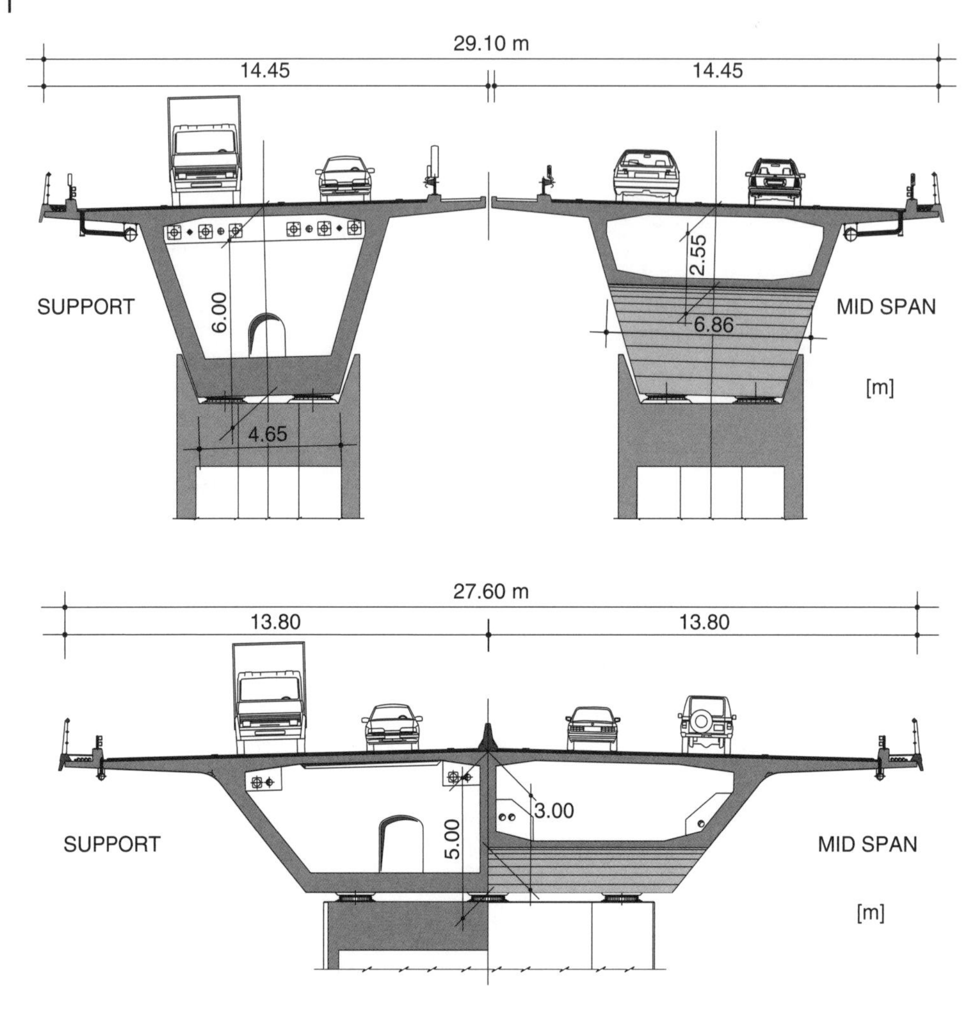

Figure 4.24 Single cell and multi-cell box girder decks.

Box girder sections, single or multiple-cell as shown on Figure 4.24, offer the following advantages with respect to a slab girder deck:

- a large compressive lower flange allowing compressive stresses at support sections under negative bending moments;
- a large torsional resistance in uniform torsion and reduced stresses due to warping torsion (Chapter 6), which is very convenient for curved bridges;
- for the same self-weight of a slab-girder deck (same h_{eq}) an increased slenderness (l/h) in the order of 20%;
- larger eccentricities for the prestressing cables at support sections subjected to negative bending moments;
- small creep deformations since the permanent compressive stresses are smaller than in an equivalent slab-girder deck.

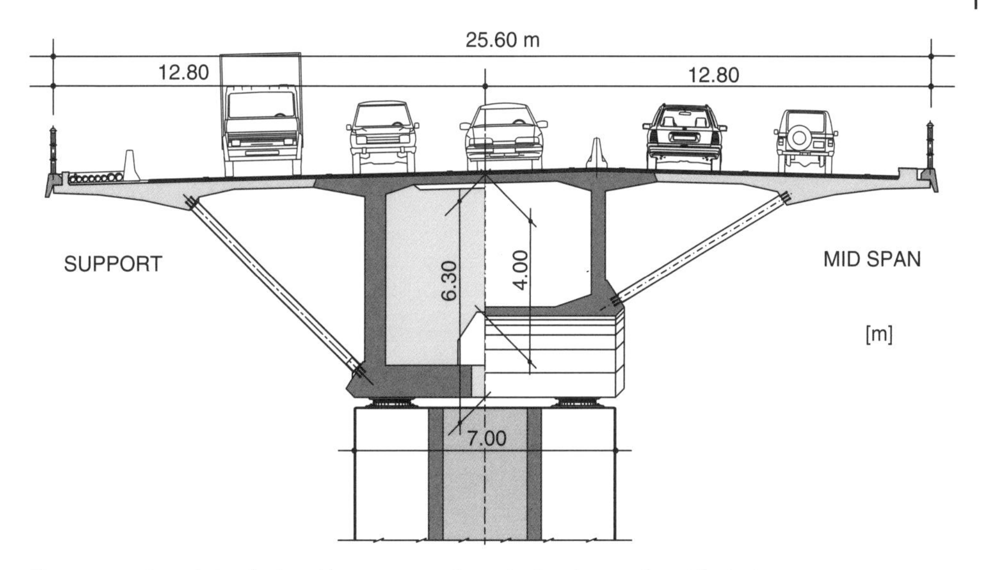

Figure 4.25 Box girder decks with transverse ribs or inclined struts for wide decks.

The most relevant disadvantage of box girder options with respect to slab-girder decks is an increased difficulty in execution, namely for concreting, formwork and installation of prestressing cables and ordinary reinforcement.

Box girder sections are the preferred options for long span bridges, namely for spans above 70 m and up to 200 m. Above this value or even less, say 150 m, cable-stayed bridges may be competitive and certainly more pleasant from an aesthetic point of view.

Nowadays, single cell box girders are preferred to multi-cell box sections, even for wide bridges. The use of transverse prestressing, the adoption of transverse ribs or even inclined struts (concrete or steel tubes) (Figure 4.25) may allow the adoption of box girder sections for deck widths with more than 20 m. Only for very limited depths, due to clearance restraints, multi-cell box girders are preferred. Even so, in these cases, it may be possible to adopt two box girder sections, independent or interconnected, through the deck slab.

Box sections in continuous spans have usually a variable thickness lower flange, with a minimum of 0.2 m at the mid-span and a thickness at support sections sufficient to accommodate the compressive stresses, while also controlling creep effects. For long spans built by the *cantilever method*, the thickness of the lower flange at the support is in the order of $l/100$, where l is the span length, but clearly this depends on the width of the lower flange. This thickness is reduced towards the mid-span in a continuous fashion along a certain length.

Typical slenderness values for box girders are indicated in the following table:

Constant depth box girders	(l/h) = 18–25 for continuous spans
	(l/h) = 17–21 for simple supported spans
Variable depth box girders	(l/h) support = 15–20
	(l/h) mid span = 30–45

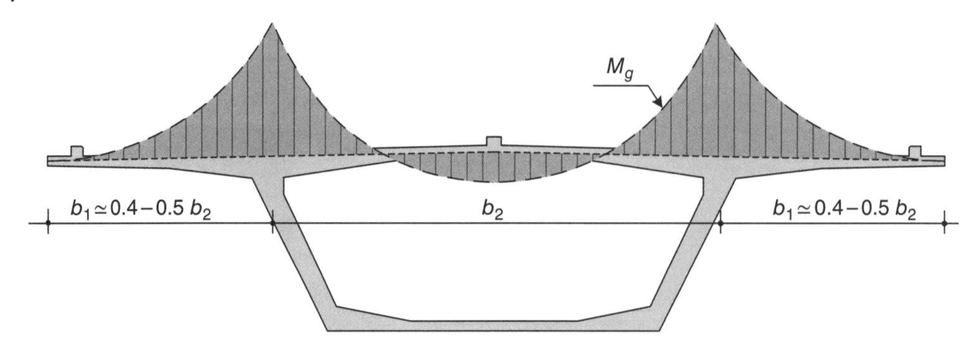

Figure 4.26 Relationship between the length of the overhang b_1 and distance between webs b_2 to avoid permanent transverse bending in the webs.

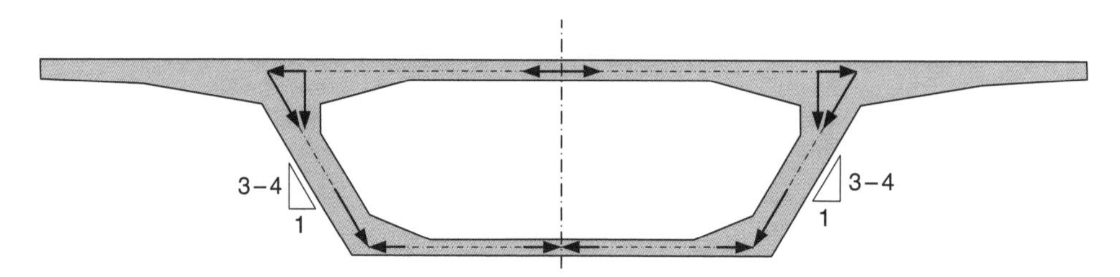

Figure 4.27 Box girders with inclined webs. Membrane forces in slab deck and webs.

To facilitate execution, a minimum depth of 2.0 m (better 2.2 m) is recommended at mid-span sections; otherwise access along the span length, very much required during construction, is more difficult.

The overhangs of the slab deck (transverse cantilevers) have a length, b_1, that is very much dependent on the distance, b_2, between webs. Of course, b_1 depends very much on the negative transverse bending moments induced by truck loading and width of the walkways. The relationship b_1/b_2 is in the order of 0.4–0.5, to reduce permanent transverse bending moments in the webs (Figure 4.26) and to avoid negative transverse bending moments at the mid-span between webs. However, a large part of the transverse bending moments due to traffic loading (eccentric live loads) are taken by transverse bending of the webs.

The webs may be vertical or inclined (order $i = 3$–4). Vertical webs are easy to execute and allow a constant width lower flange in variable depth box girders. Inclining the webs induces membrane tensile stresses at the upper flange (Figure 4.27), but has several structural and aesthetical advantages; namely, it allows bottom flange width reductions and induces an apparent decrease in the depth of the box girder.

In variable depth box girders, the width of the bottom flange is variable increasing from the support sections towards the mid span (Figure 4.27). It is possible to keep the width of the bottom flange by adopting webs with variable inclination along the span length. This may be only justified for spans above 120 m. Two examples where variable

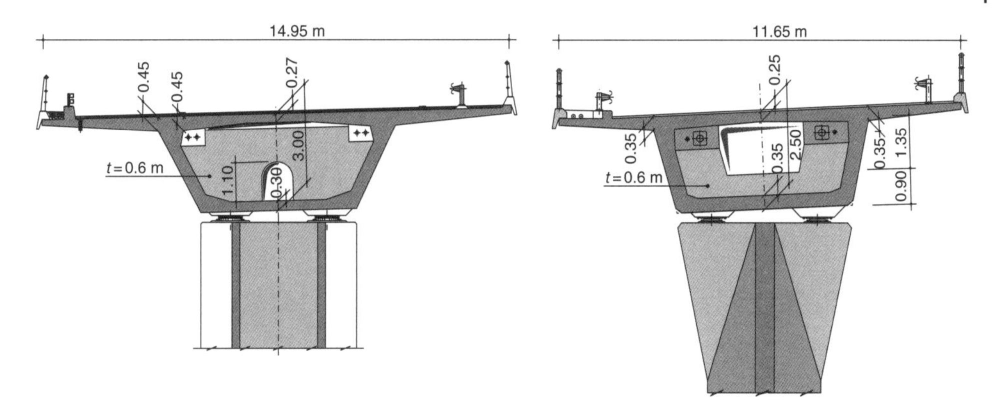

Figure 4.28 Diaphragms at support sections of box girder bridge decks.

inclination webs were adopted, are presented in Figures 5.1 and 5.9. As discussed in Chapter 5, this concept is not so difficult to implement in the execution of segmental bridges, because the gradient of the variation in web inclination along the span length is only relevant nearby the support sections for a parabolic variation of the depth (see Figure 5.10). It requires small adjustments of segment formwork specifically designed for that purpose. This concept of *webs of variable inclination* allows keeping the width of the bottom flange constant along the length of a variable box girder depth. That may be an advantage when the box is continuous on several piers, because the pier width does not need to be different at each pier to accommodate the bearings.

At support sections, diaphragms are adopted (Figure 4.28) with the following functions:

- to improve the loading path of the reactions into the webs,
- to reduce deformability by distortion of the box section.

The slab deck may be separated from the diaphragm in order to keep the slab deck with unidirectional transverse bending predominantly. If the slab is made monolithic with the diaphragm, longitudinal negative bending moments exist at the slab cross section over the diaphragm due to permanent and live loads at the slab deck, increasing and modifying the slab deck reinforcement at these zones. An exception is made, as for slab-girder decks, at the end sections where the slab should be connected to the end cross beam (diaphragm). Intermediate diaphragms along the span length are no longer adopted, although *partial diaphragms* (Figure 4.29), made by two transverse ribs along the depth of the webs, may be adopted particularly for transferring localized deviation loads from external prestressing cables.

With respect to the thickness of the diaphragms, they should be as small as possible with a minimum of 0.3 m. However, due to geometrical constraints, namely bearing dimensions, the thickness of the diaphragms is usually in the order of 0.6 m or even more with a maximum of 0.9 m near the bottom flange (Figure 4.30). The main inconvenience of having very thick diaphragms is due to the induced lock in stresses in the webs from differential shrinkage of the concrete. This effect tends to induce some visible cracking in the webs at the support sections.

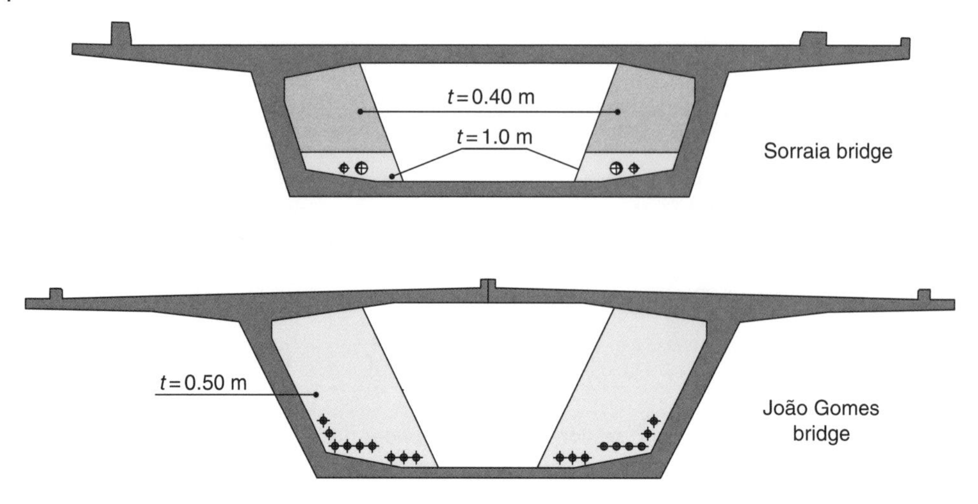

Figure 4.29 Partial diaphragms at intermediate cross sections for deviation of external prestressing cables in Sorraia Bridge and the João Gomes Bridge, Madeira Island.

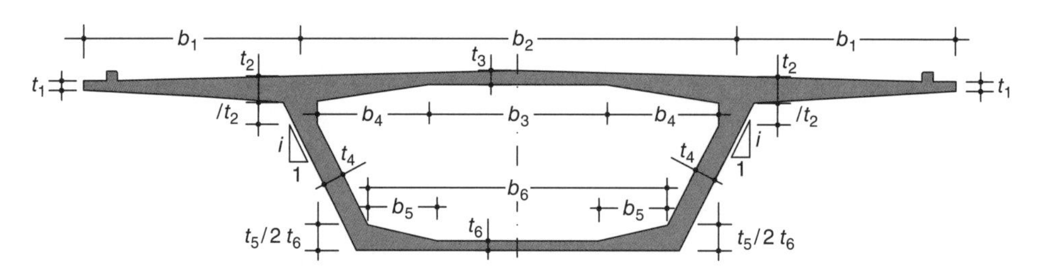

Figure 4.30 Typical geometry for the pre-design of a box girder deck.

For the preliminary design of box girder sections, one may adopt the following relationships from the design, according to notations of Figure 4.30 [6]:

- Slab widths $b_1/b_2 = 0.45$; $b_4/b_2 \leq 0.2$; $b_5/b_6 \leq 0.2$
- Minimum thicknesses $t_1 \geq 0.2\,\text{m}$; $t_3 > 0.2\,\text{m}$
  $t_4 > 0.3\,\text{m}$ or ($0.2\,\text{m} + 2 \times$ diameter of the duct)
  $t_6 \geq 0.18\,\text{m}$; $t_5 \geq 2 \times t_6$
- Flanges slenderness $b_3/t_3 \leq 30$; $b_6/t_6 \leq 30$
- Web inclination $i \geq 3\text{–}4$

4.5 Superstructure: Steel and Steel-Concrete Composite Bridges

4.5.1 Options for Bridge Type: Plated Structures

Steel-concrete composite bridges are likely to be the most competitive solutions with respect to PC bridges. For medium spans, 40–100 m, composite plate girders (Figure 4.31) or box girder bridges (Figure 4.32) offer a variety of possibilities to deal

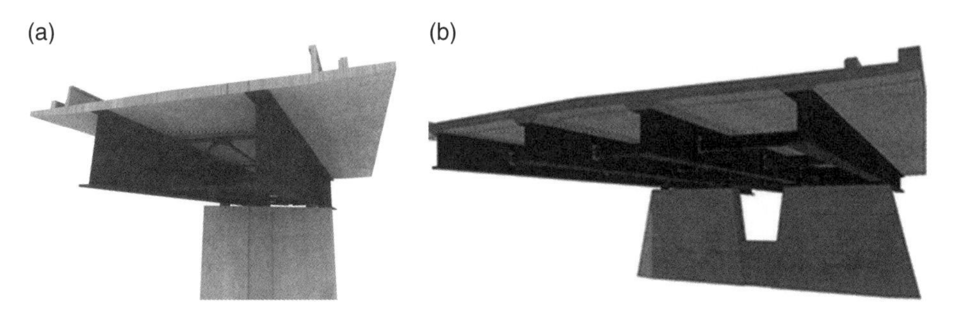

Figure 4.31 Plate girder decks: (a) A two girder bridge (Railway Sado Bridge, Portugal – viaducts, typical spans 45 m) and (b) a multiple plate girder bridge (Bridge over Cavaco River in Angola, typical spans 32 m) (*Source:* Courtesy GRID, SA).

with wide decks and curved bridges. When the span exceeds 100 m, cable-stayed (Figure 4.33) and bowstring arch bridges (Figure 4.34) may be the most suitable solutions. For two-level double decks, usually adopted for rail and highway traffic, a steel-concrete composite truss is usually required.

For large spans reaching 400–600 m, cable-stayed bridges with a steel concrete composite deck may be adopted. Above this range of spans, cable-stayed or suspension bridges are preferred with steel decks to reduce the self-weight. In place of a concrete slab deck, an orthotropic steel plate is preferred.

Steel and steel-concrete composite bridges, may offer the following advantages with respect to PC bridges:

- A reduced dead load
- More economic foundations
- Easier erection methods
- Shorter execution delay

The disadvantages may be the following ones:

- A higher cost
- Increased maintenance requirements
- A more demanding technology for execution

The longitudinal structural systems adopted for steel and composite bridges are the general ones (Chapter 1), namely simply supported or continuous girder bridges, frame and arch bridges, cable-stayed or suspension bridges. The main typologies for cross section decks of steel and composite bridges are plate girder bridges, box girder bridges and truss bridges. What makes the difference between steel and composite bridges is the deck type. In the former, the slab deck is a steel orthotropic plate (Figure 4.33) while in the last, steel concrete composite bridges, the deck is made of a RC or partial PC slab (Figures 4.31, 4.32 and 4.34).

For span lengths between 30 and 120 m, composite steel-concrete bridges tend to be the preferred solution for economy and low maintenance costs. For large spans, savings in deck self-weight, reducing the permanent load of the superstructure and consequently the amount of steelwork quantity and substructure foundation costs, may justify the adoption of an orthotropic steel deck.

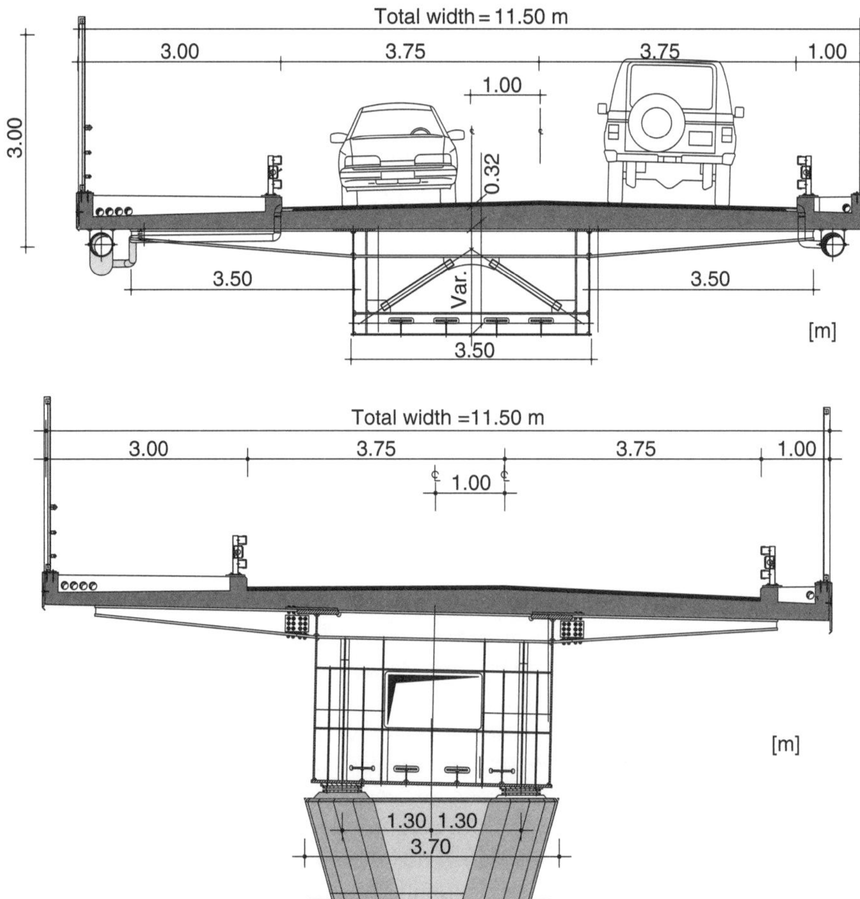

Figure 4.32 A box girder deck viaduct over the Highway IC19 in Lisbon. Main span 54 m. Cross sections over the piers and at mid-span (*Source:* Courtesy GRID, SA).

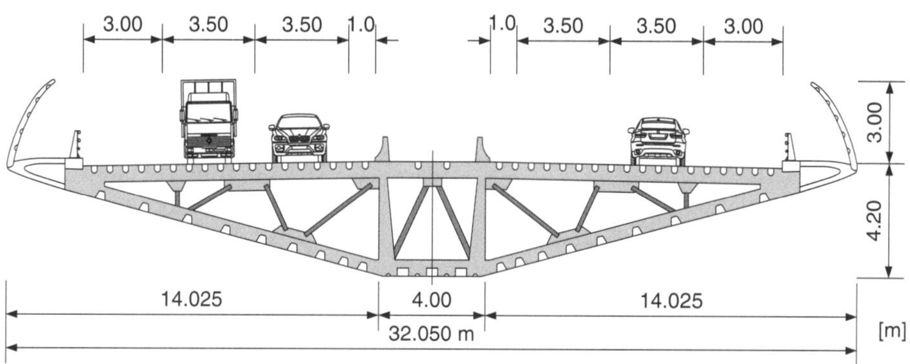

Figure 4.33 A steel box girder deck for a cable-stayed bridge – Millau Viaduct, France. Typical spans 342 m (*Source:* Photograph: Mike Switzerland/ https://fr.wikipedia.org).

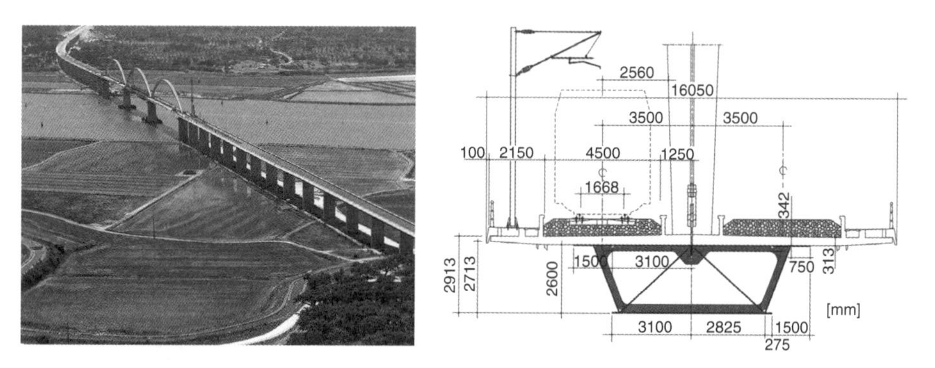

Figure 4.34 The railway bridge over river Sado, Portugal. A bowstring arch bridge with main spans of 160 m. A continuous composite box girder deck 480 m length. Approach viaducts – plate girder deck, 45 m typical spans (Figure 4.31a) (*Source:* Courtesy GRID, SA).

The conceptual design of steel or composite bridge decks should consider the following conditions:

- Deck dead load/cost of deck steelwork;
- Fabrication, transport and erection cost of the steelwork;
- Limiting conditions for transport of the steelwork;
- Conditions requiring the increase of deck slenderness (span length, l/depth, h of the superstructure) like aesthetic requirements or vertical clearances in urban bridges;
- The geometry of bridge plan alignment, namely straight, skew or curved bridges.

Particular attention in conceptual design to transport and erection conditions, at the bridge site, is required. Some important constrains are maximum segment lengths, vertical clearances of existing overpasses in the access roads and limits in navigation channels in existing bridges to access the bridge site. For ship transport, maximum container-allowable dimensions are another issue, as well as maximum width of segments for transport by trucks.

Two main typologies may be adopted for the cross section of road or rail bridges: plate girder bridges (Figures 4.31 and 4.32) and box girder bridges (Figures 4.33 and 4.34). The last one is the preferred option for long spans, curved bridges and urban bridges with special requirements for aesthetics. Box girders allow an increased slenderness but their fabrication costs are usually higher than plate girder bridges. Besides, due to the width of the box section, transport is usually limited to 3.5 m width boxes (Figure 4.35); one possibility is to transport two half boxes, which requires a longitudinal site weld along the lower flange. For urban bridges, small boxes (Figure 4.36) may be adopted that are easy to transport and avoid longitudinal site welds. Transport of plate girder bridges is made in segments of separated girders with 16–20 m maximum lengths, assembled on site.

For most cases of road bridges with spans up to 120 m, when the deck width does not exceed 15–20 m, it is possible to adopt a composite deck with two plate girders only.

Figure 4.35 Special transport of a box girder deck (*Source:* Courtesy GRID, SA).

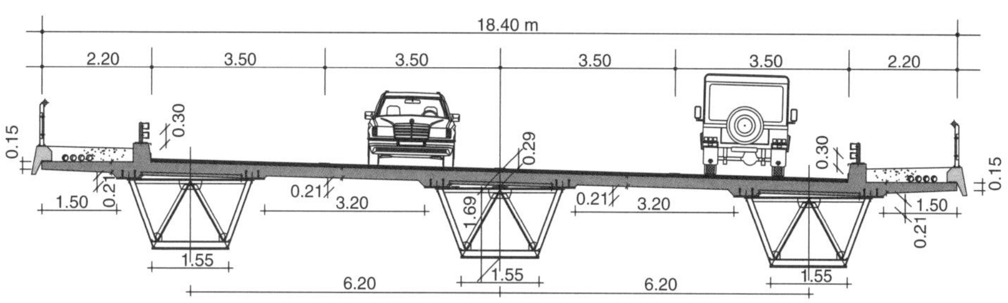

Figure 4.36 Small box girders for a composite continuous deck viaduct, two spans 43 + 45 m, in a viaduct over highway A5 in Lisbon, with 'double composite action' (*Source:* Courtesy GRID, SA).

Particular attention is given in the next sections to this composite bridge deck typology, quite common today in Europe. For rail bridges, decks with only two girders are also adopted for single or double track cases (Figures 4.5 and 4.52).

Plate girders and box girders for steel and composite bridges are *plated structures*. For pre-design of these structures, the reader should be aware of the basic aspects of buckling and post-buckling behaviour of plates under in-plane loading. The fundamental aspects are dealt with in Annex A of this book. The elastic buckling stresses of plates under compressive or shear stresses are designated σ_{cr} or τ_{cr}, respectively. Plate elements may fail by plasticity or by elastic instability under compressive in-plane loading (compressive stresses) followed by yielding. Depending on the slenderness of the plate element, collapse is mainly controlled by plasticity (stocky plates) or by plate buckling (slender plates). For intermediate plate slenderness, the collapse is always governed by the interaction of elastic buckling and plasticity as in a compressed column. The design parameter governing plate behaviour is the non-dimensional slenderness parameter of the plate element, given by

$$\sqrt{f_y/\sigma_{cr}} \tag{4.3}$$

where f_y and σ_{cr} are the yielding and elastic critical stresses of the plate. For shear, one takes (in place of f_y and σ_{cr}), $\tau_y = f_y/\sqrt{3}$ to define the nondimensional slenderness, the yield shear stress according to the von-Mises criterion, and τ_{cr} the elastic critical stress under pure shear. Plate elements with low $\bar{\lambda}_p$ values (say < 0.2) are governed by plasticity, while for large slenderness values (say > 1.2) the behaviour is governed by elastic instability. Slender plates under compression or shear loading have a considerable reserve of resistance after reaching the elastic critical stresses – post-buckling resistance, as may be seen in Annex A.

4.5.2 Steels for Metal Bridges and Corrosion Protection

4.5.2.1 Materials and Weldability

Steel materials for metal bridges should have sufficient resistance, ductility and good weldability characteristics. The ultimate resistance of steel materials may be reached by plasticity, brittle fracture or fatiguering.

Plasticity occurs after a steel specimen reaches the elastic limit strength associated with *yielding*, while *brittle fracture* (inducing a crack like in fatigue failure, Figure 4.37) may occur at low stress levels usually associated with low temperatures and high strain rates induced by impact loading. *Lamellar tearing* (Figure 4.38) is a kind of brittle fracture occurring across the thickness of a steel specimen induced by deformation in the direction perpendicular to the surface of a steel plate.

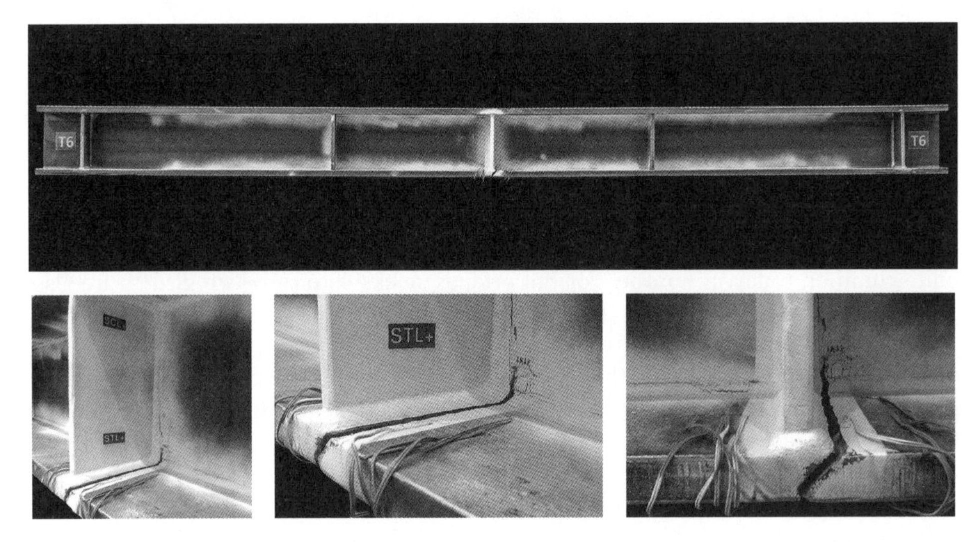

Figure 4.37 Fatigue crack in a steel girder (*Source:* Courtesy of University of Stuttgart – Ref. [7]).

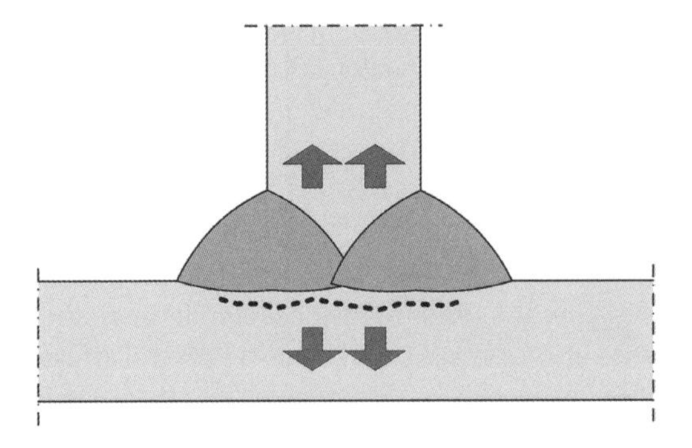

Figure 4.38 Lamellar tearing in a T welded joint.

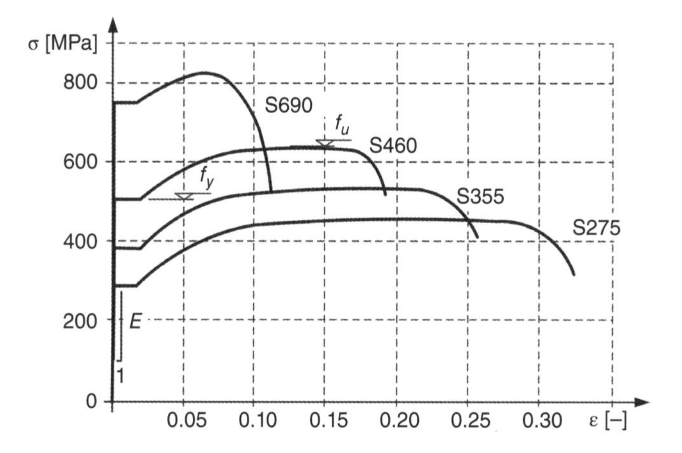

Figure 4.39 Stress–strain diagram from a tensile steel test. Yield stress f_y, ultimate stress f_u and modulus of elasticity E.

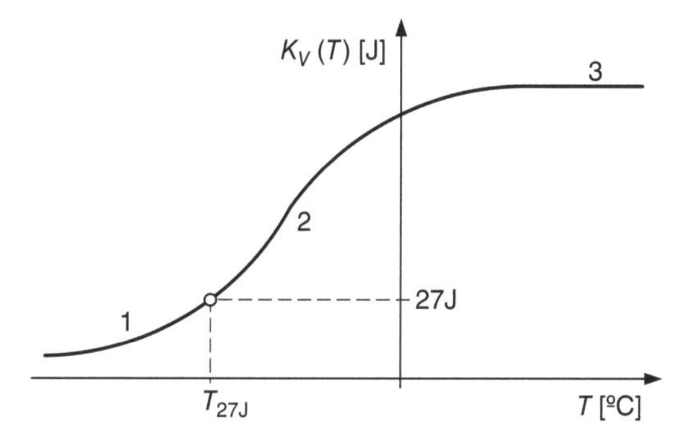

Figure 4.40 Relationship between impact energy (Joules J) and temperature (T) from Charpy tests.

The plastic resistance is characterized by *yielding and ultimate failure stresses* defined from *tensile tests* (f_y and f_u, respectively). Resistance to *brittle fracture* is defined by the Charpy test [8] where the *toughness* of the steel is defined as its capacity to resist to crack propagation, usually associated with high tensile stresses or multiaxial tensile stresses. Lamellar tearing occurs in the direction perpendicular to hot rolling and is characterized from specific tensile tests according to [9].

From tensile tests, one defines stress–strain diagrams (Figure 4.39) and from the Charpy test, one defines the *steel toughness*. This last mechanical characteristics, is defined in terms of capacity to absorb energy required to propagation of a crack. The *stress intensity factor K* at the vicinity of a certain crack depends on the geometry and dimensions of the crack and stress level. Under increasing stresses, a critical value K_V is reached at a certain temperature giving rise to a brittle fracture. From low to high temperatures, the steel behaviour passes from brittle to ductile. A nominal transition temperature may be defined (Figure 4.40). The toughness is measured by impact bending tests on a specimen with a V-notch, usually by using a Charpy Pendulum machine. The impact energy is measured by a certain mass M with weight P falling from a certain height, h. The impact energy is measured as $A_v = P\,h$ in Joules (1 J = 1 N × 1 m). On the

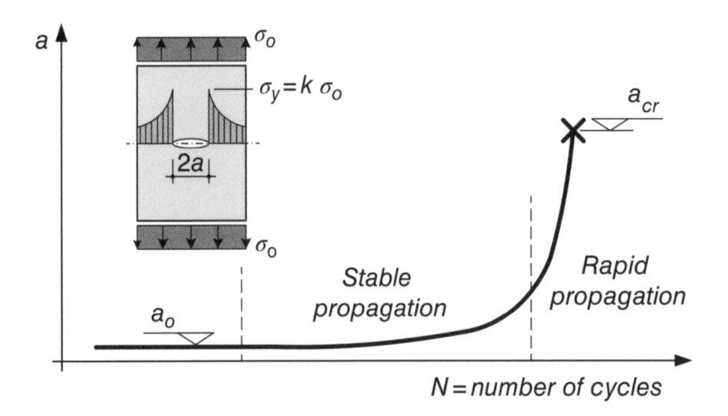

Figure 4.41 Crack propagation as a function of the number of fatigue cycles.

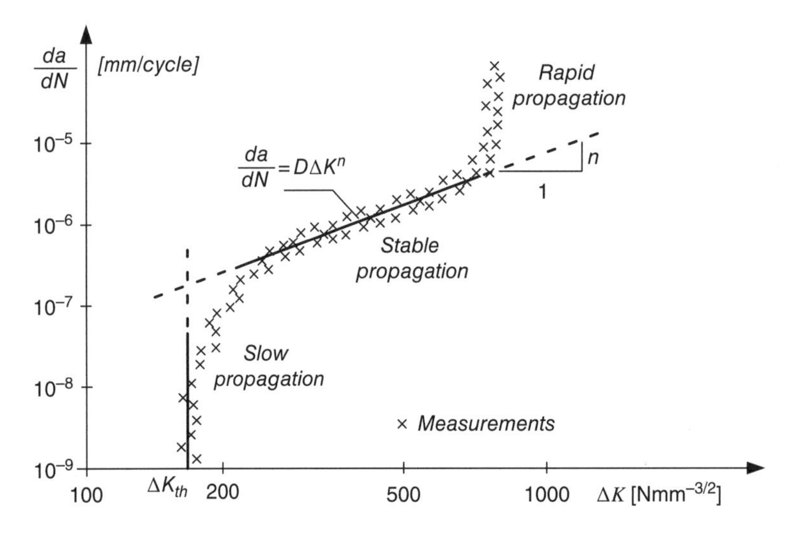

Figure 4.42 Relationship between crack propagation ratio and variation of the stress intensity factor (*Source:* Adapted from Ref. [10]).

other hand, it is possible to define the temperature associated to a nominal *brittle fracture impact energy* specified as 27 J in Part 1-10 of EN 1993 [10]. Figure 4.40 shows a diagram where the K_V is defined (Charpy V-Notch) as the value of impact energy $A_V(T)$ in Joules (J) required to fracture a Charpy V-Notch specimen at a given test temperature, T. Toughness tends to reduce under decreasing temperatures. Sufficient toughness is particularly required for welded bridges to prevent cracks from appearing or to increase rapidly under fatigue loading. Fracture mechanics adopts the concepts of *stress intensity facture* K [11] for a certain nominal crack (Figure 4.41) and *toughness* at certain temperature. If a crack amplitude is measured by a certain geometrical parameter a (Figure 4.41) and if that crack exists at service conditions of a steel bridge, the propagation of the crack dimension a has a typical diagram as shown in Figure 4.41. The propagation ratio (gradient) of the crack dimension da/dN may be correlated with the variation of the stress intensity factor ΔK. Paris law [12, 13], represented in Figure 4.42, gives the simplest correlation. The constants D and n defining the linear correlation in the stable range of the crack are material dependent only. If the variation of the stress intensity factor does not exceed a certain value defined in Figure 4.41 by ΔK_{th}, the crack does not tend to increase or may increase in a very slow fashion during the life of the structure. This is important to know for cracks detected during maintenance inspections of steel bridges. From this limit ΔK_{th} on, the crack tends to progress in a stable fashion according to Paris law. However, if a certain limit is exceeded, the propagation gradient da/dN is very high and the crack increases rapidly. It is possible [11] to define the critical dimension of a fatigue crack not inducing local failure by unstable crack propagation. In this case, the stress intensity factor remains smaller than its critical value K_c. This critical value is called *toughness* and is approximately dependent on the material only and independent of the geometry of the detail. However, it is temperature and strain rate (impact loading) dependent. Brittle fracture and fatigue crack is in steel bridges are usually induced by weld details. The detail of the plate welded joint to the lower flange in Figure 4.37 induces a high stress intensity factor inducing a crack in the flange propagated to the web.

Yield strength is specified in design by the *steel grade* while toughness is fixed by *steel quality*. Hence, steel for bridges may be specified in the design according to EN 1993-2 and EN 1993-1-10. The following steels are usually adopted for bridges [14]:

1) *Non-alloy steels*, according to EN10025-2 – Steel grades S235, S275 and S355 with qualities J0, J2 e K2 the last one is usually for S355 only.
2) *Fine grain weldable steels* according to EN10025-3 – Steel grades S275, S355, S420 and S460 with qualities N or NL.
3) *Thermomechanical steels* according to EN10025-4 – Steel grades S275, S355, S420 and S460 with qualities M or ML.

Qualities J and K are associated with 27 and 40 J. The numbers 0 and 2, associated with J and K, represent a temperature of 0 and −20°C, respectively. Hence, a steel of quality J2 has a toughness of 27 J at −20°C while a steel of quality K2 has 40 J at −20°C. In EN 10025-2 [14], there is also a lower quality steel designated JR (27 J at +20°C) that should only be adopted for secondary elements. Steels for the main structural bridge elements, specified according to EN 10025-2, should be at least S355 J2 or S355 K2.

Steels may be subjected to a *normalizing treatment*, a thermal treatment during steel fabrication improving its qualities. The *normalized delivery condition* should be required for steel for bridges and steel specification is written as +N, as, for example, S355 J2 + N or S355 K2 + N. However, for fine *grain weldable steels* specified according to EN10025-3 [14], the normalizing delivery condition is already included and steels are denoted just as N. For 'N steels', the toughness is defined as 40 J at −20°C or 27 J at −50°C for NL steels (L – stand for low temperature).

Thermomechanical steels ('steels M') are subjected specific thermal treatments during the rolling process and for the same mechanical characteristics of other steels; they have less carbon in its chemical composition. For *weldability*, it is important to define the *equivalent carbon*, C_e, defined as:

$$C_e = C + \frac{Mn}{6} + \frac{Cr + Mo + V}{5} + \frac{Ni + Cu}{15} \quad (4.4)$$

where the several terms *C*, *Mn*, *Cr* and so on define the chemical composition of the steel material as a percentage of weight. Structural steels always have less than 0.25% of carbon (*C*), as, for example, a maximum of 0.18% for S355 steels. The remaining elements of the chemical composition defining the C_e value are manganese (*Mn*), chromium (*Cr*), molybdenum (*Mo*), vanadium (*V*), nickel (*N*) and copper (*Cu*). Other elements, such as sulfur (*S*), phosphorus (*P*) and hydrogen (*H*), have an adverse effect on the mechanical characteristics of steel materials.

The C_e value has an important role in the weldability of steel components. In design specifications for welded structures, this value should be specified. Steels with C_e values less than 0.4% may be welded with fewer or even no risks of cracking. Cracks appearing after are due to the appearance of martensite or cracking may be due to presence of hydrogen in the designated Heat Affected Zone (HAZ) at the welding zone.

The cooling velocity is an important factor to avoid cracking development due to welding. Thick plates are more susceptible of cracking due to hydrogen than thin plates. To reduce susceptibility to cracking, preheating before welding is usually required, at least for thick welds. One advantage of thermomechanical steels ('M steels') is reducing or even totally avoiding preheating for welding. These steels, developed in last decade of the last century, or fine grain weldable steels ('N steels'), are nowadays the preferred ones for bridges. They are less susceptible to brittle fracture at least if the thickness of the plates exceeds 30 mm. Flange plates for bridge girders nowadays may reach 150 mm or even more, if necessary.

Resistance to brittle failure is considered in Part 1-10 of EN 1993 [10]. Minimum steel quality in bridge design may be defined from *Fracture Mechanics* but a simplified procedure is predicted in EN 1993-1-10 by defining maximum allowable plate thickness to be welded at a minimum design service temperature. An accidental combination of actions is defined where the leading action is the reference temperature, T_{Ed} in EN 1993-1-10. The temperature T_{Ed} is obtained from the minimum air temperature at the bridge site, which is defined as T_{md} associated to a certain return period (usually 50 years) defined in EN 1991-1-5, and from ΔT_r an adjustment for radiation lost, usually −3 or −5°C. For most cases $T_{Ed} = T_{md} + \Delta T_r$.

The stress level at the referred load combination, as a ratio of the service load stress σ_{Ed} to the yield stress f_y, is taken in EN1993-1-10 with its frequent load value. Table 4.3 present the maximum thicknesses as a function of steel qualities, T_{Ed} and $\sigma_{Ed}/f_y(t)$.

Additional values are given in EN 1993-1-10 [10] for other steel grades. If one takes, for example, $T_{Ed} = -10°C$ and $\sigma_{Ed} = 0.5f_y$, the allowable maximum thicknesses for a steel plate are, from Table 4.3: $t_{max} = 65$ mm for steel S355JO; $t_{max} = 110$ mm for steels S355 K2 or S355 N and $t_{max} = 155$ mm for steel S355NL.

Finally, concerning lamellar tearing (Figure 4.38), this is dependent on:

- susceptibility of the steel material
- deformation across the thickness of the steel plates induced by shrinkage due to welding
- welding orientation
- loading type (cycle load or impact load)

Lamellar tearing, when it occurs, is usually, but not always, in the base material and outside of the HAZ, as a result of a stress state perpendicular to the plate thickness.

Table 4.3 Maximum allowable thicknesses t_{max} (mm).

		Charpy energy CVN		Reference temperature T_{Ed} [°C]						
				10	0	−10	−20	−30	−40	−50
Steel grade	**Sub-grade**	T [°C]	J_{min}	$\sigma_{Ed} = 0.75\ f_y(t)\ /\ 0.50\ f_y(t)\ /\ 0.25\ f_y(t)$						
S355	JR	20	27	40/65/110	35/55/95	25/45/80	20/40/70	15/30/60	15/25/55	10/25/45
	J0	0	27	60/95/150	50/80/130	40/65/110	35/55/95	25/45/80	20/40/70	15/30/60
	J2	−20	27	90/135/200	75/110/175	60/95/150	50/80/130	40/65/110	35/55/95	25/45/80
	K2,M,N	−20	40	110/155/200	90/135/200	75/110/175	60/95/150	50/80/130	40/65/110	35/55/95
	ML,NL	−50	27	155/200/210	130/180/200	110/155/200	90/135/200	75/110/175	60/95/150	50/80/130
S420	M,N	−20	40	95/140/200	80/120/185	65/100/160	55/85/140	45/70/120	35/60/100	30/50/85
	ML,NL	−50	27	135/190/200	115/165/200	95/140/200	80/120/185	65/100/160	55/85/140	45/70/120
S460	Q	−20	30	70/110/175	60/95/155	50/75/130	40/65/115	30/55/95	25/45/80	20/35/70
	M,N	−20	40	90/130/200	70/110/175	60/95/155	50/75/130	40/65/115	30/55/95	25/45/80
	QL	−40	30	105/155/200	90/130/200	70/110/175	60/95/155	50/75/130	40/65/115	30/55/95
	ML,NL	−50	27	125/180/200	105/155/200	90/130/200	70/110/175	60/95/155	50/75/130	40/65/115
	QL1	−60	30	150/200/215	125/180/200	105/155/200	90/130/200	70/110/175	60/95/155	50/75/130

Source: Adapted from EN 1993-1-10 [10].

For design, it is important for thick plates (say above a thickness of 30 mm) to specify properties across the thickness of the plate element. These properties are defined according to EN 10164:2004 [9]. A tensile test of a specimen according to EN 10002-1 [12], where tensile stresses are applied perpendicular to the surface of the specimen with an initial area S_o. At ultimate stress, the area S_o is reduced to S_u and the relative deformation is defined by the Z parameter

$$Z = \frac{S_o - S_u}{S_o} \times 100\% \tag{4.5}$$

The quality classes $Z15$, $Z25$ and $Z35$ are defined from minimum values of $Z = 15\%$, $Z = 25\%$ or $Z = 35\%$, respectively, obtained from average values in three tests. The minimum Z value for the steel plates is defined in the steel specification of welded structures with thick plates. The design verification may be done according to EN 1993-1-10 [10] by defining a 'design value' for Z_{Ed}

$$Z_{\mathrm{Ed}} = Z_{\mathrm{a}} + Z_{\mathrm{b}} + Z_{\mathrm{c}} + Z_{\mathrm{d}} + Z_{\mathrm{e}} \tag{4.6}$$

where the several terms are taken from Table 3.2 of EN 1993-1-10. Lamellar tearing may be neglected if $Z_{\mathrm{Ed}} < Z_{\mathrm{Rd}}$ where Z_{Rd} is defined from the quality class, that is:

$$Z_{\mathrm{Ed}} \le Z_{\mathrm{Rd}} = Z15, Z25 \text{ or } Z35 \tag{4.7}$$

The terms Z_{a} to Z_{e} in Eq. (4.6) may have positive or negative values depending, for example, on connection detail and conditions for restrained or free shrinkage due to welding. Another aspect is the beneficial effect of pre-heating for thick welds that is associated with Z_{e}. For example, $Z_{\mathrm{e}} = 0$ without pre-heating and $Z_{\mathrm{e}} = -8$ with preheating (>100°C). Hence, pre-heating reduces the Z_{Ed} and consequently the required Z_{Rd}. An example of calculation of Z_{Ed} is given in Figure 4.43 for a cross-girder welded connection to a bottom chord of a truss girder bridge.

4.5.2.2 Corrosion Protection

With respect to maintenance requirements in steel and composite bridges, one of the main issues is usually corrosion protection. The protection schemes have been very improved in the last decades. Nowadays, it is usual to specify a corrosion protection scheme with a durability of at least 20 years for steel bridges. The improvement of modern coating systems may guarantee bridges in excess of 30 years without requiring the first major maintenance. The atmospheric corrosion categories are defined according to EN ISO 12944-2 [15] as C1 very low, C2 low, C3 medium, C4 high, C5I very high (industrial) and C5M very high (marine).

The coating scheme may be defined according to EN ISO 12944, but the minimum total dry film thickness of paint system (primer, intermediate and finish coat) may vary from 200 up to 1000 μm. The primer is applied on steel after abrasive blasting cleaning defined by a standard grade of cleanliness from EN ISO 8501-1 as SA1, SA2 SA2 ½ and SA3, from light blast cleaning to blast cleaning to visually clean steel. Grades Sa2 ½ and SA3 are usually specified for bridges.

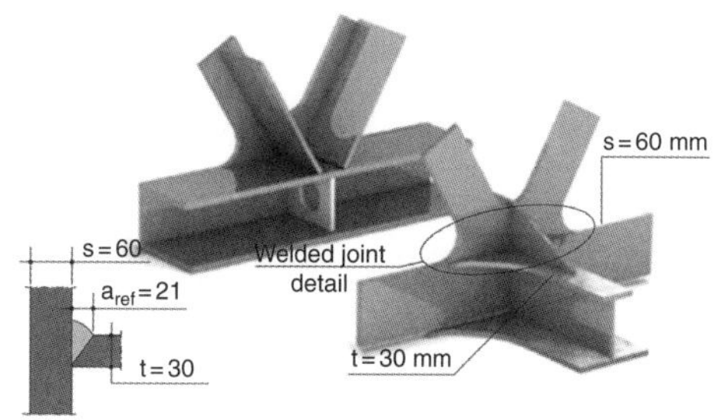

$Z_{Ed} = Z_a + Z_b + Z_c + Z_d + Z_e \leq Z_{Rd} = \{15,25,35\}$ with:

Weld depth relevant for straining from metal shrinkage	$a_{eff} = 21$ mm	$Z_a = 9$
Shape and position of welds	Full penetration weld with appropriate welding sequence to reduce shrinkage effects	$Z_b = 3$
Effect of material thickness s on restraint to shrinkage	$s = 60$ mm	$Z_c = 12$
Remote restraint of shrinkage after welding by other portions of the structure	Low restraint: Free shrinkage possible	$Z_d = 0$
Influence of pre-heating	Without pre-heating	$Z_e = 0$
Bottom chord plate specification: EN 10025-3 S355NL Z25 (EN 10164)	**TOTAL**	$\mathbf{Z_{Ed} = 24}$

Figure 4.43 Example of Z_{Ed} evaluation for a welded detail of a steel truss girder bridge: full penetration T welded joint between the top flange of a floor beam 30 mm thick and the lateral plate of the bottom chord 60 mm thick.

For bridges in environments of high corrosivity, a 'duplex' scheme for coating may be adopted were paints are applied over thermal metal spraying coatings.

Weathering steels ('Corten'), avoiding any type of coatings, is a common practice in several European countries (Figure 4.44).

The influence of design detailing on corrosion is quite meaningful. The handling of structural steelwork to bridge site may induce severe damage to the corrosion protection system. Site connections and splices are potential aspects to be taken into consideration, also because they may induce corrosion problems. The reader is referred to references [15, 16] for further information on corrosion protection systems.

4.5.3 Slab Deck: Concrete Slabs and Orthotropic Plates

The slab deck of metal bridges may be made of a concrete or partial PC slab or a steel *orthotropic plate* where a thin steel plate is stiffened by longitudinal and transverse stiffeners. A composite steel-concrete slab may also make the slab deck by adopting a thin steel plate (8–10 mm thick) in composite action with the concrete (70–110 mm) cast on top of the steel plate adopted as formwork. In short, one may have the following three types of slab deck:

- A concrete slab, reinforced or PC, weighing approximately 5.0–7.5 kN m^{-2};
- A composite slab weighing approximately 2.5–3.5 kN m^{-2};
- A steel orthotropic plate weighing a maximum of 2.0 kN m^{-2}.

Figure 4.44 Example of a bridge made with 'corten' steel. Bridge 'sur la Dala entre Varone et Loèche' Switzerland (*Source:* Photograph by Björn Sothmann/ https://commons.wikimedia.org).

4.5.3.1 Concrete Slab Decks

Concrete slab decks are adopted in general with thicknesses between 0.20 and 0.30 m, connected to deck steelwork through one of the types of *shear connectors* shown in Figure 4.45. The longitudinal shear flow, q_s, between concrete and steel (Figure 4.46) is resisted by the shear connectors and, if so, the concrete slab works in composite action with the steelwork. An equivalent homogeneous cross section in steel is taken by considering a reduced slab width, b_h, working in composite action with the steel girder, $b_h = b/n$ where $n = E_s/E_c$ is the ratio between steel and concrete elastic modulus. The shear flow, q_s, is a force per unit length dF/dx, and is determined by the well-known formula from Strength of Materials:

$$q_s = \frac{V_z S_{hy}}{I_{hy}} \tag{4.8}$$

where V_z is the vertical shear force at the cross section, S_{hy} is the static moment of the homogenized concrete part as shown in Figure 4.46 and I_{hy} is the moment of inertia with respect to the bending neutral axis (yy) of the homogenized transverse cross section.

Shear connectors of one of the types shown in Figure 4.45 may be classified as *rigid or flexible*. In rigid connectors, the shear flow q_s (a force per unit length) is considered to be taken at each section, even at ULS. If a is the longitudinal distance between shear connectors, the force acting at each group of connectors located at the same cross section is $q_s a$. *Flexible connectors* may redistribute the longitudinal forces between them at ULS and longitudinal equilibrium is resisted by a set of connectors between two transverse cross sections of the composite girder. If the forces due to bending stresses in the concrete part are denoted as F_A and F_B, at cross sections A and B, the longitudinal equilibrium requires $\Delta F = F_B - F_A = n\ F_R$ where F_R is the resistant shear force of each connector. Flexible connectors, *head stud* type, are the most adopted ones and are specified

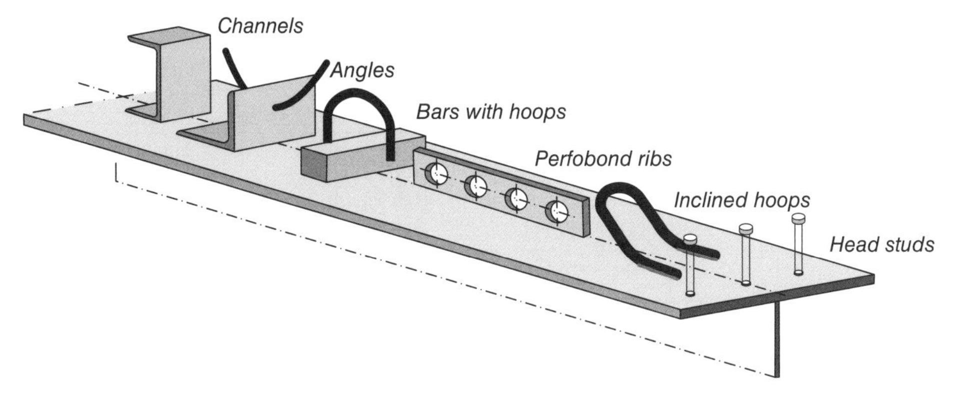

Figure 4.45 Types of shear connectors – flexible and rigid.

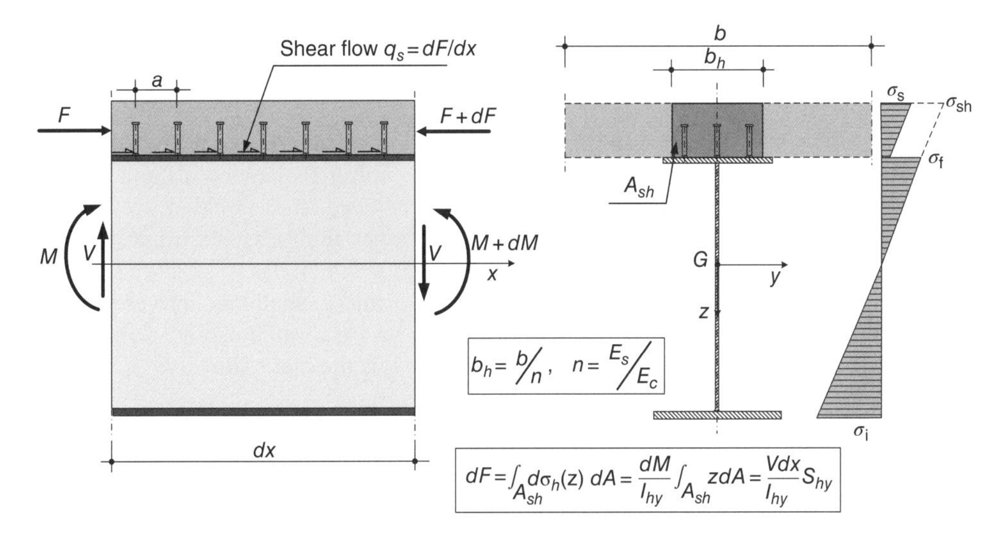

Figure 4.46 Shear flow in composite bridge decks.

in Europe by ISO 13918 [17]. The most common diameters for head studs in bridges are 16, 19, 22 and 25 mm. Steels adopted in Europe for head studs usually have:

- yield strength f_y – minimum value of 350 N mm^{-2};
- ultimate strength f_u – minimum value of 450 N mm^{-2};
- minimum strain of 15%, measured in a normalized specimen with a length $L_o = 5.65\sqrt{A_o}$, where A_o is the initial cross section area of the specimen.

The geometrical characteristics of head studs (Figure 4.47a) should satisfy the following relationships – $D > 1.5d$ and $H > 0.4d$. The height of a shear stud is usually $h_{sc} = 75$, 100, 150 or 175 mm and in any case should not be less than three times the nominal diameter, d.

In general, head studs are welded to girder flanges by arc welding, applied at the head of the stud by inducing an electrical discharge by a 'pistol'. By this reason, the height L of the head stud cannot be specified arbitrarily. The lower end section has a ceramic ring

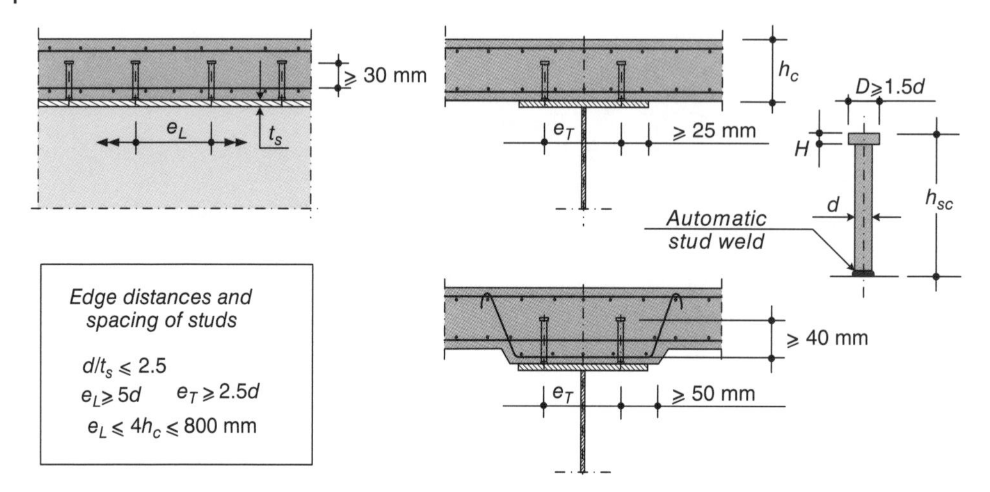

Figure 4.47 Geometrical characteristics of head studs, edge distances and spacings to be respected by head studs in composite slab decks.

to protect the weld during fusion of the material. The nominal diameter d should not be greater than 2.5 times the plate thickness to which the head stud is welded.

The geometrical arrangement of head studs in flanges should satisfy minimum distances specified, for example, in [18] as shown in Figure 4.47b.

Maximum longitudinal and transverse distances e_L and e_T shall take into consideration local plate buckling between shear connectors and are approximately $18t_s$ and $28t_s$, respectively, for e_L and e_T for S355 steel grade where t_s is the plate thickness.

4.5.3.2 Steel Orthotropic Plate Decks

Main advantage of steel orthotropic plate decks, are reduced weight, as previously mentioned. They are the preferred option for long span bridges, say spans above 300 m, like cable-stayed and suspension bridges. However, in cable-stayed bridges even with long spans, say 500 m or even more, steel-concrete composite decks are possible options taking into consideration high axial forces induced by stay action in the deck.

Increasing the span length above 600 m, steel orthotropic plate decks are generally the preferred option, even if they have a high initial cost and increased maintenance cost. However, the steel slab deck may represent a saving in self-weight of 2.5–4 times the self-weight of a concrete slab deck and this has a large impact on savings for main deck steel structure, either a plate girder or a box girder.

The design examples of steel orthotropic plate decks, shown in Figure 4.48, are integrated as slab deck of an open section superstructure made of two truss girders (Figure 4.48a) and in a closed section (box girder) of a modern suspension bridge (Figure 4.48b). Another example of an orthotropic plate deck is presented in Figure 4.33 for the Millau cable-stayed Viaduct, in France.

An orthotropic plate deck is composed by a plate with longitudinal stiffeners (stringers) and *transverse cross* beams supporting the stringers, as shown in Figure 4.49. The deck plate should have a minimum thickness of t 14 mm for durability and resistance to local concentrated loads due to traffic and a maximum of 20 mm. Minimum thickness

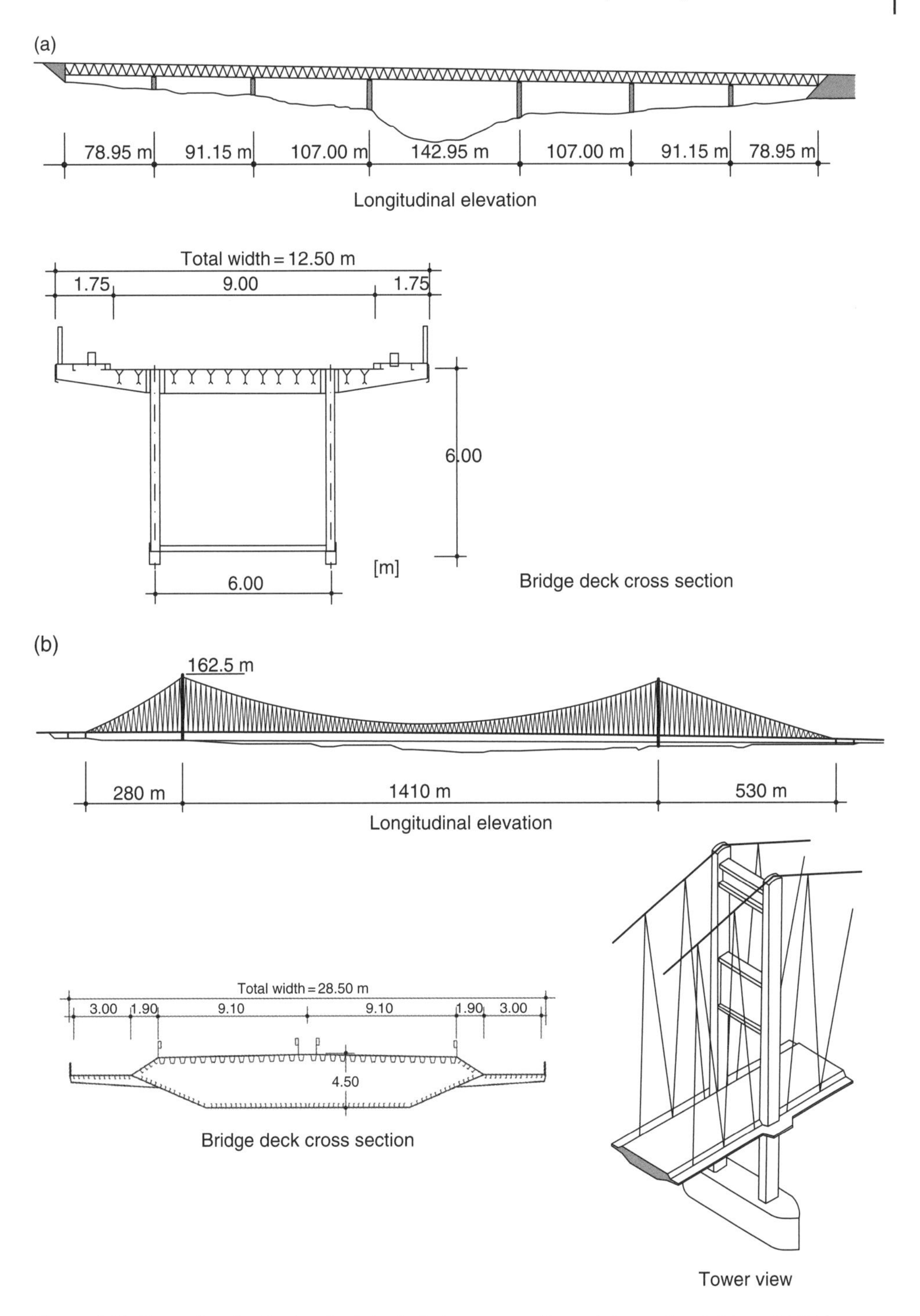

Figure 4.48 Examples of orthotropic plate decks in open and box girder section bridges: (a) Bridge over river Tulda, Germany and (b) Humber Bridge, UK.

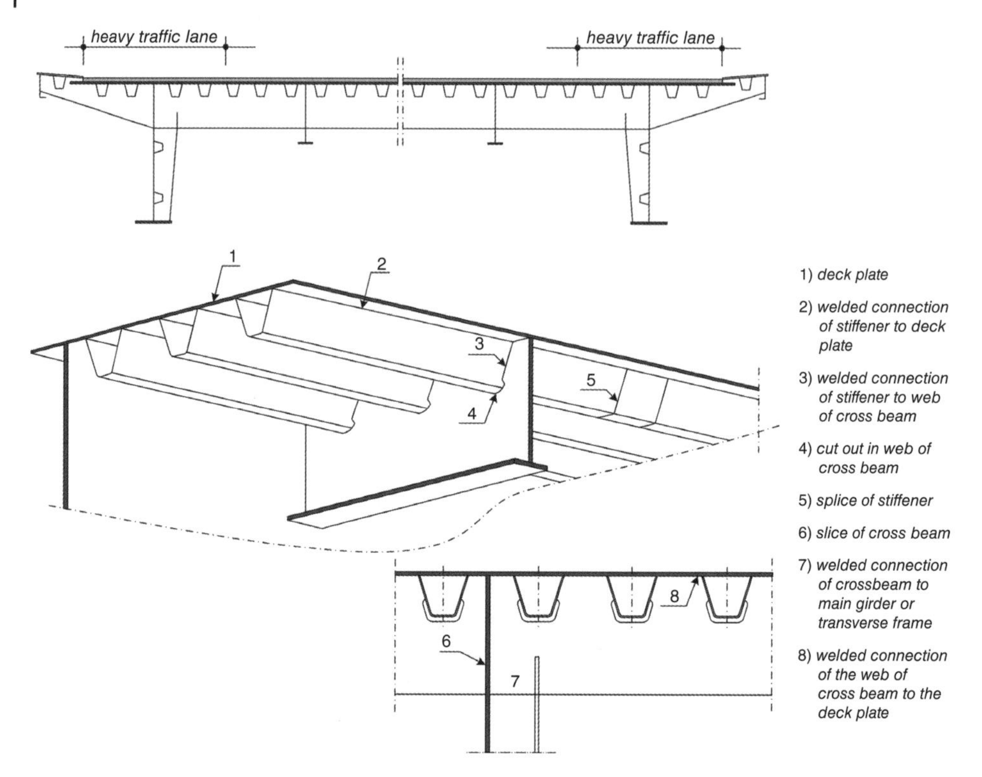

Figure 4.49 Basic scheme of an orthotropic plate deck in an open section superstructure made of two main plate girders, according to EN1993-2 [18]. (*Source:* Eurocode 3-2).

is defined by bending deflections under local load effects, usually restrained to $e/300$, where e is the transverse distance between stringers.

Stringers may be of open or closed cross section (Figure 4.50) but its transverse distance e should not be greater than 25 t for open section stiffeners and 50 t for closed sections, for durability of surfacing. This yields 350 and 700 mm for maximum distance between stringers, respectively, for open and closed cross sections e measured between stringer axes. For pedestrian bridges, one may adopt $t \geq 10$ mm and $e/t \leq 40$ with $e \leq 600$ mm.

Cross beams, spaced apart 1–2 m for open section stringers and 1.5–4.5 m for closed section stringers, transfer to longitudinal main girders or to webs of box girders, the vertical forces induced by permanent and traffic loads on the slab deck. The behaviour of the slab deck as an orthotropic plate is discussed in Chapter 6.

Open section stiffeners should be adopted with a depth $b_s < 10t_s$ to avoid *torsional buckling* (Figure 4.50). Torsional buckling is a column-like buckling mode involving torsional rotations of the cross sections and transverse displacements in the plane of the cross sections (see Annex A). In general, only open sections elements are susceptible of torsional buckling modes. A stiffener is attached to the plate where transverse cross sections are restricted but not torsional rotations. Open stringers may be adopted with flat section angles (L sections) or T sections.

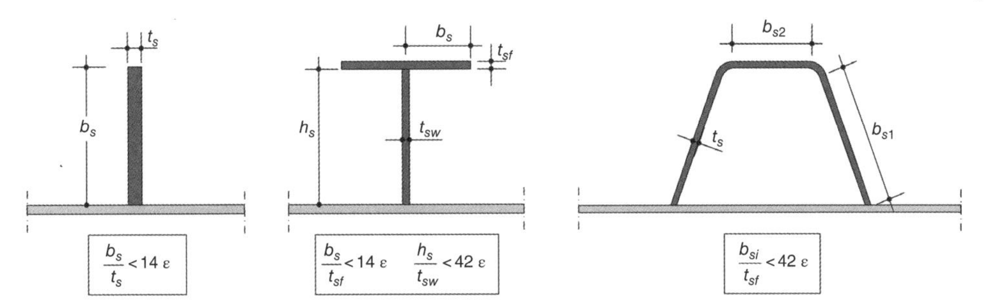

Figure 4.50 Open and closed shape stiffeners. Slenderness limits for preliminary design [19]. (*Source:* PPUR editions).

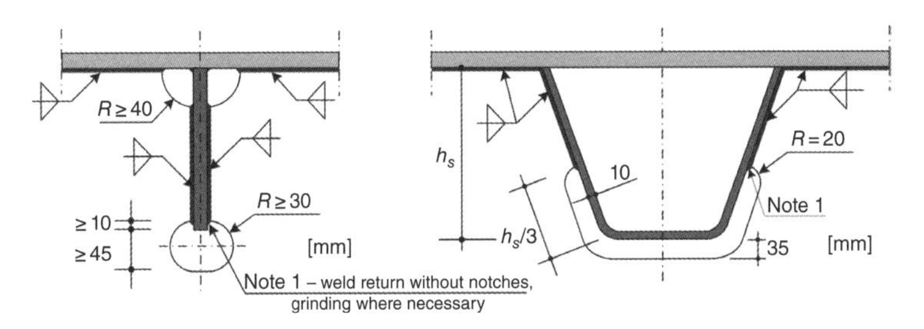

Figure 4.51 Details for crossing stringers in cross beams and diaphragms according to EN1993-2, to avoid fatigue problems [18]. (*Source:* Eurocode 3-2).

Closed section stringers, are of a trapezoidal shape made by cold forming a plate with a thickness of 6–10 mm. Stringer dimensions (Figure 4.50) should not exceed $42\varepsilon\ t_s$, where t_s is the stringer thickness. Nowadays, the trend of adopting stringers with large cross sections and slender plates may require the need to consider *local buckling* effects in the stringer. Minimum bending stiffness of stringers needs to be respected to avoid large deflections between the cross beams or diaphragms. The minimum stringer stiffness can be found in Ref. [19] as a function of cross beam spacing.

Finally, a main aspect in designing steel orthotropic plate decks is fatigue resistance. As a rule, the welding connection details of Figure 4.51 should be respected for open and closed shape stringers. The lower flange of the stringer is not welded to the web of the cross girder, which is done by *copes holes* as shown in Figure 4.51.

4.5.4 Plate Girder Bridges

4.5.4.1 Superstructure Components

A plate girder bridge superstructure (Figure 4.52) integrates the following components:

- the slab deck, usually a concrete slab in reinforced or PC;
- two or more (four at the most nowadays) steel plate girders – the *main girders*;
- a *vertical bracing* system between the main girders composed by cross girders or steel trusses with a diaphragm function;
- an *horizontal bracing* system nearby the lower flange level, composed by an horizontal truss between main girders and cross girders or diaphragms.

(a)

(b)

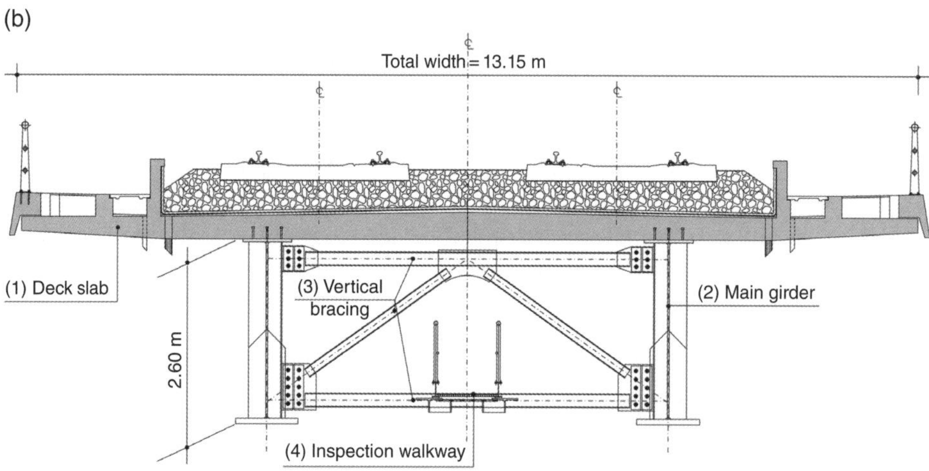

Figure 4.52 A plate girder bridge superstructure: (a) Example of the Sado railway bridge approach viaducts (*Source:* Courtesy GRID, SA); (b) Deck cross section: (1) Slab deck, (2) Main girder, (3) Vertical bracing (diaphragms) and (4) Inspection walkway.

The last component, the horizontal bracing system, is usually adopted only for rail bridges and curved road bridges. The overall system works as an equivalent box girder, the lower flange being an equivalent thin plate to the truss. The torsional stiffness is increased and consequently it reduces torsion deformability under eccentric loads and torsion vibration frequencies. This aspect is quite relevant in High Speed Railway bridges.

The slab deck has a variable transverse thickness, varying between a minimum of 17 cm at the tip of the overhangs and a maximum of 40 cm over the girders; at transverse

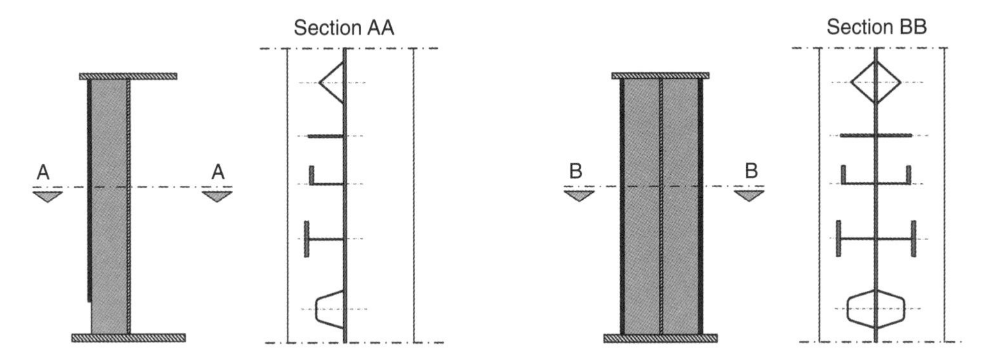

Figure 4.53 Bridge plate girders with longitudinal and transverse stiffeners placed asymmetrically or symmetrically at the cross section.

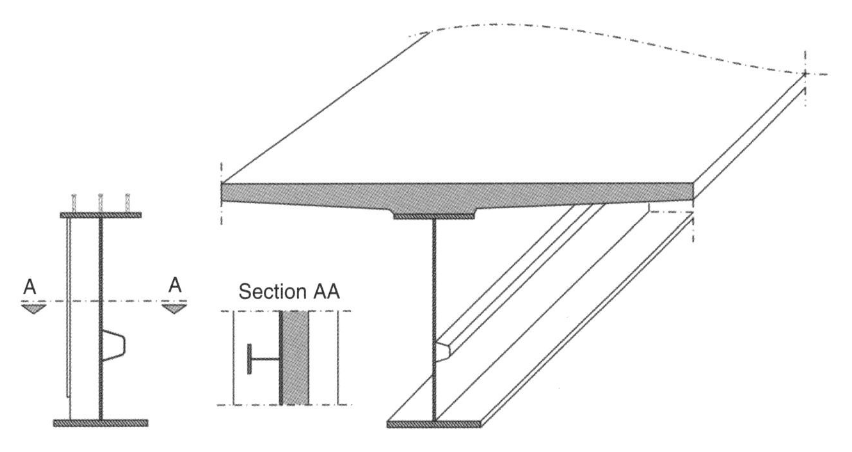

Figure 4.54 Plate girder bridge with one longitudinal stiffener at the outside and a transverse stiffener at the inside of the girder.

mid-span between main girders, the slab thickness is generally 20–25 cm. For rail bridges, the minimum allowable slab thickness is generally 30 cm (better 35 cm).

The plate girders may have only transverse stiffeners with different shapes (Figure 4.53), or transverse and one or two longitudinal stiffeners, as shown in Figure 4.54. At support sections, the vertical transverse stiffeners are placed symmetrically at the girder cross section (Figure 4.53). Hence, a symmetrical horizontal deck cross section, composed by the vertical stiffeners and parts of the web, is obtained to resist the vertical reaction from the bearings. At span sections, the vertical stiffeners are placed at one side only of the webs of the main girders. In case one or two longitudinal stiffeners are adopted, they may be placed at the outside of the cross section (Figure 4.54). If they are placed at the inside, they cross the vertical stiffeners and should adopt a fatigue detail of Figure 4.51.

The vertical bracing is made of hot rolled or welded composed sections, as is usually the case for road bridges. In the case of straight road bridges, the vertical bracing is generally reduced to a cross beam (Figure 4.55). Trusses are, also adopted for the vertical bracing system or plate diaphragms (Figure 4.56), with a *manhole* at the mid-depth to allow an inspection walkway in many railway bridges (Figure 4.52).

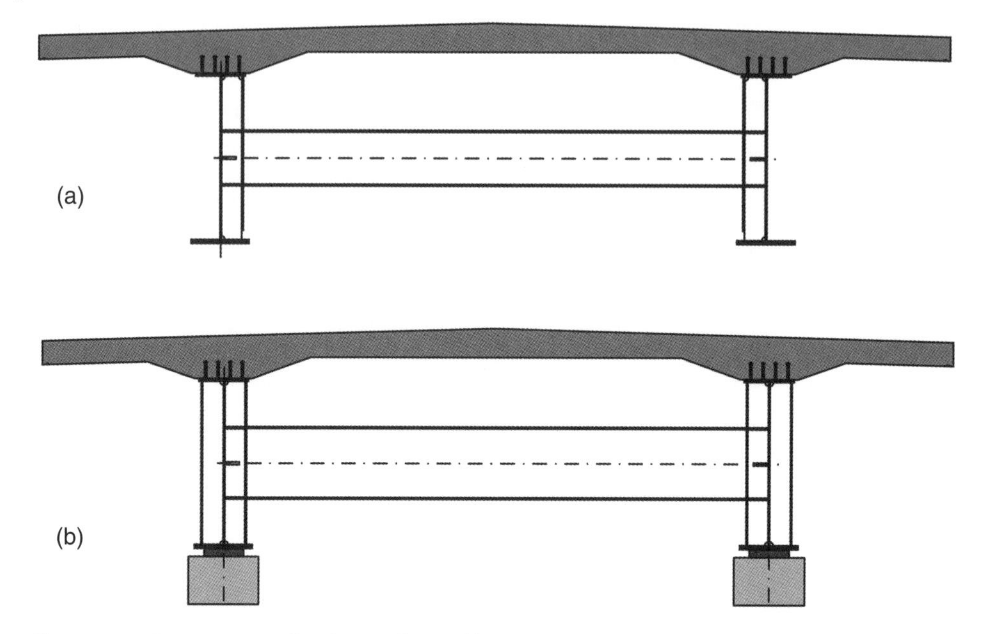

Figure 4.55 Cross sections of a plate girder road bridge at span (a) and at pier supports (b).

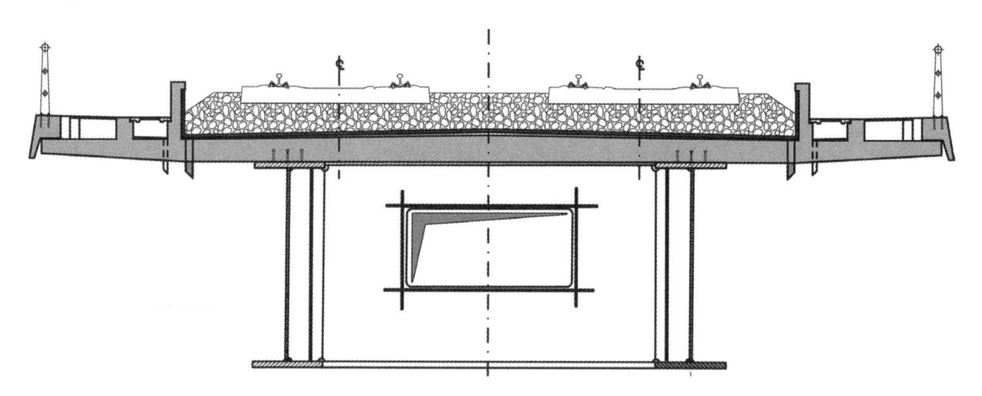

Figure 4.56 Diaphragms in railway bridges made by a solid plate.

4.5.4.2 Preliminary Design of the Main Girders

Taking into consideration resistance at ULS and elastic behaviour at SLS, the following geometrical characteristics (Figure 4.57) for main plate girders are adopted.

4.5.4.2.1 Deck Slenderness

- For constant depth girders, $l/h = 22–28$ for continuous girders of road bridges with internal span lengths, l.
- For variable depth girders, $l/h = 20–25$ at support sections and 40–50 at mid-span sections.
- The lower limits of slenderness ratios, referred to previously, may be adopted for the preliminary design of simply supported girder decks or for the case of rail bridges in general.

Figure 4.57 Geometry of a plate girder.

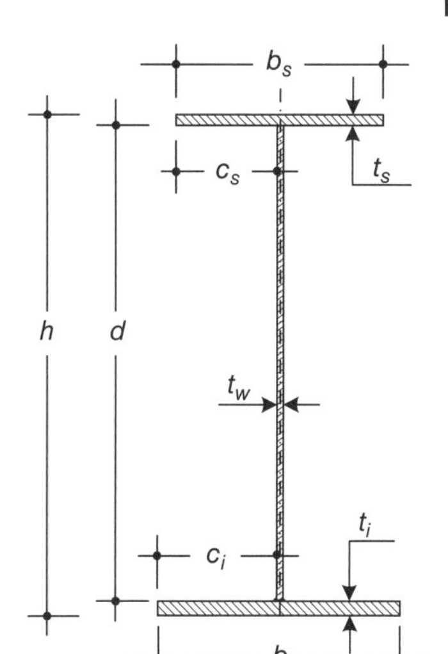

4.5.4.2.2 Slenderness of Flanges and Webs

Flanges are generally designed for being *totally effective* at ULS (see Annex A) allowing its full plastic resistance $N_{f,pl}$ to be mobilized. For a preliminary design of the cross section, one assumes the bending moment is taken by flanges and the shear force is taken by the web.

A plate girder bridge section (Figure 4.57) is considered, with unequal flanges of areas A_{fs} and A_{fi}, respectively, for the upper and lower flanges and a section depth, h. If the depth h is assumed, for the time being, to be approximately equal to the distance between centroids of the flanges, the resistant bending moment of the cross section is reached when the stress at the most loaded flange (the upper flange in the figure) reaches the yield stress:

$$M_{Rd} = N_{fs,pl}\ h = A_{fs}\ f_y\ h = b_s\ t_s\ f_y \tag{4.9}$$

where f_y is the flange yield stress of width b_s and thickness t_s. To reach $N_{fs,pl} = A_{fs} f_y$ *local plate buckling* in the compressed flange (upper flange in the figure) should be avoided (Figure 4.58).

As shown in Annex A, the flange is fully effective if it has a non-dimensional slenderness $\bar{\lambda}_p \leq 0.748$. For flanges, this yields:

$$\frac{c}{t} \leq 0.466\sqrt{\frac{E}{f_y}} \tag{4.10}$$

If this limit value is exceeded the flange should be considered reduced to its effective width (Annex A), as shown in Figure 4.59.

By taking $E = 210\,\text{GPa}$, one has a fully effective flange if approximately $c/t < 14$ for S235, $c/t < 13$ for S275 and $c/t < 11$ for S355. For pre-design of a plate girder in a bridge composite deck, one may take the following plate dimensions:

- upper flanges $500\,\text{mm} < b_s = 2c_s < 800\,\text{mm}$ with $30\,\text{mm} < t_s < 80\,\text{mm}$
- lower flanges $600\,\text{mm} < b_i = 2c_i < 1200\,\text{mm}$ with $40\,\text{mm} < t_i < 120\,\text{mm}$

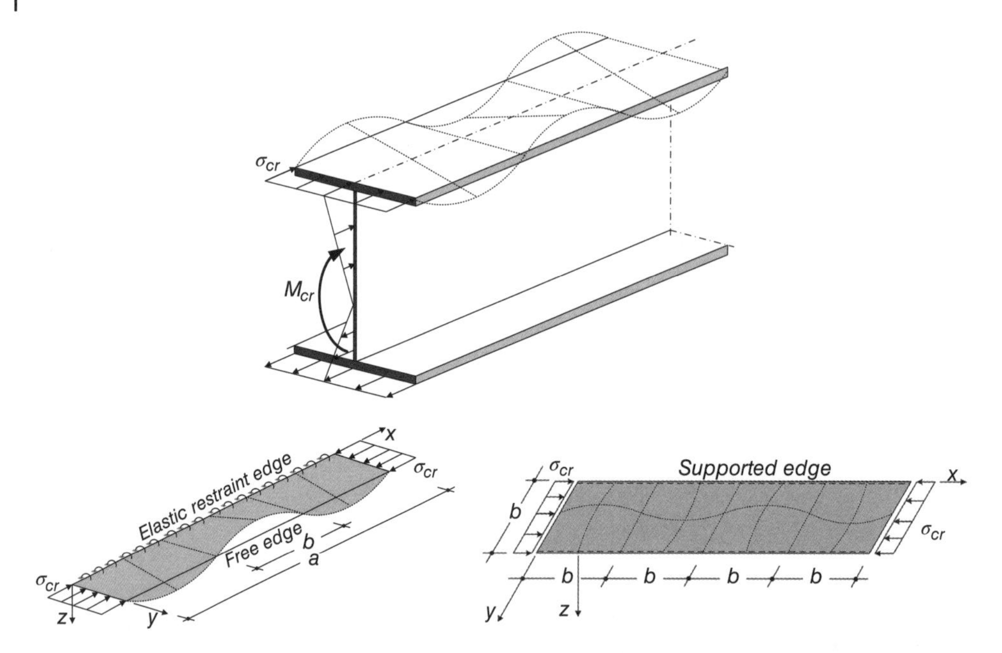

Figure 4.58 Plate buckling modes of long plates (a >> b) elastically built in at one longitudinal edge and free at the other as the case of a plate girder flange.

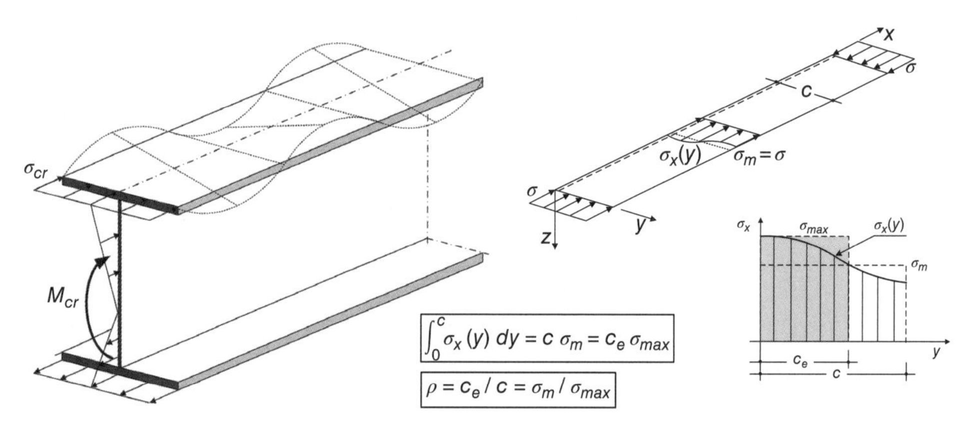

Figure 4.59 Effective width concept for a flange of a plate girder in the post-buckling range.

The plate thicknesses vary along the span length to accommodate the tensile or compressive stresses due to bending moment variation.

In Figure 4.60, one shows the post-buckling behaviour of web panels under normal stresses due to bending and under shear stresses. The elastic critical stress of a plate girder web, under in-plane bending only, is determined as a function of the stress ratio $\psi = \sigma_2/\sigma_1$ at top and bottom longitudinal edges, with compressive stresses taken as positive. For an equal flange girder (quite unusual in plate girder bridges) one has $\psi = \sigma_2/\sigma_1 = -1$. Slender webs do not allow for taking the full depth as effective under the bending moment. As shown in Figure 4.60, the compressed zone of a web under bending goes into the postbuckling range and its effective width is taken as shown in Figure 4.61.

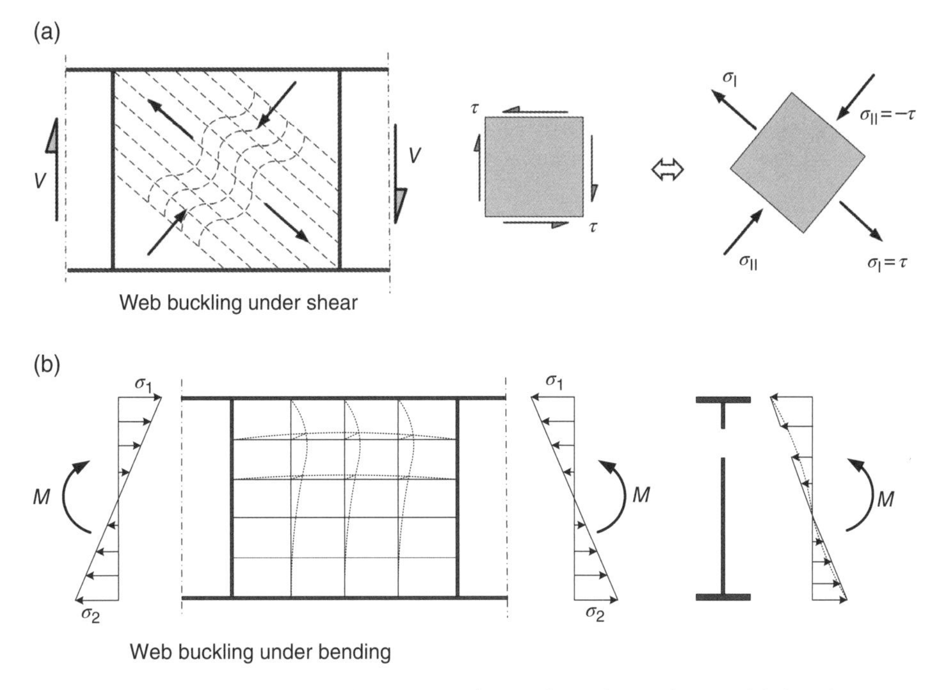

Figure 4.60 Web post-buckling behaviour in a plate girder under (a) shear and (b) bending moment.

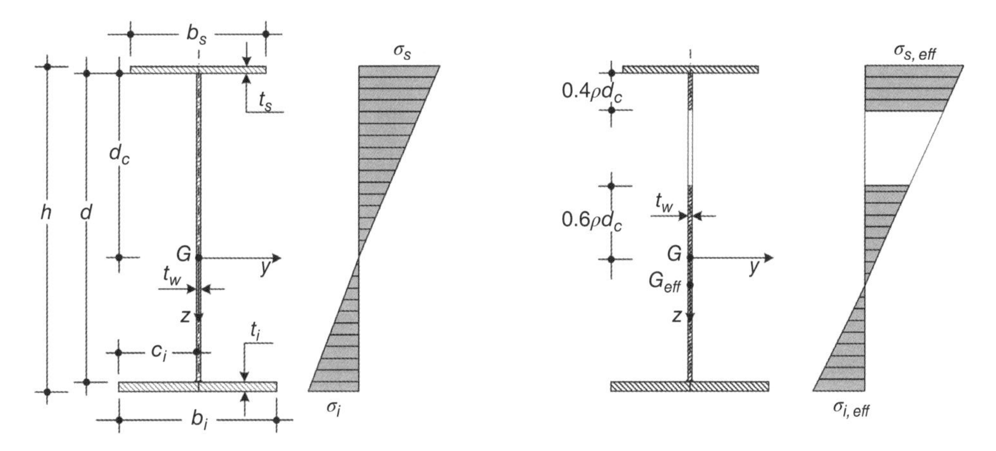

Figure 4.61 Effective section of a plate girder under bending.

Thus, the resistance of the section under bending is taken for the *effective section* composed by the flanges and the effective parts of the web.

Assuming the flanges are fully effective, the ultimate moment in Eq. (4.9) is always smaller than the resistant bending moment of the reduced section in Figure 4.61.

For the webs (Figure 4.60a), assuming they are going to predominantly resist shear force, they reach the elastic shear critical stress (Annex A) determined as a function of the geometry of the web panels between transverse stiffeners, a/d.

Slender webs have a considerable post-buckling resistance under shear forces. They reach the ultimate shear force resistance at values much higher than the elastic critical value for the shear force $V_{cr} = \tau_{cr}\, d\, t_w$. Nowadays, steel bridge design takes advantage of the post-buckling strength of web panels, designing webs with a slenderness d/t_w such that, in general, for webs with only transverse stiffeners one has,

$$70 < \frac{d}{t_w} < 200 \tag{4.11}$$

The usual web thicknesses in plate girder bridges are as follows:

- At span sections – $t_w = 14$–$18\,\text{mm}$
- At support sections $t_w = 16$–$25\,\text{mm}$

For pre-design, webs may be taken with a thickness allowing a maximum shear stress for the characteristic load combination (permanent + live loads) less than $100\,\text{N}\,\text{mm}^{-2}$ for steel grade S355.

If one restricts the bending resistance to flange action, as previously referred, the webs are assumed in the post-buckling range under shear and a *diagonal tension field* is developed at ULS, as shown Figure 4.60a.

Reducing the distance a between transverse stiffeners, taking in general as $0.5d < a < 1.5d$), increases τ_{cr} and consequently reduces its slenderness, $\overline{\lambda}_p$, thus increasing its post-buckling strength. The reader is referred to Chapter 6 for the detailed evaluation of ultimate strength of webs working in the post-buckling range and to specific literature [19, 20]. Adding one or two longitudinal stiffeners to the web may be justified in deep plate girders, say $d > 3.0\,\text{m}$. The elastic critical stress, τ_{cr}, is still obtained in the same way, but is now applied to each sub-panel. The diagonal tension field for post-buckling shear resistance is not significantly increased by adding longitudinal stiffeners. However, they may help, if located nearby the flanges (say approximately $d/3$ from the flanges) in bending resistance and for resistance under concentrate loads as it occurs during launching operations ('patch loading' – Figure 4.62).

The slenderness of the web panels cannot be increased above a certain limit due to *flange induced buckling*. This instability problem is associated to buckling of the flanges in the plane of the web as shown in Figure 4.63. The web is subjected to compressive stresses along the longitudinal edges, coming from the deviation forces induced by flanges, due to overall bending of the cross section. Taking the web as a long plate simply supported and loaded along the longitudinal edges, its critical load should always

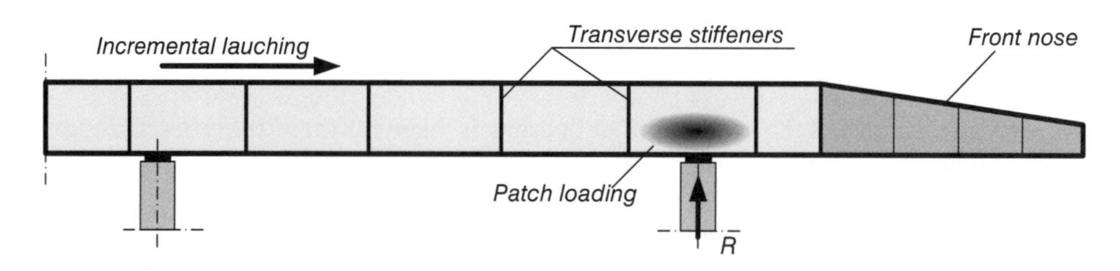

Figure 4.62 Patch loading in webs of plate girder bridges under launching.

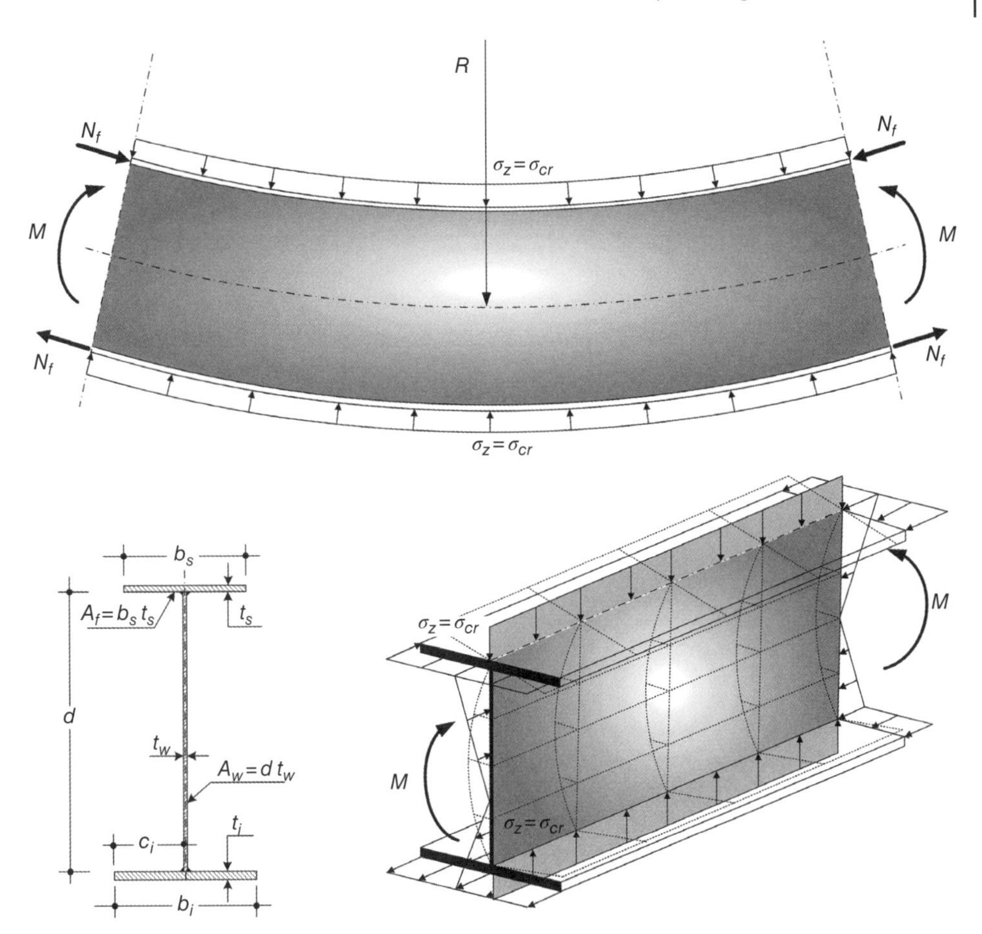

Figure 4.63 Flange induced buckling in a plat girder.

be higher than the deviation force coming from the flanges when the ultimate resistance is reached. Flange induced buckling does not occur (see Chapter 6) if the slenderness of the web of height h_w, thickness t_w and area $A_w = h_w t_w$, is limited to:

$$\frac{h_w}{t_w} \leq k \frac{E}{f_{yf}} \sqrt{\frac{A_w}{A_{fc}}} \tag{4.12}$$

where f_{yf} is the yield stress of the flange of area A_{fc} (first reaching yielding) and k a factor, taken as 0.55 or 0.40, respectively, for an elastic or plastic design. For different steel grades and web and flange area ratios, the expression 4.12 yields the limits for web slenderness of Table 4.4.

The nominal stresses in the section at SLS also need to be controlled when designing a plate girder bridge, and the web slenderness should be restricted by *web breathing*

Table 4.4 Web slenderness limits for S235 and S355 grades to avoid flange induced buckling (elastic design).

	A_w/A_{fc}				
Steel grade	**0.2**	**0.4**	**0.6**	**0.8**	**1.0**
S235	219	310	380	439	491
S355	145	205	252	291	325

phenomena. This is a problem at SLS inducing of transverse vibrations of the web under traffic loading, which may yield cracks at the welding between the web and the flanges (Figure 4.64). The limits to avoid checking web breathing are given in [21], for road and railway bridges. Designating the depth of the web as d and its thickness as t_w, one has

$$d/t_w \leq 30 + 4.0\ l \quad \leq 300 \text{ for road bridges}$$
$$d/t_w \leq 55 + 3.3\ l \quad \leq 250 \text{ for rail bridges}$$

where l is the span length.

4.5.4.2.3 *Preliminary Design of Stiffeners*

Preliminary design of stiffeners for plate girders considers conditions of resistance and minimum stiffness. These limits are defined in [22]. The slenderness of stiffeners shall be limited to avoid local buckling phenomena or torsion buckling in open section stiffeners. The most frequently adopted types of stiffeners are represented in Figure 4.50 with the recommended limits for the slenderness of plate elements.

4.5.4.3 Vertical Bracing System

The main function of the vertical bracing system is to limit distortion of the superstructure cross section in its plane (Figure 4.65) under eccentric loading due to traffic and

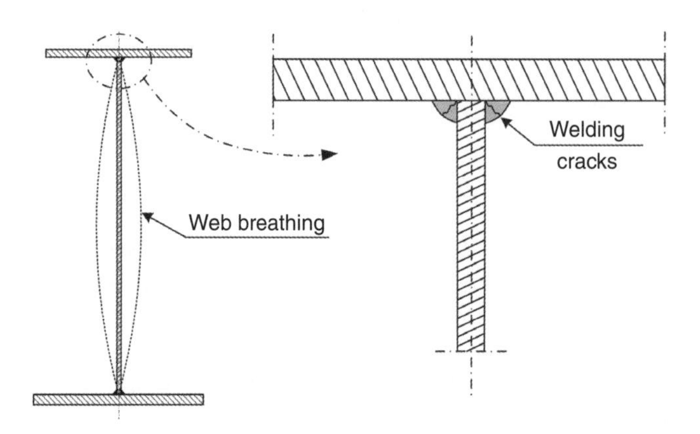

Figure 4.64 Web breathing in plate girder bridges.

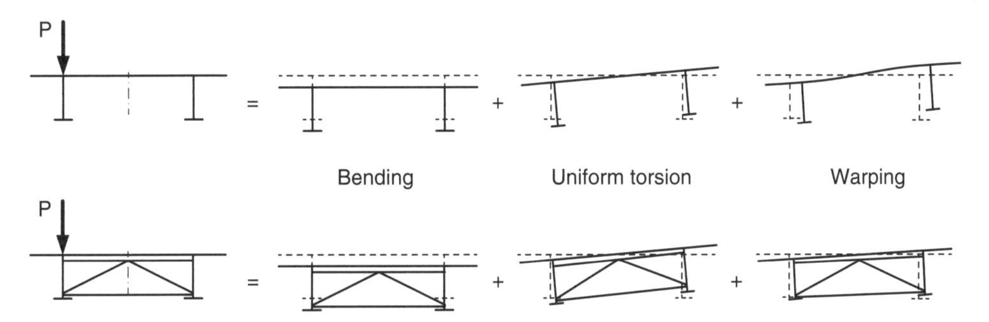

Figure 4.65 Effects of the vertical bracing system in open section decks.

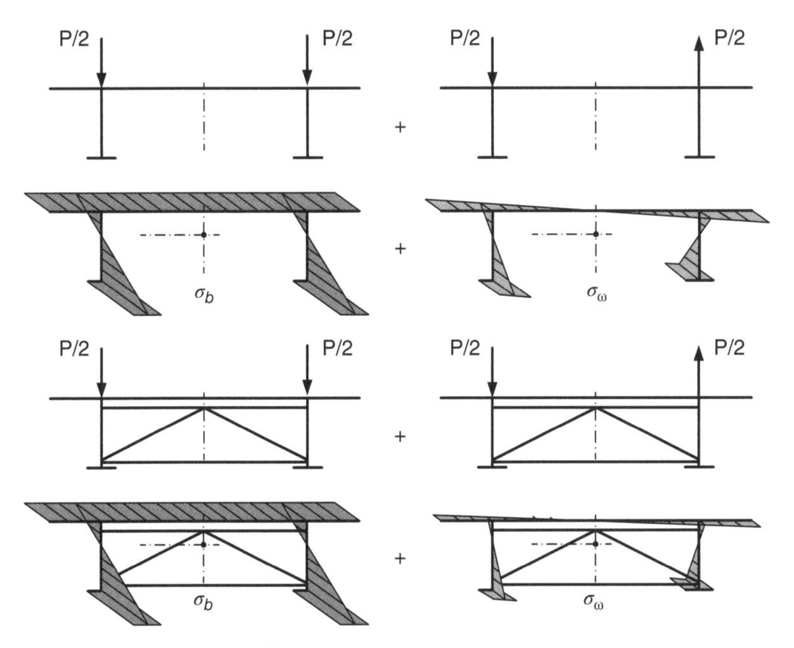

Figure 4.66 Bending and warping stresses in open section decks with eccentric loading, with and without vertical bracing.

lateral horizontal loads due to wind or earthquakes. It is composed of cross girders or diaphragm trusses as solid plates, as previously referred to. The vertical bracing system benefits transverse load distribution between main girders under torsion loads.

An open section superstructure, composed by two main girders and a vertical bracing system, under eccentric loading due to traffic or (and) lateral transverse horizontal loads, is subjected to warping stresses σ_ω that should be added to the normal stresses σ_b due to global bending of the cross section (Figure 4.66). An asymmetric vertical loading may always be decomposed into a symmetric component, inducing overall bending only, and a skew symmetric load case, inducing torsion only. The last component induces a differential bending between main girders.

To reduce distortion and warping effects, cross girders are introduced at approximately mid-height of the main girders, or even at the upper flange level as shown in

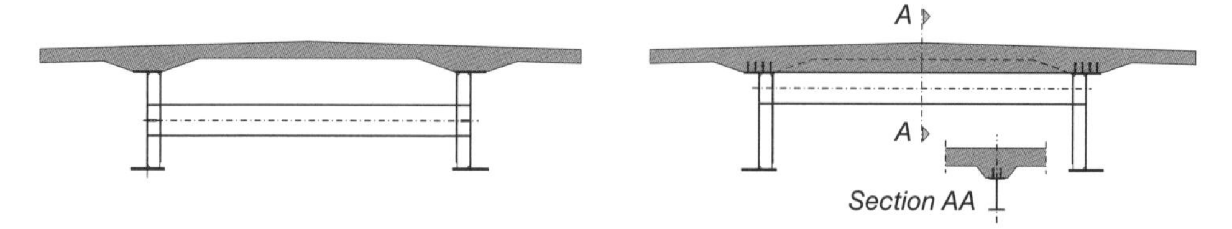

Figure 4.67 Cross girders in plate girder bridge decks: at below or at underneath of the deck slab.

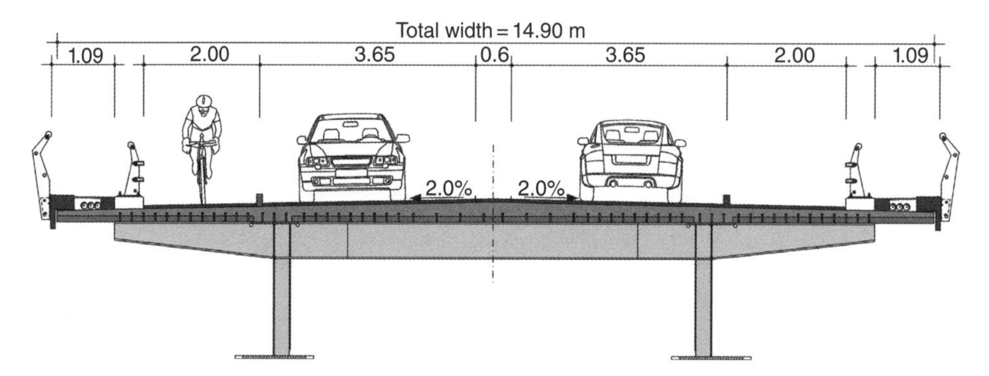

Figure 4.68 Cross girders underneath a wide deck. Bridge over the River Zapotal, Ecuador.

Figure 4.67. In this last case – cross girders at upper flange level – one may benefit from them to support the slab deck. Since the distance of cross girders is always less than 5–8 m in road bridges with a maximum of $4h$, h being the height of the main girders, the deck slab has bi-directional bending reducing the transverse bending moments. In this case, it is usual to locate the cross girders at approximately 4 m maximum for main girders at 7–8 m distance apart. The cross girders are usually connected to the deck slab by shear head studs and may be extended to the transverse cantilevers of the deck slab (the overhangs) allowing wide decks with two girders only. Wide decks (Figure 4.68) with only two girders may be adopted with the mentioned cross girders, with distances reduced to 2–3 m only. Precasted slabs, supported by the cross girders, allow to cast the in situ slab deck. The precasted slabs may work just as a lost formwork or may work together with the cast in situ slab deck.

The height of the cross girders is usually 1/15–1/10 of the height of the main girders. The cross girders and diaphragms also have a function of avoiding *lateral torsional buckling* of the compressed flange. This global buckling mode, discussed in Chapter 6, is particularly relevant in continuous girders nearby the support sections, where important negative moments induce a high compressive stress in the lower flange. During the erection stages, flanges may need to be braced against lateral buckling. After casting the deck slab, lateral buckling of the upper flange cannot occur due to the connection with the slab. A temporary bracing of the upper flange in composite bridge decks, where the width of the upper flange is usually smaller than the lower flange, may be necessary for stability.

Lateral buckling effects in the lower flange, may be reduced by limiting the longitudinal distance between the vertical bracing elements (l_b) to values of approximately $9b_f$ for S355 steel, where b_f is the flange width. In fact, if $l_b < 9b_f$ one can show [20] the *non-dimensional slenderness* of the compressed flange for lateral torsional buckling [19, 20, 23] $\bar{\lambda}_{LT}$ does not exceed 0.4, which allows a reduction, due to flange lateral buckling effects, of not more than 10% of f_y on maximum compressive stresses. Hence, the preliminary design of the cross section of the main girders may be performed by allowing a compressive stress at ULS in the order of $0.9 f_y$ at the compressive flange.

4.5.4.4 Horizontal Bracing System

As previously referred to in Section 4.5.4.1, in rail bridges and curved road bridges open section decks need certain torsion stiffness, which may be achieved by adopting a horizontal bracing system at the level of the lower flange of the main girders (Figure 4.69). The horizontal cross bracing may be made by hot rolled sections (e.g. Type 1/2 HEA 400) with a cross (X) configuration, a diamond configuration and a K configuration, as

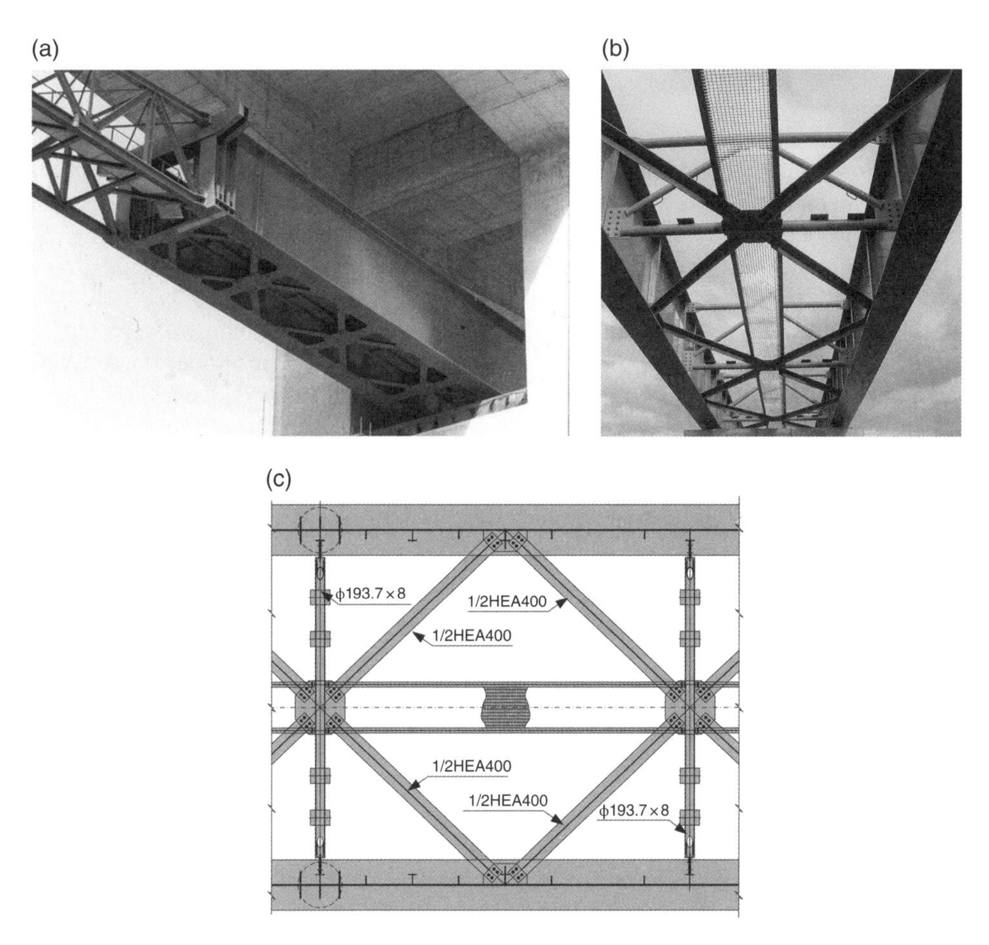

Figure 4.69 Horizontal bracing at lower flange level for railway bridges – approach viaducts to bridges over (a) the Tagus river, (b) and (c) the Sado river (Figures 5.22, 5.23 and Figure 4.52) (*Source:* Courtesy GRID, SA).

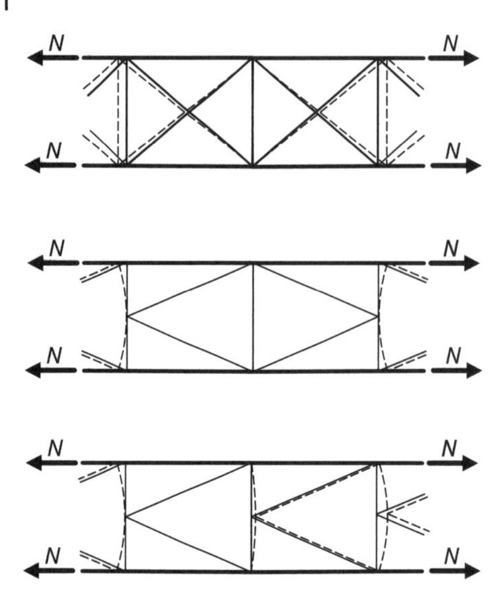

Figure 4.70 Horizontal Bracing in plate girder bridges: X, diamond and K configurations.

shown in Figure 4.70. The horizontal bracing may participate in the resistance to global longitudinal bending moments at the overall cross sections. However, its main function is to yield an equivalent horizontal plate, closing the section and working as a horizontal girder with a depth approximately equal to the distance between webs of the main girders. The in-plane forces from the lower flanges are transferred to this horizontal girder and then to the bearings at support sections.

The thickness of the equivalent plate to the horizontal cross bracing is evaluated by strain energy consideration of a segment between two diaphragms under torsion. For different horizontal bracing configurations, a different value of the equivalent thickness is obtained as presented in [19, 24].

One relevant design aspect of the horizontal bracing configuration is the effect on the thermal locking stresses induced by the bracing. These stresses in the flanges are induced because thermal deformations are not free to occur. The configuration in X induces higher lock-in stresses, while the K configuration is the less rigid one, inducing very light lock-in stresses. This may be easily understood by comparing the deformation of the X configuration with the K configuration (Figure 4.70) under an axial force, N, at the flanges.

4.5.5 Box Girder Bridges

4.5.5.1 General

Box girder decks, as referred to in Section 4.5.1, are adopted for long span bridges, curved bridges (Figure 4.71) and bridges with specific requirements for increased slenderness (span/depth ratio) like urban bridges (Figures 4.32 and 4.36). Box sections are also adopted for axially suspended decks [25] due to requirements of torsional stiffness for static and aerodynamic stability. Examples of these cases are presented in Figures 4.33 and 4.34.

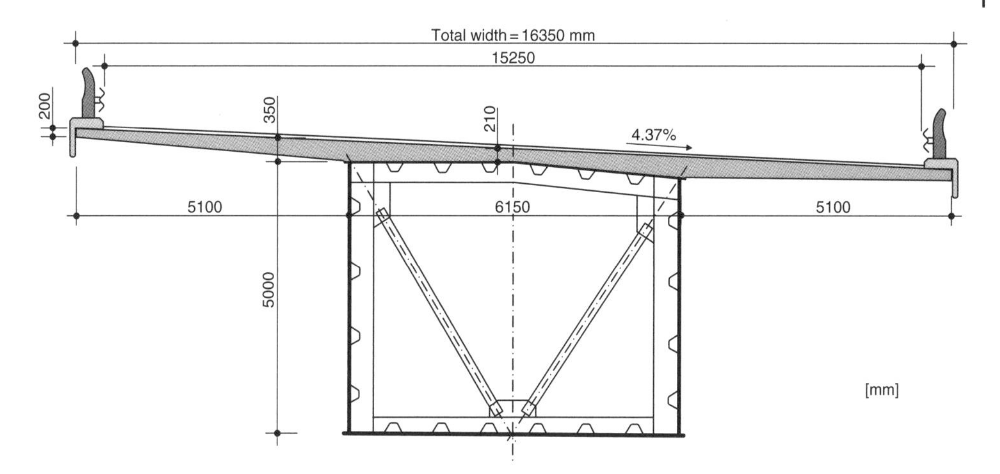

Figure 4.71 Box girder span section of the bridge over Veveyse (Switzerland): curved bridge deck with a main span of 129 m.

Box girder cross sections may have vertical (Figures 4.32 and 4.71) or inclined webs (Figure 4.33) and may be single cell or multicell box girders (Figures 4.34 and 4.72).

In cable-stayed bridges and suspension bridges, steel box girders with an aerofoil cross sectional shape (Figures 4.33, 4.48b and 4.72) are adopted for aerodynamic stability reasons. One of the first suspension bridges built with this type of aerodynamic shape and with inclined hangers was the Humber Bridge in the UK (Figure 4.48b), with a main span of 1410 m: this was the record in Europe for some years before the Storebelt Bridge in Denmark. The recently built third Crossing of Bosporus, Turkey, is a more recent example (Figure 4.72). The deck of this modern steel suspension / cable-stayed bridge has a main span of 1408 m with 1360 m being a steel deck and the remainder a PC box girder as the side spans with the same external configuration. The total deck width is 50 m and accommodates eight road traffic lanes and two rail tracks at the axis. Some stay cables complementing the main suspension cables stiffen the deck.

In the following sections, the main design concepts for the pre-design of box girder decks are discussed. The analysis and design of box girders are discussed in Chapter 6.

4.5.5.2 Superstructure Components

Box girder decks integrate the following structural elements:

- A deck slab made of a steel orthotropic plate (steel box girders) or by a reinforced or PC slab (composite box girders), as happens in plate girder bridges.
- A complete steel box section or a U-shaped steel girder, respectively, in steel decks or composite decks: in both cases the webs being vertical or perpendicular to the lower flange, or inclined as in trapezoidal boxes. The flanges and webs of box girders are steel plates with longitudinal and transverse stiffeners.
- Vertical transverse cross frames or (and) diaphragms, with or without web and flange transverse stiffeners between them.

Figure 4.73 shows the basic typologies for single cell box girder decks with vertical or inclined webs. In composite bridges, if the deck is large, the slab in RC or PC may

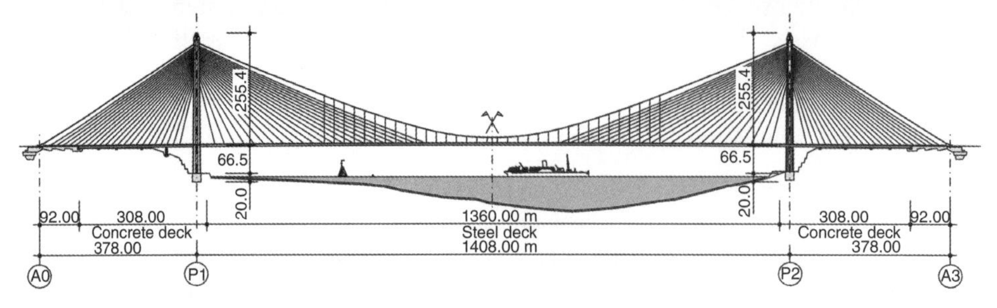

Figure 4.72 The third Bosporus crossing – Yavuz Sultan SelimBridge (*Source:* Photograph by VikiPicture/ https://commons.wikimedia.org). A steel box girder deck in the major part of the main span 1408 m long (*Source:* Courtesy T Engineering, SA).

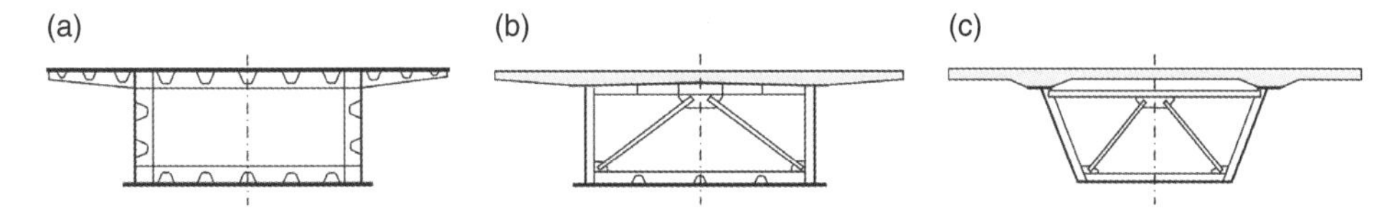

Figure 4.73 Typology of (a) steel box girder and composite box girder decks with (b) vertical or (c) inclined webs.

Figure 4.74 Model of a box girder deck with double composite action near the support sections (design study for the ring road in Antwerp). Main span 108 m, deck height 4.0 m and deck width 14.85 m. (*Source:* Courtesy GRID, SA).

be supported by cross beams extending to the overhangs. Another option is to adopt a longitudinal central beam supported the top flange and transferring its load to the diaphragms as shown in Figure 4.74. The lower flange in box composite girder decks is usually made of a steel reinforced panel (Figures 4.32 and 4.72) transferring its load to the diaphragms or cross frames. However, to save steel, one may adopt a *double composite action* (as shown in Figures 4.36 and 4.74), nearby support sections where high negative bending moments induce high compressive stresses in the lower flange.

The deck slab may also be cast on top of a steel plate (Figure 4.71). In this case, the top steel plate should have sufficient rigidity to avoid transverse deformations of the slab. It may have shear connectors transferring the longitudinal and transverse shear flows due to transverse bending and torsion. The steel box girder may be erected as a complete steel box (Figure 4.75) and the slab is cast after closing the superstructure. In negative bending sections, nearby supports, the steel plate is in tension but at the span sections the longitudinal compressive stresses due to bending requires a stiffened plate as in the lower flange section near the supports.

One of the main differences between plate girder bridges and box girder bridges is the existence of a stiffened plate at the bottom flange. The longitudinal high compressive

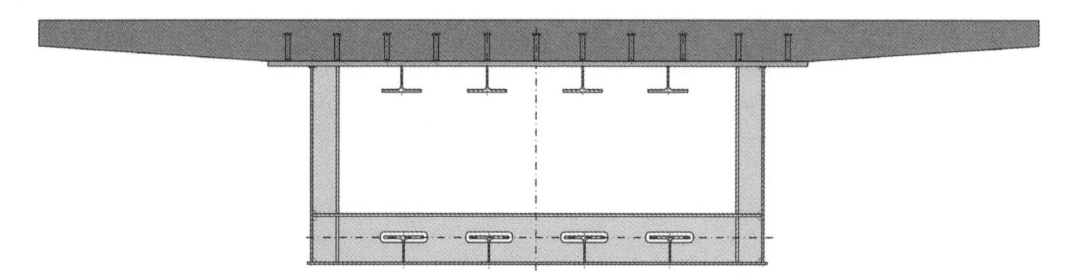

Figure 4.75 Top slab casted on a steel plate included in a complete steel box section.

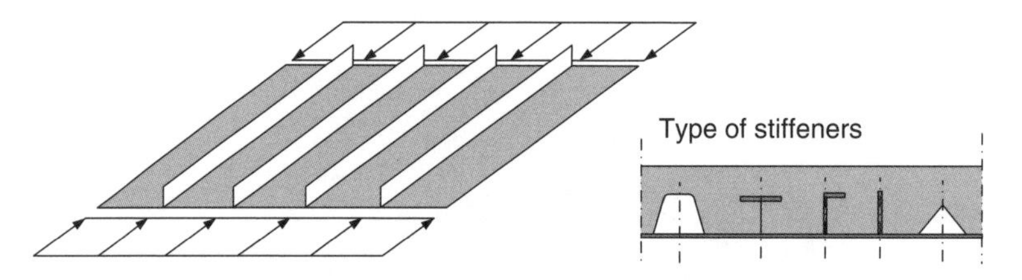

Figure 4.76 Lower flange of a box girder with a stiffened plate. Open or closed stiffeners.

stresses at this panel requires longitudinal stiffeners, which may be open or closed section stiffeners as shown in Figure 4.76.

4.5.5.3 Pre-Design of Composite Box Girder Sections

A composite box girder section, with vertical or inclined webs, is generally a U or trapezoidal shape. Two top flanges, welded to the webs, are adopted and the bottom flange is made a stiffened plate (Figure 4.76). Since it is an open section during erection and before casting the slab deck, a top horizontal bracing is required for stability and torsion stiffness at the construction stages.

4.5.5.3.1 Deck Slenderness

Box sections generally have an increased slenderness with respect to plate girder superstructures for the same span length. Designating l as the internal span length of a continuous girder and h the depth of the box girder, values in the order of 1/25–1/30 are typical ones in practice. Of course, the width, B, of the deck influences the slenderness to be adopted. Empirical expressions have been proposed in Ref. [4.26] for pre-design, like

$$h/l = 1/36\left(B/12\right)^{0.7} \tag{4.13}$$

For example, for a wide deck with $B = 18\,\text{m}$, one has $h/l = 27$ for a continuous span.

4.5.5.3.2 Flange and Web Slenderness

The upper flanges may be pre-designed as follows (Figure 4.77):

- $b_s = 600–1000\,\text{mm}$
- $t_s = 30–120\,\text{mm}$

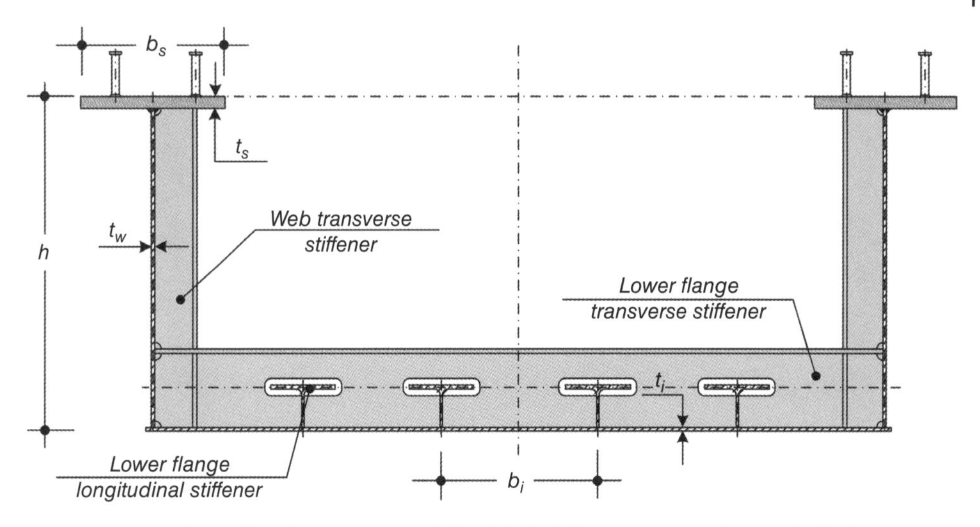

Figure 4.77 Steel 'open box' before casting of the slab deck for a composite box girder deck.

With $c_s/t_s < 10$, in general, ($c_s \approx b_s/2$) for the same reasons as plate girder bridges.

The webs are pre-designed in a similar way to the webs of plate girder bridges. One generally has $12 < t_w < 16$ mm at span sections and $15 < t_w < 25$ mm at support sections. To take into account the width of the deck, some empirical expressions, taken from practice may be adopted (at least for support sections), such as the following [26]:

$$t_w = 10 + Bl/100\,[\text{mm}] \text{ with } B \text{ and } l \text{ in metres} \quad (4.14)$$

For example, for B = 18 m and l = 50 m, one has t_w = 20 mm at support sections; at span sections thinner webs (min. t_w = 12 mm) may be adopted.

The lower flange is pre-designed taking the three main buckling modes of the compressive steel panel into consideration (Figure 4.78)

- The *local plate buckling mode* between longitudinal stiffeners;
- The *global plate buckling mode* where the stiffened panel buckles as an orthotropic plate supported at the longitudinal edges by the webs and at end sections by the transverse cross frames or diaphragms;
- A *column buckling mode,* typical for a wide panel with stiffeners deflecting with the plating between transverse cross frames or diaphragms.

The interaction between the plate mode and the column mode may exist, as discussed in Chapter 6. The plate thickness is generally $12 \le t_i \le 60$ mm. At span sections, one generally has $12 \le t_i \le 30$ mm and at support sections, $22 \le t_i \le 50$ mm.

The slenderness of the plate between longitudinal stiffeners should be pre-designed following: $b/t_i < 60$ in compressed regions and $b/t_i < 120$ in tension regions.

The longitudinal and transverse stiffeners should satisfy conditions of strength and minimum rigidity. The transverse stiffeners, cross frames or diaphragms supporting the end sections of the longitudinal stiffeners, should constitute rigid supports for the longitudinal stiffeners. Its slenderness (l_s/h_s), measured for the free length, l_s, between webs, for a transverse stiffener of height, h_s (Figure 4.49), should be limited to $l_s/h_s \le 15$.

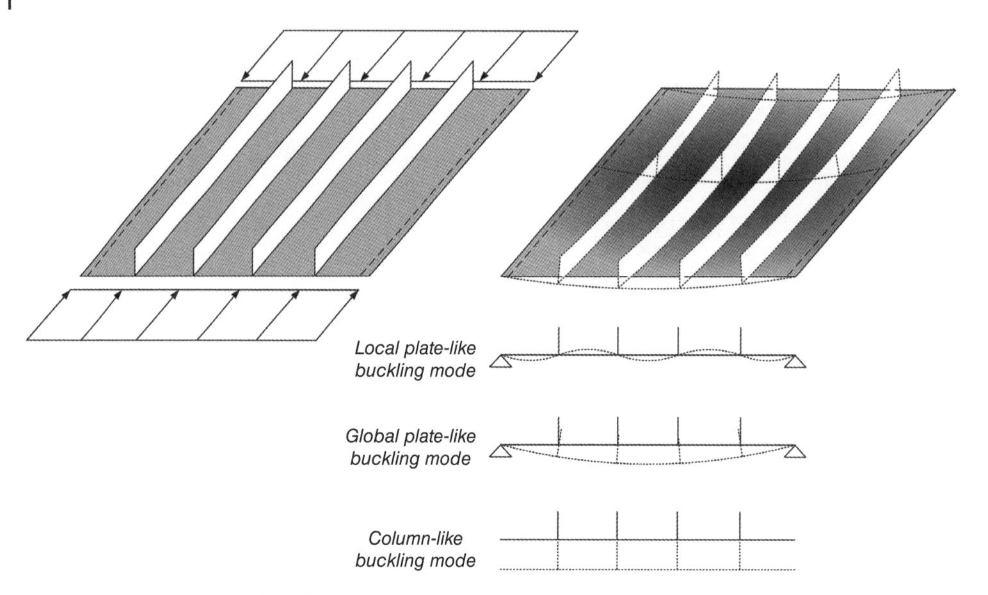

Figure 4.78 Buckling modes of stiffened panels under compression (*Source:* Adapted from Ref. [19]).

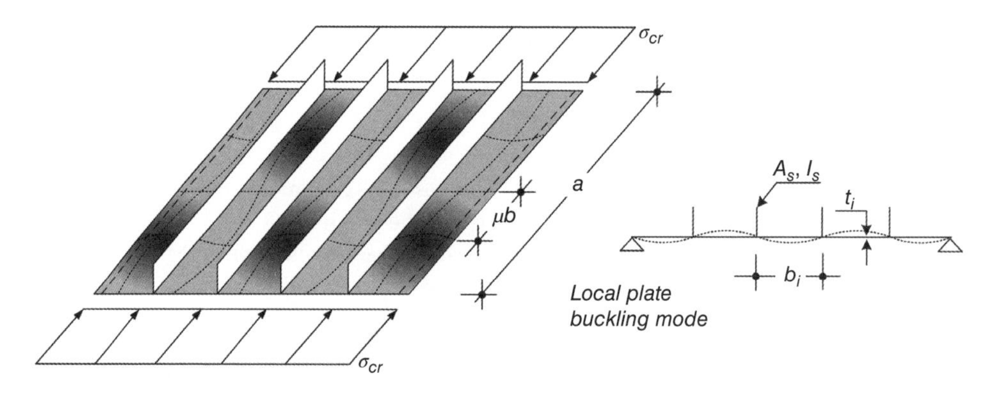

Figure 4.79 Local buckling mode of a steel panel under compression (*Source:* Adapted from Ref. [20]).

If the longitudinal stiffeners are very rigid, local buckling modes control the design. However this condition is difficult to fulfil in design practice and the longitudinal stiffeners, characterized by area A_s and inertia I_s, will deflect between transverse stiffeners with the plate (Figure 4.79). Even assuming the longitudinal stiffeners are designed as flexible, slenderness should be limited to $a/h_s \leq 25$, where a is the distance between transverse cross beams and h_s is the depth of the longitudinal stiffener.

The behaviour of a stiffened panel under compressive stresses, like at the bottom flange of a box girder, is considered in Chapter 6. However, it is anticipated the simplest model is the so-called *strut approach*. The ultimate load of the panel is evaluated as a series of independent columns (struts) under compression, which is always a

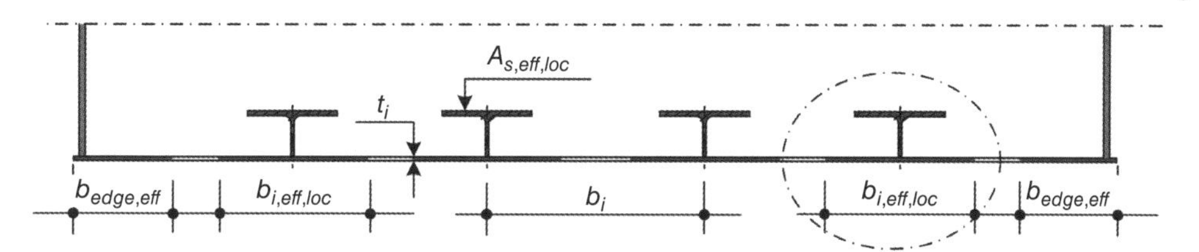

Figure 4.80 Bottom flange of a box girder deck: a stiffened plate under compression. Individual stiffeners and associated effective plating $b_{i,eff\,loc}$ (*Source:* Adapted from Ref. [20]).

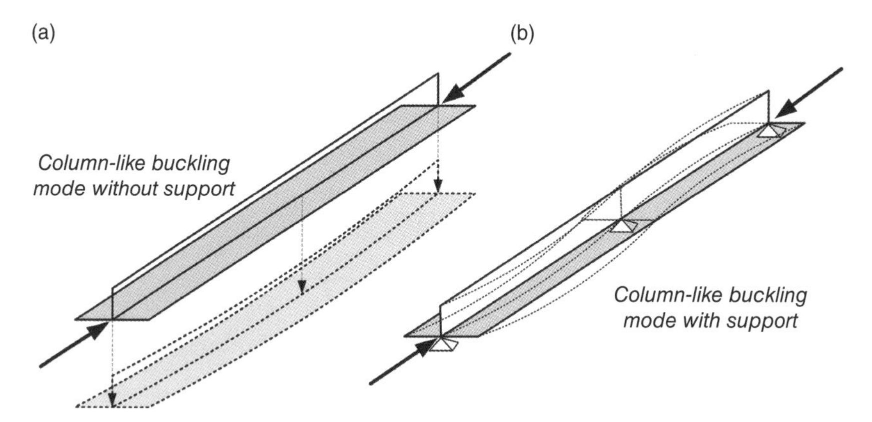

Figure 4.81 Strut approach to evaluate the ultimate compressive load of a stiffened plate: (a) without support at the transverse stiffeners and (b) with multiple supports at transverse stiffeners (*Source:* Adapted from Ref. [20]).

conservative assumption for pre-design. The ultimate load is evaluated from the buckling loads of the individual columns composed by a stiffener and its associated effective plate widths to each side of the stiffener, as shown in Figure 4.80. The struts are assumed to be simply supported at the diaphragms (Figure 4.81) because the bottom flange is considered continuous over multiple supports.

4.5.5.4 Pre-Design of Diaphragms or Cross Frames

Box girders need to have diaphragms or cross frames of two types: (i) at support sections and (ii) at intermediate sections.

The aim of diaphragms, made by solid plates with manholes for inspection and maintenance, cross frames or truss systems (Figure 4.82), is to reduce distortional effects leading to increased normal stresses. Slightly different schemes, such as the one shown in Figure 4.83, may be adopted. Distortional effects are associated with the one of the components of torsional loading as shown in Figure 4.84. Distortional effects lead to *folded plate action* in box sections [6, 24, 28] inducing axial normal stresses that should be added to the primary bending normal stresses. To avoid this effect, the maximum distance between intermediate diaphragms, generally between 5 and 8 m, should not exceed four times the web girder depth ($l_d \leq 4d$).

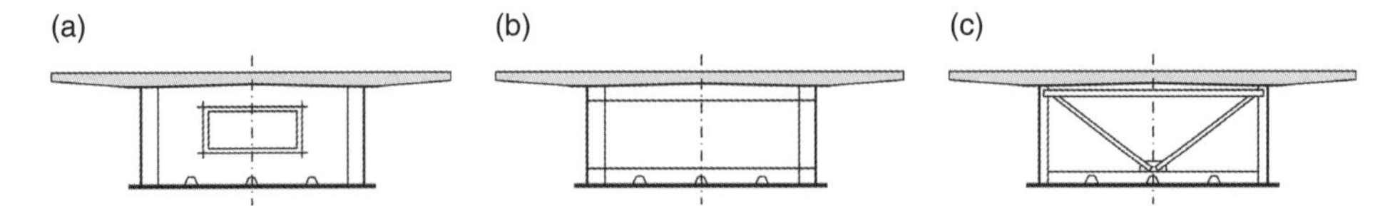

Figure 4.82 Intermediate diaphragms in box girders made of (a) solid plate with manholes and (b) cross frame and (c) truss system.

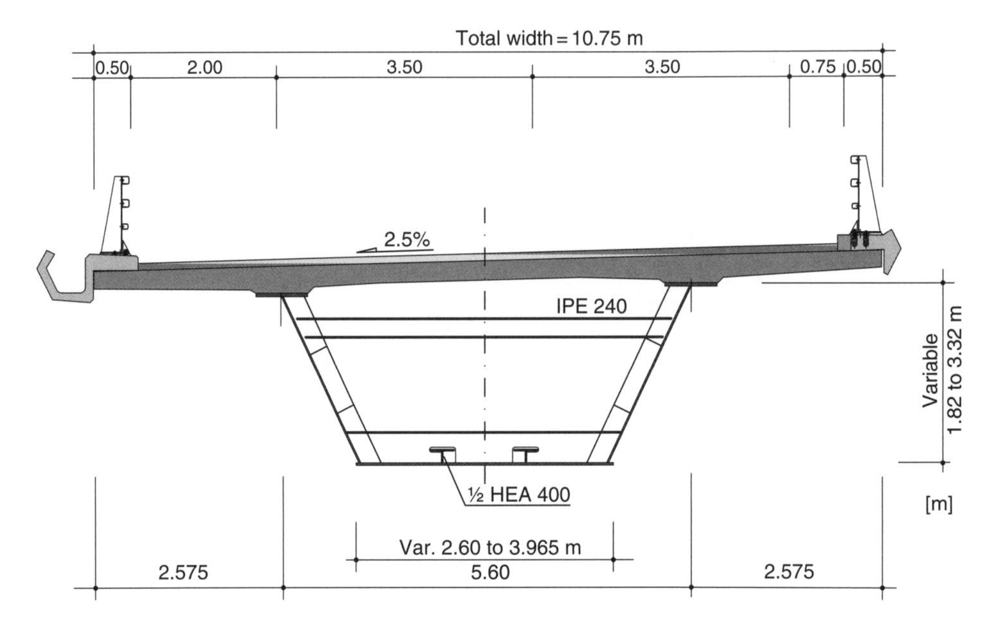

Figure 4.83 Intermediate cross frames: deck cross section of a composite box girder bridge. Bridge over Ante Valley, France (*Source:* Adapted from Ref. [27]).

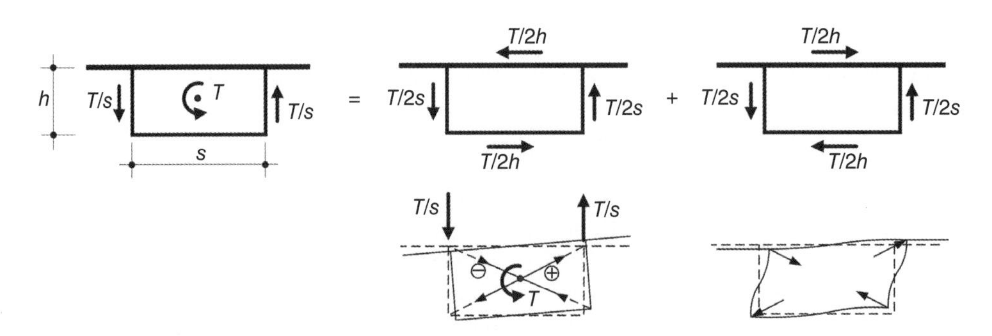

Figure 4.84 Torsion of a box section: torsion and distortional components.

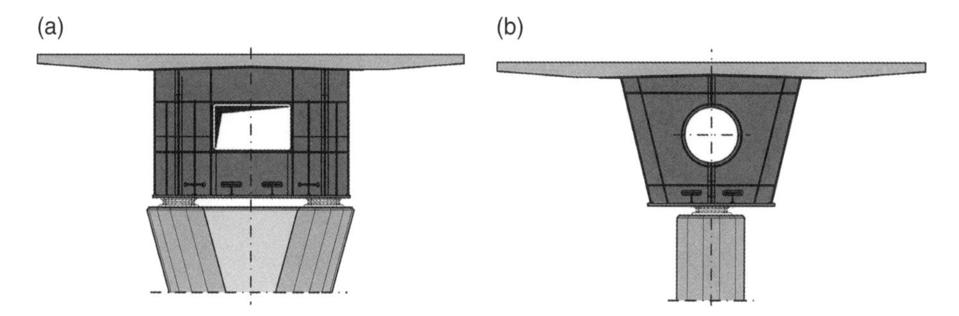

Figure 4.85 Diaphragms at support sections (a) with double bearings and (b) with a single bearing.

Besides, the stiffness of the diaphragms shall be sufficient to reduce distortional effects. During erection of 'open box sections' (U shape) of composite girder decks, a horizontal top bracing is necessary to 'close' the section (Figure 4.36) because otherwise there is no sufficient resistance to torsion under wind loading and eccentric construction loading.

Diaphragms at support sections (piers or abutments) are generally made of solid plates (Figure 4.85) since, apart from avoiding distortional effects, they have to transfer vertical and horizontal bearing reactions to the webs of the box girder. In case of Figure 4.85a, a double bearing exists at pier sections and torsion moment reactions are taken by a couple of vertical forces in the bearings. In Figure 4.85b there is no resistance to torsion moments at support pier sections. The plate diaphragms, generally 18–22 mm thick, have vertical and horizontal transverse stiffeners; the slenderness of each subpanel should avoid plate-buckling effects as a basic criterion. For maintenance requirements, these diaphragms are designed to accommodate, at the bottom flange of the box girder, concentrated forces due to hydraulic jacks required to replace bridge bearings. That requires clearances and specific stiffeners in the diaphragm.

4.5.6 Typical Steel Quantities

Empirical formulas to estimate at pre-design steelwork quantities have been proposed [26, 29] based on statistical analysis of many design cases (Figure 4.86):

$$\text{Plate girder decks} \qquad \text{Steel quantity}\left(\text{kg/m}^2\right) = 0.105x^{1.6} + 100$$

$$\text{Box girder decks} \qquad \text{Steel quantity}\left(\text{kg/m}^2\right) = 2.85x + 45$$

In these expressions, x (m) is the internal span l of a continuous deck or $x = 1.4l$ for simply supported decks or two-span continuous decks. Since the steel quantity is referred to as a unit area (m^2) of the bridge deck, this is defined here as the area of the whole traffic platform, including shoulders, plus 60% of the walkway surface.

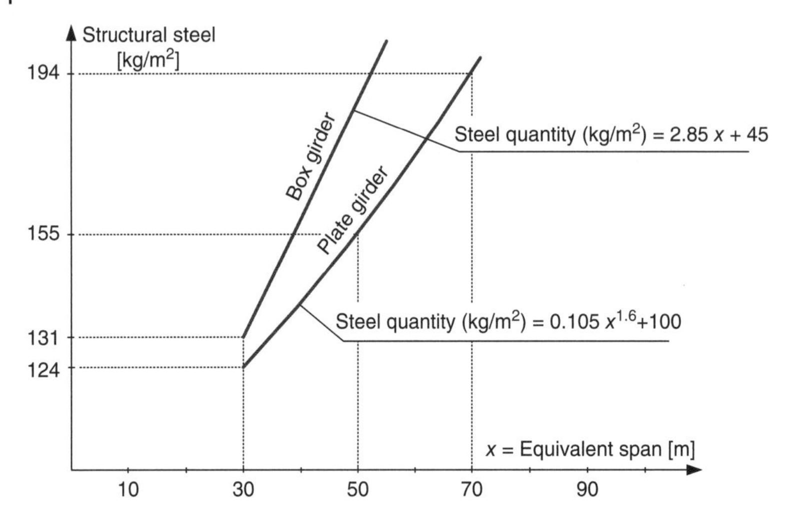

Figure 4.86 Typical steel quantities for composite plate girder and box girder bridges.

4.6 Superstructure: Execution Methods

4.6.1 General Aspects

Both concrete and steel bridge decks, discussed in the last sections, may be executed by a variety of construction methods. The construction method cannot be selected independently of the concept design of the bridge deck, as already shown. The construction method is part of design and influences the geometry of the deck, namely the layout of cross sections.

Concrete bridge decks may be cast in place or precasted. The erection of precasted longitudinal elements (precasted girders) has similar issues to erection of steel bridge girders. Precasted or steel girders may be erected from the ground with cranes, both of them can also be erected with *launching girders*. The basic concepts of some execution methods, as well as their designations, are applicable to both concrete and steel or steel-concrete composite bridges, such as erection by incremental launching or by the cantilever method. There are, however, significant differences in weight of elements to be erected, steel elements are lightest allowing reduction in capacities for the erection devices or increased span lengths for a certain capacity.

To select the erection method for both concrete and steel bridges, the following aspects should be taken into consideration:

- The height of the deck with respect to the ground
- The in-plane alignment of the bridge axis
- The span lengths and the total bridge length
- The shape of deck cross sections

Although the span length is not the unique parameter, it is one of the most important ones. A synthesis of application of construction methods is presented in Table 4.5 as a function of typical span lengths for concrete and steel and composite bridges.

Table 4.5 Main execution methods for concrete and steel bridges – influence of the span length.

Execution method	Concrete bridges cast in place	Concrete bridges Precasted girders or segmental construction	Steel or composite bridges
Scaffolding supported from the ground	30–60 m	–	–
Erection with cranes supported from the ground	–	30–50 m	30–80 m
Formwork launching girders	30–60 m	–	–
Launching girders	–	30–45 m – girders 40–80 m – segments	(rare)
Incremental launching	–	30–60 m	30–70 m[1]
Cantilever construction	60–250 m	60–150 m[2]	70–300 m

Notes
[1] Spans well above 70 m, for steel and composite girders, may be erected by incremental launching if temporary piers or stays are adopted;
[2] Cantilever construction of concrete bridges with precasted segments.

4.6.2 Execution Methods for Concrete Decks

4.6.2.1 General

Concrete decks are executed by one of the following methods:

- Scaffolding or stationary formwork
- Formwork launching girders
- Incremental launching
- Cantilever construction

As already mentioned in the previous section, concrete decks may be cast in place or precasted. In this latter case, one adopts precast girders (Section 4.4.6) or a precast segmental scheme, where entire cross section deck elements, 2–3 m long in general, are assembled in place as will be discussed in 4.6.2.6.

The range of span lengths for the main execution methods for concrete decks is represented in Figure 4.87, showing optimum typical spans and exceptional span lengths for each method. The reader is referred to specific literature [4, 30, 31] for a general discussion on execution methods for concrete bridges.

4.6.2.2 Scaffoldings and Falseworks

For cast in place decks, one has two options – *classical scaffolding* or *stationary falsework*. The first is a 'continuous' falsework (Figure 4.88) made with multiple small posts, adequately braced and supported from the ground. The stationary falsework is a 'discrete' scheme (Figure 4.89) where temporary piers are adopted at certain sections to support temporary girders, generally trusses. Classical scaffolding or stationary falsework are generally adopted for small or medium spans between 30 and 50 m and usually for shorter bridges. Classical scaffolding is not convenient when the height of the deck to the ground is greater than approximately 20 m. The same happens when the bridge is located in a valley or when the ground has relevant slopes making foundations of the scaffolding difficult. Standardized steel posts or towers (Figure 4.90) are adopted in

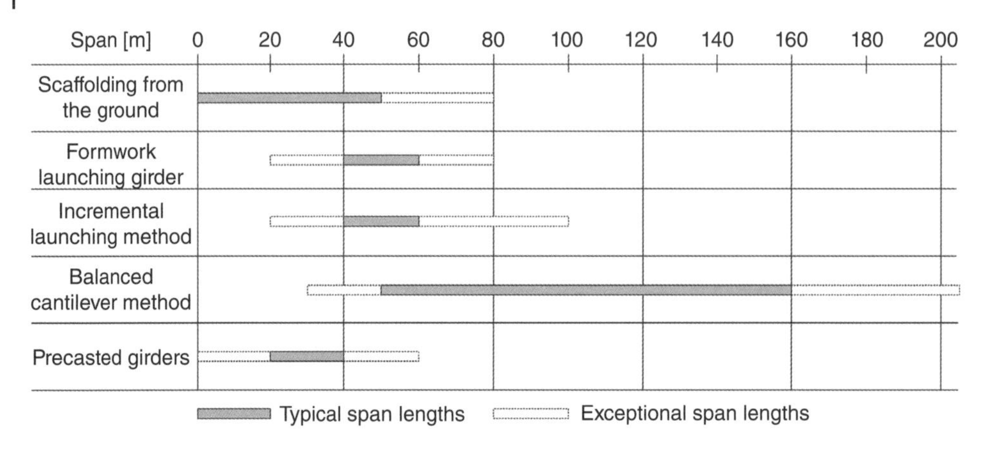

Figure 4.87 Execution methods for concrete bridge decks: typical and exceptional span lengths.

Figure 4.88 Classical scaffolding for the Quintanilha Bridge approach spans (*Source:* Courtesy GRID, SA).

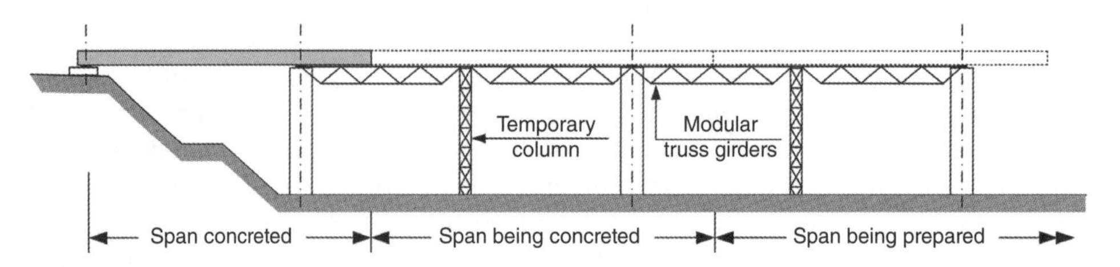

Figure 4.89 Stationary falsework.

Figure 4.90 Columns and modular truss girders for stationary falsework made of modular elements (*Source:* Photograph by José O. Pedro).

both schemes, with hot rolled profiles as supporting beams to the formwork panels. The posts and towers, as well as truss girders adopted in these systems, are very often made of modular elements reutilized in a sequence of construction stages of the deck.

The main aspects for safety of scaffoldings or stationary falsework are adequate bracing and adequate foundations. Settlements of foundations should be considered. The foundation of the posts on the ground is generally through wood or steel plates. Accidents during construction associated with inadequate stability of scaffoldings have generally occurred during casting stages of the deck.

For classical scaffolding, a main issue is to understand the concept for the bracing system. The posts are small diameter tubes, with approximately 50 mm, braced between them to resist buckling effects. The connections between bracing elements and posts are through standardized node devices. The scaffolding, as a global structure, should have an overall bracing system to withstand lateral loads due to wind, construction loads and instability effects of the vertical members. When the height of the deck to the ground exceeds approximately 20 m, the stationary falsework is easier to control than a "continuous" falsework, with respect to overall stability. Tall towers, composed of multiple standardized elements braced between them, support main truss girders at its top. These girders, composed by modular elements connected by special devices, support the secondary elements of the formwork. The truss girders may generally reach spans of approximately 15–20 m, exceptionally spans up to 30 m.

Classical scaffolding or stationary falsework, except for very small bridges, are not economic for adoption without a sequence of construction phases allowing the reutilization of the construction equipment. In Figure 4.91, one represents the most adopted scheme for these bridges: a span by span construction with construction joints approximately one-fifth of the internal span lengths. At theses sections, the bending moments for the final static scheme (a continuous girder) are very small for the permanent loads. The prestressing stages are also span-by-span allowing the scaffolding or the falsework to be moved to the next span at each stage. A set of prestressing cables at each stage are tensioned. The remaining cables of the span may be stressed at the next construction joint. At each construction joint, it is necessary to adopt continuity anchorages (couplers) for the cables stressed at that section.

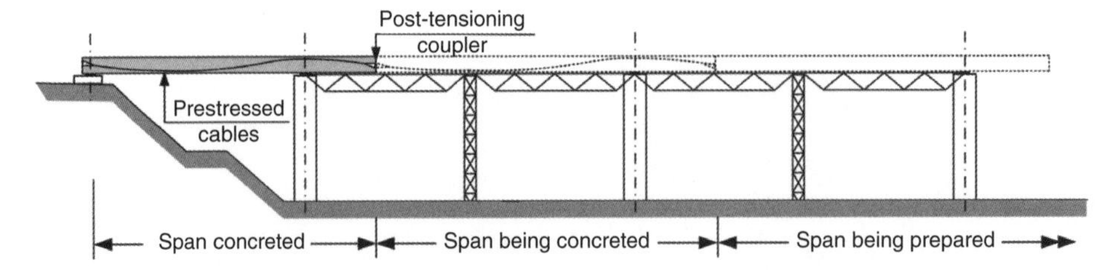

Figure 4.91 Construction phases- concreting and post-tensioning for a cast in place bridge deck.

4.6.2.3 Formwork Launching Girders

Formwork launching girders are construction equipments consisting of a moving formwork supported by launching girders usually made by plate, box or truss girders. These girders may work from below (Figure 4.92) or from above (Figure 4.93) with respect to the deck level. In the latter case, they may designated by *launching gantries.*

Launching girders are adequate for long bridges (viaducts) with straight or curved in-plane geometries and with multiple spans between 30 and 60 m. The deck cross section may be a ribbed slab, a slab girder or a box girder. The launching girders generally have a length corresponding to two spans of the deck or have a support at the part of the deck already built, as shown in Figure 4.92. They are equipped with one or two launching noses (one at the front and one at the back) to reach the next pier during launching operations (Figure 4.92). They are equipped with hydraulic devices to move the equipment forward and to install and re-install the formwork as exemplified in Figure 4.92 and 4.93. The most convenient sections to be built with formwork launching girders are slab twin girder decks, with span lengths up to 45 m, allowing building of one span per week. For box girders, with spans between 45 and 60 m, the progress of construction is generally lower and approximately equal to one span every 10–12 days. Formwork launching girders working from below are usually more costly than formwork launching girders working from above, because the available space for the structural depth for launching girders working from above may be larger allowing a lightest girder. Truss girders may be adopted in this case. The steelwork required for launching girders is controlled by maximum deformability requirements. A maximum deflection of the launching girders in the order of 1/600 of the span length is usually required in design specifications. The main inconvenience of launching girders working from above is the constraints resulting from the hangers suspending the formwork (Figure 4.93) not allowing prefabrication of the reinforcing cages. Moving forward operations are usually more difficult than with formwork launching girders working from below. But, supporting the launching girders in the pier shafts (Figure 4.93) is usually more difficult for formwork launching girders working from below. Launching girders may also be adopted for *precasted segmental construction* (section 4.6.2.6) as shown in Figure 4.94.

The cost of formwork launching girders is an investment that cannot be recovered without reutilization of this equipment. The adaptation to a new bridge cross section is usually possible but the span lengths cannot be much larger than that adopted for the original design of the launching girder.

4.6.2.4 Incremental Launching

Incremental launching is one of the most recent methods for the execution of concrete decks. It was initially developed last century in the 1960s [30, 32] and has been one of the most adopted methods in Europe in last decades.

(a)

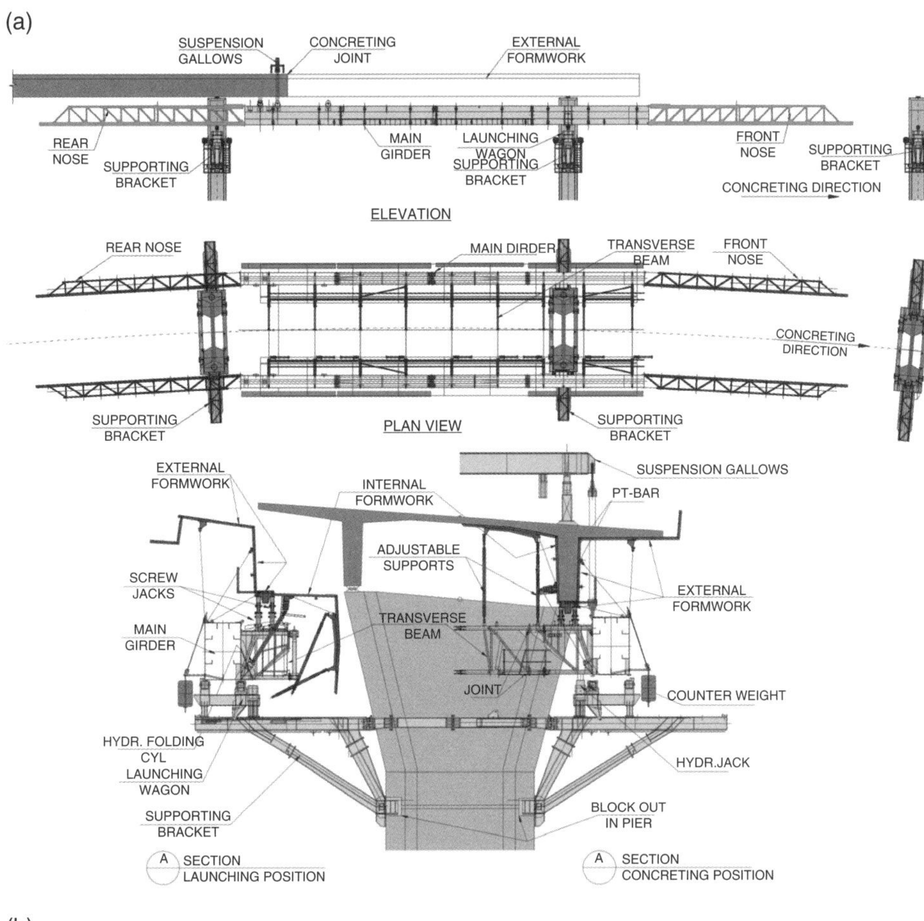

(b)

Figure 4.92 (a) Formwork launching girder working from below – Elevation view and cross sections of the working stages (Ref. Strukturas Catalogue – Movable Scaffolding System) and (b) The Sorraia Viaduct, A13 Highway, Portugal.

The method consists of producing deck segments, usually between one-third and half of the bridge span length, to the back of an abutment. After a certain bridge deck segment is completed, this is pushed or pulled forwards by hydraulic devices as shown in Figure 4.95. The reaction forces to move the deck forwards are induced at the abutment.

(a)

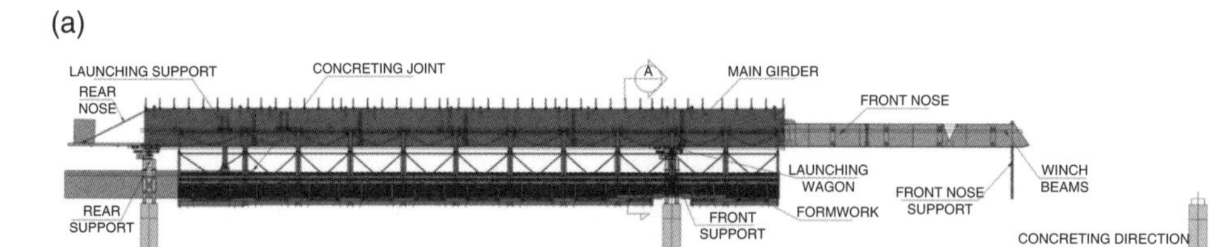

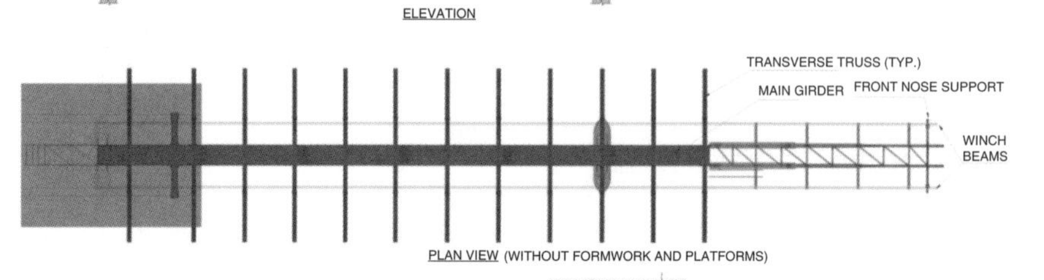

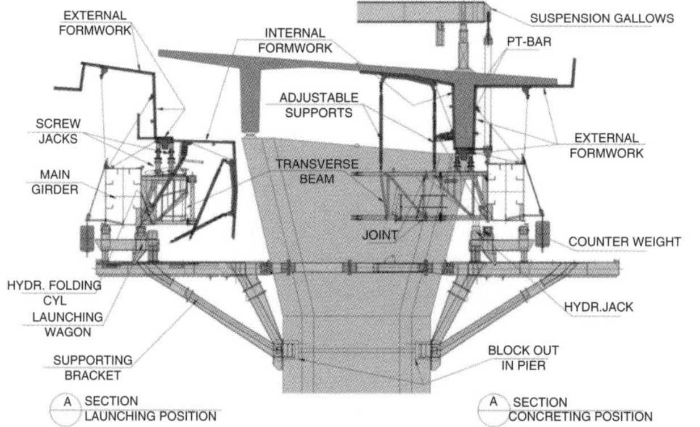

(b)

Figure 4.93 (a) Formwork launching girder working from above – Cross sections of the working stages (Ref. Strukturas Catalogue – Movable Scaffolding System); (b) The A15 Highway viaducts, Portugal.

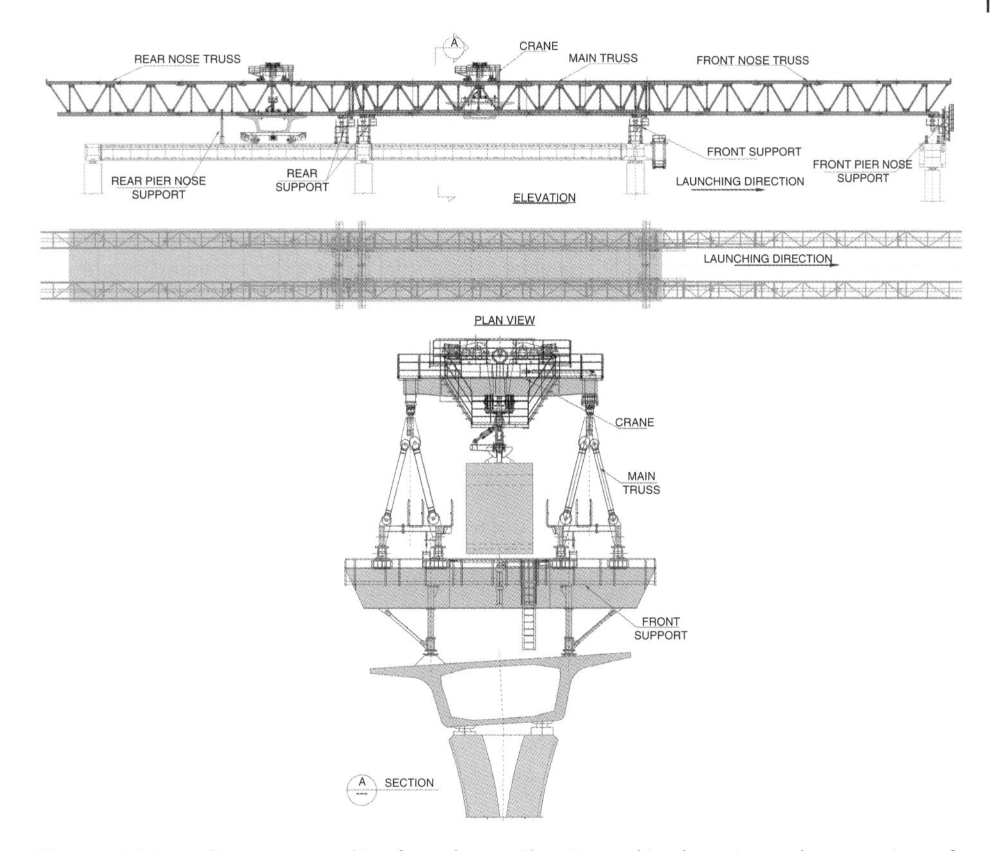

Figure 4.94 Launching gantry working from above – Elevation and in-plane view and cross sections of the working stages (Ref. Strukturas Catalogue – Launching Gantries for Segmental Bridges).

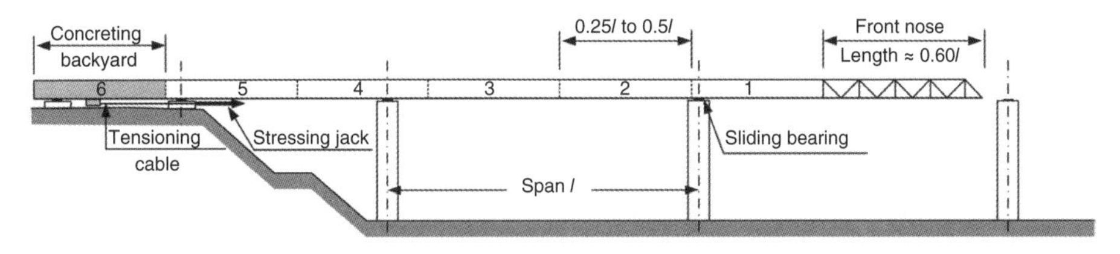

Figure 4.95 Incremental launching method for a concrete deck.

The method is usually applicable for long bridges with an in-plane straight or constant curvature alignment (circular – R = const.) and with medium span lengths between 30 and 60 m. To reach one pier when pushing the deck forward at the rear section, a constant radius of the in-plane alignment is necessary. Otherwise, a longitudinal movement induces a transverse deviation from the in-plane alignment at the front section and the new pier support cannot be reached. However, the in-plane alignment may be composed, for example, by straight and a circular alignments, but the deck should be pushed or pulled from both sides by executing one platform at the back of each one of the abutments. Intermediate launching platforms can be adopted to overcome the problem of in-plane alignments made of more than two constant curvature bridge segments.

A front nose made of a plate or truss steel structure is necessary to reduce the bending moments at the front pier section as shown in Figure 4.95. The length of the nose is usually in the order of 50–60% of the span length. The same segment passes during the movement over support sections (negative bending moments) and at mid-span locations (positive bending moments). The evolution of the static scheme during launching phases induces an envelope of bending moments as shown in Figure 4.96. The critical section for negative bending moments is of course at the last pier section before the nose landing at the pier. The nose may be equipped with a curved 'landing device' to facilitate support at the front piers.

In order to avoid tensile stresses in the concrete sections, the prestressing for the execution phase is likely to be centred and the cross section should have similar lower and upper section modulus. The ideal cross section deck is a single cell box girder. Span lengths of 30–60 m are the most adopted ones although longer spans are possible with intermediate piers or with a temporary staying scheme (Figure 4.97).

Temporary prestressing is required for the construction phases that may be done by external cables, some of them dismantled at the end before additional cables are added to support positive and negative deck bending moments at final stage.

Long bridges (more than 200 m length), or bridges with main spans crossing a river where no support from the ground is possible, may be executed by incremental launching at a construction progress of at least one segment per week. For segment lengths of half the span, one has at least one span made every two weeks.

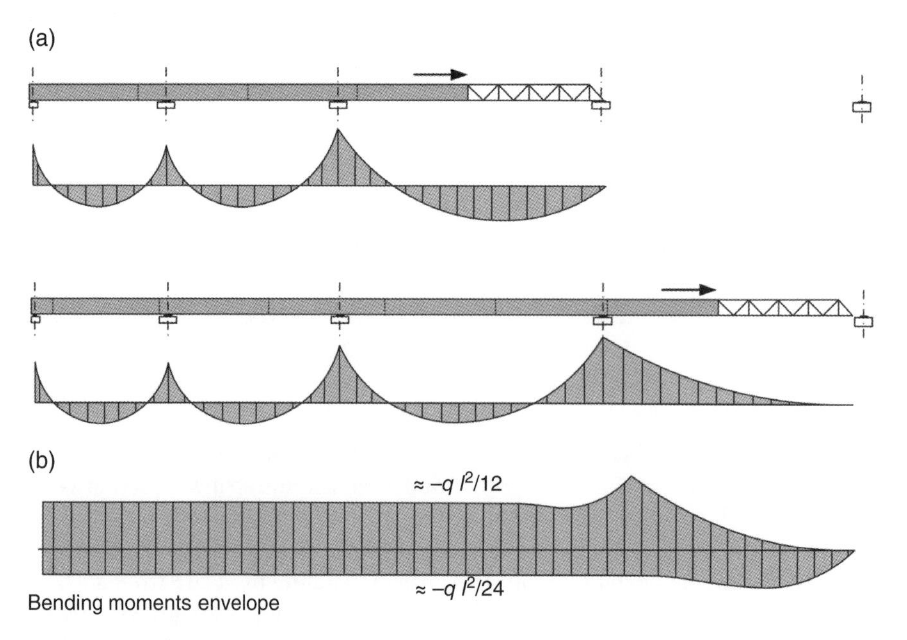

Figure 4.96 (a) Evolution of the static scheme and (b) bending moments envelope during deck incremental launching.

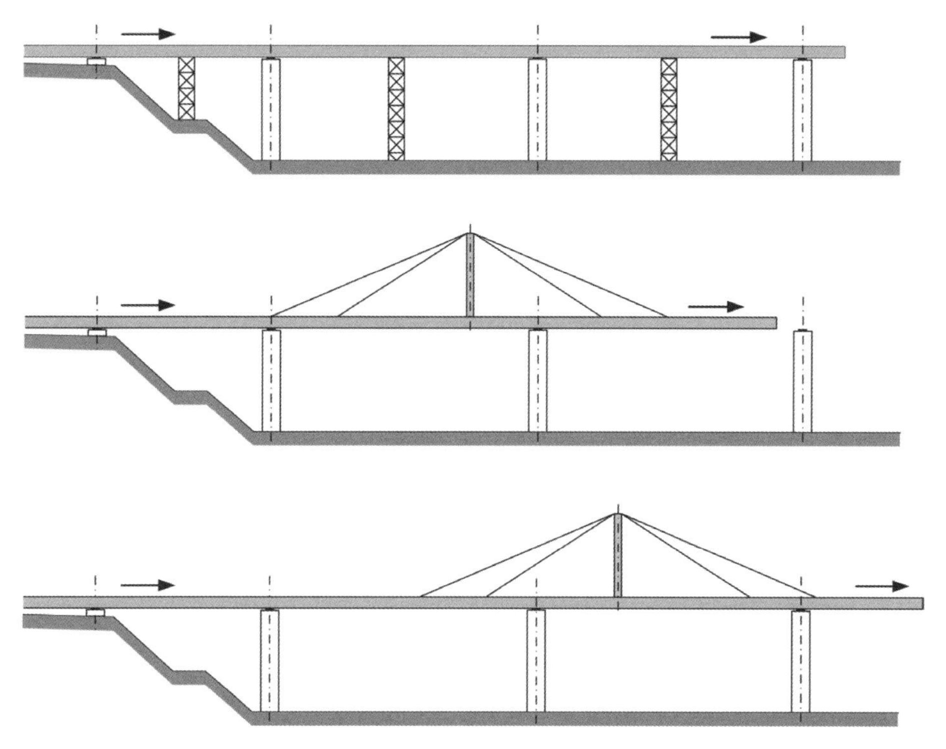

Figure 4.97 Temporary piers or staying device, reducing bending moments at front cross section during launching of the deck.

For pre-design, one may take the following guidelines:

- spans – 20–70 m, usual 35–55 m
- slenderness of box girder cross sections are l/h = 14–17 for road bridges and l/h = 12–15 for railway decks
- steel reinforcement quantity – 130 to 150 kg m^{-3}
- prestressing quantities – 40–70 kg m^{-3}

The length of the launching plant is at least one or two span lengths. The segments are cast in a formwork supported from the launching platform (Figure 4.98). Hydraulic jacks lower the formwork before the deck is moved forwards. Sliding supports under the webs are adopted as shown in Figure 4.98. The capacity of the launching equipment to move the deck is calculated on the basis of a friction coefficient (μ) of at least 5% and if the longitudinal profile has a positive slope i (%), which has to be taken into consideration when evaluating the required capacity of the hydraulic jacks, that is, $\mu > 5\% + i$. When the slope is negative, retaining cables have to be added to guarantee stability during launching operations. The most common jacking system to push the deck forwards is as shown in the scheme of Figure 4.99. Two jacks are adopted – a lifting jack reacting by friction at the deck underneath and a horizontal jack reacting against the abutment to induce the movement. The total dead load of the deck to be launched may reach some thousand tonnes (65 000 tons for the ship-canal Bridge in Belgium [4]).

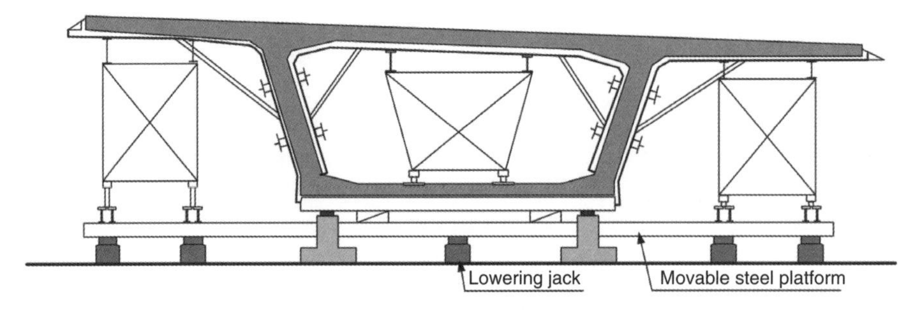

Figure 4.98 Formwork and falsework at launching platform.

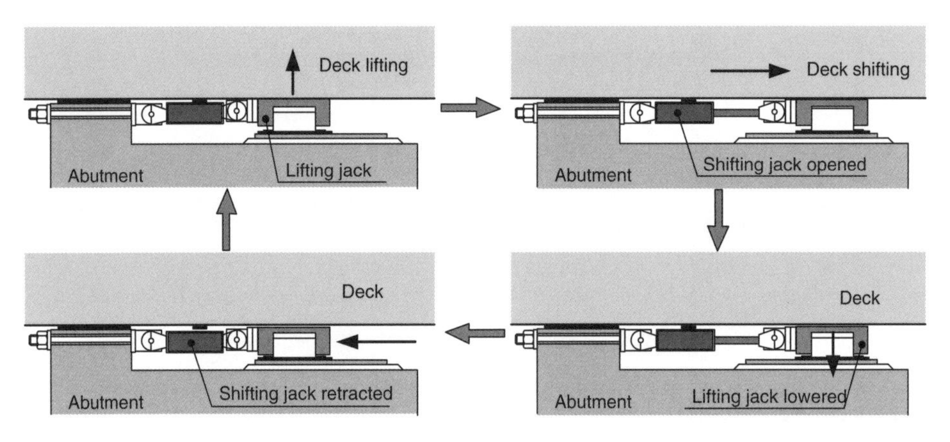

Figure 4.99 Example of a hydraulic jack system for moving a concrete deck by incremental launching (Ref. Eberspächer launching system – site http://www.eberspaecher-hydraulik.de).

Temporary sliding guided-bearings are adopted at each pier for launching operations and made of PTFE plates with neoprene pads to accommodate rotations.

4.6.2.5 Cantilever Construction

4.6.2.5.1 Basic Aspects

Historically, the cantilever construction method, developed initially for wood bridges, has taken its place in bridge engineering with cantilever construction of steel bridges. With the development of RC bridges, the method has attracted attention from bridge builders after the Herval Bridge over River Peixe in Brazil was built in 1930 by a construction technique similar to that adopted nowadays. Although Freyssinet had previously (1945–1950) adopted the cantilever method to some extent, it was only with Finsterwalder in Germany (1950–1951) that the cantilever method for PC bridges was similar to modern technology.

Cantilever construction is a segmental construction method where small segments, generally in the order of approximately 2.5–5.0 m length, are built and assembled from the end section of an already-built deck cantilever. The segments may be cast in place or precasted. In the former case, a moving scaffolding and formwork (Figure 4.100a), attached to the last segment already built is needed. In the last case (Figure 4.100b), precasted segmental cantilever construction, the segments are positioned at the end

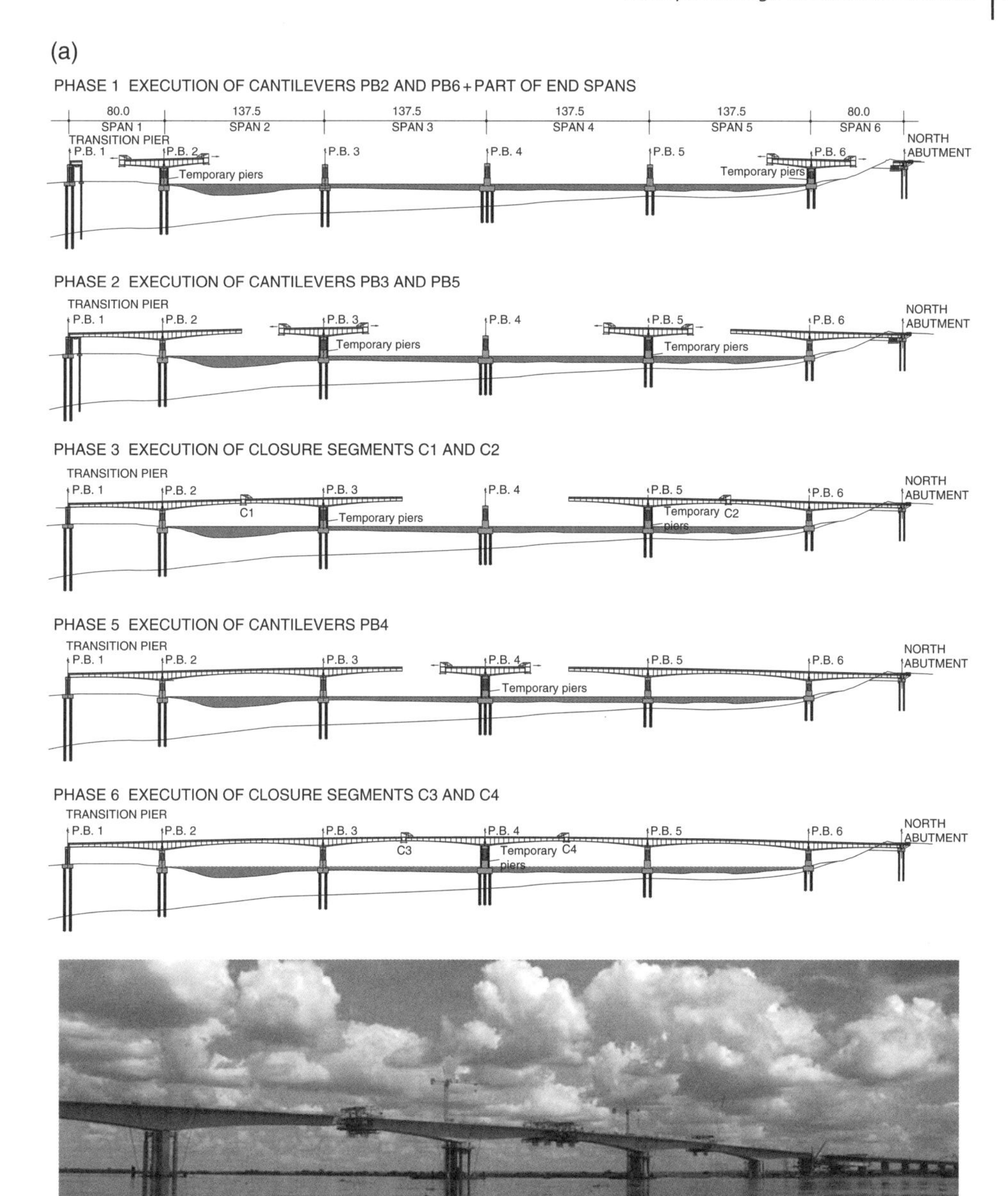

Figure 4.100 (a) Bridge built by in-situ concrete balanced cantilever system – The Zambezi River Bridge, Mozambique (137.5 m typical spans) (*Source:* Courtesy GRID, SA) or by (b) precasted segments – Oléron Bridge, in France (*Source:* Photograph by Mutichou/ https://commons.wikimedia.org).

(b)

Figure 4.100 (Continued)

section of the cantilever. For precasted segmental cantilever construction, the segments are positioned with a launching girder, lifted from a barge or from the ground with a derrick. The cantilever construction scheme may be executed:

- Symmetrically to each side of a pier – *balanced cantilever scheme*
- Asymmetrically, from a span already built
- Asymmetrically, from an abutment

Usually, at the end span, a mixed technique is adopted as shown in Figure 4.3. The cantilever construction scheme (Figures 4.3 and 4.4) is adopted from the end pier for part of the end span and the remaining area is made with scaffolding supported from the ground.

In the cantilever construction method, cast in place or precasted segments are prestressed at each phase after casting or assembling a new segment (as shown later in Figure 4.101).

The connection between the end sections of each one of the cantilevers coming from two piers of one span is made through a closing segment with a general length

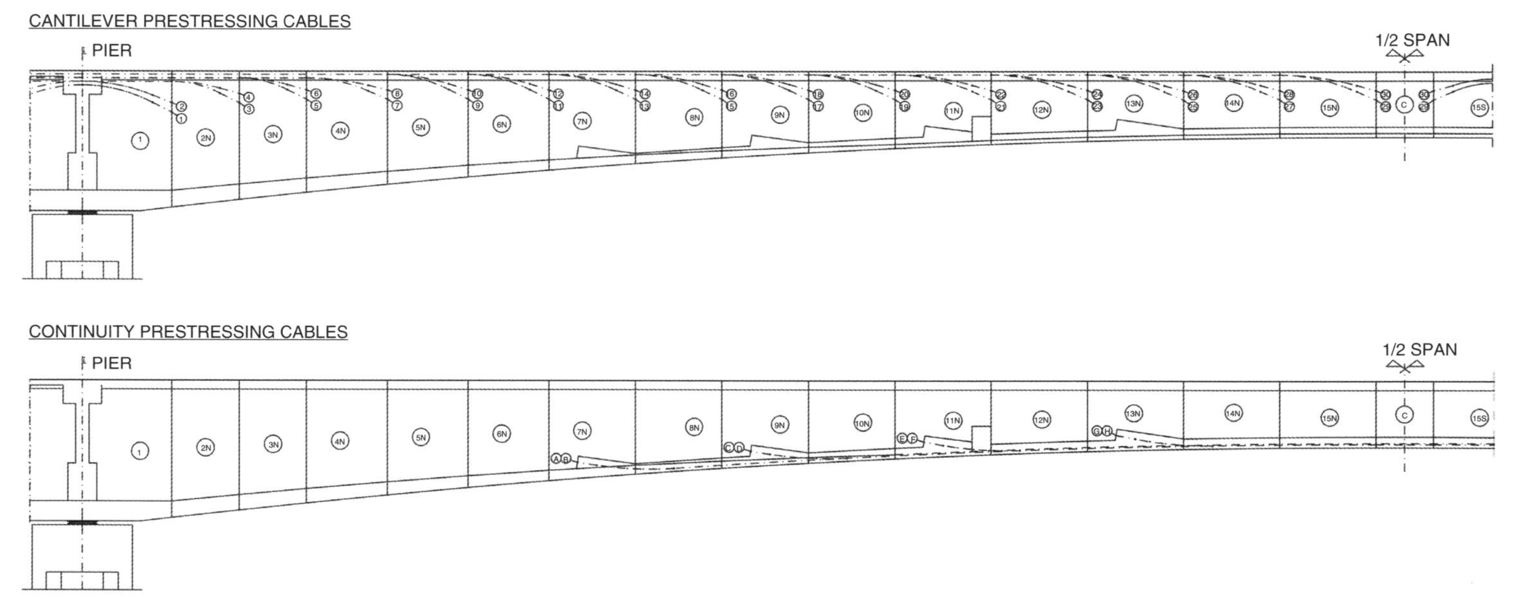

Figure 4.101 Cantilever and continuity prestressing cable layout.

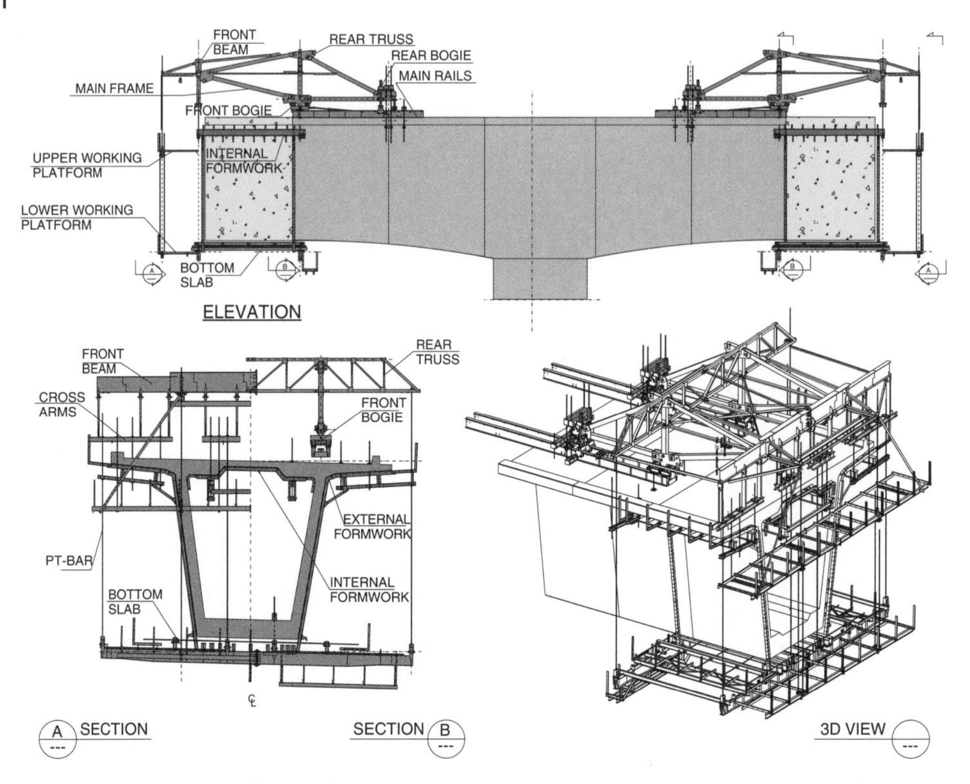

Figure 4.102 Moving scaffolding for cantilever construction with cast in place segments working from above (Ref. Strukturas – Form traveller for cast in situ bridge decks).

between 2 and 3 m. Continuity prestressing cables (Figure 4.101) are applied after casting of the closure segment to face the induced bending moments at mid-span sections, developing in the continuous structure as explained in Section 4.3. Part of this prestressing may be made by external cables.

The dead weight of each cantilever and the weight of the moving equipment (scaffolding and formwork) is supported in bending by the already completed part of the cantilever. Figures 4.102 and 4.103 show moving scaffoldings for cantilever construction. The moving scaffoldings are usually designed for capacities between 250 and 750 kN and standard maximum segment lengths of 5.0 m.

Large negative bending moments exist during the cantilever scheme. The most adequate section to resist these bending moments is, of course, a box girder section due to its large lower flange. The box girder is in most cases a single cell box girder with vertical or inclined webs. For very large width sections, a double cell box girder may be adopted, or a single box with inclined struts or with transverse ribs.

The depth of single cell box girder at pier sections for cantilever construction is between 1/20 and 1/17 of the span length. A maximum slenderness of 1/22 at the support sections may be adopted for aesthetics (Figure 5.14 – Alcacer do Sal Bridge at IP1 over the River Sado, Portugal) when the vertical alignment is at a low level.

In balanced cantilever schemes, the piers (Figure 4.104) resist the differences between the moments coming from each cantilever. These differences result from deviations in

Figure 4.103 Moving scaffolding for balanced cantilever construction in Quintanilha Bridge (multiple spans of 80 m), for maximum weight segments of 1700 kN with 5.0 m lengths (*Source:* Courtesy GRID, SA).

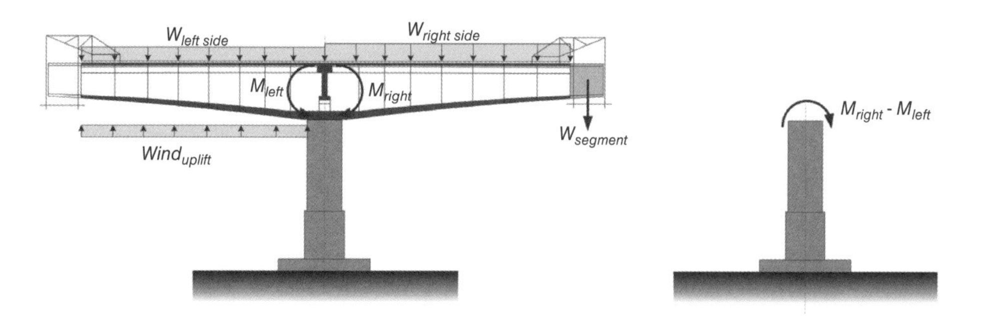

Figure 4.104 Bending moments to be resisted by piers in balanced cantilever schemes.

weight of the segments, and possible asymmetry in concreting the segments, construction loads and wind loading. The quantification of these loadings is addressed in Chapter 6.

The connection between the piers and the deck to resist these moments during construction may be monolithic in frame bridges, connected through anchorages by prestressing cables or supported by temporary piers (Figure 4.105) in the case of continuous girder decks supported on bearings.

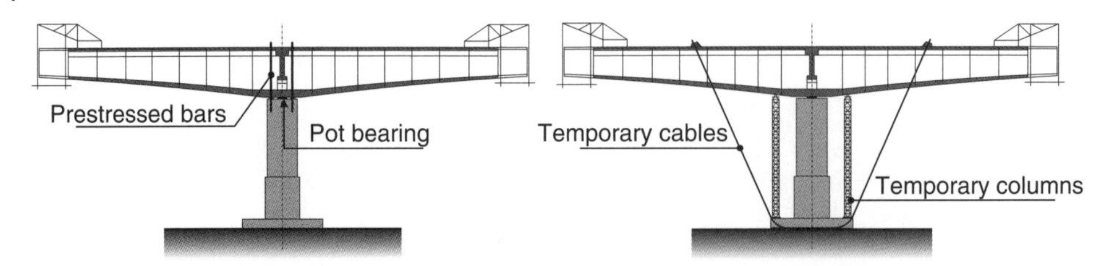

Figure 4.105 Possible schemes to resist unbalanced moments from the cantilevers.

4.6.2.5.2 Span Arrangement for Balanced Cantilever

In continuous multiple span decks, the easiest solution for span arrangement is to make, as far as possible, multiple internal span lengths (l) and two end spans with lengths in the order of $0.6l$ to $0.8l$. Some examples are shown in Figure 4.106a. In the Porto Novo Viaduct (1988), 80 m typical internal spans with a constant depth (4 m) and two side spans of 60 m with a reducing depth towards the abutments were adopted. For the internal spans, the balanced cantilever was executed from each pier and a sequence of closure segments from the side spans to the centre was executed. At the end piers, 38 m deck lengths to each side of the pier were executed and the remainder 20 m of the side spans (20 m) between the abutment and end cantilever were executed on a scaffolding from the ground. Then closure segments were executed in the first and last spans, followed by the closure segments at the second and fourth span and, finally, the centre closure segment. The evolution of the static scheme is similar to that already discussed in Section 4.3. In the bridge over the Zambezi River in Caia, Mozambique, with 137.5 m internal typical spans, the end spans are 80 m and a sequence of closure segments was executed from the side spans to the centre, see Figure 4.106b. Details on this long bridge (2376 km) and sequence of construction may be seen in [33]. When different internal spans are adopted as in the Rosso Bridge over the River Senegal (Figure 4.106c), with a main span of 120 m over the main channel, the balanced cantilever scheme can also be adopted but the span distribution should be established taking into consideration the sequence of closure segments. In a sequence of internal spans, l_1, l_2, l_3,..., for a general symmetrical arrangement as shown in Figure 4.106c, a balanced cantilever scheme from piers P1, P2 and P3 requires, for example, $l_2 = (l_1 + l_3)/2$.

4.6.2.5.3 Asymmetric Cantilever Construction

Site data, like topography, geotechnical conditions or a main span over a river, may justify adopting an asymmetric (unbalanced) cantilever scheme. An example of a design case [34] of two parallel bridge decks is shown in Figure 4.107 The topography and geotechnical conditions of the valley have shown the best option to be a deck for each of the bridges built from a single pier in a balanced cantilever scheme and the remaining part of the deck from the tunnel as a single cantilever in an unbalanced scheme. The closing segment was then executed at the centre of the main span of 104 m. Site conditions showed the location of the single pier for each of the decks to be skew symmetric, as shown in Figure 4.107. The bridge is curved in-plan and the pier is made of two thin RC walls. The deck is made monolithic with the pier, which allows the imposed deformations (thermal, shrinkage and creep) by flexibility of the two shafts of the deck, rigidly

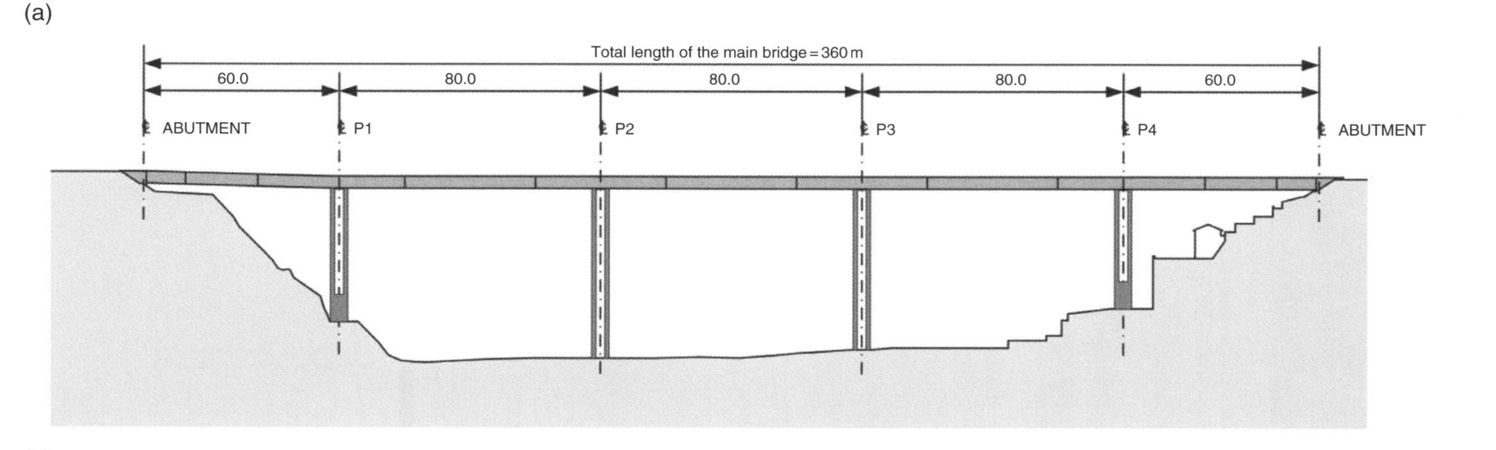

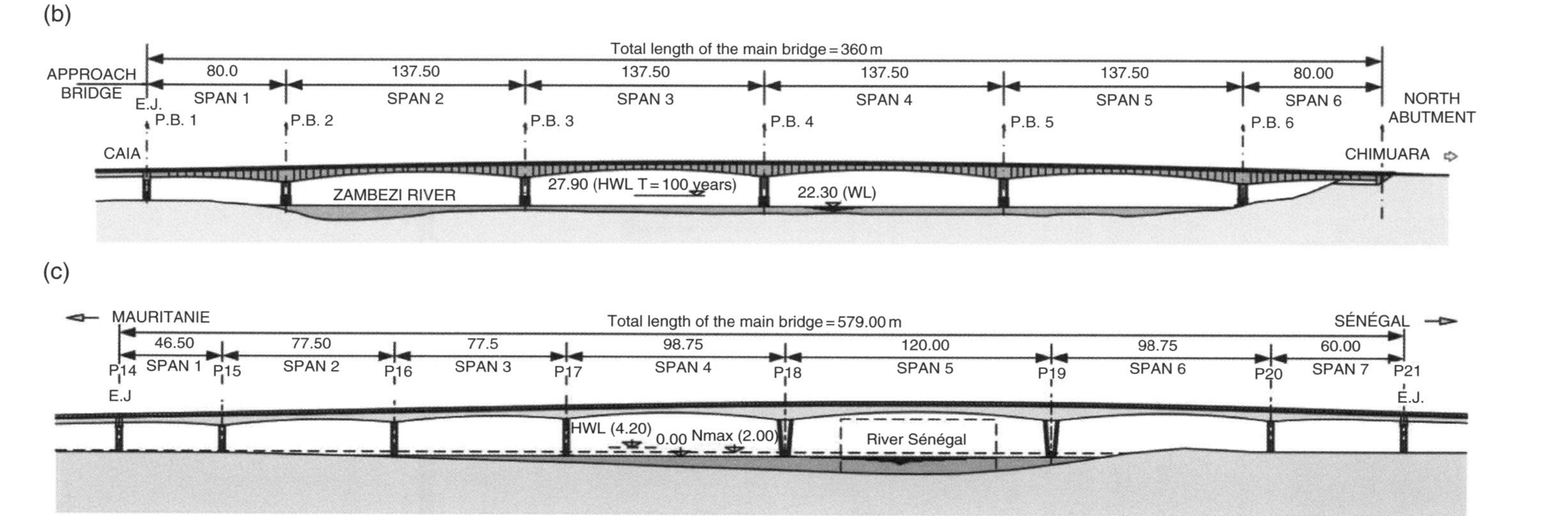

Figure 4.106 Span arrangement for balanced cantilever construction with equal internal spans (a) Porto Novo Viaduct in Madeira Island with 80 m spans, (b) Bridge over Zambezi River in Caia Mozambique with 137.5 m spans) and with different spans internal spans and (c) Rosso Bridge over river Senegal (tender design), with a main span of 120 m for the main channel).

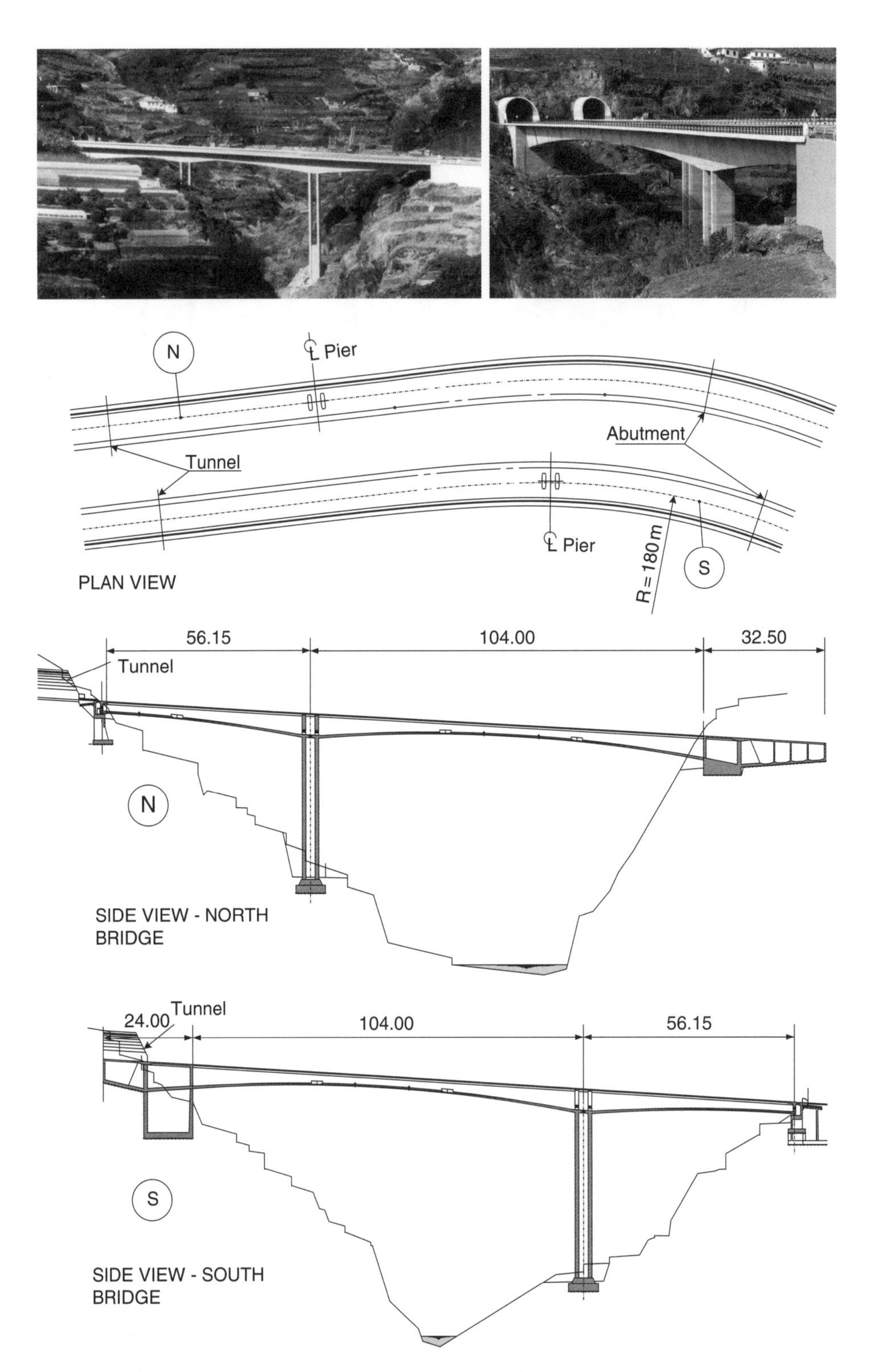

Figure 4.107 An asymmetric cantilever construction. The Bridge over Vigario Creek in Madeira Island (*Source:* Courtesy GRID, SA).

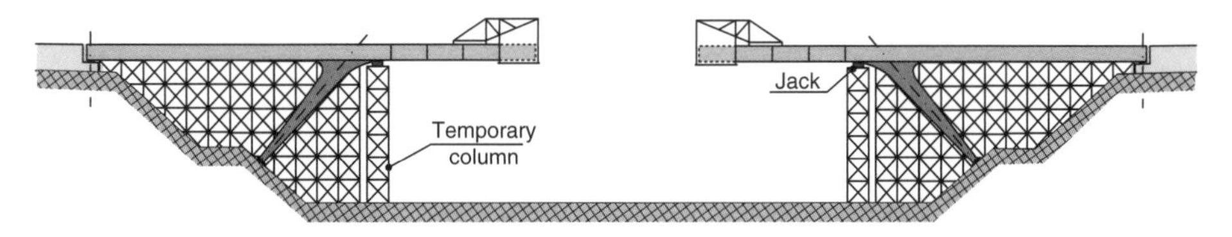

Figure 4.108 Cantilever construction scheme for a frame bridge with inclined piers.

connected at the tunnel and having unidirectional bearings at the abutment. The bridge, with two spans for each of the decks, from the front view looks like a classical three-span bridge. The reader is referred to Ref. [34] for details.

4.6.2.5.4 Frame Bridges with Inclined Piers and Arches

For frame bridges with inclined piers, usually hydraulic jacks are inserted between the deck and the temporary piers (Figure 4.108) for removing the temporary supports after closing the deck at the mid-span section. The cantilever method may be also adopted for the construction of arch bridges. The reader is referred to [35] for developments on the cantilever construction method.

4.6.2.6 Precasted Segmental Cantilever Construction

It has been already mentioned that precasted concrete segments can be adopted with launching girders and cantilever construction. This is a very convenient alternative for multiple span long bridges, because segments are being executed and at same time others are being assembled. The rate of construction may reach an average of 10 m of deck per day. If one compares it, for example, with a cast in situ cantilever scheme, with two pairs of movable false work where it is possible to build 20 m week^{-1} (2×2 segments of 5.0 m length), the speed of construction with precasted segments is at least twice the one achieved with a cast in-place cantilever scheme.

The length of segments is up to 3.5 m in general, limited by transport, lifting and assembling capacities of the equipment and the amount of prestressing required at each assembling stage. They are precasted on site or in a specific facility and transported to the bridge site. The most adopted method for precasting is 'match cast' segmental construction, where the segments are cast one against the other in order to achieve a coupled joint (Figure 4.109). The segments are then assembled on the alignment of the bridge deck, by cranes (if the height of the bridge deck allows), via a launching girder or the cantilever method.

An important issue in precasted segmental bridges concerns the joints between segments. The three possible options are cast in situ joints (requiring approximately 200 mm for the joint and a formwork), mortar joints allowing a reduction in the gap between segments to approximately 5 cm and, finally, coupled match cast joints with 'epoxy' glue. The last procedure is nowadays the most adopted one due to the quality usually achieved and the rate of construction allowed. In coupled match cast joints, the 'epoxy' allows lubrication of the contact surfaces of the segments facilitating the assembling and, most important, seals the joint allowing waterproofing, thus avoiding the water ingress into the ducts of the prestressing cables crossing the joint. The joints are designed to be permanently compressed even under live loads and

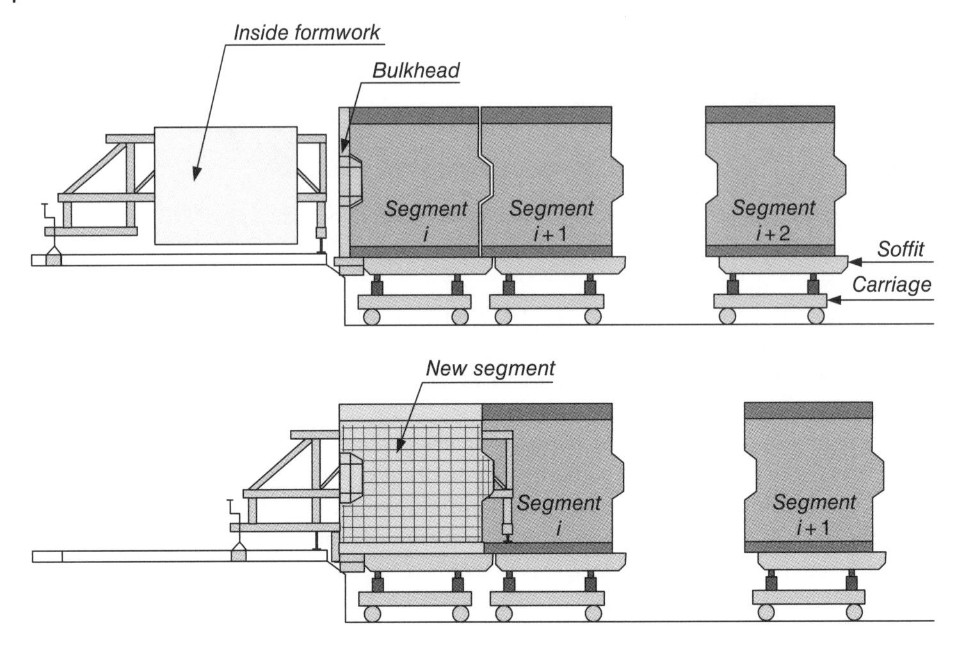

Figure 4.109 Precasted segmental construction with match cast joints.

thermal gradients. The shear resistance is achieved nowadays by a multiple 'shear key' detail, as shown in Figure 4.109, improved by the presence of the 'epoxy'. The reader is referred to Refs. [4, 26, 35, 36] for details and specifications on precasted segmental construction.

4.6.2.7 Other Methods

A brief reference is here made to other possible options for execution of concrete bridge decks. Three additional methods are:

- Deck erection by transverse displacement
- Deck erection by rotation
- Deck erection by heavy lifting equipment

One is the possibility of executing the bridge deck on a site parallel and adjacent to the in-plane bridge alignment and moving the deck laterally after completion, previously erected on sliding bearings, by hydraulic equipment. The transverse sliding method is adopted for small bridges and usually when replacing an existing deck where severe traffic disruption is a main constraint.

Rotation of the deck built on site is, perhaps, the most spectacular bridge erection method. The deck may be executed, for example, along a river bank on scaffolding supported from the ground and then rotated to the final bridge alignment. It is adopted for the same span range spans of the cantilever method since during rotation the static system is similar. Yet, this method has been already adopted for cable-stayed bridges like Ben-Ahin bridge, Belgium [4].

Heavy lifting erection methods have been developed in the last few decades for long bridges offshore or crossing long estuaries like the Great Belt Bridge in Denmark, Vasco da Gama Bridge in Portugal, and the Prince Edward Island Bridge in Canada. Precasted entire spans may be transported on barges to the site and lifted with special equipment: gantry cranes, reaching capacities of several thousand tonnes. In the Vasco da Gama Bridge, entire 80 m box girder spans were transported and lifted and for the Prince Edward Bridge, with multiple spans of 250 m, segments reaching 7800 tons were erected. The construction rate with heavy lifting is the largest, generally reaching 250 m per week for the Prince Edward Bridge [31].

4.6.3 Erection Methods for Steel and Composite Bridges

4.6.3.1 Erection Methods, Transport and Erection Joints

The most adopted methods for the erection of the superstructure of steel and composite bridges are:

- Erection from the ground with cranes
- Incremental launching of the steelwork
- Cantilever construction

The last method, cantilever construction, is adopted for long spans as for concrete bridges, namely for box girder bridges and cable-stayed bridges. Specific methods, such as the ones referred to for concrete bridges in Section 4.6.2.6, like transverse displacement and rotation, are also adopted in a more limited range of applications. The fundamentals of the methods are similar to the ones already mentioned for concrete bridges. One of the specific aspects of steel and composite bridges is transport to the site of the steelwork. After fabrication, a main issue is how the steelwork should be transported to the bridge site. This aspect has been already mentioned in the introductory section on steel bridges (Section 4.5) but is also related to the erection procedures. The most common transportation is by trucks, but without special permissions a segment limit in the order of 30 m length and 3 m width by 4 m height is imposed in most countries. The weight of the segment is usually limited to approximately 50 tons. Rail and ship or barge transport is a possible option in some cases, with segment lengths limited to approximately to 12 m for railway transport and container limitation for ship transport.

Plate girder steelwork decks are generally transported in independent girders, the number of segments being limited to a maximum total weight of approximately 40 tons for the truck. Box girders are transported in two halves (Figure 4.35) requiring a longitudinal site joint as previously referred to.

The segments after transport to the site may be assembled in longer segments by execution of full penetration welding joints at flanges and web sections that are subjected to Non Destructive Testing (NDT) on site. The quality welding control on site is currently done by ultrasonic or magnetoscopic tests and less frequently by RX tests. Of course, basic geometric quality control, visual inspection of the welding and dye penetrant tests are always done.

The segments are then erected and positioned at the deck alignment and bolted or welded at erection joints. The tendency nowadays is to prefer welding erection joints for the benefit of durability of the structure but, of course, this requires the appropriate means for site execution and inspection. The number of NTDs to be carried out on-site inspection depends on the Execution Class (EXE) of the structure according to EN 1090 [37]. For bridges, EXE 4 is required in most cases and for site welding:

- Transverse butt weld subjected to tensile stresses 100%
- Transverse fillet welds, in general 10%
- Longitudinal welds and welds to stiffeners 10%

In a span-by-span erection, it is possible to have erection joints at intermediate span sections, for example, at one-fifth span sections, and to execute complete penetration site welding on webs and flanges. For most cases, bolted temporary erection joints are needed, which could be done according to the scheme of Figure 4.110. After final adjustment of the new segment position the definitive site welding is executed.

4.6.3.2 Erection with Cranes Supported from the Ground

If the height of the deck with respect to the ground is relatively low (order of less than 15 m) and access to bridge site allows, it is possible to erect the steelwork with cranes from the ground. The crane capacity is dependent on the lever arm required and on the weight of the segments of the steelwork. A reference value of 100 tons at maximum distances of 10–20 m may be considered for standard crane erection of the steelwork. The deck may be erected by a span-by-span scheme with erection joints as discussed in Section 4.6.3.1.

For a main span on a river, it is possible to erect the central part of the steelwork by cranes on barges, or even with two cranes located on each river bank as in the example of Figure 4.111.

4.6.3.3 Incremental Launching

The incremental launching method for the erection of the steelwork is, concerning its basic aspects, constraints, advantages and disadvantages, quite similar to the incremental launching method for concrete bridge decks discussed in Section 4.6.2.4. The erection

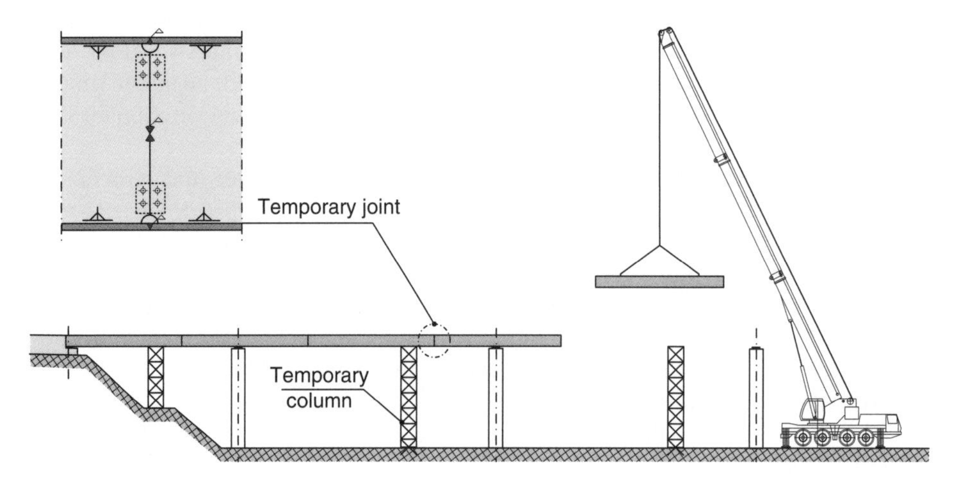

Figure 4.110 Erection of the steel work with cranes on the ground, temporary supports and erection joints.

Figure 4.111 Erection by cranes of the main segment of the central span of Lanaye bridge in Belgium with a main span of 72 m (Depth of the steel box girder 2.1 m) (*Source:* Courtesy BEGreisch, SA).

scheme is as represented in Figure 4.95 for concrete decks. The segments of the steel deck coming from the manufacturer are positioned along a launching platform, welded at the end section of the previously launched segment and moved forward, by pushing or pulling, by hydraulic devices.

Long straight alignments or curved with constant curvature steelwork decks, with spans up to 150 m, may be incrementally launched. Longer spans are possible with intermediate temporary piers or staying schemes during launching (Figure 4.96). The span length is one of the main differences with respect to the incremental launching of concrete bridge decks, where spans above 60 or 70 m are very difficult to launch due to the deck dead weight inducing too large internal forces during erection. The incremental launching method of steel and composite decks (Figure 4.112) requires:

- a launching platform with a minimum of two span lengths
- an hydraulic device system for pulling or pushing the steelwork
- temporary bearings at all piers and guiding devices
- a launching nose equipped at the front with a 'landing device' or with the possibility for lifting the end section from the pier

The capacity of the hydraulic devices system is usually between 3 and 10% of the weight of the steelwork. Temporary sliding bearings at piers are shown schematically in Figure 4.113. The friction coefficient is for design purposes at least 5%. If the geometric

Figure 4.112 Incremental launching of the railway deck of (a) the access viaduct of Tagus Suspension Bridge in Lisbon (launching nose) and (b) Sado Viaducts (typical spans of 76 m and 45 m, respectively) (*Source:* Courtesy GRID, SA).

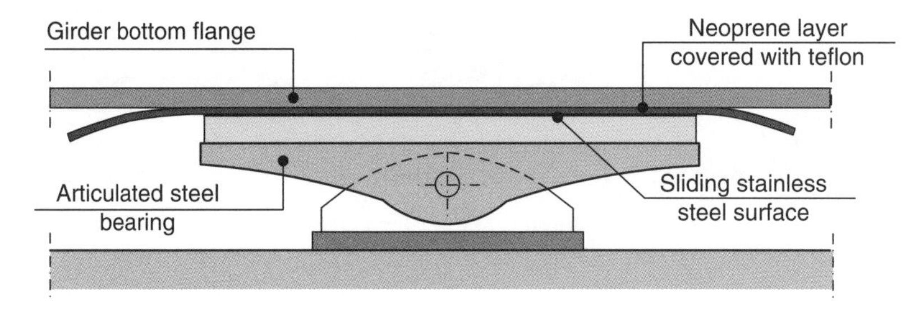

Figure 4.113 Temporary bearings scheme, for incremental launching (*Source:* Adapted from Ref. [19]).

slope of the longitudinal profile is, for example, 2% as in case of Figure 4.112a, an equivalent friction to 7% is considered. Launching stages in the order of 2 h for 20 m segments are typical, as in the case of Figure 4.112a.

When the deck approaches a new pier, the vertical deflections at the front section of the launching nose requires either a special type of device – a *landing device* – or the possibility of lifting the end section of the nose upwards. Introducing upward movements in the order of 200–300 mm is usual. All these deflections induce internal forces and the final precamber to adopt for the deck should be determined at design stages. Patch loading in the webs (Figure 4.63) under combined shear and bending moments should be checked at all erection stages and critical sections.

The deck is usually launched at its final level or at level of the lower flange and then moved downwards to its final position. In a twin girder deck, both girders are launched simultaneously, in general and if needed, with a temporary transverse horizontal bracing, usually at level of the upper flanges. The transverse stability of both flanges should be checked at the erection stages taking the distance between diaphragms restraining horizontal transverse buckling of the compressed flanges into consideration.

4.6.3.4 Erection by the Cantilever Method

The cantilever method, already discussed for concrete bridges, is the most adaptable method for erection of steel decks of large spans (more than 100 m) and with or without a variable curved in-plane alignment. The segments are erected by derrick devices at the end section of the cantilever, from the ground or from a barge in a river. The length of the segments may be much longer than in concrete bridges, reaching 15–20 m in length. The method is adopted based on a balanced cantilever scheme or by an asymmetric scheme as shown in Figure 4.114a. At closure of the segment section (Figure 4.3), the bending moment due to dead loads is zero when the erection joint is done by site butt welding of flanges and webs, and it remains zero contrary to what happens to concrete bridges, as explained previously in Section 4.3.

For very long spans, intermediate temporary piers or staying schemes are adopted. Cable-stayed bridges with steel, concrete or steel-concrete composite decks can also be erected by the cantilever method (Figure 4.114b).

4.6.3.5 Other Methods

Other methods, like the ones referred to for concrete bridges, are possible. In particular, transverse sliding [19] is very much adopted for replacement of an existing steel deck when, in railway bridges (in particular), traffic interruption is a main constraint. Another option is erection of a central part of a long main span (Figure 4.115) from a barge with heavy lifting equipment or of a bowstring arch in three segments, as in Figure 4.116.

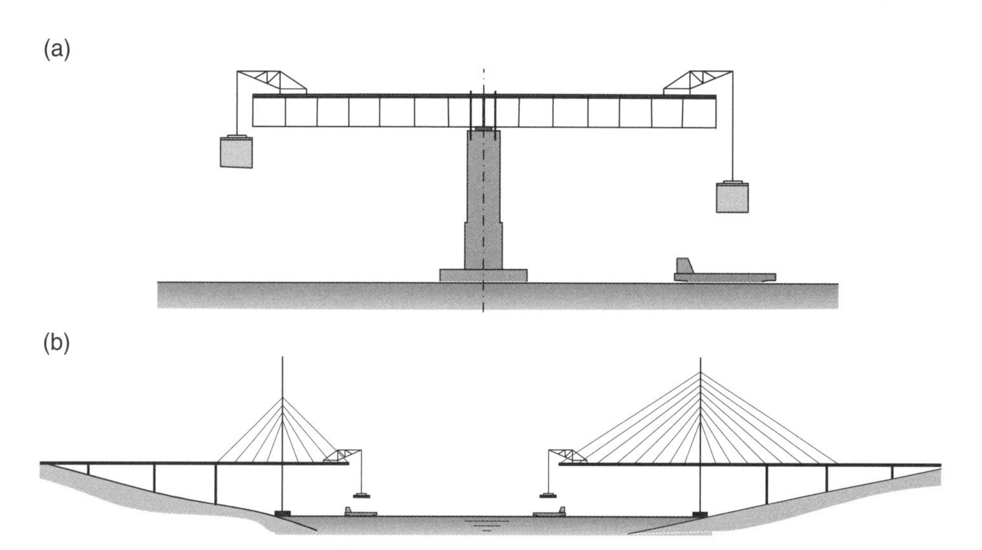

Figure 4.114 (a) Balanced cantilever schemes for the erection of a prestress concrete decks and (b) asymmetric scheme for the erection of steel and composite cable-stayed decks.

Figure 4.115 Erection from a barge of a 162 m length segment, of the main span (242 m) of Cheviré Bridge over Loire River in France (*Source:* Adapted form Ref. [38]).

Figure 4.116 Erection of the arches in the bowstring railway bridge over river Sado, in Portugal (*Source:* Courtesy GRID, SA).

4.7 Substructure: Conceptual Design and Execution Methods

4.7.1 Elements and Functions

A bridge substructure integrates piers, abutments and foundations. Its main function is to transfer to the ground the loads of the superstructure. The substructure should constitute, along with the superstructure, an integrated stable and resistant system complying with the imposed displacements of the superstructure due to thermal deck movements, creep and shrinkage effects of concrete decks, prestressing actions and settlements of foundations.

The abutments make the transition with the earthfill of road or railway approaches. Bridge bearings are introduced, in general, between the superstructure and the abutments, but small bridges may be made with *integral abutments* monolithic with the superstructure [39].

From an architectural point of view, piles and abutments should be carefully studied due to their importance on bridge aesthetics and surroundings integration, as discussed in Chapter 5.

4.7.2 Bridge Piers

4.7.2.1 Structural Materials and Pier Typology

Piers may be made in masonry, wood, concrete or steel. Masonry piers have been adopted for a long time, since bridges were first made. The lack of tension resistance is one of their main limitations. Even in the middle of last century, beautiful bridges (Figure 4.117) were

Figure 4.117 Bridge in Caniçada dam with masonry tubular piers (0.30–0.40 m thickness) reaching 60 m height as designed by E. Cardoso in the 1950s. (*Source:* Photograph by José O. Pedro).

made with masonry piers and engineers should be aware of masonry as structural material. Effectively, retrofitting of old bridges very often requires replacement of the deck keeping the existing piers for economy but, most importantly, for respecting historical heritage. It should be borne in mind the resistance of a hard stone may reach 50–100 MPa, while standard masonry due to the influence of the mortar may be limited to values 5–10 MPa. However, this depends very much on the case, and values for compressive stress resistance from tests of prismatic specimens of masonry for the piers of the bridge in Figure 4.117, with joints of 3–4 cm, can reach minimum values as high as 70 MPa [40].

Wood piers are adopted in wood and temporary bridges. It is also adopted as a structural material for scaffoldings in bridge construction. However, the majority of bridge piers nowadays are made in RC. Some prestressing is adopted in some cases but currently compressive stresses due to permanent loads are enough to balance any tensile stresses due to variable actions. In any case, if tensile stresses are induced due to live loads or imposed displacements, the problem is overcome, in most cases, with just ordinary reinforcement. RC is an excellent structural material for piers, since the main actions induce compressive stresses, and reinforcement steel is required to limit crack widths and for ULS resistance due to induced tensile stresses under the combination of compression and bending due to permanent and variable actions.

A *pier shaft* with its *foundation* is shown in Figure 4.118. If the pier has a double shaft, or more than two shafts, a transverse cross beam, *capping beam*, between pier shafts may be needed for transverse stability (Figure 4.119). Although not very aesthetically pleasant, the cross beam yields a transverse frame action to the pier.

The geometry of cross section of bridge piers may be quite diverse and it is possible to define the following typologies:

- solid section piers or tubular piers – *column piers* (Figure 4.120)
- wall piers – *leaf piers* (Figure 4.121)

In *column piers*, cross sectional dimensions *a* and *b* (Figure 4.120) are of the same order of magnitude, usually in the order of 0.5–1.0 m. For pre-design of the cross

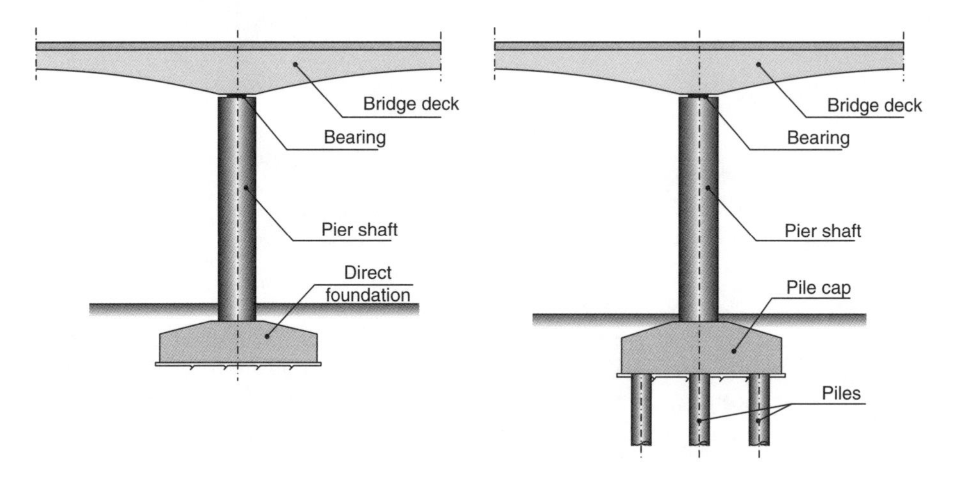

Figure 4.118 Elements of concrete pier with a single shaft.

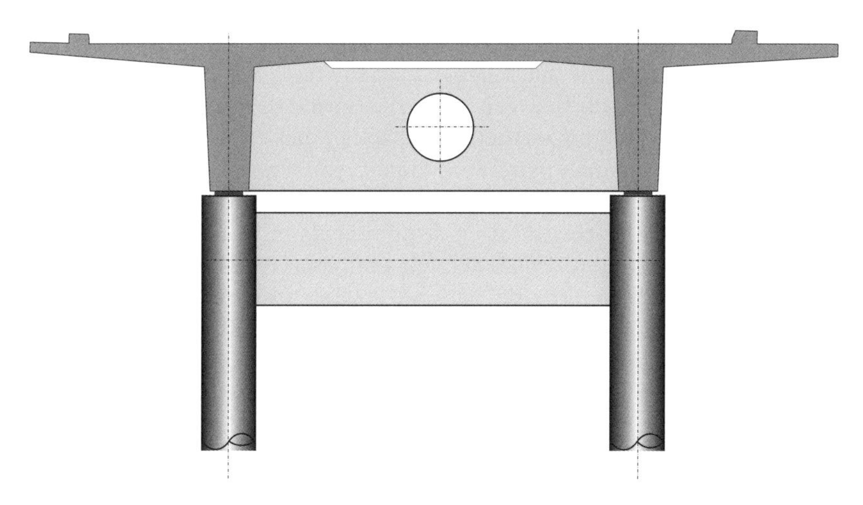

Figure 4.119 A pier with a double shaft.

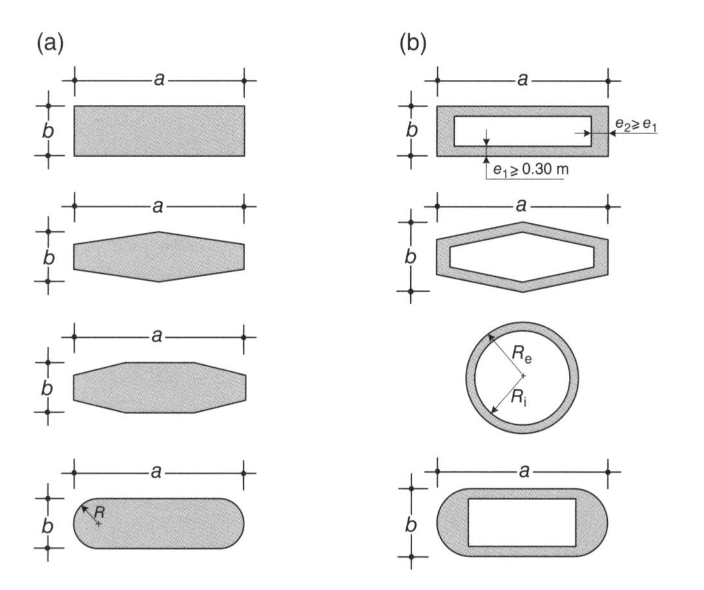

Figure 4.120 Column Piers: (a) solid section and (b) tubular piers.

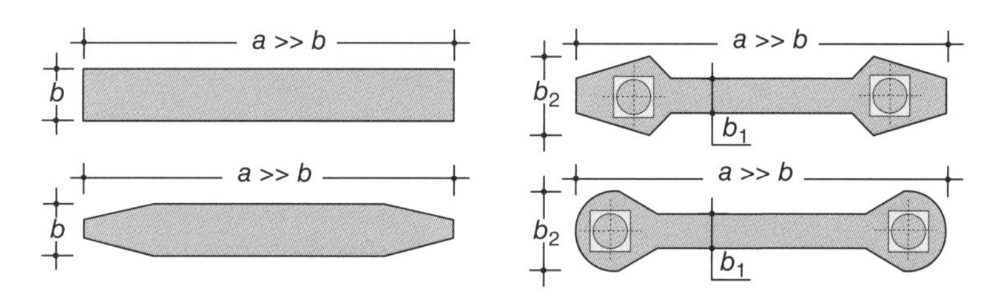

Figure 4.121 Leaf (wall) piers.

section, an average medium compressive stress at SLS in the order of 4–5 MPa may be adopted. In *leaf piers*, the width a is much larger than the thickness b, usually at least four times. In this case, b is generally between 0.4 and 1.0 m and the width a is very often reduced towards the base section for aesthetics and sometimes to increase horizontal clearances at the ground level, namely in urban viaducts. An average compressive stress at SLS in the order of 2–3 MPa may be adopted. At the top section, cross sectional dimension a is dependent on bearing location requirements in most cases. For piers with a single shaft, and heights up to 15–20 m in general, *column piers* (Figure 4.120a) are usually preferred. For medium piers, say between 20 and 30 m, solid or tubular box sections may be adopted. Above 30 m and sometimes reaching 200 m height or even more (Figure 4.33 – Viaduct the Millau), *tubular piers* (Figure 4.120b) are the preferred option. The wall thickness is in general at least 30 cm for easiness of formwork and placement of reinforcement.

4.7.2.2 Piers Pre-Design

Designing pier shaft cross sections should take into consideration:

- *resistance*
- *stability*
- *aesthetics*

Stability should be considered at the final static model of the bridge and also at construction stages. Its overall cross section dimensions are kept constant for short and medium piers, say up to 30–40 m. In tubular tall piers, at least one of the cross sectional dimensions, usually the transverse dimension, is variable (see Figure 5.17).

The slenderness of a pier *for aesthetics* is usually defined as l/b or l/a, l being the pier height of its shaft and b or a the overall cross section dimensions (Figure 4.120). For stability, slenderness is defined by:

$$\lambda = l_e / i \tag{4.15}$$

where l_e is the buckling length and i is the relevant radius of gyration of the pier cross section. The radius of gyration $i = \sqrt{I / A}$, where I is the relevant moment of inertia and A the cross sectional area. For a rectangular section, $i = b\sqrt{12}$ or $i = a\sqrt{12}$. This means the slenderness for aesthetics is approximately 28% of slenderness for stability. Assuming b is the smaller dimension of a constant rectangular cross section, a pier with l/b = 20 has λ = 70 if $l_e = l$. At the final static model of the bridge, one may have from $l_e = 0.5\,l$ if the pier is monolithic to the deck, up to $l_e = 2\,l$ if a sliding bearing is adopted between the deck and the top of the pier shaft. At the construction stage, one very often has $l_e = 2\,l$, or even more if rotation at connection with foundations is considered, as discussed in Chapter 7.

A basic pre-design criterion for selecting the cross sectional dimension of a pier section is to limit its slenderness to 70 for the final static scheme and 100 during construction. A limit of 70 at the final static scheme is adopted to avoid relevant creep effects of the concrete on the ultimate buckling resistance of the pier. For tubular piers, if one compares its cross section with a solid section pier with same cross section area, one has a larger radius of gyration i. That is why tubular sections are preferred for high piers.

In tubular piers one has, compared to a solid section pier with the same cross sectional area, A, a more stable shaft (smaller λ) and the same compressive stresses due to axial loads, which is relevant to reducing tensile stresses coming from imposed permanent actions due to creep and shrinkage and variable actions.

The slenderness of a concrete pier is generally smaller than in a steel pier, and second order effects due to instability effects govern this. RC piers are always adopted in concrete bridges and are the main option for steel and composite bridges. In piers in rivers, it is usual to adopt wall piers, with hydrodynamic details as shown in Figure 4.121. If the risk of ship impact exists, the cross sectional dimension b of a wall pier may be larger than 3 m. In those cases, ship impact defences ('fencers') may be adopted. A similar case occurs in viaducts over railways where wall piers are usually required and minimum distance to tracks should be respected [41]. In short, whenever an accidental impact event in pier shafts due to traffic needs to be considered, possible options for design are:

- to locate the pier at a sufficient distance to minimize risk of impact
- to adopt pier protections (fences in rivers, walls nearby tracks, metal fences near road traffic lanes)
- to design the pier shaft for impact loading

The risk of ship impact in piers depends on the dimensions, tonnage and speed allowed of the vessels in the navigation channel. Ship impact, reaching thousands of kN, tends to be the critical action for the design of piers in rivers.

Design of pier shaft cross sections should take into consideration resistance, stability and aesthetics. Overall cross sectional dimensions in piers are kept constant for short piers, say $l < 10$–15 m, and medium piers, say $15\,\text{m} < l < 40\,\text{m}$. In tubular tall piers, at least one of the cross section dimensions, usually the transverse dimension, is variable.

One important aspect when designing pier cross sections is the dimension required for the bearings as indicated, for example, in a leaf pier in Figure 4.121. Besides, the space required for hydraulic jacks for replacement of bearings is another aspect to be taken into consideration.

4.7.2.3 Execution Method of the Deck and Pier Concept Design

The execution method of the deck may influence the shape of the pier shaft. In incremental launching girders working from below, the upper part of the pier shaft should be designed to accommodate the fixation devices of the launching girders as shown in Figure 4.92. In the incremental launching method, the pier top cross section requires sufficient clearance to accommodate temporary bearings and transverse guiding devices to avoid deviations of the in-plane alignment during launching operations. Besides, the pier shaft should be designed to withstand the horizontal forces due to bearing friction during launching or even due to the longitudinal slope of the bridge deck. However, from all execution methods for the bridge deck, the cantilever method for deck execution generally has the largest impact on the concept design of piers. By this reason, it is dealt with specifically in the following.

4.7.2.3.1 Piers for Bridges Built by Cantilever Construction

Actions induced in bridge piers built by balanced cantilever during construction are due to a multiplicity of actions (Figure 4.122), namely:

- dead loads (pier and deck)
- unbalanced segments during construction

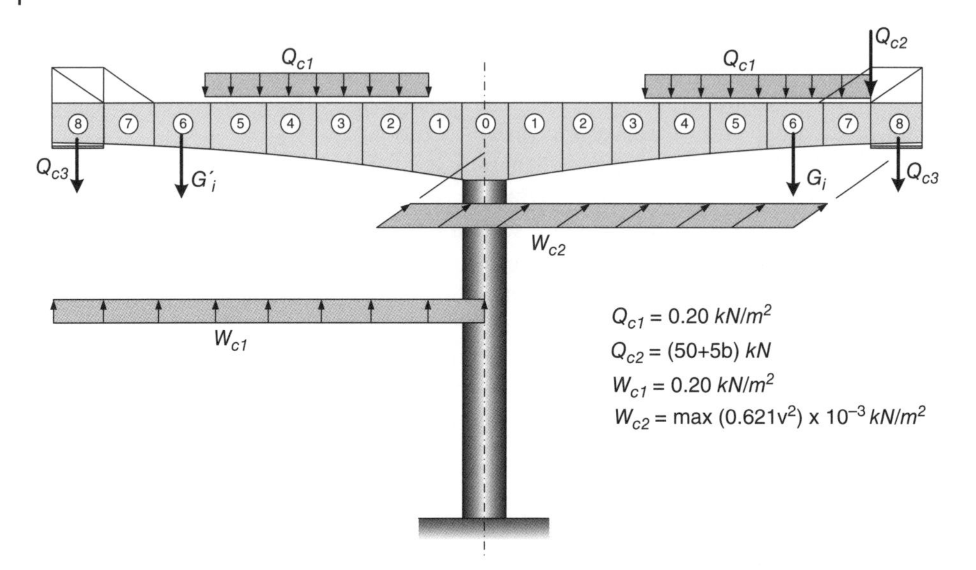

Figure 4.122 Example of design loads for piers during cantilever construction of the box girder deck (*b* – deck width, *v* – design construction wind speed).

- dead loads of moving travellers (scaffoldings)
- construction live loads
- wind loads

All these loads induce large axial loads and bending moments (in two directions) and are easy to resist with tubular column piers. In tall piers, second order effects due to instability come into play and may be relevant, as will be discussed in Chapter 7.

The connection between the pier and the deck may be rigid (frame bridges) or through bearings (continuous girder bridges), as shown in Figure 4.123. Bending moment resistant of the piers is usually adopted for tall piers to avoid bearings, and temporary devices for stability during construction and to benefit from frame action at the final static scheme.

Short piers are usually connected to the deck through one single row of bearings. Temporary anchorage, stay cables or temporary piers (Figure 4.124) are needed to guarantee a stable system during the erection stages.

Medium height piers are usually provided with one or two rows of bearings reducing the stiffness of the pier for imposed deformation (creep, shrinkage and temperature) of the deck; in long continuous bridges, the central piers are usually rigidly connected to the deck, while the remainder have fixed pot bearings or unidirectional movable pot bearings. In some cases, a pier made of two flexible concrete shafts (Figure 4.107 and 4.125) is adopted, which are stabilized during the erection stage by a temporary steel structure, as will be discussed in the following.

In short, there are three main possible options for typology and schemes of pier-deck connections:

- tubular monolithic piers with the deck
- tubular or solid section piers (depending on height) supporting the deck by bearings
- piers made by two independent solid section shafts monolithically connected to the deck or eventually hinged at the deck

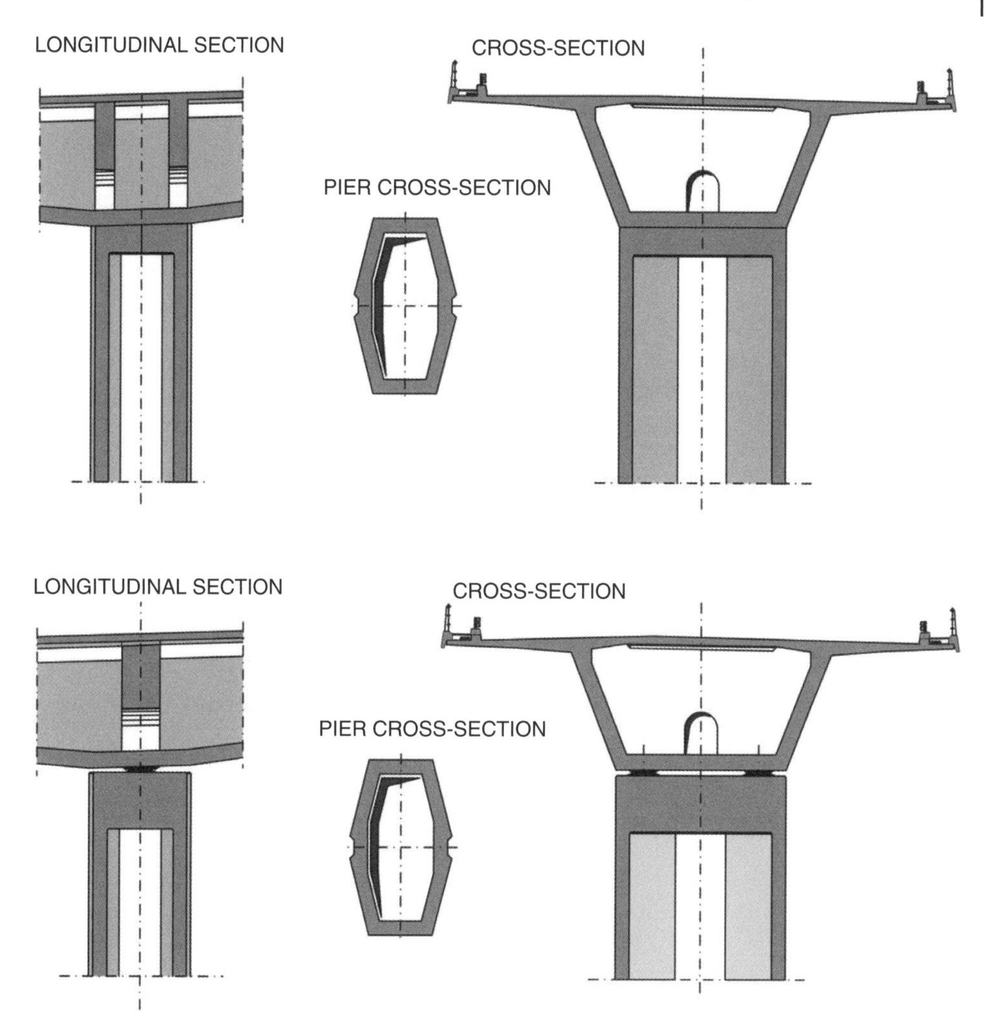

Figure 4.123 Monolithic and simply supported decks of box girder bridges built by cantilevering – support sections. (*Source:* Courtesy GRID, SA).

Stability in the first case – monolithic tubular piers – adopted for tall piers, may guarantee stability during construction phases by resistance of the tubular section shaft. The second case, a deck being articulated to the pier shaft, requires specific devices of course, as said before, guaranteeing overall stability of the deck. There are the following options:

- adopt temporary high strength prestressing bars or cables connecting deck and pier sections
- adopt one or two temporary piers adjacent to the definitive pier shaft
- adopt temporary stability cables (stays) anchored to the deck and to pier foundation

Finally, the option of adopting a pier with two slender pier shafts (Figure 4.125) is a possibility when the imposed displacements by the deck tend to induce large forces. In a pier made of two flexible shafts, any unbalanced moment M during construction is

Figure 4.124 Temporary stability systems – anchorages, temporary piers and stay cables for overall stability of the deck during cantilever construction stages. Sorraia River Bridge. Main span 120 m (*Source:* Courtesy GRID, SA/José Pedro).

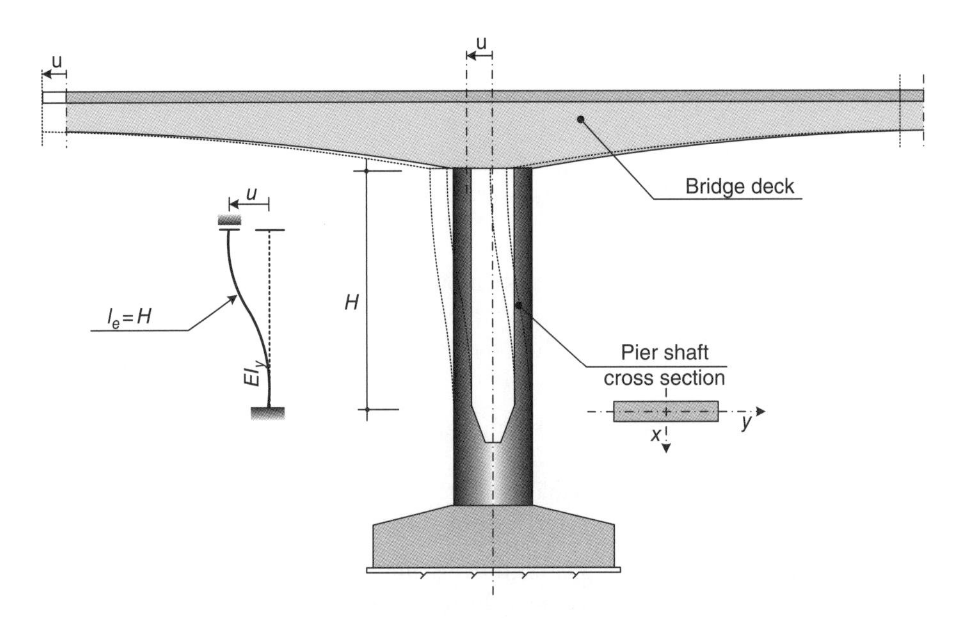

Figure 4.125 Pier made with two slender pier shafts.

taken by a couple effect; no tensile stresses are usually accepted in the section at any construction stage, except for accidental load combinations. The shafts shall be sufficiently braced between them in such way that the full section works as a global section during construction stages. After closing the deck, the bracing system is removed and the shafts can deform as independent piers towards the centre of stiffness, as shown in Figure 4.125. The buckling length of the shafts at the built structure tends to be half the free length. During construction, the two pier shafts work together and the buckling length is taken for the pier with the overall cross section inertia.

The temporary stability systems referred to here – anchorages, temporary piers and stay cables – should be designed to guarantee the necessary overall stability of the deck and to have sufficient stiffness for good control of deflections during construction. Any rotation of the deck at support section has a large impact on deflections of the cantilevers, and precamber control becomes difficult to achieve. Sometimes, the three schemes (as before – anchorages, temporary piers or stay cables) are combined (Figure 4.124). It is usual to adopt anchorages and temporary piers or anchorages and stay cables due to lack of space at the top section of the pier in order to introduce a sufficient number of anchorages to guarantee overturning stability of the deck until the casting of the closure segment. Stay cables are very often sufficient for stability of the deck during execution but are not very efficient concerning deflection control (camber) because the deformability of the stays is usually too large. The use of temporary supports, if possible, is usually simpler and more efficient. The anchor bars for temporary connections between segment 0 and the end section of the pier are always required.

All these stability devices are designed for forces coming from overturning moments of the deck. Temporary bearings or wood shims are usually adopted in the anchorage scheme system between the underneath of the deck and top of pier section. At the end, after the closure segments are executed and stability of the deck is guaranteed, the anchorages and any temporary devices are removed.

4.7.2.3.2 Design Safety Checks of Piers During Deck Cantilever Construction

Actions during construction stages, referred to in Section 3.19, have been mentioned in the present chapter. For designing bridge piers built by the cantilever method, or the stability devices previously referred to, some specific construction loads and load combinations need to be specified. The basic system is represented in Figure 4.122, where G_i and G_i' denote the weights of segments i and i'. The weights of each cantilever are $G = \Sigma\, G_i$ and $G' = \Sigma\, G_i'$. For G_{max} and G_{min}, one usually takes a ±2% deviation on the nominal weight, G, of one of the cantilevers; that is, $G_{max} = 1.02\, G$ and $G_{min} = 0.98\, G$. Construction loads (equipment, formwork travellers, construction live loads etc.), are denoted by Q_{c1}, Q_{c2} and Q_{c3}, in Figure 4.122. Load Q_{c1} denotes a live load during construction, applied anywhere, Q_{c2} is associated with a construction load at the end segment due to weight of prestressing equipment (cables, anchorages, etc.) and Q_{c3} denotes the weight of the form traveller (500–800 kN in general). W_c designates an uplift differential wind action between the cantilevers.

Safety should be checked involving stability systems, for a load combination:

$$\gamma_g G + \gamma_q \left[\sum_{i=1}^{n} Q_{ci} + W_c \right] \tag{4.16}$$

where G are permanent loads, Q_{ci} construction loads and W_c unbalanced wind loading. Partial safety coefficients $\gamma_g = 1.1$ or 0.9 as most unfavourable and $\gamma_q = 1.25$ have been proposed in [26, 42, 43].

An accidental load combination should also be checked replacing Eq. (4.16) by

$$G + \sum_{i=1}^{n} Q_{ci} + W_c + F_a \tag{4.17}$$

where F_a represents the action induced by the falling of a form traveller during a manoeuvring or the falling of a precasted segment during erection, with a dynamic coefficient of 2.

In cast in-place balanced cantilever bridges, sometimes in Eqs. (4.16) or (4.17) one considers a construction load Q_{ci} associated with a difference in weight between the end segment n and n' due to a differential time between castings. To take $Q_{ci} = 0.5\ G_n$ where G_n is the dead weight of segment n is, in general, enough, provided the concreting of the slab deck in one side is not initiated before the bottom flanges and webs at both sides are concluded. This control, to be made by the Engineer, is sufficient because in box girder bridges the deck slab weighs about 50% of the full weight of the segment. Even with $Q_c = 0.5\ G_n$ applied at the end section of only one of the cantilevers, the associated load combination tends to be the most critical for the pier, at least for long span balanced cantilever bridges.

During the balanced cantilever scheme of a curved box girder bridge (Figure 4.126), the dead load of the deck induces transverse bending moments in the piers, unless some provisional prestressing, internal or external, is adopted. It is possible to balance the total applied transverse bending moments during construction by external prestressing tendons applied with an effective force P and eccentricity e (Figure 4.126).

Balanced cantilever bridges, with tall piers, present low natural frequencies in bending and torsional modes of the piers during construction. It is usual to have longitudinal and transverse bending frequencies and torsional frequencies f_x, f_y and f_t, respectively for tall piers (say between 60 and 100 m) and spans between 100 and 180 m in the range 0.1–0.4 Hz. An example was shown in Figure 4.4 where $f_x = 0.17$ Hz and $f_t = 0.29$ Hz.

Under random excitation and due to wind gusts, bending and torsional effects due to unbalanced wind buffeting should be considered in designs when checking the structural safety during the construction stages. For torsional effects, the historical Sir Benjamin Baker's rule-of-thumb, as considered in the design of the Firth of Forth Bridge, assumes a 'full mean wind' with a wind velocity v_m on one overhang and 'no wind' on the other [43] as wind pressure W_{c2} shown in Figure 4.122. This load needs not to be combined simultaneously with normal construction loads, since it is unlikely construction is being done at a strong wind event. The pioneering work of Davenport [44] in 1967 evaluates the torsional effects induced by wind gusts on a cantilever bridge due to the resonant component of the torque only. Dyrbye and Hansen in 1996 [45] presented an approach suitable for design practice to consider the problem of torsional effects of balanced cantilever bridges of constant height decks. The approach was generalized by Mendes and Branco [46] for variable depth box girder bridges, as well as for the joint analysis or bending and torsion effects. Codes and design recommendations, like ASCE 1982, ECCS 1989 and NBCC 1990, refer to the need to investigate the wind torsional effects in balanced cantilever bridges during construction.

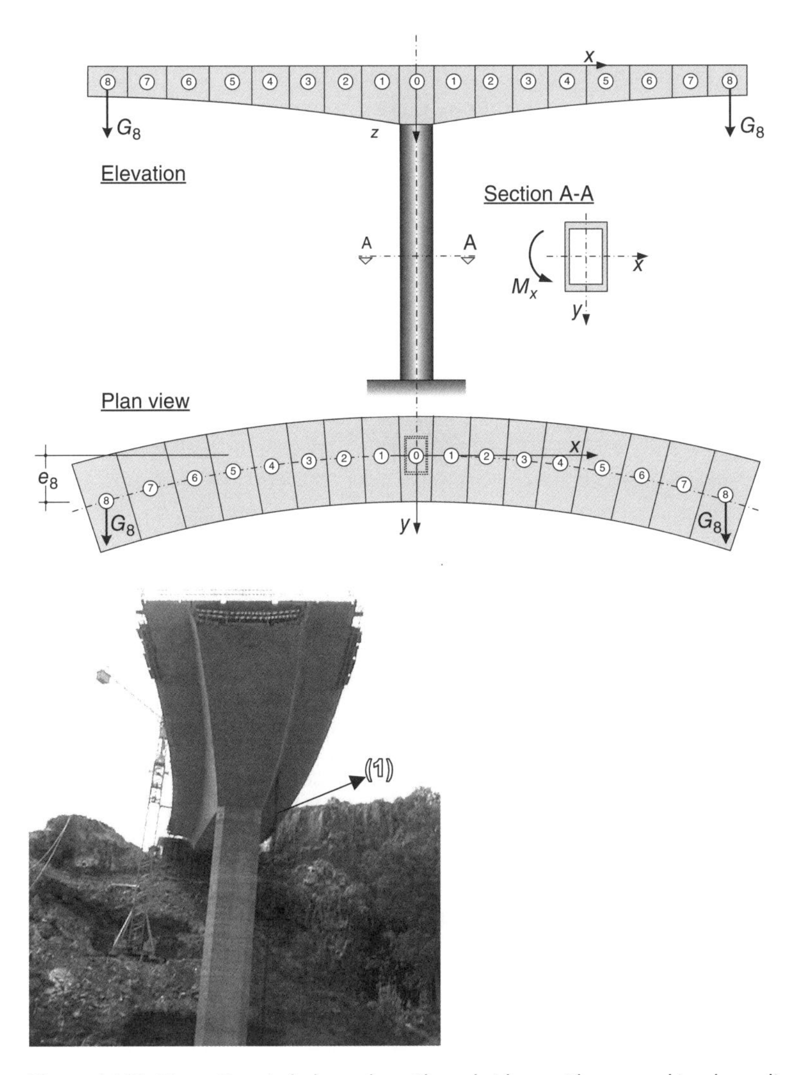

Figure 4.126 Pier actions in balanced cantilever bridges with a curved in-plane alignment. (*Source:* Courtesy GRID, SA).

The characteristic average wind speeds ($U_{k,50}$) are usually defined for time intervals of 10 minutes with a reference period of 50 years. For checking the structure during construction, one adopts a shorter reference period T, say $T = 10$ years for a construction time not greater than 1 year according to EN 1991-1-4. The associated characteristic averages wind speed yields for $T = 10$ years $U_{k,T} = 0.88\ U_{k,50}$, that is, a 12% reduction in wind velocity representing in the wind pressures (depending on U^2) a 23% reduction. The reader is referred to specific literature [45] on this subject for further information.

For the design of temporary stability devices, one should take into consideration the uplift condition at a temporary pier or the minimum prestressing force required for temporary stays. The reader is referred to references [35, 45] for additional information.

4.7.2.4 Construction Methods for Piers

Piers are usually built by griping formwork. The formwork for the shaft with 3–4 m length, is moved upwards in steps discontinuously, after concreting the previously segment of the shaft. It is fixed to the part of the pier shaft already built and moved upwards by hydraulic devices. Then, after erection of the steel reinforcement of next segment the formwork is moved upwards and the new segment is casted.

The example of the Guadiana bridge in Figure 4.127, shows a griping formwork for casting a tubular shaft of a box girder bridge built by the cantilever method.

Figure 4.127 Piers built using a gripping formwork for casting tubular shafts – Bridge over Guadiana River in Portugal – Alqueva dam in Portugal with multiple spans of 112.5 m (*Source:* Courtesy GRID, SA).

Tall piers may be executed with *sliding forms*, where forms are moved continuously. The method requires specific measures to guarantee the required position of the reinforcement, namely the required concrete cover. Sliding forms have encountered some restrictions by technical specifications and tender documents in some countries, due to the sensitivity of the method in achieving a good concrete quality control and control of geometry of the reinforcement steel. Sub-contractors specialize in sliding forms, as was the case of the tall piers of the João Gomes Bridge (Figure 5.1); with a variable pier cross section, it reached a construction progress with sliding forms of approximately 5 m per day for piers of about 100 m.

4.7.3 Abutments

4.7.3.1 Functions of the Abutments

The abutments, establishing the transition between superstructure and access roads or railways, have the following main functions:

- supporting vertical and horizontal loads from the superstructure at the end spans
- supporting soil actions transmitted by the backfill and to avoid erosion of the soil contained in the abutment
- accommodating movements of the deck, in particular at the expansion joints between deck and abutment
- accommodating small foundation settlements without inducing risks to the superstructure
- allowing the positioning of the end section bearings of the superstructure

4.7.3.2 Abutment Concepts and Typology

Options for abutment types are very much dependent on topographic and geotechnical conditions at the bridge end sections, type of bridge superstructure and aesthetical conditions. Transition of the bridge to the ground should be as smooth as possible but abutments should be at the *scale of the bridge*. A large bridge should not have a very small abutment, even if that is possible by extending the superstructure towards the slopes of the valley. On the other hand, a pedestrian bridge should not have a large and tall abutment since it is out of the bridge scale. The height of an abutment should be established leaving a minimum space of at least 1–2 m between the underneath of the superstructure and the ground. This clearance is required to avoid accumulation of debris, to allow an easy access for abutment bearing inspection and replacement and to allow a clear visual separation between the superstructure and the ground. There are two main types of abutments (Figure 4.128):

- Closed/solid abutments
- Open or skeletal/spill-through abutments

Additional abutment types can be considered, namely:

- Bankseats
- Reinforced earth abutments
- Integral abutments

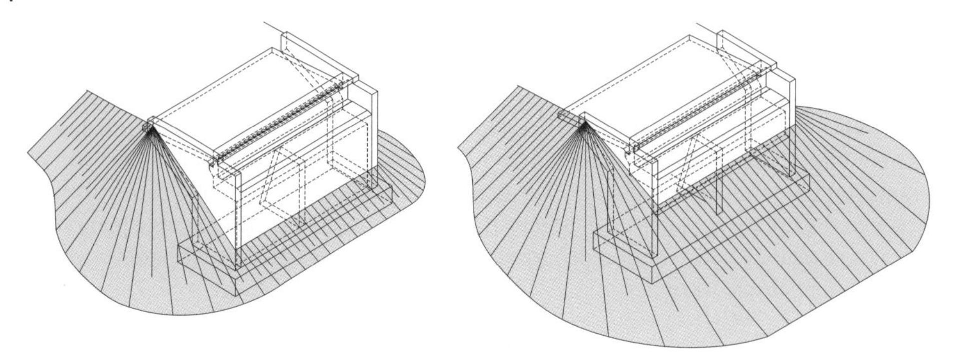

Figure 4.128 Closed (solid) and open (skeletal) abutments.

Closed abutments have a fully visible superstructure in such a way the complete structure of the abutment, apart its foundation, is apparent. They integrate a *front wall, wing walls* and a *seating beam* for bridge deck bearings.

If the topography allows, or if the end spans of the superstructure are sufficiently extended against the slopes, it is possible to avoid the front wall (at least one part of it) and reduce the wing walls to small lateral retaining walls connected to the seating beam. This yields an *open abutment*, also called a *skeletal or spill-through abutment*. These are cheaper than closed abutments but an additional cost is involved in extending the deck to allow the adoption of an open abutment. Open abutments have aesthetical advantages and do not induce, in cases of overpasses over highways, the visual impression of having front walls too close to the road platform. Viewed from a certain distance, closed abutments near the road platform induce a tunnel visual impression, which is avoided with open abutments (as shown in Figure 4.132 later). Natural slopes of the soil are retained in open abutment cases.

Wing walls in closed abutments can be parallel or inclined with respect to the bridge axis, as shown in Figure 4.129. Inclined wing walls have generally a better appearance but their geometry is very much dependent on topographic and geotechnical conditions, since they need to retain the soil at the backfill. The longitudinal horizontal free length of the wing wall (Figure 4.129) is generally limited to 5 m. The lateral soil pressure at rest conditions should be considered when designing this lateral cantilever wall.

Front and wing walls may have counterforts if needed, depending on the height of the abutment. If the height of the abutment is higher than approximately 7 m it may be economic to adopt counterforts at 3–5 m distances (Figure 4.130) in such a way to take the soil pressure by wall resistance as a slab working mainly in the transverse direction, as will be discussed in Chapter 7. If the counterforts are in the front wall, they are the main resistant elements to longitudinal horizontal forces at the abutment, namely soil pressures and seismic longitudinal forces in the case of a deck fixed at the abutment.

A *transition slab* as shown in Figure 4.130 makes the transition between the abutment and the earthfill. It reduces effects of differential settlements between soil at the backfill and the rigid superstructure of the abutment. Transition slabs have thicknesses usually of 0.25 or 0.30 m and should have sufficient steel reinforcement (min Φ12 every 0.10–0.20 m) in two layers (upper and lower faces) avoiding excessive crack widths due to differential settlements of the soil under impact of traffic loading. The transition slab should be articulated to the abutment in order to accommodate rotations due to settlement of the soil at the backfill (Figure 4.130).

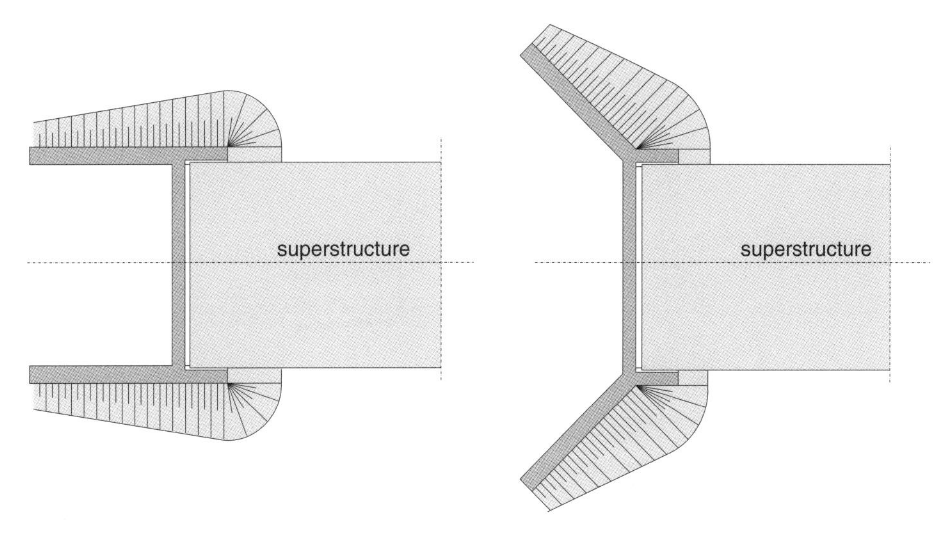

Figure 4.129 Wing walls parallel or inclined with respect to the bridge axis.

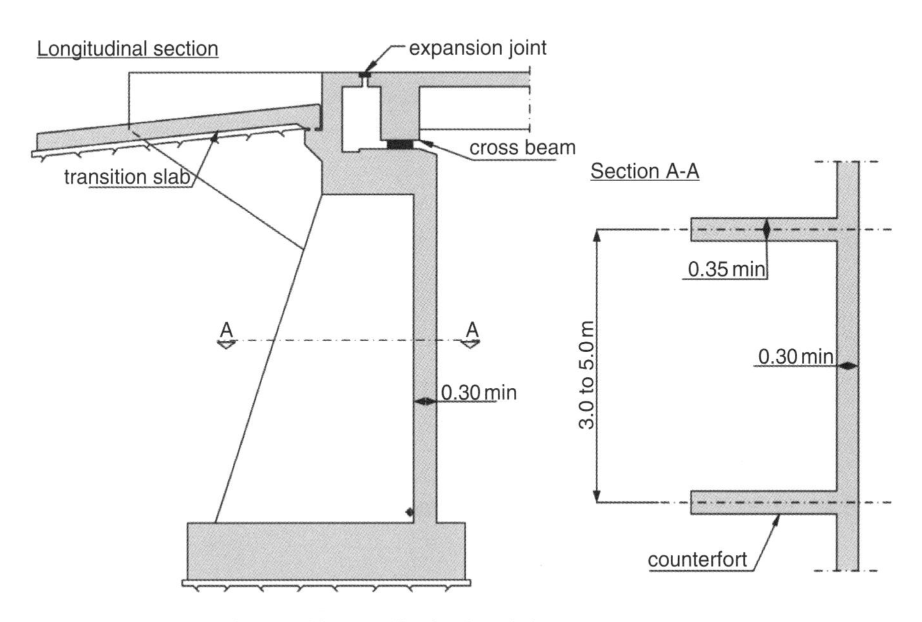

Figure 4.130 Counterforts and front walls of a closed abutment.

Abutments should integrate the drainage system at the backfill, which is presently made in most cases by a scheme as shown in Figure 4.131, including a lateral discharge as indicated.

The earthfill at the abutment shall respect the minimum distances shown in Figure 4.132. In some cases, it may be justified to reduce the abutment to a simple *bankseat*. The superstructure of the abutment is reduced to the seating beam supported directly by a footing or by pile foundation as in Figure 4.133.

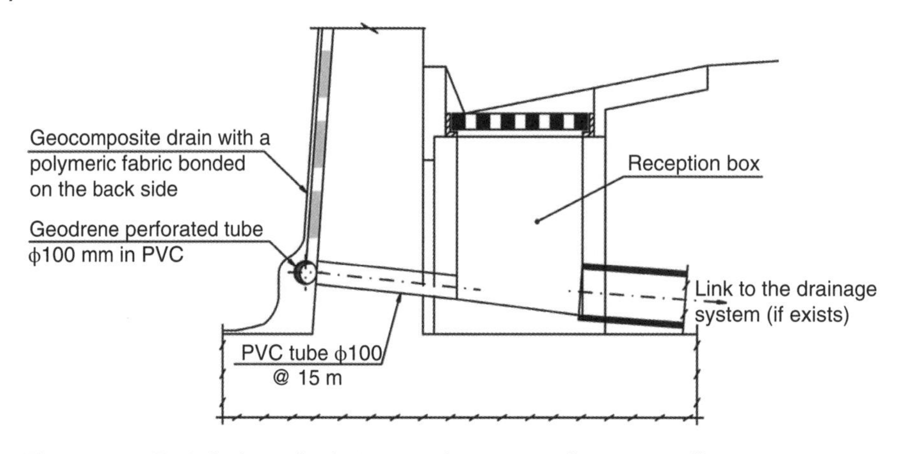

Figure 4.131 Typical scheme for drainage at abutments and retaining walls.

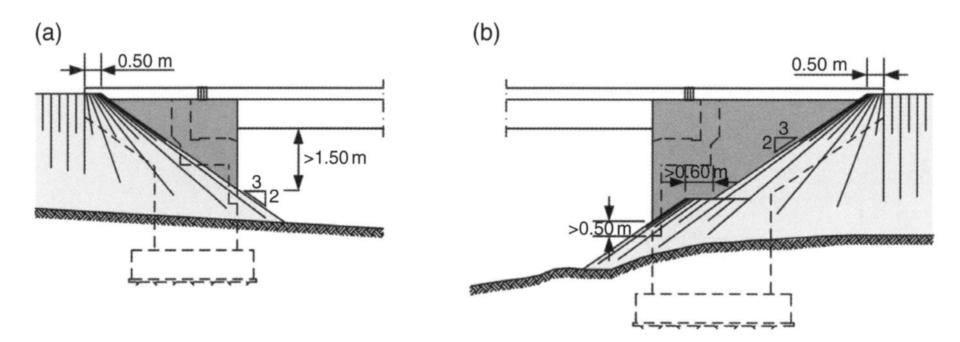

Figure 4.132 Earth fill layout and typical minimum distances at abutments.

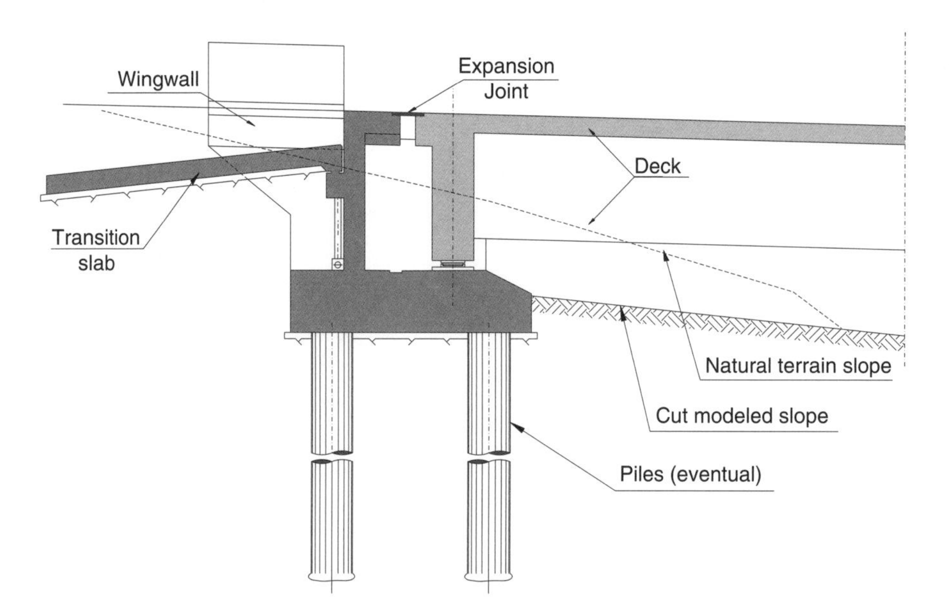

Figure 4.133 Scheme of a bankseat abutment with piles.

Reinforced earth abutments integrate reinforcement earth walls, replacing RC wing walls or (and) front walls. Earth walls are made of precasted concrete panels and steel reinforcement bars inside the earthfill. A back seat is adopted integrated with the *reinforcement earth abutment.*

Integral abutments are a possible option for shorter bridges, say in the order of 60 m at a maximum, avoiding any expansion joint and bearings between the superstructure and the abutments. The superstructure is monolithic with the abutment, reducing problems of maintenance of expansion joints. In an integral abutment, the structure moves against or away from the backfill depending on induced movements due to temperature. Reference is made here to specific literature [39] on bridges with integral abutments, a subject of many investigations in recent years.

4.7.4 Bridge Foundations

4.7.4.1 Foundation Typology

Bridge foundations follow typologies and design rules of structural foundations in most cases. For this reason, the reader is referred to references on foundation engineering [47]. Only specific aspects are dealt with in the present chapter. The main bridge foundation types are:

- Direct foundations by footings
- Indirect foundations by piles
- Special foundations

The last case includes foundations by digging shafts, pile walls, micropiles, jet grouting and caissons. Another aspect specific to bridge foundations is how to execute foundations in rivers, a subject discussed here.

A direct foundation is a shallow foundation where load is transferred at a depth h_f by direct soil pressures at the underneath of a footing with a base dimension characterized by a typical dimension, b. What differs from an indirect foundation made by piles, with a cross section dimension b (diameter in a circular pile), is the ratio, h_f/b. One has, in general, $h_f/b < 5$ for shallow foundations and $h_f/b > 10$ for deep foundations. Sometimes, one has direct foundations quite deep as represented in Figure 2.2. In pile foundations, the load is transferred at bearing pressures at the base, but also to some extent by friction along the length of the pile.

4.7.4.2 Direct Foundations

Direct foundations of bridge piers are made by footings at a depth h_f at least 1.5–2 m in order to avoid any erosion effects of the soil above the footing due to expansion effects of soil or erosion due to superficial waters. After execution of the footing, refilling, as shown in scheme of Figure 4.134, is done. The inclination i of excavation is a function of soil type and excavation method. Distance a may be reduced to zero if lateral formwork is avoided, mainly for hard/compact soils. The best choice, if possible, is always to cast the footing against the soil.

The height h of bridge footings is at least 0.70 m and may reach more than 3.0 m in very large footings. Minimum footing height h is established as a function of the horizontal length c of the free cantilever measured from the pier shaft. A *rigid footing* (say, $h > 0.4c$)

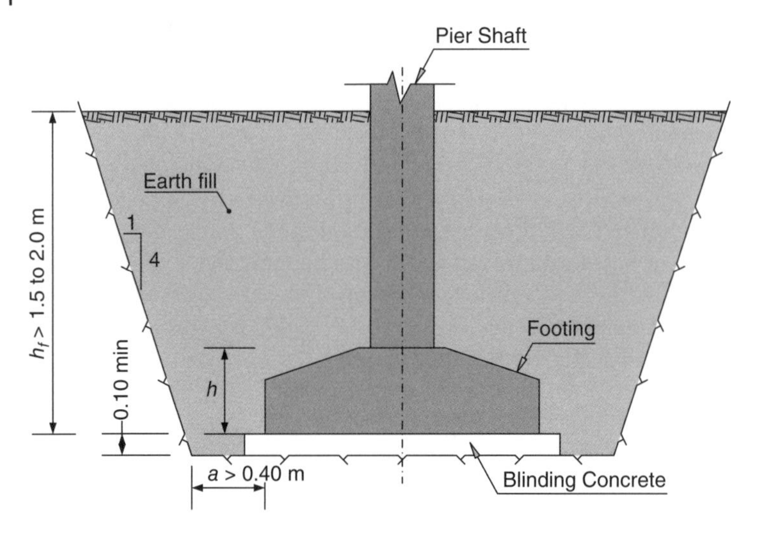

Figure 4.134 Direct foundation of a bridge pier.

is usually the best choice allowing assumption of a linear distribution of bearing pressures on soil. The upper surface of the footing may be tapered, in large footings, as shown in Figure 4.134, but care should be taken to avoid the need for an upper formwork for concreting the footing. In general, slopes up to 10% are acceptable without upper formwork.

Design of bridge footings are made for soil resistance at ULS and deformability at SLS. Differential settlements between pier bridge footings needed to be evaluated and considered in design of the superstructure.

4.7.4.3 Pile Foundations

Bridge piles are the most current type of deep foundations, made by *driven piles* in steel (I, H, circular tubular sections) or precasted concrete small diameter piles and cast in situ concrete piles. The latter have diameters in the range 0.6–3.0 m.

Risk of corrosion in driven steel piles in rivers, for example, is ruled out, apart from the pile length above the water table. Thickness of tubular piles is established taking into consideration long-term corrosion effects, which is done by adopting an over thickness of some millimetres. The upper part of the steel tube above the water level needs to have adequate corrosion protection.

Cast in place piles may be executed with a recovered or lost steel tube. In piers in rivers, the steel tube is generally left in the pile to guarantee quality of execution. If soil permits, cast in place piles are executed with bentonite if that is allowed by environmental protection and bridge site constraints. Piles should penetrate the bed-rock between 0.5–2 diameters, in general, for load diffusion at the end sections. If required, a trepan is adopted to excavate the bed-rock. An execution scheme for concrete cast in situ piles is shown in Figure 4.135.

Distance between pile axis is at least 2.5 diameters to avoid pile interaction effects at the foundation. This distance is generally what defines the in-plan dimensions of the pile cap, as shown in the example of Figure 4.136 for piles of 2 m diameter. The height

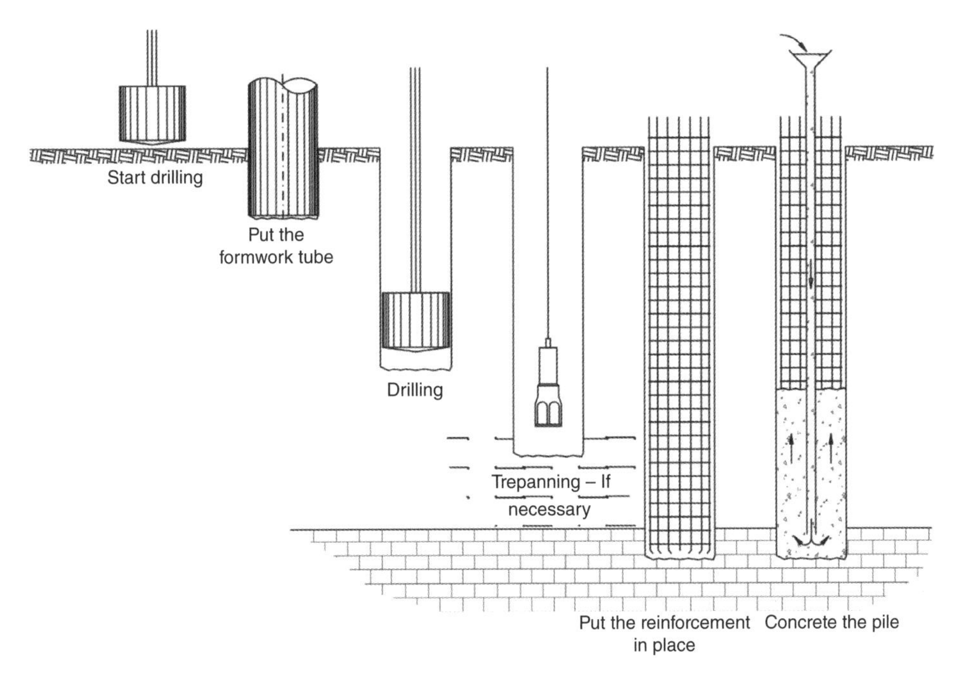

Figure 4.135 Execution scheme for cast in situ concrete piles (adapted from ref. [48]).

of the pile cap (between 1.0 and 3.0 m in general) should be in the order of 1.5–2Φ, where Φ is the pile diameter. In large diameter piles, with 2–3 m diameter, the pile cap depth can be reduced to a minimum of 1.0–1.5 pile diameter. A rigid pile cap allows designing based on a linear distribution of pile forces due to axial load and longitudinal and transverse bending moments at the base of the pier shaft.

4.7.4.4 Special Bridge Foundations

Special foundations in bridges with caisson shafts are not so common nowadays. They are similar to large diameter cast in place piers, executed by excavation and adopting an external and internal formwork to make a RC tube that is filled up with concrete. The depth reached is moderate but allows to reach a soil layer with sufficient load capacity.

Pile walls are rectangular foundation elements in cast in place RC, executed by a similar technology of cast in place piles made with bentonite. They may be associated to yield T or U sections presenting a high bearing capacity.

Micropiles have been adopted in some cases, sometimes when standard pile diameters are not easily available at the bridge site or adopted as a technique for improvement of existing bridge foundations. Micropiles have been adopted as standard equipment for bore holes in soil site investigations, for perforating small diameter holes. Steel tubes in the order of 150 or 200 mm, with a 5–10 mm thickness, are adopted as structural elements for the foundations. Steel reinforcement bars are installed at the inside of the tubes and afterwards the tubes are injected with cement grout. Micropiles reach load capacities in the order of 500–1000 kN, depending on the tube diameter and thickness as well as the soil proprieties. One of the advantages of micropiles is the capacity to resist to tensile forces, which for some accidental load cases may be considered.

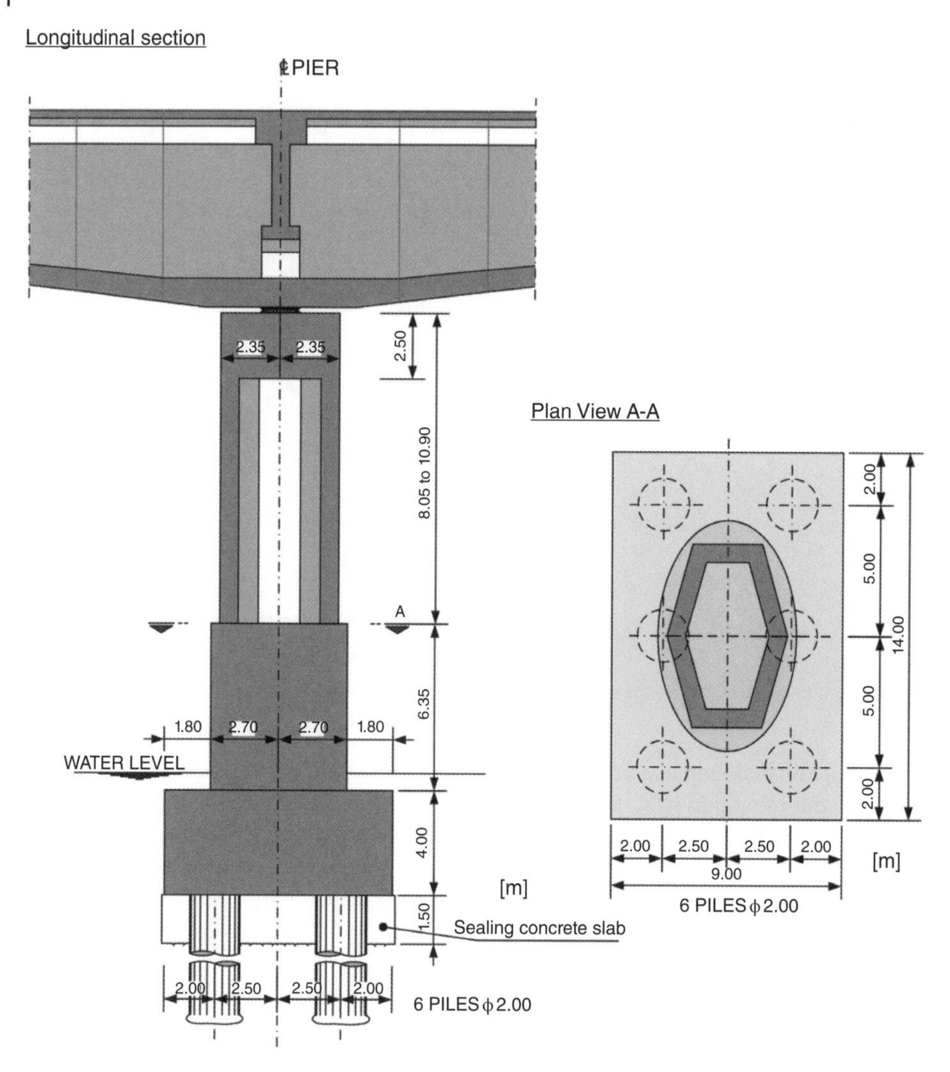

Figure 4.136 Pile foundations in piers of the Zambezi river Bridge (*Source:* Courtesy GRID, SA).

Jet grouting is a soil improvement reinforcement technique done by adopting high pressure water to disaggregate the soil, then cement grout injections are made. A bridge foundation example using this technique is shown in Figure 4.137.

Finally, in the case of *caissons*, a foundation type adopted only for large foundations in water, usually of long span bridges like cable-stayed and suspension bridges. A caisson is a RC prefabricated structure that is driven to a certain depth according to the scheme shown in Figure 4.138. Caissons are driven by a technique called 'havage': progressively excavating the soil at the inside. Soil may be excavated in the open air or by using pressurized air allowing excavation below the water level. Working health requirements limit this technique usually to depths of 25 m, which, even so, requires air pressures of 2.5 atm (0.25 MPa). The reader is referred to specific literature for further information [48] on this type of foundation.

Figure 4.137 Foundation of a pier of a balanced cantilever bridge (typical span lengths of 115 m) in Alqueva dam in Portugal. Direct foundations of piers by adopting jet grouting technique (*Source:* Courtesy GRID, SA).

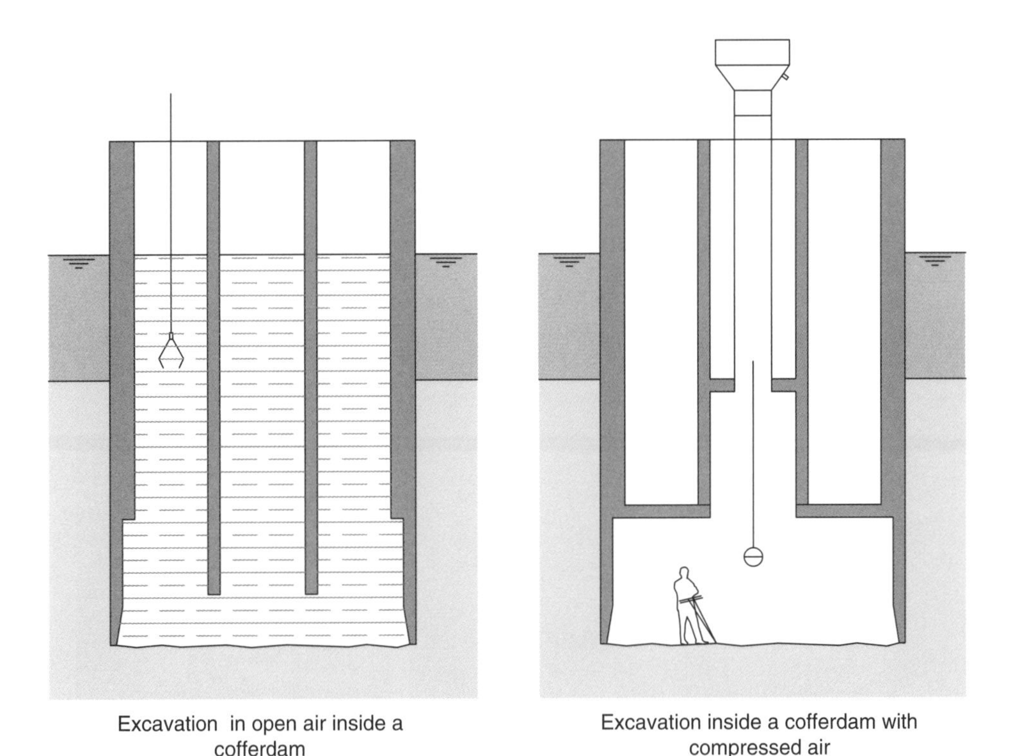

Figure 4.138 Caissons for bridge foundations of piers in water executed by 'havage' (adapted from ref. [48]).

4.7.4.5 Bridge Pier Foundations in Rivers

The main issues for pier foundations in rivers are:

- Foundation levels and scour effects
- Cofferdams for pile foundations in rivers
- Protection against scour effects

The level of a direct foundation of a pier in a river can only be established after a hydraulic study is performed determining the design level against *scour effects* (Figure 2.4), discussed in Chapter 2. The main approaches to determine the referred maximum scour level are discussed in specialized bridge hydraulic manuals [49, 50]. In Figure 4.139, a scheme is shown for the direct foundation in a river after the maximum scour level has been determined.

Execution of pier foundation in rivers generally requires a cofferdam execution, by driving sheet piles to allow to work inside and below water levels. A cofferdam is shown in Figure 4.140, where the bracing system to withstand water pressures may be observed. At the bottom of the cofferdam, a solid RC sealing slab to resist water pressures from

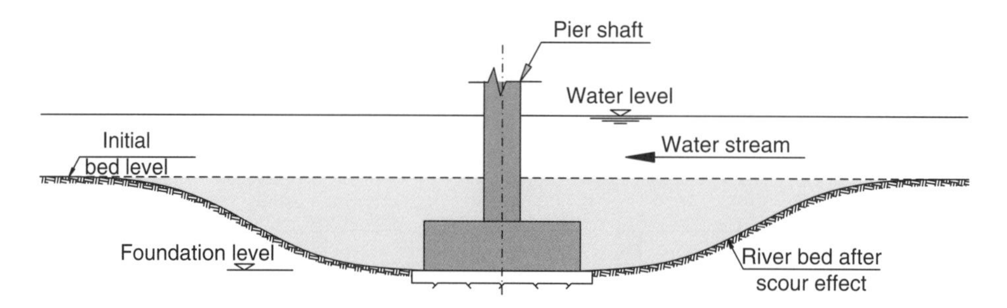

Figure 4.139 Foundation level of a bridge pier accounting for scour effects.

Figure 4.140 Foundations of the Zambezi river Bridge piers executed inside cofferdams. (*Source:* Courtesy GRID, SA).

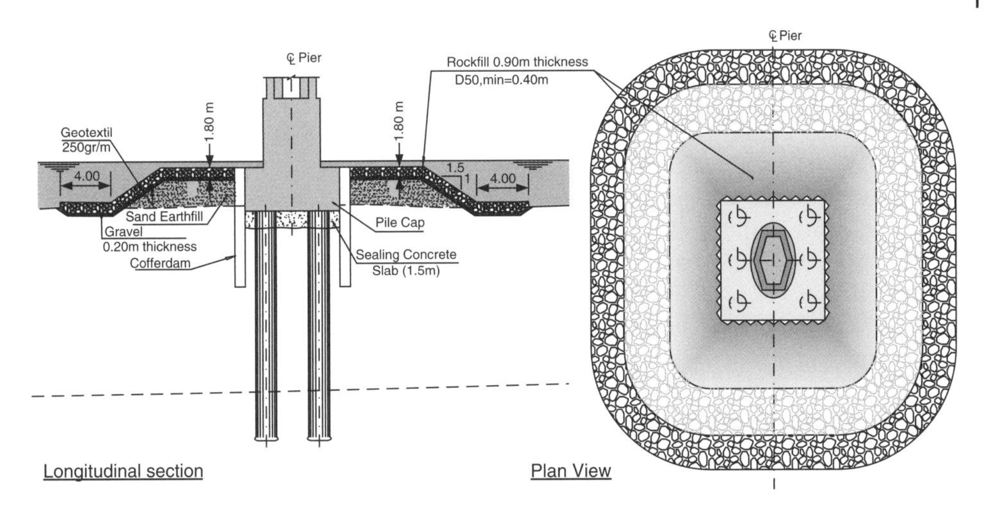

Figure 4.141 Example of a pier foundation protection – Zambezi river bridge design.

the underneath is required. Self-weight of this slab should be sufficient, with a certain safety factor, to withstand upwards water pressures to the underside. Conditions for erecting the steel reinforcement of the pile cap at the inside of the cofferdam do exist. Hence, the pile cap is cast at the end.

For executing piles in rivers, if distance to the riverside is not too large, the simplest scheme is to make a land reclamation, an *artificial peninsula*, usually using sand, and to do the piles from the top of it. If this possibility does not exist, the cofferdam should be driven directly to the riverbed.

After execution of the pile caps, riverbed protection is very often required to avoid local scour around the foundations. Rock materials with a variety of diameters, the upper ones with sufficient weight to resist water flow hydrodynamic forces, are adopted. This is again a bridge hydraulics problem and reference is made to particular literature. In Figure 4.141, an example is shown of the geometry and specification of a foundation protection of a bridge pier in a river.

References

1 UIC Code 774-3-R (2001). *Track/Bridge Interaction – Recommendations for Calculation*, 2e, 70. International Union of Railways.

2 EN 206-1 (2007). *Concrete – Part 1: Specification, Performance, Production and Conformity*. Brussels: CEN.

3 EN 10138-3 (2005). *Prestressing Steels – Part 3: Strand*. Brussels: CEN.

4 fib-CEB-FIP (2000). *fib Bulletin 9 – Guidance for good bridge design*, 180. Stuttgard: Sprint-Druck.

5 SETRA (1996). Ponts à poutres préfabriquées précontraintes par post-tension – Guide de concepcion. 262 pp. France.

6 Schlaich, J. and Scheef, H. (1982). *Structural Engineering Documents SED 1: Concrete Box-Girder Bridges*. Zurich, Switzerland: IABSE, International Association for Bridge and Structural Engineering.

7 Kuhlmann, U. and Breunig, S. (2017). Report on experimental results of beams. OptiBri Research Project RFSR-CT-2014-00026, 57 pp. Institute of Structural Design, University of Stuttgart, Germany.
8 EN ISO 148-1 (2009). Metallic materials – Charpy pendulum impact test – Part 1: Test method. ISO 2009, Switzerland.
9 EN 10164 (2004). Steel products with improved deformation properties perpendicular to the surface of the product. Technical delivery conditions.
10 EN 1993-1-10 (2005). *Eurocode 3 – Design of Steel Structures – Part 1-10: Material Toughness and through-Thickness Properties*. Brussels: CEN.
11 Hirt, M., Bez, R., and Nussbaumer, A. (2006). *Traité de Génie Civil, Volume 10: Construction métallique*. Lausanne: PPUR Presses Polytechniques.
12 EN 10002-1 (2001). *Metallic Materials – Tensile Testing – Part 1: Method of Test at Ambient Temperature*. Brussels: CEN.
13 Paris, P. and Erdogan, F. (1963). A critical analysis of crack propagation laws. *Journal of Basic Engineering* 85: 528–534.
14 EN 10025 (2004). *Hot Rolled Products of Structural Steels: Part 2: Technical Delivery Conditions for Non-Alloy Structural Steels; Part 3: Technical Delivery Conditions for Normalized/Normalized Rolled Weldable Fine Grain Structural Steels; Part 4: Technical Delivery Conditions for Thermomechanical Rolled Weldable Fine Grain Structural Steels*. Brussels: CEN.
15 EN ISO 12944 (2007). Paints and varnishes – Corrosion protection of steel structures by protective paint systems: Parts 1 to 6. ISO 2007, Switzerland.
16 Corus (2009). Steel Bridges – Material Matters: Corrosion Protection.
17 ISO 13918 (2008). Welding – Studs and ceramic ferrules for arc stud welding. ISO 2008, Switzerland.
18 EN 1994-2 (2005). *Eurocode 4 – Design of Composite Steel and Concrete Structures – Part 2: General Rules and Rules for Bridges*. Brussels: CEN.
19 Lebet, J.-P. and Hirt, M. (2009). *Traité de Génie Civil volume 12: Ponts en acier – Conception et dimensionnement des ponts métalliques et mixtes acier-béton*. Lausanne: PPUR presses polytechniques.
20 Reis, A. and Camotim, D. (2012). *Estabilidade e dimensionamento de estruturas*. Lisbon: Orion.
21 EN 1993-2 (2006). *Eurocode 3 – Design of Steel Structures – Part 2: Steel Bridges*. Brussels: CEN.
22 EN 1993-1-5 (2006). *Eurocode 3 – Design of Steel Structures – Part 1-5: Plated Structural Elements: Steel Bridges*. Brussels: CEN.
23. EN1993-1-1 (2005). *Eurocode 3: Design of Steel Structures – Part 1-1: General Rules and Rules for Buildings*. Brussels: CEN.
24 Kollbrunner, C.F. and Basler, K. (2013). *Torsion in Structures: An Engineering Approach*. Springer Science & Business Media.
25 Reis, A. J. and Pedro, J. O. (2011). Axially suspended decks for road and railway bridges. 35th IABSE International Symposium on Bridge and Structural Engineering. London.
26 Bernard-Gély, A. and Calgaro, J.A. (1994). *Conception des Ponts (Design of Bridges)*, 361. Presses de l'Ecole Nationale des Ponts et Chaussées ENPC.
27 Corfdir, P., Kretz, T., Leclerc, G., et al (1994). Pont à béquilles sur l'Ante. Bulletin Ponts Métalliques N°17, pp. 75-91. Ed. OTUA, France.
28 Viñuela Rueda, L. and Salcedo, J. (2009). *Proyecto y construcción de puentes metálicos y mixtos*. APTA.

29 SETRA (2010). Ponts mixtes acier-béton – Guide de conception durable. 196 pp. ISBN: 978-2-11-099163-8.
30 Leonhardt, F. (1979). Construções de Concreto - Vol. 6 - Princípios Básicos da Construção de Pontes de Concreto. Editora Interciência. 241 pp. ISBN: 9788571933378.
31 Podolny, W. Jr. and Muller, J.M. (1982). *Construction and Design of Prestressed Concrete Segmental Bridges*, 543–545. New York, NY: Wiley.
32 AFPC (1999). *Guide des ponts poussés*, 240. Presses des ponts.
33 Reis, A.J., Pedro, J.O., and Dalili, B. (2013). Design and construction of Zambezi River bridge at Caia, Mozambique. *Proceedings of the ICE – Bridge Engineering* 166 (2): 104–125. doi: 10.1680/bren.11.00028.
34 Reis, A.J. (1996). Designing post-tensioned concrete bridges for innovation. *FIP Symposium, Post-Tensioned Concrete Structure* 2: 963–970.
35 Mathivat, J. (1983). *Construction Par Encorbellement des Ponts en Beton Precontraint*, 341. Wiley.
36 Post-Tensioning Institute/Prestressed Concrete Institute (1978). *Precast Segmental Box Girder Bridge Manual*, 116. PTI-PCI.
37 EN 1090 (2008). Execution of steel structures and aluminium structures. Part 1: Requirements for conformity assessment for structural components; Part 2: Technical requirements for the execution of steel structures.
38 Bouchon, E., Boutonnet, L., Epinoux, J.P., et al (1992). La travée métallique du Pont de Cheviré sur la Loire. Bulletin Ponts Métalliques N°15, pp. 115–132. Ed. OTUA, France.
39 Parke, G. and Hewson, N. (2008). *ICE Manual of Bridge Engineering*, 2e. Thomas Telford Lim.
40 Lousada Soares, L. (2003). *Edgar Cardoso: Engenheiro Civil*, 370. FEUP.
41 UIC Code 777-1 (2002). *Measures to Protect Railway Bridges Against Impact from Road Vehicles, and to Protect Rail Traffic from Road Vehicles Fouling the Track*, 2e, 16. International Union of Railways.
42 SETRA (2007). Prestressed concrete bridges built using the cantilever method – Design guide. Translation of the work 'Ponts en béton précontraint construits par encorbellements successifs' published in June 2003 under the reference F0308.
43 Reis, A. J. (2006). Safety of balanced cantilever and cable stayed bridges during construction. 3rd IABMAS Conference on Bridge Maintenance, Safety and Management. Porto.
44 Davenport, A.G. (1967). Gust loading factors. *Journal of the Structural Division* 93 (3): 11–34.
45 Mendes, P.A. and Branco, F.A. (2001). Unbalanced wind buffeting effects on bridges during double cantilever erection stages. *Wind and Structures* 4 (1): 45–62.
46 Dyrbye, C. and Hansen, S.O. (1996). *Wind Loads on Structures*. Wiley, Chichester.
47 Bowles, J. (1997). *Foundation Analysis and Design*, 5e. The McGraw-Hill Companies, Inc.
48 Calgaro, J.A. (2000). *Projet et construction des ponts – Généralités, Fondations, Appuis et Ouvrages courants*. Paris: Presse des Ponts et Chaussées.
49 Neil, C.R. (1973). *Guide to Bridge Hydraulics*. Roads and Transportation Association of Canada, University of Toronto Press.
50 Matias Ramos, C. (2005). *Drenagem em infra-estruturas de transportes e hidráulica de pontes*, 262. LNEC.

5

Aesthetics and Environmental Integration

5.1 Introduction

The roles of aesthetics and environmental integration in bridge design are recognized nowadays as key parameters in any design or design and build tender. Cost and execution methods are no longer adopted as unique parameters for comparing different bridge solutions. Aesthetics is taken into consideration like other criteria related to structural, execution and durability aspects.

Design options must reflect equilibrium between functional requirements, structural safety and durability, cost, planning for execution, aesthetics and environmental integration. If most of these aspects may be compared on basis of objective criteria, aesthetics, and environmental integration may be quite subjective. Apart from this, an even more difficult aspect to define is the 'weight' of an aesthetic parameter to compare bridge solutions.

The 'cost of aesthetics' in a bridge project is not easy to define even if some attempts have been made. Minimum values in the order of 10% have been mentioned [1] as very often accepted in design-built competitions. However, this aspect can not be accepted without taking in to consideration the bridge site and the way the bridge is seen. An urban bridge requires more aesthetics requirements than, for example, a highway viaduct integrated in a site where bridge views are rare. Bridge design should therefore be the outcome of scientific and technological progress respecting integration in the environment and aesthetic perception of structural forms. Engineering sensitivity discussing and taking into consideration environmental issues has very much increased in last few decades.

A series of basic concepts on architectural aspects of bridge design may be formulated from design practice. Even if considered as common sense rules of aesthetics and landscape integration, they may be useful as design guidelines. Design cases to illustrate aesthetical design rules are taken only from authors design experience. This avoids any aesthetic appraisal on bridge designs from other authors, since aesthetics is a subjective aspect. The reader is adressed to the excellent references [2–4] on the subject, for many other design examples.

Bridge Design: Concepts and Analysis, First Edition. António J. Reis and José J. Oliveira Pedro.

As a basic rule, designing a bridge should not be a simple structural exercise; bridges should reflect the art of structural engineering [1, 5] and environmental integration. The reader is referred to more general aspects on architecture [6] and texts on specific projects [7–11] and some outstanding works on the "art of structural design" [12].

5.2 Integration and Formal Aspects

The development of the design concept requires understanding *environment integration* aspects from *formal aspects*. The former includes the scale of the bridge in the landscape and the way span arrangement, location of the piers and geometry of structural elements are integrated with the landscape and the environment.

The last, formal aspects are concepts of slenderness of the deck, 'transparency' of the substructure, namely the impact on the landscape of forms and proximity of the piers, or even more detailed aspects such as fascia beam forms, handrails and lighting posts.

5.3 Bridge Environment

Understanding the bridge environment requires understanding how the space/landscape is occupied by the bridge. A main issue is: what is the relationship between the 'scale' of the bridge and the 'scale' of the bridge site?

The horizontal and vertical alignments, after the bridge location decision is taken, are the most important aspects for site bridge integration. The level of the bridge deck is a key parameter when the bridge is located near a small village. A deck at a high level will require large spans and the general layout of the bridge results in a 'scale' unbalanced with the 'scale' of the village. On the contrary, when the bridge is located in a deep valley (Figure 5.1) a high level deck requires a reduce number of piers and a good balance with the landscape.

Urban bridges are often more difficult for environment integration. The location of the piers is usually constrained by existing roads, railways or any other type of existing construction. The vertical alignment may induce reduced vertical clearances and a deck of 'high slenderness' defined as the ratio l/h, where l is the span length and h the depth of the bridge deck. The situation is even more difficult when the in plan alignment is curved due to local constraints, as already discussed in Chapter 2 (Figure 2.14).

Simple sketches on photographs of the bridge site may be useful tools to approach the main issue about the landscape integration of the bridge. In Figure 5.2, the scale of a suspension pedestrian bridge, with triangular towers and a single main cable suspending a thin precasted concrete deck, designed to be built on a sensitive environment, is approached in this way. By adopting a self-anchored suspension bridge the impact of the abutments and cable anchorages were avoided. The deck is suspended from the main cable through inclined hangers with a triangular geometric arrangement such as a Warren truss. From Figure 5.2 it is possible to assess the scale ratio and affinity between towers and trees and from span deck length and scale of the river. New computer technologies may be adopted to develop renderings of the bridge inserted on topography and

Figure 5.1 A bridge located in a deep valley. The João Gomes Bridge in Funchal, Madeira Island (main span of 125 m, deck at 140 m height from the river bed) (*Source:* Courtesy GRID, SA).

(a)

(b)

Figure 5.2 Bridge scale and integration with the environment. A sketch and a rendering for a pedestrian bridge proposed for a design competition in Serbia Republic (*Source:* Courtesy GRID, SA).

landscape. Some examples have been shown in previous chapters (Figure 2.1). These 3D models allow study of the 'bridge scale', its volumetric occupancy in the landscape and the integration of the super and substructures on the site.

A basic issue is, what is the impact of the structure on the landscape/environment? There is no simple answer for this question since, in some cases, it may be justified to design a bridge 'aesthetically detached from the contest' to be built as a landmark; in other cases there is no need for it – *the structure holds by itself* [5].

5.4 Shape and Function

A basic concept for bridge aesthetics [2, 3] is 'shape follows function'. In bridge engineering, the bridge geometry is quite often the result of structural requirements and the execution method. Some examples are:

- The parabolic depth variation of continuous bridge decks (Figure 5.3a), consistent with its function to large bending moments at support sections and at mid-spans;
- In a frame bridge with inclined piers (Figure 2.1), articulated at the foundations and monolithic with the deck, the piers cross section should be reduced from the top to bottom;
- Bridge structures may be adopted with constant height decks (Figure 5.3c), even with large span lengths, if the height (h) of the deck above the ground is large enough to avoid a negative appearance.

Shape and function may result, as stated, from the execution method. For example, when the deck of a segmental prestressed concrete bridge is built by a balanced cantilever scheme, large bending are developed at support sections during erection; the mid-span sections have much lower permanent bending moments and may have a height less than half of the depth at support sections. However, when a concrete bridge deck is executed by an incremental launching scheme or by a formwork launching girder, a constant depth girder deck is the most convenient one. Of course, the adoption of one of the previously mentioned execution schemes is dependent of the span lengths adopted. For the balanced cantilever scheme, the spans are usually in the order of 60–200 m, while for incremental launching or for formwork launching girders, spans lengths are cost-effective for 30 and 60 m spans.

The longitudinal layout of the bridge is a key aspect for aesthetics quality. Taking the three span bridge in Figure 5.3, a variable parabolic depth is quite often the most convenient solution. A bridge with a linear variable deck near support sections may also be adopted. Comparing the three solutions in the figure, solution (c) is the most easy to execute and solution (a) the most complex one; option (b) represents a compromise. For the case of constant bending stiffness decks ($EI = \text{const}$) and $l_1 = 0.7\,l$ one has $M_1 = 0.4 M_0$ and $M_2 = 0.6 M_0$. For option (a) it is quite convenient to adopt a depth d_0 of order $1.3\,d_1$–$2.5\,d_1$ and the parabolic transition at the end span should be made at a distance $c = l/2$. If the deck is built by a 'balanced cantilever scheme', the rule $c = l/2$ is relevant for the equilibrium of the cantilevers during execution. All these examples show how the execution scheme may affect the layout in elevation of the bridge influencing its appearance.

For a bridge over a river (Figure 5.4) with a deck not too high from the water the appearance improves for large ratios d_0/d_1, say between 2 and 2.5, yielding an 'arch' shape to the superstructure.

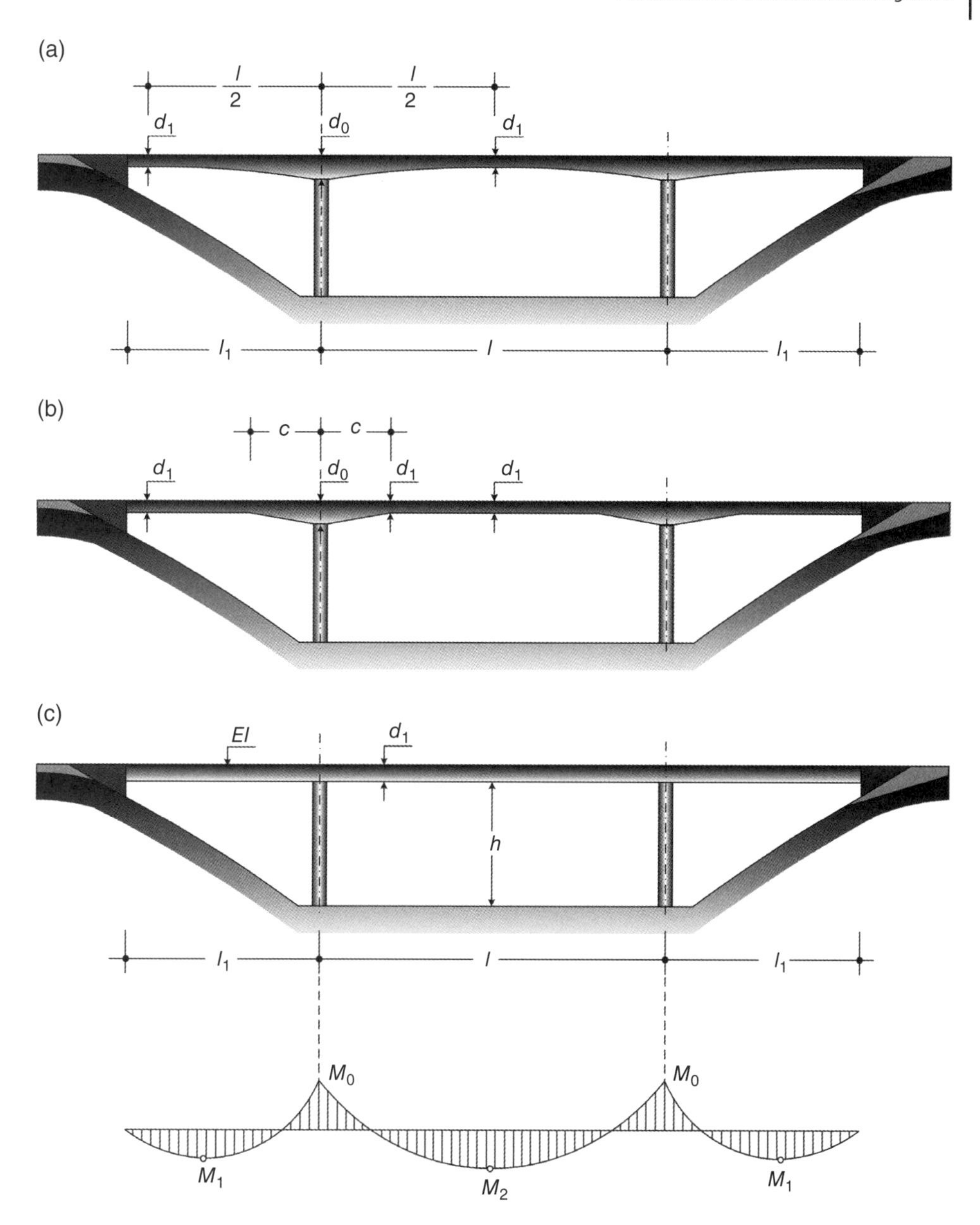

Figure 5.3 Shape and function. Geometry of continuous bridge decks consistent with the shape of the permanent bending moment diagram: (a) Parabolic, (b) linear and (c) constant deck height solutions.

For the case of low longitudinal alignments with decks with large depths at the support sections (Figure 5.4), the width *b* of the piers should be much larger than the width strictly necessary from structural requirements.

A large pier width yields a stable appearance for the bridge and a quite good aesthetic equilibrium with the deck at support sections. If the bridge deck is not continuous with an approach viaduct (Figure 5.4), the apparent 'continuity' of the depth of the deck, at the transition piers, is quite important for aesthetics. A discontinuous variable depth at transition piers (Figure 5.5) should be avoided.

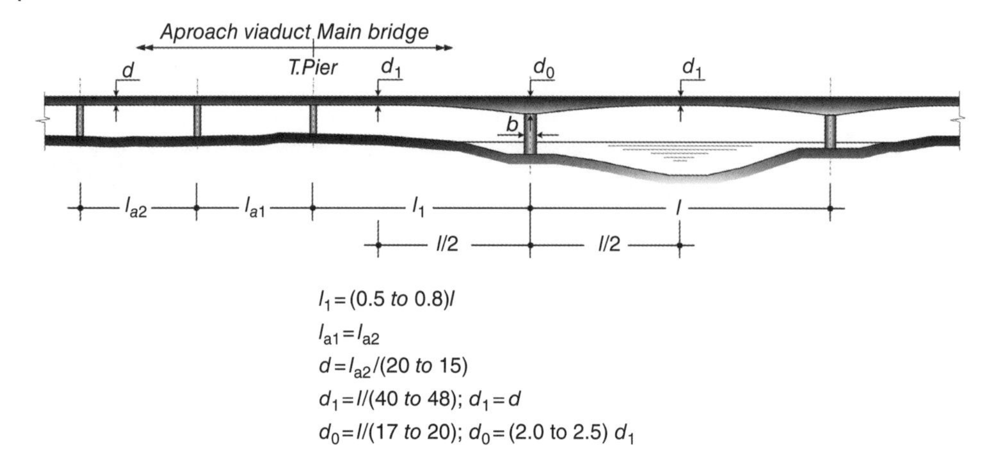

Figure 5.4 Bridges with continuous decks and approach viaducts.

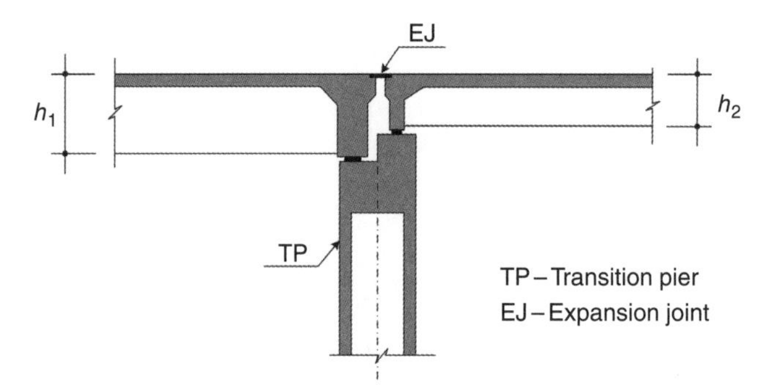

Figure 5.5 Discontinuities at the transition piers.

The concept of 'form follows function' is not restricted to bridge decks, as already exemplified for the piers of the portal frame in Figure 2.1. The shape of the piers influences the bridge appearance [1]. Vertical, inclined or V-type piers may be adopted as required by the span length or for aesthetics. In Figure 5.6 the V-shaped pier at the roundabout has been adopted to reduce maximum free span lengths to 41 m to improve the visual slenderness of the deck. This has also been made to reduce the weight of the precasted girders. The deck is transversely made of three precasted concrete bridge girders with a U shape cross section. The central V shaped pier turns out to be the landmark for the roundabout.

In Chapter 2 (Figure 2.2) another example was shown where V shaped piers have been adopted to allow a slender box girder deck, reusing the formwork of another bridge of the same roadway project, with 42 m span lengths. The V shape piers at a 60 m distance (at the foundation levels) allowed the free span lengths to be reduced to 42 m.

5.5 Order and Continuity

The concepts of 'order and continuity' plays a relevant role in bridge aesthetics, namely for the layout and span arrangement. The case of Figure 5.4, where two different sequences of span lengths are adopted, shows the importance of 'visual continuity' of

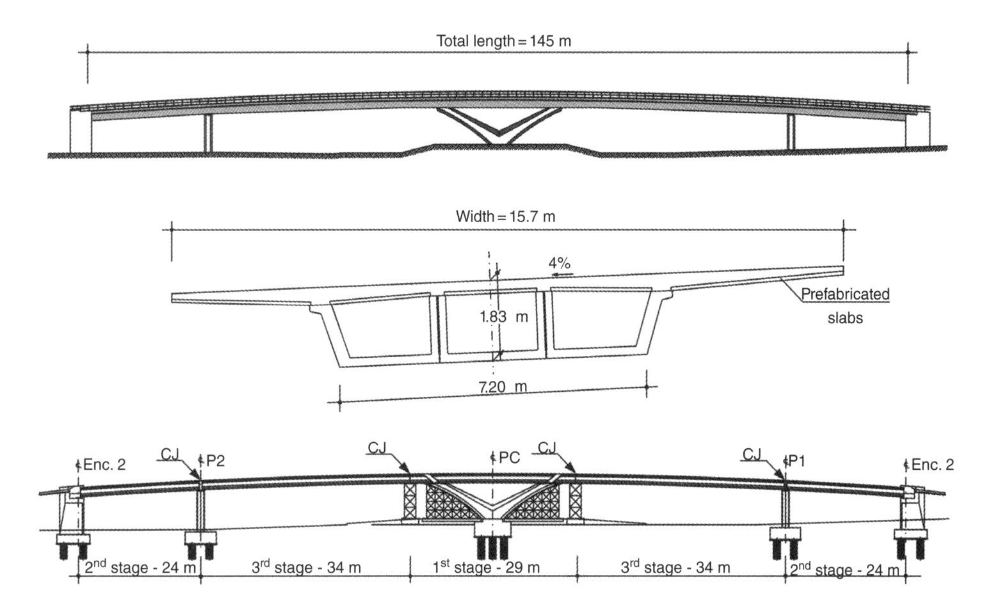

Figure 5.6 The use of V shaped piers in a viaduct in a roundabout (approach to Lisbon airport) to reduce the span lengths of concrete precasted U girders (*Source:* Courtesy GRID, SA).

the bridge deck, that is, a bridge deck for the approach spans with the same depth d_1 of the main bridge at the end span. Of course, d and d_1 depend on the execution methods adopted for the approach bridge (or viaduct) and the main bridge. An expansion joint is adopted between the two bridge decks but a smooth transition, like in Figure 5.4, is always the best option.

The span arrangement should be selected as a function of the execution method, but also as a function of the longitudinal alignment. In a deep valley, span lengths should decrease (Figure 5.7) from the centre to the end spans. The span length/pier height ratio (l_i/h_i) should remain approximately constant at all spans. This condition should be applied for the views of the bridge under its normal service conditions. That is the case of the long

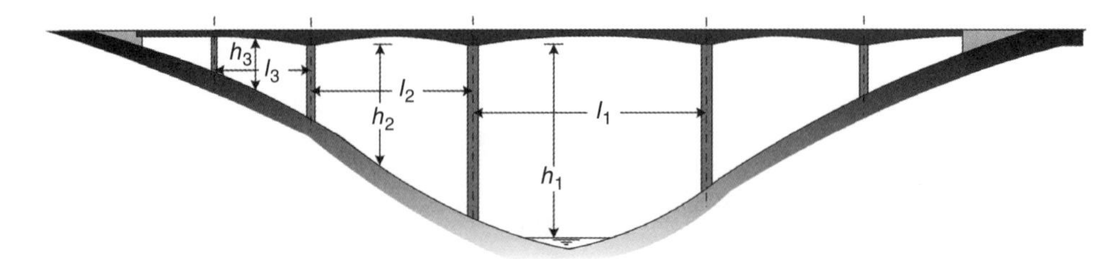

Figure 5.7 Span arrangement for multiple span bridges in deep valleys, from the centre to the end spans keeping the ratios l_i/h_i approximately equal.

bridge (1230 m) in a dam shown in Figure 5.8, where (l_i/h_i) ratios at the bridge sites before the dam was built, were very much different from the 'service' conditions. The water table yields an approximately constant (l_i/h_{ic}), where h_{ic} is the vertical clearance at usual water table levels.

5.6 Slenderness and Transparency

The concept of slenderness may be tied in to the deck or to the piers. Slenderness may be defined as the (span length/depth of the deck) ratio or the (height/cross section dimension) ratio of a pier. Hence, one refers in bridge aesthetics to the slenderness of the superstructure (deck) or of the substructure (piers and abutments).

For roadway superstructures, l/h ratios at mid spans of continuous box girder decks may be as large as 40 or even 50 at mid span sections, provided at support sections, over the piers, lower values are adopted like 17–22. Constant height decks have l/h ratios in the order of 15–27, values depending on the method of execution. For example, for incremental launching of a concrete box girder of constant depth, low values (14–16) may be needed while for the same bridge deck type when executed with a moving scaffolding, a value l/h between 17 and 20 may be adopted.

However, if the incremental launching method is adopted for a composite bridge deck, the slenderness of the deck steelwork may reach 22–27. One relevant aspect when defining the deck slenderness is the height of the deck above the ground.

If the height of the bridge deck to the ground is too low, as is usually the case for urban bridges, the deck slenderness should be increased, as much as possible, for aesthetics. When the height is quite large, for example, a bridge in a deep valley, the slenderness of the deck is not so relevant. The concept of 'apparent slenderness', that is, the slenderness as detected by the viewer, is related to the apparent depth of the bridge deck. The apparent slenderness increases with the inclination with respect to the vertical of the webs of a box girder; larger inclination, say slopes (v/h) from 3 to 1 may be reached for the purpose of reducing the apparent height of the deck. The inclination of the webs of a box girder may be adopted in variable depth girders, by keeping the slope as constant and varying the width of the lower flange, or by varying the web slopes and keeping the width of the lower flange. This last concept yields a more complex formwork but may be interesting when dealing with large spans, say above 120 m. For this option (variable web inclination) one has more vertical webs near the support section where the shear force is higher, and more inclined webs at span sections where the shear force is lower. The gradient of the inclination, that is, the variation (per unit length) of the inclination

(a)

(b)

Figure 5.8 Bridge over the Guadiana River at the Alqueva Dam, Portugal: renderings before and after the dam has been built (*Source:* Courtesy GRID, SA).

Figure 5.9 Freixo Bridge over Douro River in Oporto. Main span of 150 m, width of the deck 36 m, integrating two independent box girders with varying slope webs (*Source:* Courtesy GRID, SA; Photograph by José Moutinho/https://commons.wikimedia.org).

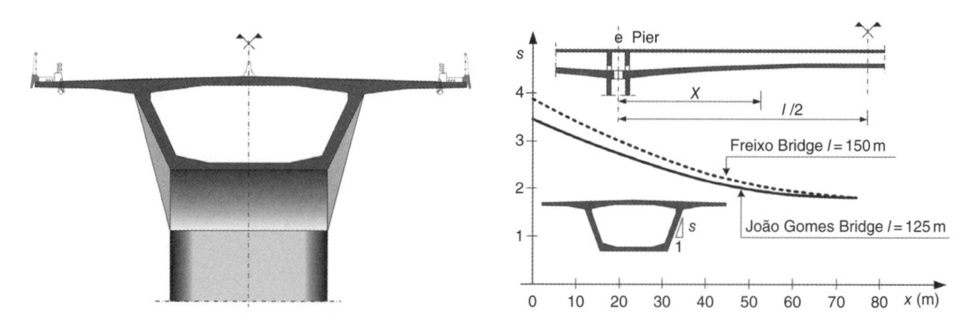

Figure 5.10 Box girder decks with webs of variable inclination. Value of s varies with the distance x of the cross section to the pier for the bridges of Figures 5.1 and 5.9.

of the web along the span length, is larger near the support sections and has a very small variation at the central part of the span. Box girders with webs with varying inclinations were designed for the bridge cases shown in Figures 5.1 and 5.9, with 125 and 150 m main span lengths, respectively. The variation of the inclination of the webs is represented in Figure 5.10 for the mentioned box girder bridges.

The transparency of the substructure is associated with the number and dimensions of the pier shafts. The best solution is of course to adopt a single pier at each bridge support. The transverse dimension of the pier section should be no more than one-eighth of the span length (Figure 5.11) and no more than one third of the deck width to increase the transparency when the bridge is observed from skew angles.

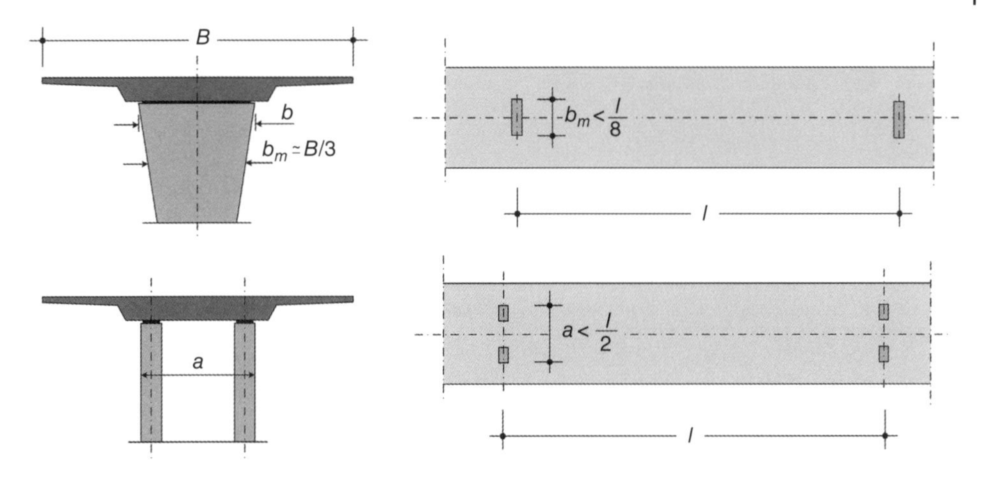

Figure 5.11 Width of single pier shafts and distance between the pier shafts in decks with two piers in the transverse direction.

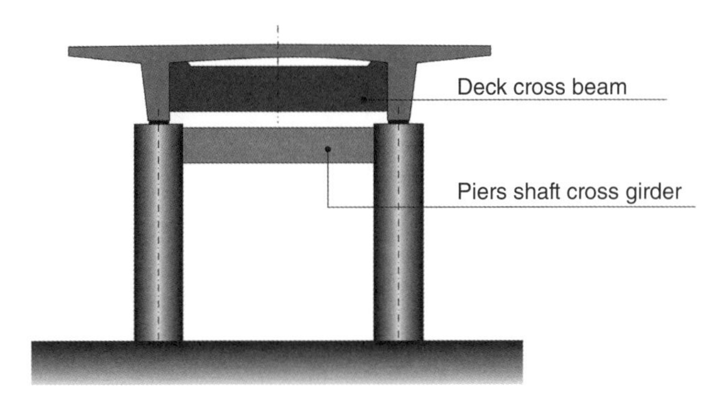

Figure 5.12 A cross girder underneath the deck between the piers shafts always has a negative impact on the transparency of the substructure.

However, situations exist where at least two pier shafts are required at each cross section, as is the case for bridge decks integrated in highways where two independent decks are adopted. In the case shown in Figure 5.9, a bridge deck of 36 m width is made of two independent box girders (18 m width). The ratio between widths of cross sections of the elliptical pier shafts and span lengths was studied to improve the transparency of the bridge substructure.

As a basic rule, for bridge with single wide decks (25–30 m width) where two piers shafts are adopted, the distance, *a*, in the transverse direction between the inner faces of the pier shafts (Figure 5.11), should always be less than one half of the span length, *l*. The adoption of cross beams between pier shafts underneath the lower level of the bridge deck should be avoided because it has a negative impact on the transparency of the substructure for bridge skew views. That is the case of bridges with 'pile-piers' solutions, with a cross girder underneath the deck between piers shafts (Figure 5.12). This creates, for skew views, an *apparent depth* in the order of twice the height of the deck.

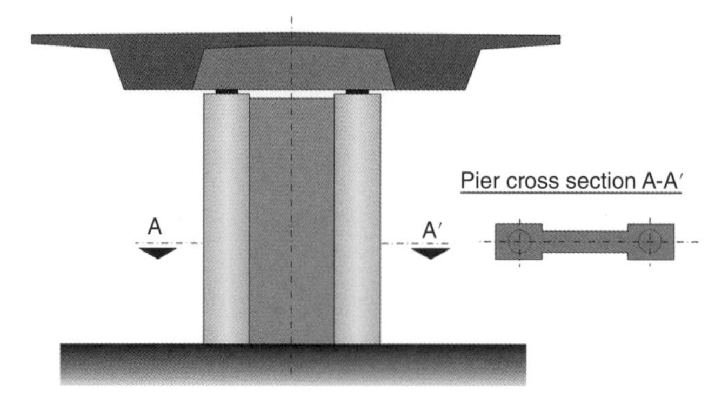

Figure 5.13 A single shaft to support a two-ribbed prestressed concrete bridge deck.

In urban viaducts with ribbed slab decks or made from two precasted small box girders, a single pier shaft with a reduced width may be adopted by inserting a transverse cross girder (Figure 5.13) underneath the slab deck. The transverse bending of this cross girder requires usually transverse prestressing of the beam supported on pier bridge bearings. It is more complex for bridge execution but it is much better for aesthetics compared to a solution with a wide pier shaft to accommodate the bridge bearings.

5.7 Symmetries, Asymmetries and Proximity with Other Bridges

Symmetric structural solutions, for example as three span bridges with a main span and two equal end spans, are simple and aesthetically pleasant in deep valleys with an approximately symmetric topography (Figure 5.1). For these bridge sites, arch or arch type solutions like a frame bridge with inclined legs (Figure 2.1) are also good bridge options, more difficult to execute but with increased aesthetical value for integration at the bridge site. These two last solutions are quite dependent on geological and geotechnical conditions at the foundations of the arch or of the frame pile legs.

However, if the geometry of the valley is not approximately symmetric, a symmetric bridge option may not be the most adequate solution for bridge site integration. The concept of an asymmetric solution may result for aesthetics and integration. Typical examples are two unequal span bridges, as shown in Figure 4.107, where the roadway with four traffic lanes is inserted in two separated box girder decks – North and Southern alignments. In longitudinal cross section, due to topographical and geotechnical conditions, the most convenient solutions are two span frame bridges with a single pier located for each bridge at east and west slopes. One of the cantilevers was built from the tunnel end section and the other one in a symmetric balanced cantilever scheme from the pier. This option explores the asymmetric conditions of the valley for each of the bridge structures, but a symmetric appearance may be observed for the general layout of the valley due to visual composition of both structures.

A final issue is raised for the design of a bridge located nearby an old bridge. As referred to in Ref. [11] 'anything complicated or dramatic would compete with the existing bridge'. An example is given in Figure 5.14 – the Alcácer do Sal Bridge, over river Sado.

Figure 5.14 Alcaçer do Sal Bridge, over the Sado River, designed close to an old arch rail bridge (*Source:* Courtesy GRID, SA).

This road bridge was designed close to an old arch rail bridge of the beginning of last century. Similarity of bridge typologies should be avoided because the new structure should not compete with the old one. They are quite separated in 'time distance', in technological availability of execution methods, but not in physical distance. It is better to idealize a very different structure, reflecting advances in technology and different structural materials. The example in Figure 5.14 shows how two bridge structures well separated in time, with different typologies, can be integrated in the same environment. The new bridge has a very slender box girder deck ($L/h = 22$ at support sections and $L/39$ at span sections) with a main span $L = 85\,\text{m}$ built by a balanced cantilever scheme.

5.8 Piers Aesthetics

Piers play an important role in bridge aesthetics. Its shape and dimensions should be considered from aesthetic and structural points of views. Increasing slenderness does not always improve aesthetics. Bridges with decks at low levels from the ground or from the river, slender piers will induce a feeling of instability and an apparent lack of equilibrium with the superstructure. That is why the main piers are not very slender in the main bridge of Figure 5.14.

As a basic criterion, the cross section dimension of the piers at support sections in the longitudinal direction should be, for aesthetics, no more than 0.6–0.8 of the depth of the deck sections.

For beam and slab decks, the transverse dimension of the piers should be enough to accommodate the bearings at the support sections; otherwise a pier cap (Figure 5.15) is needed or the dimension of the pier shaft needs to increase (Figure 5.16) towards the top. Pier caps have a relevant visual impact on bridge aesthetics and a detailed architectural study should be developed.

Tall piers should have, for aesthetics, a tubular variable section at least in the transverse direction, as shown in Figure 5.17.

For asymmetric valleys, it is better to adopt piers with a constant cross sectional dimension in the longitudinal direction only. For three span bridges in asymmetric deep

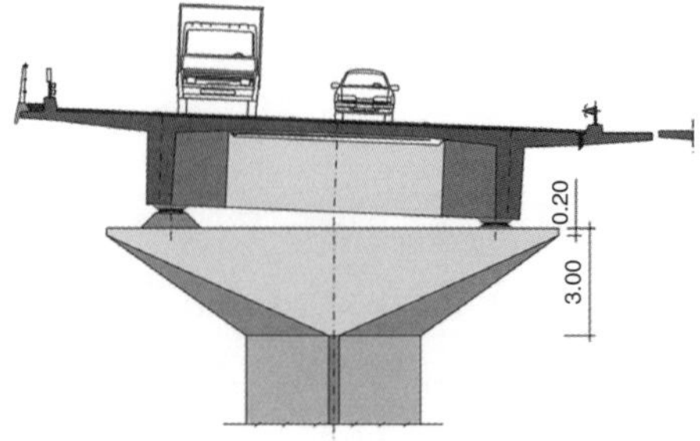

Figure 5.15 Piers caps for slab-girder decks. Integrated in a highway with four traffic lanes, the viaduct in A2 Lisbon-Algarve has two independent superstructures made of prestressed concrete slab and girder decks (*Source:* Courtesy GRID, SA).

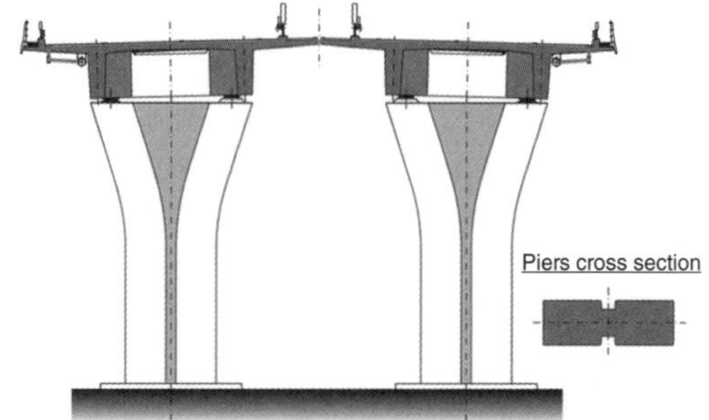

Figure 5.16 Piers shaft increasing the width towards the support deck section for the approach spans of the Sorraia River Bridge in Portugal (*Source:* Courtesy GRID, SA; Photograph by J. Pedro).

valleys, it is better to adopt two piers with the same cross section dimension in the longitudinal direction adopting a variable section in the transverse direction if necessary for stability. An example of this option was adopted for the bridge in Figure 5.1.

For very tall piers, say with more than 80–100 m height, piers with variable cross section in both directions (longitudinal and transverse) may be adopted for the benefit of aesthetics. The variation of the piers shafts cross sections should be done, taking into account the need to avoid warping of the surfaces of the formwork.

5.9 Colours, Shadows, and Detailing

Colour may be the result of assuming structural materials in their natural colour (the grey colour of concrete depending of the type of aggregates used, the dark-grey colour of steel) or artificial colours to identify different materials as in a composite bridge deck where the steel is painted in a different colour from the grey colour of the concrete of the deck slab.

The aim of the colour to be adopted in a bridge, or parts of it, is usually determined by one or more of the following objectives:

- To integrate the bridge with natural colours of the environment, not necessarily avoiding a bridge colour contrasting with colours of the bridge site.
- To increase the impact of the bridge on the site, such as for urban bridges.

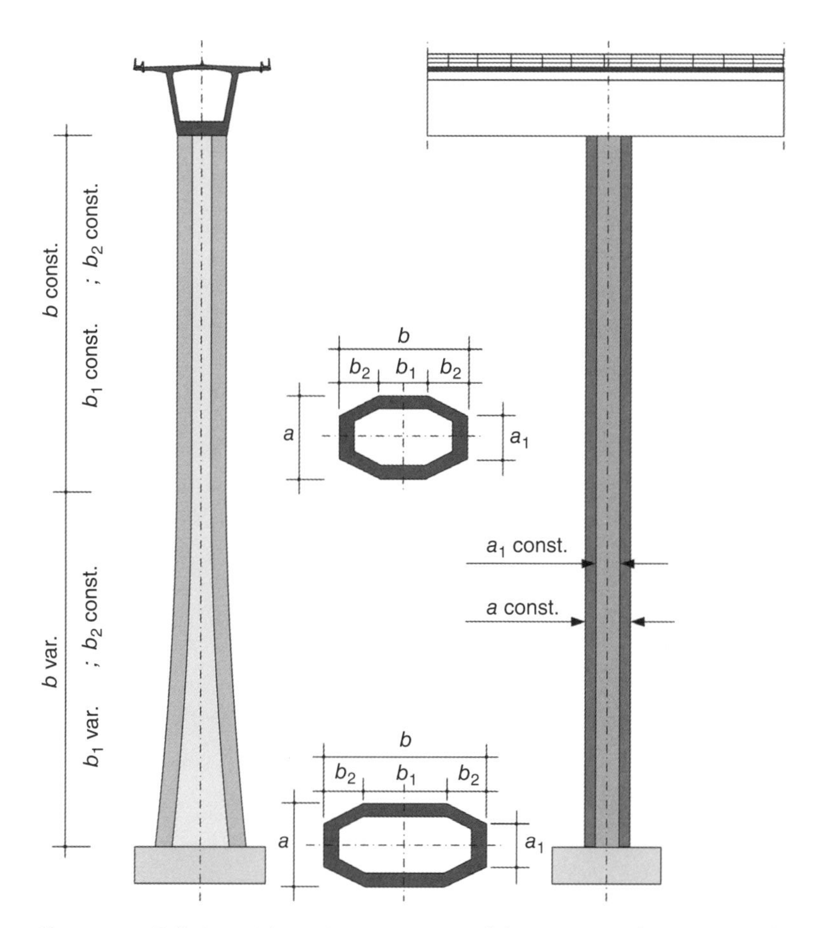

Figure 5.17 Tall piers with varying cross sectional dimensions in the transverse direction only.

- To express the function of some structural elements, for example, some colours adopted for the ducts of stay cables in cable-stayed bridges.

To select bridge colours, design engineers should be aware of some basic concepts about colours [2], namely:

- *Reflexion and absorption* of light in a colour spectrum ranging from white to black; the grey scale in which the white absorbs 0% and reflects 100%; A light grey colour concrete may absorb 20% reflecting 80% while a dark-grey concrete may absorb 30% reflecting 70%.
- *The basic natural colours* (red, yellow and blue) and natural secondary colours (orange, green, violet) obtained by mixing the basic natural colours, for example, 'green = yellow + blue' and 'orange = red + yellow'.
- *The complementary colours*, such as, red/green, yellow/violet and blue/orange, usually identified at 180° positions from each other in a colour wheel; when located one adjacent to the other, these complementary colours induce the most contrasting effects.

Other concepts include colour intensity, adjusted by mixing the basic colour with black, and *colour temperature*, allowing a differentiation on the psychological sensation of the colour, for example, the difference between warm colours, such as red and yellow, and cool colours like green and blue. Natural colours at the bridge site are of course the colours on which the bridge is observed. The orientation of the bridge alignment, north–south to east–west, make quite a difference concerning colour reflection; southern oriented bridge parts have a much lighter appearance under sunshine. The bridges in Figures 5.1, 5.9 and 5.18 are concrete bridges that have been painted in white-blue for

Figure 5.18 The Amieira Bridge over the Alqueva Dam before and after the filling of the dam (*Source:* Courtesy GRID, SA).

protection of the concrete, but also for better integration with the basic colours of the landscape and lighter appearance of the bridges. Painting of concrete has been supported for durability in some environments and as a means to improve the appearance of the bridge. A lighter appearance results from white or light grey colours.

If the aim is to place the bridge to be *aesthetically detached from the context*, complementary colours to the natural ones should be adopted for the bridge; on the other hand, when the idea is to insert ('hide') the bridge in the landscape, the dominant natural colour of the landscape, such as the green of the trees or the blue of the sea, should be adopted.

Details play also a relevant role in bridge appearance. Starting from fascia beams, they may be executed with standard colour grey concrete or white concrete when a lighter bridge appearance is sought. In some cases, a steel fascia beam painted in the same colour as the steel bridge deck may contribute to a uniform appearance of the bridge elements. The depth d of the fascia beam should be at least 0.4 m (Figure 5.19) increasing with the height h of the bridge deck. The apparent surface of the fascia beam may be a single plane or made of two planes; different options may also be adopted as shown in Figure 5.19.

The dominant direction of the elements in handrails should be horizontal like the dominant direction of the bridge superstructure. The colour of hand railings is quite relevant for bridge appearance. Concrete handrails are rare and should be avoided due to their weight, heavier appearance and difficulty to repair.

Shadow effects induced by slab overhangs in box girder bridge decks may be explored to the benefit of an apparent reduced depth of the bridge deck. The apparent slenderness in box girder bridge decks increases with the shadow effect of the overhangs (Figure 5.20), reducing the apparent depth and by adopting inclined webs as previously discussed. Large overhangs should be adopted in vertical web box girders.

Details in piers surfaces are quite relevant for the appearance of the bridge. Large dimension concrete pier surfaces should be avoided since a uniform finish of concrete is difficult to achieve.

Details for the concrete pier surfaces, as shown in Figure 5.21 may be adopted as already presented for the bridge design cases in Figures 5.15 and 5.16. The width c and the depth t_c, of the transition detail should be sufficient to induce a shadow effect. The

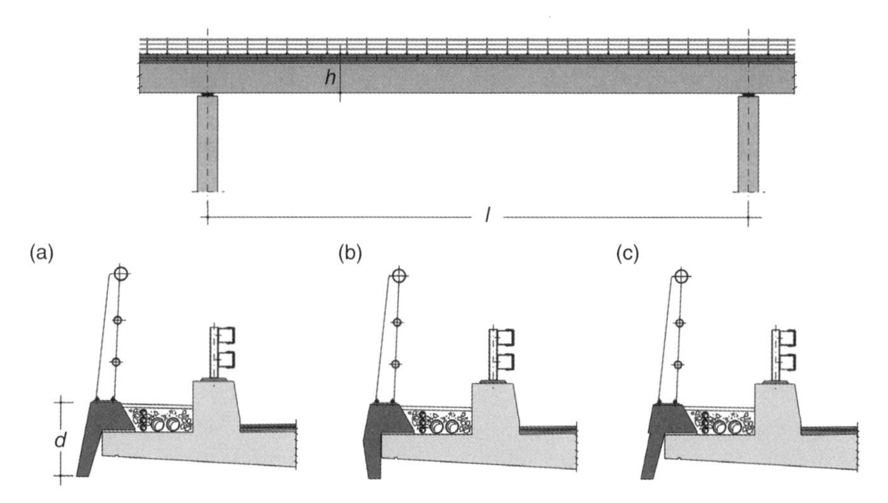

Figure 5.19 Details to improve the appearance of the fascia beams.

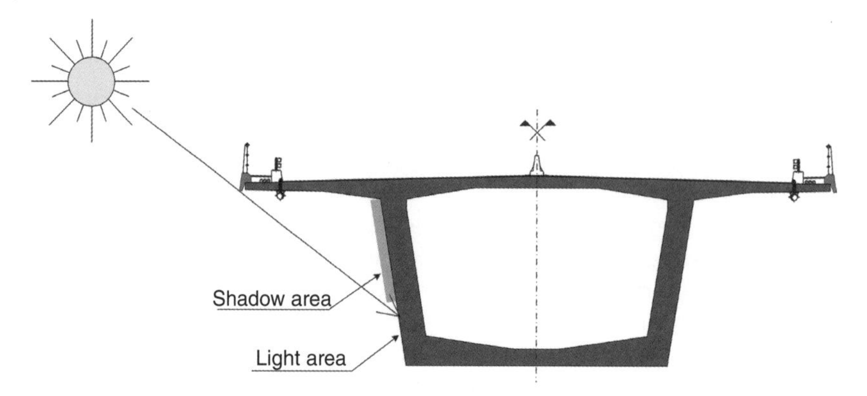

Figure 5.20 Shadow effects induced by slab overhangs in box girder decks.

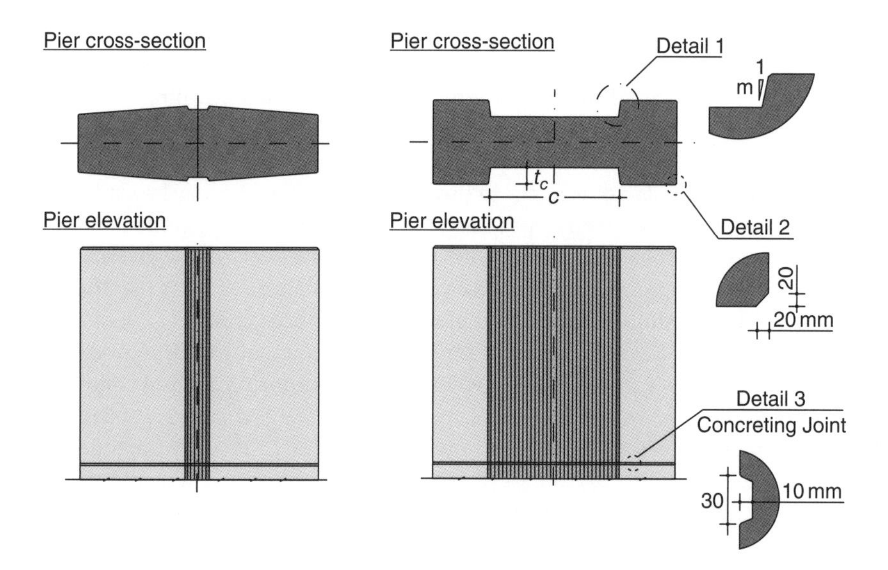

Figure 5.21 Details in pier shafts to improve the appearance of concrete surfaces.

inclination $1/i$ of the detail should be as small possible, but not zero due to the difficulty of taking off the formwork without damaging the concrete finish. Construction joints in pier shafts should include details as shown in Figure 5.21 under the same principle of avoiding large apparent concrete surfaces. The depth of the detail should be sufficient for aesthetics, but small enough to avoid complex detailing of the horizontal steel reinforcement bars. The finish of concrete surfaces in abutments is also relevant for bridge aesthetics, as discussed in Chapter 7.

5.10 Urban Bridges

The design of bridges in urban spaces requires particular attention to aesthetics and environmental issues [9, 10]. The first difficulty is the impact of the bridge in the urban environment, not only concerning aesthetics, but also the impact of the bridge on human life. An urban architect should work with the bridge design team.

Apart from aesthetics, some specific problems concerning the design of urban bridges are as follows:

- The environmental noise impact, very often requiring noise impact barriers, a main issue for bridge aesthetics;
- The execution method to be adopted, very often depending on traffic maintenance requirements during bridge construction or urban occupancy underneath the deck;
- The need to adopt long spans, in some cases due to crossing over roads or railways.

Some bridge design cases are presented next to illustrate these aspects. Figure 5.22 presents the Tagus Suspension Bridge in Lisbon (25th of April Bridge). Approximately 30 years after the bridge and the approach viaduct were built, the railway deck was installed underneath the upper road deck. For the approach viaduct, this required special consideration of:

- noise impact associated with trains rolling on the lower bridge deck;
- an incremental launching scheme to install the steel structure due to the occupancy of the urban space.

For noise impact, it is important to distinguish between *structural noise*, due to vibration of the superstructure under traffic loads, from *aerial noise*, due to the propagation of sound waves induced by the trains rolling on the tracks, which could be controlled by noise barriers. For the case under discussion, a *ballasted track* option on a noise isolation membrane on the top of the deck concrete slab of the superstructure was preferred to a *direct fixation track* on concrete stringers integrating the deck slab. The preferred option resulted in much higher permanent loads (the ballast represents approximately 30% of the total permanent loads of the deck), but the noise impact was much lower. The only way to reduce structural noise impact in a direct fixation solution was to build a separated concrete slab (30 cm thick) isolated by a noise isolation membrane from the structural slab. The top slab, being independent of the structural slab, could not participate in composite action with the steel structure. This represents an additional permanent load similar to ballast with additional inconvenient for maintenance. An option with an *embedded rail* (Figure 2.34) was possible to avoid direct fixation and ballast, but due to the length of the viaduct (approximately 1 km long) and special maintenance requirements, the embedded track solution was discarded. The ballasted option was finally retained.

Concerning the execution method, due to urban occupancy underneath the bridge deck, an incremental launched composite steel-concrete superstructure was the preferred option (Figure 4.112a). The steel superstructure of the viaduct was transported to the site in segments 12–16 m long and erected to two platforms – one located at the north abutment and the other one adjacent to the transition pier with the suspension bridge.

Hence, the superstructure of the viaduct was incrementally launched 600 m from the north abutment in an in-plan straight alignment and 400 m from the opposite side (the platform at the transition pier) in a constant curved (1000 m radius). The launched operations took about 2 hours each ($10\,\mathrm{m\,h^{-1}}$), after which a new segment was welded at the end sections of the steel structure at the platforms, welding control was made and a new launching operation performed.

Traffic maintenance requirements during construction very often require deck solutions built from precasted concrete girders or steel-concrete bridge decks. The classical

Figure 5.22 The Tagus River suspension Bridge in Lisbon after the second phase of construction for the addition of railroad deck at the lower level of the existing truss deck and the northern approach viaduct (*Source:* Courtesy GRID, SA; Photographs by J. Pedro).

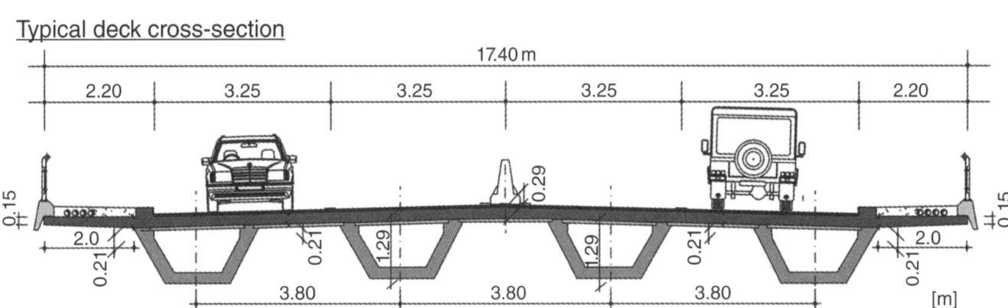

Figure 5.23 Concrete decks made from precasted U girders. The cross beam is inserted at the pier cross section for aesthetics. (*Source:* Courtesy GRID, SA.)

solution with concrete I shape concrete girders is the most commonly adopted, with usually poor results for bridge appearance. The slenderness of the girders is usually low due to the construction phase were the beams have to support the slab deck dead weight cast on the top of the girders and the beams are quite often supported at the piers by cross girders. A much better solution for aesthetics, but more difficult to execute, is to adopt small box girders supported by an embedded cross girder at the piers (Figure 5.23). The concrete slab may be executed on precasted slabs supported by the U prestressed concrete girders. Another option is to adopt small steel box girders, made from U shaped welded sections supported directly on piers that could be installed during the night (Figure 5.24); the deck slab is then cast, as for the previous solution, with concrete precasted U girders. In all these solutions, the end results a multiple box girder, very often far more slender than equivalent solutions made from I girders. The complexity of the construction may be slightly higher, but that is the price to pay for aesthetics.

Finally, the case of very long spans requiring traffic maintenance in urban viaducts is discussed. Two main issues may be mentioned:

- The erection/execution method
- The slenderness of the superstructure

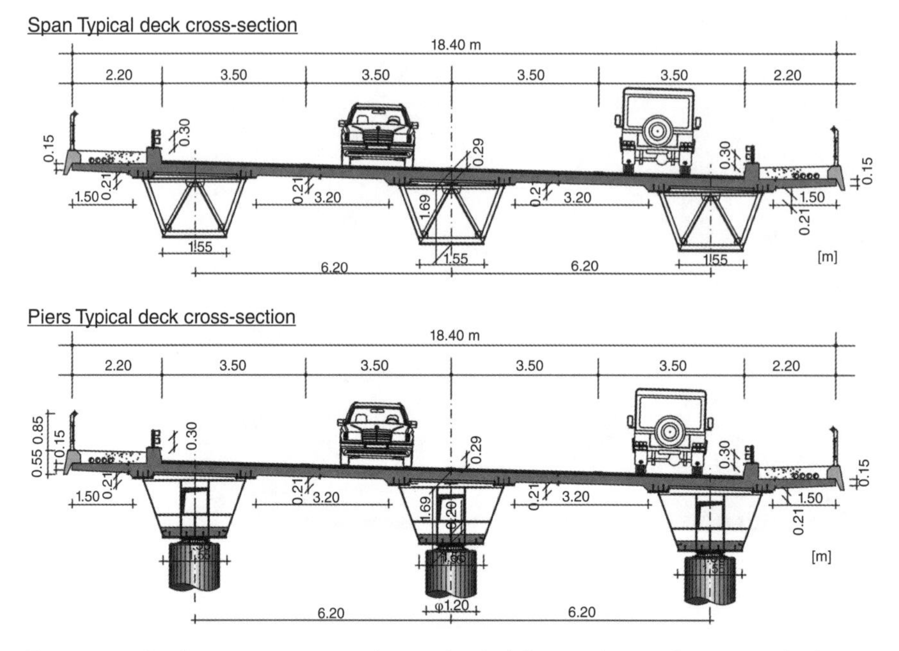

Figure 5.24 Steel concrete composite box girder deck for an urban viaduct over an highway made from U steel girders erected directly over the piers during the night.

If main spans in the order of 100 m are under discussion, a prestressed concrete box girder or a steel box girder are usually the preferred options.

For slenderness reasons, aesthetics associated with inclined webs, reduced pier widths and adequacy of execution methods, box girder options are preferred. The main issue with box girders is the need for a sufficient depth at support sections, and minimum free depth (order of 2 m) at span sections. The deck built by a cantilever scheme would require a support cross section depth of at least 5 m, yielding in most cases a negative aesthetic impact at the urban environment. It should be pointed out that, quite often, the deck stands at a low level from the roads below, with a minimum vertical clearance of 5.5 m.

Cable-stayed solutions or bowstring arches, even for moderate span lengths as usually required in urban environments, may help in improving aesthetics. The first case is illustrated by a cable-stayed viaduct in Oporto (Figure 5.25a) where a 120 m main span was the required length to overcome several traffic maintenance requirements in the underneath roads and rail tracks. The solution, retained in a bridge design competition, was a cable-stayed bridge with a single mast and a 120 m main span built by a cantilever scheme from the mast support sections. The deck was a prestressed concrete box girder with steel diagonals (Figure 5.25b), since the staying scheme in the main deck was made with a single plane of stays located along the central pedestrian walkway, 5 m width. A 3D arrangement for the staying scheme was adopted, with two arrangements for the backstays. The mast was designed as an iconic element for the site with the pedestrian walkway passing thought it. The reader is referred to [10] for further details on this urban bridge design.

(a)

(b)

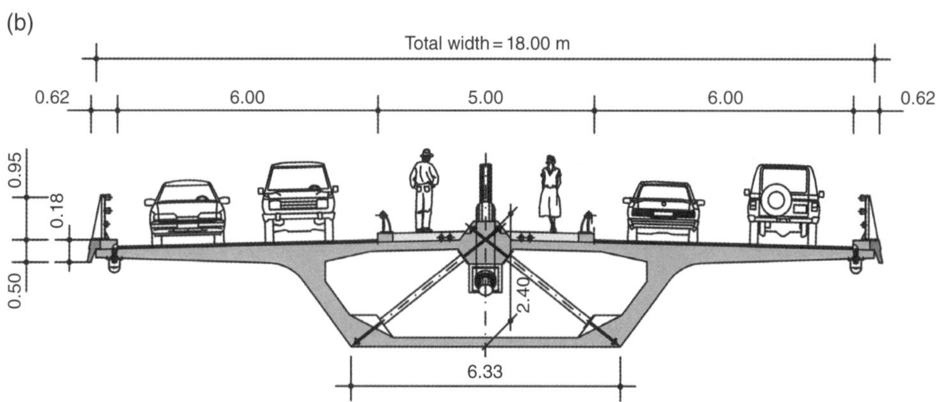

Figure 5.25 A cable stayed viaduct in Oporto with a main span of 120 m: (a) viaduct as built (*Source:* Courtesy GRID, SA) and (b) deck cross section.

References

1 Menn, C. (1998). Functional shaping of piers and pylons. *Structural Engineering International* 8 (4): 249–251.

2 Leonhardt, F. (1982). *Bridges – Aesthetics and Design*, 308 pp. London: The Architectural Press Ltd.

3 Strasky, J. (1994). Architecture of Bridges as Developed from the Structural Solution. Proceedings, XII FIP Congress/40th PCI Annual Convention, Washington, D.C.

4 Gottemoeller, F. (1988). *Bridgescape: The Art of Designing Bridges*, 288 pp. Wiley.
5 Reis, A. (2010). Two large bridge projects in environmentally constrained areas. In *IABSE Symposium Report*, 97(17), 61–68. International Association for Bridge and Structural Engineering.
6 Melvin, J. (2006). *Découvrir l'architecture*, 160 pp. Eyrolles.
7 Calzón, J.M. (2006). *Puentes, estructuras, actitudes*, 373 pp. Turner.
8 Tzonis, A. and Donadei, R. (2005). *Calatrava Bridges*, 272 pp. Thames & Hudson.
9 Reis, A. (2001). Urban bridges: design, environmental and construction issues. *Structural Engineering International* 11 (3): 196–201.
10 Reis, A., Pereira, A., Pedro, J. and Sousa, D. (1999). Cable-stayed bridges for urban spaces. In Proceedings of the IABSE Conference on Cable-Stayed Bridges: Past, Present and Future. International Association for Bridge and Structural Engineering.
11 Bennett, D. (1997). *The Architecture of Bridge Design*, 200 pp. Thomas Telford.
12 Billington, D. (2330). *The Art of Structural Design: A Swiss Legacy*, 211 pp. Yale University Press.

6

Superstructure

Analysis and Design

6.1 Introduction

Analysis and design of bridge superstructures is made, after the concept design stage, and is made on the basis of simplified models. Concept design may need modifications as a consequence of detailed analysis, but, no matter what, the complexity of the structure concept, design should be made on the basis of simple structural models highlighting bridge structural behaviour. The complexity of the structural models should progress with design evolution from concept to final design stages.

Structural analysis and design integrates:

- Definition of structural models for the complete structure and evolution of static schemes under construction;
- Analysis of internal forces, stresses, deformations and displacements;
- Design verifications of prestressing steel reinforcement in concrete superstructures or steelwork in steel and composite bridges;
- Design verifications for detailing and bridge equipment, like bearings, dampers and expansion joints.

In this chapter, specific aspects of structural analysis of bridge decks are presented and discussed. The aim is not to cover general aspects of structural analysis and prestressed structures or steel structures, but main specific aspects of analysis and design of bridge superstructures. It is assumed the reader is familiar with the general aspects referred to above.

One specific aspect of bridge structural analysis is the need and possibility of performing transverse and longitudinal structural analyses on the basis of two different models. Of course, this approach may be replaced by an integrated 3D shell model analysis where transverse and longitudinal behaviours of the superstructure are jointly considered. However, this is not recommended before transverse and longitudinal models have been developed and interpreted separately.

Bridge Design: Concepts and Analysis, First Edition. António J. Reis and José J. Oliveira Pedro.

6.2 Structural Models

Design should be made on global internal forces at deck transverse sections or even at independent girders connected by the deck slab. Box girder decks can be analysed by a 3D beam analysis with local effects on the slab deck and cross section deformations considered on the basis of transverse models.

Local and global effects in bridge decks may be explained on the basis of bridge deck models from Figures 6.1 and 6.2. Both cases are longitudinally three-span continuous decks. The former case is a solid slab deck under a concentrated load at transverse section AA′. Longitudinal bending moments in the slab (kNm m^{-1}), at a transverse cross section AA′, are defined per unit width as stress resultants

$$M_x(y) = \int_{-h/2}^{h/2} \sigma_x(y)dz \tag{6.1}$$

If a beam analysis is done, a positive bending moment induces a linear stress distribution along the depth of the slab, uniformly distributed along the cross section AA′. At the bottom fibres, one has stresses $\sigma^i_x(y)$ evaluated by a strength of materials beam model, with an average value $\sigma^i_{xm}(y)$. However, real stresses should be evaluated by $\sigma^i_{xm}(y) = M_x(y)\frac{z}{I}$ where $I = 1 \times h^3/12$ and M_x is the non-uniform bending moment distribution along the cross width as shown in the figure. The average bending moment at section AA′ is:

$$M_{xm}(y) = \frac{1}{b}\int_0^b M_x(y)dy \tag{6.2}$$

This value is equal to M/b where M is the bending moment evaluated for a beam model. If one now takes the slab-girder case of Figure 6.2, preventing transverse cross section deformations by diaphragm action, the deck has a longitudinal beam behaviour with bending, shear and torsion moments at cross sections evaluated by a beam linear analysis. Vertical transverse displacements v and torsional rotations θ are beam

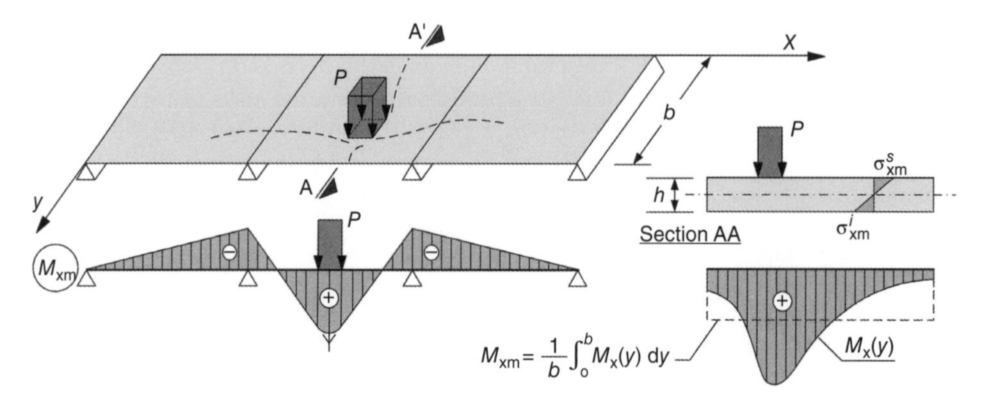

Figure 6.1 Longitudinal bending moments $M_x(y)$ of a transverse section of a three span continuous deck slab, under a concentrate load P, and average (beam) moments M_{xm}.

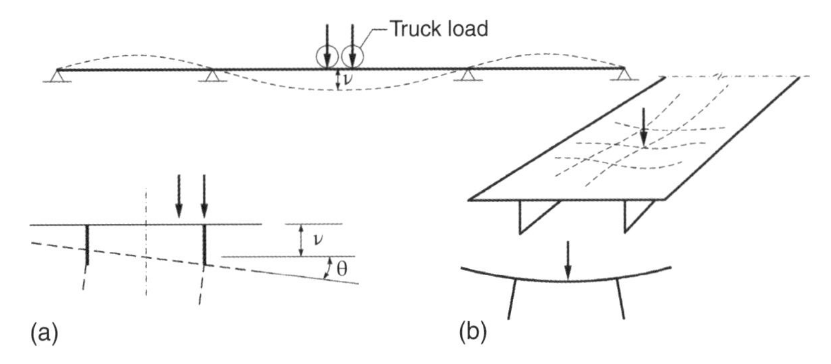

Figure 6.2 Global and local behaviours of a slab-girder deck under a truck load.

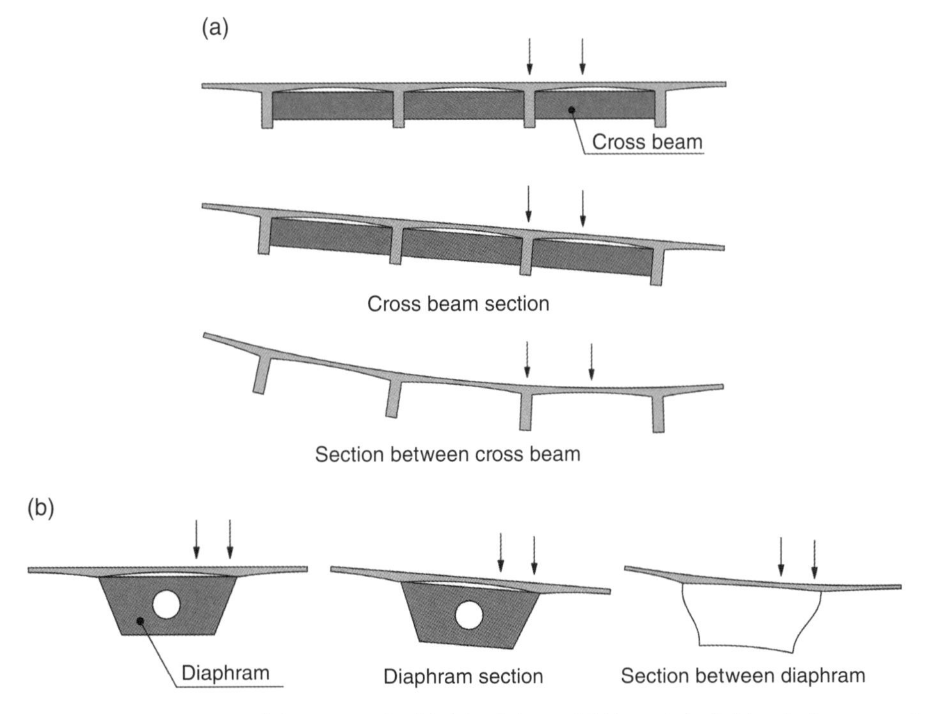

Figure 6.3 Cross section deformations for (a) slab-girder and (b) box girder bridge deck cross sections.

deflections associated with the *global behaviour* of the deck. Vertical displacements of the deck slab between girders can only be evaluated by a slab analysis. These deflections $w(x, y)$ – x and y being local coordinates at the mid-surface of the slab and w the slab vertical deflection, are associated with *local behaviour*. Local transverse deflections at level of the cross section contribute to a transverse load distribution between girders in a slab girder deck (Figure 6.3a). Nearby, if the cross beams existent, the deck has a global beam behaviour allowing a transverse distribution of load effects on the basis of a simplified model. This will be shown in one of the following sections. In a box girder deck, in particular, a prestressed concrete box girder, transverse rigidity of the cross section allows to assume in general, a beam behaviour under bending and uniform torsion.

A concentrated load, in a section of a single cell prestressed concrete box girder, is distributed approximately uniformly between the webs. In Figure 6.3b) one shows the behaviour of slab girder decks and box girder decks taking into consideration cross girder/diaphragm action.

The global behaviours of bridge superstructures yields a 3D behaviour that can be modelled by a shell Finite Element Model (FEM), webs and flanges of the deck having membrane internal forces (N_x, N_y and N_{xy}) and bending and torsion moments (M_x, M_y, $M_{xy} = M_{yx}$). All these internal forces are defined as stress resultants (Figure 6.4) and can be evaluated by an shell model of the superstructure. However, the design in most cases is based on global internal forces at the bridge deck and membrane and bending moments for local effects. These internal forces are the ones adopted for designing, for example, the overall prestressing cables in a PC bridge and local bending moments to design transverse prestressing in a deck slab. If analysis is done on the basis of a FEM, it is necessary to determined overall cross section stress resultants (internal forces for a beam model) for prestressing design or to check Ultimate Limit States (ULS) shear capacity in webs of box girders. Local effects on slab decks are usually analysed on the basis of a slab model elastically restrained by the main girders or by the webs of box girders. Deck-slab, when not connected to diaphragms, may be analysed by modelling as an infinite slab as shown in Figure 6.5.

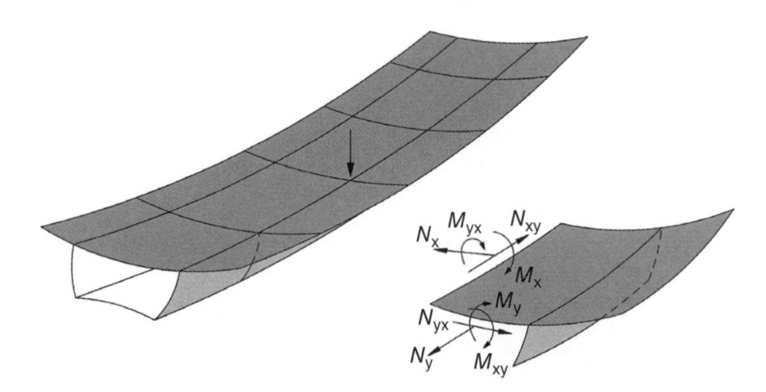

Figure 6.4 Membrane, bending and torsion moments in plate elements of box girder decks.

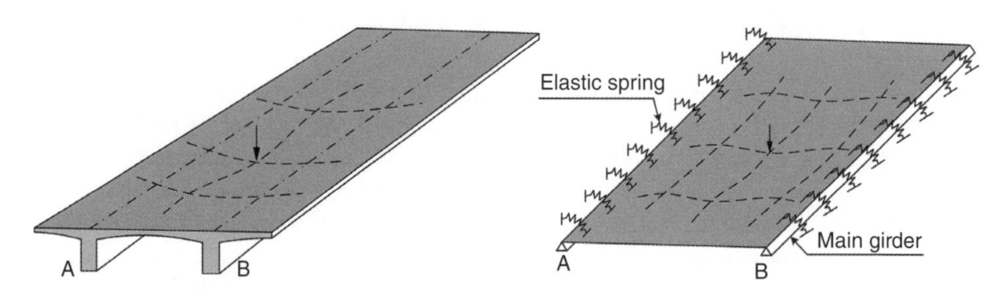

Figure 6.5 Deck-slab of a bridge deck. Model for local effects.

6.3 Deck Slabs

6.3.1 General

Main functions of deck slabs are to support direct load effects due to traffic loading and to connect transversely main structural elements of the deck. Deck slabs connect main girders in girder slab or ribbed slab decks, and webs of box girders as well, allowing a transverse load distribution. Besides, the slab deck works as a compressed or tension flange of the superstructure. Hence, it is subjected to:

- In-plane loading due to overall bending actions in deck transverse cross sections.
- Out of plane loading inducing local bending and torsion internal forces in the slab.

As discussed in Chapter 4, deck slabs may be made by orthotropic plates in steel bridges (Figure 4.49) or in reinforced or prestressed concrete in concrete or composite bridges. Several examples have been given of deck slab typologies in Chapter 4 (see, e.g. Figures 4.13, 4.17 and 4.24). Concrete deck slabs will be dealt with specifically in this chapter. A minimum thickness of 20 cm, better at 22 cm, is usually adopted for durability reasons. Nearby the main girders or the webs of box girders, the thickness is increased to withstand local transverse bending moments and local transverse shear forces. At the tip of the overhangs the thickness is generally decreased to a minimum of 15 cm, better at 20 cm if transverse prestressing is adopted (Figure 4.17). Deck slabs may be, between girders or webs of box girders, a constant or variable thickness (linearly segmented or parabolic) as discussed in Chapter 4.

6.3.2 Overall Bending: Shear Lag Effects

Overall bending of cross section decks induces normal stresses in deck slabs with a non-uniform transverse distribution because the slab is more rigid, for in-plane shear, near the main girders. The non-uniform distribution of normal stresses due to overall bending of the cross section is known as *shear lag*. This is a problem associated with a wide flange as shown in Figure 6.6. In-plane shear deformations of the slab deck should be compatible

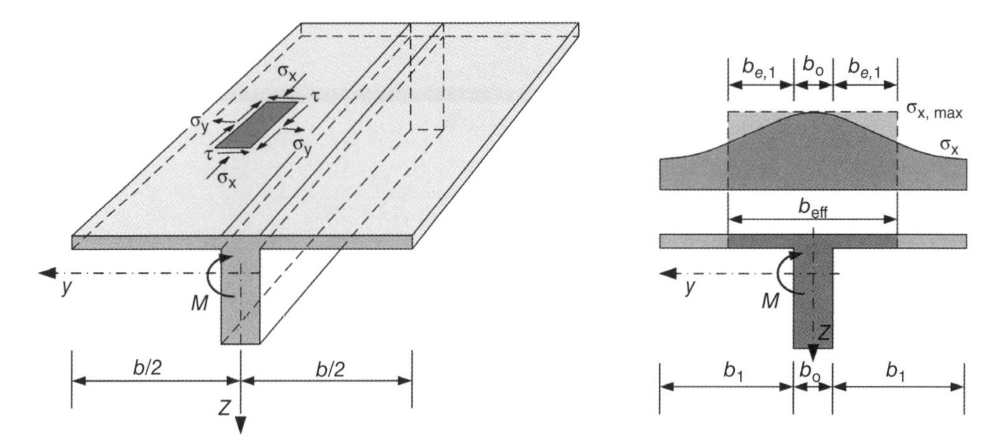

Figure 6.6 Shear lag and effective width concepts.

with longitudinal deformation of the girder at plate girder junction. The increased in plane stiffness at the slab-girder junction yields a non-uniform distribution of the longitudinal stresses σ_x as shown in the figure, compared to the average stress, given by strength of materials, $\sigma_x = M/W_s$, where W_s is the section modulus with respect to the middle surface of the deck-slab. Elastic shear lag is a linear effect that can be dealt with on the basis of an *effective width* concept illustrated on Figure 6.6. Analytical series solutions and based on energy methods for elastic shear lag in slab decks are available [1, 2] but nowadays a finite element model of slab and beam elements, or with shell elements, may be easily developed yielding the non-uniform distribution of normal bending stresses represented in Figure 6.7. Maximum bending stresses σ_{max} occur on the girders or on the webs of a box girder and an *effective width*, b_{eff}, may be defined according to Figure 6.6. Hence, the effective width is defined as the flange width of an equivalent T section with uniform flange stress equal to σ_{max}. Knowing $\sigma_x(y)$, b_{eff} is determined from

$$\int_{-b/2}^{b/2} \sigma_x(y)dy = b\,\sigma = b_{eff}\sigma_{x,max} \tag{6.3}$$

An *effective width* of the deck slab works together with the main girders, this effective width being always less than the transverse distance between girders or between webs

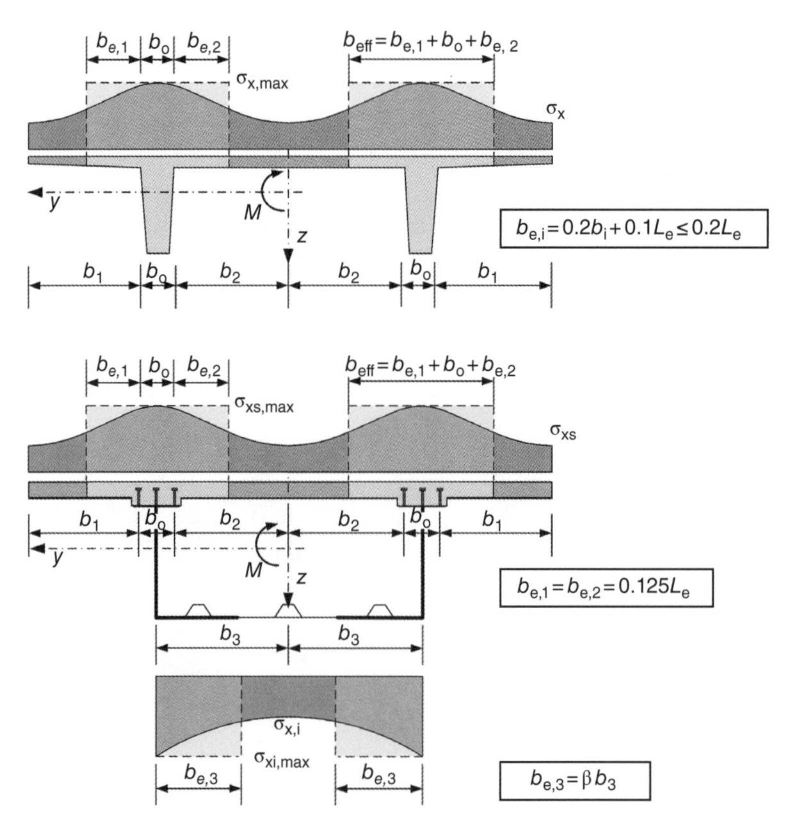

Figure 6.7 Shear lag in concrete and steel-concrete composite decks.

of a composite plate girder. In Figure 6.7, the same concept is shown for the steel lower flange. The main parameters affecting effective widths for shear lag are the width of the flange, more precisely the *width/span ratio* defined here by the parameter $\kappa = b/L_e$ and loading. The length L_e is the length of an equivalent simply supported span, that is, the distance between zero moments (inflexion points) in a continuous deck. In concrete decks, shear reinforcement also plays a role.

Shear lag effective width at the ULS is not necessarily equal to the effective width for SLS. At the ULS a plastic redistribution may be accepted [3–6] and the effective width increases. This effect is taken into account in steel bridges (elastic and plastic effective widths) where shear lag effects are particularly important for checking fatigue resistance [69]. For concrete and composite bridge decks a single effective width is defined for ULS and SLS [7, 8]. The length L_e is defined for intermediate spans as $L_e = 0.7L$, where L is the internal span of a continuous girder. For definition of b_{eff} over support zones, L_e is taken as shown in Figure 6.8 with different values for steel-composite bridge decks and concrete decks. These different effective widths values are adopted for stress checking and ultimate moment cross section resistance. For structural analysis of decks, a uniform effective width along the entire deck may be taken.

The effective width may be defined in a unified way for any bridge type cross section and material, in terms of the *width/length ratio* parameter, κ

$$b_{eff} = \beta b \tag{6.4}$$

where $\beta = \beta(\kappa)$ is the *effective width parameter*. In concrete and composite decks, the width of the web or the distance between shear connectors at the flange level,

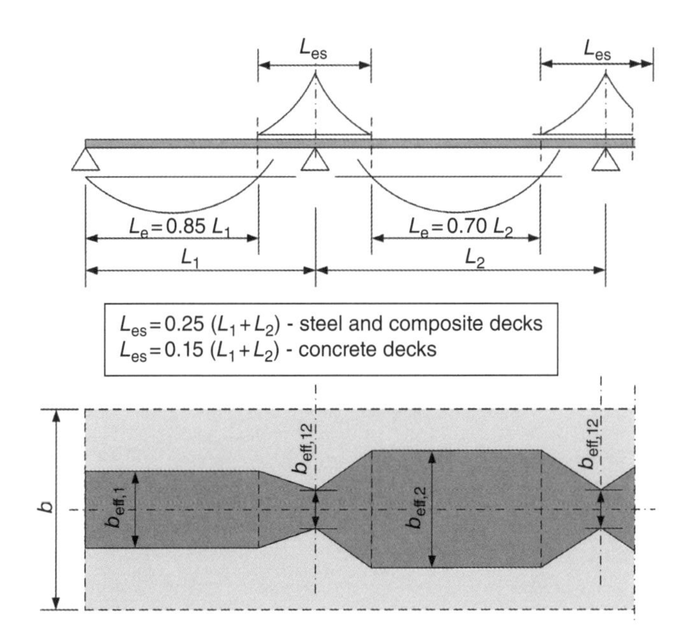

Figure 6.8 Definition of equivalent length for concrete, steel and composite bridges.

respectively, is always considered fully effective. By adding the remaining effective width parts of the flanges b_{ei} one has

$$b_{eff} = b_0 + \sum b_{ei} \qquad \text{with } b_{ei} = \beta b_i \tag{6.5}$$

$$\beta = 0.2 + \frac{0.1}{\kappa} \leq \frac{0.2}{\kappa} \qquad \text{for concrete decks} \tag{6.6}$$

$$\beta = \frac{0.125}{\kappa} \qquad \text{for composite decks} \tag{6.7}$$

Of course, the effective width of a flange cannot be larger than the real width (b); for widths b less than 1/16 of the length L (concrete and composite decks) and $L/25$ in steel decks, the effect of shear lag may be neglected ($\beta = 1.0$). Expressions for shear lag effective widths of concrete and composite decks are given in [7, 8].

For composite decks the effective widths b_{ei} are equal to $L_e/8$. For steel decks, effective widths are defined for positive and negative bending moments in EC3-1-5 [5] yielding the shear lag parameters β of Figure 6.9.

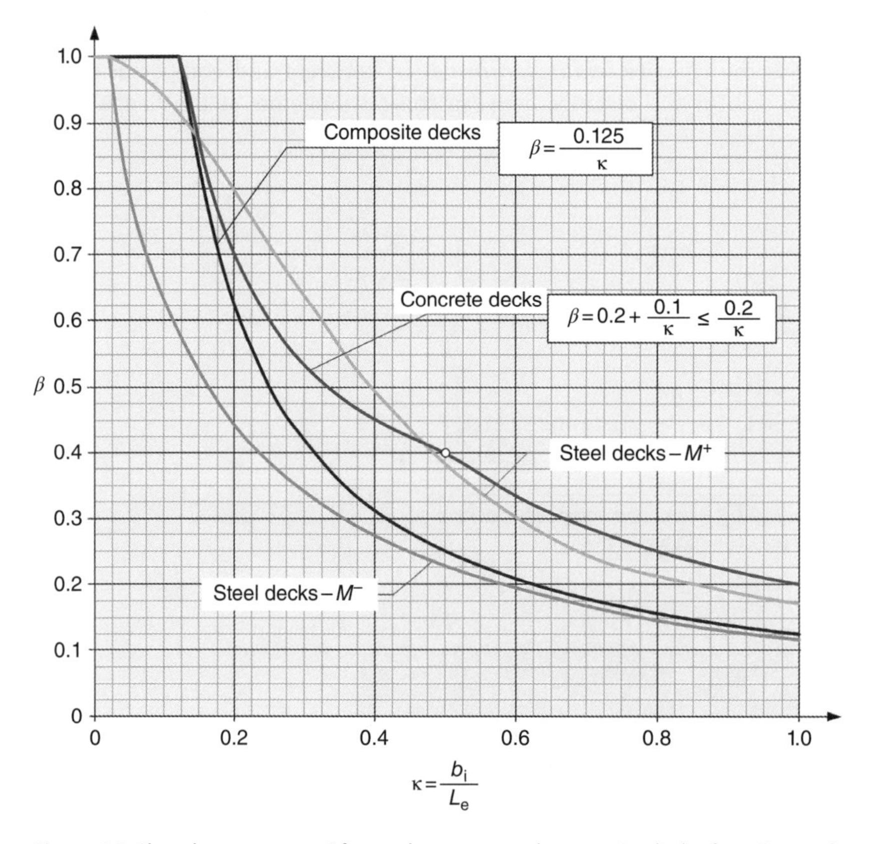

Figure 6.9 Shear lag parameter β for steel, concrete and composite decks, from Eurocodes.

6.3.3 Local Bending Effects: Influence Surfaces

As referred to in Section 6.1, deck-slabs are modelled as a long plate elastically restrained at both longitudinal edges. Boundary conditions at transverse edges depend on the slab being supported by diaphragms or cross beams. For determining local effects in deck slabs due to truck loading, first wheel loads from vehicles are transferred to the mid-surface of the slab as shown in Figure 6.10. The wheel load at the mid-surface is transformed into a uniformly distributed load on a small area, as shown in the figure in the case of tandem load of EN 1991-2 [6]. If internal forces from a point load $P = 1$ are denoted by $\xi = \xi\,(x, y)$, for a distributed load $p = p(x, y)$ in an area A, the action effects are calculated by integration as shown in Figure 6.11, which may be done numerically for constant p, by

$$M = \int_A p(x,y)\xi \mathrm{d}A = p\sum \xi_k \Delta A_k = pV \tag{6.8}$$

where ξ_k are ξ values at centre of elementary areas ΔA_k, in which A is decomposed and V is the volume under the influence surface with projected area A.

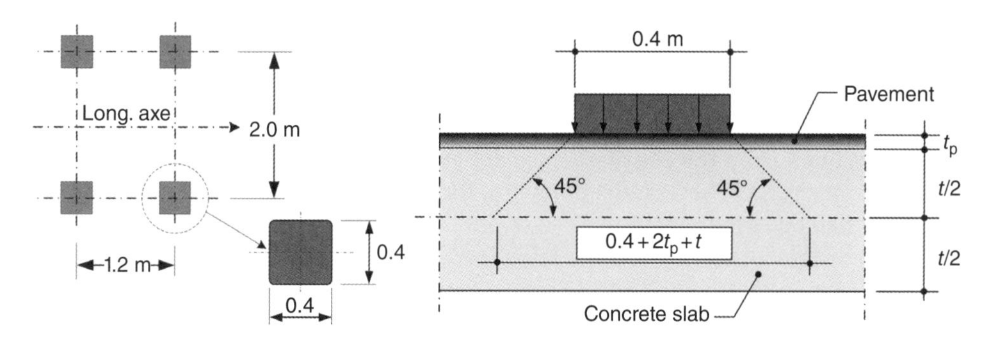

Figure 6.10 Wheel load actions of LM1 vehicle from EN 1991-2 [6], transferred to the mid-surface of the deck slab.

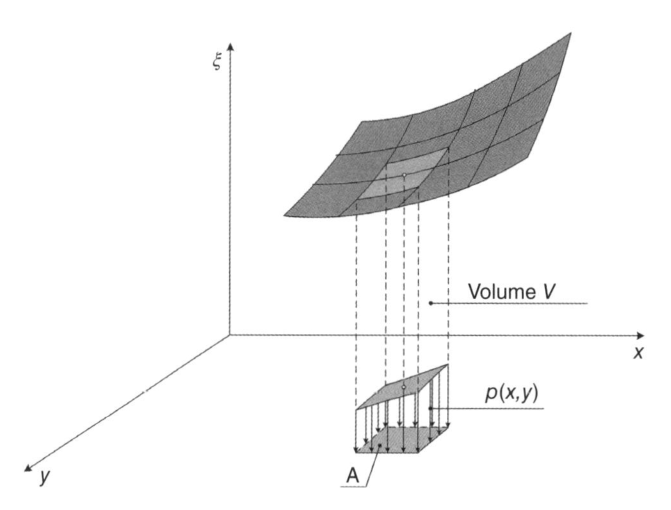

Figure 6.11 Evaluation of an action effect for a distributed load $p = p(x, y)$ form the influence surface values $\xi = \xi\,(x, y)$.

If the gradient of the influence surface is not very large, for wheel loads Q_k (not the point where the action effect is to be calculated) the action effect is $M = \Sigma Q_k \xi_k$. For a uniformly distributed load, p, applied at an area A, point loads Q_k at points i ($i = 1$ to n) and line load $q = q(s)$ applied along a curve c, one has

$$M = \Sigma_i Q_i \xi_i + \int_c q \xi \, ds + \int_A p \xi \, dA \quad (6.9)$$

For design applications, the Eq. (6.9) is transformed in

$$M = \Sigma_i Q_i \xi_i + \Sigma_j q_j \xi_j \Delta s_j + \Sigma_k p_k \xi_k \Delta A_k \quad (6.10)$$

Internal forces at deck slabs are usually evaluated on the basis of elastic analysis of isotropic slabs. The most common and direct approach are *influence surfaces*. The influence surfaces for internal forces in a slab, for example, the bending moments M_x and M_y at a general point (x_0, y_0) are the internal forces due to a point load $P = 1$ applied at a general point with coordinates (x,y). The vertical deflection at (x_0,y_0) due to $P = 1$ at (x,y), designated by $w(x_0,y_0,x,y)$ is equal, by the reciprocity theorem of structural analysis [9], to the vertical deflection at (x,y) due to $P = 1$ at (x_0,y_0) as represented in Figure 6.12: $w(x_0, y_0, x, y) = w(x, y, x_0, y_0)$.

Note that $w(x_0,y_0,x,y)$ represents the *influence surface* for deflections at (x_0,y_0) that can be obtained from the elastic surface deflections of the slab by applying a point load $P = 1$ at (x_0,y_0). These deflections are, for isotropic elastic slabs, governed by Lagrange slab equation:

$$D\nabla^4 w = p \quad (6.11)$$

where D is the bending rigidity $D = Eh^3/12(1-\nu^2)$, $p = p(x,y)$ the distributed load perpendicular to the plate and

$$\nabla^4 w = \frac{\partial^4 w}{\partial x^4} + 2\frac{\partial^4 w}{\partial x^2 \partial y^2} + \frac{\partial^4 w}{\partial y^4} \quad (6.12)$$

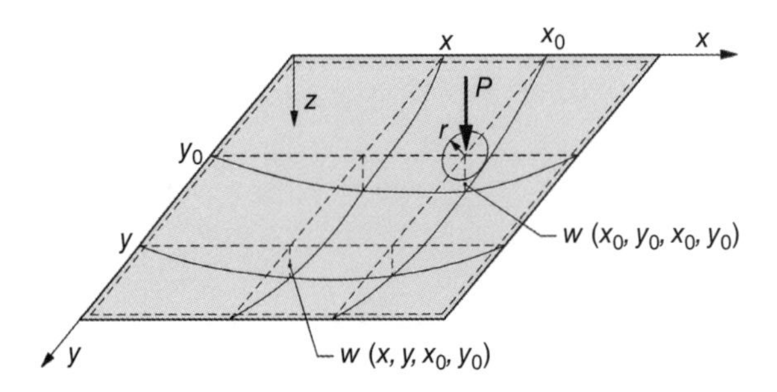

Figure 6.12 Basic model for a slab under a point load.

Deflections $w(x,y)$ are obtained by integration of Eq. (6.12) for a point load $P = 1$ at (x_0,y_0) and $p(x,y) = 0$. The solution of Eq. (6.12) is of the form

$$w(x,y) = w_0(x,y) + w_1(x,y) \tag{6.13}$$

where $w_0(x, y)$ is a particular solution for $P = 1$ loading and $w_1(x, y)$ is the solution of the homogeneous differential equation, that is, for $p(x, y) = 0$, satisfying slab specific boundary conditions along the supported edges. One has (Figure 6.12):

$$w_0 = \frac{1}{8\pi D} r^2 \ln r \tag{6.14}$$

where $r = \sqrt{[(x-x_0)^2 + (y-y_0)^2]}$ is the distance to (x_0,y_0). For $r = 0$, w_0 is indeterminate. However, the bending moments M_x and M_y, and torsion moments $M_{xy} = M_{yx}$, are obtained from moment-curvature relationships

$$M_x = -D\left(w,_{xx} + v\, w,_{yy}\right); \quad M_y = -D\left(w,_{yy} + v\, w,_{xx}\right); \quad M_{xy} = -D\, w,_{xy} \tag{6.15}$$

Here, partial derivatives are denoted by commas (,). Influence surfaces for internal forces have been developed and published in tables and diagrams by Pucher [10] for constant thickness slabs. Homberg [11] published influence surfaces for deck slabs, with linear or parabolic variation, including continuous slabs.

In general, these surfaces are obtained for the case $v = 0$; displacements and moments for $v \neq 0$ can be obtained by:

$$M_x = \bar{M}_x + v\,\bar{M}_y \quad \text{and} \quad M_y = \bar{M}_y + v\,\bar{M}_x \tag{6.16}$$

where $\bar{M}_x$ and $\bar{M}_y$ are the values for $v = 0$. Hence, from integration of Eq. (6.12) one has $w = w\,(x_0,y_0,x,y)$ and the influence surface, for example, for bending moments M_x at point (x_0,y_0) are obtained from the moment–curvature relationships:

$$\bar{M}_x = -D \frac{\partial^2 \bar{w}(x_0, y_0, x, y)}{\partial x_0^2} \tag{6.17}$$

One should note, $\bar{w}(x,y,x_0,y_0) = \bar{w}(x_0,y_0,x,y)$ where $\bar{w}\,(x,y,x_0,y_0)$ is the elastic deflection of the slab $\bar{w} = \bar{w}(x,y)$ for a point load applied at (x_0,y_0) for $v = 0$. According to elastic theory of plates, at the point of load application P, one has from the solution of Eq. (6.17) finite vertical displacements w but infinite moments (Figure 6.13). This is the result of the elastic infinite curvatures (second derivatives of w) at the point where the load is applied.

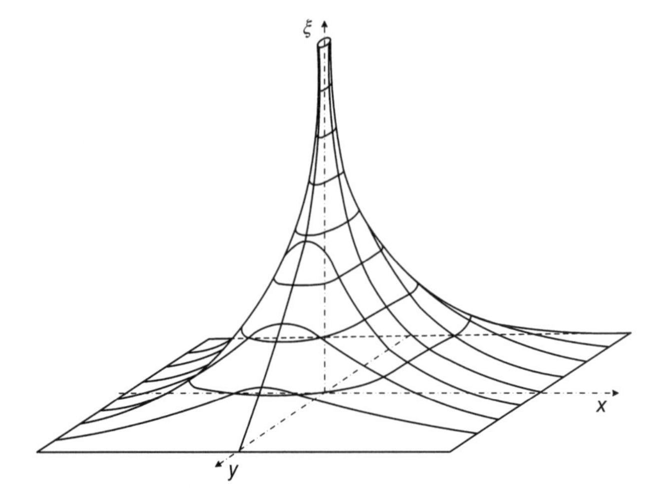

Figure 6.13 Influence surface for bending moments M_x at centre of a simply supported plate.

Nowadays, influence surfaces may be generated for any type of boundary conditions and geometry of the slab; skew slabs, curved slabs and so on by adopting the FEM. The principle for doing this is the Muller–Breslau Principle [9] from beam structures, extended here to plate structures:

> the ordinates of influence surfaces of any effect (generalized displacement or internal force) in a plate structure may be obtained from the deformed elastic surface of the plate by releasing the restraint to referred effect and introducing a corresponding unit displacement in the remaining structure.

This approach has been specifically developed for this book and results are presented in Figures 6.14–6.17. Hence, to obtain from a FEM the influence bending moment m_x at a point (x_0,y_0), a unit rotation $\theta_y = 1$ rad is applied at the point after a hinge at the element boundary has been introduced; the ordinate of the influence surface for m_x are the displacements $w(x,y)$ in the FEM. Based on this principle and adopting a mesh of an eight node FEM, a series of influence surfaces have been generated. The values have been checked against published results. The advantage of this procedure is to give the reader a powerful method for generating influence surfaces based on standard existing FEM software packages. Cases of stiffened plates, any type of variable thickness or in-plane geometry such as skew and curved bridge decks, may be dealt with by this numerical approach. On it, Figures 6.14 and 6.15 represent influence surfaces for span bending moments, m_x and m_y, for constant thickness isotropic long slabs, simply supported or restrained along two long longitudinal edges. Figure 6.16 presents the usual case of deck overhang (transverse slab cantilever) with linear variable thickness. And Figure 6.17 presents the influence surface for the support transverse bending moment of a two-restraint panel and support longitudinal bending moment of a three-edged restraint panel.

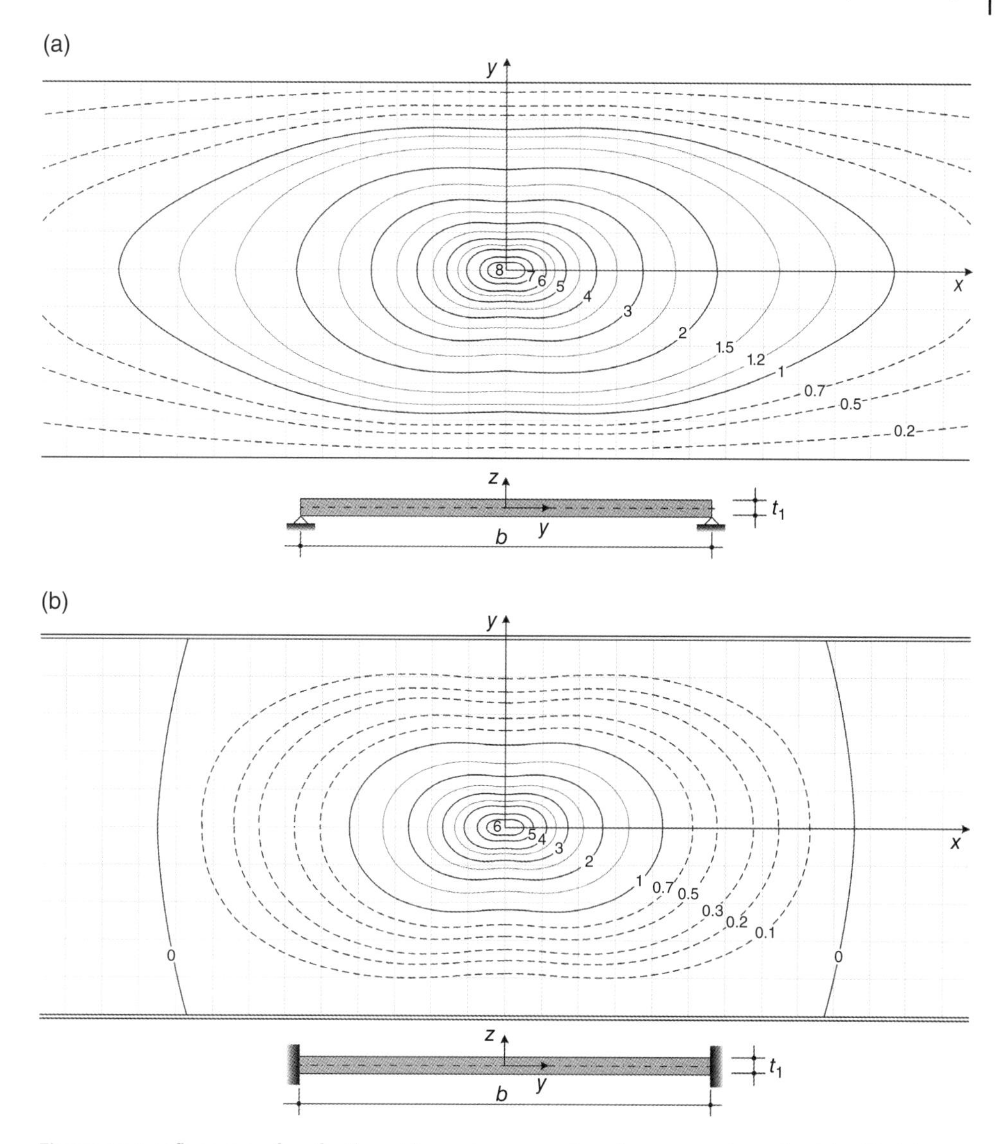

Figure 6.14 Influence surface for the mid-span transverse bending moment m_y of a slab with: (a) two supported edges and (b) two restrained edges (t_1 = const.; $\nu = 0.2$, multiplied by 8π).

For overhangs, a design simple approach is to adopt the expression and charts developed by Bakht and Jaeger [12] and reproduced in Figure 6.18. The transverse bending moment, for a constant or linear variable thickness slab cantilever, is given by

$$M_y = -\frac{P}{\Pi}\left[\frac{k}{\cosh\left(\dfrac{kx}{c-y}\right)}\right] \quad for \;\; y < c \tag{6.18}$$

where k is a constant given in charts and presented in Figure 6.18 for $t_2/t_1 = 1$, 2 or 3.

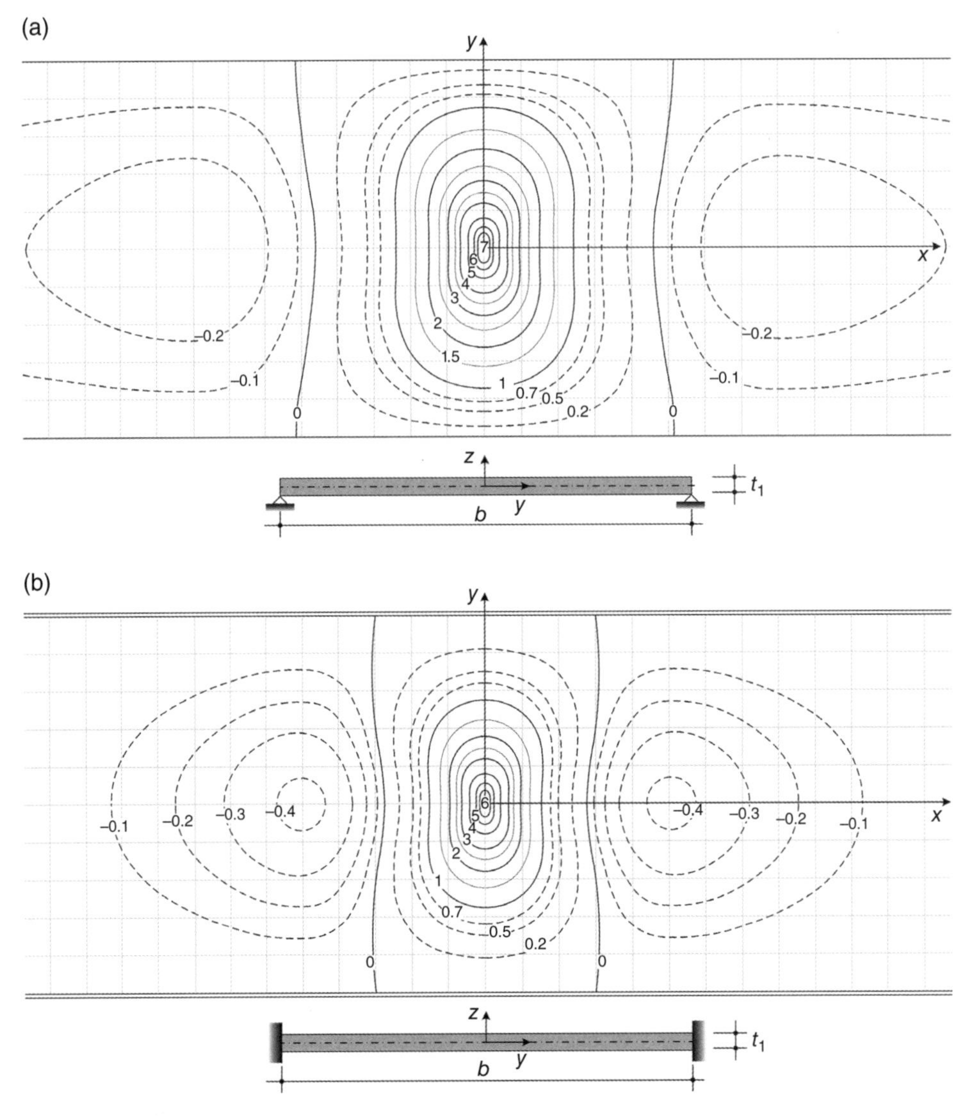

Figure 6.15 Influence surface for the mid-span longitudinal bending moment m_x of a slab with: (a) two supported edges and (b) two restrained edges (t_1 = const.; $\nu = 0.2$, multiplied by 8π).

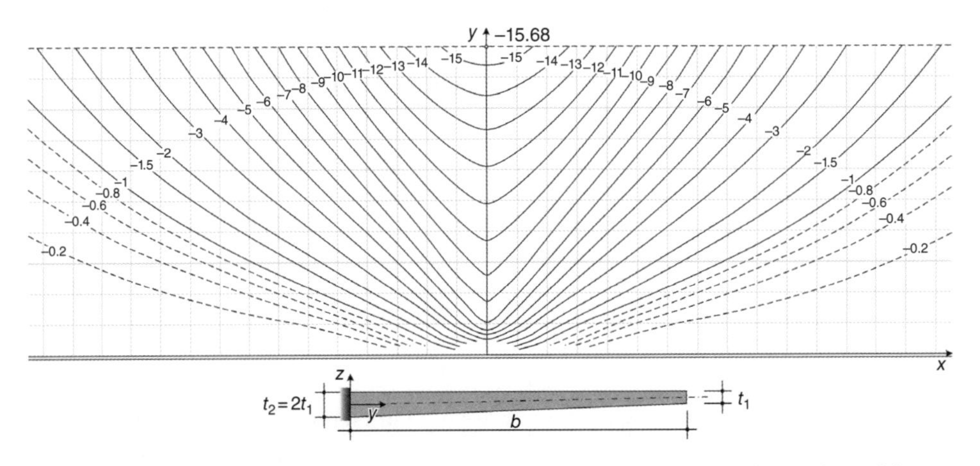

Figure 6.16 Influence surface for the support transverse bending moment m_y of a cantilever slab ($t_1/t_2 = 1/2.$; $\nu = 0.2$, multiplied by 8π).

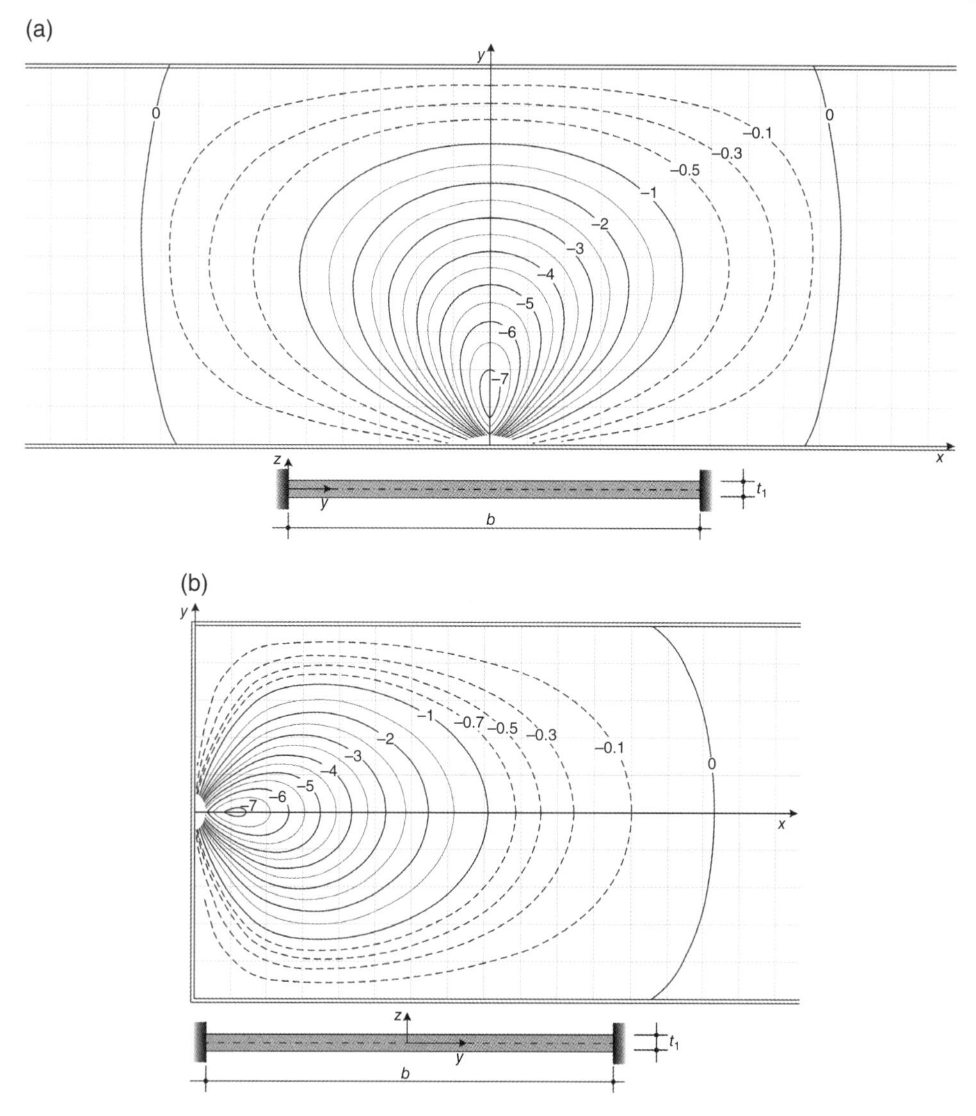

Figure 6.17 Influence surface for: (a) the support transverse m_y with two restrained edges, and (b) the longitudinal m_x bending moments of a slab with three restrained edges (t_1 = const.; $v = 0.2$, multiplied by 8π).

The transverse bending moment distribution along the built in edge is obtained from Eq. (6.18) for $y = 0$, for the cases $t_2/t_1 = 1$ ($k = 1.54$) and $t_2/t_1 = 2$ ($k = 1.87$) as shown in Figure 6.19.

For a load P applied at the free edge, one has for maximum transverse bending moment at built in edge ($c = b$, $x = 0$ and $y = 0$) $k = 1.54$ and $M_y = 0.49P$. From the influence surface, one has $M_y = 12.65/(8\pi) = 0.497P$. Note this value is also obtained by 'spreading' at the point load in the slab plane at 45°, because $M_y = -P/(2c) = -0.5P$. For a variable thickness slab $t_2/t_1 = 2$, one has $k \simeq 1.87$ and $M_y = -0.6P$, compared to the

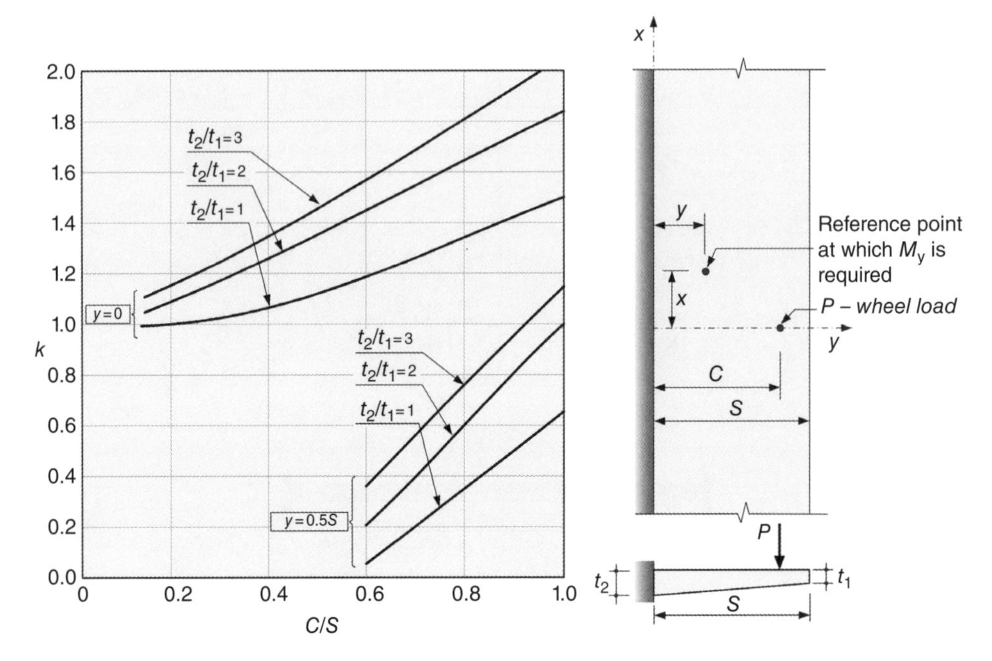

Figure 6.18 Charts for transverse bending moments in constant or variable thickness overhangs (*Source:* Adapted from Ref. [11], National Academy of Sciences).

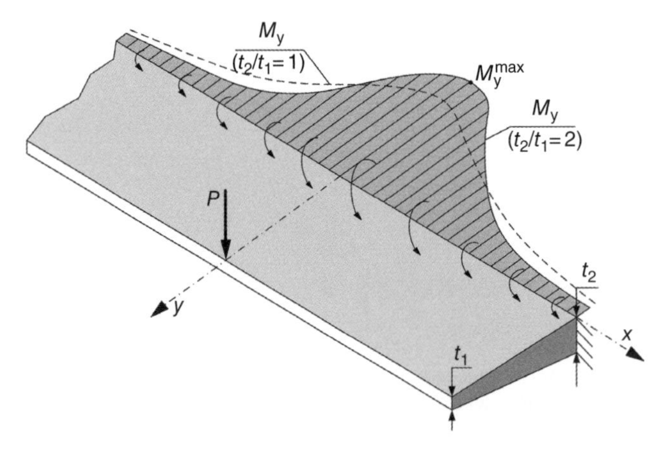

Figure 6.19 Distribution of the bending moment M_y along the built in edge of a cantilever slab for a point load at the free edge.

influence surface value $M_y = 15.68/(8\pi) = -0.62P$. This means that doubling the thickness of the plate at the built-in edge, with respect to the free edge, induces a 20% increase in the maximum bending moment. Increasing the stiffness of the slab induces an increased action effect, which is always the case for statically indeterminate structures. On the contrary, for a cantilever beam instead of a slab, the bending moment is independent of the stiffness distribution.

A numerical example of the application of influence surfaces to determine local bending moments in a slab deck due to tandem loading is presented in Chapter 8. It should be noted that wheel loads of tandems are transformed in distributed loads ($q_i = Q_i/(a \cdot b)$) in small areas $a \cdot b$ at the mid surface of the slab. Near a singularity point of influence surfaces, the distributed loads should be decomposed in small areas ΔAi and by knowing the contour lines (constant levels of the influence surface ordinates ξ_i) at the centre if each element, obtained by, that is, intersecting the influence surface by horizontal planes, the wheel effect is $\Sigma\ \xi_i\ \Delta Ai$. This approach is exemplified in Figure 6.20 for the central point of a simply supported slab. The other approach, usually giving a more precise result (due to the high gradient of the influence surface at singularity points), is to intersect the influence surface by vertical planes and from the section cut (see Figure 6.19) the integral that is numerically calculated by using, for example, Simpson's rule as shown in the figure.

6.3.4 Elastic Restraint of Deck Slabs

As referred to in Section 3.1 for the case of slab-girder decks, rib-slab decks or box girder decks, the slab may be considered elastically restrained in the girders, ribs or webs. For the last case, the elastic degree of restraint may be determined from a cross frame analysis, as will be discussed in Section 6.7. For the girder and rib cases (Figure 6.21) neglecting the effect of the overhangs, the degree of restraint may be evaluated by compatibility of torsion rotations of the girders/ribs and bending rotation of the slab at the supported edge. The load case is assumed associated with a uniform distributed load q in the slab. By adopting moment X as a hyperstatic unknown, EI_T as the bending stiffness of the slab per unit length and GJ the torsional stiffness of beam cross section, and using the Force Method of Structural analysis, the end rotation at the slab longitudinal edges θ_0 and the rotations of slab and beam sections at mid span, θ_1 and θ_2, are given in Figure 6.21.

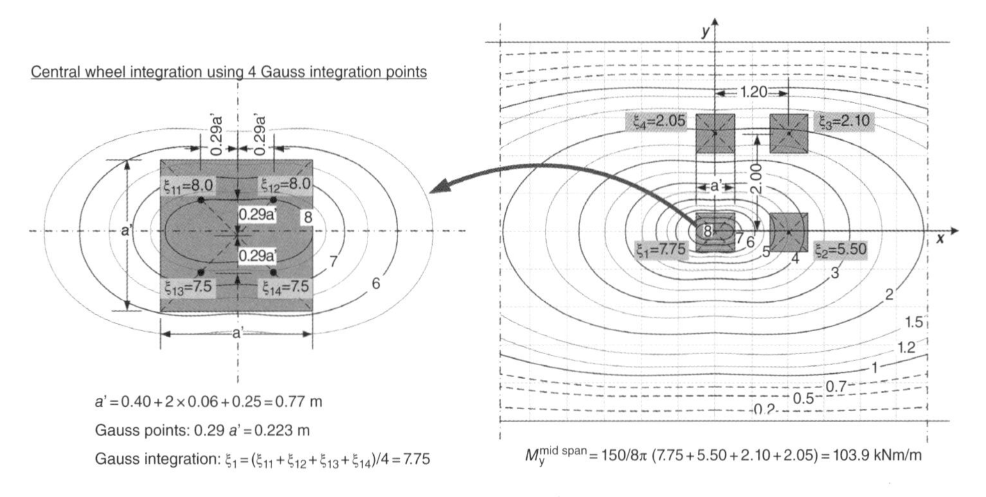

Figure 6.20 Calculation of the wheels load effect (m_y) at the mid span of a simply supported slab by an influence surface, adopting a numerical integration with four Gauss points for the central wheel effect.

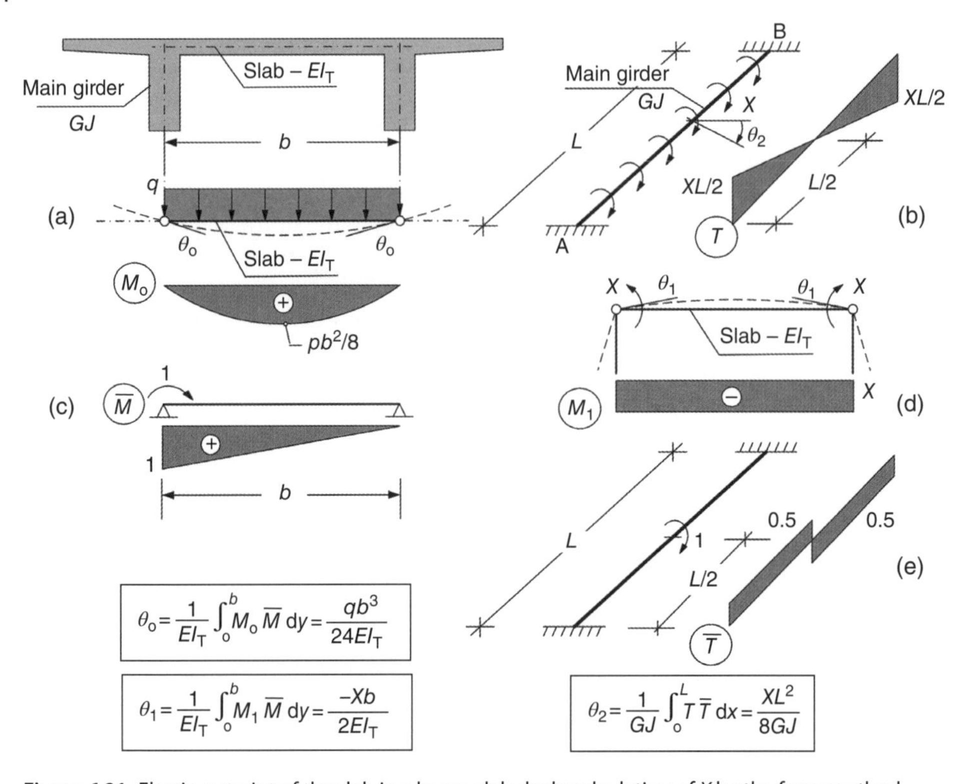

Figure 6.21 Elastic restraint of the slab in a beam-slab deck; calculation of X by the force method.

For compatibility, the slab and main girder rotations are equal and, for $q = 1$, one defines the degree of restraint α, with respect to the built in moment, that is, $\alpha = X/(qb^2/12)$, yielding for concrete slabs $G = E/[2(1+\nu)] = E/2.4$:

$$\alpha = \frac{1}{1 + \frac{EI_T}{GJ} \cdot \frac{L^2}{4b}} = \frac{1}{1 + 0.6\frac{I_T}{J} \cdot \frac{L^2}{b}} \tag{6.19}$$

As expected, $\alpha = 1$ for J tending to infinite. If $J = 0$, $\alpha = 0$.

Equation (6.19) gives a precise degree of restraint only at mid-span section of the deck slab panel between cross girders or diaphragms (where torsion rotations are restricted). Besides, it has been obtained for a uniformly distributed load q and assumes a linear uncracked stiffness (state I) for slab bending stiffness and torsion stiffness of the girder. The first limitation is easy to remove by deriving a similar formula for α for intermediate sections; however, it is acceptable in practice to assume a linear variation for α between 1 at the diaphragm sections and α obtained at the mid-span section. The limitation associated with uncracked properties is particularly relevant for torsion stiffness. It may be removed by adopting GJ at ULS, designated by GJ_{II}, which is no more than 50% of GJ at state I [13]. The limitation associated with the assumption of a uniformly distributed load is not easy to remove. If one derives the formula for truck loading, it depends on the location of the truck on the slab. Besides, there are two aspects to be considered:

- local loads, compared to uniformly distributed load, induce a more localized deformation of the deck slab (Figure 6.5), with bidirectional bending, tending to increase the degree of restraint;
- the values of the bending moments at the centre of the slab under truck loading are not so different for the built-in case (M_{y1}) and the simply supported one (M_{y0}). In a beam, there is a large difference influence of boundary conditions; that is not the case for a slab.

The bending moment M_y in function of the α value is finally obtained by interpolation:

$$M_y = M_{y0} - \alpha\left(M_{y0} - M_{y1}\right) \tag{6.20}$$

For example, for $\alpha = 0.3$ and $M_{y1} = 0.4 M_{y0}$, one has $M_y = 0.82 M_{y0}$; if $\alpha = 0.4$ one has $M_y = 0.76\ M_{y0}$. Near the diaphragm sections, one has at the centre of the slab, $\alpha = 1.0$ and therefore, $M_y = M_{y1}$.

Since influence surfaces allow only values at specific points for longitudinal and transverse bending moments in slabs, a numeric integration using four Gauss points, as shown in Figure 6.20, is often adopted for assess the effect of the wheels where the influence surface has important gradients.

By proper transverse location of the girders, the negative bending moments induced at the support section of the slab are balanced between the bending moment coming from the internal span and from the cantilever. If no transverse rotation exists, the negative bending moment due to dead load q is $qb^2/12$. The design moment is very much influenced by the transverse moment coming from the cantilever under tandem loading. The cantilever is usually assumed to be completely built in and no torsion rotation of the girder is considered to redistribute this moment, which is exact near a diaphragm or cross beam. At the mid-span of the slab, the bending moments are usually positive. However, due to simplified assumptions adopted in deriving α, bending moments at the centre of transverse spans are recommended to be taken in the envelope, approximately 20% of the negative bending moment at the support (Figure 6.22).

A final aspect to be considered in decks without cross beams is differential bending between main girders (Figure 6.23). This induces positive transverse bending moments at the underneath of the deck slab. A minimum reinforcement (typically bars of 12 mm diameter each 150 mm) should be adopted.

6.3.5 Transverse Prestressing of Deck Slabs

Slab-girder decks may usually be designed without transverse prestressing, for small transverse spans of the slab, say $b_2 < 6\,\text{m}$ with overhangs $b_1 < 2.5\,\text{m}$. This means a total deck width $b = 2b_1 + b_2$ of less than 11–12 m. The ratio of steel reinforcement of the deck slab is usually quite large, in the order of 120–140 kgm^{-3}, and can represent up to 50% of the total quantity of steel reinforcement in the deck. If the deck is transversely

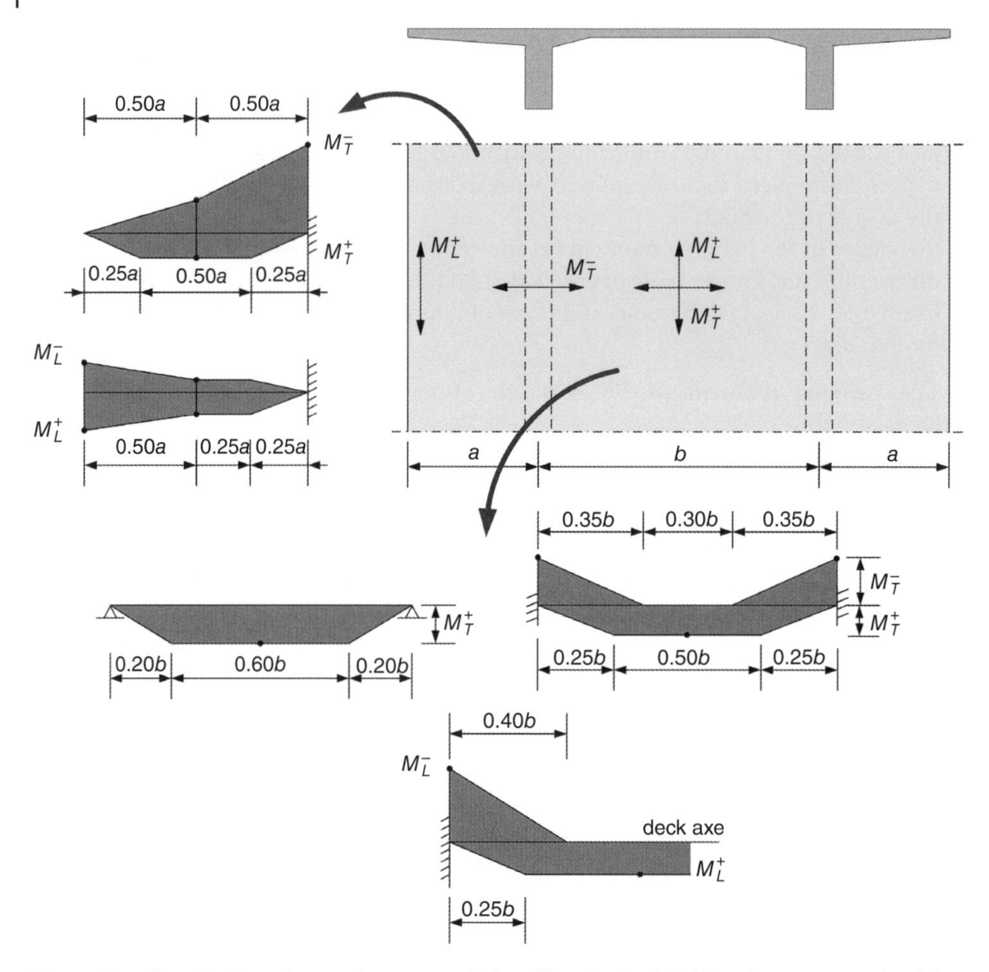

Figure 6.22 Simplified envelopes of transverse (M_T) and longitudinal (M_L) bending moments for slabs in slab-girder or rib slab decks.

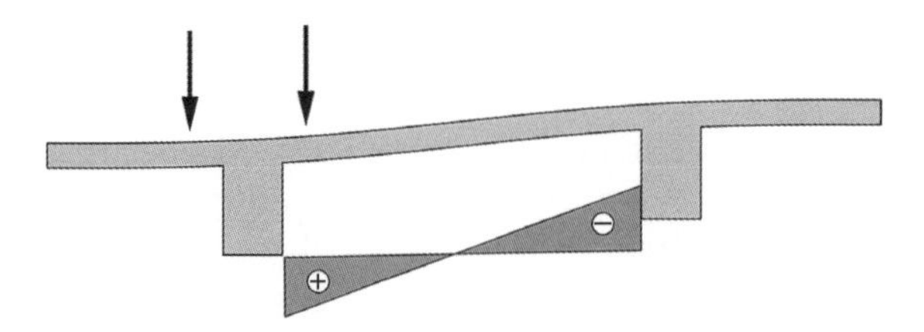

Figure 6.23 Differential bending between main girders in a beam-slab deck inducing transverse positive bending moments at the slab.

prestressed (Figure 6.24), reinforcement may be reduced to small diameter bars of 10–12 mm diameter if resistance to ULS is warranted.

Transverse prestressing cables are preferably made by small capacity cables (order of 500–700 kN of effective force each), made in general with 3–4 strands of 15 mm each, spaced

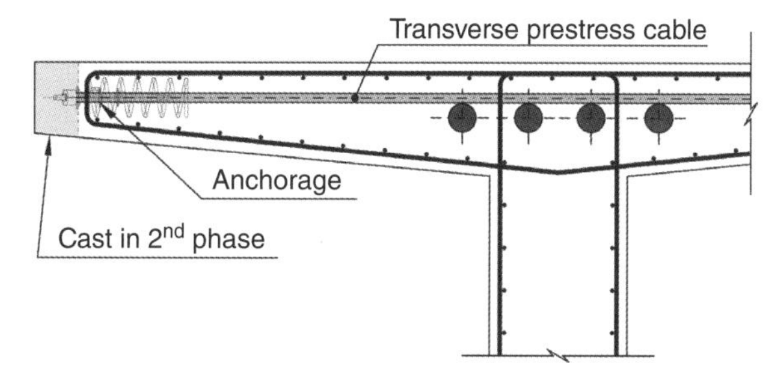

Figure 6.24 Transverse prestressing in deck slabs.

between 0.5 and 1.0 m apart. In this way, local bending moments' effects due to wheel loads are covered. Passive and active anchorages may be alternated at free longitudinal edges of the slab. Since the cables are short, effect of the loss of prestressing due to re-entrance of edges is someway compensated each side. The anchorages should be located at the free edge of the slab or nearby if the depth is not sufficient to locate the anchorage; the remaining part cast in a second phase. The best approach is to always adopt a minimum of 0.20–0.25 m thickness to be able to locate the anchorages at free edges of the slab. Rectangular anchorages allow for a reduction of the thickness of the slab at the free edges.

One of the drawbacks of the transverse prestressing cables is the reduced eccentricity of the cables with respect to the slab mid-surface. Small ducts should be adopted but with standard cables the diameter will be always of 50–60 mm at the least. Due to concrete covering, effective eccentricity e of transverse prestressing cables is quite reduced. The effect of cables curvature inducing transverse deviation forces on the concrete and balancing dead loads and part of traffic loads is small.

The SLS bending moment for prestressing design is denoted here by $M = M_{DL} + \psi M_{LL}$ at one section of the slab (mid-span or support section). The required effective prestressing force is designated by P_∞ in the long term, after all instantaneous losses (edge effects re-entrance and friction forces along the cable) and long-term losses (creep, relaxation and shrinkage). Besides, any secondary moment effects (hyperstatic effects) induced by transverse prestressing cables is generally small and may be neglected by the time being. The decompression limit state is verified under the condition $P_\infty/(1 \times h) = (P_\infty\, e)/(1 \times h^2/6) = M/(1 \times h^2/6)$ yielding $P_\infty = M/(e + h/6)$, being h the thickness of the slab and e the eccentricity of transverse prestressing cables.

One important action to be taken into account when defining transverse prestressing is the thermal gradient in the thickness (Figure 6.23). At slab span sections and at the mid-span length between cross girders, torsion stiffness rotation of the girder is low. However, approaching the cross girders, for a positive linear thermal gradient ΔT, a positive bending moment $M_{\Delta T}$ is induced at span sections of the slab, $M_{\Delta T} = D\alpha_T\, \Delta T/h = E\, \alpha_T\, \Delta T\, h^2/[12(1\text{-}\nu^2)]$, in (kNm m^{-1}) being α_T the linear coefficient for thermal expansion of concrete.

Part of this moment should be added to previously referred value of M, say $M = M_{DL} + \psi_1 M_{LL} + \psi_2 M_{\Delta T}$, where $\psi_i < 1.0$ coefficients dependent on the load combination defined for verifying decompression limit states (DLS) in concrete. In general, $\psi_1 M_{LL}$ is the quasi-permanent or the frequent value of the live load; $\psi_2 M_{\Delta T}$ is in most cases the quasi-permanent value of the thermal gradient.

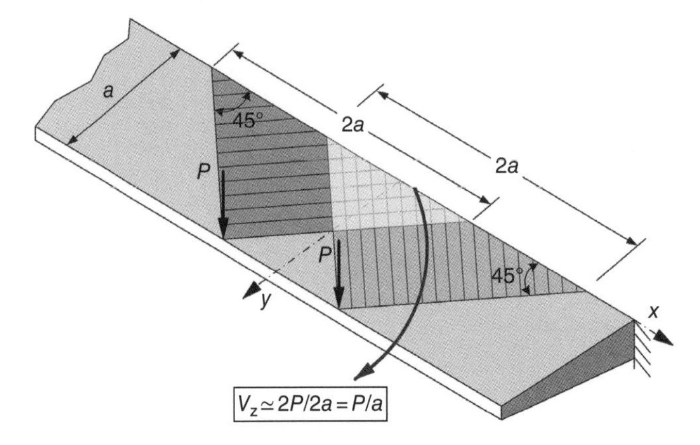

Figure 6.25 Shear force estimation in slab decks due to wheel loads.

Deck slabs usually have no specific reinforcement for vertical shear force, particularly when they are transversely prestressed. For shear force calculation either influence surfaces or FEM are adopted. However, it is generally sufficient to dissipate in-plane the point loads at 45° as shown in Figure 6.25, but of course this approximation depends on the transverse location of the point loads. For wheel loads near the girder, say at $<1.2\,h$, the loads are transferred directly by inclined struts to the girders.

6.3.6 Steel Orthotropic Plate Decks

The concept design of steel orthotropic deck slabs was discussed in Chapter 4, Section 4.5.3.2. *Orthotropic* means anisotropic in two orthogonal directions: the longitudinal direction along the stringers and the transverse direction along the cross beams and diaphragms.

The plate with thickness t is elastically built in by the stringers allowing its deflection under a wheel load Q_i to be evaluated from local uniform pressure $q = Q_i/(a\ b)$, as one-sixth of a simply supported transverse span length e between the stringers:

$$f = \frac{1}{6}\frac{5}{384}\frac{qe^4}{EI} < \frac{e}{300} \tag{6.21}$$

Here, $I = 1 \times t^3$, $E = 210\,000\,\mathrm{N\,mm^{-2}}$ and q in $\mathrm{N\,mm^{-2}}$, yielding $t > 0.0034\ e\ q^{\ 1/3}$.

Designating by $w(x,y)$ the transverse deflections and by $q(x,y)$ the applied distributed load perpendicular to the plane of the plate, the plate equation is:

$$D_x \frac{\partial^4 w}{\partial x^4} + 2H \frac{\partial^4 w}{\partial x^2 \partial y^2} + D_y \frac{\partial^4 w}{\partial y^4} = q(x,y) \tag{6.22}$$

where D_x is the longitudinal bending stiffness, D_y is the transverse bending stiffness and H is a torsion parameter. The classical Pelikan–Esslinger Method [14, 15] may be adopted to analyse this type of deck based on a Fourier series solution of Eq. (6.22) yielding design charts for influence lines. For open stringers, $D_x = H = 0$, Eq. (6.22) is reduced to the beam equation

$$D_y \frac{d^4 w}{dy^4} = q(x,y) \tag{6.23}$$

Analysis is reduced to a simple stringer, with associated plating, on elastic supports (cross girders). Orthotropic plates with closed section stringers can only be analysed bringing torsion effects into play by means of torsion parameter H in the orthotropic slab equation (Eq. (6.23)). The solution is expressed in terms of a Fourier series for the displacements $w(x,y)$. The reader is referred to [14, 15] for more details on this subject.

Nowadays, the analysis of orthotropic plate decks may be done on the basis of a FEM with shell elements. However, the amount of data for modelling and interpretation of output results is a main issue, if the model is not restricted to a short deck segment between cross beams. Deck continuity with adjacent panels should be modelled or at least three panels of the deck slab need to modelled to interpret local effects at the internal panel. Truck loads should be considered at different locations to obtain maximum stresses at stringers and plating. These models may be adopted for detailed fatigue studies. But, in regular design situations, they may be avoided if design rules and fatigue detailing are respected (Figure 4.49).

6.4 Transverse Analysis of Bridge Decks

6.4.1 Use of Influence Lines for Transverse Load Distribution

Transverse load distribution in bridge decks is a classic problem in bridge design and was introduced in Chapter 3 (Section 3.3) when traffic loading was discussed. At that stage, a statically determinate transverse load distribution was adopted for a twin girder composite deck (Figure 3.8). If the deck has more than two main girders, this simple approach is not possible and account for deformability of the deck should be considered. A variety of methods, tables and charts for transverse load distribution are available. In this section, one assumes transverse load influence lines are known and their use is explored.

The example of a continuous bridge deck in Figure 6.26a is considered. When the deck is loaded by a point load on girder Q moving transversely across the deck width, there is a global torsion rotation and an overall bending of the deck. Due to the torsion effect, load Q is distributed between girders. Besides, since the slab is deformable, there is also a distorsion effect of the cross section and the load on girder (rib) 1 varies according to one of possible influence lines shown in Figure 6.26b or c. These different influence lines are obtainable at different transverse cross sections. In the case of Figure 6.26(c) the influence line refers to a transversely more rigid cross section with less distortion yielding, for example, an uplift of girder 1 when the load is at the right-hand side of the deck. When the load is on the girder, 90% of the load is taken by the girder in case Figure 6.26(c) and 75% in case Figure 6.26(b). In Figure 6.26c, more, 100%, is taken by girder 1 if the load acts at the left-hand side. When a transverse distributed load acts on

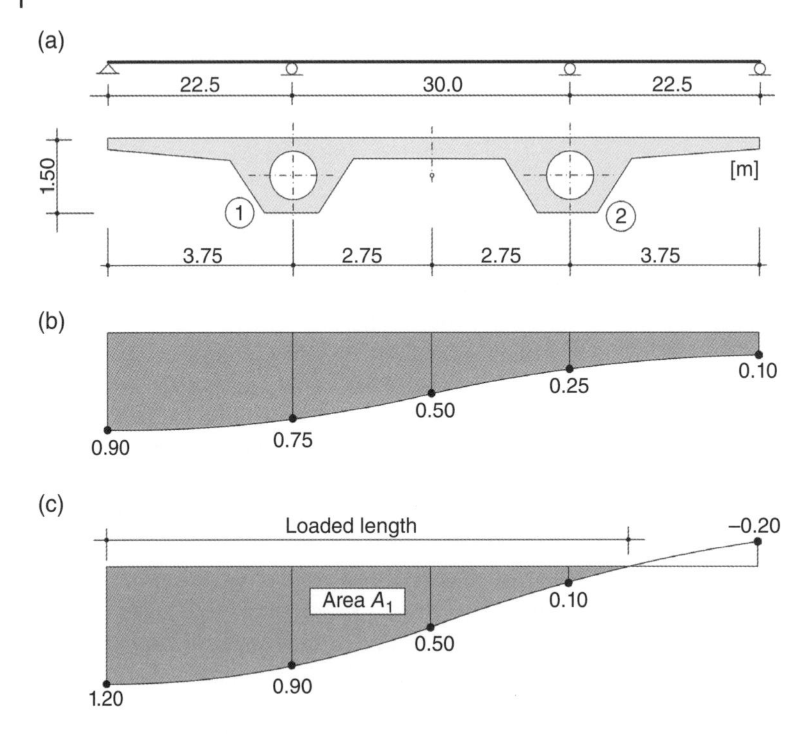

Figure 6.26 Influence lines for transverse load distribution for girder 1 a continuous rib-slab deck.

the deck, the most adverse effects for girder 1 are obtained in case 6.26(b) by loading a full deck width, while in case 6.26(c) only the positive part of the transverse influence line should be loaded.

Once the deck transverse influence line is obtained, the traffic load resultant for one girder is evaluated by loading the deck transversely and using the standard procedure of influence lines. Hence, for point loads Q_i (truck wheel loads) and distributed traffic loads q_0 are assumed to be constant in the transverse direction y if the ordinates of the influence line for the girder where load effects are to be obtained are of type (c) and denoted by $\xi(y)$, for loading girder 1 one has:

$$Q_1 = \Sigma Q_i \ \xi_i \qquad q_1 = \int_a^b q(y)\ \xi(y)\ dy = q_0\ A_1 \tag{6.24}$$

where A_1 is the area limited by the influence line as shown in Figure 6.26c. Loads Q_1 and q_1 are transported to the longitudinal model of the girder and the maximum adverse load effects (internal forces or deflections) may be obtainable by standard influence line methods of beam structures, as exemplified in Figure 3.8, Chapter 3.

6.4.2 Transverse Load Distribution Coefficients for Load Effects

Since load effects in deck elements, in linear elastic analysis, are proportional to loads, usually one says there is a *transfer function f(x)* for determining load effects from general loads in the longitudinal model. For example, in a simply supported beam with a span L, under a uniformly distributed load, at a general section x, one has $M = f(x)\, q_1$ with $f(x) = (Lx/2 - x^2/2)$. In Figure 6.26, the girder 1 bending moment M_1, for a uniformly distributed load q, is given by:

$$M_1 = f(\text{x})q_1 = f(\text{x})(A_1 q) = f(\text{x})q\; b(A_1/b) = \alpha_1 M \quad \text{with} \quad \alpha_1 = A_1/b \quad (6.25)$$

where M is the global bending moment in the full deck width, calculated as a beam loaded with (qb) and α_1 is the *transverse distribution coefficient* for girder 1. For the case of Figure 6.26b, Area A_1 is the total area under the influence line, and should be equal to $0.5\,\text{m}^2\,\text{m}^{-1}$. Hence, each girder takes one half of the global moment.

Example 6.1 Taking the case of influence line of Figure 6.26c and spans l_1 = 22.5 m and l_2 = 30.0 m, the maximum positive bending moment in one of the girders at the internal span for a traffic load uniformly distributed q = 5 kN/m^2 is obtained by:

$$b = 13.0\ \text{m};\ A_1 = 6.75;\ \alpha_1 = 6.75/13 = 0.52$$

$$l_2/l_1 = 30.0/22.5 = 1.33;\ M^{+}{}_{\max} = 0.123 \times (13.0\ q)22.5^2 = 4047.5\ \text{kNm}$$

For girder 1 $M^{+}{}_{1\max} = 0.52 \times 4047.5 = 2105$ kNm.

For a different load case, say 'q_2' (distributed, concentrate or line load) one has a different load distribution coefficient, α_2, and for load combinations '$q_1 + q_2$', the superposition principle is applicable.

6.4.3 Transverse Load Distribution Methods

Transverse load distribution in bridge decks may be dealt with on the basis of analytical and/or numerical methods. The most elementary method is by Courbon and is based on a rigid transverse cross section assumption. Analytical methods have been developed for a long time for considering the deformability of deck cross sections and improving the transverse load distribution obtained by the Courbon method. These analytical methods are based on a Fourier series solution of orthotropic plate equations yielding a series of design charts for obtaining transverse load distribution in bridge decks. One of the most popular adopted methods was the Guyon–Bares–Massonnet method [16, 17] initially proposed in 1946. A series of charts published in 1969 by Cusens and Pama [18] improved practical application of the method. With the development of numerical methods, in particular FEM, classical analytical approaches have been progressively less adopted. However, classical methods are still useful tools for pre-design and design checking. That is why an overview of analytical methods is presented here.

A simply supported bridge deck loaded perpendicular to its plane is considered. The deck is assumed to be composed of a series of longitudinal and transverse cross beams monolithically connected to the slab (Figure 6.27). The centre of gravity of the slab elements and of the beams is not at the same level.

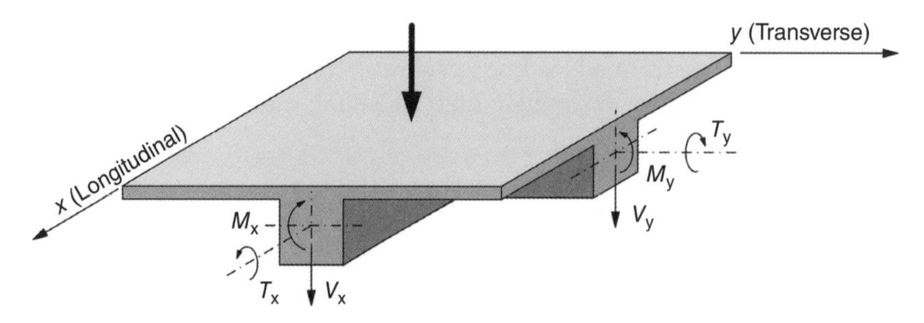

Figure 6.27 General case of a beam slab deck element with longitudinal (*x*) and transverse (*y*) cross beams.

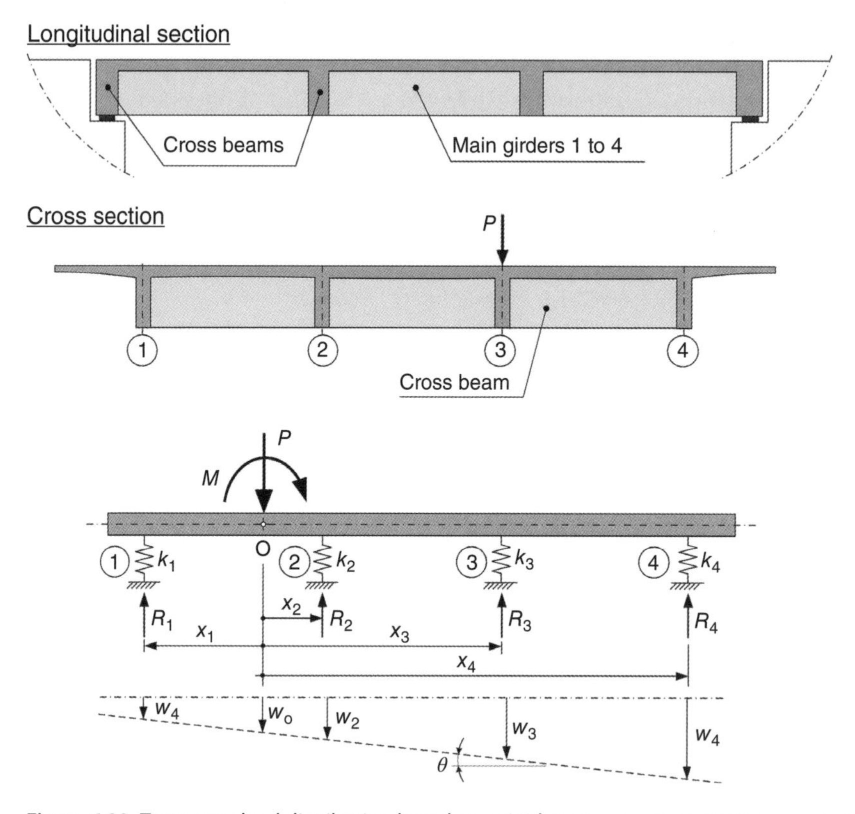

Figure 6.28 Transverse load distribution based on a rigid transverse cross section assumption.

6.4.3.1 Rigid Cross Beam Methods: Courbon Method

A transversally rigid cross section is assumed, with rigid cross beams elastically supported by longitudinal girders with torsional stiffness assumed negligible (Figure 6.28). Vertical stiffness of the girders, at the section where transverse load distribution is to be obtained, is denoted by k_i. Vertical deflections of girder i is w_i, the reference point at cross section is O and transverse coordinates of girders location are x_i. The cross section deck has a rigid movement defined by two degrees of freedom relative to point O – deflection w_o and rigid rotation θ_0. Loads resultant on the deck referring to point O are P and M. Girder reactions at spring supports are R_i. One has

$$\text{Equilibrium } \sum R_i = P \ \sum R_i x_i = M \tag{6.26}$$

$$\text{Constitutive law } R_i = K_i \ w_i \tag{6.27}$$

$$\text{Compatibility } w_i = w_e + \theta \ x_i \tag{6.28}$$

By replacing compatibility and constitutive laws in equilibrium equations, one obtains:

$$w_o \sum K_i + \theta \sum K_i \ x_i = P \tag{6.29}$$

$$w_o \sum K_i \ x_i + \theta \sum K_i \ x_i^2 = M \tag{6.30}$$

This is a system of two linear equations with two unknowns – the degrees of freedom w_o and θ. As point O is arbitrary, one may select O at the *centre of stiffness of the elastic supports* k_i, (the centre of gravity of stiffness's K_i), which means $\sum K_i \, x_i = 0$. Hence, from these equations one has

$$w_0 = \frac{P}{\sum k_i} \tag{6.31}$$

$$\theta = \frac{M}{\sum k_i x_i^2} \tag{6.32}$$

The reaction at girder j is from equilibrium Eq. (6.31)

$$R_j = K_j \left(\frac{P}{\sum K_i} + \frac{x_j M}{\sum K_i x_i^2} \right) \tag{6.33}$$

Thus, the *transverse influence line* of reaction at girder j is obtained for $P = 1$ and $M = Px$ by (Figure 6.29):

$$R_j(x) = K_j \left(\frac{1}{\sum k_i} + \frac{x_j}{\sum k_i x_i^2} x \right) \tag{6.34}$$

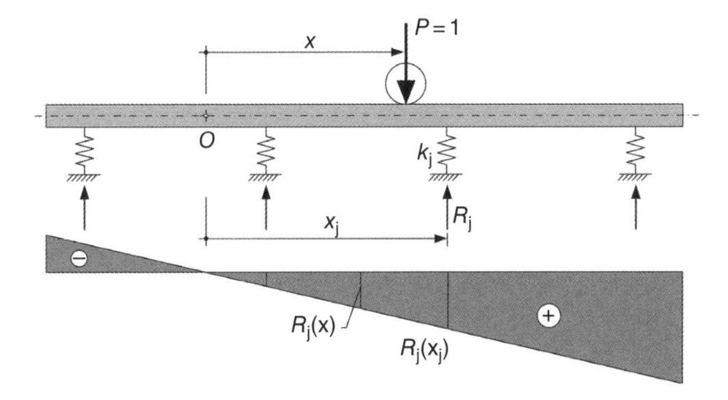

Figure 6.29 Influence line for the reaction on girder *j*.

The influence line is a straight line shown in Figure 6.29 as a consequence of a rigid cross section assumption. For the usual case of n equal girders, $k_1 = k_2 = \ldots k_n$, Eq. (6.34) yields,

$$R_j(x) = \frac{1}{n} + \frac{x_j}{\sum x_i^2} x \tag{6.35}$$

Example 6.2 A multi girder beam-slab deck has the geometry of Figure 6.30 and deck width is $b = 12\,\text{m}$; longitudinally, it is a simply supported span with a length of 30 m. Let the maximum bending moment at girder 4 under a traffic uniformly distributed load of $q_1 = 5\,\text{kN}\,\text{m}^{-2}$ be determined. With respect to point O one has:

$$x_1 = -\frac{b}{2};\; x_2 = -\frac{b}{6};\; x_3 = +\frac{b}{6};\; x_4 = +\frac{b}{2} \rightarrow \sum x_i^2 = \frac{5}{9} b^2$$

Hence, from Eq. (6.35), for girder 4 one has:

$$R_4(x) = \frac{1}{4} + \frac{b/2}{5b^2/9} x = 0.25 + \frac{0.9x}{b}$$

When load $P = 1$ is on girder 4, 70% of the load is taken by this girder; when load is at $x = -b/2$ there is an uplift of girder 4 and $R_4 = -0.35$, while for $x = +b/2$, 85% is taken by girder 4.

For the distributed load $q_1 = 5\text{kN}\,\text{m}^{-2}$ on the deck, the maximum load for longitudinal model of girder 4, and maximum (mid span) bending moment is (Figure 6.30) is:

$p = \frac{1}{2} \times 11.33 \times 0.85 \times 5\ \text{kN}\ \text{m}^{-2} = 24.08\ \text{kN}\ \text{m}^{-1}$ and $M_{\max} = 24.08 \times 30^2/8 = 2709.4\ \text{kNm}$.

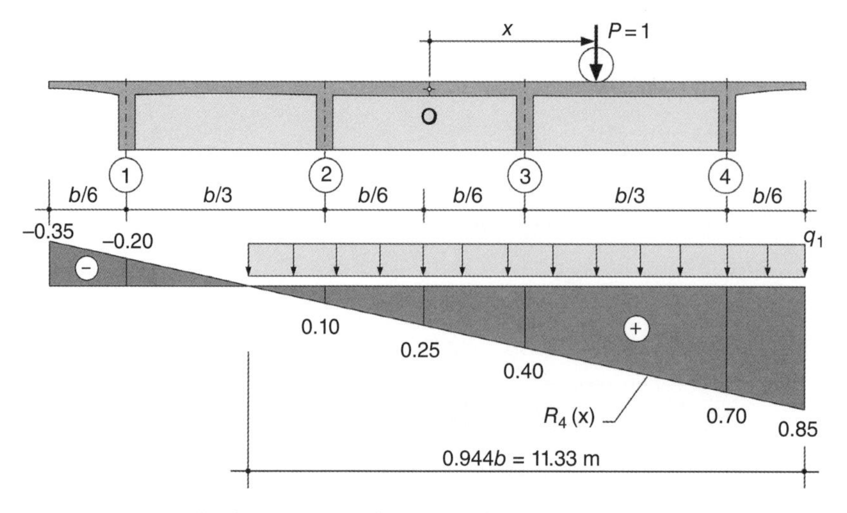

Figure 6.30 Example of a transverse influence line for load at girder 4 by the Courbon method.

6.4.3.2 Transverse Load Distribution on Cross Beams

The load distribution on the cross beams or load effects may be obtained from the previous simplified method. The influence line for maximum bending moment at cross beams of Example 6.2 is considered here. For the mid-section of cross beams, adopting the Muller–Breslau principle [9], a hinge is introduced and a relative unit rotation $\theta = 1$ is applied between the two adjacent sections. The deformed shape is the influence line (Figure 6.31). Two points are enough, due to symmetry, to define it numerically. Let the load at positions (pos.) 1 and 2 be considered. Load reactions on girders 1 and 2 are determined from Eq. (6.35), and bending moments at the mid-span section determined accordingly:

$$Pos.1 \rightarrow x = -\frac{b}{2} \rightarrow R_1 = 0.7; R_2 = 0.4; M_A = 0.7 \times 6.0 + 0.4 \times 2.0 - 1 \times 6.0 = -1.0$$

$$Pos.2 \rightarrow x = +\frac{b}{6} \rightarrow R_1 = 0.1; R_2 = 0.2; M_A = 0.1 \times 6.0 + 0.2 \times 2.0 = 2.0$$

6.4.3.3 Extensions of the Courbon Method: Influence of Torsional Stiffness of Main Girders and Deformability of Cross Beams

The influence of torsion rigidity of longitudinal girders may be taken into account by an extension of the Courbon method [16, 18], by modelling the spring supports of the cross beams on main girders with two springs at each girder. The distance between the springs is determined by the torsional stiffness of the girder. Expressions for $R_j(x)$ are more complex and practical simplicity of the method is nowadays questionable due to the common use of numerical models.

A more exact method consists of idealizing the deck as a set of deformable cross beams supported on springs (stiffness of the main girders) has been developed and published by Homberg [19] with a series of charts for design practice. It should be noted

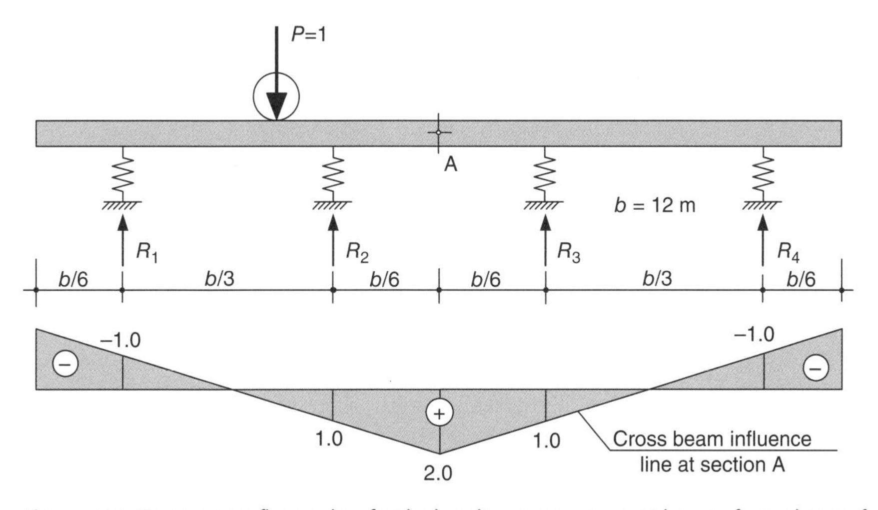

Figure 6.31 Transverse influence line for the bending moments at mid span of cross beams for the example of Figure 6.29.

that, while in previously discussed methods transverse influence lines are straight, in Homberg's method influence curved lines are obtained very close to more exact ones obtained using the orthotropic plate method or numerical models.

6.4.3.4 The Orthotropic Plate Approach

When the deck has several longitudinal girders, it may be modelled by an *orthotropic plate* (Figure 6.32) loaded in the transverse direction by a distributed load $q = q(x, y)$ (note that orthotropic means anisotropic in two perpendicular directions – longitudinal (xx) and transverse (yy) beam directions).

Deflections $w = w(x, y)$ are governed by an orthotropic slab, Eq. (6.22), already referred to in Section 6.3.6 for orthotropic steel plate decks. The behaviour is governed by partial differential Eq. (6.22), where D_x and D_y are bending stiffness (slab + beams) in (x,y) directions and H is a torsion stiffness parameter. For a beam-slab deck, the deck is idealized as a uniform thickness plate with orthotropic properties defined per unit of length from bending and torsion stiffness of beams and slab. For the deck geometry of Figure 6.32, one has

$$D_x = Ei_x = EI_v / l_v \tag{6.36a}$$

$$D_y = Ei_y = EI_c / l_c \tag{6.36b}$$

$$2H \cong Gj_x + Gj_y + \frac{Et^3}{6} = G\frac{J_v}{l_c} + G\frac{J_c}{l_c} + \frac{Et^3}{6} \tag{6.37}$$

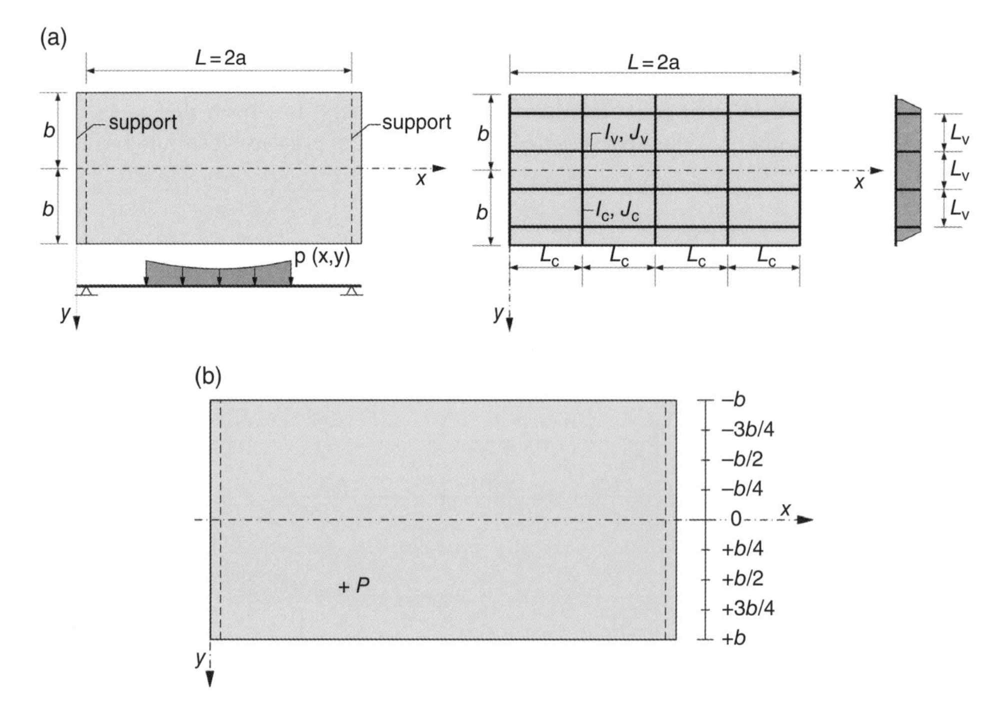

Figure 6.32 (a) Geometry of beam and slab deck for the orthotropic plate approach and (b) correspondent model for determining transverse influence lines by charts.

where i_x and i_y are equivalent orthotropic slab inertias per unit width, obtained from the longitudinal and transverse beam inertias with effective slab widths. Torsional stiffness parameter H is obtained per unit width from torsion stiffness constants J_v and J_c with slab contribution (2x $Et^3/12$).

For determining transverse influence lines of a bending moment in longitudinal or transverse directions, the model of Figure 6.33 is considered. In a continuous deck, this model refers to a deck slab between two cross sections of zero DL bending moments. A point load P is applied at section $x = c$. The aim is to determine longitudinal bending moments $M_x(y)$ at a certain section $x = x_0$, at different y coordinates. In general, the influence line distribution at mid span section $x = L/2$, for loading at that section ($c = L/2$) is adopted for design purposes. A Fourier series solution of Eq. (6.22) for displacement field $w = w(x, y)$ may be obtained and from moment curvature relationships for orthotropic plates [17, 19, 20] longitudinal bending moments M_x are determined. The method was initially developed by Guyon (1946), Massonnet (1950) and Bares and Massonnet (1968). Charts were produced initially (1956) by Morice–Little and Rowe. More recent developments have been made by Cusens and Pama [18] obtaining a series of design charts numerically by taking nine terms of the Fourier series solution and expressing bending moments by:

$$M_x = \frac{PL}{\pi^2 b} \sum_{n=1}^{9} \frac{1}{n^2} \sin^2 \frac{n\pi}{2} k_1' \tag{6.38}$$

where k_1' is a coefficient dependent on n, on the y coordinate and on load eccentricity e at the section.

The average value M_{xm} at full width $2b$ of the cross section can be obtained from the previous expression with the result:

$$M_{xm} = \frac{PL}{\pi^2 b} \sum_{n=1}^{9} \frac{1}{n^2} \sin^2 \frac{n\pi}{2} \tag{6.39}$$

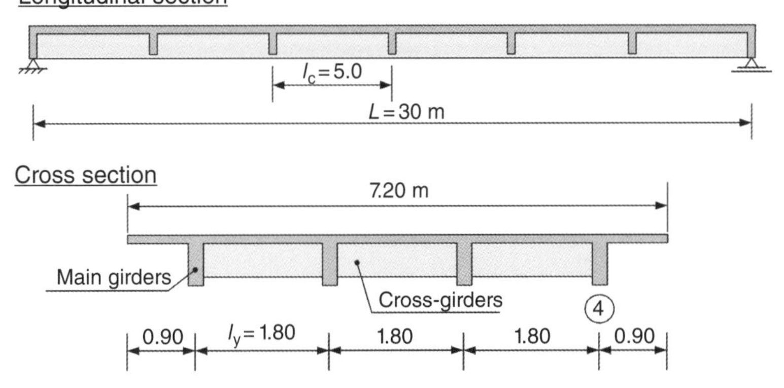

Figure 6.33 Example of application of the orthotropic plate approach to determine transverse influence line at girder 4 of the deck cross section.

Of course, M_{xm} converges to the average value in $2b$ of the bending moment at mid span section evaluated as a beam, that is, $M_{xm} = PL/(8b)$. The transverse distribution coefficient k_{mx} is defined from

$$M_x(y) = k_{mx}\, M_{xm} \tag{6.40}$$

Since the integral of $M_x(y)$ in $2b$ is equal to M and $M_{xm} = M/(2b)$, one has the following condition to check distribution coefficients K_{mx}

$$\frac{1}{2b}\int_0^{2b} k_{mx}dy = 1 \tag{6.41}$$

Charts are available for k_{mx} for different values of the torsion and bending parameters

$$\alpha = \frac{H}{\sqrt{D_x D_y}} \tag{6.42}$$

$$\theta = \frac{b}{L}\sqrt[4]{D_x/D_y} \tag{6.43}$$

Similarly, charts are given for k_{mx} to determine M_y in terms of M_{xm}

$$M_y = k_{my}\sqrt{D_y/D_x}\, M_{xm} \tag{6.44}$$

Coefficients k_{mx} and k_{my} are given by charts for nine *station points* represented in Figure 6.34, for values of $\alpha = 0$, $\alpha = 1$ and $\alpha = 2$ and $0 < \theta < 2.5$. Charts give $k_{mx} = k_{mx}(y,e)$ and $k_{my}(y,e)$ at $y = b, -3/4b, \ldots, +b$, when load $P = 1$ is at each of the nine station points $y = b, 3/4b, \ldots, +b$ in terms of θ. These coefficients define the influence line for transverse load distribution when the load is at the mid-span section. Two of these charts are presented in the example at the end of this section.

In general, $0 < \alpha < 1$ and interpolation for k_{mx} and k_{my} is possible by adopting:

$$k_\alpha = k_0 - (k_0 - k_1)\sqrt{\alpha} \tag{6.45}$$

where k_0 and k_1 are k values for $\alpha = 0$ and $\alpha = 1$. For $1.0 < \alpha < 2.0$ the following interpolation formula is recommended between the values of k for $\alpha = 1$ and $\alpha = 2$, respectively, k_1 and k_2

$$k_\propto = k_1 + \frac{\sqrt{\alpha} - 1}{\sqrt{2} - 1}(k_2 - k_1) \tag{6.46}$$

The charts have been developed for determining the transverse influence line at the mid-span section $x = a$. If the load is not at that section but at a section $x = c$, at mid-span the load induces an overall bending moment $M = Pc/a$. It is possible to use the charts replacing load P at $x = c$, by an equivalent load $Q = Pc/a$ at mid span ($x = a$) inducing the same overall bending moment M.

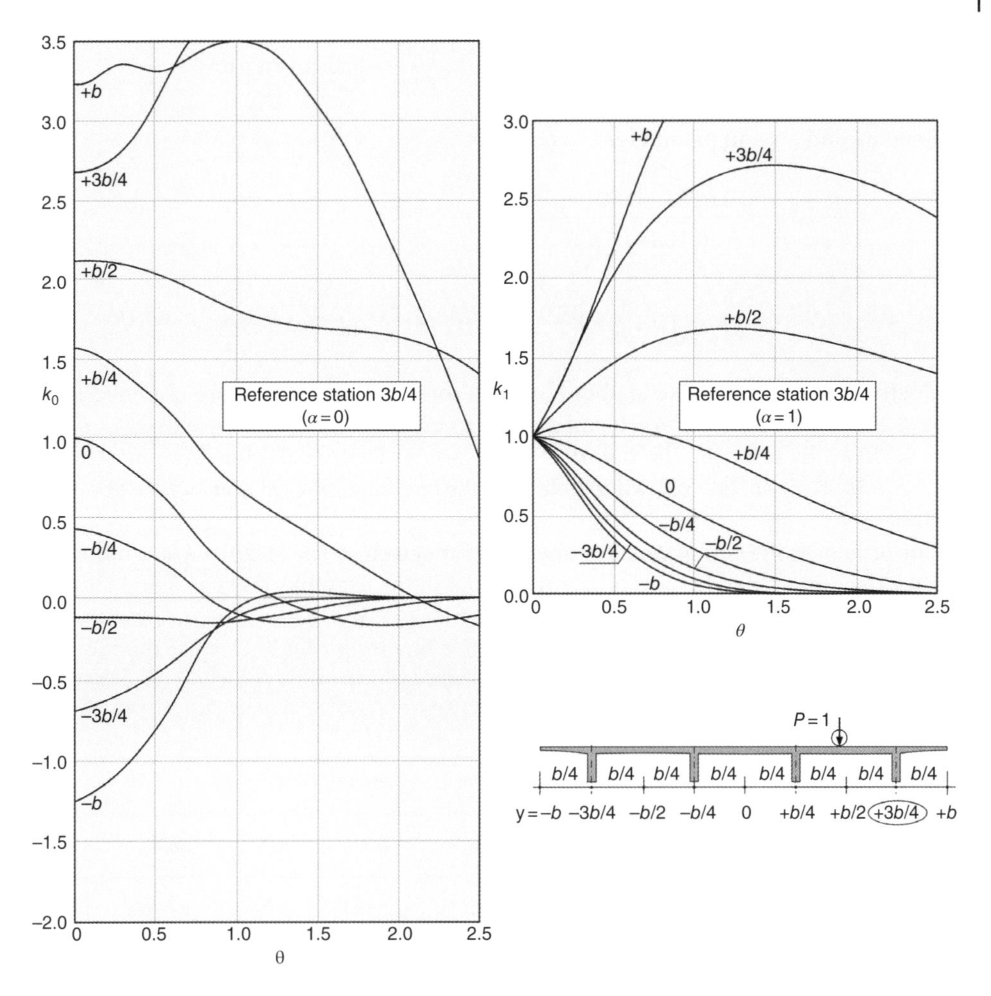

Figure 6.34 Charts for k_0 and k_1 at station $+3b/4$ (i.e. at girder 4), for $\alpha = 0$ and $\alpha = 1$, and load positions $y = e = -b, -3b/4, \ldots, +b$.

Example 6.3 For the bridge deck in Figure 6.33, simply supported and with a span length of 30 m, the aim is to determine, by orthotropic plate approach and the charts of Cusens and Pama, the transverse influence line at mid span section for unit load $P = 1$ acting at that section. The deck has four main girders and a series of five cross beams 5.0 m distance apart.

Resolution:

Deck geometry: $L = 2a = 30\,\text{m}$; $2b = 7.2\,\text{m}$; $l_v = 1.8\,\text{m}$; $l_c = 5.0\,\text{m}$

Section properties Main girders $I_v = 0.303\,\text{m}^4$; $J_v = 0.024\,\text{m}^4$

Cross beams $I_c = 0.113\,\text{m}^4$; $J_c = 0.007\,\text{m}^4$

Properties for the orthotropic equivalent slab:

$$i_v = \frac{I_v}{l_v} = \frac{0.303}{1.80} = 0.168\ \text{m}^4/\text{m} \qquad j_v = \frac{J_v}{l_v} = \frac{0.024}{1.80} = 0.0133\ \text{m}^4/\text{m}$$

$$i_c = \frac{I_c}{l_c} = \frac{0.113}{5.00} = 0.0226\ \text{m}^4/\text{m} \qquad j_c = \frac{J_c}{l_c} = \frac{0.007}{5.00} = 0.0014\ \text{m}^4/\text{m}$$

Bending and torsion parameters:

$$\alpha = \frac{j_v + j_c}{4\sqrt{i_v \cdot i_c}} = \frac{0.0133 + 0.0014}{4\sqrt{0.168 \cdot 0.0226}} = 0.06$$

$$\theta = \frac{b}{L} \cdot \sqrt[4]{\frac{i_v}{i_c}} = \frac{3.6}{30} \cdot \sqrt[4]{\frac{0.168}{0.0226}} = 0.198 \approx 0.20$$

Coefficients for transverse load distribution for girder at $x = 3b/4$ are obtained from Curve 4 for $\theta = 0.20$ and for $\alpha = 0$ and $\alpha = 1.0$. Interpolation Eq. (6.46) yields $k_{mx} = 0.755\ k_0 + 0.245\ k_1$. By selecting the points at the charts of Figure 6.34, for the nine stations $e = -b, -3b/4; \ldots; +b$, the following table with the coefficients k_0, k_1 and k_{mx} are obtained in Table 6.1.

The influence line coefficients for one girder are given by $K = (1/2b) \times k_{mx} \times$ influence the width of the girder. In particular, for girder 4 in the example, one has $K = (1/2b)\ k_{mx}\ (b/2) = k_{mx}/4$.

In Figure 6.35, one shows transverse influence line for girder 4. Despite that, for the present case, due to the stiffness of the cross girders approaching to a rigid cross section

Table 6.1 Values k_0 and k_1 at station $+3b/4$ (i.e. at girder 4) for load positions $y = e = -b, -3b/4, \ldots, +b$.

$e=$	$-b$	$-3b/4$	$-b/2$	$-b/4$	0	$b/4$	$b/2$	$3b/4$	b
k_0	−1.10	−0.62	−0.12	+0.38	+0.90	+1.50	+2.10	+2.75	+3.35
k_1	+0.75	+0.78	+0.80	+0.84	+0.90	+1.06	+1.20	+1.35	+1.35
k_{mx}	−0.65	−0.28	+0.11	+0.49	+0.90	+1.39	+1.88	+2.41	+2.86

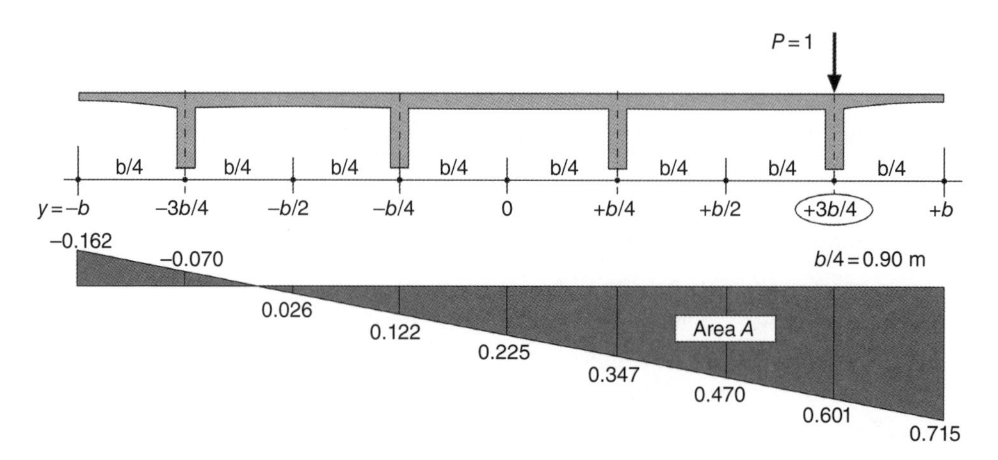

Figure 6.35 Transverse influence line for girder 4 (station + 3b/4).

behaviour (as in the assumption of Courbon's method), the transverse influence line looks straight, it is a curve. If one seeks to find, for example, the maximum load effects for girder 4 due to a uniform traffic load q (kN m^{-2}) on the deck, the load is $q' = q\,A$ where A is the area indicated on Figure 6.35.

6.4.3.5 Other Transverse Load Distribution Methods

Methods based on the *folded plate approach*, where the deck is modelled as a series of longitudinal plate strips between the nodes of the cross section, have been developed by Hambly [21] and charts provided for direct determination of transverse influence lines in decks. The compatibility conditions in folded plate approaches are established along the longitudinal connections between the plates in which the cross section is decomposed.

Hambly's charts may be adopted for a variety of deck cross sections. The reader is referred to the original reference [21] for further information.

Another method based on warping torsion analysis of twin-girder decks, in particular for composite decks, allowing a direct determination of transverse influence lines, is discussed in Section 6.6.5.3.

6.5 Deck Analysis by Grid and FEM Models

Transverse load distribution analysis, discussed in Section 6.4, may also be obtained by numerical approaches based on a 3D beam element model or on a FEM with shell elements. These models allow a complete deck analysis, transversally and longitudinal and are the preferred methods nowadays for advanced design stages. For preliminary design and for checking results of numerical methods, simplified approaches as previously discussed are most valuable. Besides, they allow an understanding of deck behaviour and to obtain the loads for the longitudinal model. Fundamentals of 3D beam models and FEM for structural analysis are covered in structural analysis references [17, 22–24]. Only specific aspects of application of these methods to bridge deck analysis are dealt with in this section.

6.5.1 Grid Models

6.5.1.1 Fundamentals

In *grid* models the deck is considered as an in-plane bar system, rigidly connected at the nodes as shown in Figure 6.36. For permanent and traffic loads, the grid model is loaded perpendicular to its plane. At each node there are three generalized displacements – one

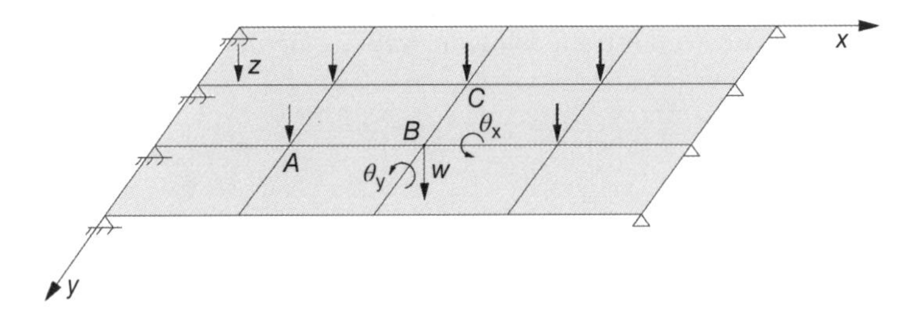

Figure 6.36 Grid model of a bridge deck.

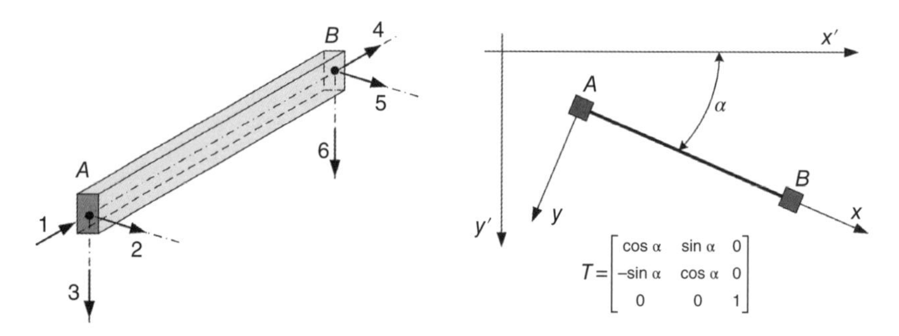

Figure 6.37 Grid element bar model and local coordinate system rotation matrix for transformation of local into global axes system.

vertical deflection w and two rotations θ_x and θ_y. For a beam element θ_x is a torsion rotation while θ_y is a bending rotation. For a general bar element of *grid model* generalized displacement and nodal forces are represented in Figure 6.37 for structural analysis by the displacement method. The stiffness matrix of a grid bar element is

$$k = \begin{bmatrix} k_{AA} & k_{AB} \\ k_{BA} & k_{BB} \end{bmatrix} \tag{6.47}$$

Here, k_{AA}, $k_{AB} = k_{BA}$ and k_{BB} are given in terms of bending EI and uniform torsion stiffness GJ bar properties, where $I = I_y$ and J are the moment of inertia and uniform torsion constant of the cross section, respectively, and E and G are the the modulus of elasticity and distortional modules of the material. The sub-matrices in (6.47) are:

$$k_{AA} = k_{BB} = \begin{bmatrix} GJ/l & 0 & 0 \\ 0 & 4EI/l & 6EI/l^2 \\ 0 & 6EI/l^2 & 12EI/l^3 \end{bmatrix} \tag{6.48}$$

$$k_{AB} = k_{BA} = \begin{bmatrix} -GJ/l & 0 & 0 \\ 0 & 2EI/l & 6EI/l^2 \\ 0 & 6EI/l^2 & 12EI/l^3 \end{bmatrix} \tag{6.49}$$

If shear deformation is taken into account, the following replacements are adopted:

$$\frac{4EI}{l} \rightarrow \frac{4+\beta}{1+\beta}\frac{EI}{l} \quad \frac{2EI}{l} \rightarrow \frac{2-\beta}{1+\beta}\frac{EI}{l} \tag{6.50}$$

where $\beta = 2EI/(l^3 A_r)$ and A_r is the *reduced shear area* ($A_r = A/\chi$, with χ the *shear factor*, equal to 1.2 e.g. for a rectangular section). Stiffness matrices of bar elements in local coordinates are reduced to the global axis by a rotation of coordinates defined by matrix T (Figure 6.37) in terms of the angle α, with the result in matrix form $k_g = T^T kT$. After assembling the stiffness matrices in the global coordinate system, the global stiffness matrix K (dimensions $3n \times 3n$) is obtained. Equilibrium equations, in terms of vectors of generalized displacements q and generalized forces, with dimensions ($3n \times 1$) where n is the number of nodes, are in the matrix form $Kq = Q$. After introducing displacement boundary conditions, the linear system of equations is solved for the $3n$ unknowns q_i.

6.5.1.2 Deck Modelling

Modelling the deck by a grid requires establishing the geometry of the mesh (Figure 6.38), bar properties and nodal generalized force vector $Q = \{Q_i\}$. At each node i, one has $Q_i = \{M_{ix}, M_{iy}, F_{zi}\}$. These nodal loads result from distributed loads on the longitudinal and transverse bars transferred by the deck slab. For straight decks, longitudinal bars of the grid are adopted equally spaced in the transverse direction as shown in Figure 6.38. The number of longitudinal bars depends on the deck width b and shape of the deck cross section as shown in Figure 6.39. A minimum of five longitudinal elements should be adopted, with elements passing on bearings at piers and abutments. In some cases, if the computer program allows the consideration of beams with rigid segments within, it is possible to reduce the longitudinal beam number as in Figure 6.39c. For decks with multiple girders, it is usually sufficient to associate one element per girder with the effective part of the slab within yielding a T beam longitudinal element for the grid. In the case of twin deck girders (beam and slab decks with two girders only), the slab should be modelled by transverse and longitudinal beams as shown in Figure 6.39e.

Transverse beams in the grid model should be adopted at cross beams alignments (if existing) and they should always be considered as intermediate beams to simulate the slab stiffness as well. The relationship a/b (Figure 6.38) between transverse and longitudinal beam spacing should be in the order of 1.5; lower relationships than 1.0 are not needed because they unnecessarily increase the number of elements with only a small improvement in the results. When large gradients of variation of internal forces, such as at deck intermediate supports (Figure 6.40), the number of elements should be increased.

Skew slab decks analysed by grid models with small skew effects (usually for $\varphi > 70°$) may be done with longitudinal beams parallel to longitudinal edges and transverse beams parallel to support lines. The criterion is to model the longitudinal beams as far as possible according to the direction of *principal bending moments* in the slab. For large skew effects, $\varphi < 70°$, and for moderated deck widths, b, transverse and longitudinal beams should be perpendicular and parallel to longitudinal free edges respectively.

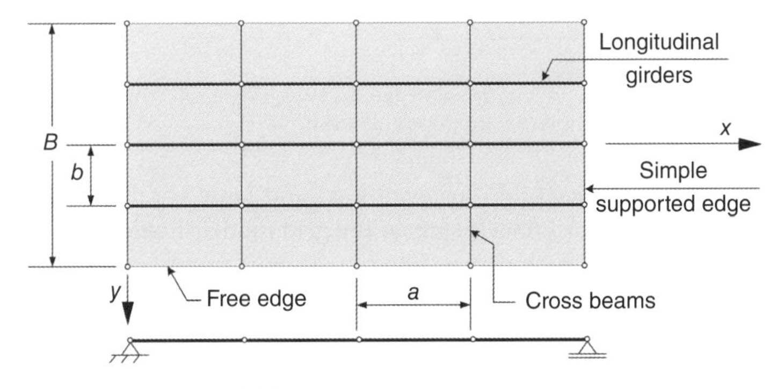

Figure 6.38 Grid model for a deck.

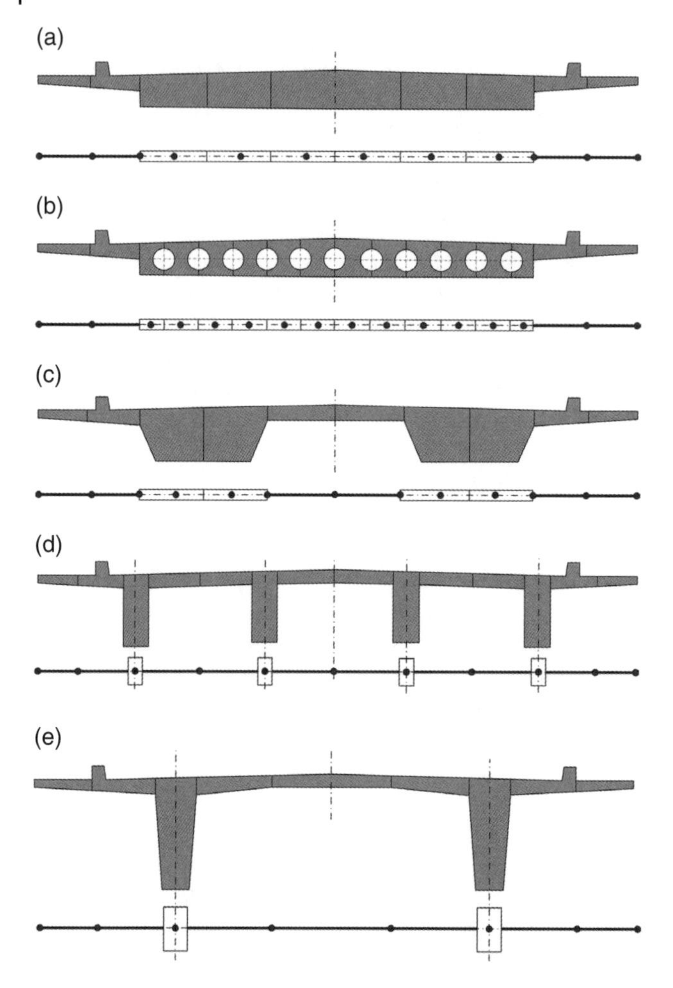

Figure 6.39 Grid models for different cross section decks.

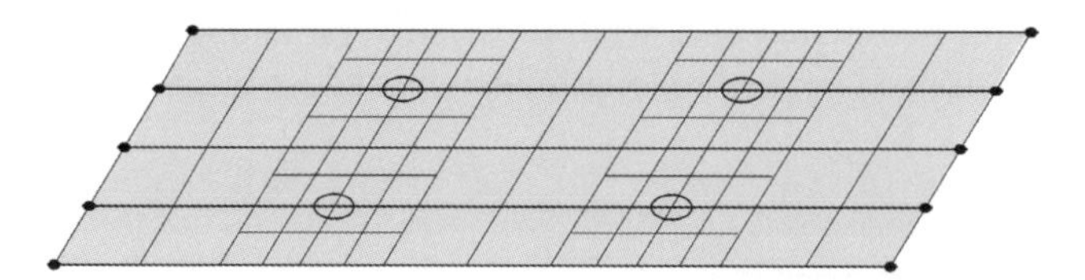

Figure 6.40 Grid mesh for a continuous deck. Internal supports and mesh.

For very large skews (say φ in the order of 45° or even less) and large widths b (say, widths in the order of the length), longitudinal beams in the grid model should be perpendicular to support line directions as shown in Figure 6.41 and transverse beams perpendicular to the longitudinal ones; otherwise, a large part of the slab stiffness will not be taken into account.

Modelling skew slab decks by grid models should consider the existence of large negative moments and large bearing reactions near obtuse corner angles. Besides, variation

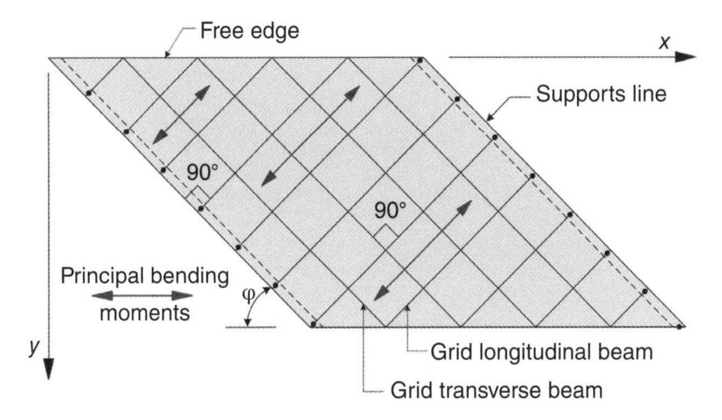

Figure 6.41 A very skew slab deck of large width. Principal bending moment directions and grid mesh modelling.

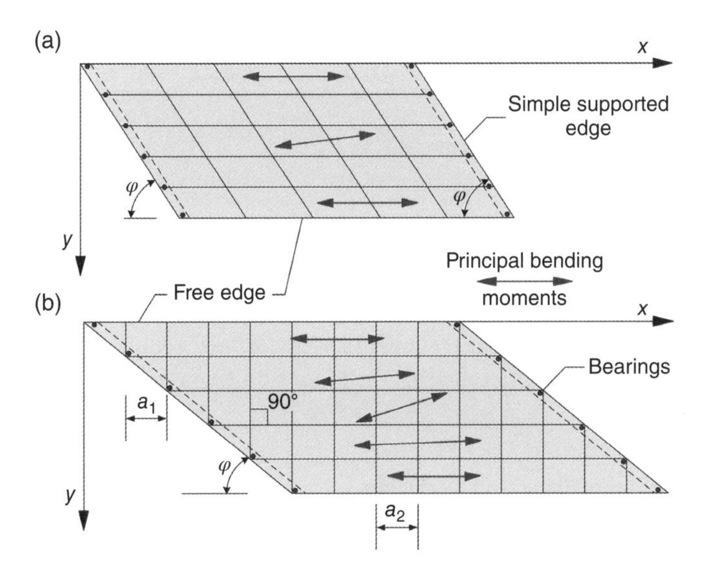

Figure 6.42 Skew slab decks. Principal bending moment directions and grid meshes.

of vertical bearing reactions along supported edges with possible negative reactions (uplift effects) nearby acute corner angles should be taken into account. Positive maximum span bending moments are oriented according to the direction perpendicular to the support lines Figure 6.42. Finally, one should note principal bending moments M_I are along free edges (moment vector perpendicular to the free edge) where $M_{II} = 0$, as shown in Figure 6.43.

6.5.1.3 Properties of Beam Elements in Grid Models

For the beam element properties definition, the section properties of the deck modelled by the bars should take into account the beam or ribs and effective associated part

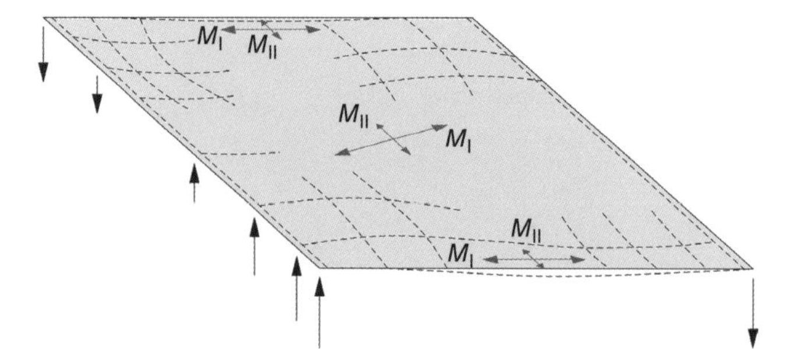

Figure 6.43 A skew slab deck. Variation of principal moment directions and support reactions.

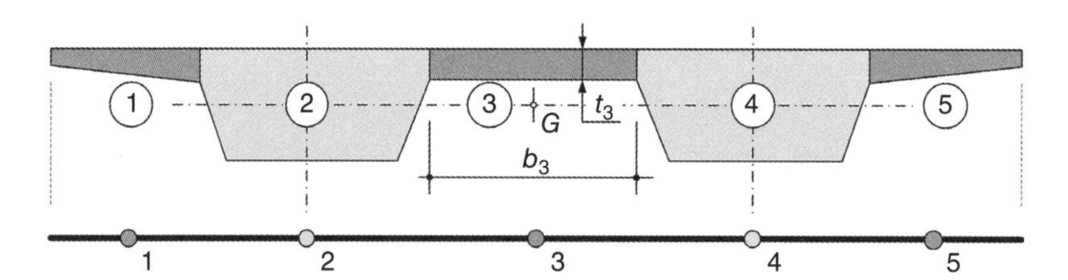

Figure 6.44 Idealization of section properties for a grid model.

of the slab deck. Besides, the bending stiffness about the local or global axis of the section should be taken into account for different elements as illustrated in Figure 6.44. For element 3, the inertia is taken in terms of its own axis (i.e. $I = b_3 t_3^3/12$), while for elements 2 and 4 inertia should be taken about the global bending axis. Voided slabs should be modelled with longitudinal beam elements with a box section as in Figure 6.39b). The reader is referred to references [17, 21] for additional information on deck grid models.

6.5.1.4 Limitations and Extensions of Plane Grid Modelling

One of the main limitations of grid modelling is a two-dimensional idealization of the superstructure. Membrane effects are not taken into account because the elements are considered in a single plane only. Idealization of properties may overcome some aspects of this limitation but grid modelling requires some skill as, for example, in multiple box girder sections. Extension of plane grid models to spatial grid models (3D bar model) is sometimes adopted, where, for example, slab and overall level differences of centres of gravity may be taken into account. However, this grid modelling has been progressively replaced by integrated finite element models with shell and beam elements. In these models, membrane effects are directly taken into account as discussed in the next section.

6.5.2 FEM Models

6.5.2.1 Fundamentals

The basic aspects of FEM are discussed in structural analysis texts [17, 22–24]. Only specific aspects related to bridge decks are dealt with in this section. FEM may be seen

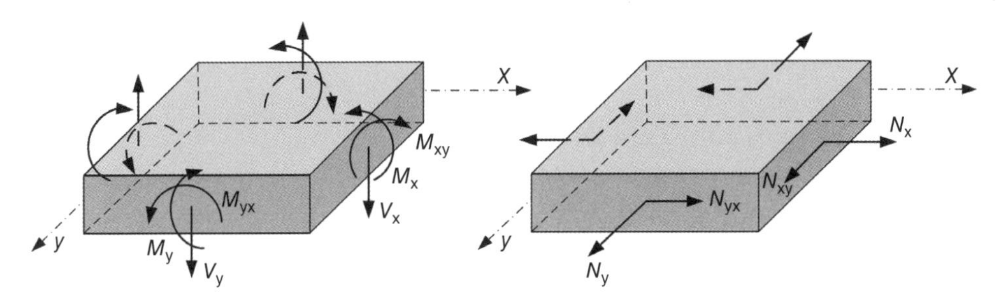

Figure 6.45 Moment and membrane forces in a shell type element.

as a generalization of the Ritz Method for finding a stationary solution for a functional [25]. In structural analysis, the functional is the Total Potential Energy $V[w]$ where $w = w(x, y)$ denotes, for example, the vertical displacement field in a deck. For the slab deck element of Figure 6.45, generalized displacements, generalized strain and stress vectors are,

$$\underline{u} = \begin{bmatrix} w \\ \theta_x \\ \theta_y \end{bmatrix} = \begin{bmatrix} w \\ \partial w / \partial y \\ \partial w / \partial x \end{bmatrix}; \tag{6.51}$$

$$\underline{\varepsilon} = \begin{bmatrix} -\partial^2 w / \partial x^2 \\ -\partial^2 w / \partial y^2 \\ -2\partial^2 w / \partial x \partial y \end{bmatrix}; \tag{6.52}$$

$$\underline{\sigma} = \begin{bmatrix} M_x \\ M_y \\ M_{xy} \end{bmatrix} \tag{6.53}$$

The generalized stress-strain relationship, for an isotropic plate of thickness h, Poisson ratio v and bending stiffness $D = Eh^3/[12\,(1 - v^2)]$, is

$$\underline{\sigma} = \underline{H}\,\underline{\varepsilon} \tag{6.54}$$

$$\text{with } \underline{H} = D \begin{bmatrix} 1 & v & 0 \\ v & 1 & 0 \\ 0 & 0 & \frac{1-v}{2} \end{bmatrix} \tag{6.55}$$

At the internal domain, A, one has distributed forces p perpendicular to the plane of plate. The total potential energy functional is

$$V[w]=\frac{1}{2}\int_A \underline{\varepsilon}^T \underline{H}\,\underline{\varepsilon}\,dA-\int_A p^T w\,dA \tag{6.56}$$

The equilibrium equation for the displacement field $w = w(x, y)$ is among all kinematically admissible functions the one that makes stationary the total potential energy function, requiring $\delta V[w] = 0$. By introducing *geometric (kinematic) boundary conditions* and using *Variational Calculus* [25] one obtains the equilibrium partial differential equation to be satisfied by the exact solution $w = w(x, y)$ and *forced (static) boundary conditions*. Approximate solutions may be obtained by transforming the continuous system in a discrete n degrees of freedom system, by the Ritz Method [26], selecting the displacement field in the form:

$$w(x,y)\approx\sum_{j=1}^{n} q_j\psi_j(x,y) \tag{6.57}$$

Here, q_j are the n degrees of freedom (*generalized coordinates*) and ψ_j are kinematic admissible *shape functions*. The shape functions are not required to satisfy the forced boundary conditions, but the kinematic ones only. The continuous system of Ritz's method is transformed in *discrete system* with n degrees of freedom (nDOF). The *potential energy functional* is transformed in a *function*

$$V(q_j)=1/2\,\underline{q}^T\,\underline{K}\,\underline{q}-\underline{Q}^T\,\underline{q} \tag{6.58}$$

where, $\underline{q} = \{q_i\}$ and $\underline{Q} = \{Q_i\}$ are generalized coordinate and force vectors and $\underline{K} = [k_{ij}]$ is the stiffness matrix, with $i, j = 1$ to n:

$$k_{ij}=\int_A \underline{B}_i^T\,\underline{H}\,\underline{B}_j\,dA \quad Q_i=\int_A \underline{N}_i^T\,\underline{p}\,dA \tag{6.59}$$

where vectors $\underline{B}$ and $\underline{N}$ are defined from the shape functions by

$$\underline{B}_i^T=\left[-\frac{\partial^2\psi_i}{\partial x^2},-\frac{\partial^2\psi_i}{\partial y^2},2\frac{\partial^2\psi_i}{\partial x\,\partial y}\right] \quad \underline{N}_i^T=\left[\psi_i,\frac{\partial\psi_i}{\partial y},-\frac{\partial\psi_i}{\partial x}\right] \tag{6.60}$$

For equilibrium, $V(q_i)$ should be stationary with respect to admissible variations of q_i and the following equilibrium equations for discrete n DOF system are obtained:

$$\frac{\partial V(q_i)}{\partial q_i}=0 \text{ for } i=1,2,\ldots,n \tag{6.61}$$

Introducing Eq. (6.58) for V in Eq. (6.61), yields in matrix form

$$[K]\{q\}=\{Q\} \tag{6.62}$$

This is a system of linear n algebraic equations at n unknowns, with K being a positive definite matrix, det $[K] > 0$. After determining q_i, the displacement field $w = w(x, y)$ is known as is the displacement vector $\underline{u}$ as a consequence. Generalized stress and strain vectors (moments and curvatures) are obtained.

In a plate analysis or general bridge deck, it is not easy to find a complete set of admissible shape functions ψ_j, that is satisfying all kinematic boundary conditions. These conditions are, for example, for a supported built-in edge at $y = b$, $\psi_j(x, y) = 0$ and $\psi_{j,y}(x, y) = 0$ at $y = b$. It is quite important to distinguish kinematic and static boundary conditions, when applying Ritz or FEM. A kinematic boundary condition and a static boundary condition along a simply supported edge $y = b$, are, respectively, $w = 0$ and $M_y = 0$ yielding for the shape functions $\psi_j(x, y) = 0$ and $\psi_{j,yy}(x, y) = 0$. Only the former need to be satisfied by admissible shape functions.

In FEM the domain is discretized in sub domains, the elements e that are for bridge deck analysis, in most cases, triangles and quadrilaterals shell type elements, that is, with bending and membrane forces (Figure 6.45). The elements should be easily adapted to deck boundaries, and usually this is done with triangular elements. In singular points, like an intermediate support, as shown in Figure 6.46, a combination of triangles and quadrilaterals is possible. Within the element, the displacement field is discretized as is done in the Ritz method for the entire domain.

The generalized coordinates in the FEM are identified with *nodal generalized displacements*, that could be node i deflections or rotations as in a slab element ($q_e = \{q^e{}_i\} = \{w_i, w_{i,x}, w_{i,y}\}^e$) or two displacements per node as in a plane stress plate element ($q_e = \{q^e{}_i\} = \{u_i, v_i\}^e$. For example, in a FEM for a slab deck, adopting slab elements with three degrees of freedom (w, θ_x, θ_y) per node, one has a discrete system with $3m$ unknowns for a FE mesh with m nodes. The stiffness matrix K is $(3m \times 3m)$ before kinematic boundary conditions are imposed at the boundary. The stiffness matrix is

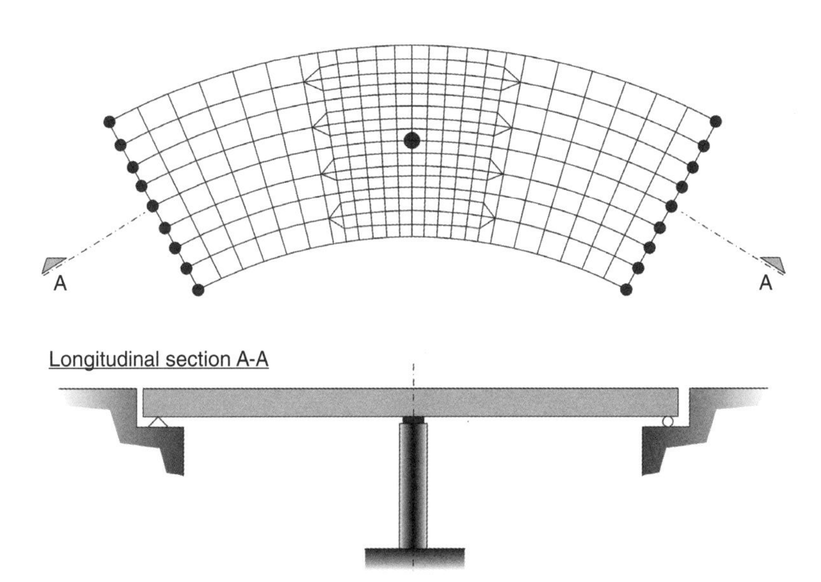

Figure 6.46 Finite element mesh for a continuous slab deck with a single pier support at mid-span.

then *condensed* [9] by eliminating the nodal degrees of freedom associated with kinematic boundary conditions.

One specific aspect of the FEM compared to the Ritz method is the choice of shape functions; in the FEM the shape functions are selected as displacement functions taking unit values associated with degrees of freedom at each node and zero for all remaining degrees of freedom. These functions for a generic element *e* are designated in the following by $N^e(x, y)$.

The displacement field at element level is written, as in the Ritz method,

$$w^e(x,y) = \sum_{i=1}^{m} N_i^e(x,y) q_i^e = \underline{N}^e \, \underline{q}^e \tag{6.63}$$

where N_i^e and q_i^e are, respectively, for a slab element, (1×3) and (3×1), and N^e and q^e are $(1 \times 3\text{m})$ and $(3\text{m} \times 1)$ vectors.

From this point on, stiffness matrix k_e and equivalent force vector P_e for the element are obtained as in the Ritz method. By the standard *assembly process* of structural analysis, the overall stiffness matrix K ($3n \times 3n$ for a slab element mesh) is obtained and equilibrium equations for linear elastic static analysis are like Eq. (6.62) of the Ritz method. These equations are solved for $3n$ unknowns $\{q\}$ and stress and strain generalized vectors are obtained.

These equilibrium equations are extended to *linear stability analysis* for determining buckling modes and critical loads by introducing the *geometrical stiffness matrix G*, including second order effects due to axial loads in bar elements or nonlinear effects induced by compressive membrane stresses in plates or shell elements. The reader is referred to structural stability references [26, 27] for specific aspects on this subject. For linear buckling analysis, the loads are assumed proportional to a load parameter λ, displacements q_i are *additional small displacements* from the *fundamental equilibrium path* $q_f = q_f(\lambda)$ and matrix G is written as $G = -\lambda G'$. Equilibrium states in the neighbourhood of the fundamental path are investigated under the equilibrium conditions

$$[K - \lambda G']\{q\} = \{0\} \tag{6.64}$$

This equation represents an *eigenvalue problem* in mathematics, with non-zero solutions (*buckling modes*) for *n* different roots (λ_{cri} – *critical loads*) of the characteristic equation

$$det[K - \lambda G'] = 0 \tag{6.65}$$

In dynamic analysis [28, 29], a similar eigenvalue problem is obtained for determining *vibration modes* and *natural frequencies* with the characteristic equation

$$det\left[K - \omega^2 M\right] = \{0\} \tag{6.66}$$

where M is the *mass matrix* and ω are the circular natural frequencies. By solving this algebraic equation, *n* natural frequencies (ω_i) are determined, each one correspondent to a vibration mode obtained for each frequency ω_i from the linear system of equations

$$\left[K - \omega_i^2 M\right]\{q\} = \{0\} \tag{6.67}$$

6.5.2.2 FEM for Analysis of Bridge Decks

Modelling and element types – The large applicability of FEM in bridge deck analysis results from its adaptability to complex geometries, variety of applied loads or imposed displacements, like prestressing, concrete creep and shrinkage, thermal actions and imposed settlements. A large potential of FEM results also form the possibility of adopting a variety of integrated elements in the same model. Shell and 3D bar elements are the most adopted ones for bridge deck analysis. Connecting elements may be modelled by *rigid links* between the nodes and specific elements to model bearings and seismic dampers are available in standard software for bridge analysis. Static, dynamic and stability analysis are performed with this standard software. Phase by phase analysis of a bridge superstructure modelling the evolution of the static system during execution phases can be also performed. Finite element meshes may also be generated automatically from concept design drawings and graphical outputs to simplify design checking.

Seismic analysis is easily done by a modal dynamic approach defining seismic spectrum response. Dynamic analysis under moving loads, namely for High Speed Railway bridges, allow evaluation of maximum vertical accelerations of the deck for ULS and SLS verifications. FEM is nowadays the most adopted method for the analysis of bridge structures. However, its use should be done in a step by step approach, starting by modelling the structure with simplified FEM models, namely a bar model, before a complete modelling is performed. It is the only way of controlling the results to predict the correct structural response and get an in-depth knowledge of bridge behaviour. The use of powerful FE Models does not transform a bad concept design in a better one. Simplified models are key instruments for design and FEM powerful tools for checking. If a structural behaviour, no matter the complexity of the structure, cannot be predicted by a simplified model, it is not reliable.

Use of FEM requires a good knowledge of its fundamentals and potential limitations of the method, namely concerning the type of element to be adopted and convergence conditions towards the exact solution. Near singularity points like 'concentrated' loads induced by a bearing (Figure 6.46) or zones where large gradients of stress variations are expected require refining FE meshes. For example, if a bearing supporting a slab deck is simulated by a nodal point, since it is being modelled as a concentrated load in a slab the exact elastic solution yields infinite bending moments at this singularity point. Of course, when a FEM is adopted, the dimensions of the elements will not yield the exact elastic response. If the bearing dimensions are in the order of the slab thickness, modelling the bearing as a rigid support yields acceptable results at a distance of the referred dimensions.

Convergence towards the exact solution is not necessarily monotonic and improved by increasing the number of elements. One should distinguish between *conform* and *non-conform elements*. For example, using slab elements and polynomial functions for the shape functions N^e, compatibility conditions for w_i, θ_{ix} and θ_{iy}, are satisfied necessarily at the nodes but not necessarily rotations θ_x along a common edge between two elements as shown in Figure 6.47. Equality of displacement vectors at common nodes between two adjacent elements is a minimum requirement for compatibility; however, this condition does not assure equality of displacement vectors along a common edge of two adjacent elements. If compatibility is satisfied at the node only, the elements are designated as *non-conforming elements* and convergence for the exact solution is not monotonic when increasing the mesh. However, better results are very often obtained with non-conforming

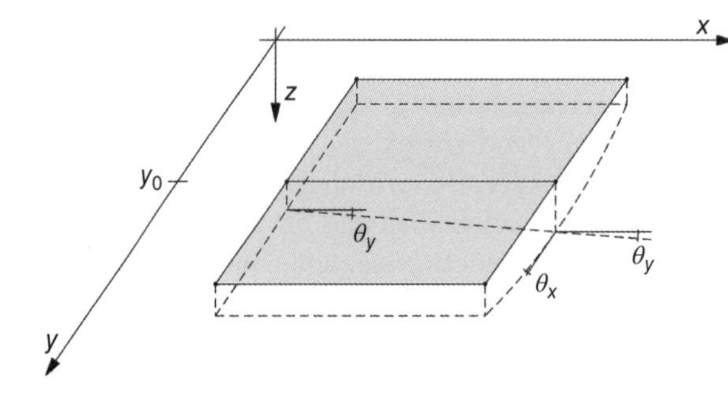

Figure 6.47 Compatibility of generalized displacements between boundaries of slab finite elements.

elements, reducing the required mesh to reach a sufficiently precise solution. In conforming elements, if one compares convergence towards the exact solution for the displacement field, say $w = w(x, y)$, for two meshes, one with n elements yielding a solution w_n and another with $n' > n$ yielding $w_n{}'$, one has for the total potential energy, V,

$$V(w_n) \geq V(w_n{}') \geq V(w) \tag{6.68}$$

In non-conforming elements, convergence is not monotonic, but it is very often quicker. What is important when selecting element types is to check if the displacement field of the element:

- allows displacement fields correspondent to constant strain fields;
- allows rigid body motions, that is, displacement fields associated with zero strain fields.

Modelling bridge decks very often requires the use of different element types. A variety of finite elements are available for bridge deck analysis, the most common ones for plate analysis are 12-DOF with 3-DOF per node (w, $w_{,x}$ and $w_{,y}$) quadrilaterals and triangles for plane stress analysis. Curved boundaries may be dealt with *isoparametric elements* in which geometry boundaries of the element and displacement fields adopt the same shape functions. The reader is referred to specific references on the FEM for additional information.

FEM of slab-girder or rib-slab decks – The case of slab-girder decks (Figure 6.48) is a good example: the slab is modelled by slab or shell elements and the girders are modelled by 3D bar elements. There is an eccentricity, e, between the centres of gravity of the slab elements and of the beam elements. If the eccentricity is small, as happens in rib-slab decks, it may be acceptable to adopt an equivalent depth h_{eq} for a beam with the same moment of inertia with respect to the global centre of gravity of the section. A better modelling introduces the eccentricity by rigid links between the plane of centres of gravity as shown in Figure 6.48c. Finally, another possibility is to model the slab and the beams by shell elements, with the inconvenience of obtaining stress resultants per unit length in the shell elements of the girders that requires integration to obtain internal forces. In a complete shell modelling of a slab-girder deck, longitudinal membrane forces N_x obtained in the slab elements act with an eccentricity with respect to the overall centre of gravity of the section, being the main components for the overall bending moments M_y. Besides, one obtains local bending moments in the slab $M_{y,slab}$ yielding a component for the overall bending moment as well. The variation of the membrane

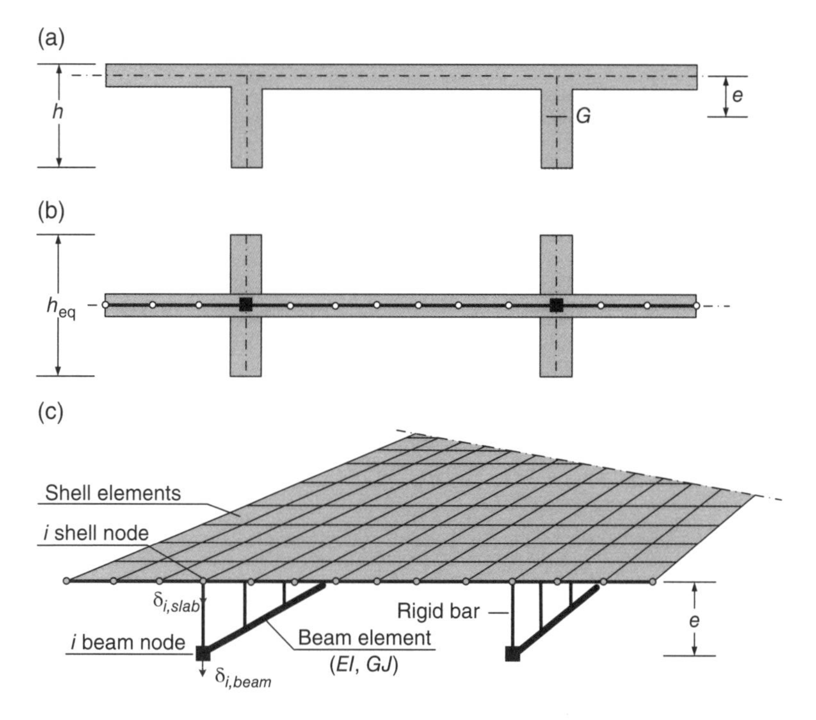

Figure 6.48 FE modelling of slab-beam decks (a) with shell and beam elements centred with mid-surface of the slab (b) or with eccentric bar elements (c).

forces N_x along the width of the slab also reflects the shear lag effect. From a design point of view, it is easier and more efficient to model the slab and the girders with slab and beam integrated elements connected by rigid links.

FEM of box girder decks – Modelling of single- or multi-cell box girder decks requires two types of approaches:

- 3D bar elements for global overall analysis, that is, for displacement and internal forces in box girder sections, for static and dynamic analysis;
- Shell FEM for local analysis and diaphragms.

In design practice, sometimes global bar models and shell models are combined by adopting a sufficient long deck segment between two transverse cross sections where beam stresses are applied and modelling the segment, including support diaphragms and any intermediate diaphragm with shell elements. An example of a FEM to model a segment over a pier of a single cell box girder deck is shown is shown in Figure 6.49. Diaphragms should be modelled by plane stress elements connected to webs and flanges, by the nodes. Shell elements are usually adopted for slab decks, webs and lower flanges. Webs are subjected to out of plane bending effects due to rigid connection with the upper flange. In plane shear stresses, due to global vertical and horizontal shear forces, interact with transverse bending effects. In prestressed concrete box girders, at the upper parts of the webs these local bending stresses, in particular due to live loading also inducing distortional cross section effects, are usually relevant for checking crack

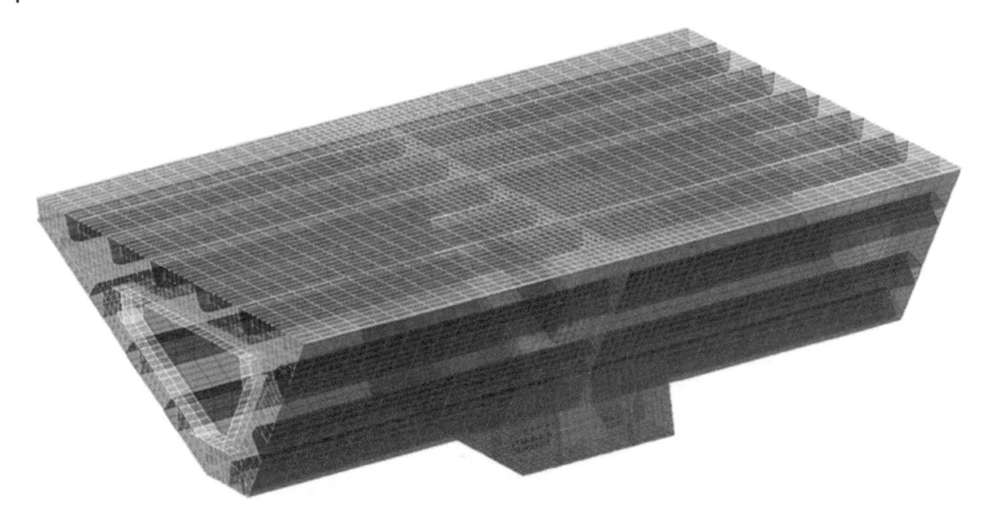

Figure 6.49 Modelling of a box girder deck segment over a pier (Bluewaters pedestrian bridge).

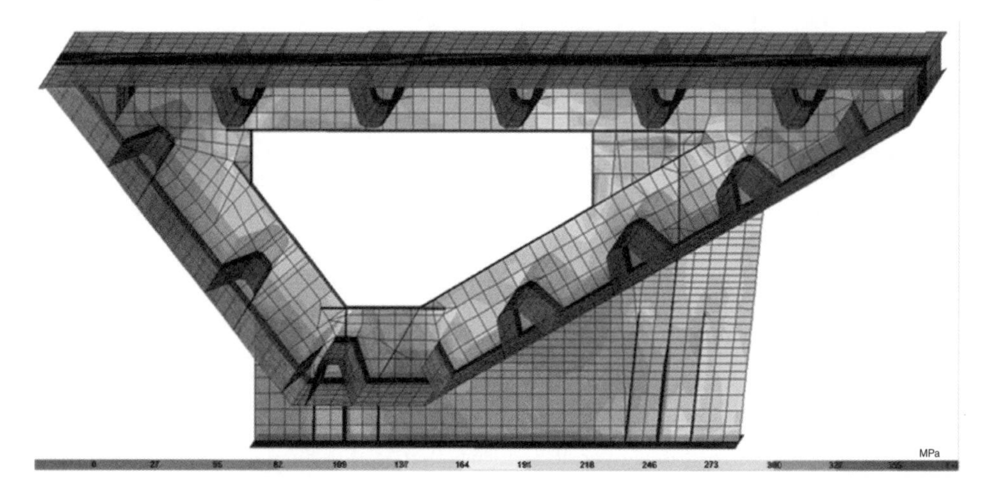

Figure 6.50 Diaphragm von Mises stresses in a steel box girder (Bluewaters pedestrian bridge).

limits. In long span box girder bridges near intermediate supports, vertical prestressing of the webs may be required for crack width limitation.

Support diaphragms are subjected to important in plane stresses due to the effect of bridge bearing reactions. Figure 6.50 shows the results of ` steel box girder deck with the effect of different reactions at the bearings due to eccentricities of the loads along the deck. Von Mises stresses are taken from the FEM to check safety at SLS.

For box girder diaphragms, in particular for prestressed concrete decks, two types of analysis may be done in design practice: elastic SLS stress analysis and a rigid plastic ULS analysis by modelling flow of forces by a strut and tie approach, as shown for a single cell box girder deck in Figure 6.51. A concrete strut is verified under compressive

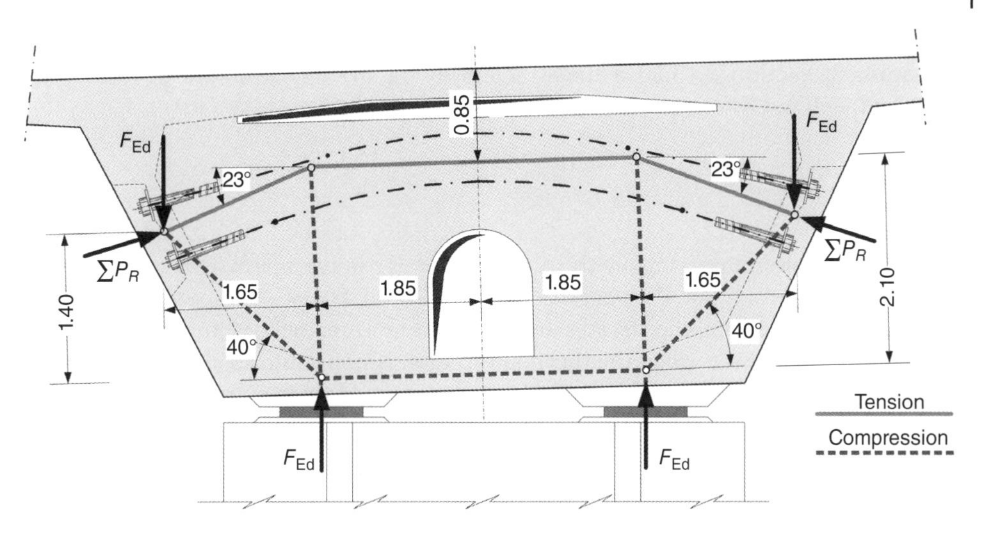

Figure 6.51 Strut and tie modelling of a diaphragm in a prestressed concrete box girder.

stresses and steel reinforcement and eventual prestressing in the tie are designed for the tensile stresses. This rigid-plastic model should not avoid a linear elastic stress analysis for SLS based on a shell FEM as previously referred to. Crack with limitation and ordinary steel reinforcement are designed according to reinforced concrete rules based on this SLS model.

Equivalent nodal loads, imposed deformations and prestressing effects – In FEM permanent and traffic loading in bridge decks are simulated by *equivalent nodal loads*, Q_i, consistent with the *element shape functions* and evaluated from distributed loads $p = p(x,y)$ by similar expressions to the ones referred to for the Ritz method. Imposed deformations are due to:

- *Thermal loading*, uniform and differential actions;
- *Concrete shrinkage*, uniform and differential shrinkage;
- *Foundation settlements*, instantaneous and long term.

Thermal loading and concrete shrinkage, if uniform through the deck, can only be analysed by a FEM of the superstructure by considering shell elements in the deck with membrane and bending effects. However, consideration of slab elements only, that is, without membrane effects, is possible if, for example, the deck is not rigidly connected to the piers. In this case, a separate analysis for the piers adopting a beam model is possible by considering the thermal uniform deformations of the deck and imposing on the piers. If the piers are monolithic with the deck, local bending effects are introduced by uniform thermal strains or shrinkage. Of course, when analysing effects on piers, time dependent deformations of the concrete, under imposed uniform temperature variation or uniform shrinkage (equivalent to a temperature drop) should be taken into account. This is usually done in an elastic FEM, by adopting an *effective modulus of elasticity* (Chapter 3, Section 3.14) for concrete. As $E_{c,eff}$ is usually less than 50% of the elastic value, thermal internal forces are reduced with respect to elastic values, obtained with an elastic FEM, in the same proportion.

For differential thermal actions in the deck, for the linear component ΔT_l (Chapter 3, Section 3.13.2.) a linear strain along the depth of the deck may be imposed, defined by:

$$\varepsilon_x^0 = \varepsilon_y^0 = \frac{\Delta T_l}{h} z \tag{6.69}$$

Usually, FE programmes allow this deformation (or linear thermal variations) to be considered directly. If not, for a deck slab these thermal strains are imposed on a totally restricted slab and resulting stresses are evaluated by Duhamel's Method of Theory of Elasticity [28, 70]. Even when the thermal variation is non-uniform along the depth of the deck, the methodology is the same. First, the restricted stresses are evaluated assuming a plane stress state on the slab ($\sigma_z = 0$) for $\varepsilon_x^0 = \varepsilon_y^0 = \alpha_T \, \Delta T(z)$, and plane strain-stress relationships are solved for σ_x^0 and σ_y^0, with the result:

$$\sigma_x^0 = \sigma_y^0 = -\frac{E\varepsilon^0}{1-\nu} \tag{6.70}$$

Then, these stresses are applied in the deck contour (a line boundary) and a FEM for these applied stresses. Elastic stresses σ_x^e and σ_y^e are obtained from the FEM and the final stresses induced by the thermal action are:

$$\sigma_x = -\frac{E\varepsilon^0}{1-\nu} + \sigma_x^e; \; \sigma_y = -\frac{E\varepsilon^0}{1-\nu} + \sigma_y^e; \; \tau_{xy} = \tau_{xy}^e \tag{6.71}$$

Imposed foundation settlements are defined in the FEM by imposed displacements in the nodes connected to bearings or directly at piers. If they are instantaneous settlements, internal forces are reduced with time according to *concrete relaxation function* $r(t, t_0)$ (Chapter 3, Section 3.14) and the elastic stresses and stress resultants (e.g. moments) are reduced according to:

$$\sigma_{\alpha\beta}(t,t_0) = r(t,t_0)\sigma_{\alpha\beta}^e \;\text{ with }\; \alpha\beta = x, y \tag{6.72}$$

where $\sigma_{\alpha\beta}^e$ is the stress tensor obtained from the elastic FEM. A similar expression holds for stress resultants; for example, for moments $M_{\alpha\beta} = r(t, t_0)\, M^e_{\alpha\beta}$. When settlements are time dependent, such as foundations on clay, the elastic results cannot be so easily extrapolated to take viscoelastic effects into account. Superposition effects according to Boltzmann's Principle of Linear Viscoelasticity should be adopted as discussed in Chapter 3, Section 3.14.

The case of prestressing effects can be considered in deck FEM by adopting an equivalent loads approach. Bi-axial prestressing in a slab deck is imposed by families of cables with eccentricities e_x and e_y with respect to the centre of gravity of slab cross sections by the equivalent loads approach, that is, by considering upward or downward line loads defined by

$$p_{eq} = P_x \frac{\partial^2 e_x}{\partial x^2} + P_y \frac{\partial^2 e_y}{\partial y^2} \tag{6.73}$$

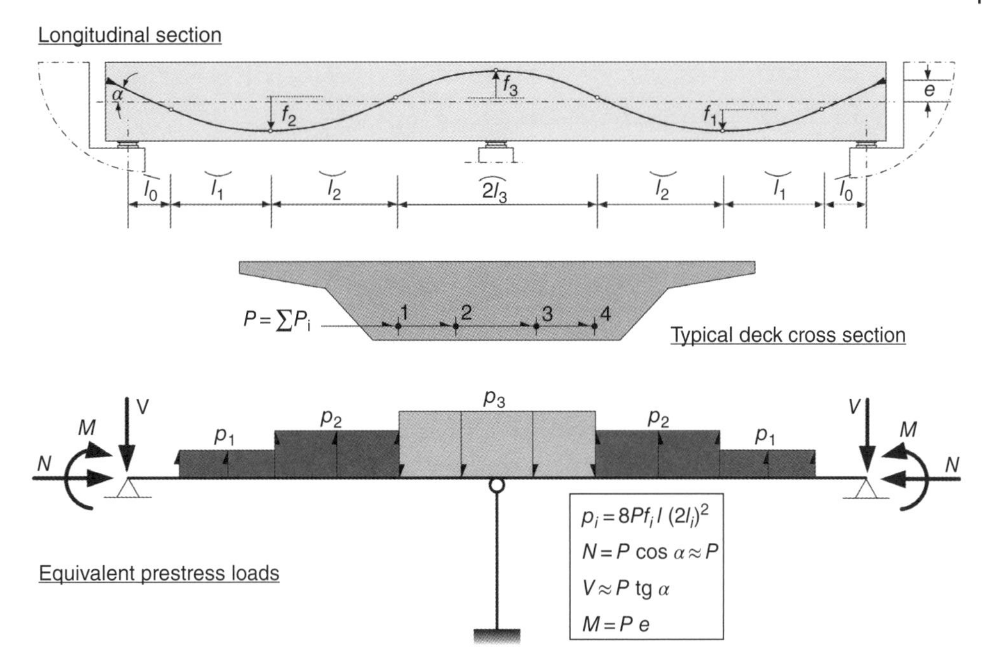

Figure 6.52 Prestressing equivalent loads for a rib-slab deck.

where P_x and P_y are the effective (after all losses) prestressing loads. At the boundary, the cables have eccentricities that should be modelled as well with the applied prestressing forces. Then, the FEM may be analysed for the equivalent prestressing loads. Most available FEM software allows also a direct definition of prestressing cables for beam structures; this is not so usual for FEM programmes with shell elements. In practice, transverse prestressing and longitudinal prestressing are often considered separately, allowing, for example, modelling longitudinal prestressing (Figure 6.52) by a bar model.

6.6 Longitudinal Analysis of the Superstructure

6.6.1 Generalities – Geometrical Non-Linear Effects: Cables and Arches

Longitudinal analysis of superstructures is usually made by bar element models – girder, frame models and arch models. Even cable-stayed bridges and suspension bridges are analysed for overall action effects, based on 3D beam models.

Geometrical nonlinear effects are illustrated in Figure 6.53 for a cable with a parabolic configuration and a shallow arch, both with a rise/span ratio with $f/L = 1/10$, both under a uniform load q. The stiffness of the cable increases under loading while the stiffness of the arch decreases due to increased bending moments induced by the thrust H.

Cables under self-weight take the configuration of a *catenary* (cos *h* shape). However, when the tensile force increases, the shape is very close from a second degree parabola under a uniform distributed load q (Figure 6.53a). Non-linear effects of cables – stays or main cables in suspension bridges – may be modelled by specific cable elements to take into account non-linear geometrical effects on the force-axial displacement

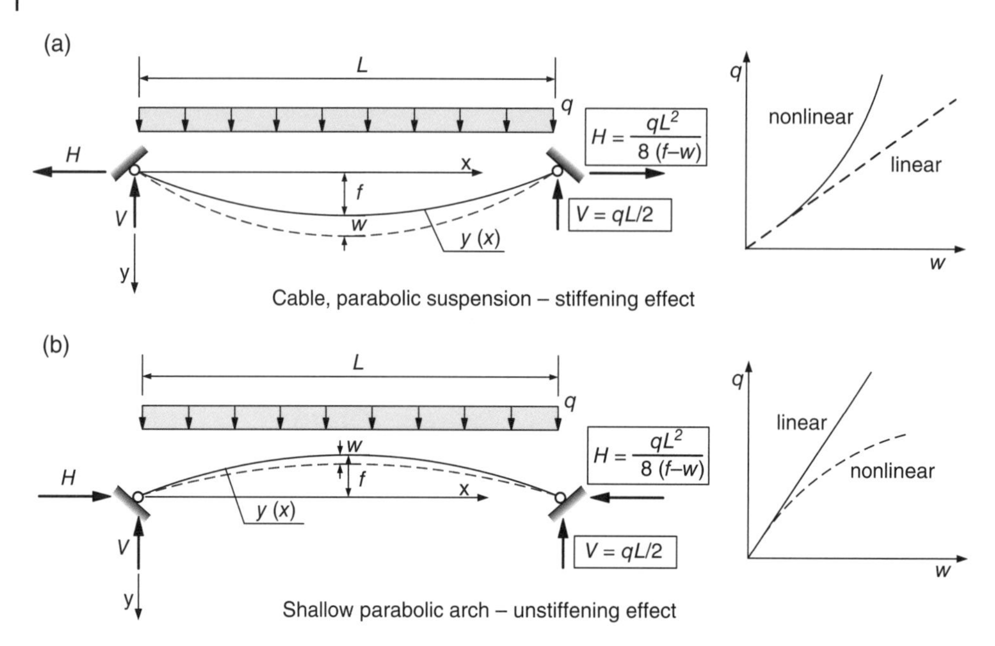

Figure 6.53 Geometrical non-linear effects in cables (stiffening effect) and arches (unstiffening effect).

relationship depending on *catenary* effects. Even so, it is possible by segmenting the cable in short elements to work with a linear stiffness matrix K_e for the element. Geometrical non-linear effects may be dealt with in the overall structural modelling where the effect of node displacements and geometrical configuration changes and evolution during erection stages is taken into account.

For the catenary effect on stay cables, the simplest approach is to model the stay as a bar element with an *equivalent modulus of elasticity* –Ernst's modulus E_t. This is not a material nonlinearity but a simplified approach to take geometrical non-linear effects into consideration. Only for long stay cables or stays under very low tensile forces is the nonlinear effect relevant, as may be concluded from Figure 6.54.

If the arch has a geometrical configuration according to the pressure line as discussed in Chapter 1 (see Figure 1.9), only axial forces are induced (Figure 6.53b) and the thrust $H = qL^2/(8f)$ is obtainable by imposing, for example, $M = 0$ at the mid-span. To find the geometrical shape of the arch, one imposes a $M = 0$ condition for a general section of coordinates (x, y), with the result $y = q(x^2 - xL)/(2H)$. This ideal arch is subject to axial forces only, as shown in Figure 6.55.

Arches with a low rise/span relationship (f/L) are subject to nonlinear geometrical effects yielding instability phenomena. An example is shown in Figure 6.56 where one plots qualitatively for a shallow sinusoidal arch under a sinusoidal distributed load the central deflection with respect to load. The fundamental equilibrium path is non-linear and two instability phenomena may occur. The arch may deflect under a symmetrical configuration but a certain load level the arch changes to an inverted configuration and starts behaving like a cable with a stiffening effect. The deflection shape of the arch may be modelled as a 2-DOF model with generalized coordinates q_1 and q_2 and sinusoidal

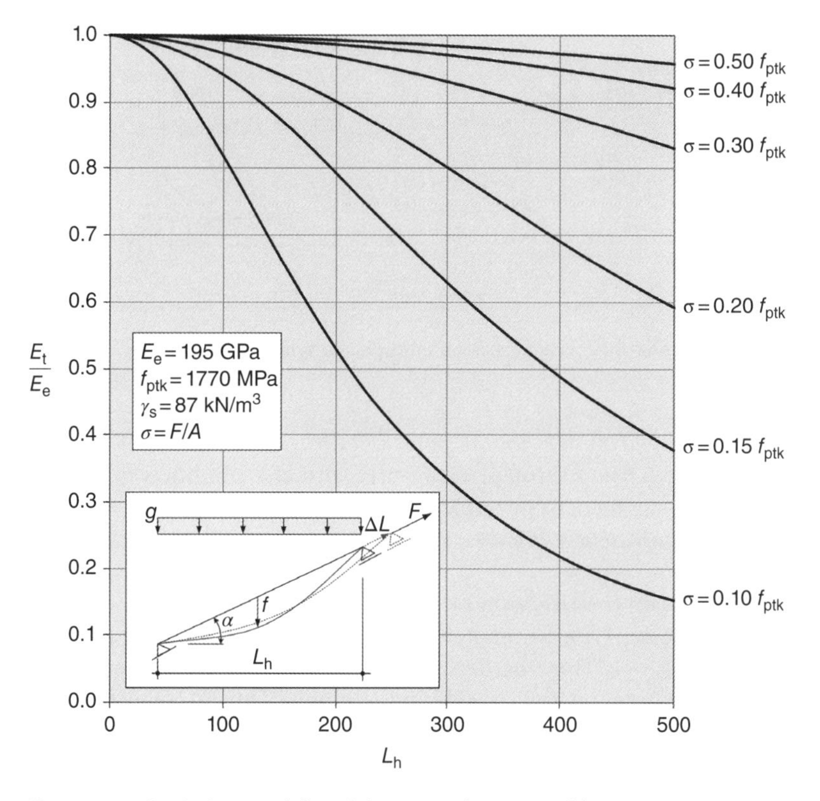

Figure 6.54 Equivalent modulus of elasticity E_t for a stay cable.

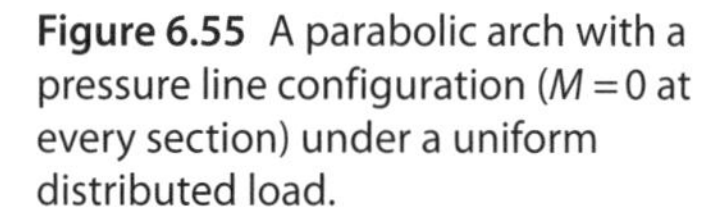

Figure 6.55 A parabolic arch with a pressure line configuration ($M = 0$ at every section) under a uniform distributed load.

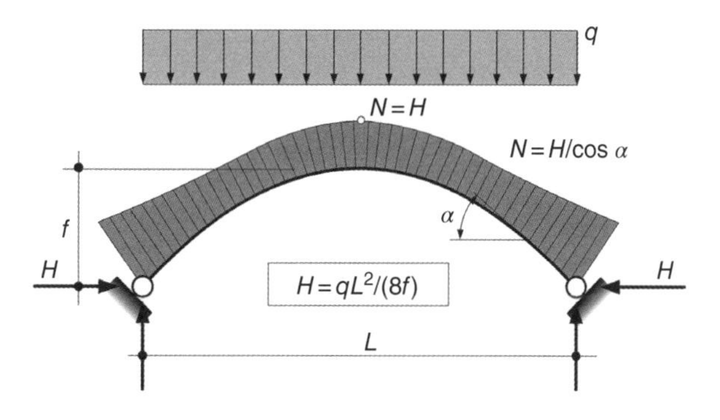

shape functions with a single semi-wave and two semi-waves. Under imposed displacements, it is possible to follow the equilibrium path along the unloading part of the path. However, under increasing loading, at a certain critical load level instability by *snap through* at a *limit point* occurs and a dynamic change of configuration occurs as shown in Figure 6.56. However, this instability mode occurs only in very shallow arches, since for usual (f/L) and before the arch 'snaps', bifurcation instability at a critical load $p_{cr} = p_b$, in a symmetric mode (two semi-waves) occurs. That is usually the lowest elastic buckling load of a shallow arch.

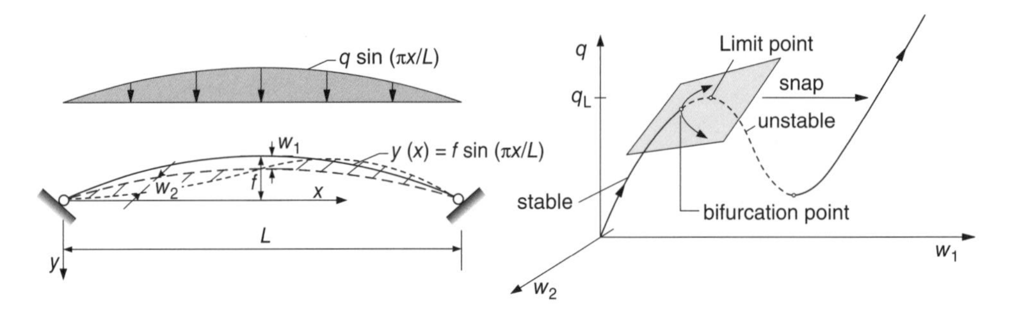

Figure 6.56 Nonlinear behaviour of shallow arches. Snap through and symmetric unstable bifurcational instabilities.

For a parabolic arch hinged at the foundations (Figure 6.53b) and subject to a uniformly distributed load, q, snap through is not critical if the condition given, for example, in EN 1993-2 [6] is satisfied. By introducing section radius of gyration $i = \sqrt{I_y/A}$ and the *slenderness of the equivalent column* $\lambda = L/i$ the expression in EN 1993-2 may be rewritten as $> \sqrt{12}\,k$.

The parameter k depends on the rise/span ratio (f/L), that is, the main parameter to roll out snap through buckling. The parameter k is almost linear with (L/f) and the relation $k = k_0\ (L/f)$ with $k_0 \approx 1.7$ may be derived by linear interpolation from table values [30]. Hence, this expression for considering the possibility of snap through buckling as rolled out in two hinged arches under a uniformly distributed load is:

$$\left(\frac{f}{L}\right) > \frac{\sqrt{12}\,k_0}{\lambda} \tag{6.74}$$

This means, for $k_0 \approx 1.7$, snap through buckling is rolled out if approximately $(f/L) > 5.9/\lambda$. This condition is regularly satisfied for arch bridges, since in general (f/L) is in the order of 1/6 to 1/5, and thus λ should be higher than 36, which is usually the case. This means bifurcation instabilities are the only ones relevant for in-plane buckling of *arch bridges*.

Geometrical non-linear effects in frame bridges are usually relevant for pier actions and that will be considered in Chapter 7. From transverse analysis of the superstructure, loading patterns for the longitudinal model are obtained. This is assumed as known for the development of longitudinal modelling. Of course, if a global FEM is developed both effects, transverse and longitudinal, are taken simultaneously into consideration. However, that is not usually the best approach for bridge structures until the final design stage.

6.6.2 Frame and Arch Effects

When piers are monolithic with the superstructure, a frame longitudinal model is adopted. However, one may take consideration the fact the bending stiffness of the connecting section of piers is usually much lower than the stiffness of the deck, that is, $(EI/h)_{pier} \ll (EI/l)_{deck}$ (Figure 6.57).

In a concrete bridge, only in the case of very short piers, (EI/h) of the piers may be of the same order of magnitude of (EI/l) of the deck. However, for these short piers,

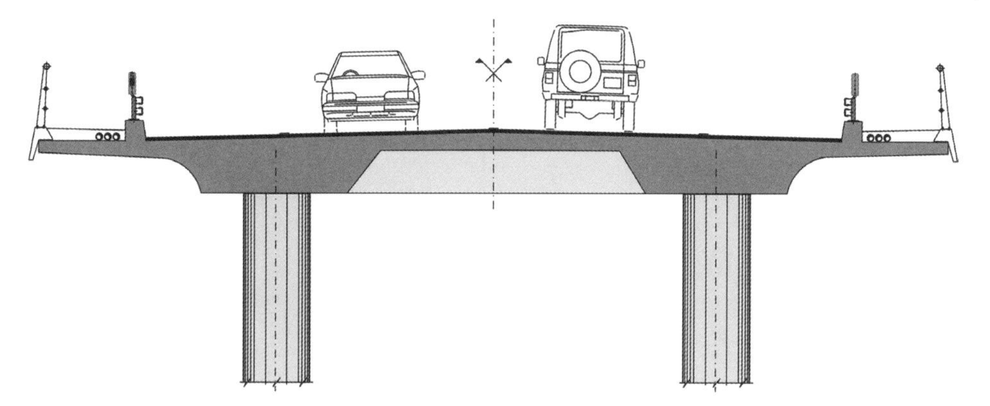

Figure 6.57 Longitudinal and transverse frame action due to monolithically connection between bridge piers and the deck.

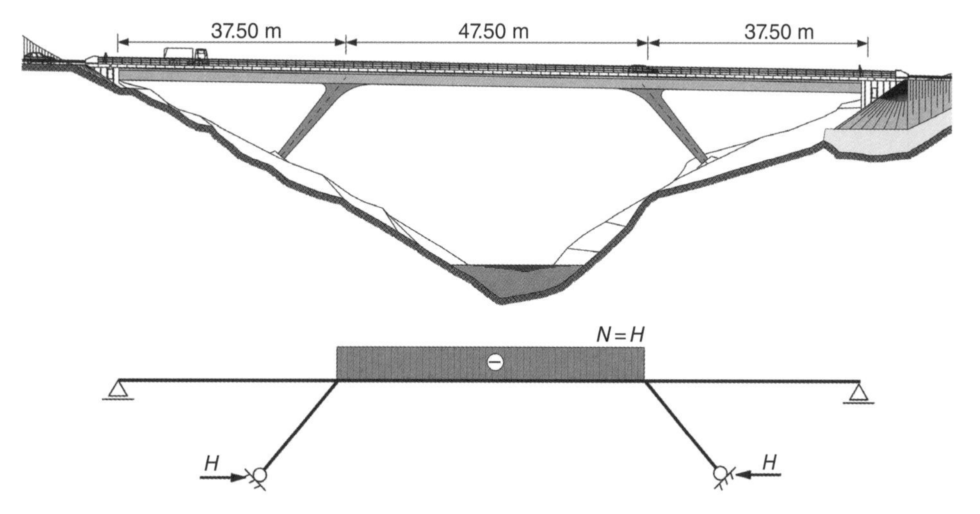

Figure 6.58 A frame with inclined piers. Horizontal reaction H and axial force in internal span of the deck.

bearings are usually adopted. In general, for vertical permanent and traffic loading, deck behaviour and internal forces in a frame bridge with vertical piers may be determined for a continuous girder bridge.

Frame bridges with inclined piers (Figure 6.58) have an arch effect due to horizontal reaction H (impulse) reducing bending moments in the deck under vertical loading compared to a frame bridge with vertical piers. The deck between inclined piers is subject to a compressive force as shown in the figure. Foundation capacity to withstand the horizontal reaction H is a basic requirement to design a frame with inclined struts. The compressive force induced in the central span of the deck is like a natural prestressing, in that it can be accounted for in the design of prestressed concrete bridges. However, it should be noted this H force is dependent on the evolution of the static system during construction. For example, in a cantilever scheme (Figure 4.108)

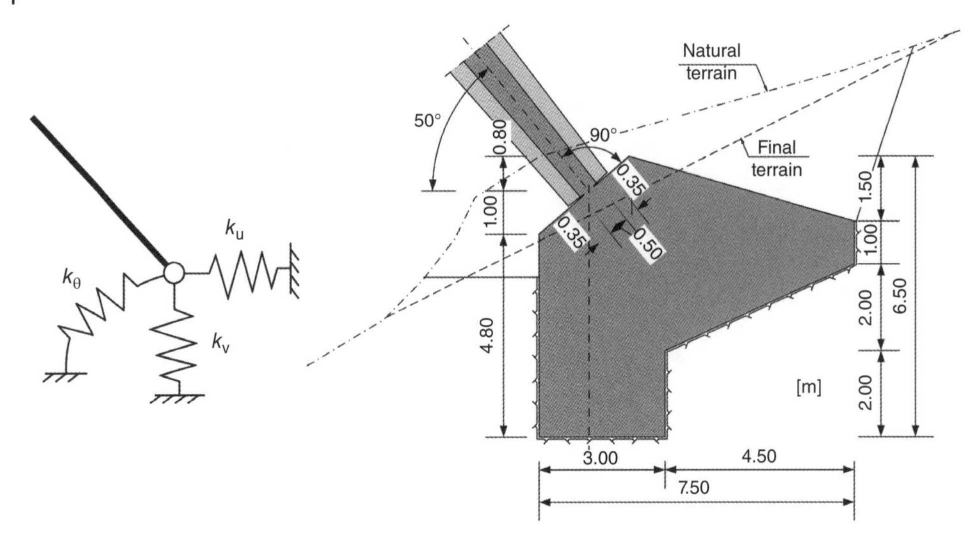

Figure 6.59 Elastic stiffness coefficients at foundations, for the analysis of frame bridges with inclined legs. Foundation of the Reis Magos bridge in Madeira Island (Figure 6.58), with a Freyssinet pin connection.

sections connecting deck and legs are supported by temporary piers. If the deck is cast, closure segment at the mid-span executed and, at last, temporary supports removed, the axial force N at the deck mid-span is induced. However, this force is time dependent due to relaxation effects in the concrete as discussed in Section 6.8. The same does not happen in a steel frame, where relaxation effects do not occur.

One important aspect of analysing frame bridges with inclined legs is how to account deformability of foundations tending to reduce the H reaction at the pier base. This may be done by considering Vogt coefficients [31] for shallow foundations defining support elastic stiffness coefficients (Figure 6.59) k_u, k_v, k_θ for the longitudinal bridge model.

6.6.3 Effect of Longitudinal Variation of Cross Sections

In girder and frame bridges, variation of the depth of cross sections is very often adopted, at least for medium and long spans. The axis of the slab-girder or box girder deck is curved along the longitudinal alignment. Cross sections are defined by planes perpendicular to the bridge axis. However, it is usual to work with 'vertical' cross sections as shown in Figure 6.60. Differences with respect to axial forces and bending moments obtained by both approaches are very small. However, for shear forces the difference between section values obtained by the two approaches should be taken into consideration. Part of shear force is equilibrated by the resulting compressive stresses at the lower flange, designated by N_b, as exemplified in Figure 6.60. Taking the example of a box girder section in Figure 6.61, part of the vertical forces are equilibrated by shear stresses in the webs and the other part by the compressive stresses at the bottom flange. The resulting shear force for the vertical cross section is:

$$V_R = V' - N_b \sin\beta \tag{6.75}$$

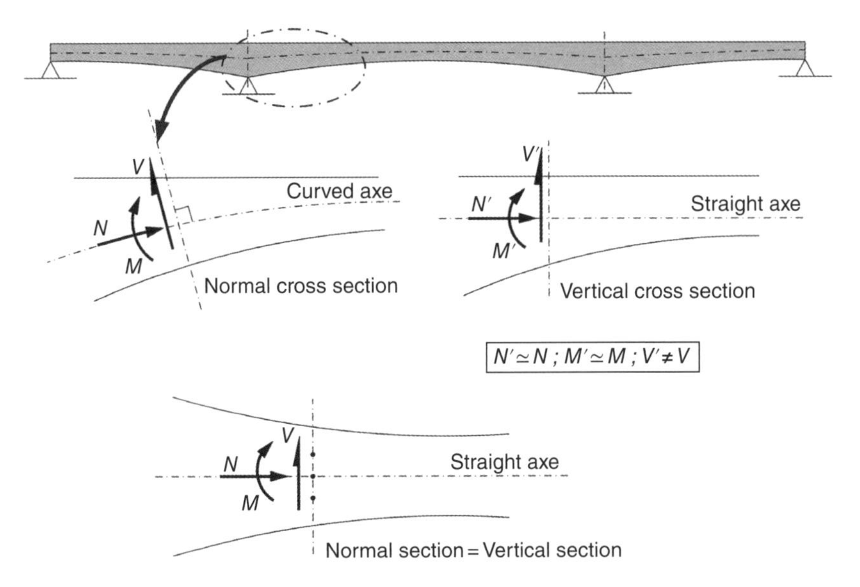

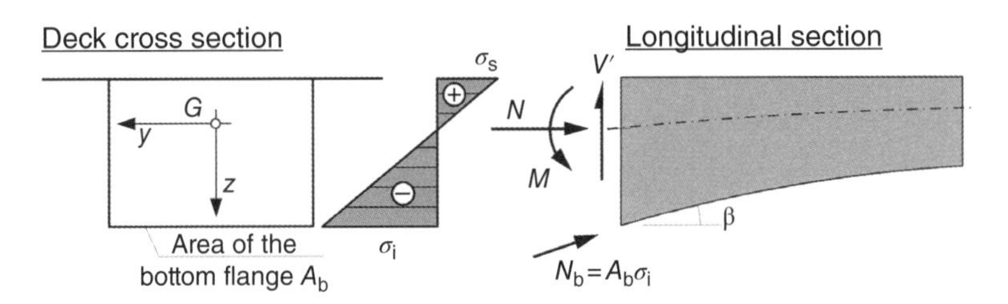

Figure 6.60 Effect of longitudinal variation of deck cross sections.

Figure 6.61 Effect of longitudinal variation of deck cross sections. The Resal effect.

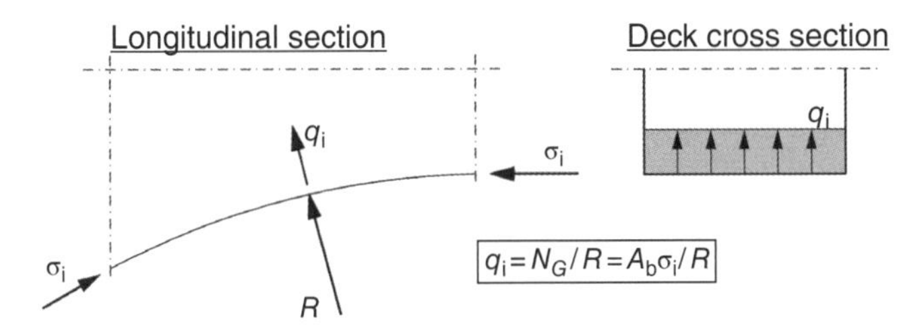

Figure 6.62 Deviation force on the compressed bottom flange of a box girder due to longitudinal variation of deck cross section depth.

The shear force at the section V' is reduced by $N_b \sin\beta$. This is known as Resal's effect.

In variable section box girders, the compressive stresses at the bottom flange induce a deviation force q_i dependent on the curvature radius R as shown in Figure 6.62. The bottom flange, subjected to compressive stresses σ_i, should be verified as a transverse slab supported by the webs and subjected to a transverse distributed load q_i.

6.6.4 Torsion Effects in Bridge Decks – Non-Uniform Torsion

Bridge decks subject to eccentricity of live loadings or horizontal transverse loads are subject to torsion. Bearings at piers and abutments should be designed to equilibrate vertical reactions (uplift if exists) taking torsional effects into account. At the abutments, torsion rotations are usually restricted by the bearings; at pier sections, these rotations may be restricted by a couple of bearings but in some cases, as in urban viaducts, a single bearing exists or the pier should resist to a transverse bending moment resulting from restricted torsion of the deck (Figure 6.63).

Bridge decks are subjected to non-uniform torsion, inducing shear stresses and normal stresses at the cross sections due to warping [32, 33]. These torsion effects are governed by the equation

$$EI_{\omega}\phi^{IV} - GJ\phi'' = -m_t \tag{6.76}$$

where ϕ is the torsion rotation of the cross section, EI_{ω} and GJ are the warping and uniform torsion (Saint Venant) stiffness's of the cross section, respectively, where E and G are the elastic modulus and the distortional modulus of the material. The constants I_{ω} and J are the warping and uniform torsion constants of the cross section, respectively, in mm^6 and mm^4. The distributed torsion moment along the deck is $m_t = -dT/dx$. For an open cross section (Figure 6.64) the torsion and warping constants are defined by:

$$J = \frac{1}{3}\sum_{i} b_i t_i^3 \quad ; \quad I_{\omega} = \int_{A} \omega^2\, dA \tag{6.77}$$

Here, ω is the *normalized sectorial coordinate* defined from the sectorial coordinate $\Omega(s)$ in the cross section (Figure 6.64) by $\omega(s) = \Omega(s) - \bar{\Omega}(s)$ where $\bar{\Omega}(s)$ is the average value of $\Omega(s)$ along the cross section, that is, between $s = 0$ and $s = s_t$. By definition $\int_{A} \omega(s)dA = 0$.

Considering a deck section where torsion rotations are prevented by the support system, one has $\phi = \phi' = 0$ as boundary conditions. For free warping at this section, one has, in addition, $\phi'' = 0$.

The general solution of linear differential Eq. (6.76) is obtained as the sum of the solution of the homogeneous equation and a particular solution, $\phi = \phi_h + \phi_p$, being [1]

$$\phi = C_1 + C_2\frac{x}{\ell} + C_3\cosh\left(\chi\frac{x}{\ell}\right) + C_4\sinh\left(\chi\frac{x}{\ell}\right) + \phi_p \tag{6.78}$$

1 General solution $\phi = A + Bx + Ce^{\lambda x} + De^{-\lambda x} + \phi_p$, where λ are the roots of the algebraic characteristic equation $EI_{\omega}\lambda^4 - GJ\lambda^2 = 0$; The exponential functions are written in terms of hyperbolic functions $e^{\lambda x} = \sinh(\lambda x) + \cosh(\lambda x)$ and $e^{-\lambda x} = \cosh(-\lambda x)\text{-}\sinh(-\lambda x)$.

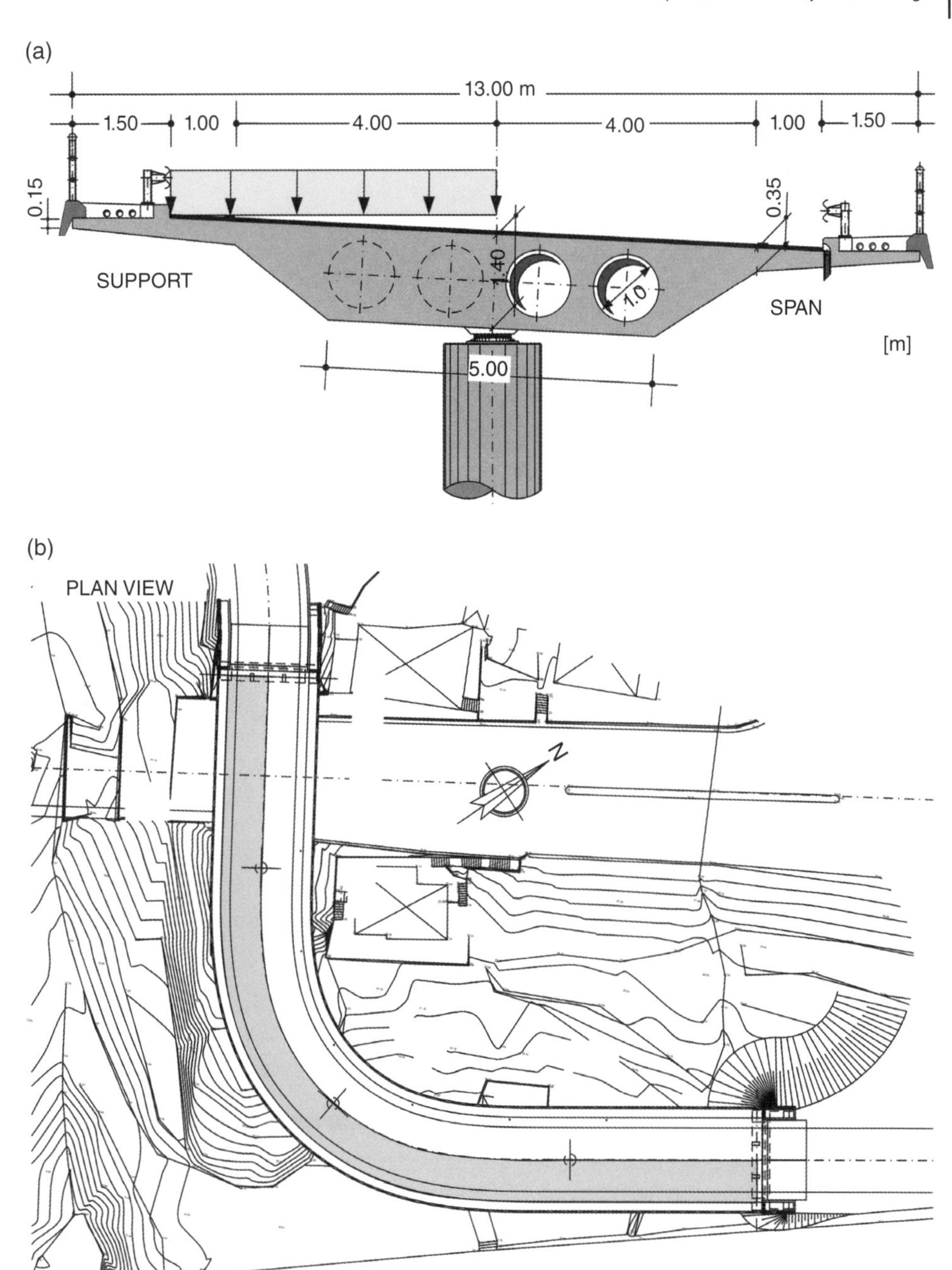

Figure 6.63 Cross section (a) of a bridge deck subjected to torsion effects due to asymmetric loading and (b) in plan curvature effects.

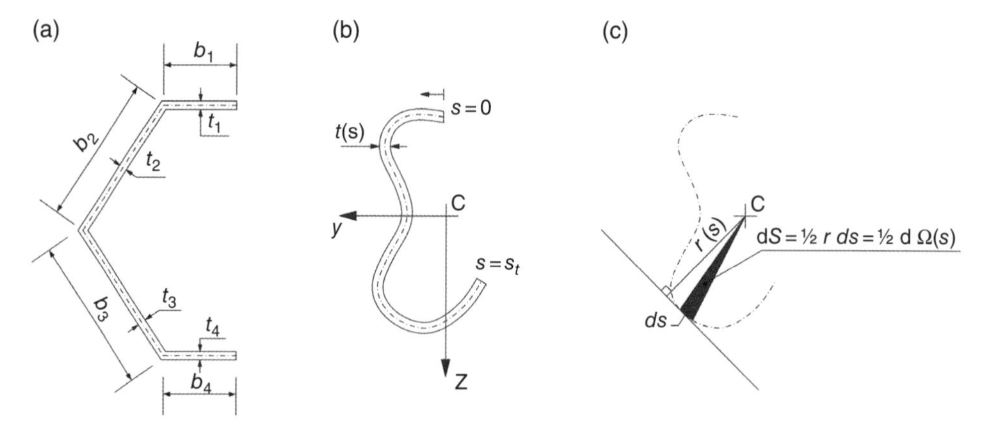

Figure 6.64 Open cross section. Definition of sectorial coordinate.

Here, C_1, C_2, C_3 and C_4 are constants of integration and χ is a non-dimensional parameter dependent on the span length and on the relationship between uniform and warping stiffness of the cross section:

$$\chi = \ell \sqrt{\frac{GJ}{EI_{\omega}}} \tag{6.79}$$

From the solution of Eq. (6.76) the torsional moment is obtained

$$T = T_{\mathrm{u}} + T_{\mathrm{w}} = GJ\,\phi' - EI_{\omega}\,\phi''' \tag{6.80}$$

Open section decks mainly resist to an applied torsional moment by non-uniform torsion (warping) while box sections resist mainly by uniform torsion. In Figure 6.65, the effect of parameter χ on the relationship $T_{\mathrm{w}}/(T_{\mathrm{u}} + T_{\mathrm{w}})$ is represented.

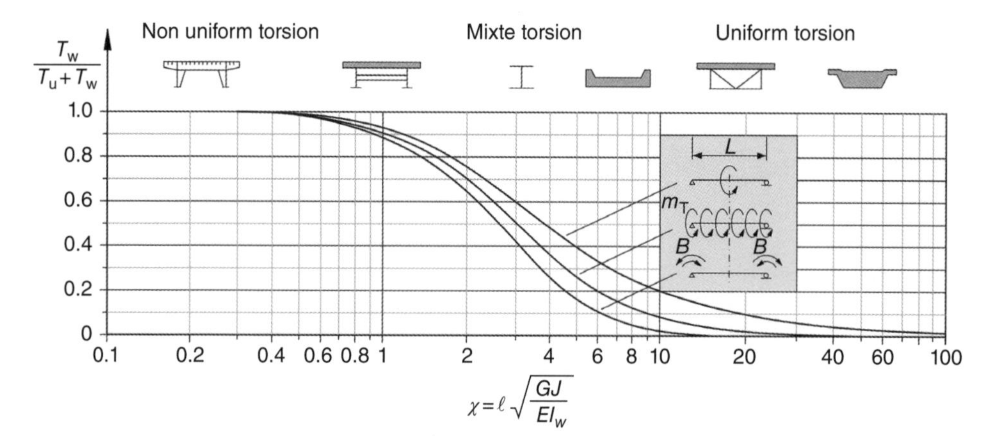

Figure 6.65 Bimoment at mid span section of a simply supported bridge deck with free warping at end sections under torsion, as a function of section parameter χ.

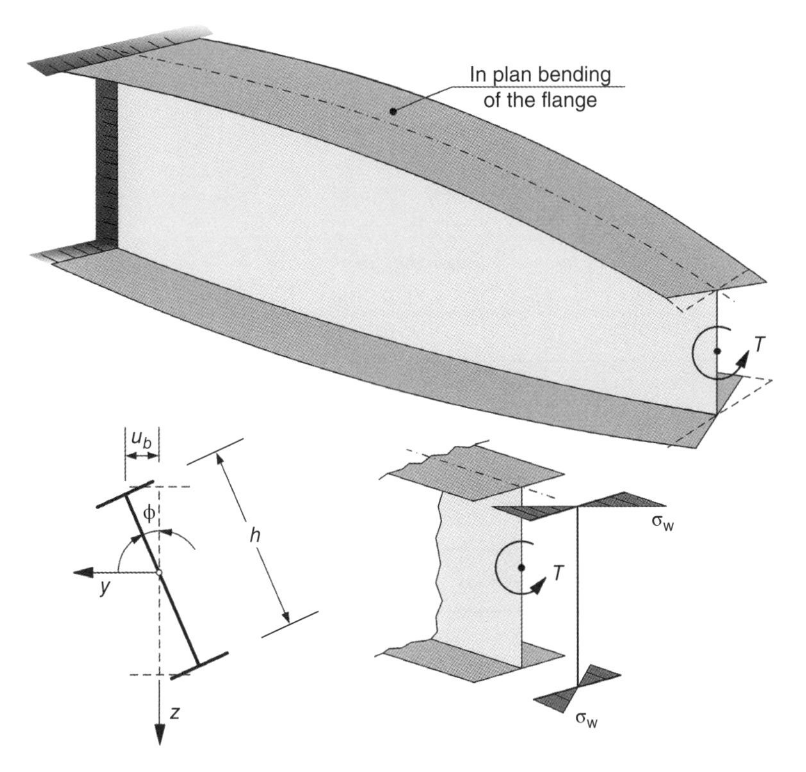

Figure 6.66 Warping stresses in a I section under torsion.

For $\chi < 1$ and $\chi > 10$, non-uniform (warping) and uniform torsion governs deck behaviour, respectively. Warping stresses for an open I section are represented in Figure 6.66. The warping effect is associated with in-plane bending of the flanges inducing longitudinal normal stresses at the cross sections. These stresses are self-equilibrated and they do not induce any internal force at the section but a couple of bending moments in the flanges equivalent to the *bimoment B*. The bimoment is defined for any type of cross section from the *torsion curvature* $B = -EI_w \phi''$.

For bridge decks where uniform torsion can be neglected (small χ values) Eq. (6.76) is reduced for GJ=0, to $EI_w \phi^{IV} = -m_t$. This equation is similar to the basic beam equation for bending effects $EI\, w^{IV} = -p$. Hence, the beam solutions for deflections w are equal to the torsion rotation ϕ, if one takes a distributed load $p = m_t$. In Table 6.2, a complete analogy between bending and warping torsion is established.

Knowing the solution w(x) for bending, one has the torsion rotations $\phi(x)$ and the bimoment is equal to the bending moment in the equivalent beam in bending; the shear force associated with warping torsion V_w is equal to the shear force in the beam. Table 6.3 shows the solutions for warping torsion taken from the bending analogy.

The warping normal stresses due to non-uniform torsion can be evaluated for any cross section from the torsion curvature and *sectorial coordinate* ω by

$$\sigma_\omega = \frac{B_\omega}{I_w} \qquad (6.81)$$

Table 6.2 Analogy between bending and warping torsion phenomena.

	Warping torsion	**Bending**
Loading	$m_t\,(x)$	$p(x)$
Generalized displacements	$\phi(x)$	$w(x)$
Load-deflection equation	$EI_w\,\phi^{IV} = -m_t$	$EI_z w^{IV} = -p$
Internal forces	$B = -EI_w\phi''$	$M = -EI\,w''$
	$T_w = B' = -EI\phi'''$	$V = M' = -EIw'''$
Stresses	$\sigma_\omega = B\omega/I_w$	$\sigma = My/I_z$

Table 6.3 Solutions for warping torsion taken from the bending analogy.

	Bimoment		
Loading case	**At end sections**	**At mid-span**	**Torsion rotation φ**
T; ℓ/2, ℓ/2	0	$-T\dfrac{\lambda}{4}$	$\dfrac{T\lambda^3}{48\,EI_\omega}$
m_t	0	$\dfrac{m_t\lambda^2}{8}$	$\dfrac{5}{384}\,\dfrac{m_t\lambda^4}{EI_\omega}$
T	$\pm T\dfrac{\lambda}{8}$	$T\dfrac{\lambda}{8}$	$\dfrac{T\lambda^3}{192\,EI_\omega}$
m_t	$-\dfrac{m_t\lambda^2}{12}$	$\dfrac{m_t\lambda^2}{24}$	$\dfrac{m_t\lambda^4}{384\,EI_\omega}$

These warping stresses are self-equilibrated at the entire cross section, that is, $\int_A \sigma_\omega dA = 0$. Equation (6.81) is similar to the basic bending formula for normal stresses $\sigma = My/I_z$.

Figure 6.67 shows warping stress distributions for open and closed box sections. In the following, an example is presented to illustrate how to calculate these stresses.

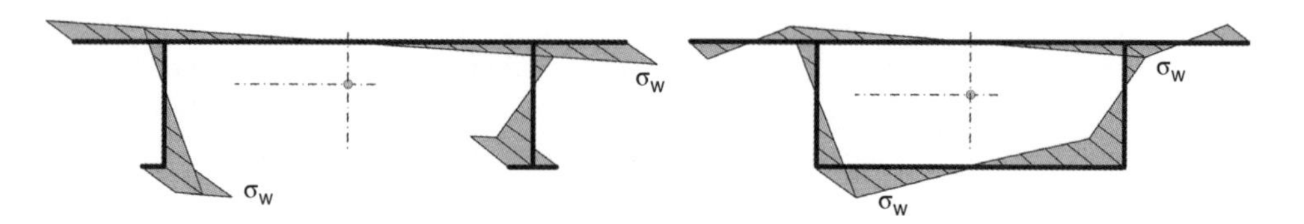

Figure 6.67 Warping stresses in an open section twin girder deck and in a box girder deck.

Example 6.4 Continuous Three-Span Slab-Girder Deck

Let a slab girder deck be considered with a cross section with simplified geometry shown in Figure 6.68. The deck is subject to a uniformly distributed live load q acting at half width ($a/2$) of the internal span l_1 of the deck. One assumes $l_2 = 0.7l_1$ (Figure 6.69) and torsion prevented at support sections with free warping at the abutment sections only.

1) Shear centre

   The shear centre, S, is located at a distance defined from the resultant of shear stress distribution for an horizontal shear force, V_y

$$F_z = \int_o^h \frac{V_y}{I_z} b_1 s \frac{b}{2} ds = \frac{V_y b_1 b h^2}{4 I_z} \tag{6.82}$$

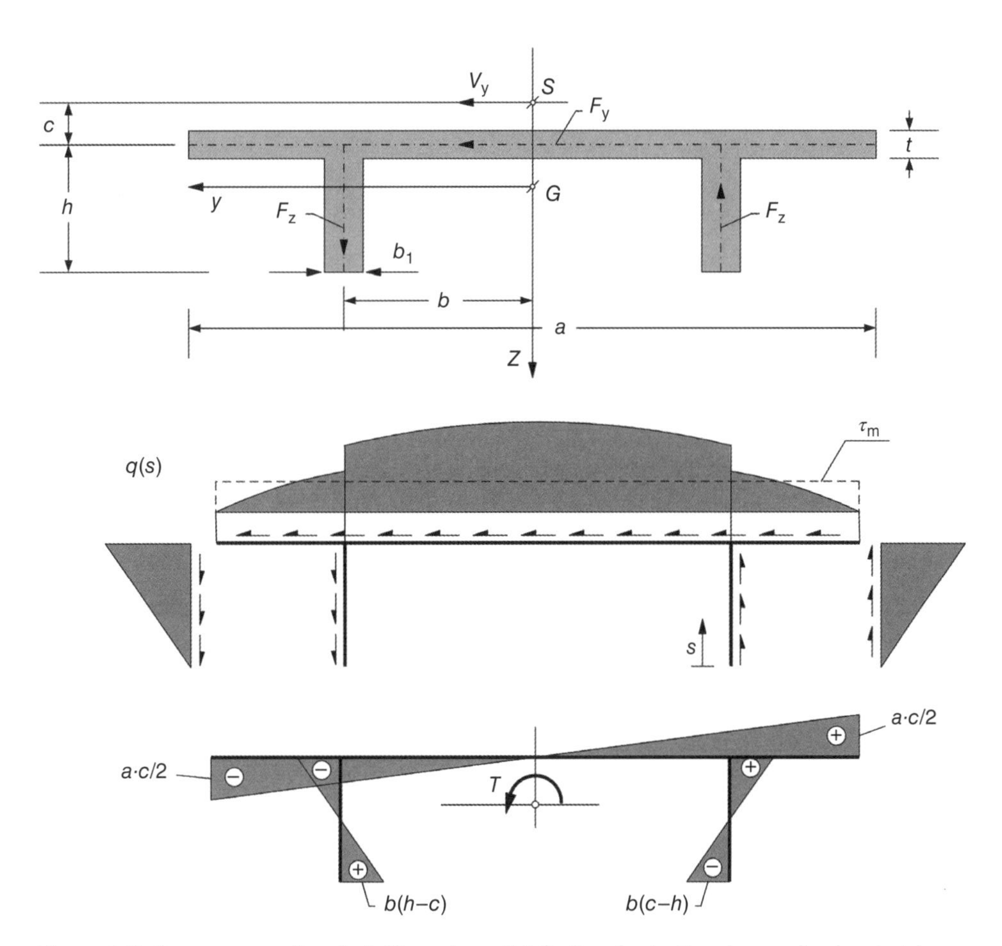

Figure 6.68 Open cross section deck. Shear stress distribution due to V_y and normalized sectorial coordinate.

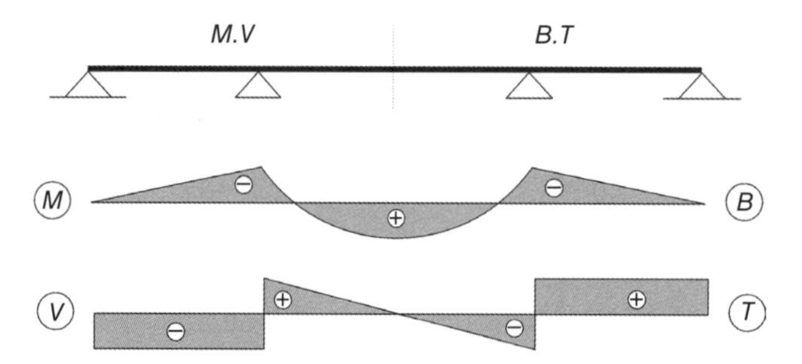

Figure 6.69 A three-span continuous slab girder deck (Figure 6.68) under torsion.

At S, the resultant of the shear flow should be a single force and by equilibrium one has $F_y\, c = F_z\, b$, with the resultant for the location of the shear centre

$$c = b_1 \frac{b^2 h^2}{4 I_z} \tag{6.83}$$

2) Normalized coordinate
   The normalized sectorial coordinate ω is defined by the coordinate s_1 measured from the point where $s = 0$ (axis of symmetry)

$$\begin{aligned} &\text{slab}: r = c\,,\ \omega = \int_0^s r ds = c\, s \\ &\text{girders}: r = b\,,\ \omega = \int_0^{b/2} r ds + \int_0^s r ds = (c - s_1) b/2 \end{aligned} \tag{6.84}$$

   where s_1 is the coordinate s measured from point A. At the end points of the sections, one obtains the values of ω shown in Figure 6.68.

3) Loading

$$p = q\frac{a}{2}\ (\text{kNm}^{-1}) \qquad m_t = q\frac{a}{2}\frac{a}{4}\ (\text{kNm m}^{-1}) \qquad M^- = 0{,}11 p l_1^2\ (\text{kNm}) \tag{6.85}$$

   Assuming the deck has a parameter $\chi < 5$, torsion is resisted by warping effects only and the diagram of the bimoment B is similar to the bending moment diagram. The analogy is made in Figure 6.69. Maximum B is over the intermediate supports with the value taken from the bending moment diagram and with $p = m_t$

$$B_{\max} = 0.11\ m_t l_1^{\,2} = 0.014\ q a^2 l_1^{\,2} \tag{6.86}$$

   An interesting aspect to notice is the existence of torsion moments in the end spans that would be zero under uniform torsion but are not when warping effects are taken into consideration.

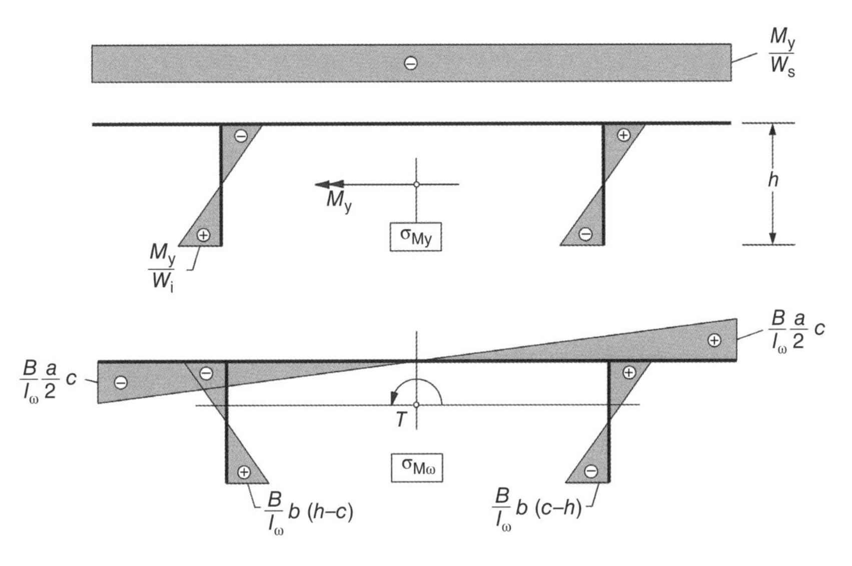

Figure 6.70 Bending and torsion warping normal stresses in the cross section deck of Figure 6.68.

4) Normal stresses

   The normal stresses due to bending and warping torsion are defined in the diagrams of Figure 6.70, with maximum values

$$\sigma_{max}^{sup} = -\frac{M_y}{W_s} - B\frac{a}{2I_\omega}c(\text{compression}) \qquad \sigma_{\max}^{\inf} = \frac{M_y}{W_i} + \frac{B}{I_\omega}(b/2)(h-c)(\text{tension}) \tag{6.87}$$

   Maximum stresses occur at intermediate support cross sections where M and B have simultaneous maximum values:

$$\sigma_{max}^{sup} = -0.555\, qa(l_1)^2\left[1/W_s + a^2c/(8l_\omega)\right] \tag{6.88}$$

$$\sigma_{max}^{inf} = -0.055\, qa(l_1)^2\left[1/W_i + ab/2(h-c)/(4I_\omega)\right] \tag{6.89}$$

6.6.5 Torsion in Steel-Concrete Composite Decks

In steel-concrete composite decks uniform torsion in the deck slab cannot be neglected, and a general case of non-uniform torsion (uniform + warping torsion) should be considered.

For box girder sections, warping effects can usually be neglected. In plate girder composite decks, non-uniform torsion should be taken into consideration.

6.6.5.1 Composite Box Girder Decks

In a box girder section, the shear flow f is constant along the middle line of the cross section defined by the coordinate s; at each point, one defines the distance $r(s)$ to the

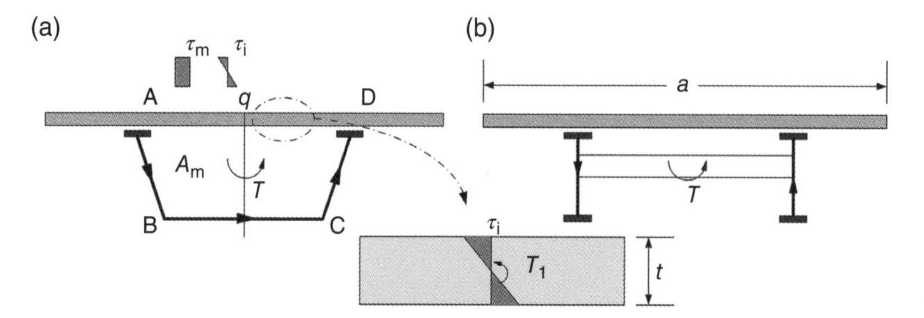

Figure 6.71 Torsion in (a) a box girder and (b) a plate girder steel-concrete composite deck.

shear (torsion) centre (Figure 6.71) and the torsion moment, torsion shear flow and average torsion stress are defined by:

$$T = \oint \tau r(s)t(s)ds \quad f = \frac{T}{2A_m} \quad \tau_m = \frac{f}{t} \tag{6.90}$$

where A_m is the area defined by the middle line of the section as, for example, ABCD in Figure 6.71. The uniform torsion stiffness GJ is defined from the distortional modulus G and torsion constant J from Bret's formula

$$J = \frac{4A_m^2}{\oint \frac{ds}{t(s)}} \tag{6.91}$$

If one has a composite box girder section with a concrete slab of thickness t_1, the torsional moment is resisted by both parts of the section (the closed part and the open part (slab) as shown in Figure 6.71), that is, $T = T_0 + T_1$, distributed proportionally to the torsion stiffness $k = GJ$ of each part

$$T_i = \frac{k_i}{k_0 + k_1} \quad i = 0,1 \text{ and } k_0 = \frac{4A_m^2}{\frac{t_0}{b_0} + 2\frac{t_w}{s_w} + n\frac{t_1}{b_1}} \quad k_1 = \frac{1}{3}\frac{at_1^3}{n} \tag{6.92}$$

where $n = G_s/G_c$ is the modulus factor between the shear steel modulus and concrete shear modulus. An equivalent homogeneous section in steel only was considered for k_0 by replacing the slab by and equivalent steel slab (Figure 6.72) with a thickness $t_{eq} = t_1/n$ obtained from compatibility of the torsion rotations of the global section and of the slab.

The additional stress τ_1 due to uniform torsion of the slab (Figure 6.71) is

$$\tau_1 = \frac{3T_1}{bt_1^2} = \frac{T_1 t_1}{J_c} \tag{6.93}$$

$$J_c = \frac{1}{3}at_1^3 \tag{6.94}$$

The total shear stress in the slab is $\tau_0 + \tau_1$, where $\tau_0 = T_0/(2A_m t_1)$ as shown in Figure 6.71.

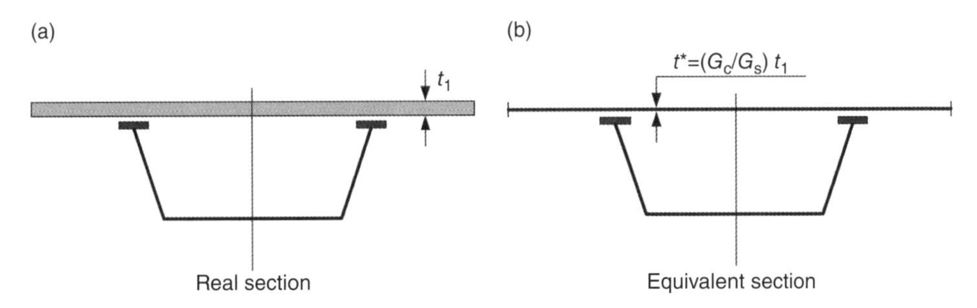

Figure 6.72 A composite box girder section under uniform torsion. Equivalent thickness of the deck slab.

6.6.5.2 Composite Plate Girder Decks

Composite plate girder decks resist mainly by mixed torsion, one could say the steel part in purely warping torsion and the deck slab in uniform torsion. This makes the analysis more complex, but some simplifications are possible as discussed in the following.

Let the deck of Figure 6.73 be considered under a uniform distribution of torsion moments m_t decomposed under an antisymmetric load case each girder working under a bending moment of opposite sign. The warping stresses σ_ω at the flange width of the girders may be considered constant and it is possible to evaluate an effective width for the slab b_1, associated with each girder under a uniform average normal stress σ_m due to warping. This equivalent section will work under bending for one of the bending moments in which warping is decomposed. The bending moment in the plane of the slab (M_z) induced by each of the stress distribution in the equivalent flange widths is

$$M_z = \int_{-a/2}^{a/2} \sigma_\omega y\, t\, dy = \frac{1}{2}\sigma_m \frac{a}{b}\frac{a}{2}\, 2\text{x}\frac{2}{3}\frac{a}{2}t = \sigma_m \frac{a^3 t}{6b} \tag{6.95}$$

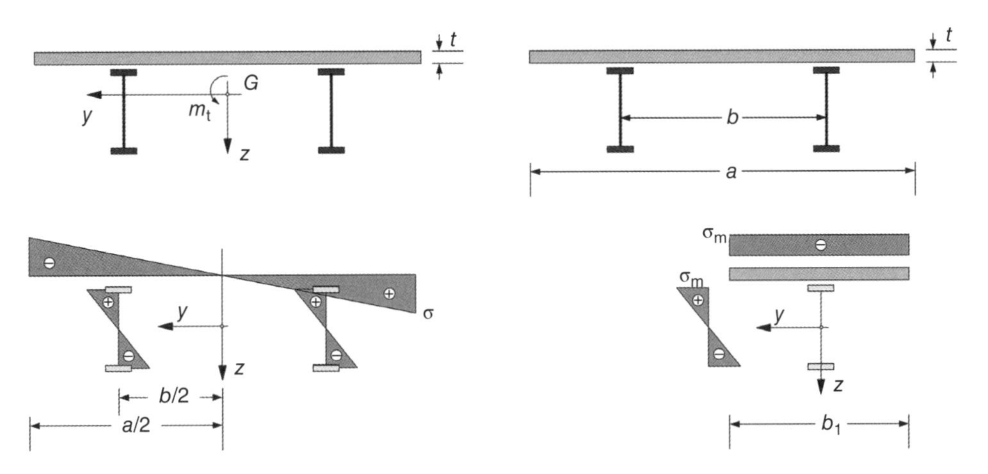

Figure 6.73 Composite open section deck under torsion.

This bending moment is equivalent to two horizontal longitudinal forces F_x at the slab-girder connection, that, on the other side should be equal to the stress resultant at the equivalent flange width b_1 (Figure 6.74)

$$F_x = \frac{M_z}{b} = \sigma_m \frac{a^3 t}{6b^2} \quad \text{and} \quad F_x = \sigma_m b_1 t \tag{6.96}$$

From these equations, one obtains $b_1 = \frac{a^3}{6b^2}$.

The warping constant I_ω may be obtained by considering the section is working in purely warping torsion governed by the differential equation

$$E I_\omega \phi^{iv} = m_t \tag{6.97}$$

Since under antis symmetric bending of the girders the vertical deflections are governed by

$$E I_{1y} w^{iv} = p_1 \tag{6.98}$$

where I_{1y} is the moment of inertia with respect to y of the girder width a flange width b_1.

The vertical deflection w is related to ϕ by $w = \pm\phi b/2$. By replacing in Eq. (6.98):

$$E I_\omega \left(\frac{2}{b} w\right)^{iv} = p_1 b \tag{6.99}$$

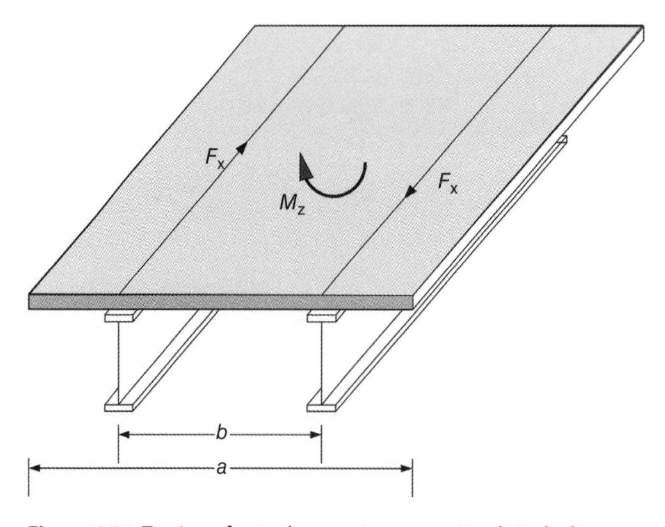

Figure 6.74 Torsion of a steel-concrete composite plate deck.

from which, by comparing with Eq. (6.98), one has

$$I_\omega = I_{1y}\frac{b^2}{2} \tag{6.100}$$

The bending moment M_{1y} to evaluate the warping stresses is determined from

$$M_{1y} = -EI_{1y}\frac{b}{2}\phi'' \tag{6.101}$$

Uniform torsion, restricted to the slab, is now considered and the overall section is working under mixt torsion governed by

$$E\,I_\omega\,\phi^{\mathrm{lv}} - GJ\,\phi'' = m_\mathrm{t} \quad \text{with} \quad GJ = G_\mathrm{c}\frac{1}{3}at^3 \quad \text{and} \quad EI_\omega = EI_{1y}\frac{b^2}{2} \tag{6.102}$$

From Eqs. (6.101) and (6.102), and taking $m_\mathrm{t} = p_1 b$, one has

$$M''_{1y} - \left(\frac{2\alpha}{\ell}\right)^2 M_{1y} + p_1 = 0 \tag{6.103}$$

Where

$$\alpha^2 = \frac{GJ}{2\,EI_{1y}}\left(\frac{\ell}{b}\right)^2 \tag{6.104}$$

with ℓ being the span length of the deck. This linear differential equation (6.104) with constant coefficients can be integrated to determine the bending moment M_1 at each girder as shown in the following example.

Example 6.5 Let the deck of Figure 6.75 under uniform eccentric distributed loading be considered. By decomposing into two load cases, load case A induces bending only

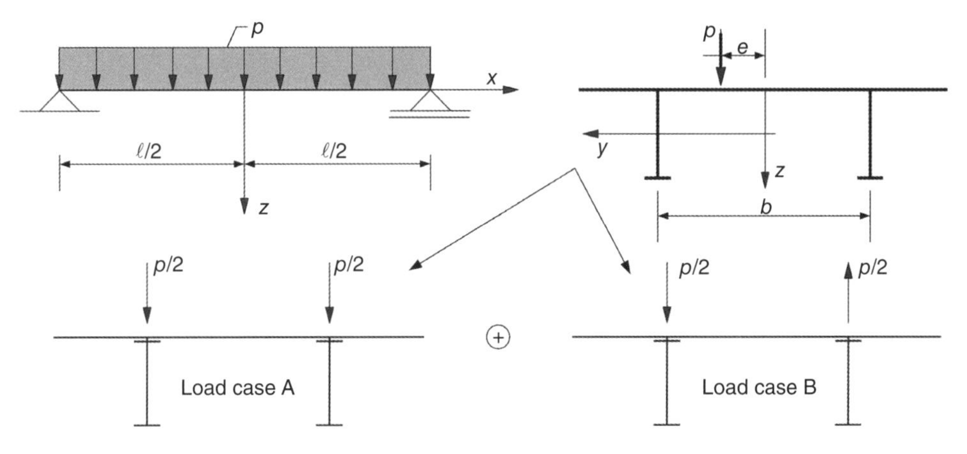

Figure 6.75 A composite plate deck under uniform eccentric distributed loading p, yielding an antisymmetric load case $p_1 = p \cdot e/b$.

and load case B torsion only. The solution of the differential Eq. (6.103) for M_y may be obtained by the standard procedure from the solutions of the homogeneous equation and from a particular solution ($M_{1y} = M_{1y}{}^h + M_{1y}{}^p$). Taking the boundary conditions $M_{1y} = 0$ *for* $x = \pm \ell/2$, one has

$$M_{1y}^{p} = cp_1 \quad \text{with} \quad c = \left(\frac{\ell}{2\alpha}\right)^2 \tag{6.105}$$

$$M_{1y}^{h} = C_1 \cos h\left(\frac{2\alpha}{\ell}x\right) + C_2 \sin h\left(\frac{2\alpha}{\ell}x\right) \tag{6.106}$$

and the final solution, after introducing the boundary conditions and evaluation the integration constants C_1 and C_2, is

$$M_{1y} = \frac{p\ell^2}{4\alpha^2}\frac{e}{b}\left[1 - \frac{\cos h(2\alpha x/\ell)}{\cos h\alpha}\right] \tag{6.107}$$

6.6.5.3 Transverse Load Distribution in Open Section Decks

The non-uniform torsion theory can be applied to determine the transverse load distribution in slab-girder decks. The total normal stresses due to bending and warping are

$$\sigma_s = \frac{M_y z_s}{I_y} + \frac{B\omega_s}{I_\omega} = \sigma_{s0} + \sigma_{s\omega} \tag{6.108}$$

where σ_{s0} is taken as the reference bending stress and all the other variables were previously defined. Hence,

$$\sigma_s = \sigma_{s0}\left[1 + \frac{B}{M_y}\frac{I_y}{I_\omega}\frac{\omega_s}{z_s}\right] \tag{6.109}$$

As B/M_y is a function of the eccentricity e, say $B/M_y = f(e)$, one may define an influence function [32, 34]

$$\eta(e) = \frac{\sigma_s}{\sigma_{s0}} = 1 + f(e)\frac{I_y}{I_\omega}\frac{\omega_s}{z_s} \tag{6.110}$$

This function defines the total stress at any point of the section as a function of the reference normal bending stress, σ_{s0}. Taking the example of Figure 6.75, one has at the mid-span section

$$M_y = \frac{p}{2}\frac{\ell^2}{8}; \quad M_{1y} = \frac{pl^2}{4\alpha^2}\left[1 - \frac{1}{\cosh\alpha}\right]\frac{e}{b} \tag{6.111}$$

Hence, assuming for W_i the same value for bending and torsion effects (i.e. for the girder with the effective slab part or for the equivalent girder under warping torsion, i.e. section $W_{yi} \approx W_{yi1}$, which is acceptable because $b_{eff} \approx b_1$) one has

$$\sigma_i = \frac{1}{W_{yi}}\left[\frac{p\ell^2}{16} + \frac{p\ell^2}{4\alpha^2}\frac{e}{b}\left(1 - \frac{1}{\cosh\alpha}\right)\right] \tag{6.112}$$

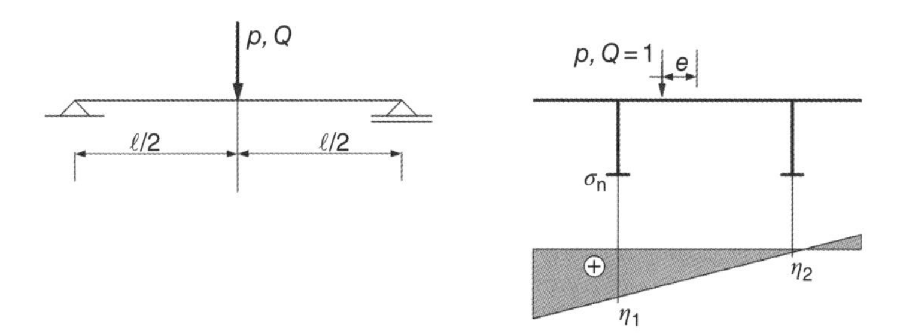

Figure 6.76 Transverse influence lines for composite decks loaded by uniformly distributed line loads $p = 1$ or concentrated loads $Q = 1$ at the mid-span sections.

The stress at the bottom flange is

$$\eta_1 = \frac{\sigma_i^{máx}}{\sigma_{io}} \quad \text{with} \quad \sigma_{io} = \frac{p\ell^2}{8W_{yi}} \tag{6.113}$$

From the above equations one has

$$\eta_1 = \frac{1}{2} + \frac{2}{\alpha^2}\frac{e}{b}\left(1 - \frac{1}{\cos h\alpha}\right) \tag{6.114}$$

This is the influence function to evaluate the stress at the bottom flange of an eccentric uniformly line load deck. This influence function represents a straight line as a function of the load eccentricity e as shown in Figure 6.76. It is defined by two points for example for $e = 0$ one has $\eta_1 = \eta_2 = 1/2$ (one half of the load goes to each girder as it should be) and $e = +-b/2$ (over girders 1 or 2) where one has respectively

$$\eta_1 = \frac{1}{2} + \frac{1}{\alpha^2}\left(1 - \frac{1}{\cos h\alpha}\right) \qquad \eta_2 = \frac{1}{2} - \frac{1}{\alpha^2}\left(1 - \frac{1}{\cos h\alpha}\right) \tag{6.115}$$

Although this influence line has been defined for a uniformly distributed line load $p = 1$ acting with a variable eccentricity e, and for simply supported span, it may be adopted for a continuous deck by choosing an equivalent simply supported span with a span length between zero moment points of the permanent bending moment diagram. Besides, usually influence lines are obtained for a concentrated load $Q = 1$ at the mid-span section and with a transverse variable eccentricity e. This problem can still be solved from differential Eq. (6.103), with the result [34] for the points defining the influence line

$$\eta(e) = \frac{\sigma_s}{\sigma_{s0}} = 1 + f(e)\frac{I_y}{I_\omega}\frac{\omega_s}{z_s} \tag{6.116}$$

$$\eta_1 = \frac{1}{2} + \frac{1}{2\alpha}\tan h\alpha \qquad \eta_2 = \frac{1}{2} - \frac{1}{2\alpha}\tan h\alpha \tag{6.117}$$

These results are compared in Figure 6.77 with the previous one for a line load. For pre-design, one may take for steel-concrete composite decks the approximation $\eta_1 = 0.9$ and $\eta_2 = 0.1$, as shown in Figure 6.77.

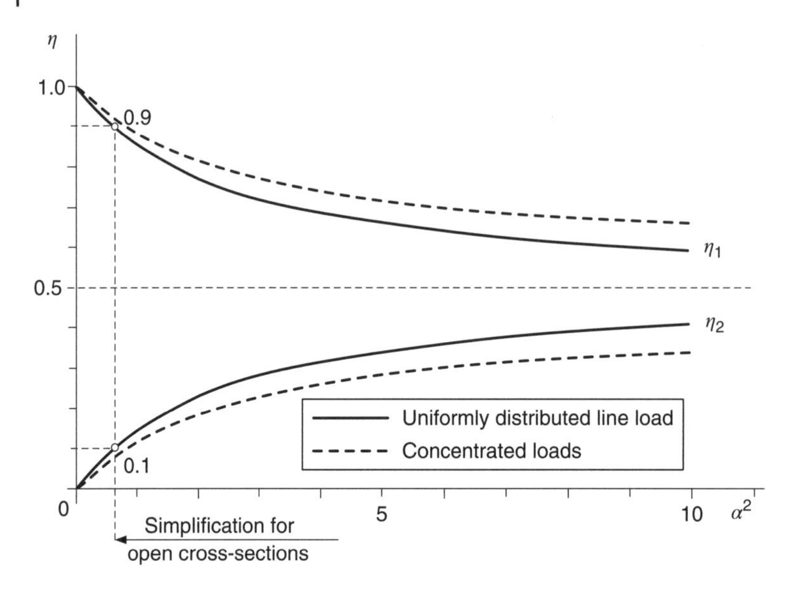

Figure 6.77 Transverse influence functions for steel-concrete composite decks loaded by uniformly distributed line loads $p = 1$ or concentrated loads $Q = 1$ at the mid-span sections as a function of the deck parameter, α^2.

6.6.6 Curved Bridges

Road and rail in-plane alignments very often require the adoption of curved bridge decks. Basic aspects of curved bridges are referred to here and the reader is referred for additional information to specific literature on this subject [21].

6.6.6.1 Statics of Curved Bridges

Equilibrium equations between internal forces and applied distributed loads are established by statics by taking an element of a curved deck with an infinitesimal length $ds = Rd\theta$ between two cross sections, as shown in Figure 6.78, under a vertical load and torsion moment distributions p and m_t. Neglecting terms in $(d\theta)^2$ and taking $\sin d\theta \approx d\theta$ and $\cos d\theta \approx 1$, one has from $\Sigma F_y = 0$, $\Sigma M_x{}^B = 0$ and $\Sigma M_z{}^B = 0$,

$$dV_y + pRd\theta = 0 \tag{6.118}$$

$$dM_x - V_y Rd\theta - Td\theta = 0 \tag{6.119}$$

$$dT + M_x d\theta + m_t Rd\theta = 0 \tag{6.120}$$

These equations may be rewritten (omitting indices for simplicity, $V \equiv V_y$, $M_x \equiv M$) in the form

$$\frac{1}{R}\frac{dV}{d\theta} + p = 0 \quad \frac{1}{R}\frac{dM}{d\theta} - V - \frac{T}{R} = 0 \quad \frac{1}{R}\frac{dT}{d\theta} + \frac{M}{R} + m_t = 0 \tag{6.121}$$

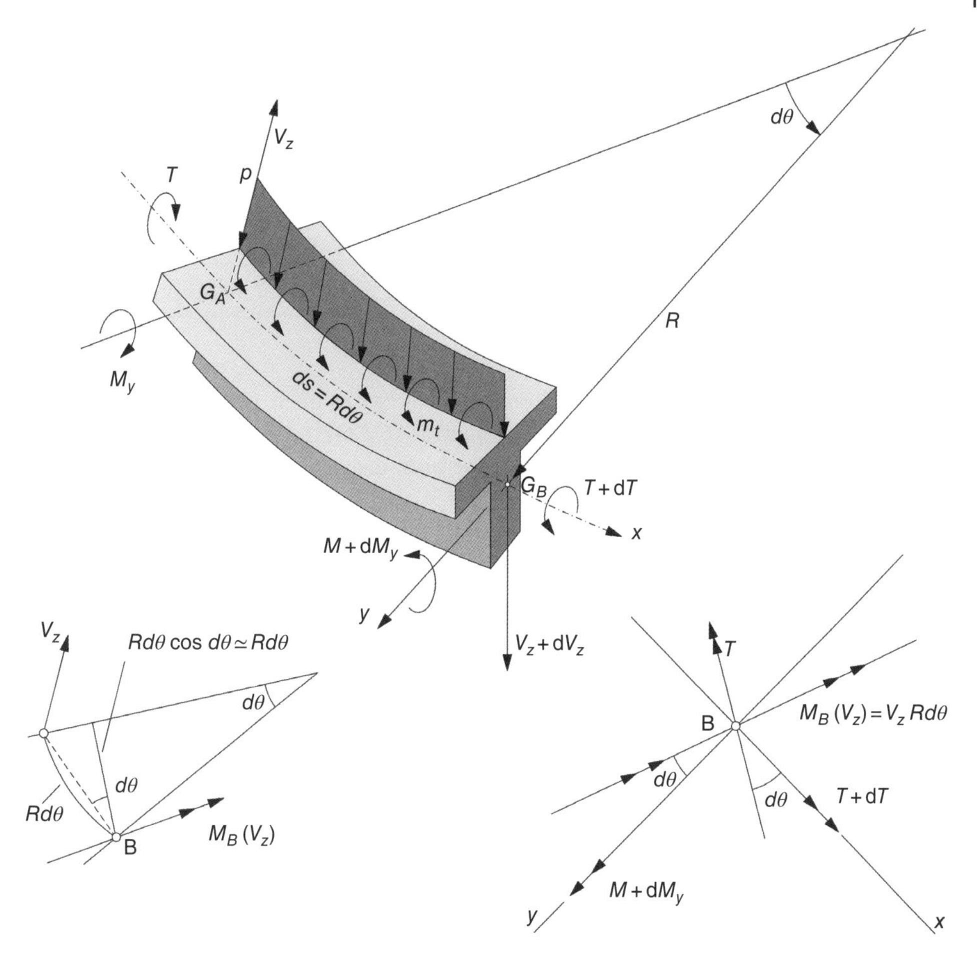

Figure 6.78 Equilibrium of a curved bridge deck segment under distributed loading perpendicular to its plane.

Since $Rd\theta = ds = dx$ one has for $R = \infty$ the well-known equilibrium equations for a straight bar ($dV/dx = -p$, $dM/dx = V$, $dT/dx = -m_t$). When $m_t = 0$, one has from the Eq. (6.121) M = $-dT/d\theta$ and

$$\Delta T = T(\theta_1) - T(\theta_0) = -\int_{\theta_0}^{\theta_1} M\, d\theta \tag{6.122}$$

Hence, the variation of the torsion moment between two sections is equal to the area of the bending moment diagram between those two sections. From Eq. (6.122), when the T is known at one section $\theta = \theta_0$ (e.g. if $T = 0$ due symmetry conditions or any other reason) one may evaluate the torsional moment at any other section by

$$T(\theta) = T(\theta_0) - \int_{\theta_0}^{\theta_1} (M + m_t R)\, d\theta \tag{6.123}$$

By differencing the second of these equations and using the other two, one has

$$\frac{d^2M}{d\theta^2}+M=-R\left(pR+m_t\right) \tag{6.124}$$

The general solution of this linear differential equation is

$$M\left(\theta\right)=C_1\sin\theta+C_2\cos\left(\theta+\theta_p\right) \tag{6.125}$$

where C_1 and C_1 are constants of integration and θ_p is a particular solution.

6.6.6.2 Simply Supported Curved Bridge Deck

For the case of Figure 6.79 of a single span deck with an in-plan radius R, under uniformly distributed load p, one has from equation (6.125) the following conditions for determining the integration constants from the boundary conditions

$$\theta=0, M=0\rightarrow C_2R^2p=0;\quad \theta=\alpha, M=0\rightarrow C_1\sin\alpha+C_2\cos\alpha-R^2p=0 \tag{6.126}$$

Solving for C_1 and C_2 and replacing it in the general solution and with the particular solution $\theta_p=-R^2p$, one has the result

$$M\left(\theta\right)=R^2p\left(\frac{1-\cos\alpha}{\sin\alpha}\sin\theta+\cos\theta-1\right) \tag{6.127}$$

The bending moment at mid-span ($\theta=\alpha/2$) is obtained from this expression, yielding after transforming cos α and sin α in terms of cos(α/2) and sin(α/2), the result

$$M\left(\frac{\alpha}{2}\right)=R^2p\left(\tan\frac{\alpha}{2}\sin\frac{\alpha}{2}+\cos\frac{\alpha}{2}-1\right) \tag{6.128}$$

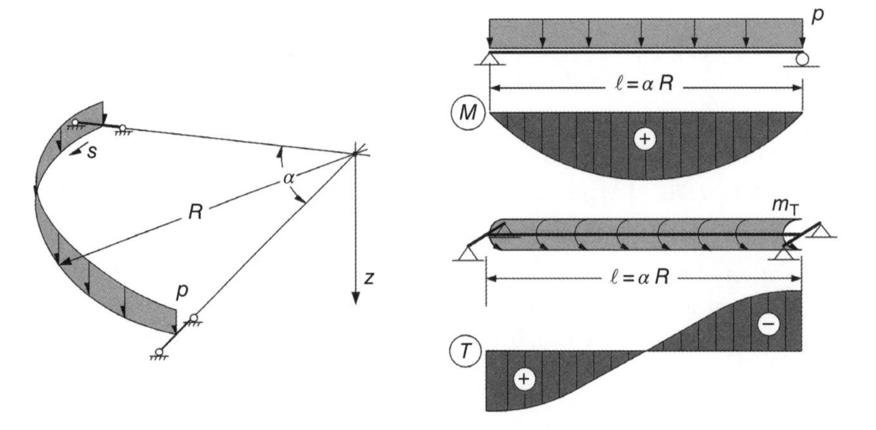

Figure 6.79 Simplified method to determine torsion moments in curved bridge decks loaded perpendicular to its plane.

This may be written as

$$M\left(\frac{\alpha}{2}\right)=M_{\mathrm{eq}}\frac{8}{\alpha^{2}}\left(\tan\frac{\alpha}{2}\sin\frac{\alpha}{2}+\cos\frac{\alpha}{2}-1\right)=M_{\mathrm{eq}}C(\alpha) \tag{6.129}$$

where $M_{\mathrm{eq}} = pl^2/8\ l$ is the equivalent bending moment in a straight deck with l equal to the span length of the curved deck ($l = \alpha R$). The value of $C(\alpha)$ is approximately 1.0 (error less than 1% for $\alpha < 0.3$ rad = 17°), which means the bending moments in the curved deck are very close to the ones obtained for a straight deck with an equivalent length. Note, for example, for $l = 40$ m and $R = 133$ m one has $\alpha = 0.3$. This means, in practice, the effect of curvature on the bending moments may be neglected.

6.6.6.3 Approximate Method

Assuming M is known (from the straight deck) Eq. (6.124) yields

$$\frac{dT}{ds}=-\left(m_{\mathrm{t}}+\frac{M}{R}\right) \tag{6.130}$$

Since dT/ds is an applied torsion distributed moment, this means knowing M, the torsion moments may be obtained by loading an equivalent straight deck with a distributed moment equal to ($m_{\mathrm{t}} + M/R$). In Figure 6.79 this approximate method of analysis is illustrated for a simply supported span. When there is no applied torsion moment distribution, the equivalent straight deck is loaded with the bending moment distribution divided by R. The method is applicable for any type of bridge deck: single span or continuous.

6.6.6.4 Bearing System and Deck Elongations

Torsion moments in curved bridge decks should be taken at predetermine deck sections at piers and abutments where torsion rotations are prevented. In continuous decks, if the in plane radius of curvature is not too small, torsion rotations may be restricted at all support sections. If that is the case, independent spans may be considered for torsion analysis, built in at support sections for torsion. That is, torsion moments may be obtained independently for each span. This is true under uniform torsion but not if warping torsion is considered as discussed in Section 6.6.5 for a continuous three-span deck. Besides, sometimes the flexibility of the piers in the transverse direction does not allow the fully restrained assumption under torsion to be adopted. Elastically restrained torsion spans over support sections may be considered in these cases. When the section has sufficient torsion rigidity, like box sections and solid or voided slab sections, it is possible to restraint torsion at some section only and to adopt single supports at the remaining pier sections. A high curvature bridge example is shown in Figure 6.80. Torsion rotations are restricted only at the access ramp abutment and at the transition piers of the ramp with the main deck. All other ramp piers have single bearings yielding to free torsion rotations of the deck.

In-plane movements at bearings in curved bridges, due to uniform thermal actions and shrinkage effects of the concrete, are along the direction defined by the fixed in plane point and the bearing. This is easily shown by the Principle of Virtual Work (unit

Figure 6.80 Design example of a high curvature bridge deck in the access ramp to a cable-stayed bridge (*Source:* Bridge over River Ave at St Tirso, Portugal, courtesy of GRID, SA).

load method for displacements) from which the displacements at the bearing (Figure 6.81) are

$$u_x = \int_{AB} \bar{N}_{Fx} \varepsilon \, ds \qquad u_y = \int_{AB} \bar{N}_{Fy} \varepsilon \, ds \tag{6.131}$$

where $\bar{N}_{Fx}$ and $\bar{N}_{Fy}$ are axial forces due to unit forces $F_x = 1$ and $F_y = 1$ applied at the bearing deck section as shown in the figure and ε is the thermal or shrinkage strain. As

$$\bar{N}_{Fx} = 1 \times \cos\varphi \qquad \bar{N}_{Fy} = 1 \times \sin\varphi \tag{6.132}$$

One has

$$u_x = \varepsilon_0 \int_{AB} \cos\varphi \, ds = \varepsilon_0 \int_{x_A}^{x_B} dx = \varepsilon_0 \, l_x \tag{6.133}$$

$$u_y = \varepsilon_0 \int_{AB} \sin\varphi \, ds = \varepsilon_0 \int_{y_A}^{y_A} dx = \varepsilon_0 \, l_y \tag{6.134}$$

where l_x and l_y are the projections on the axis of the curved chord *AB*. Hence

$$\frac{u_y}{u_x} = \frac{l_y}{l_x} = \tan\beta \tag{6.135}$$

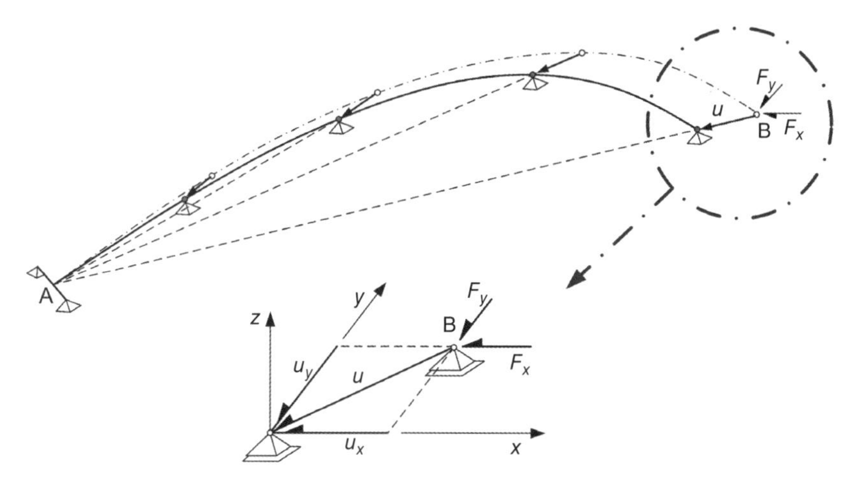

Figure 6.81 In-plane displacements at the bearings of an in-plane curved bridge.

which means the displacement is according to the radius vector $\overrightarrow{AB}$ and with an absolute value

$$u = \sqrt{u_x^2 + u_y^2} = \varepsilon_0 \sqrt{l_x^2 + l_y^2} = \varepsilon_0 \left|\overrightarrow{AB}\right| \tag{6.136}$$

Hence, the displacement is equal to the total elongation of the chord AB due to a uniform strain ε_0. It should be noted if the axial strains are not constant along the bridge length, the conclusion about the direction of the displacement vector is not valid. This is often the case and unidirectional bearings may not be installed according to the radius vector AB. For example, it may be shown creep deformations are according to the direction of the bridge axis and not following the radius vector $\overrightarrow{AB}$. In practice, unidirectional bearings have small transverse clearances in the guides and the bearings are very often installed along the curved direction of the in-plan bridge alignment. At abutments the effects of transverse movements for the bridge expansion joints should be carefully analysed.

6.7 Influence of Construction Methods on Superstructure Analysis

As previously discussed in Chapter 4, the influence of the construction method on the analysis of the superstructure should be taken into account in the design. The evolution of the static system during erection and viscoelastic effects of concrete, namely in prestressed concrete bridges or in steel-concrete composite bridges, influences redistribution of permanent bending moments and hyperstatic effects of prestressing. Specific aspects for bridges built span by span, segmental construction by the cantilever method, prestressed concrete precasted girder bridges and steel-concrete composite bridge decks, are discussed in the following.

6.7.1 Span by Span Erection of Prestressed Concrete Decks

The static system of the superstructure is subject to an evolution changing at each construction phase, as shown in Figure 6.82, for a span by span execution of a prestressed concrete deck. Construction joints are in this case located nearby one-fifth part of the spans in order to match with sections of approximately zero bending moments at the final static system, after erection has been completed. The load at each phase corresponds to the dead weight (DW) of the concrete applied between the construction section and the free end section; this dead load acts after removing the scaffolding. The elastic bending moment at the end of construction is the sum of the bending moments at each phase. However, this bending moment differs from the actual bending moment due to viscoelastic redistribution of internal forces.

At each construction phase and before moving the scaffolding for the next span, the prestressing actions should be verified to check whether the decompression limit state is satisfied. Usually, no tensile stresses in the deck under permanent loads are allowed during any construction stage. Part of the prestressing cables are stressed at the each phase and the remainder are extended to be stressed at the next stage. Usually, when the decompression limit state is verified at a certain construction phase, safety with respect to ULS is satisfied as well but this depends on the existing ordinary reinforcement and should always be checked in any case.

With reference to Figure 6.82, valid for classical scaffolding, stationary falsework or formwork launching girders, it should be noted the final elastic bending moment distribution, including elastic prestressing effects, $M_0 = \Sigma M_{0i}$, is not very different from the

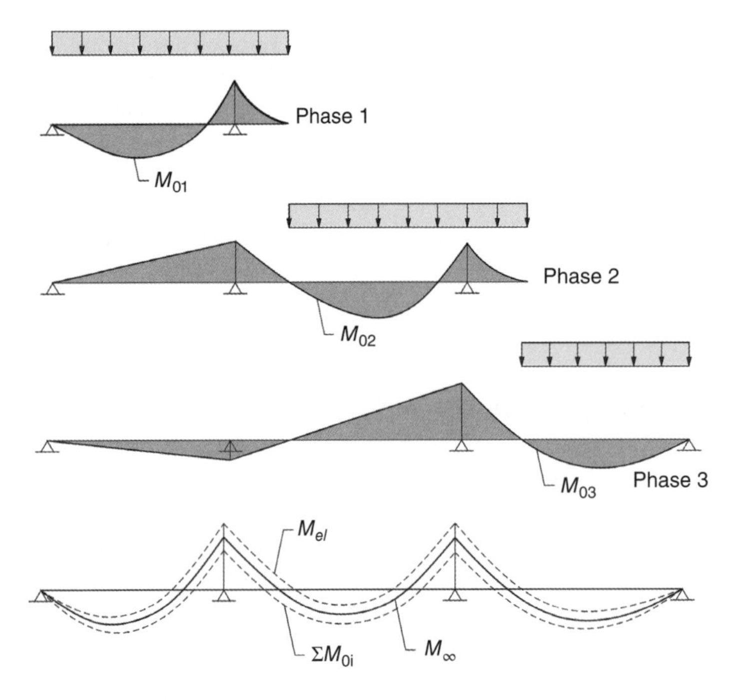

Figure 6.82 Evolution of the static moment and bending moment distribution in a span by span execution of a prestressed concrete deck.

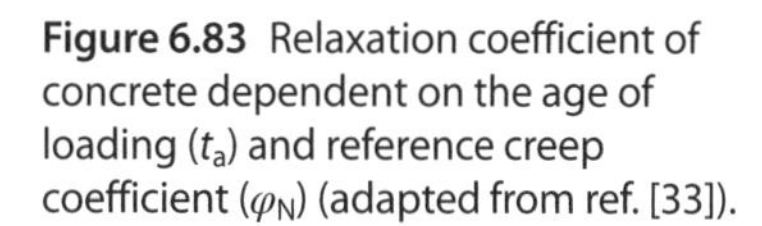

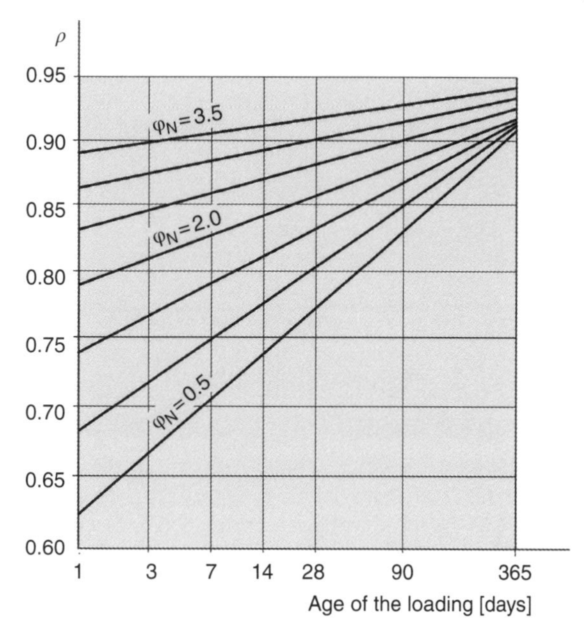

Figure 6.83 Relaxation coefficient of concrete dependent on the age of loading (t_a) and reference creep coefficient (φ_N) (adapted from ref. [33]).

bending moment distribution M_{el} obtained by applying the permanent load on the final static system. A step by step analysis, considering the evolution of the static system, ages of concrete at each prestressing phase and at each time instant when scaffolding is removed, can be done on the basis of available commercial software packages. In design practice, this may not be justified and simple redistribution formulas are available. The reader is referred to specific references on this subject [9, 33, 35, 36]. One of the simplest approaches to this problem is to evaluate M_{el} and M_0 as previously mentioned, and to estimate the long term bending moments by [37]

$$M_\infty = \Sigma M_{0i} + \frac{\varphi_\infty}{1+\rho\varphi_\infty}\left(M_{el} - \Sigma M_{0i}\right) \tag{6.137}$$

where φ_∞ is the concrete creep long term creep coefficient dependent on concrete age at loading (order of 2.5–3.0) and ρ is a relaxation coefficient for concrete (Figure 6.83) depending on the age at loading and on the creep coefficient of concrete at a reference age (usually at 28 days). If, for example, one assumes $\varphi_\infty = 2.5$ and $t_o = 14$ days, one has $\rho = 0.87$ and $M_\infty = 0.21\ M_0 + 0.79 M_e$.

6.7.2 Cantilever Construction of Prestressed Concrete Decks

The example of the cantilever construction in Section 4.3.1 is taken again here (Figure 6.84) to explain how to estimate the bending moment redistribution at the mid-span section due to creep and evolution of the static system. A simple model of two cantilevers, one of each built from each pier and modelling a complete span, is shown in Figure 6.85. Just before executing the closure segment, there is a relative rotation θ_0^e between the two end sections. Due to creep, this rotation tends to increase according to the creep coefficient φ and the concept of an adjusted effective modulus (Eq. (3.71))

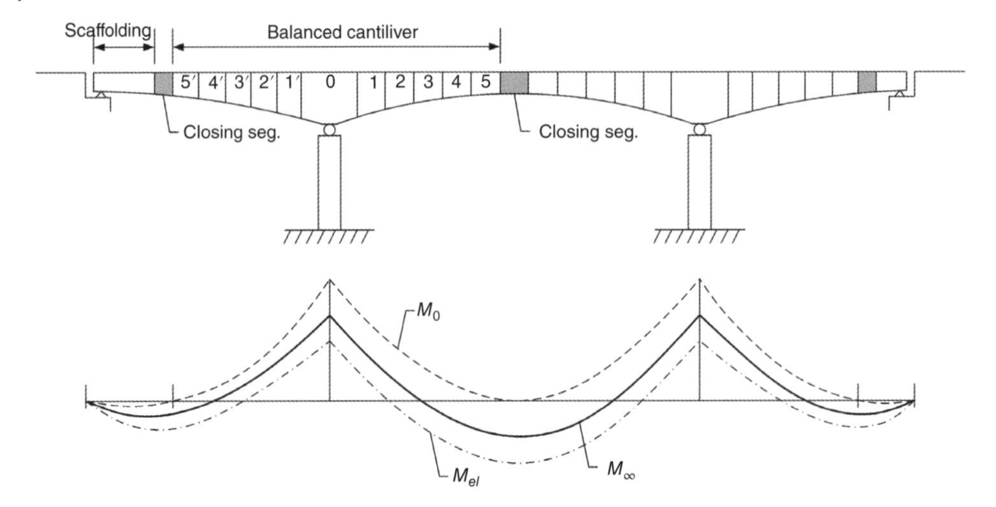

Figure 6.84 Balanced cantilever construction by cast in place segments. Continuity established at the closures' segments. Evolution of bending moment diagram, due to DL.

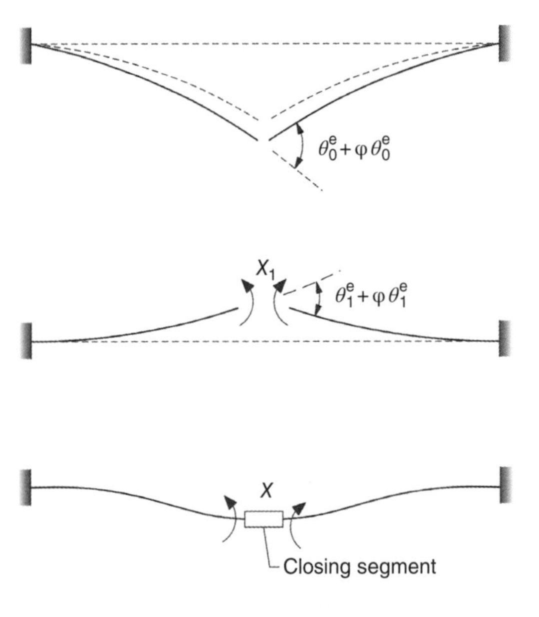

Figure 6.85 Cantilever construction by cast in place segments. Model for determining the hyperstatic redistribution bending moment due to DL.

towards a value $(1 + \varphi)\,\theta_0^{\mathrm{e}}$. This tendency is restricted by a hyperstatic moment X_1 at the end sections. Only the creep deformation $\varphi\,\theta_0^{\mathrm{e}}$ will be restricted. The relative rotation induced by the X_1, if it was elastic was $f_{11}X_1$ where f_{11} is the flexibility coefficient (rotation due to $X_1 = 1$); due to creep it tends to increase towards $(1 + \varphi)\,f_{11}X_1$. Hence, the compatibility condition from the force method of structural analysis with due account to opposite rotation signs, requires

$$(1+\varphi)\,f_{11}X_1 = \varphi\theta_0^{\mathrm{e}} \tag{6.138}$$

Hence the hyperstatic bending moment due to creep is

$$X_1 = \frac{\varphi}{1+\varphi} X_1^e \tag{6.139}$$

with $X_1^e = \theta_0{}^e / f_{11}$ that would be the hyperstatic bending moment induced if the system was elastic and the structure had been cast in a single phase on a scaffolding. The final elastic bending moment distribution is $M_e = M_0 + X_1^e$ where M_0 are the bending moments before the closure segment has been executed. Hence, one has for the final bending moments, creep and redistribution effects taking into account:

$$M = M_0 + \frac{\varphi}{1+\varphi}\left(M_{el} - M_0\right) \tag{6.140}$$

The second term on the right-hand side of the equation represents the redistribution of bending moments due to creep effects. At the mid-span and long term, one has for a φ = 2 to 3 and $M(L/2)$ = 0.67–0.75 of M_{el}. The Eq. (6.140) is of course an approximate expression for the following reasons:

- the approach of equivalent elastic modulus E_{eq} does not take into account the variation of E with time;
- rotations θ_0 are not elastic, but are also affected by creep and by the construction history with incremental DW loading at each time of casting segments and prestressing operations;
- the age difference between each segment has not been taken into account;
- at each cantilever there are prestressing cables that induce an hyperstatic effect by creep after the closure segment has been executed;
- time-dependent effects on the continuity prestressing has not been taken into account.

The limitation associated with E_{eq} may be removed by taking an *adjusted equivalent modulus of elasticity* (Section 3.14), yielding an expression of the type of Eq. (6.140) referred to for span by span deck execution on scaffoldings.

Concerning the effects of the cantilever cables, it should be noted that isostatic effects (*Pe*) are considered in the bending moment distribution M_0 (Figure 6.85). Prestressing instantaneous and deferred losses should be considered as well as shrinkage effects. It is only possible to take all effects into consideration on the basis of a numerical model available from a variety of existing software. One may rewrite Eq. (6.140) in the long term as

$$M_\infty = M_0 + \alpha\left(M_{el} - M_0\right) \tag{6.141}$$

where α represents the redistribution coefficient to take all the effects in due account.

An illustrative example of a model case of a three-span bridge (28 + 50 + 28 m) is represented in Figure 6.86; details on the prestressing layout, cross section properties and construction history are given in [37]. For this example, the redistribution coefficients obtained at the mid-span section were $\alpha = 0.67$ for the DW and $\alpha = 0.71$ for the hyperstatic bending moment due to prestressing.

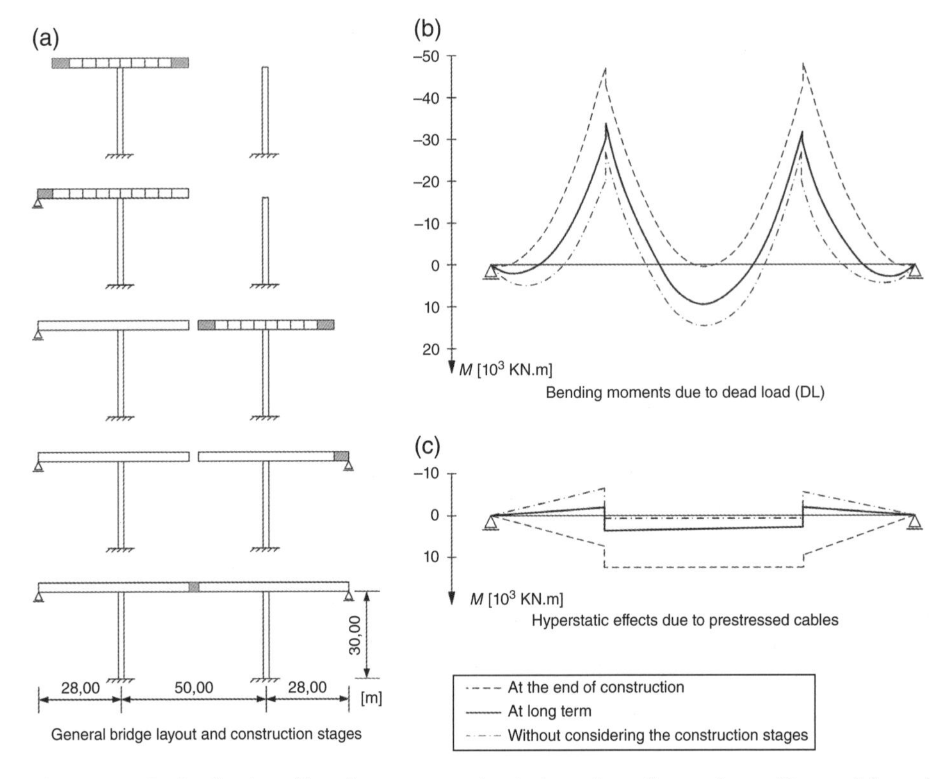

Figure 6.86 Redistribution of bending moments in a balanced cantilever scheme (*Source:* Adapted from Ref. [37]).

6.7.3 Prestressed Concrete Decks with Prefabricated Girders

In the case of decks executed with precasted girders (Figures 4.21–4.23) the continuity may be given at support sections as explained in Section 4.4.5. First, the precasted girders ($\Delta W = p$) erected on the piers work as simply supported with maximum bending moments of $pl^2/8$; second, cross beams over supports are cast followed by the slab. At this stage, the girders, with the prestressing cables for the isostatic phase, support the full DW of the slab deck. Finally, full continuity at support sections is established after tensioning the continuity prestressing cables and a continuous deck girder is obtained. The bending moment diagram M_0 starts being one of a simply supported span but a hyperstatic moment X_1 at the support section is developed after continuity is introduced (Figure 6.87). The bending moment redistribution following the evolution of static system and creep effects may be estimated according to a simple approach based on the force method. As has been done for cantilever construction, the compatibility equation is identical to Eq. (6.138). In this case, $\theta_0^{\rm e}$ is the elastic rotation at the end sections of the simply supported girders immediately after casting the slab deck; this rotation tends to increase due to creep towards values of $\varphi\theta_0^{\rm e}$. The hyperstatic bending moment X_1 induces a total (elastic + creep) strain given by $\varepsilon_{\rm total} = f^{\rm e}_{11}X_1 + \phi f^{\rm e}_{11}X_1$. Due to compatibility, Eq. (6.138) is obtained. The hyperstatic moment X_1 is given by Eq. (6.139). Hence, the redistribution coefficient is again given by $\varphi/(1+\varphi)$. In this case, φ is lower

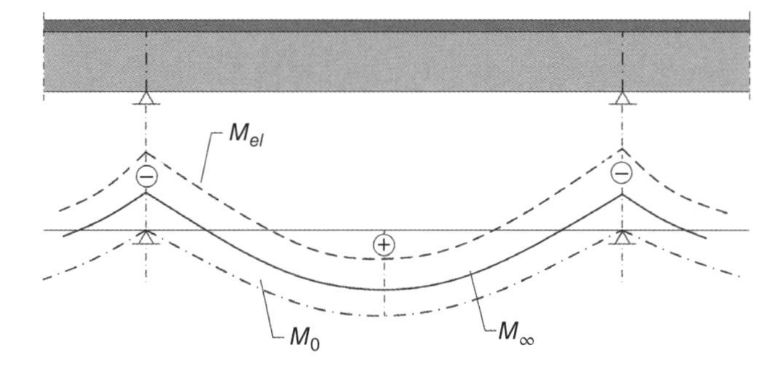

Figure 6.87 Deck executed with precasted girders, with continuity over support sections, after erection of the girders. Evolution of the bending moment diagram due to dead loads.

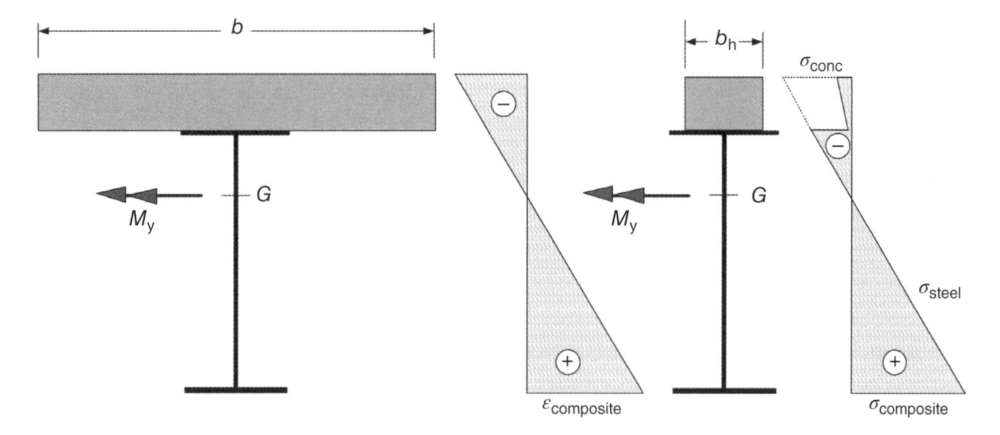

Figure 6.88 Stresses in steel-concrete composite sections. Equivalent homogeneous section in steel with $b_h = b/n$.

than previously referred to for cantilever construction. Values of $\varphi = 1$ to 2 may be taken for continuous decks with precasted girders. Redistribution coefficients α are in the order of 0.5–0.7.

6.7.4 Steel-Concrete Composite Decks

Steel-concrete composite decks are subject to a variety of redistribution effects, starting with stress redistribution between steel and concrete. This has been already refereed in Chapter 4 (Section 4.5.3.1) when dealing with how the composite action is induced. Referring to Figure 6.88, the usual approach is to adopt for concrete an *adjusted affective modulus* (Section 3.14) with an ageing coefficient χ between 0.8 and 0.9 for concrete loading ages between 10 and 100 days:

$$E_{c,adj}(t,t_o) = \frac{E_c(t_o)}{1+\chi\varphi'(t,t_o)} \tag{6.142}$$

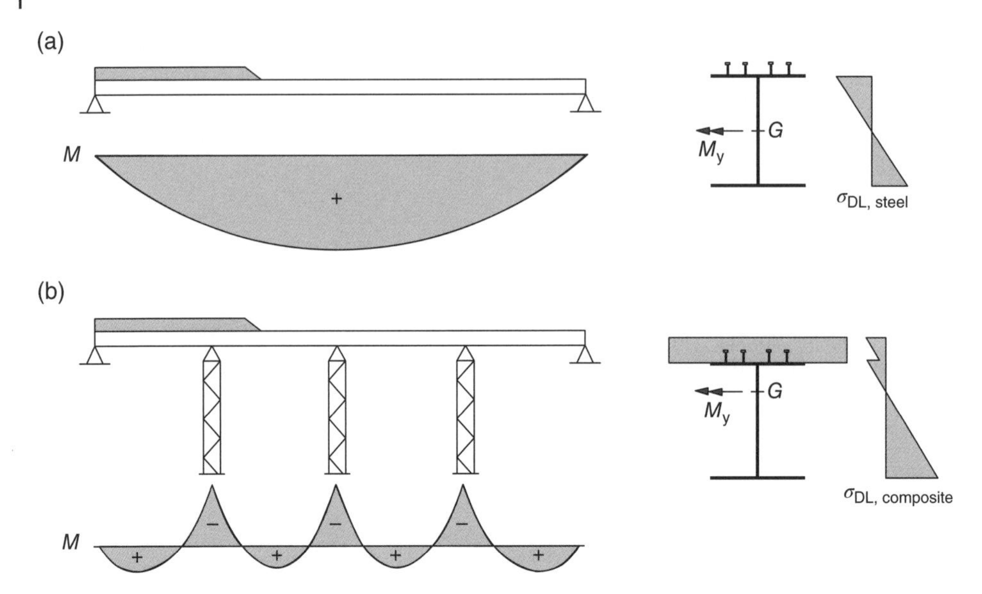

Figure 6.89 (a) Cast in place slab deck on the steel girders without temporary supports. Stresses at the steel section after casting the slab over the steel girders. (b) Cast in place slab deck on steel girders on temporary supports. Bending moments and stresses at the composite section after removing the temporary supports and a simply supported composite deck.

A *modular ratio* $n_0 = E_s/E_{cm}$ is defined, with E_{cm} being the average secant modulus of the concrete. A composite section under bending (Figure 6.89) has a linear distribution of strains according to the Bernoulli Hypothesis, and a stress distribution with a discontinuity at the steel-concrete interface. The equivalent homogeneous section has a linear continuous distribution of stresses as shown in Figure 6.89. For taking long term effects into consideration, namely stress redistribution due to creep between the concrete and the steel parts, an adjusted modular ratio $n(t,t_o)$ should be taken, where t_o is the time at loading and t is time at observed stresses. This may be done by taking a modular ratio from (3.4.12) defined by

$$n = \frac{E_a}{E_{c,eff}\left(t,t_o\right)} \tag{6.143}$$

that can be written as

$$n = \frac{E_a}{E_c\left(t_o\right)}\left[1+\chi\left(t,t_o\right)\frac{E_c\left(t_o\right)}{E_{c,28}}\varphi\left(t,t_o\right)\right] \tag{6.144}$$

In short, one has

$$n = n_o\left(1+\psi\varphi\right) \tag{6.145}$$

with

$$\psi = \frac{E_{cm}}{E_{c,28}}\chi(t,t_o) + \frac{1}{\varphi(t,t_o)}\left[\frac{E_{cm}}{E_c(t_o)} - 1\right] \tag{6.146}$$

Here, E_{c28} is the average tangent modulus of elasticity of concrete at the origin of the stress-strain diagram that can be evaluated from the characteristic stress f_{ck}, while E_{cm} is the average value of the secant modulus of elasticity of concrete for short term loading. The *creep multiplier* ψ may be taken in design practice with the following values [8]:

- $\psi = 1.10$ for permanent loads including prestressing effects after the composite action has been introduced.
- $\psi = 0.55$ for shrinkage time dependent effects (including hyper static effects) when the bending moment distribution at age t_0 is significantly changed by creep, as for continuous girders with or without composite action during the execution phases.
- $\psi = 1.50$ for prestressing effects induced by imposed displacements at the supports.

This last case is very much adopted for controlling tensile slab stresses over supports of continuous steel-concrete composite decks as shown in Figure 6.90. The structure is erected in phase 2 of the figure, then the deck slab is cast and finally the structure is moved downwards to its final position by an imposed displacement between phases 3 and 4. The elastic bending moment diagram, due to this imposed displacement, should be evaluated with $\psi = 1.50$. There is a strong reduction of the imposed elastic compressive stresses in the slab due to relaxation effects of concrete. The effect is similar as to assuming a bending moment reduction by relaxation in a viscoelastic structure, as illustrated in Figure 6.90.

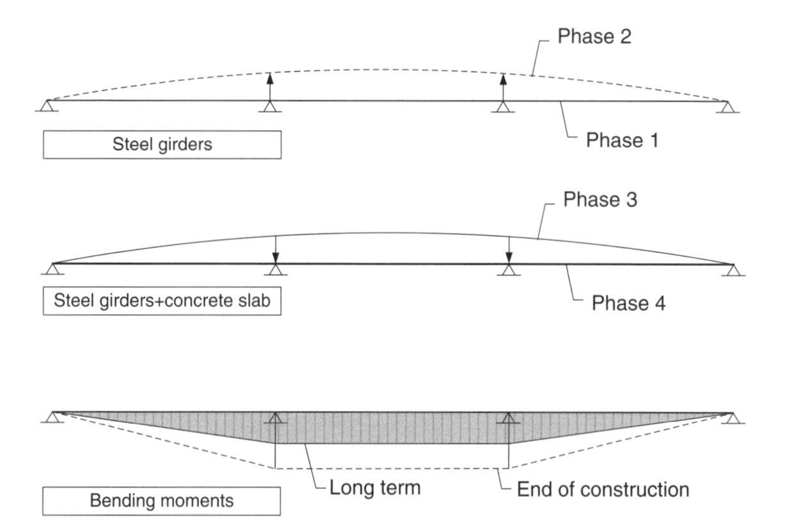

Figure 6.90 Imposed displacements in a continuous three-span steel-concrete composite deck.

6.8 Prestressed Concrete Decks: Design Aspects

6.8.1 Generalities

General aspects about the design of prestressed concrete decks have been discussed in previous chapters, namely concepts, construction and erection methods and general methods of analysis. The aim of the present section is not to discuss general aspects of reinforced or prestressed concrete design. The reader is assumed to be familiar with these aspects and for developments is referred to specific literature on the subject [13, 38]. Therefore, only specific aspects on prestressed concrete bridge design are dealt with here.

6.8.2 Design Concepts and Basic Criteria

Limit state design methodology is adopted for prestressed concrete bridge structures. Bridge decks should be verified for Serviceability Limit States (SLS) and Ultimate Limit States (ULS).

Prestressing design of bridge decks is done to satisfy *decompression limit state (*DLS*)*, defined as a stress state of zero tension at the most loaded fibres of the section. Of course, this has to be done for a certain load combination – characteristic, frequent or quasi-permanent, according to definitions in Chapter 3. Codes define the required load combination as a function of the exposure class environment.

Bridges in most cases are subjected to chlorides or freeze/thaw attack with exposure classes XD3 applicable to parts exposed to cycle wet and dry of bridges exposed to spray containing chlorides. For bridge decks exposed to de-icing agents, exposure class XF4 is applicable.

Requirements on maximum crack widths (w_{max}) and DLS are referred to in EN 1992-1 and 2; for the usual case of bridge decks with bonded tendons where an exposure class XD3 should be taken into account, the basic criteria are limited to DLS for Quasi-Permanent Load Combination and maximum crack width to w_{max} = 0.2 mm for the frequent load combination.

6.8.3 Durability

Durability of prestressed concrete bridges is basically dependent on:

- The quality of concrete – the use of high performance concrete, namely high strength concrete is a step forward towards an improved durability;
- The corrosion resistance of steel prestressing strands and in particular protection of the strands by injection of cement grout or wax products in unbounded tendons;
- The covering of the steel reinforcement and prestressed cable ducts;
- The adoption the concept of partial prestressing instead of a full prestressing concept.

6.8.4 Concept of Partial Prestressed Concrete (PPC)

Nowadays, bridges are not usually designed as fully prestressed concrete structures as they were in the early days of prestressing. The exception is made for precasted segmental bridges (Section 4.6.2.6). Ordinary reinforcement has to be added to achieve sufficient ductility under all design action scenarios, in particular after exceeding DLS and

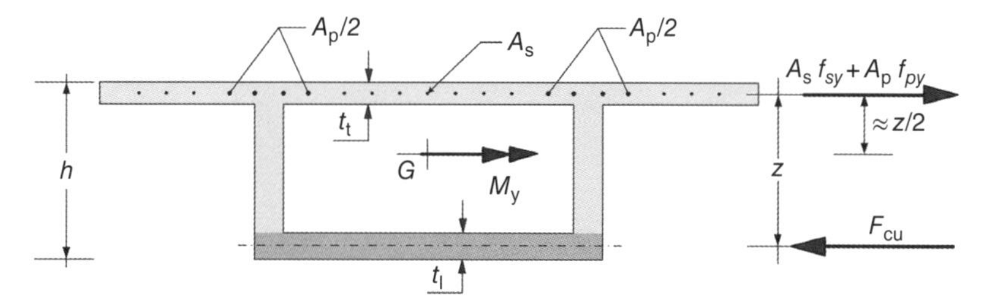

Figure 6.91 Partial prestressed concrete section under bending.

under imposed deformations due to uniform and differential shrinkage and creep effects, thermal actions and foundation settlements. In a PPC section under bending, the ULS bending moment resistant (Figure 6.91) is given by

$$M_R = \left(A_s f_{sy} + A_p f_{py}\right) z \tag{6.147}$$

where A_s and A_p are the areas of ordinary reinforcement steel and bonded prestressing tendons working at design stresses f_{sy} and f_{py}, respectively, and z is the lever arm. If the top and lower deck flanges with thicknesses t_t and t_l are considered, one may take as an approximation $z = h - 0.5(t_t + t_l)$ with h being the total height of the cross section.

Full prestressing, at least for permanent loads is always necessary. The cracking moment at some sections may be exceeded under certain load combinations provided sufficient ordinary reinforcing steel has been added for crack width control under imposed deformations inducing self-equilibrating stresses. Minimum reinforcement ratios are imposed in the codes to deal with crack width control.

6.8.5 Particular Aspects of Bridges Built by Cantilevering

An illustrative case for the amount of prestressing required in designing a bridge cross section is given for the case of cantilever construction. The area of prestressing steel at the support section is designed to satisfy DLS during cantilever construction, at service and at ULS. Of course, the bending moment due to prestressing M_p is maximized by restricting the ordinary reinforcement to a minimum for crack control. Let the ratio between the prestressing moments and permanent moments be considered, say M_p/M_g. Designating the live load moments by M_q, adopting a load and material factors 1.35 and 1.15, respectively, one has the ULS safety condition given by:

$$1.35\left(M_g + M_q\right) \leq z\left(A_s f_{sy} + A_p f_{py}\right)/1.15 \tag{6.148}$$

One may define for the support section $M_p = A_p \sigma_p z/2$, take $A_s = 0$ for maximum A_p and prestressing cables stressed at 70% of f_{py} ($\sigma_p = 0.7 f_{py}$). For road bridges M_q is in the order of magnitude of 20% of M_g (i.e. $M_q = 0.2M_g$) and this equation yields, $M_p = 0.6M_g$. Hence, to satisfy ULS requirements, moments required for prestressing are much lower than moments due to permanent load. That is why DLS usually governs the prestressing design. Normal stresses due to prestressing (N_p/A_c) in cantilever constructed bridges are quite high compared to bridges built on scaffoldings or with formwork launching girders. The prestressing force required over support sections in bridges built by

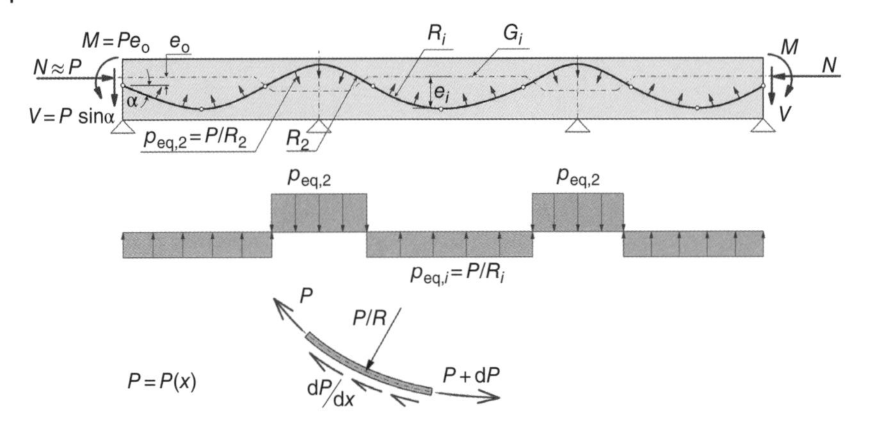

Figure 6.92 Concept of prestressing equivalent loads.

cantilevering usually yields a full prestressing concept for permanent loads and part of live loads. At span sections, it depends on the amount of continuity prestressing adopted to control redistribution of bending moments and positive bending moments induced by differential temperature effects.

Deflection control in cantilever constructed bridges is a main issue. Correct geometry cannot be achieved by compensating permanent deflections by prestressing. This yields a huge and unnecessary quantity of prestressing. Geometry control should be made by introducing the required camber to compensate at least for permanent and creep induced deflections. A good balance of permanent loads by prestressing is a very good criterion to avoid uncontrolled long term creep effects. The *concept of prestressing equivalent load* is reviewed in Figure 6.92 for a continuous beam. If the equivalent load p_{eq} to prestressing compensates the permanent load, creep deflections are usually well controlled.

6.8.6 Ductility and Precasted Segmental Construction

Ductility is a main issue particularly when designing precasted segmental bridge decks, with external cables as usual and *match cast segmental construction* with 'epoxy' joints (see Section 4.6.2.6). At the characteristic load combination full prestressing is required but joints will open at factored loads for ULS, as shown in Figure 6.93. The resistance and sufficient ductility at section zones under compression should be satisfied. Shear forces in the webs need to pass through shear keys at all decompressed joints to satisfy equilibrium at ULS.

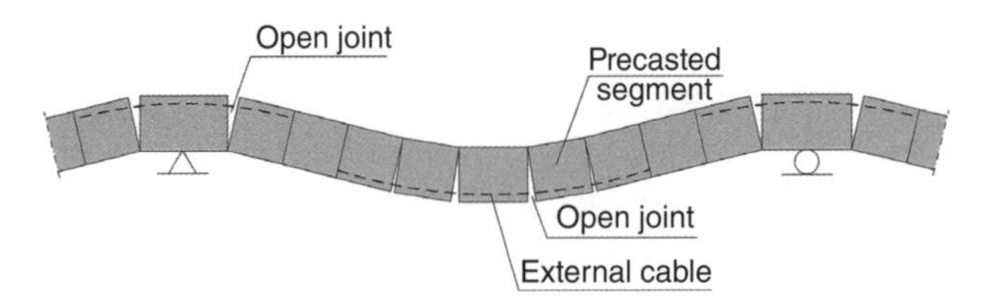

Figure 6.93 Behaviour of precasted segmental decks at ULS.

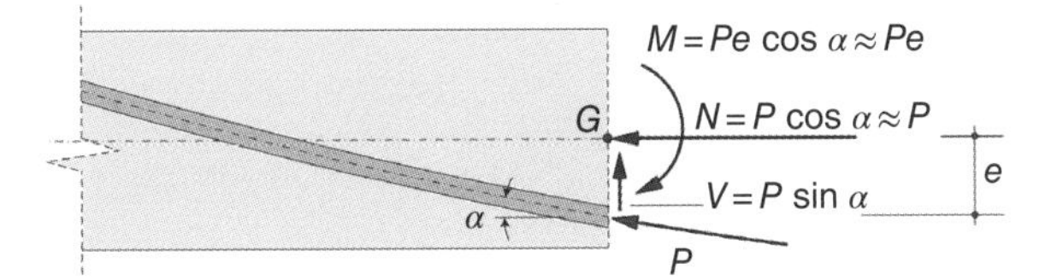

Figure 6.94 Prestressed introduced as internal forces at one section.

6.8.6.1 Internal and External Prestressing

In bridges with internal (bonded) tendons, an elastic analysis is done to estimate section forces due to prestressing. Prestressing actions may be introduced as *internal forces* (Figure 6.94) by the resultant section forces due to effective prestressing P. These are an axial force $N_p = P\cos\alpha$, shear force $V_p = P\sin\alpha$ and section moment $M_p = Pe\cos\alpha$. Alternatively, prestressing may be introduced as an *external load* through induced *equivalent forces to prestressing* (Figure 6.92), that is, a distributed perpendicular force to the axis$=P/R$ and the distributed axial force $n = dP/dx$ along the cable. In these expressions, $P = P(x)$ and $R = R(x)$ are the variable prestressing force and the cable radius. At the end section, the section forces are introduced through the stress resultants determined from the forces at the anchorages.

For *external prestressing cables,* if the cable is assumed to be blocked at the deviation saddles and the prestressing force is assumed as not being influenced by the deck deformations, the simplest approach is to adopt the induced deviation forces at each saddle and at the end sections, and to proceed with an elastic structural analysis with prestressing forces as an external load. This is, of course, an approximate analysis, valid in most practical cases at least for SLS. At ULS it is likely stress variation in the external prestressing cables do exist and the cables slide at the deviation saddles. At this stage a non-linear incremental analysis is needed to estimate the cable and internal section deck forces.

6.8.7 Hyperstatic Prestressing Effects

It is well known from prestressed concrete design due to restrained deformation in statically indeterminate structures, like a three-span continuous bridge deck in Figure 6.95, internal or external prestressing induces hyperstatic effects commonly designated *secondary effects due to prestressing.* These effects are usually evaluated by standard structural analysis programs, by introducing the prestressing forces as referred to in previous sections. Taking the example of Figure 6.95, X_1 and X_2 at the internal supports may be taken as statically indeterminate unknowns. For a general case, one has $\{X_i\}$ ($i = 1$ to n) unknowns due to prestressing. The *base case* for the Force method of structural analysis corresponds to remove the n deck internal supports. The flexibility matrix $[F_{ij}]$ and the vertical displacement vector $\{u_i\}$ at the base system, allows one to write the linear system of compatibility equations

$$\left[F_{ij}\right]\left\{X_j\right\}+\left\{u_i\right\}=\left\{0\right\} \tag{6.149}$$

from which the hyperstatic support reactions are evaluated $\{X_i\}$.

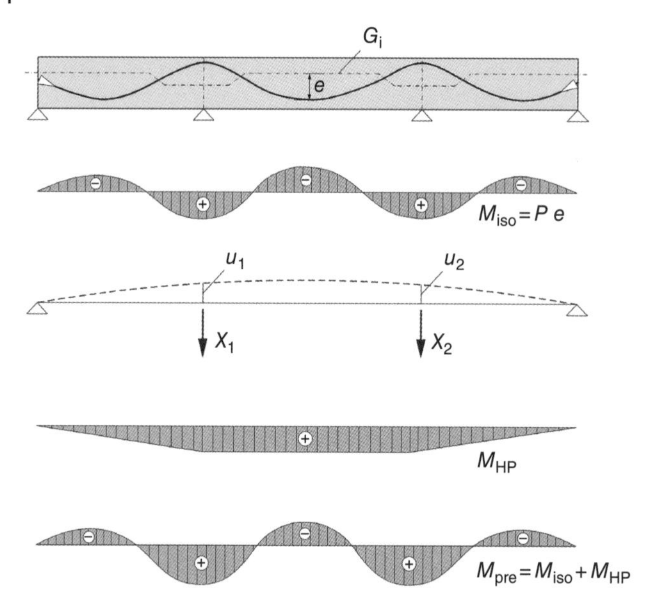

Figure 6.95 Isostatic (primary loading) bending moments M_{ISO} and hyperstatic (secondary) bending moments M_{HP} due to prestressing in a three-span continuous deck.

For the example under discussion, primary prestressing moments Pe, the secondary moments and the total prestressing bending moment diagrams are shown in Figure 6.95.

Since these secondary moments are *self-equilibrated* (they balance a zero external load) and if the bridge structure has sufficient ductility to perform a plastic analysis at ULS, the secondary moments will not play any role at ULS as is the case of imposed deformations by settlements or temperature gradients inducing eigenstresses (Sections 3.13 and 3.15). The assumption of geometrically linear structure is assumed. However, plastic analysis of bridge structures is very much limited by lack of ductility and fatigue load effects. That is why secondary moments cannot be disregarded at ULS because an elastic global analysis is required. If a rigid plastic model is accepted, secondary moments will not play any role at ULS, when geometrical non-linear effects can be negligible.

6.8.8 Deflections, Vibration and Fatigue

Deformability is not generally a critical issue for designing partial prestressed concrete (PPC) bridges. This is particularly true for road bridges where deck stiffness is usually sufficiently high enough to satisfy SLS for maximum allowable deflections. The only cases where deformability tends to be an issue are pedestrian bridges and some cases of rail bridges, in particular High Speed Railway bridges. However, that is more a problem of limiting vertical and horizontal accelerations of the deck rather than a limit deflection problem. Of course, the two aspects are associated because accelerations depend on frequency of vibrations that are associated with mass and stiffness of the structure. Let the simplest case of a 1-DOF system (Mass M and stiffness k, see Chapter 3, Section 3.10) be considered. The response is defined by $\delta(t) = \delta_{max} sin(\omega t)$ where δ_1 is the deflection amplitude and ω the natural circular frequency ($\omega = \sqrt{k/M}$).

The accelerations are obtained by double derivative with respect to time t, with the result $a_{max} = -\delta_{max}\omega^2 = -\delta_{max}k/M$. So limiting deflections and frequencies induces a limit on maximum accelerations.

Natural frequencies are evaluated by a dynamic analysis of the deck (see Section 3.10) but for simply supported, built in spans or continuous girder decks, the fundamental bending frequency may be estimated by

$$f_1 = \frac{k^2}{2\pi L^2}\sqrt{\frac{EI}{m}} \tag{6.150}$$

where k is a coefficient depending on span boundary conditions that is given below, L is the span length, EI the bending stiffness of the cross section and m the mass of the deck per unit length. One has $k = \pi$ for simply supported spans, and $k = 4.73$ for built-in end sections. For decks with two or three continuous spans, one has the k values from Figure 6.96. The fundamental bending frequency may also be calculated using the Rayleigh Method, by

$$f_1 = \frac{1.1}{2\pi}\sqrt{\frac{g}{\delta_{\max}}} \tag{6.151}$$

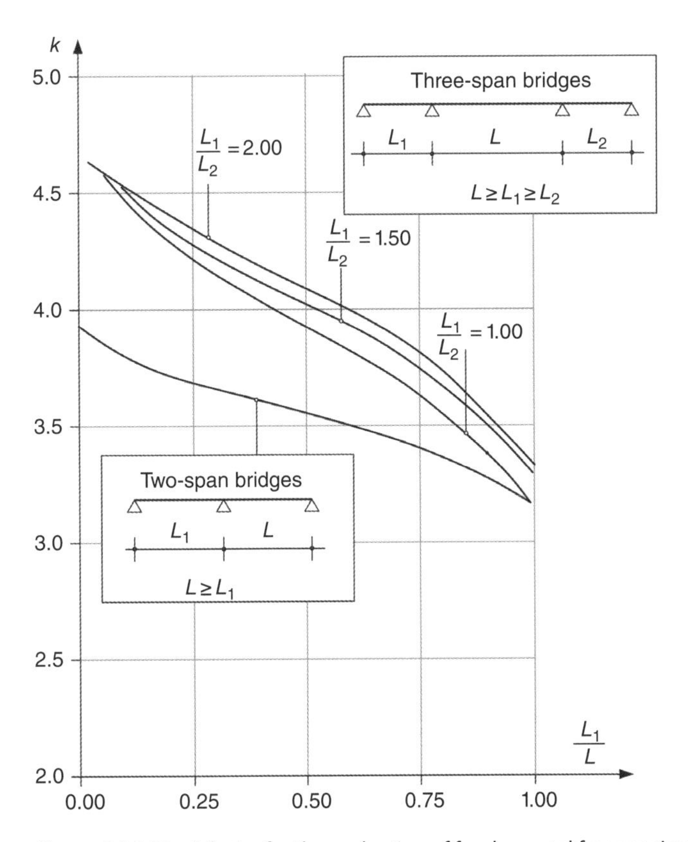

Figure 6.96 The k factor for the evaluation of fundamental frequencies of two- or three-span continuous decks (*Source:* Adapted from EN 1991-2).

where δ_{max} is the maximum vertical deflection under permanent loads, $g = 9.81$ m s^{-2} is the acceleration of gravity and the coefficient 1.1 is an adjusted coefficient to numerical results on a variety of decks. An order of magnitude of road deck vibration frequencies, for multiple span bridges with span lengths $L(m)$, may be estimated by an empirical formula:

$$f_1 = \frac{L}{100} + 0.5 \; [\text{Hz}] \tag{6.152}$$

Torsion vibration frequencies, without involving pier transverse deformability (which should be considered in some cases, in particular for tall piers) may be evaluated by considering every span as independent and built in at the pier supports if that is the case for torsion rotations. In the case of a single bearing (Figure 6.63), these torsion rotations are free and the span length should be considered between sections where torsion rotations are effectively prevented. If the total deck length between those sections is still defined by L the torsion frequency of the deck is given by

$$f_t = \frac{1}{2L}\sqrt{\frac{GJ}{I_0}} \tag{6.153}$$

where GJ is the uniform (St. Venant) torsion stiffness of the cross section and I_0 is the mass polar section moment of inertia $I_0 = \rho(I_y + I_z)$ with ρ being the mass density of the deck. Note that these formulas have been defined for constant deck cross section properties, which in practice is not usually the case. An average value or a weighted average value should be adopted to improve the accuracy of the approximate formulas. One aspect that should be taken into account is the role of the additional permanent loads contributing to m but not to EI or GJ.

Pedestrian PPC Bridges – Maximum deflections, traffic vibrations or wind induced vibrations are not usual problems in PPC road bridges. On the contrary, design of pedestrian bridges, even in PPC, is usually governed by limits on maximum vertical and horizontal accelerations of the deck. In pedestrian bridges, when the vertical fundamental frequency exceeds 5 Hz ($f_{1v} > 5$ Hz), maximum vertical accelerations are not a main issue. The same holds for horizontal vibrations when $f_{1h} > 2.5$ Hz. However, that is not usually the case, at least for medium to long span lengths because to reach a vertical frequency of 5 Hz, the superstructure usually has too heavy appearance and is aesthetically unacceptable. For medium comfort, maximum vertical accelerations are usually limited to 0.5–1.0 m s^{-2}. One may limit accelerations as a function of the natural bending vertical frequency f_v to $a_{\text{lim}} = 0.5\sqrt{f_v} < 0.7\,\text{m s}^{-2}$ [39].

This is a severe limitation on vertical maximum accelerations, very often not easily satisfied in practice. In any case, maximum vertical accelerations above 1 m s^{-2} are not usually acceptable for comfort of the users. The vertical accelerations a_{max} are evaluated under a pedestrian traffic loading case, usually inducing the most adverse effects.

As an approximation, a_{max} may be evaluated by [40]

$$a_{\max} = 4\pi^2 f_v^2 w_0 / (2\zeta) \tag{6.154}$$

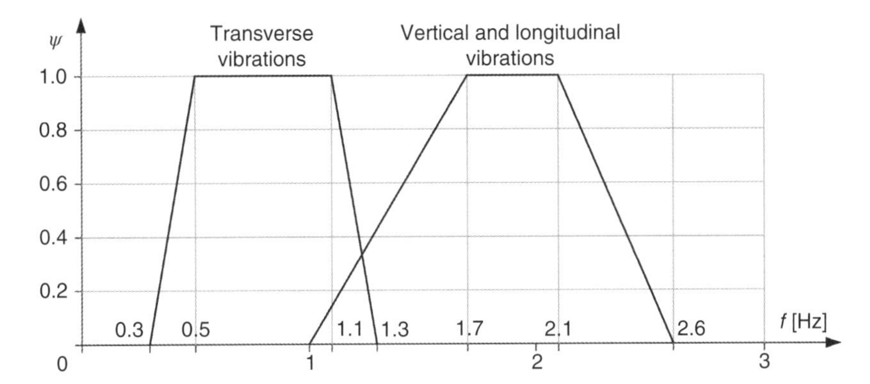

Figure 6.97 Reduction factor for evaluating the equivalent number of users inducing vibrations in a pedestrian bridge deck.

where w_0 is the maximum deck deflection under an equivalent load case inducing the most adverse effects for vertical accelerations. Pedestrian loads are composed by a static part G_0 (taken usually as 700 N) and a dynamic component made of a sum of harmonic functions with frequencies that are multiple of the action frequency. Pedestrians' vertical frequency actions tend to be between to 2 and 3.5 Hz, while imposed horizontal exciting frequencies are approximately 50% of the vertical frequencies, that is, between 1.0 and 1.75 Hz. Based on a harmonic analysis by Fourier series, it is shown [34, 40] this formula for vertical accelerations holds for w_0 evaluated for an equivalent uniformly distributed load

$$q_1 = \psi \, G_1 \, N_{eq} / L \tag{6.155}$$

where N_{eq} is the number of pedestrians on the bridge (depending on density of traffic considered), and $\psi \leq 1.0$ is a reduction factor (Figure 6.97) to take into account that not all pedestrians walk simultaneously to induce resonance effects on the deck. The load G_1 is the force amplitude of the first harmonic; usually taken as $G_1 = 0.4G_0 = 280$ N for $f_p \cong 2$ Hz – frequency of the force induced by pedestrians.

For low or normal pedestrian traffic $N_{eq} = 10.8\sqrt{(N\zeta)}$ where N is the total number of pedestrians on the bridge and ζ the damping coefficient, usually taken as 0.7 for concrete pedestrian bridges. To evaluate N, the pedestrian density is taken as d = 0.8 pedestrian/m^{-2} for dense traffic and d = 0.5 pedestrian/m^{-2} for low density traffic. For high density traffic, d = 1 pedestrian/m^{-2} and $N_{eq} = 1.85\sqrt{N}$.

As mentioned, pedestrian traffic tends to induce horizontal transverse acceleration on the bridge deck, which has been recently the cause of large unexpected vibrations in some bridges. The maximum horizontal accelerations should be limited usually to 0.2 m s^{-2} under normal traffic and 0.4 m s^{-2} for exceptional crowd conditions. Of course, the horizontal action forces only have a dynamic component $F(t) = G_1 \sin(2\pi f_p t)$ where one may adopt $G_1 = 0.2\ G_0 = 140$ kN and f_p is the frequency of the force induced by pedestrians.

Very often, slender pedestrian bridges cannot satisfy the required maximum vertical or horizontal accelerations as specified previously. One should say this is more often the case for steel or composite pedestrian bridges than prestressed concrete bridges. In any case, if that happens and if increasing deck stiffness is ruled out, the solution is to adopt a Tuned Mass Damper (TMD) [34, 40, 41]. One important aspect when checking this necessity is the real damping ratio displayed by the structure. Very often, the assumed design values are too conservative. One possibility is to consider the TMD at the design stage, but adopt it only after measuring on site the real damping ratio after construction. Often TMDs are adopted if justified after a dynamic site test has been done.

Rail PPC Bridges – Rail PPC bridges should also be checked for deformations and excessive vibrations. Excessive deflections (vertical or horizontal) induce problems in the track geometry that may cause derailment and (or) excessive rail stresses. The maximum vertical deflection along any track due to rail traffic actions, defined by the characteristic load values, is limited to $L/600$, with L being the span length. The maximum twist t (mm m^{-1}) of a track gauge s(m) measured on a standard length l, should also be checked to avoid excessive deformations. For an European track gauge $s = 1.435$ m in a standard length $l = 3$ m, $t = 3$ mm is recommended for train speeds between 120 and 200 km h^{-1} (Figure 6.98).

Excessive vibrations may cause ballast instability or discomfort for passengers. Recommended maximum values for vertical accelerations have been proposed in Annex A2 from EN1990 [39] as 3.5 m s^{-2} for ballasted tracks and 5 m s^{-2} for track direct fixation. Discussions on these limits can be found in [42].

Passenger comfort is also controlled as a function of maximum vertical accelerations b_v inside the coach. Depending on the level of comfort, desired limits for b_v are established. For an indirect approach, these limits may also be related to maximum vertical deflections as a function of span length L [6] and design speed V (km h^{-1}). As an order of magnitude for a very good level of comfort in a multiple span continuous deck, deflections are limited for $V = 200$ km h^{-1} to values between $L/1350$ and $L/600$. For example, for $V = 200$ km h^{-1}, medium span lengths in the order of 40 m, the allowable limit reaches approximately the maximum allowable value of $L/1350$, while for a $L = 100$ m the allowable deflections may reach $L/600$.

Fatigue – Fatigue is not a usual issue in prestressed concrete bridges. Fatigue in reinforced and prestressed concrete road and railway decks is, even so, considered in EN 1992-1 [7]. However, since fatigue loadings are usually associated with values of frequent live loads and not characteristic loads, stress amplitudes in prestressing cables and steel reinforcement of prestressed concrete decks is usually very low because the section is working at SLS usually in state I – uncracked. This means that under bending actions stress

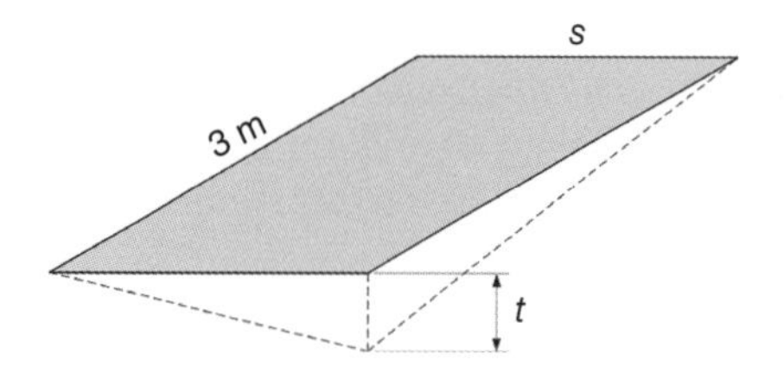

Figure 6.98 Definition of maximum twist t between track gauge (s) in railway bridge decks.

variations are mainly taken by the concrete part of the section since its contribution for the bending stiffness is much larger than that of the cables or even the steel reinforcement.

The only important issue is fatigue induced by vibration of external prestressing cables under the effect of moving loads, particularly in railway bridges. If the frequency of the free length of tendons between deviators or anchor points is sufficiently different from the fundamental frequency of the deck, the fatigue problem in external cables may be neglected. That frequency (in Hz), for a free tendon length, L_t, with mass m_t per unit length and installed force, F, is given by the usual formula for vibration of strings, that is,

$$f = \frac{L_t}{2}\sqrt{\frac{F}{m_t}} \tag{6.156}$$

It is usual to limit the free tendon length between deviators to 9–12 m.

6.9 Steel and Composite Decks

6.9.1 Generalities

Design concepts and erection methods for steel and steel-concrete composite decks have been discussed in previous chapters. General methods of analysis for determining internal forces, transverse load distribution and structural behaviour have been discussed in the present chapter. Design of steel and composite decks is based on general ULS and SLS criteria discussed in Chapter 3 with specific aspects considered in this section.

6.9.2 Design Criteria for ULS

Steel-concrete composite bridge decks may be designed elastically or according to an elasto-plastic criterion. For bridge decks, elastic structural analysis to determine internal forces should be done, although section analyses may be done elastically or based on an elasto-plastic model.

Most bridge sections are made of slender plate elements, at least for the webs, and this restrains the fully plastic design of the cross sections. Even when the slenderness of the plate elements is sufficiently low to allow a fully plastic design, care should be taken with two basic problems:

- rotation capacity of cross sections to allow redistribution of internal forces;
- fatigue resistance under moving road and rail traffic.

That is why in most bridge decks it is not possible to take advantage of full plastic resistance, allowed for Class 1 or 2 sections, according to class limits in EN 1993-1-1 [4]. Class limits defined for plate elements of the deck cross sections (flanges and webs) in EN 1993-1-1, illustrated in Figure 6.99 with a moment-curvature relationship, are related to the section capacity to reach a plastic limit state of bending section resistance without being affected by local buckling. Designing in post-buckling range is allowed, namely for plate girders and box girder webs and compressed flanges of box girders as

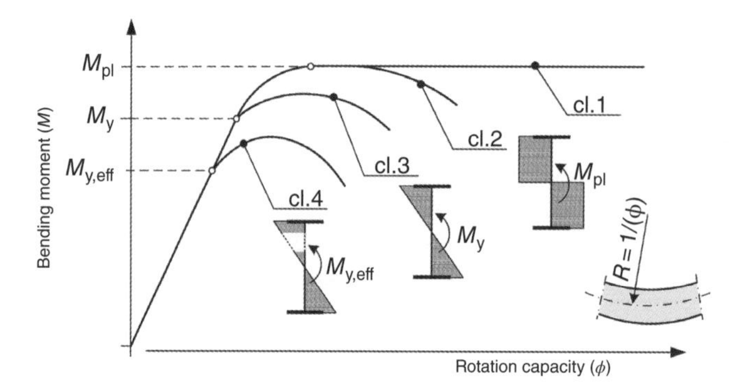

Figure 6.99 Cross section classes of steel girders under bending moment (*M*) – rotation capacity (ϕ) diagram.

well. However, the maximum normal stress at the effective section should be limited to the yield stress. Flanges of plate girder sections may also be designed to go into the local post-buckling range, but the economic advantage of this option is not usually very large. In this case, it may be more economic to reduce the width of the flange and to increase its thickness, preferably without entering in thickness values that are too large with potential brittle fracture problems under low temperature events, as discussed in Section 4.5.2.1.

The full plastic moment (M_{pl}) is feasible in Class 1 sections without any restriction in the rotation capacity, allowing redistribution of bending moments between cross sections if a rigid plastic or elasto-plastic analysis is done for global structural analysis. Class 2 sections have sufficient resistance to reach M_{pl} but limited rotation capacity; local buckling can occur at M_{pl} under increased rotations.

Class 3 sections have only capacity to reach the elastic limit moment (yield moment M_y) without any local buckling effects. When the sections integrate slender plate elements, in webs or flanges, they may have capacity to reach M_y in the post-buckling range and should be classified as Class 4 sections. Limits between section classes are defined in terms of slenderness of plate elements. Since that is dependent on the local buckling stress depending on the thickness/width ratio (t/b) of the element, class limits given in [4] are established in terms of that ratio. The reader is referred to specific steel structure design references for additional information on class section limits and elasto-plastic design [43,44].

In principle, composite sections in bridge decks should be designed according to the section model as shown in Figure 6.100 where a partial plastification at ULS in the tension zones is allowed in Class 3 sections. The strains should be limited at the lower flange that has reached the yield stress. This is acceptable in some codes without specifying any limit for the maximum strain [4] at the section or with a limit of two to four times the yield strain ε_y [43]. For Class 4 sections, for example, when the web is in the post-buckling range, a partial plastic state in the tensile zones as shown in Figure 6.100, is not explicitly allowed in the present Eurocodes. If so, maximum stresses and strains, even in tensile zones, should be limited to yield stress values. In short, and concerning composite sections, the following design criteria are referred to with recommended material safety factors from the current Eurocodes:

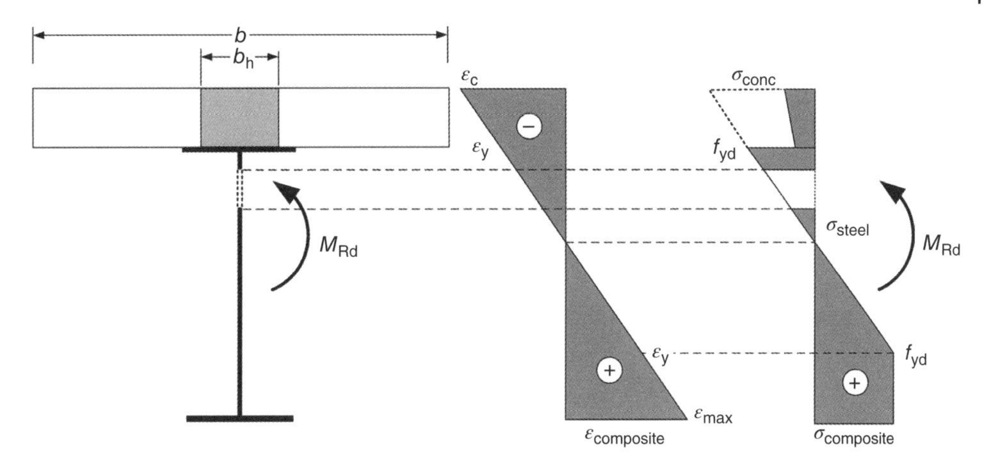

Figure 6.100 Composite bridge girder section at ULS allowing a partial plastification at the lower flange and the web at the post-buckling range.

Maximum stress in steel part of the cross section	$\sigma_{\max}^{a} = \dfrac{f_{yk}}{\gamma_{ma}} = f_{yd}; \gamma_{ma} = 1.1 \text{ or } 1.0$	(6.157)
Concrete assumed without any resistance under tensile stresses and with maximum compressive stresses	$\sigma_{max}^{c} = 0.85\dfrac{f_{ck}}{\gamma_{mc}} = 0.85 f_{yd}; \gamma_{mc} = 1.5$	(6.158)
Maximum stresses in steel reinforcement	$\sigma_{max}^{s} = \dfrac{f_{sk}}{\gamma_{ms}} = f_{sd}; \gamma_{ms} = 1.15$	(6.159)
Maximum stresses in prestressing tendons (if existing)	$\sigma_{max}^{sp} = \dfrac{f_{pk}}{\gamma_{ms}} = f_{pd}; \gamma_{ms} = 1.15$	(6.160)

Under negative bending moments, the bottom flange is under compressive stresses and it may buckle laterally. The maximum compressive stress at the bottom flange should be limited to a stress $\sigma_{LT,Rd} < f_y$ where $\sigma_{LT} = M_{LT,Rd}/W_{e,inf}$, with $M_{LT,Rd}$ being the design resistant bending moment for lateral torsional buckling and $W_{e,inf}$ the elastic section modulus for the bottom fibre. The top steel flange is restricted laterally by the deck slab and lateral buckling under positive bending moments is excluded, unless in the construction stages where it should be verified between lateral bracings. The support sections should be designed elastically with the bottom stresses restricted as referred to before and with the slab restricted to the steel reinforcement. The ultimate bending moment is then determined from the stress diagram as shown in Figure 6.101. A neutral axis should be determined first under the condition of $N = 0$ at the section.

6.9.3 Design Criteria for SLS

At SLS, steel and composite bridges should have an elastic behaviour, avoiding excessive yielding and excessive deformations. Limiting deflections may avoid unexpected dynamic (impact) load effects or fatigue damage caused by *resonance*. Natural vibration

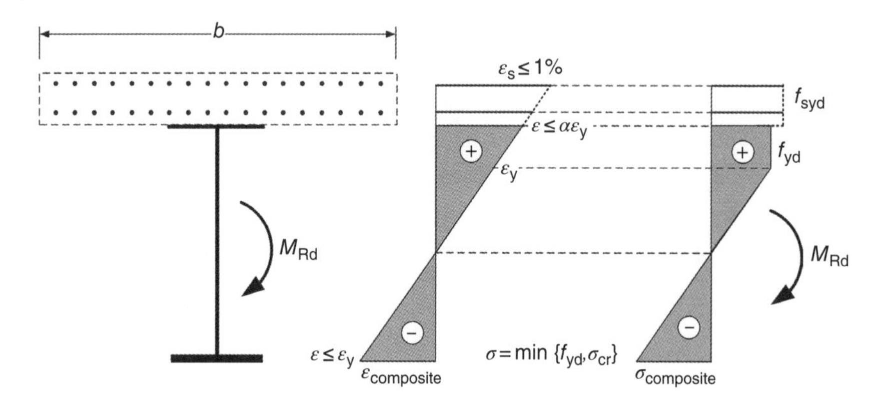

Figure 6.101 Strains and stresses at ultimate bending resistance under negative bending moment of a composite section allowing a partial plastification ($\alpha < 2$ to 4) of the tensile zone of a class 3 cross section (maximum stresses at the compressed lower flange limited by lateral torsion buckling).

frequencies of bridge elements and of the overall bridge deck should be checked at concept design and wind induced vibrations should be appraised.

The slenderness of web plates should be limited to avoid *web breathing* (see Section 4.5.4.2.2) and stiffness reduction that may increase deflections.

In composite bridge decks, crack width limitation in the deck slab to values of 0.2–0.3 mm is usually recommended at the SLS characteristic load combination. An important aspect in steel and composite bridges refers to easiness of maintenance and repair, with due account drainage, corrosion protection and replacement of bearings and expansion joints.

6.9.3.1 Stress Limitations and Web Breathing

Stresses at SLS should be limited to elastic limit values, namely for the characteristic load combination. Normal and shear stresses should be limited to the correspondent design values of the yield stresses. The material safety factors γ_M are usually taken as 1.0 for these verifications and hence $\sigma \leq f_y$ and $\tau < \tau_y = f_y/\sqrt{3}$. For combined stress states, the von Mises yield criterion is adopted with normal ($\sigma = \sigma_x$) and shear stresses ($\tau = \tau_{xz}$) evaluated for the *characteristic load combination* and usually with $\gamma_{M,ser} = 1.0$

$$\sqrt{\sigma_{Ed,ser}{}^2 + 3\tau_{Ed,ser}^2} \leq f_y/\gamma_{M,ser} \tag{6.161}$$

This expression may be rewritten as an interaction equation,

$$\sqrt{\left(\frac{\sigma_{Ed,ser}}{f_y}\right)^2 + \left(\frac{\tau_{Ed,ser}}{\tau_y}\right)^2} \leq 1.0 \tag{6.162}$$

Sometimes, normal stresses in the perpendicular direction σ_y are relevant, like at the crossing flanges of main and secondary girders (like *ribs* – see Figure 4.49) or in flange of box girders and the Eq. (6.162) should incorporate these stresses [5].

Web breathing was already discussed in Chapter 4 (Section 4.5.4.2.2). A simplified procedure limiting the slenderness (d/t) to avoid web breathing stress verifications was given for road and rail bridges. If the limits are exceeded, stresses in the webs for the *frequent load combination* should be verified from the following expression adapted from [6], and quite similar to the yield criterion expression but with the yield stress replaced by the elastic critical stresses of the web in bending and shear,

$$\sqrt{\left(\frac{\sigma_{Ed,ser}}{\sigma_{cr}}\right)^2 + \left(\frac{1.1\tau_{Ed,ser}}{\tau_{cr}}\right)^2} \leq 1.1 \tag{6.163}$$

6.9.3.2 Deflection Limitations and Vibrations

These aspects have been already discussed in Section 6.8.8 for prestressed concrete road and rail bridges. However, some specific aspects should be highlighted for steel and composite bridges. Permanent deflections, or permanent plus one part of live load deflections, should always be compensated for by *precamber*. Deflections in steel or composite bridges are usually far more relevant than in prestressed concrete bridges. An appraisal of the relevance of possible excessive deformability of the bridge, which is likely to induce associated problems of excessive vibration, induced impact loading effects due to traffic, drainage problems, visual impact of excessive deformations or, in short, any type of bridge deck problems associated with bridge performance under service, may be checked at the conceptual design phase. Deck live load deflections in the order of 1/300 of the span length due to characteristic live loads may be tolerated, specifically in large spans. As a basic rule, under frequent values of the live loads, deflections for road bridges are limited to values of 1/1000 of the span length. Rail bridges are usually designed to limit deflections to 1/600 of the span length under traffic rail loading as per [39].

When evaluating maximum deflections slipping effects at bolted connections should be taken into consideration. Usually, as a design rule, bolted connections in steel bridges should be done with prestressed bolts in order to avoid any slipping at SLS. Alternatively, fitted bolts or injected bolts may be adopted but fabrication and (or) execution is usually more difficult. In truss bridges, slipping effects at connections are usually quite relevant for medium and long spans at least and prestressed bolting should be adopted designed as slip resistance connections for the frequent load combination.

Vibrations in steel and composite pedestrian bridges should be limited based on maximum values of pedestrian induced accelerations as per the criteria discussed for prestressed concrete bridges (Section 6.8.8).

6.9.4 Design Criteria for Fatigue Limit State

Fatigue design is generally a leading ultimate limit state verification in steel and composite bridges, particularly in rail bridges. Fatigue design is a main issue from the conceptual design stage, since appropriate detailing should be envisaged in order to have as much as possible a high fatigue resistance. It is not possible to discuss all basic and main aspects of fatigue in steel structures here; the reader is referred to specific references on the subject [44, 45]. However, in short, one may say fatigue design should be considered in steel and composite bridge design on the following basis:

- define the fatigue loading;
- identify the critical details;
- define the *fatigue resistance classes* (FAT detail) for the selected details;
- check fatigue resistance by either a simplified *damage equivalent method* for constant stress amplitudes or by *damage accumulation method.*

Fatigue occurs under stress variation due to live loads in road or rail bridges. Under stress concentration effects at some locations, a crack may be initiated as discussed in Chapter 4 (Section 4.5.2.1, Figure 4.41) and propagated under certain conditions and of repeated stress variations with a given number of cycles.

Let us take the example of a continuous girder bridge, Figure 6.102 under a set of moving loads Q_i, entering the bridge at different instants t_i and leaving the bridge at time $t_i + \Delta t_i$. If one takes the stresses, for example, at the lower flange of the steel girder, each of these moving loads will induce a maximum tensile stress $\sigma_{max} = \sigma_G + \sigma_{Qi\ (S1)}$ when the load is at S1 and a minimum stress $\sigma_{min} = \sigma_G + \sigma_{Qi\ (S2)}$ when the load is at S2. The *mean stress* is the permanent stress $\sigma_m = (\sigma_{max} + \sigma_{min})/2$, and *the stress range* is $\Delta\sigma = \sigma_{max} - \sigma_{min}$. Each moving load Q_i may induce n_i cycles during the life of the structure. Fatigue resistance under *constant amplitude stress ranges* is the basic reference to evaluate *fatigue under variable amplitude stress ranges* as occurs in practice, Figure 6.103.

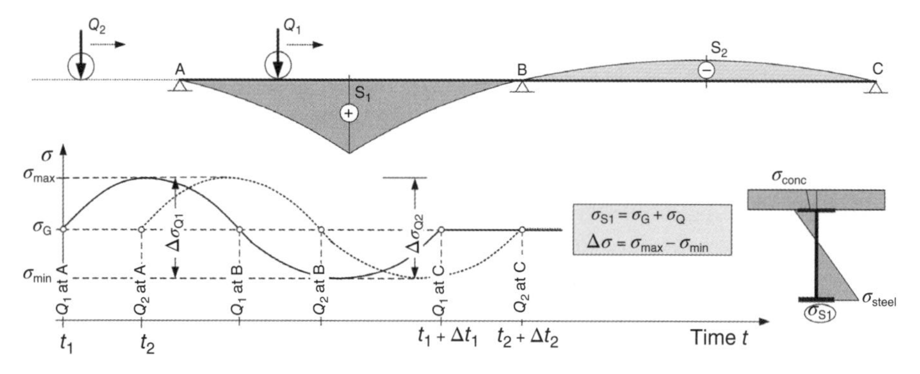

Figure 6.102 Stress amplitudes in the lower flange of a bridge girder at the mid-span section, due to moving loads Q_i entering the bridge at time t_i and leaving at time $ti + \Delta t_i$.

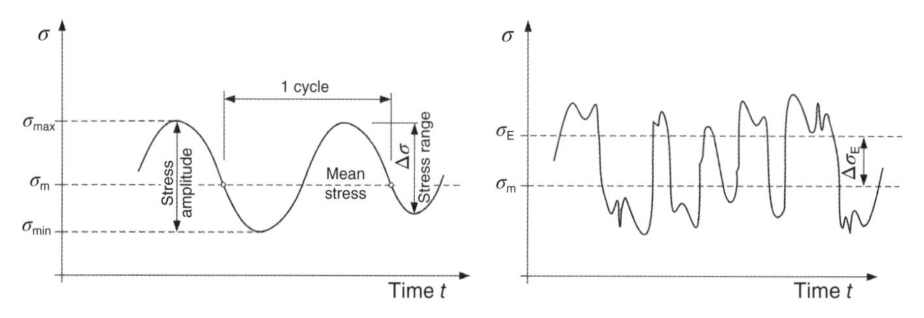

Figure 6.103 Stress amplitude, stress range $\Delta\sigma$, stress cycles and equivalent constant amplitude stress cycles $\Delta\sigma_E$ definitions.

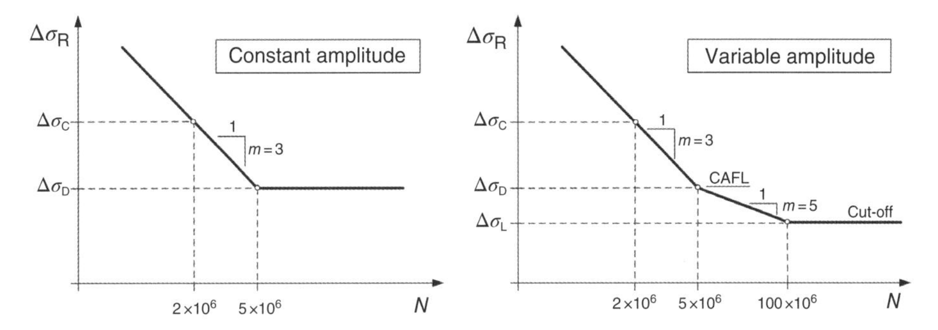

Figure 6.104 Fatigue resistance *S-N* curves defining the stress range resistance in terms of the constant amplitude cycles *N* in a log scale.

The *fatigue resistance* under constant amplitude stress ranges of a certain bridge detail (e.g. given the stress at a certain point of the lower flange) is defined by a Wohler curve (*S-N curve*) where the resistance $\Delta\sigma$ is given in terms of the total number of cycles *N*, according to the relationship

$$\Delta\sigma^{\mathrm{m}} N = Cte \tag{6.164}$$

where the exponent ^{m} is usually an integer. This relationship may be rewritten in a logarithmic form, as Log $\Delta\sigma$ = $(1/m)$ (logCte − log N) = $C - (1/m)$ log$N$, represented in Figure 6.104 for $m = 3$ up to $N = 5$ million cycles and $m = 5$ for $N <$ between 5 and 100 million cycles. The stress resistance at 2 million cycles $\Delta\sigma_C$ is defined as a reference to designate the *fatigue resistance curve* for a certain detail. Hence, Eq. (6.164) may be used to define the number of cycles (*Endurance*, N_R) for a certain stress range $\Delta\sigma_R$ as

$$\Delta\sigma_R^m N_R = \Delta\sigma_c^m 2\times10^6 \text{ with } m=3 \text{ for } N \le 5\times10^6 \tag{6.165}$$

For $N_R = 5$ million cycles, one obtains from Eq. (6.165) the *Constant Amplitude Fatigue Limit* (CAFL) as $\Delta\sigma_D = (2/5)^{(1/3)}\ \Delta\sigma_C = 0.737\ \Delta\sigma_C$. For $N = 100$ million cycles, one obtains the *cut off limit* as $\Delta\sigma_L = (5/100)(1/5)(2/5)(1/3)\ \Delta\sigma_C = 0.404\ \Delta\sigma_C$. Fatigue *S-N* curves are defined for a variety of details in the codes, such as the ones presented in Figure 6.105 for tatigue details (FAT) in [45]. A variety of details common in steel and composite decks are shown in Figure 6.106. If one takes the highest *S-N* curve, that is, FAT = 160, this curve applies only to a stress range $\Delta\sigma$ in a typical 'free' point of a flange of a hot rolled section. For a typical point of flange of a welded bridge section, as previously referred to in the example of Figure 6.102, for automatic fillet welds between the flange and the web, the FAT detail is lower (FAT 125). It should be noted that for the curve FAT 125, one has $\Delta\sigma_R = 125\,\mathrm{N\,mm^{-2}}$ at $N = 2\times10^6$ as per the *S-N* curve definition. On the other hand, if one takes a point at the girder flange nearby a transverse stiffener with flange, one has, according to Figure 6.106, FAT varying between 56 and 80 depending on the stiffener's flange length *l*. If one cuts the flange of the stiffener like the one shown in Figure 6.107, one has the maximum fatigue resistance FAT 80 with $l < 50$ mm, where *l* is the stiffener's web thickness plus weld width.

In practice, variable stress ranges occurs under traffic, as shown in Figure 6.103. If the stress cycles n_i are counted for each stress range $\Delta\sigma_i$ a *load spectrum* may be defined as

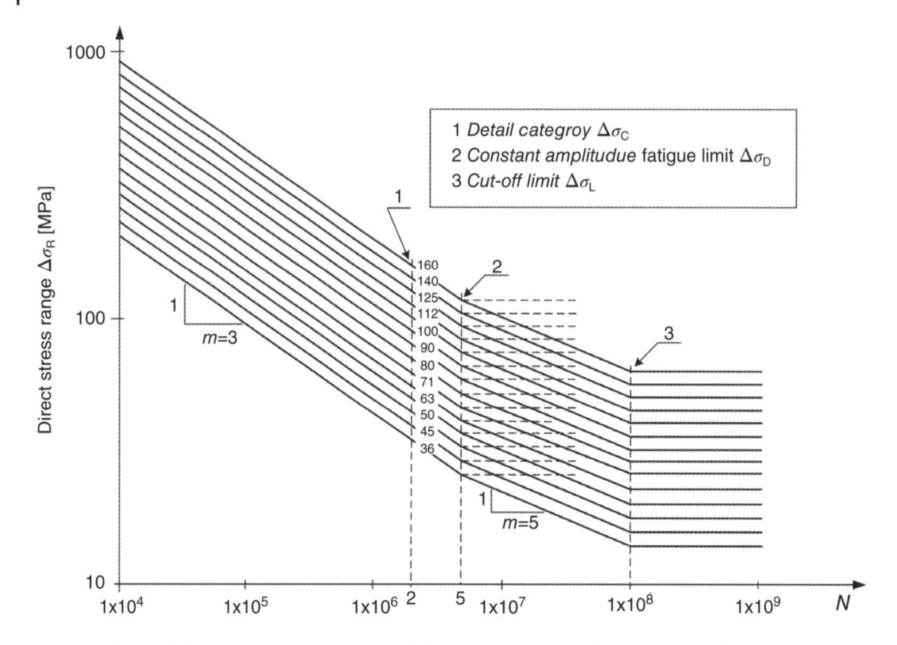

Figure 6.105 Fatigue resistance curves in EN 1993-1-9 [46] for FAT 36 to FAT 160.

shown in Figure 6.108. Each cycle stress range $\Delta\sigma_i$ can be associated in to an endurance N_i in the variable amplitude *S-N* curve as shown in Figure 6.109 and a *fatigue damage* n_i/N_i may be computed. The damage accumulation is usually evaluated by the linear Palmgren Miner rule [45]) allowing to verify fatigue resistance by the condition $\Sigma(n_i/N_i) \leq 1.0$.

To apply the damage accumulation method, a set of fatigue load models is defined for road and rail bridges [6]. A simpler procedure to avoid a damage accumulation method consists in defining a Fatigue Loading by a single vehicle as the Fatigue Load Model 3 (FLM3) defined in EN 1991-2 (Figure 6.110) and evaluate the stress range $\Delta\sigma$ for this vehicle.

Then this $\Delta\sigma$ is converted to an equivalent stress range $\Delta\sigma_{E2}$ at two million cycles by introducing a *damage equivalent factor* λ as:

$$\Delta\sigma_{E2} = \lambda\phi\Delta\sigma \tag{6.166}$$

where $\Delta\sigma$ may be amplified by a dynamic load factor ϕ (usually $\phi = 1.0$ for road bridges and $\phi > 1.0$ for rail bridges). The λ factor depends on the length of the influence line for the stress under consideration, on the yearly density of traffic, on the fatigue design life intended for the bridge and on traffic in other lanes than the one where the single loading vehicle is considered. A limit λ_{max} is established in terms of the span length, see Figure 6.111. The reader is referred to (EN 1991-2) [6] for the calculation of λ factors for road and rail bridges.

Knowing $\Delta\sigma_{E2}$ by Eq. (6.166), the fatigue load resistance verification is based on $\Delta\sigma_C$ values determined from the constant amplitude *S-N* curve for the specific FAT detail

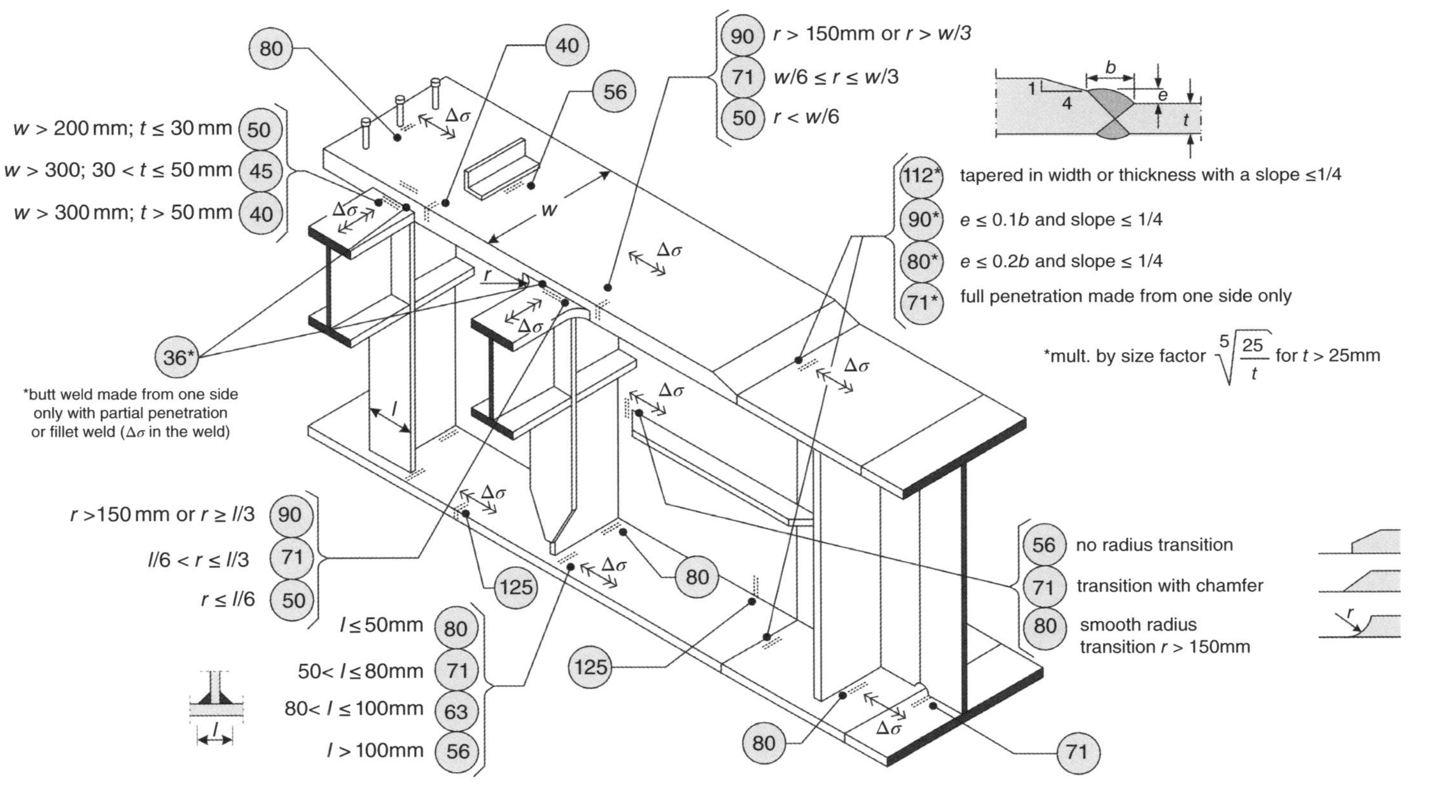

Figure 6.106 Usual FAT detail categories (in MPa) for steel bridge decks according to EN 1993-1-9 [46].

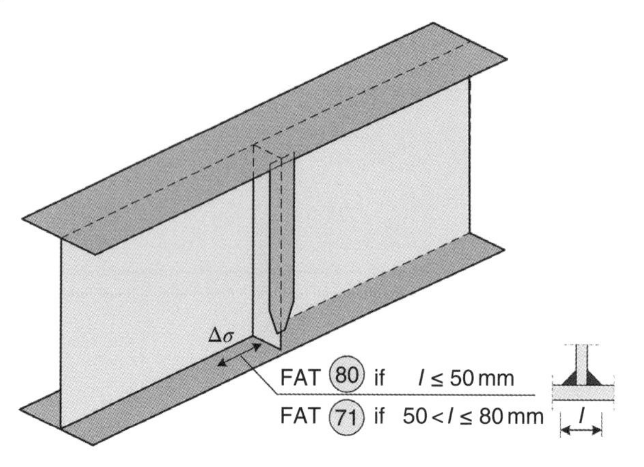

Figure 6.107 Fatigue detail for a transverse stiffener attachment to a tensile flange of a welded bridge girder.

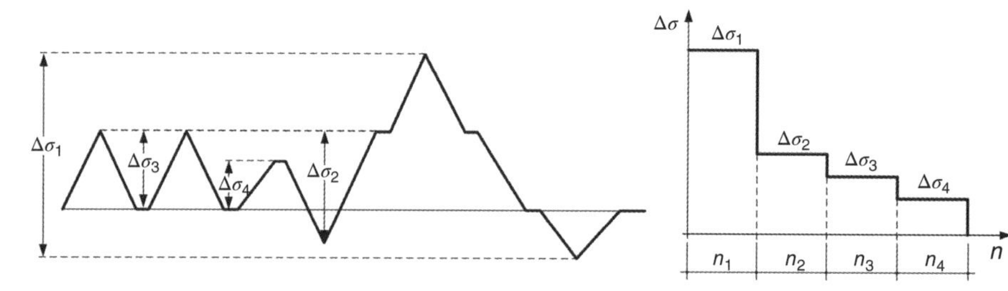

Figure 6.108 Fatigue stress ranges defined through a loading spectrum in terms of the number of cycles n_i for each stress range $\Delta\sigma_i$ (*Source:* Adapted from EN 1993-1-9 [46]).

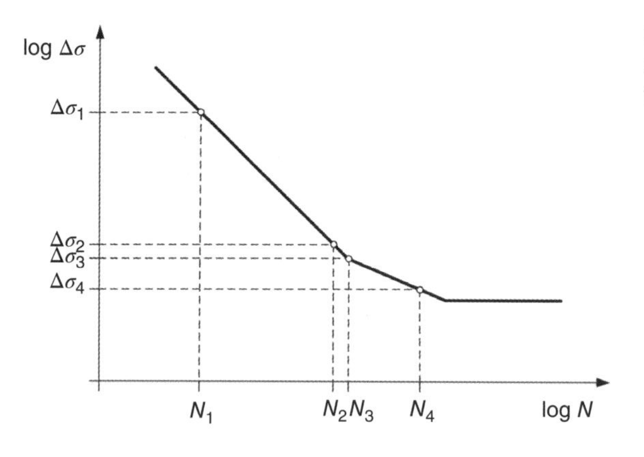

Figure 6.109 Definition of endurance N_i for each stress cycle of the load spectrum.

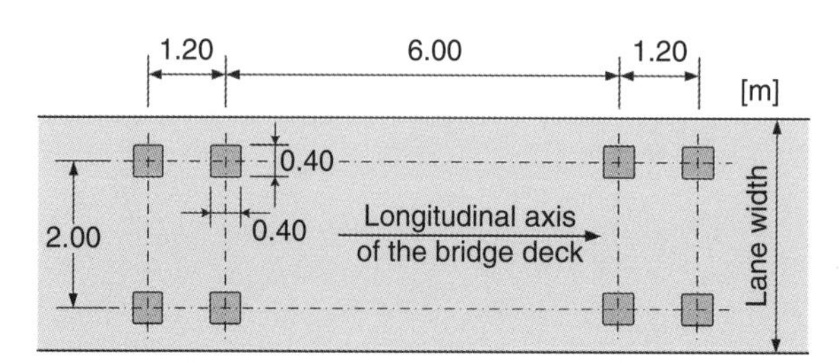

Figure 6.110 Fatigue Load Model 3 (FLM3) in EN 1991-2 [42] – load per axis 120 kN.

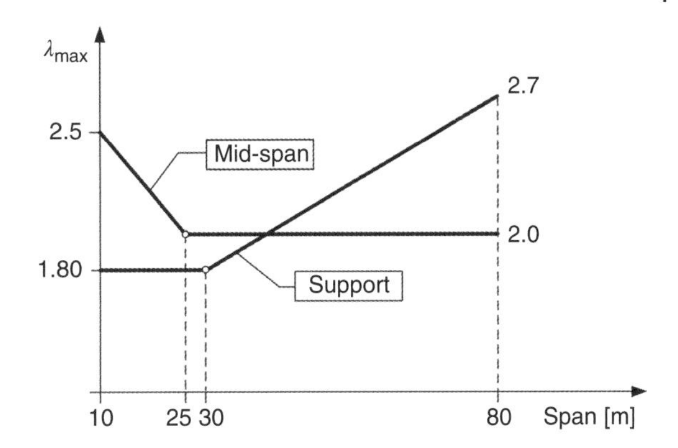

Figure 6.111 Maximum values of damage equivalent factors for road bridges in EN 1993-2 [6] in terms of the span lengths or average value at the support.

under consideration. With due account for fatigue safety factors predicted in the codes, the following condition should be verified:

$$\gamma_{Ff}\,\Delta\sigma_{E2} \le \Delta\sigma_C / \gamma_{Mf} \tag{6.167}$$

where γ_{Ff} = 1.0 in EN 1990-A2 [39] and γ_{Mf} varies in general between 1.0 and 1.35 depending on consequences of fatigue failure in the element and basically on the frequency of inspections predicted for the bridge.

6.9.5 Web Design of Plate and Box Girder Sections

Conceptual and pre-design of plate girders and box girder bridges was presented in Sections 4.5.4 and 4.5.5. Specific aspects that concern web design verifications are considered here for webs with transverse stiffeners only. The case of webs with longitudinal stiffeners is considered in the next section.

6.9.5.1 Web Under in Plane Bending and Shear Forces

Figure 4.60 shows the post-buckling behaviour of web panels under normal stresses due to bending and under shear stresses. The *elastic critical stress* of a web of a plate girder or box girder under in plane bending only is determined as a function of the stress ratio $\psi = \sigma_2/\sigma_1$ by:

$$\sigma_{cr} = k \frac{\pi^2 E}{12\left(1-\nu^2\right)}\left(\frac{t}{d}\right)^2 \tag{6.168}$$

with the buckling coefficient k given in Annex A. Webs are usually assumed conservatively hinged at the flanges and are considered as long plates with two simply supported edges and loaded at the transverse end edges only. For an equal flange girder (which is quite unusual in plate girder bridges) one has $\psi = \sigma_2/\sigma_1 = -1$ and assuming simply supported longitudinal edges, $k = 23.9$. The non-dimensional plate slenderness is defined by:

$$\bar{\lambda}_p = \sqrt{\frac{f_y}{\sigma_{cr}}} \tag{6.169}$$

and the *effective width* (Figure 4.61) of the web is obtained by $d_{\text{eff}} = \rho\ d_{\text{c}}$ using the *reduction factor* ρ [5] given by

$$\rho = \frac{\bar{\lambda}_p - 0.055(3-\psi)}{\bar{\lambda}_p^2} \quad \text{for } \bar{\lambda}_p > 0.673 \tag{6.170}$$

If $\bar{\lambda}_p \leq 0.673$ the web is fully effective, that is, $\rho = 1.0$. In Eq. (6.170) $(3-\psi) > 0$. For $\psi = 1.0$ – web under uniform compression, this is effective with the formula defined from the ρ factor and is reduced to the classical Winter formula [26, 27, 47] for plate effective widths.

The resistance of the section under bending is taken for the *effective section* composed by the flanges and the effective parts of the web (Figure 4.61) and the ultimate moment in elastic design is $M_u = W_{eff} f_y$, where W_{eff} is the section modulus of the effective cross section referred to the extreme fibre where the maximum bending stress occurs as shown in Figure 4.61.

For webs of plate or box girders, assuming they resist predominantly to shear, they reach the elastic shear critical stress at a value

$$\tau_{cr} = k_\tau \frac{\pi^2 E}{12(1-\nu^2)} \left(\frac{t_w}{h_w}\right)^2 \tag{6.171}$$

where the plate buckling coefficient is determined as a function of the geometry of the web panels between transverse stiffeners a/h_w (Annex A).

The non-dimensional slenderness of the web panels under shear is defined (similarly to the one defined for in plane bending) by:

$$\bar{\lambda}_w = \sqrt{\frac{\tau_y}{\tau_{cr}}} \tag{6.172}$$

where $\tau_y = f_y/\sqrt{3}$ is the yield shear stress according to von Mises criterion. As mentioned in Chapter 4, slender webs have a considerable post-buckling resistance under shear forces. They reach the ultimate shear force resistance at values much higher than the elastic critical value for the shear force $V_{cr} = \tau_{cr}\ h_w\ t_w$. In the post-buckling range the ultimate shear stress of the webs is given by $\tau_{ult} = \chi_w \tau_y$ where the factor χ_w represents the influence of buckling effects. If τ_{ult} is τ_{cr}, then $\chi_w = 1/\bar{\lambda}_w^{\ 2}$ by definition of the non-dimensional slenderness parameter. The ultimate shear force V_u for a web panel is lower than the plastic resistance value $V_{pl} = \tau_y\ h_w\ t_w$ but for *slender webs* (say $\bar{\lambda}_w > 1.1$), it is larger than the elastic critical value (this post-buckling stiffness of web panels has been accounted for since a long time ago in steel structure design). The reduction with respect to the plastic resistance is defined by the *web buckling factor* χ_w as a function of the web slenderness $\bar{\lambda}_w$. The buckling resistance of web panels has been the subject of numerous investigations and a variety of models have been proposed [48]. One of the latest models, adopted in [5], is the 'rotated stress field method' [47, 48]. Hence, for stiffened web panels, assuming *rigid transverse stiffeners* according to definition given in next section, the contribution of the web for ultimate design shear resistance is given by

$$V_{w,Rd} = \chi_w V_{pl}/\gamma_{M1} \tag{6.173}$$

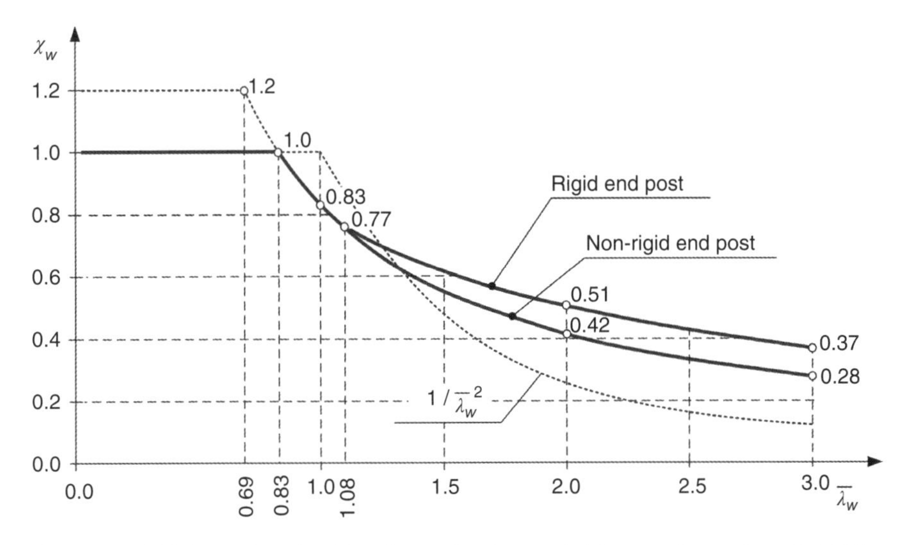

Figure 6.112 Ultimate shear buckling factor of webs panels (*Source:* Adapted from Ref. [47], reproduced with permission of Orion). Elastic critical shear forces V_{cr} and ultimate shear forces V_{ult} for web panels with non-rigid or rigid end stiffeners.

where the *web buckling factor* χ_w may be conservatively taken as $0.83/\bar{\lambda}_w$ for $\chi_w \geq 0.83$. This value is always higher than the elastic critical value $V_{cr} = \tau_{cr}\, h_w\, t_w$ as shown in Figure 6.112. For $\chi_w \leq 0.83$ the web may reach its yield stress in shear ($\chi_w = 1.0$) or even a slightly higher value than τ_y (in the order of 1.2 τ_y) due to steel hardening, as occurs in stocky webs for some steel grades. In Eq. (6.173), γ_{M1} is safety coefficient (1.1 in EN 1993-1-5 [5]). The reduction coefficient χ_w is represented in Figure 6.112 for webs with *non-rigid* or *rigid end posts* [3, 4, 48]. To allow the consideration of a rigid end posts, the end transverse stiffeners should be designed as such (apart from taking the end reactions) to allow sufficient capacity to resist the membrane stresses in the post-buckling range of the web.

Apart from the web resistance, a plate girder in shear has also an additional resistance from the contribution of the flanges working as a plastic resistance mechanism, but this is not usually quite relevant in design practice and can be conservatively neglected. In short, the ultimate resistance of plate girders to shear forces is illustrated in Figure 6.113, and may be described as

$$V_u = V_{cr} + V_t + V_f \tag{6.174}$$

where V_{cr} represents the shear force resistance associated with the elastic critical stress τ_{cr}, V_t is the component associated with the *diagonal tension field* associated with the post-buckling stiffness of the web panel, and V_f is the component associated with the plastic frame mechanism. The resistance $V_{cr} + V_t$ is the one considered in Eq. (6.173), which may be taken conservatively as the design shear resistance of the plate girder, since V_f is usually small as mentioned before.

6.9.5.2 Flange Induced Buckling

The slenderness of a web cannot be increased above a certain limit because the compressed flange may buckle in the plane of the web (flange induced buckling) as explained

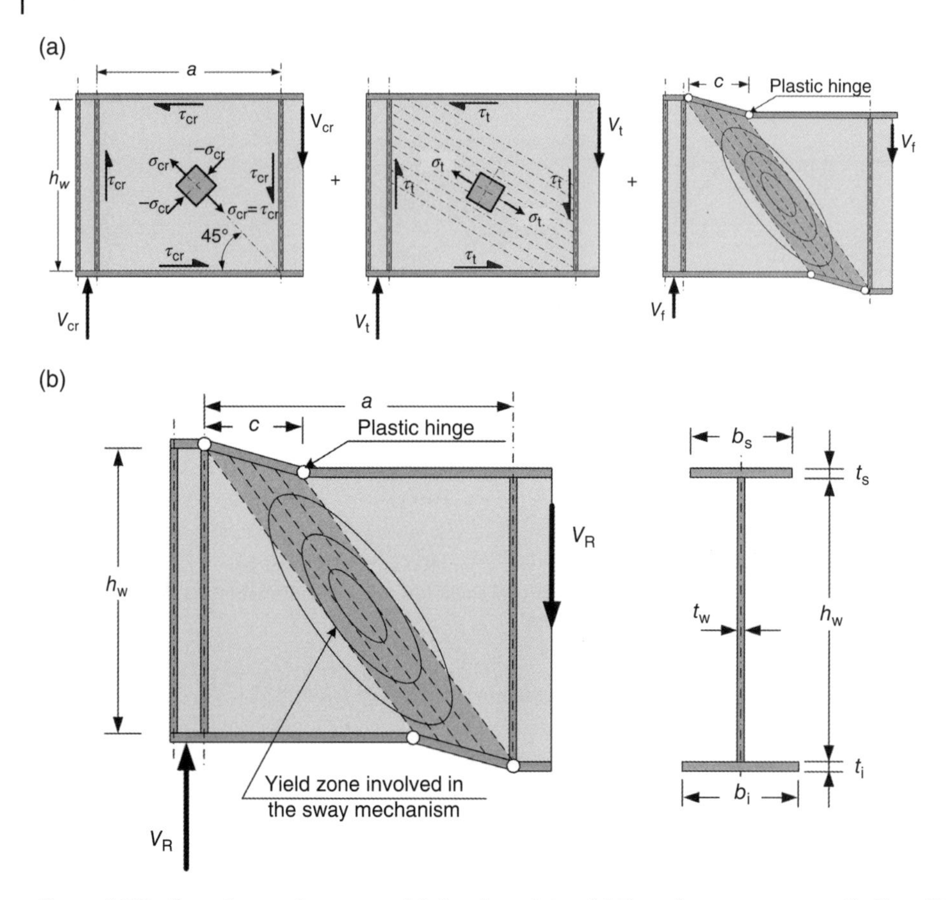

Figure 6.113 Shear force resistance models for plate girders (a) Shear force components V_{cr}, V_t and V_f. (b) Plastic mechanism at the end panel.

in Chapter 4, Section 4.5.5, Figure 4.63. The maximum curvature of the a symmetric girder is taken by assuming a compressive stress correspondent to a total strain in the flange of 1.5 f_y to take into consideration residual stresses, namely due to welding between the web and the flange,

$$\frac{1}{R}=\frac{2}{h_w}\cdot\frac{1.5f_{yf}}{E_s}=\frac{3f_{yf}}{E_s h_w} \tag{6.175}$$

Hence, assuming the flange attained yielding maximum vertical stresses in the web are given by

$$\sigma_{z,max}=\frac{3f_{yf}}{E_s h_w}\cdot\frac{A_f\, f_{yf}}{t_w}=\frac{3A\, f_{yf}^2}{E_s h_w t_w} \tag{6.176}$$

that should be less than the elastic critical stress $\sigma_{z,cr}$ of the web with a plate buckling coefficient $k = 1.0$ (buckling of a long plate – Figure 4.63, as a column) given by

$$\sigma_{z,cr} = \frac{\pi^2 E}{12(1-\nu^2)}\left(\frac{t_w}{h_w}\right)^2 \tag{6.177}$$

Under the condition $\sigma_{z,max} \le \sigma_{z,cr}$ and taken as an approximation for the area of the web $A_w = h_w\, t_w$, one obtains the limit for h_w/t_w given by Eq. (4.12) in Section 4.5.4.2 and Table 4.4. However, for web dimensions adopted in practice, this limit is not usually a design constraint, at least for steel grades up to S355.

6.9.5.3 Webs Under Patch Loading

Web panels are also subject to the effect of concentrated forces, namely during launching stages (Figure 4.62) as discussed in Section 4.5. A web panel under a 'concentrated force' (Figure 6.114) may collapse by pure plasticity if it is stocky, by local buckling ('web crippling') if it is slender, or in a general buckling mode for intermediate slenderness, h_w/t_w.

Let the case of a plate girder web between transverse stiffeners in a launching stage be considered, as in Figure 4.62. The 'concentrated' load distributed along length s_s (Figure 6.115) collapses by an interaction of plasticity and buckling, with the effective loaded length depending on the plastic bending resistance of the loaded flange and of the web. Plasticity is reached at a force F_y with internal plastic hinges at the flange only under a moment M_i and two other plastic hinges with a plastic moment M_0 involving both the flange and the web as shown in the figure. By simple equilibrium of the flange segment between an internal and external plastic hinge (Figure 6.115) the length s_y is determined:

$$s_y = \sqrt{\frac{2(M_0 + M_i)}{f_{yw} t_w}} \tag{6.178}$$

where f_{yw} and t_w are the yield stress and the thickness of the web, respectively.

The total length l_y associated with plastic resistance the web is obtained from $l_y = l_y^0 + 2t_f$ where $l_y^0 = s_s + 2s_y$ to take into consideration the spread (assumed at 45°) of the load through the flange thickness, t_f. If the contribution of the web for the plastic

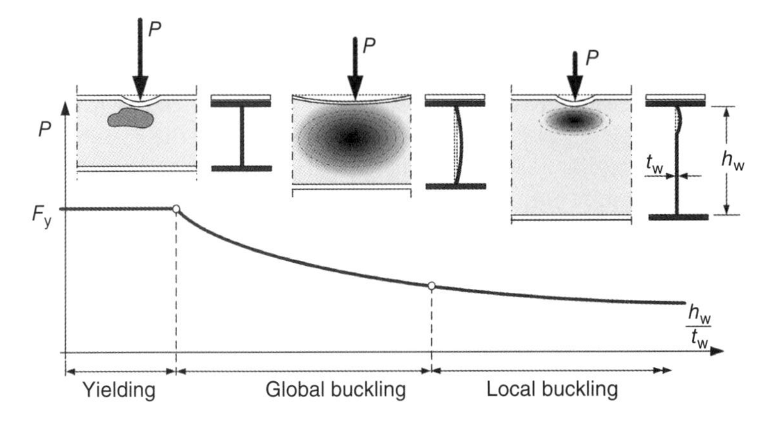

Figure 6.114 Collapse of plate girder webs under concentrated forces.

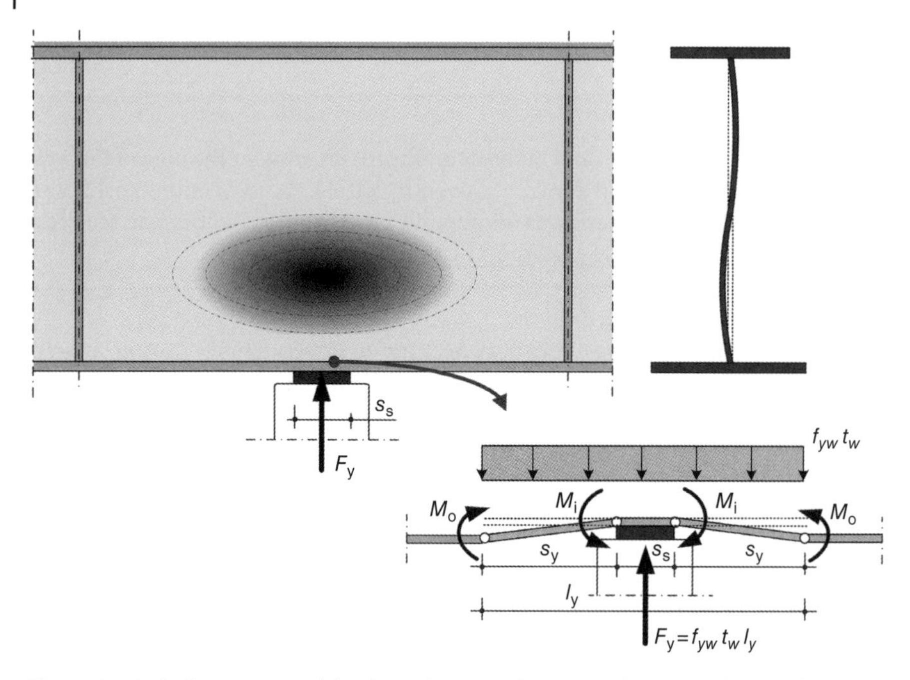

Figure 6.115 Resistance to patch loading: plastic mechanism to determine the yield force F_y.

moment M_0 is neglected, one has $M_0 = M_i = b_f t_f^2 f_{yf}/4$, where f_{yf} is the yield stress of the flange, with the result for the distance s_y,

$$s_y = t_f \sqrt{\frac{f_{yf} b_f}{f_{yw} t_w}} \tag{6.179}$$

The plastic resistance load is $F_y = f_{yw}\, t_w\, l_y$, finally given by

$$F_y = f_{yw} t_w \left[s_s + 2t_f \left(1 + \sqrt{\frac{f_{yf} b_f}{f_{yw} t_w}} \right) \right] \tag{6.180}$$

If the contribution of the web had been taken in M_0, an additional term would appear under the radical and the expression given in EN1993-1-5 [5], also adjusted by experimental tests, would be obtained. In any case, l_y should be limited to the distance a between transverse stiffeners and the width of the flange, particularly in box girders to $15\varepsilon t_f$ where $\varepsilon = \sqrt{235/f_y}$ with f_y in N mm^{-2}, to each side of the web. The elastic critical load of the web under a patch loading is given by:

$$F_{cr} = k_F \frac{\pi^2 E}{12(1-\nu^2)} \left(\frac{t_w}{d} \right)^2 d\, t_w \tag{6.181}$$

where k_F is the plate buckling coefficient for which several results are given or referred to in the literature [5, 47], and that can also be obtained by numerical approaches such

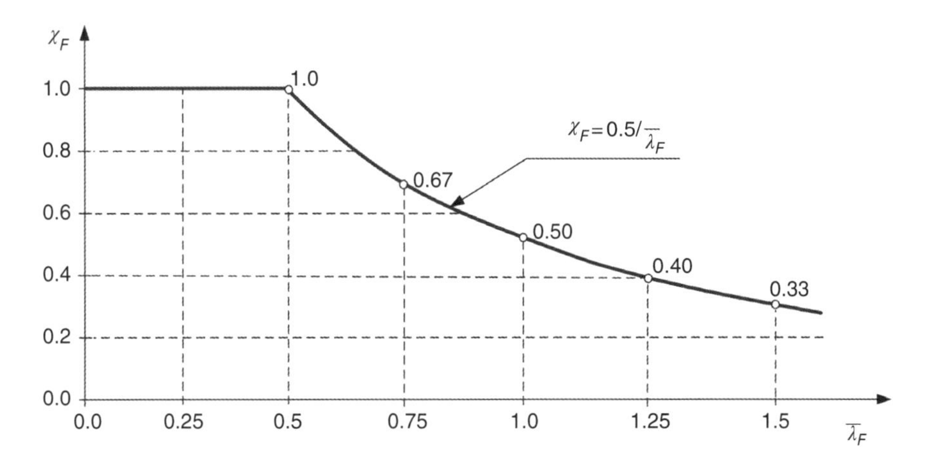

Figure 6.116 Resistance to patch loading of plate girder webs. Buckling factor χ_F according to the expression in EN 1993-1-5 [5]

as FEM or by a specific plate buckling software such as EBPlate [49]. A simple expression adapted from [5] is $k_F = 6 + 2(d/a)^2$ where F is the distance between transverse stiffeners. Knowing F_{cr} and F_y for patch loading, the non-dimensional slenderness is evaluated by:

$$\bar{\lambda}_F = \sqrt{F_y / F_{cr}} \tag{6.182}$$

The ultimate load is $F_u = \chi_F F_y$, where χ_F is, as usual, the *buckling factor for patch loading* defined as function of the nondimensional slenderness $\bar{\lambda}_F$. The ultimate load is influenced by several parameters, namely by the non-linear behaviour of the web under geometrical imperfections and residual stresses. Empirical expressions to evaluate F_u may be calibrated by tests and numerical studies. If one takes, for example, the classical von-Karman approach from which $F_u = \sqrt{F_y \, F_{cr}}$ in terms of the nondimensional slenderness this expression yields $F_u = \chi_F \, F_y$ with $\chi_F = 1/\bar{\lambda}_F$. However, this value is an unsafe value compared to, for example, values given in a variety of codes and recommendation calibrated by experimental tests, as illustrated in Figure 6.116 where $\chi_F = 0.5/\bar{\lambda}_F$. The design ultimate load of the web under patch loading, considering the resistance safety factor γ_{M1}, is finally evaluated by:

$$F_{Rd} = \chi_F \, F_y / \gamma_{M1} \tag{6.183}$$

6.9.5.4 Webs under Interaction of Internal Forces

Interaction of internal forces are usually dealt with in codes and design recommendations on the basis of interaction equations, in terms of design values of bending moments M, shear forces V and 'concentrated forces' F, of the form:

$$f\left(\frac{M_{Ed}}{M_{Rd}}, \frac{V_{Ed}}{V_{Rd}}, \frac{F_{Ed}}{F_{Rd}}\right) < 1.0 \tag{6.184}$$

Interaction between bending moments and axial forces is usually considered by defining an internal force ratio ($N_{Ed}/N_{Rd}+M_{Ed}/M_{Rd}$) with due account for eccentricity of the axial force at the effective cross section. The interaction between bending moments and shear forces is usually neglected if $V_{Ed} < 0.5\ V_{w,Rd}$. The interaction between bending moments and transverse 'concentrated' forces is usually neglected when $M_{Ed} < 0.5\ M_{Rd}$ or when $F_{Ed} < 0.5\ F_{Rd}$. If these conditions are not satisfied interaction should be considered, according to the design interaction equations, for example, as predicted in [5].

6.9.6 Transverse Web Stiffeners

Transverse web stiffeners should be designed to satisfy resistance and stability conditions. A minimum transverse stiffness is required to be considered a *rigid stiffener*. Otherwise the stiffener deforms with the web panels and the web buckling modes should take this effect into consideration. To have a rigid stiffener, the moment of inertia of the stiffener cross section (Figure 6.117), I_{st}, with respect to a cross sectional axis parallel to the plane of the web, should satisfy a minimum value. The reader is referred to design codes [5, 43, 50] for minimum rigidity conditions for stiffeners.

Transverse web stiffeners are subject to direct force actions like stiffeners at sections at the bearing location, transverse forces induced, for example, by bracings or wind actions on the webs and forces induced at the diagonal tension field in the post-buckling range of the web.

For an intermediate (internal) web panel, the induced force N_{st} in the transverse stiffener by the diagonal tension field, is usually taken conservatively as the design shear force

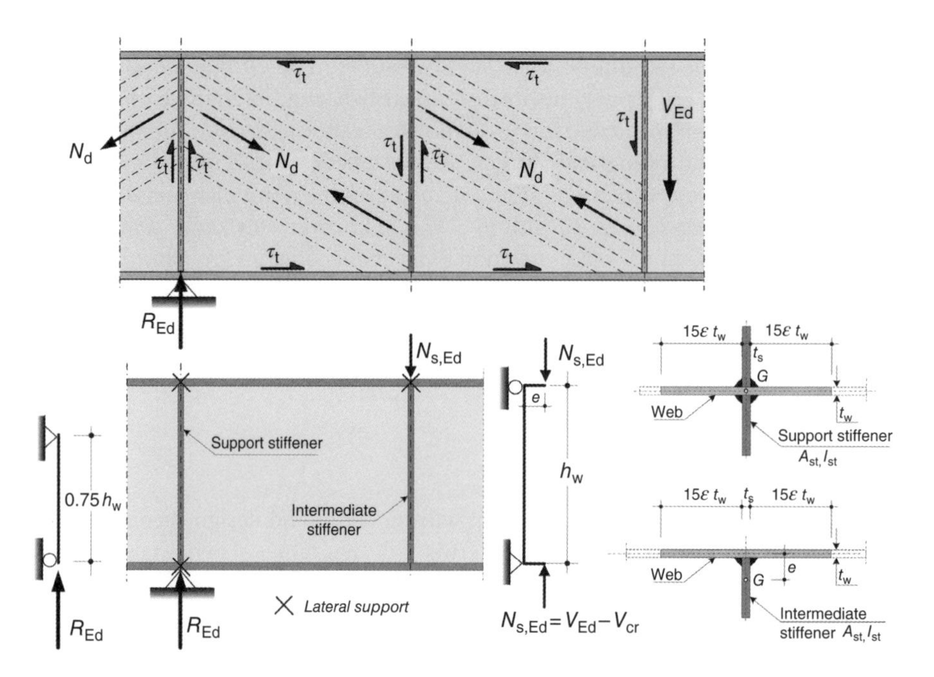

Figure 6.117 Design of internal web transverse stiffeners in plate and box girders.

reduced by the elastic critical force shear force of the web panel, that is, $N_{st} = V_{Ed} - V_{cr}$. The stiffener is considered working with an associated plate, coming from the contribution of the web, as shown in Figure 6.117. It is designed as a *beam column* (Figure 6.117), with a buckling length equal to the web depth h_w, assuming the compressed flange is laterally supported. The eccentricity e at the end sections, due to the eccentricity of the centre of gravity of the stiffener to the plate, and the *geometrical equivalent imperfection* of curve c, should be used in the design. A second order analysis may be avoided by defining an *equivalent transverse deviation force* induced by the web panels as predicted in several codes [5, 43]. That is usually the design approach. The symmetrical support stiffeners should be designed for the vertical reaction force using the buckling length $0.75\, h_w$ (Figure 6.117).

6.9.7 Stiffened Panels in Webs and Flanges

Stiffened panels are one the main structural elements of webs in bridge plate girders and flange panels in box girder bridges. The effect of transverse stiffeners in webs was already dealt with in previous sections. Plate girders have very often longitudinal stiffeners as discussed in Chapter 4 (Section 4.5). The lower flange of a box girder is stiffened by longitudinal stiffeners supported by transverse diaphragms; for example, as shown in Figures 4.75 and 4.76.

A stiffened plate in the web of a plate girder under positive in plane bending is subjected to compressive stresses at the upper part, and the longitudinal stiffener may or may not buckle together with the plating (Figure 6.118). If the stiffener is sufficiently rigid, the local buckling mode at the sub-panels between transverse and longitudinal stiffeners is induced. If not (Figure 6.119a) the plate buckling mode involves transverse deflection of the stiffener and the elastic critical stress should be evaluated for the plate mode of the stiffened panel. The same happens in a compressive flange of a box girder, where local buckling is induced at the sub-panels (Figures 4.78 and 4.79) if the longitudinal stiffeners are sufficiently rigid. According to classical elastic theory of stability, two parameters are defined for stiffened plates, the ratio between the inertia (I) and cross section areas (A) of the stiffener and of the plating:

$$\gamma = I_{sl}/I_p \quad \text{and} \quad \delta = A_{sl}/A_p \tag{6.185}$$

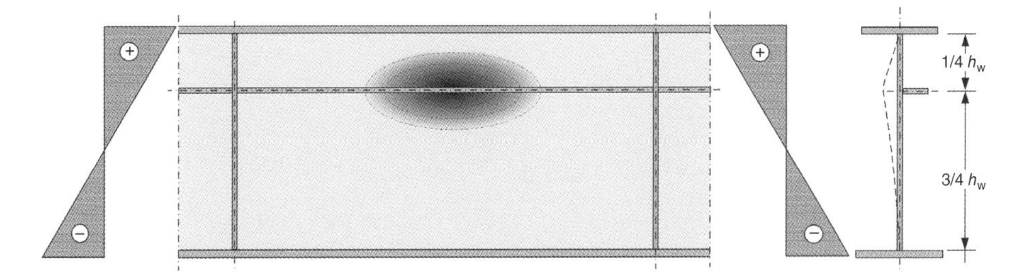

Figure 6.118 A web panel with a non-rigid longitudinal stiffener under in plane bending.

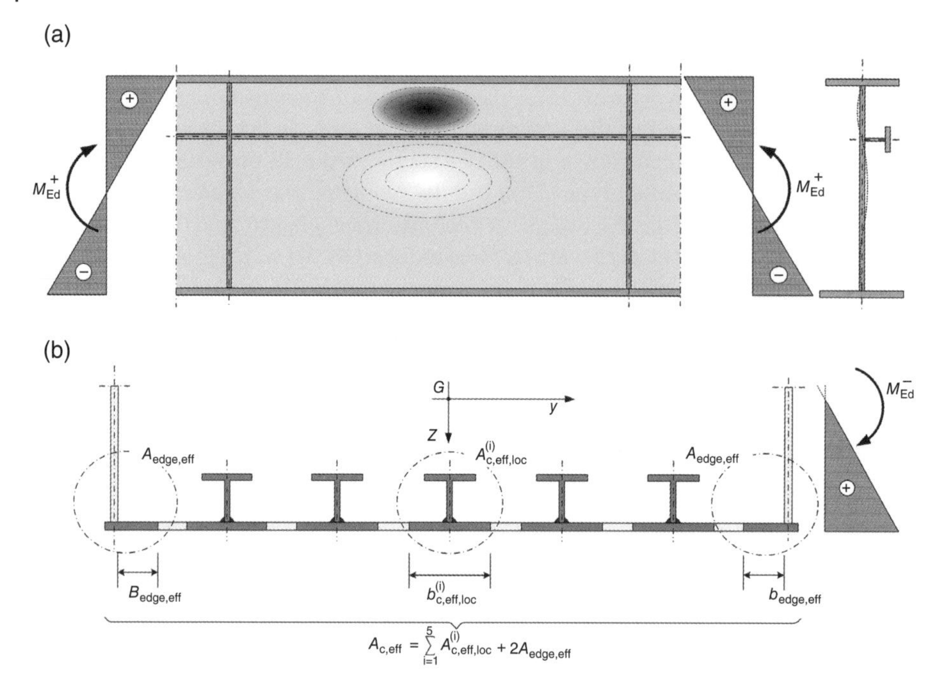

Figure 6.119 Stiffened panels under compression: (a) at webs of plate girders or (b) at the bottom flange of a box girder.

where I_{sl} and I_p refer, respectively, to the overall inertia of (stiffeners + plating) and I_p is the inertia of the plate only, that is, $I_p = bt^3/\,[12(1-\upsilon^2)]$; for the relative area parameter, $A_{sl} = \Sigma A_{sli}$ is the sum of the individual areas of the stiffeners and $A_p = bt$.

A stiffened plate under compression (Figure 4.78) has an elastic critical stress $\sigma_{cr,p}$ for the *plate mode* involving deflection of plate and of the stiffeners. For the case of a stiffened panel with three or more stiffeners, equally spaced and considered to be simply supported at all edges and loaded at the transverse edges with a linearly varying stress defined by $\psi = \sigma_2/\sigma_1$ (compressive stresses as positive), given for the case $\alpha = a/b > 0.5$, the elastic critical stress of the stiffened plate may be approximately determined by:

$$\sigma_{cr,p} = k_p \frac{\pi^2 E}{12\left(1-\nu^2\right)}\left(\frac{t}{b}\right)^2 \tag{6.186}$$

where

$$k_p = \begin{cases} \dfrac{2\left[\left(1+\alpha^2\right)^2+\gamma-1\right]}{\alpha^2(\psi+1)(1+\delta)} & if \quad \alpha \le \sqrt[4]{\gamma} \\[2ex] \dfrac{4\left(1+\sqrt{\gamma}\right)}{(\psi+1)(1+\delta)} & if \quad \alpha > \sqrt[4]{\gamma} \end{cases} \tag{6.187}$$

These expressions are valid for open stiffeners and are conservative for closed stiffeners, assuming the stiffeners are eccentric with respect to the plane of the plate.

According to the classical elastic stability of plates, the stiffener will remain rigid if the parameter γ is greater than a certain 'optimum' value γ^*. This is a theoretical and ideal value because, due to nonlinear behaviour of the plate under the presence of geometrical imperfections, the stiffener tends to deflect from the unset of the loading process unless γ is well above γ^*. In practice, rigid stiffeners require $\gamma > m\gamma^*$ generally with $m > 3$ for closed stiffeners and $m > 5$ for open section stiffeners. One may say $3 < m < 7$ depending on the type of stiffener (lower values valid for closed stiffeners) and its location on the panel. That is why the classical elastic stability theory cannot be applied directly to stiffened panels design without due account for plate non-linear behaviour and detrimental effect of geometrical and material (namely residual stresses) imperfections. Linear elastic critical stresses and optimum γ^* values for stiffeners, reducing the problem to plate buckling between sub-panels has been known to be for a long time a very unsafe design concept that yielded a series of accidents in box girders in the 1960s. Due accounts for plate buckling non-linear behaviour, geometrical and material imperfections and possible detrimental effects between plate and column buckling modes are taken into consideration in modern bridge and steel structures design codes.

The elastic critical stresses, $\sigma_{cr,p}$, for plate modes of stiffened panels in webs or flanges of plate and box girder bridges, can be evaluated by approximate formulas such as the one mentioned before, by charts [51, 52], numerical FEM or simpler methods such as the one adopted for the available software EBPlate [49]. The elastic critical stress $\sigma_{cr,p}$ should not be confused with the local plate critical stress $\sigma_{cr,l}$ associated only to buckling of the plate sub-panels between stiffeners and evaluated conservatively for plates under uniform compressive stresses with a buckling coefficient $k_l = 4.0$ and width b_l between longitudinal stiffeners. Apart from these two critical stresses, the panel (Figure 6.119b) can buckle in an overall mode like a column (column *buckling mode*) between transverse stiffeners or diaphragms. The associated critical stress is

$$\sigma_{cr,c} = \frac{\pi^2 E I_s}{A_s a^2} \tag{6.188}$$

where I_s and A_s are the inertia and cross sectional area of *a* single stiffener and associated plating. This overall mode can also be envisaged as the induced buckling mode for the plate, with removed supports at the longitudinal edges. In the column buckling formula I_s and A_s may be taken for the overall transverse cross section of the panel.

In stiffened panels of webs or flanges of plate girder or box girder bridges, the simplest and conservative approach to evaluate the ultimate load is to consider the *strut approach* as discussed in Chapter 4. In this approach, the buckling problem of the panel is reduced to a single strut or column under compression. This corresponds to take for the ultimate stress such as the one associated with the column buckling mode with an elastic critical stress $\sigma_{cr,c}$ and a non-dimensional slenderness $\bar{\lambda}_c = \sqrt{f_y / \sigma_{cr,c}}$. The ultimate load of this column is determined from the buckling factor χ_c by $N_{u,c} = \chi_c f_y$. The reader is referred to the specific literature on stiffened plates [3, 47] and design recommendations or codes [5] to define the imperfection factors to be adopted when evaluating χ_c from column buckling curves.

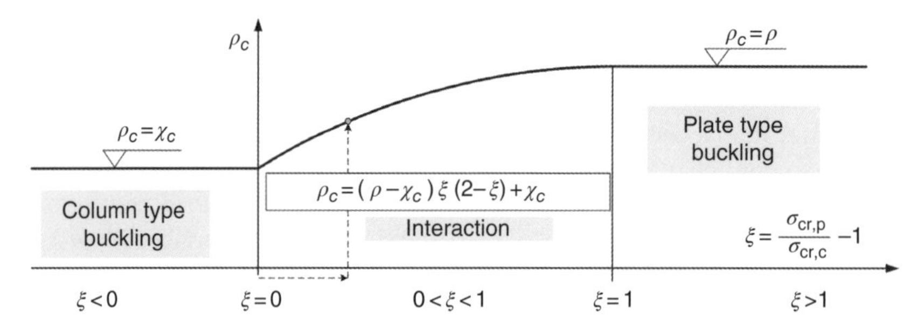

Figure 6.120 Interaction between plate and column buckling modes in stiffened plates.

In practice, a stiffened plate reaches its ultimate load in an *interactive mode* defining on the proximity of the two critical stresses for the plate mode and for the column mode, as discussed in Chapter 4. It should be noted that we are in presence of an interactive buckling problem that is been the subject on a variety of investigation since several decades ago [52–54]. For this specific case, the two elastic buckling modes (the plate mode and the column mode) are both stable modes, with the column mode being almost without any post-buckling stiffness ('neutral' mode). However, the interactive mode is unstable and quite sensitive to geometrical modal imperfections in each of the competing modes.

The ultimate load of a stiffened plate under compression, with due account for the interactive mode, may be defined from the effective plate areas associated with each stiffener (Figure 6.119b) by

$$N_u = f_y \rho_c \left(A_{c,eff,loc} + \Sigma b_{edge,eff}\, t \right) \tag{6.189}$$

where the reduction factor ρ_c is estimated from the plate reduction factor ρ and the column buckling factor χ_c. The interaction may be considered in a simplified approach, by adopting, for example, an interaction formula such as the one proposed in EC3-1-5, Figure 6.120. In Eq. (6.189) $A_{c,eff,loc}$ is the effective area of all sub-panels and associated stiffeners with the exception of the area of the end sub-panels with effective plate width ($b_{edge,eff}\,t$). These effective areas are evaluated by the plate effective width formula $b_{eff}=\rho\, b$ with ρ given by Eq. (6.170) yielding the Winter's formula for uniform compression:

$$\rho = \frac{1}{\lambda_p}\left(1 - 0.22\frac{1}{\lambda_p}\right) \tag{6.190}$$

6.9.8 Diaphragms

Diaphragms made by a vertical bracing system in plate girder bridges were discussed in Chapter 4, Section 4.5.4.3; the case of horizontal bracing systems was dealt with in Section 4.5.4.4. The design verifications of diaphragms require an analysis for strength/stability and deformability.

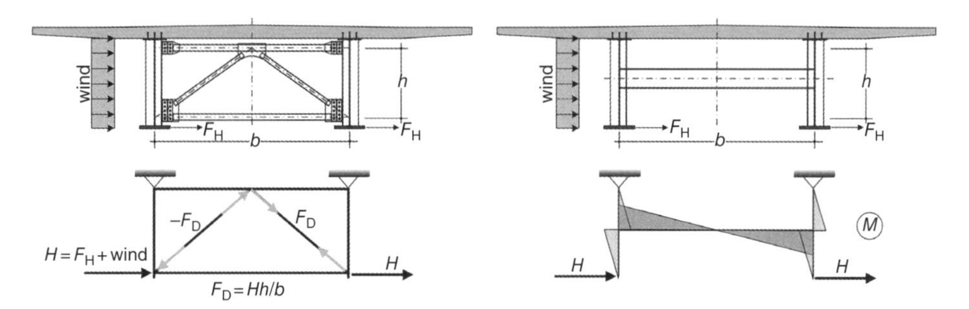

Figure 6.121 Forces induced in diaphragms (vertical bracings – trusses or cross beams) of plate girder decks.

The forces induced in intermediate vertical bracings (diaphragms) of plate girder bridges, spaced at equal distance a along the deck, are mainly due to the symmetric and asymmetric components of vertical forces on the deck slab (Figure 4.65 and 4.66) and transverse wind loads (Figure 6.121) and horizontal transverse forces (F_H) at the lower flange induced by lateral buckling effects of the compressed flange. The total force H is equilibrated by forces F_D at the diagonals as shown in Figure 6.121 for the case where the diaphragm is made of a truss. When vertical bracing is made by cross girders only, transverse bending moments are induced by the forces H as shown in Figure 6.121. For the diaphragms at the supports (piers and abutments), due account should be taken by the force reactions that should also be transferred to elements of the diaphragm [34].

In box girder decks, intermediate diaphragms are usually of one of the types represented in Figure 4.82. The action effects induced in the intermediate or support diaphragms are determined from the superstructure analysis, namely due to direct forces like bearing reactions, normal stresses (vertical and horizontal), shear flow and torsion effects. If the diaphragm is made of a stiffened plate, as the case in Figure 4.32, the sub-panels should be verified for strength and stability. Evaluating the stress state for example by a FEM and von Mises equivalent stress, plastic resistance is verified. For stability, it is recommended to keep non-dimensional slenderness values $\bar{\lambda}$ below a certain limit (say 0.7), to avoid plate buckling effects at the sub-panels. The equivalent slenderness for the multiple stress state $\bar{\lambda} = \sqrt{\alpha_{pl}/\alpha_{cr}}$ may be evaluated from the load parameters associated with yield α_{pl} and elastic instability α_{cr}.

6.10 Reference to Special Bridges: Bowstring Arches and Cable-Stayed Bridges

6.10.1 Generalities

In this last section, a reference is made to the analysis and design of superstructures of some special bridge types. The similarities of bowstring arches and cable-stayed bridges, shown schematically in Figures 1.11 and 1.14, are the following:

- both superstructures are suspended by cables/bars, specifically *hangers* in the case of bowstring arches and *stays* for the case of cable-stayed bridges.

- relevant axial forces are induced in the superstructures of both bridge types, specifically a tensile force induced by the arch in the deck of bowstring arch bridges and a compressive force induced by the stays in case of cable-stayed bridges.
- both superstructures may have a lateral or an axial suspension type from a double or a single plane of hangers/stays, respectively.
- the hangers or stays are attached to the superstructures by anchorages that in practice behave as pin connections.
- the superstructures types, in reinforced or prestressed concrete, steel or composite construction, may be a slab-girder type or box girder type.
- the superstructures work for live load vertical forces on the deck as beams on elastic foundation.
- hangers and stay cable adjustments and force control during construction stages, to achieve the correct final geometry, may be done according to similar procedures.

Of course, there are differences in the behaviour of superstructures of bowstring arches and cable-stayed bridges that will be highlighted in the following. The flow of forces for cable-stayed bridges has been highlighted in Figures 1.11 and 1.14 of Chapter 1.

6.10.2 Bowstring Arch Bridges

6.10.2.1 Geometry, Slenderness and Stability

In bowstring arch bridges, the rise-span ratio f/L is in the order of one-sixth to one-fifth for both road and rail bridges (Figure 6.122). The flow of forces from the transverse girders of the deck supporting the slab to the arch trough tensile forces N_h in the hangers induces a compressive force N_a in the arch equilibrated at the support levels by the vertical reactions and by the tension force in the deck working as a *tie member* (Figure 6.123). The stiffness of the arch and of the deck may be chosen at concept design

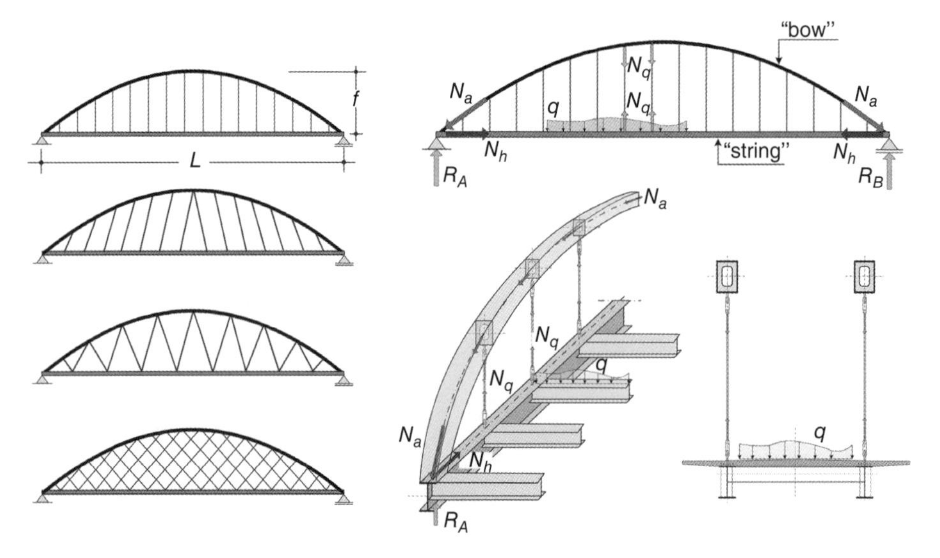

Figure 6.122 Possible options for *bowstring arch bridges* with vertical hangers, inclined hangers and with a network of hangers, typical cross section and flow of forces between deck, hangers and arch.

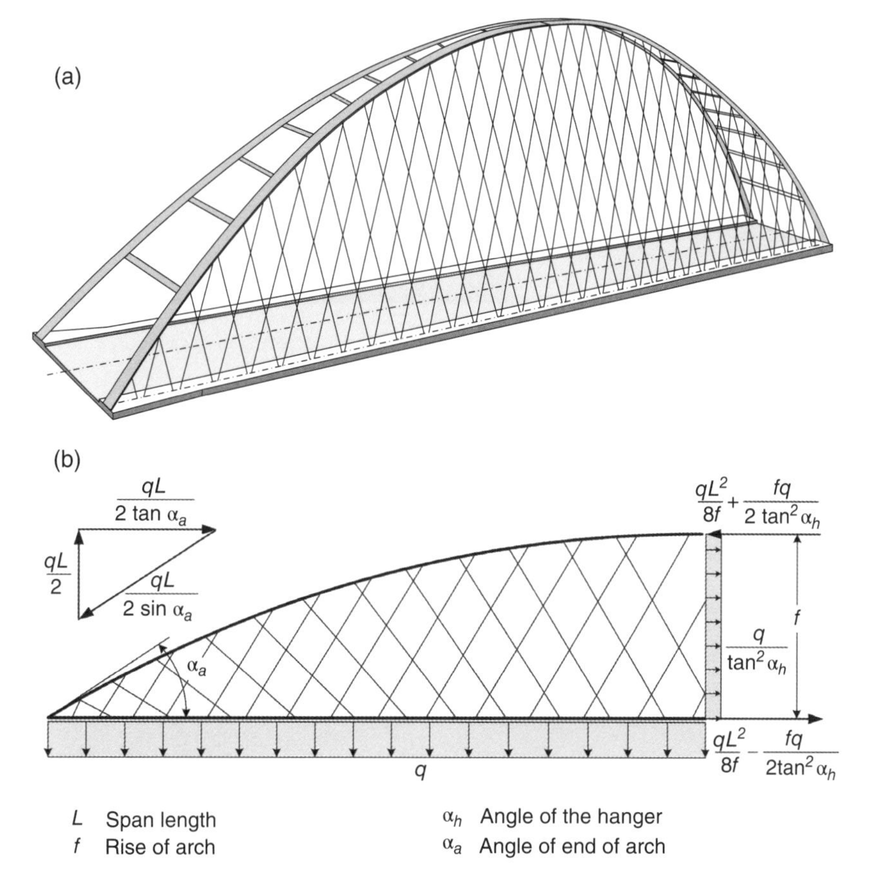

Figure 6.123 Network arch bridges proposed by Tveit [55] in 1959 and axial forces in the arch (span *L*, rise *f*), in the deck and in the hangers due to a uniformly distributed load on the deck.

between a *rigid arch option* with a *slender deck*, and a *slender arch* with a *rigid deck*. The deck may be a simple solid prestressed concrete slab (Figure 6.124) or made by a single or double prestressed concrete box girder, such as in the recent third Millennium Bridge in Spain, with a prestressed concrete deck reaching 43 m width. Alternatively, a steel plate girder (Figure 6.129), a steel box girder deck or a steel concrete composite deck (Figure 4.34) may be adopted. For the arches, steel is usually the preferred option, particularly box girder sections made of simple tubular hot rolled sections for small bow string arches (Figure 6.124) or large welded sections made of steel plates with or without stringers (Figure 4.34).

By denoting the height of the arch section by h_a and the height of the deck by h_d, one may define a *slenderness parameter* $\alpha = (h_a + h_d)/L$. In general, α lies between 1/45 and 1/30 the lower limit for double plane of hangers (lateral suspension) and the higher limit for single plane of hangers (axial suspension). In the design example of the railway bridge over the Sado River in Portugal (Figure 4.34), with $f/L = 1/5.4$ and a single plane of hangers [56], one has $h_a = 1.8$ m and $h_d = 3.0$ m and hence $\alpha = 33$. This bridge has two concepts that are not usual in bowstring arch bridges and are in some way innovative: a

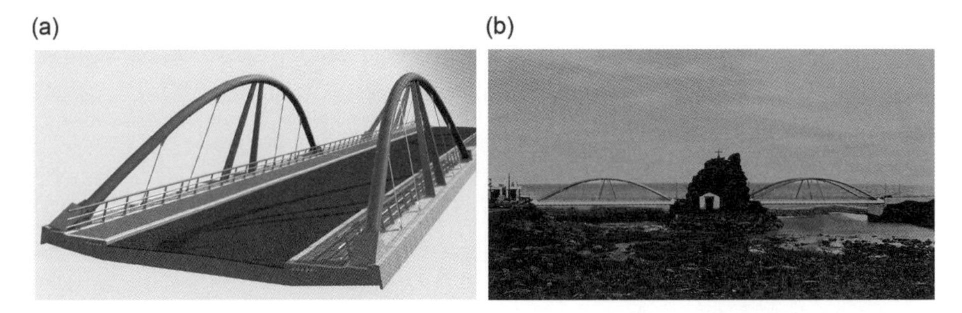

Figure 6.124 A small span bowstring arch bridge with a prestressed concrete slab deck (*Source:* São Vicente bridge in Madeira Island; Courtesy of GRID, SA).

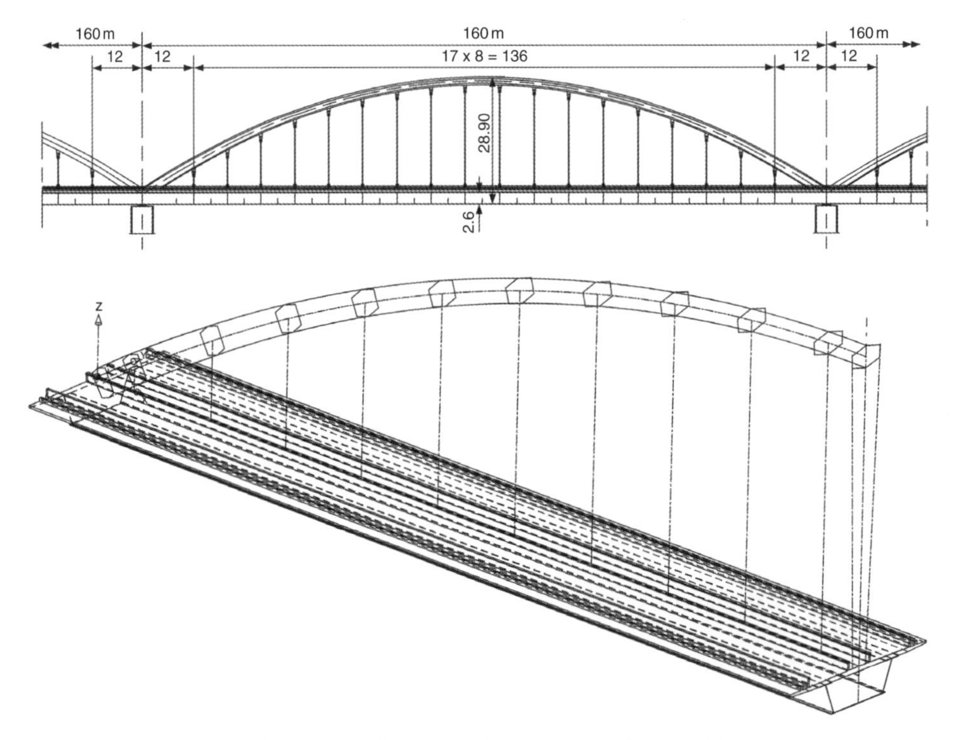

Figure 6.125 The Sado railway bridge (Figure 4.34) – longitudinal layout of the central spans (*Source:* Adapt from Ref. [56] with permission of John Wiley & Sons).

multiple span continuous deck (3 × 160 m) and a single arch-axial suspension for a double track railway bridge. The deck is a single cell trapezoidal box girder working in composite action with the reinforced concrete slab (Figure 6.125). The arch has a variable welded tubular hexagonal cross section, with plate thicknesses varying between 60 and 120 mm, increasing width towards the crown in order to achieve transverse stability and limiting the width as much as possible at the deck level to reduce the need of increasing the deck width. This last issue is similar to what happens with cable-stayed bridges with axial staying.

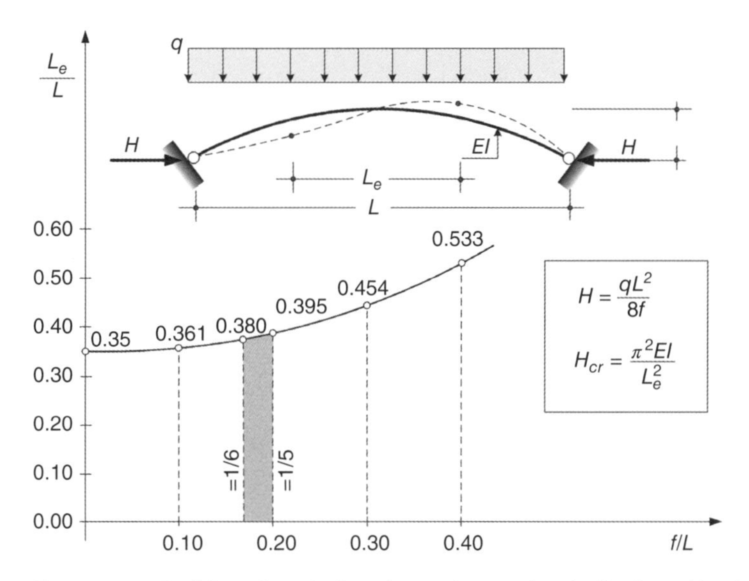

Figure 6.126 Buckling of parabolic arches under a uniformly distributed load.

Among the most slender type of bowstring arch bridges are the so called *network arch bridges* to which a variety of design and research contributions, since 1959, have been made by Tveit [55]. The slenderness $\alpha = (h_a + h_d)/L$ in network arch bridges, for narrow deck road bridges with two traffic lanes, may reach 1/100.

As discussed previously (Section 6.6.1, Figure 6.56), *snap through buckling* is rolled out for usual f/L ratios and slenderness L/i of the arch (being i the relevant radius of gyration of the deck section). The fundamental equilibrium path has a *symmetrical bifurcation point* (Figure 6.56) with arch buckling in an asymmetric configuration. For a parabolic arch, under a uniformly distributed load q, the buckling load may be referred to the elastic critical value of the thrust H with an equivalent buckling length L_e defined in Figure 6.126. For rise/span values adopted in practice (f/L between 1/6 and 1/5), the buckling length is $L_e \approx 0.39L$. In bowstring arches, the stiffness of the hangers and of the deck play a role in the in buckling load of the arch as shown in Figure 6.127. If these stiffnesses are neglected, the arch workings as a *tie member* only and expressions for buckling load in terms of f/L and the number of hangers are given in [69]. However, if arch and hanger stiffness effects are considered, no simple expressions are available for determining the buckling load. For this case, one may define elastic critical loads $H_{cr}^{(n)}$ depending on the relative bending stiffness of the deck cross section and axial stiffness of the hanger's cross section to the bending stiffness of the arch. By defining the following parameters (Figure 6.127):

$$\mu = \frac{f}{L} \qquad \lambda_v = \frac{E_d I_d}{EI_a} \qquad \lambda_h = \frac{E_h A_h}{EI_a} \tag{6.191}$$

the buckling loads may be obtained as roots of the equation

$$H_{cr}^{(n)} - \frac{4EI_a}{L^2} \frac{J_1 + \lambda_v J_2 + \lambda_h J_3}{J_4 - J_5 - J_6} = 0 \tag{6.192}$$

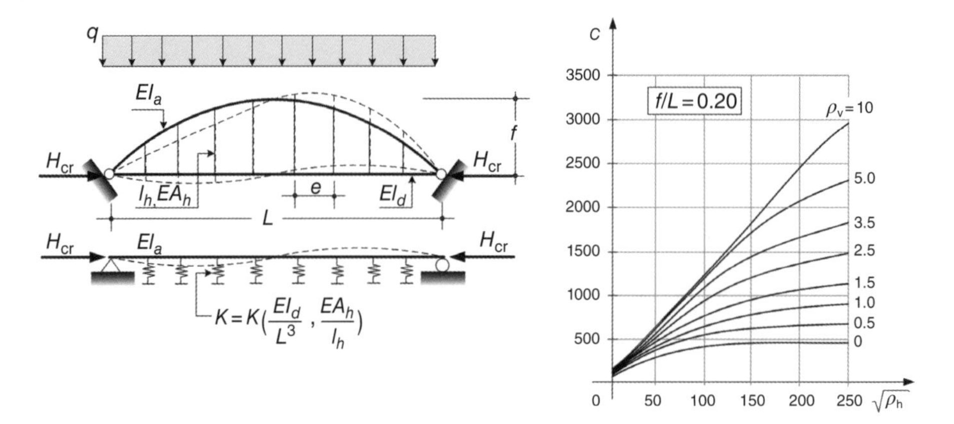

Figure 6.127 Buckling of bow-string arches $H_{cr} = C\,EI_a/L^2$ with $\rho_v = (E_d I_d)/(EI_a)$ and $\rho_h = [(m+1)\,L^2/8]$ $[(E_h A_h)/(EI_a)]$, m – number of hangers (*Source:* Adapted from Refs. [32, 57]).

with

$$J_2 = \frac{\pi^4 n^4}{2}, J_5 = \frac{\pi^2 n^2}{2} \text{ and } J_i = J_i(\mu, n, H_{cr}) \text{ for } i = 1,3,4 \text{ and } 6 \tag{6.193}$$

The expressions for J_i are given in [30] based on the original work due to Pflüger [58]. The J_i expressions are dependent on the integer values $n = 1, 2, 3,...$ and on $H_{cr}^{(n)}$ as well rendering the Eq. 6.192 nonlinear algebraic equation. The lowest critical load should be investigated for different n values, associated with the buckling modes, by solving numerically Eq. 6.192. For a specific geometry (f/L) it is also possible to use the diagrams in Figure 6.127 [57] (e.g. $f/L = 0.2$) to obtain the critical value H_{cr}. The evaluation of buckling loads of bowstring arches, taking all these referred to parameters into account, particularly when variable sections of the arch and deck are involved as is usual in practice, is nowadays more easily investigated using standard FE software. In network arch bridges the buckling modes have to be investigated numerically but, in spite of the slenderness of these structures, buckling is not usually a governing phenomenon for design practice.

Out of plane buckling of bowstring arches should be investigated in design with particular relevance for single arch bridges with axial suspension or bridges with double arches without bracing between them. The elastic critical load may be written in terms of the thrust H and this depends on the out of plane bending rigidity of the arch cross section EI_z and on the uniform (St Venant) torsional stiffness (GJ) of the arch cross section. By conservatively neglecting the stabilizing effect of the hangers ($N \sin\alpha$ in Figure 6.128), a distributed vertical load q is the only one induced to the arch assumed to be fixed in position at the end sections. The out of plane buckling load is given by [30]

$$H_{cr} = \frac{EI_z}{L^2} \frac{k_1}{1 + k_2 \Upsilon} \tag{6.194}$$

where $\Upsilon = (EI_z)/(GJ)$ and k_1 and k_2 are constants only depending on the (f/L) ratio. The in plane bending stiffness of the deck is usually such that the arch may be considered

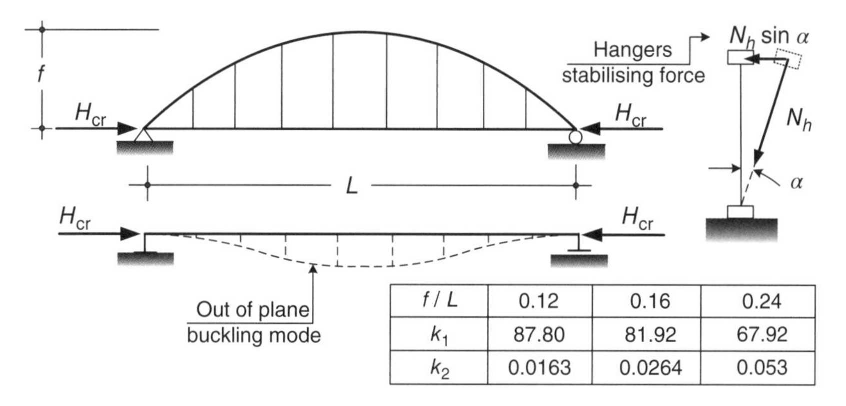

f/L	0.12	0.16	0.24
k_1	87.80	81.92	67.92
k_2	0.0163	0.0264	0.053

Figure 6.128 Out of plane buckling of a parabolic arch.

built in the deck for out of plane buckling deformations and the mode shape is as represented in Figure 6.128. For free torsion rotations of the arch at the end sections the values of k_1 and k_2 in Eq. (6.194) are given in the figure. In single arch bridges (Figure 4.34) to improve stability (H_{cr}), it is quite important to increase the torsional stiffness of the arch section GJ because the Υ factor will decreases. In the limit for very large torsion stiffness, Υ tends towards zero and one may obtain Eq. (6.194) for the out of plane buckling length of $L_e = (\pi/\sqrt{k_1})\ L$. This yields, for example, $L_e = 0.34L$ for $f/l = 0.12$ and $L_e = 0.38L$ for $f/l = 0.24$. Hence, if torsional stiffness of the arch section is great (e.g. a box section), the stability is not very dependent on the rise/span ratio and the buckling length may be taken as $L_e \approx 0.36L$.

To avoid out of plane buckling, a lateral bracing is usually adopted between the two arches. Inclining the arches towards the inside is usually a good design concept to improve out of plane arch stability reducing the length of bracing bars. The arches and hangers lie in two inclined planes, as shown in Figure 6.129 for the Fehmarn Sound Bridge with a main span of 248 m; one of the first and most impressive bowstring network arch bridges.

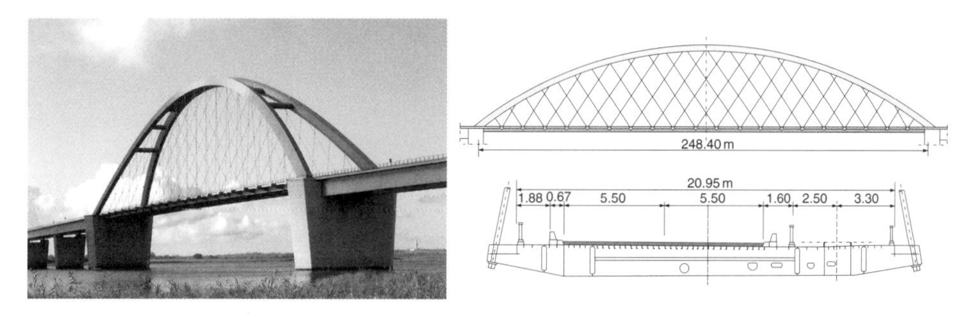

Figure 6.129 The Fehmarn Sound bridge in Germany, 1964 (*Source:* Photograph by S. Möller/ https://commons. http://wikimedia.org / Public Domain).

6.10.2.2 Hanger System and Anchorages

The hanger system may be made by *solid steel bars, locked coil cables (ropes)* or *cable strands* (with independent protected strands as adopted in cable-stayed bridges), as shown in Figure 6.130c.

For standard cases, cast steel pieces are available to make the anchorages and the hangers may be made by *high strength stainless steel bars*. Locked coil cables and high strength bars are usually fabricated with standard anchorages made of cast steel pieces (Figure 6.130a,b). These may require, in the case of bars, an intermediate coupler to join both segments directly connected to the deck and the arch.

In the plane of the arch, the hangers should be preferably connected by pin connections, allowing free rotations to improve fatigue life resistance of the hanger (Figure 6.131a). For railway bridges, both free rotations, in-plane and out of plane, are convenient to prevent fatigue problems in the hangers at the connections with the deck and with the arch. In the case of the Sado railway Bridge, there are 18 hangers in each span, 8 m apart, made of 200 mm diameter bars of steel grade S355NL, with a maximum length of 22.8 m (Figure 6.131c). The option of this hanger type was due to the need to

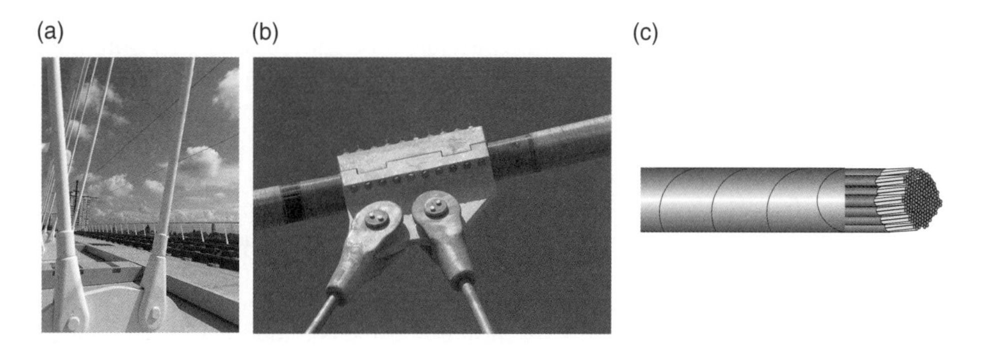

Figure 6.130 Hanger types: (a) high strength steel bars, (b) locked coil cables and (c) multiple strand cable (*Source:* (a) Photograph by José O. Pedro; (b) Courtesy GRID, SA).

Figure 6.131 Examples of use of different hanger types: (a) High strength bars (Beatus Rhenanus bridge, Strasburg), (b) locked coil cables (Zambezi river bridge, Tete) and (c) solid large diameter bars (Sado river bridge, Alçacer do Sal – Figure 4.34) (*Source:* (a) and (c) Photographs by José O. Pedro; (b) Courtesy of GRID, SA).

increase hanger stiffness under train loading. The hangers are fitted with spherical hinges where they connect to the deck and the arches. The spherical hinges are inserted into cast steel pieces, steel grade G20MN5 (EN10340). The maximum weight of these pieces is 1.5 tons and they are designed for maximum forces (permanent + variable actions) of 4830 kN.

6.10.2.3 Analysis of the Superstructure

Structural analysis of bowstring arches may usually be done on the basis of a standard elastic linear analysis. It requires, as usual, definition of loading cases for live loads based on influence lines. In Figure 6.132, some typical influence lines are shown. For example, to obtain the maximum axial force at the crown of the arch, all decks should be loaded while for the maximum bending moment at the mid-span of the deck, only the central part should be loaded.

In bowstring arch bridges with steel-concrete composite decks, an important aspect for structural analysis is how to account for the cracking effects of the concrete slab.

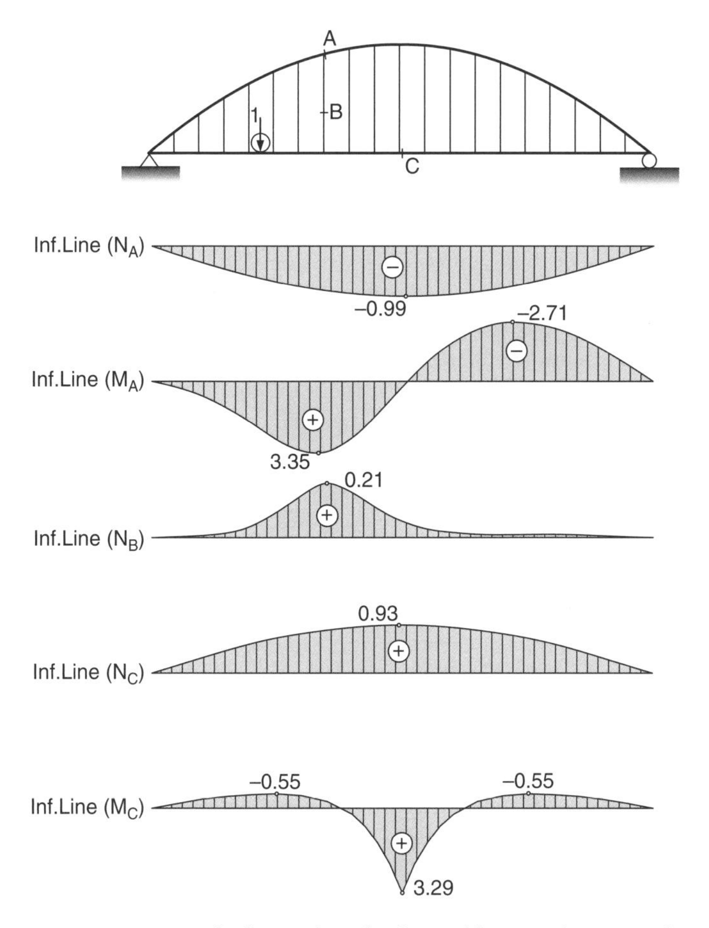

Figure 6.132 Typical influence lines for the axial force in a hanger, and axial force and bending moments in the arch and deck of a bowstring arch bridge.

This slab is usually cast on the top of steel deck (Figure 4.34). Afterwards, due to shrinkage effects and live load effects, it is likely cracking will develop in the slab, at least at ULS. Hence, the *tension stiffening effect* [7, 8] will induce an axial stiffness for the tie effect of the deck in addition to the steel structure stiffness. For structural global analysis of a bowstring arch bridge with a steel-concrete composite deck, a simplified and conservative rule is to assume, for the design of the deck slab, the concrete as uncracked with its full elastic stiffness, reduced by shear lag effects. However, for designing the steel elements, the arch, the steel girder deck and the hangers, the slab should be assumed as cracked with its *tension stiffening effects*. If the slab is restrained by shear connectors to the steel deck at the end sections only, it must is considered an *isolated tension member*, and its effective stiffness $(EA_s)_{eff}$ should be considered for designing the steel structure. To quantify the effective stiffness of the tension member the reader is referred, for example, to [8].

6.10.3 Cable-Stayed Bridges

6.10.3.1 Basic Concepts

Cable-stayed bridges have been discussed in Section 1.5 of Chapter 1 in the general frame of cable supported bridges – suspension bridges, bowstring arch bridges and cable-stayed bridges. The flow of forces was represented in Figure 1.14 and the deck may be, as referred to, a prestressed concrete deck, a steel deck or a steel-concrete composite deck. Cable-stayed structures are elegant and efficient solutions for bridges. In the past 50 years, the range of these bridges has been steadily increasing and they are currently the most used for medium and long spans (Figure 6.133). For spans over 1000 m, cable-stayed bridges can at present compete with suspension bridges.

Depending on the transverse staying scheme, two planes of stays in lateral suspension cable-stayed bridges, or a single plane of stay cables for axial suspension scheme, the deck typology is selected at concept design phase. Although single cell box girder is the most adopted typology for axial staying schemes, box girders are also often adopted for laterally suspended decks. That is the case of long span bridges with narrow decks where aerodynamic stability may be a critical issue. In Figure 6.134, one shows one of world longest span concrete deck bridges in Norway. The deck is only 13 m wide, 2.25 m depth and the span is 530 m long. For axial staying schemes, box girder decks (Figure 1.15) are required in most cases to achieve torsion resistance under eccentric live loads and aerodynamic stability. However, for small spans it is possible to adopt an axial staying scheme with a voided slab deck (Figure 4.19) [59]. Concerning stay arrangement from the three options – fan, semi-fan and harp system; the semi-fan is adopted nowadays in most cases by reasons of economy and execution easiness (Figure 1.13). The distance of stay cables along the deck is usually nowadays from 6 to 9 m for concrete decks and 10–20 m for steel-concrete composite and all steel decks.

Longitudinally, the basic schemes for a cable-stayed bridge are represented in Figure 6.135. The three-span symmetric scheme is one of the most adopted for long and medium span bridges where two masts are adopted. A typical span arrangement is represented in Figure 6.135. For road bridges with a three-span symmetric layout the end span length L_2 is in the order of 0.4–0.45 L_1. For rail bridges, it is quite important the deformability of the main span and it is usual to reduce L_2 to values in the order of 0.2–0.25 L_1. The height h of the masts is usually in the order of 0.2–0.25 L_1 for two

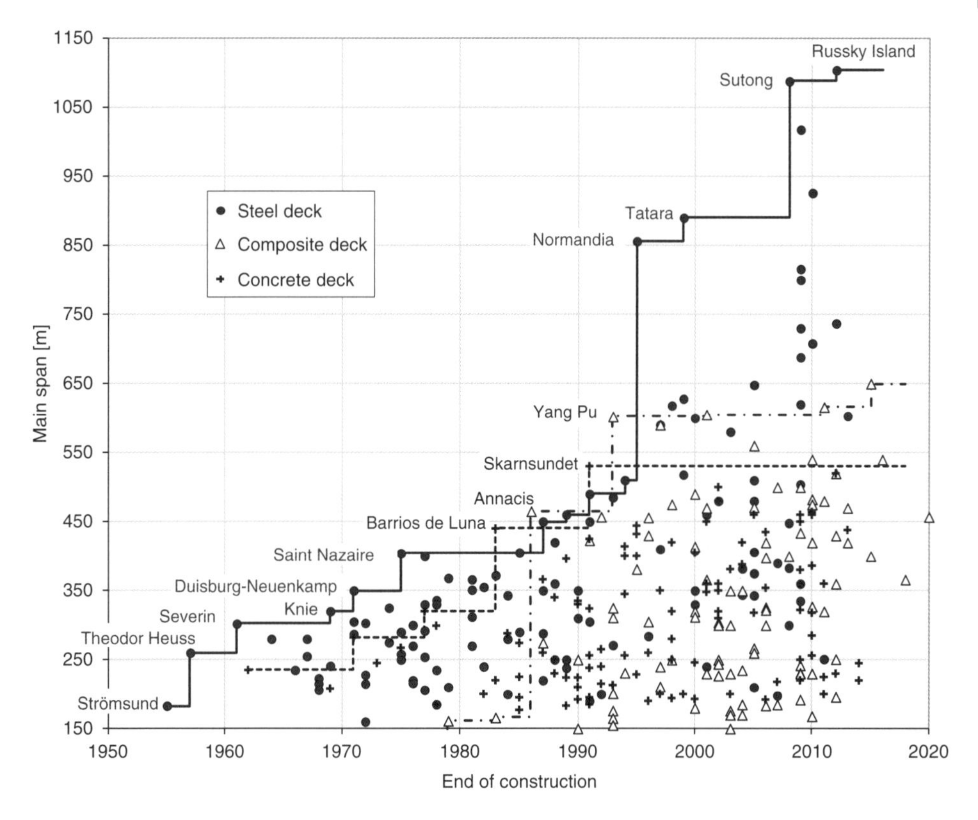

Figure 6.133 Cable-stayed bridge main span length evolution over the last 50 years.

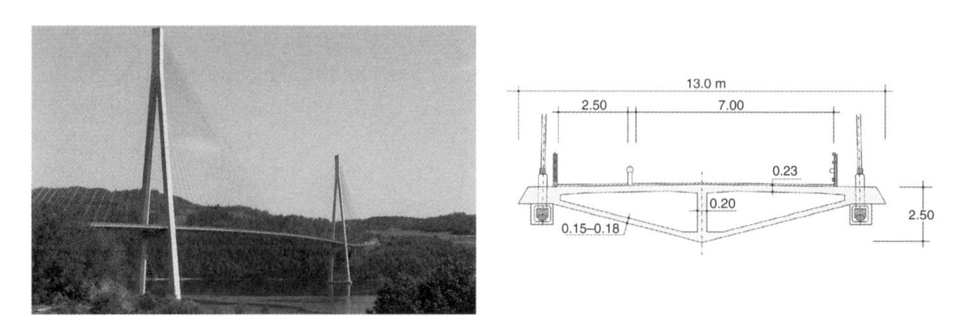

Figure 6.134 Skarnsundet bridge, Norway – a cable-stayed bridge with main span of 530 m and a concrete box girder deck with lateral suspension (*Source:* Photograph by Michael/https://commons.http://wikimedia.org / Public Domain).

mast bridges to keep the angle α between the stays and the deck at a minimum value of 20–25°. Otherwise, the staying effect is not very efficient for taking the vertical component (Figure 6.136, $N_i = q\ a\ /\sin \alpha_i$). The *effective span length* in a single mast bridge is approximately twice the span for comparison with a two mast symmetric cable-stayed bridge, that is, $L_e \approx 2L_1$, with $(L_e/h) = 0.20–0.25$. The approximate maximum

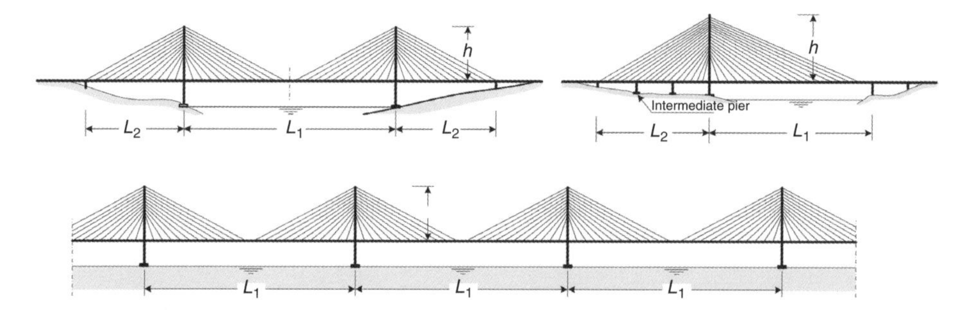

Figure 6.135 Cable-stayed bridges, longitudinal layout. Symmetric, asymmetric and multiple span staying schemes.

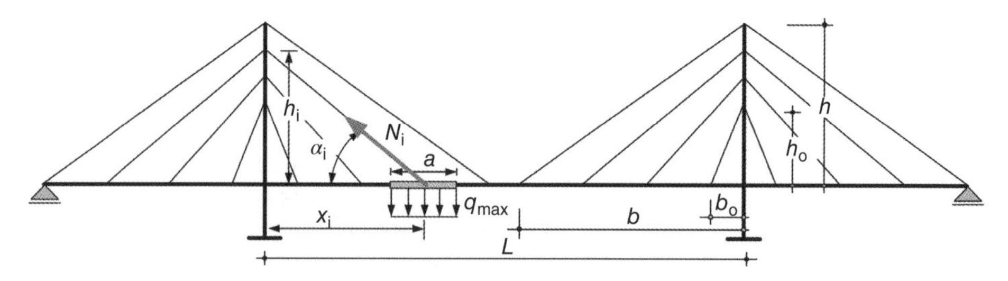

Figure 6.136 Deck transfer forces for stay cables (axial forces N_i, $i = 1$ to m each side of the mast).

axial forces in the masts and in the deck due to stay actions (m stays at each mast Figure 6.136) are

$$N_{mast} = \Sigma 2N_i \sin\alpha_i; \quad N_{deck} = \Sigma N_i \cos\alpha_i \quad \text{for } i = 1 \text{ to } m \tag{6.195}$$

The *back stays* play a key role in the transference of the forces from the mid-span to transition piers or abutments, see Figure 6.137. For end span loadings, there is a tendency for decompression at the back stays. The back stays are also quite relevant in reducing the deformability of the main span, since they control the horizontal displacements at the top of the mast.

Still, there are bridges located on sites where internal piers at the adjacent spans are unacceptable from site constraints, aesthetics or execution point of views. As exemplified in Figure 6.138 for a single mast bridge, the deformability of the main span is quite well controlled by the anchoring effect of the back stays at the adjacent span.

The back stays may be anchored at the end sections at piers or abutments – externally anchored bridges, or they may be anchored directly to the deck – a self-anchored cable-stayed bridge. These concepts come from suspension bridges as shown in Figure 6.139 for comparison. In cable-stayed bridges it is usual to anchor the back stays at the deck, even in single mast asymmetric bridges (Figure 6.140a). In suspension bridges it is possible to use the self-anchoring concept but it is not so common since the level of axial force induced by the main cable in the deck is quite large. For comparison, one shows two schemes [60] of self-anchored bridges – one suspension bridge and one cable-stayed

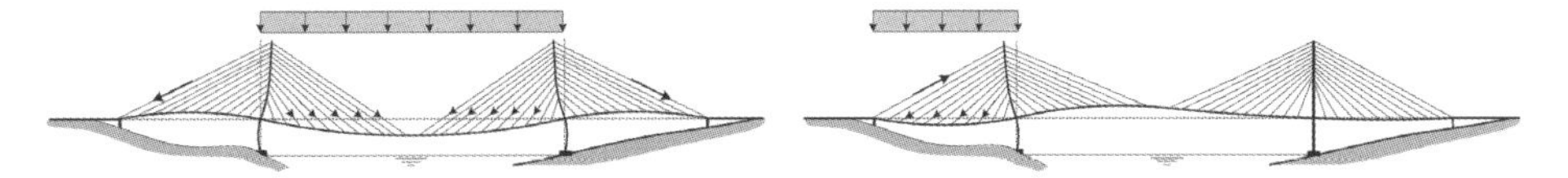

Figure 6.137 Effect of the back stays for the transference of mid-span and end span loadings.

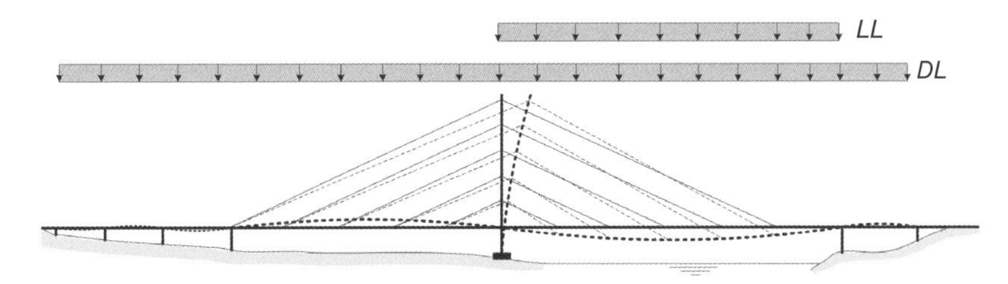

Figure 6.138 Harp stay system without intermediate side span piers.

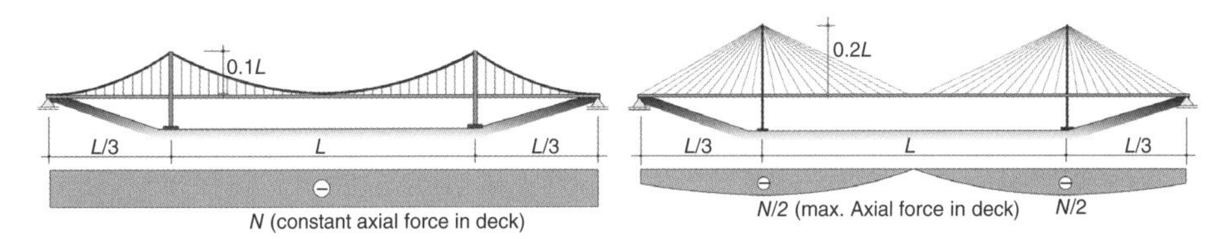

Figure 6.139 Comparison of a self-anchored suspension bridge with a cable-stayed bridge.

(a) (b)

Figure 6.140 Two examples of a 3D concept of stay arrangement – (a) Viaduct over VCI in Oporto and (b) cable-stayed bridge over Mondego River in Coimbra, Portugal (*Source:* Courtesy of GRID, SA).

bridge with the same layout. If N is the total axial force induced in the deck by the main cables, in the suspension bridge that compares with $N/2$ in a self-anchored scheme for a cable-stayed bridge. In single mast bridges, if the back stays are arranged in two planes this yields, if a single plane of stays is adopted for the main span (axial suspension), what may be called a 3D arrangement of the stay cables, as shown for two design examples in Figure 6.140. In a stay cable harp system, if no intermediate piers are adopted in the side

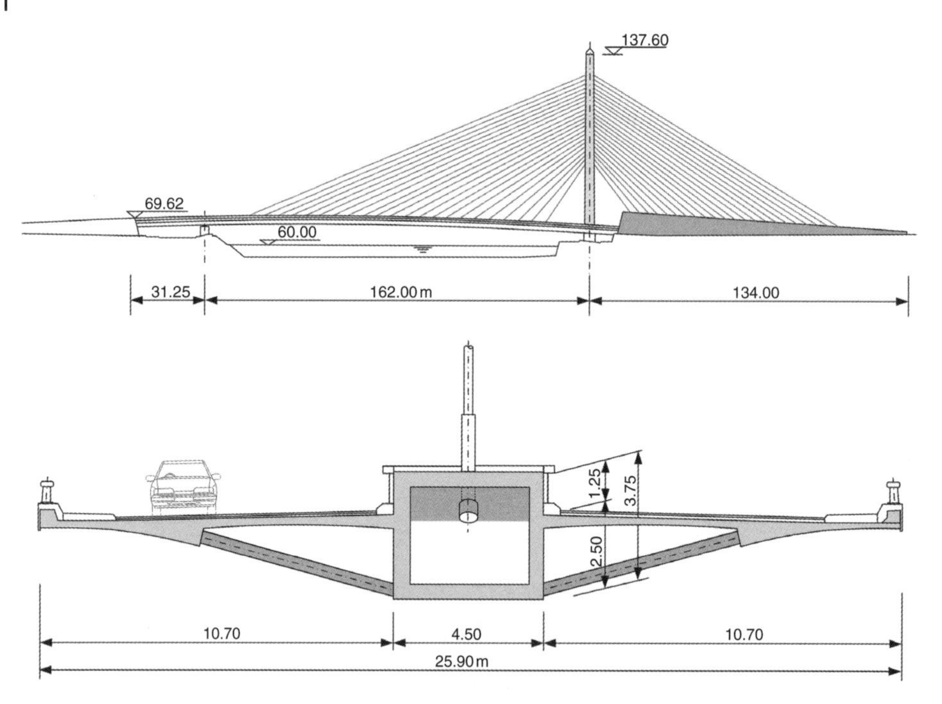

Figure 6.141 'Pont du Pays de Liège'. An asymmetric staying scheme with externally anchored back stays over the tunnel structure.

span (Figure 6.138), when the live load acts at the main span the stability of the system is controlled by the rigidity of the side span and of the mast as shown in Figure 6.137.

In most cable-stayed bridges, the deck at the lateral suspended span is continuous with at least one more spans supported on piers (Figure 1.15 and Figure 6.138). This continuity is quite important to balance internal forces at side spans and to reduce deformability of the main span because the side span tends to increase its rigidity reducing the lateral deformability of the mast and, consequently, of the main span. The deck may also be fully built in at an abutment as adopted in the Mondego River Bridge (Figure 6.140b) and in the 'Pays de Liège' Bridge (Figure 6.141) where the deck is built-in at the transition tunnel structure. This bridge has one of the most impressive examples of prestressed concrete decks with an axial staying scheme, as shown in the figure. The deck overhangs are laterally supported by stainless steel tubes.

6.10.3.2 Total and Partial Adjustment Staying Options

To introduce the concepts of staying adjustments, the deck of a cable-stayed bridge is firstly envisaged as built on fixed scaffolding. Then the stays are adjusted to meet the geometry as *passive bars* only. If the scaffolding is removed, the deck will deform (Figure 6.142) under permanent loads. At each stay anchorage point at the deck, there is a vertical deflection $u_i^0 = u(x_i)$ at point x_i, as shown in the figure. This vertical deflection is due to an elongation $\delta_{ie} = N_i l_i / E_{eq} A_i$ of the stay cable (Figure 4.143a) and a rotation of the deck due to the elongation of the back stays inducing an additional deflection $u_{ir} = (\delta_{hm}/h)\, x_i$, where δ_{hm} is the horizontal displacement of the top of the

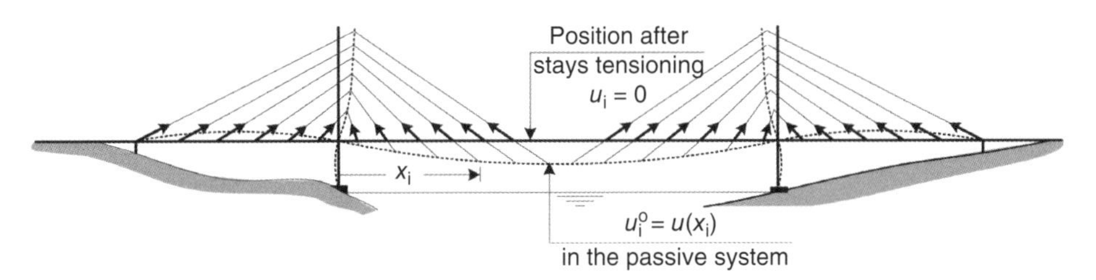

Figure 6.142 Deck deflections in a cable-stayed bridge assuming passive stays.

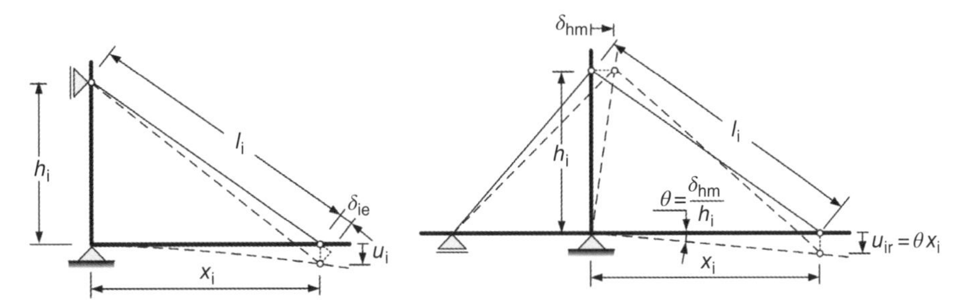

Figure 6.143 Effect of deformations of stay cables on vertical deflections of the deck.

masts, illustrated in Figure 6.143b. To obtain the correct geometry, the stay cables should be installed with a shorter length $l_{i0} = (l_i - \Delta l_i)$ to compensate for the elongations after removing the scaffolding. The other possible way of stay adjustment is to install the stays with the original length to meet the geometry of the deck on the scaffolding and after removing it, in order to tension each stay cable to compensate for the total deflection u_i^0. In the bridge model of Figure 6.144, the final bending moments may be envisaged as the bending moments in the passive system plus the induced bending moments by the stay adjustment. In the final stage of the real structure, the stay cables act as rigid supports of a multiple span continuous girder and the bending moments in the deck, under permanent loads, would be the ones shown in Figure 6.144b. This means the total permanent load at each stay cable, without considering the bending rigidity of the deck, is obtained from the support reaction of the multi span girder as $N_i = R_i/\sin \alpha_i$. Considering the deck bending stiffness, the stay cable force, in terms of the bending moments at anchor points in the deck, is

$$N_i = \left[pa + \frac{1}{a}\left(M_{i-1} - M_i \right) + \frac{1}{a}\left(M_{i+1} - M_i \right) \right] / \sin \alpha_i \tag{6.196}$$

This concept of *total adjustment staying option* can be achieved, in the structural model, by introducing the appropriate end shortening at each stay cable to completely compensate for the deflections in the passive system. This can be done by an equivalent temperature load case (temperature drop $\Delta T_{i,eq}$) at each stay *i*. Since the imposed elongation at one stay influences the deflections at all anchorages stay points, it is necessary to define an influence matrix for the effect of end shortenings at each stay and to use superposition principle.

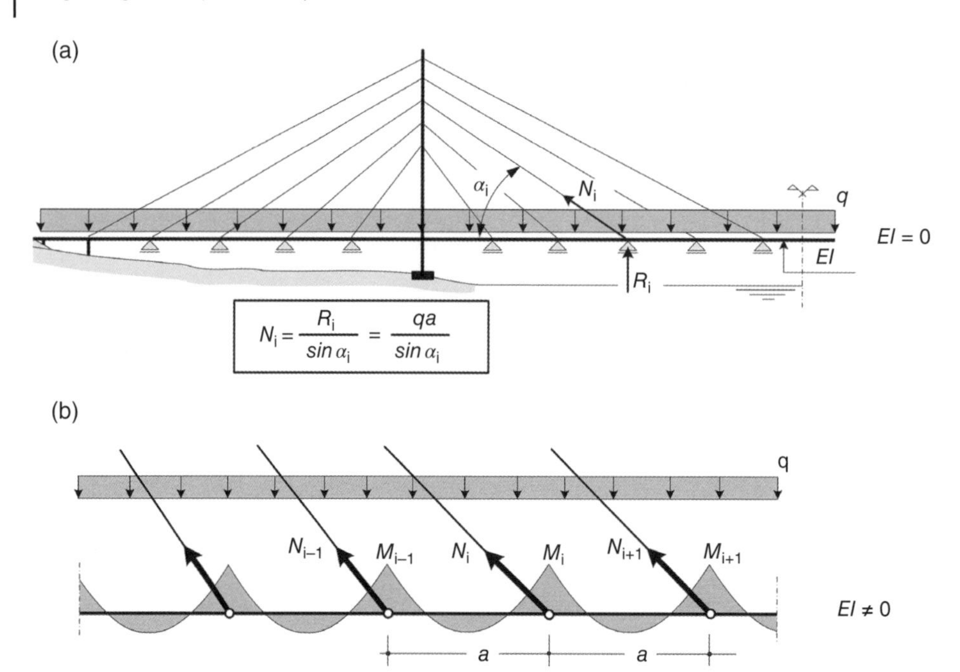

Figure 6.144 Deflections of the deck with 'passive' stay cables and after tensioning for total compensation of permanent deformations. Stay forces considering a deck with or without bending stiffness.

By determining, the *influence matrix* $[f_{ij}]$ of the vertical displacement induced at each anchor point j due to an unit axial shortening $\Delta\varepsilon_i = \alpha\,\Delta T_{i,eq}$ in stay i, the conditions for zero vertical deflections at the anchor points are, in matrix form,

$$\left[f_{ij}\right]\left\{\Delta\varepsilon_j\right\}+\left\{u_j\right\}=\left\{0\right\} \tag{6.197}$$

Here, the vertical displacements of the anchor point j of the deck, in the passive system were denoted by u_j. The horizontal displacements at the mast should also be compensated for by the effect of the imposed stay forces. At the final permanent stage, the mast should be free of bending moments as much as possible. Hence, the influence matrix f_{ij} should be extended to the deflection effects at these points. After determining the cable shortenings, $\Delta\varepsilon_i$ (or the equivalent $\Delta T_{i,eq}$), the stay forces are easily evaluated by

$$F_i = F_{i0} + EA_i\,\Delta\varepsilon_i \tag{6.198}$$

where F_{i0} are the forces in the passive system.

Apart from the influence of the erection method in concrete and steel-concrete composite decks, the deflections under permanent loads are dependent on shrinkage and creep effects. This procedure for stay forces adjustment can only be adopted as a tool for preliminary design of the stay cables.

The deck usually has bending moment resistances to allow stay forces for a partial compensation of the deflections in the passive system. The remaining deflection can be compensated by camber deflections during erection stages. This *partial adjustment staying option* is usually more economic for small and medium cable-stayed bridges. In some way, it is similar to the adopted in an extradosed *prestressed concrete bridge*

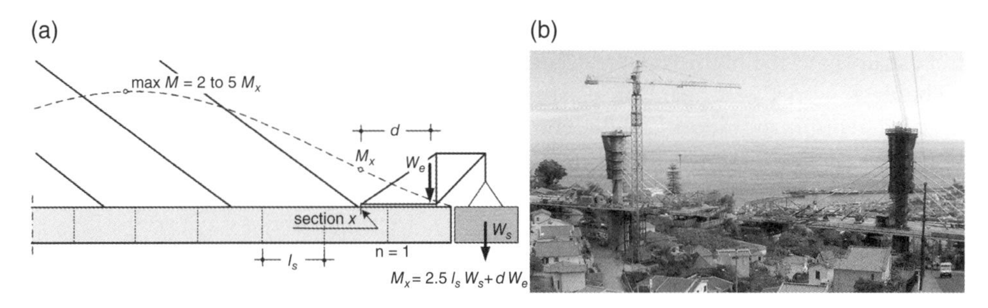

Figure 6.145 (a) Cantilevering scheme for cable-stayed bridges and (b) Comboio Viaduct during construction (*Source:* Courtesy of GRID, SA).

where stay cables are added to a box girder bridge built by cantilevering as external prestressing cables with large eccentricity effects or inserted in a concrete case [61]. As previously referred to, stay forces may be adjusted to the allowable bending moments selected for the deck. After fixing the desired final stay forces, it is necessary to determine the initial values for cable installation. During stages these forces are being modified and safety at each erection stage should be checked. Of course, the forces are dependent on the erection scheme selected. Very often cantilevering construction is the preferred scheme, at least for medium and long span bridges. During cantilevering of prestressed concrete decks, since the segments are usually approximately 5 m long and the stays are usually 5–10 m apart, the sequence of installation is as shown in Figure 6.145. The maximum bending moments in the deck during erection stages are much larger than the moment M_c at the last stay anchor point. As an order of magnitude, the maximum bending moments may reach two to five times the moment at C (Figure 6.145) given by $M_c = W_e\, e + 2.5\, a\, W_s$ where W_e is the load of the equipment (travelling form) and W_s the dead load of the segment of length a.

To find out the installation forces at the stay cables to meet the specified final dead load stay forces, usually the 'dismantling' of the final structure is necessary consisting of the consideration for the opposite direction of the sequence of the erection scheme. This starts by dismantling the closure segment (and introducing the internal forces at the tip of the two cantilevers) and then removing the cantilever segment dead loads and the stays stage by stage. For details on this subject, the reader is referred to specific literature on cable-stayed bridges [60, 62, 63].

6.10.3.3 Deck Slenderness, Static and Aerodynamic Stability

The DW of the deck is a determining parameter for the stay cable system. In general, for different decks one has:

$DW = 2.5 \text{ to } 3.5\,\text{kN/m}^2$	for steel decks with	$600 < L < 1100\,\text{m}$
$DW = 6.5 \text{ to } 8.5\,\text{kN/m}^2$	for steel concrete composite decks with	$400 < L < 600\,\text{m}$
$DW = 10 \text{ to } 15\,\text{kN/m}^2$	for concrete decks with	$L < 500\,\text{m}$

The deck slenderness of cable-stayed bridges, defined as the span/deck height ratio (L/h) may be in the order of 100–200, but theoretically it may reach much higher values. It depends very much on the type of suspension, lower values of slenderness being usually suitable for axial staying. Lateral suspended slab decks developed by Walther [62] usually have the highest values of slenderness.

The deck works as a beam on elastic foundations (BEF) under live loads and the installed bending moments (M) depend very little on the deck height. In general, one may say M is proportional to $(EI/\beta)^{1/2}$ where β is the foundation modulus per unit length depending on the section location (x) along the span (Figure 6.151). At a certain location where the spacing between stays is a, the angle of the stay (axial stiffness $E_s A_s$) with the deck is α, the foundation stiffness due to the stays only and assuming the mast as rigid is obtained from ($\beta = k/a = F/\delta a$) where k is the vertical stiffness due to the stay at location x and δ the vertical deflection induced by a vertical force F. The result is

$$\beta(x) = \frac{E_s A_s \sin^2 \alpha \cos\alpha}{ax} \tag{6.199}$$

The increased slenderness of the deck depends only on its static and aerodynamic stability. For static stability a simplified model of a column or a beam column on elastic foundation may be adopted. Under increased axial forces N in the deck, the elastic buckling load (in bending) is given by the classical Engesser formula

$$N_{cr} = 2\sqrt{EI\beta} \tag{6.200}$$

The application in practice of Eq. (6.200) has two limitations:

- the axial force N and the equivalent foundation stiffness β are not constant along the length of the deck;
- the mast and deck lateral span are not rigid as assumed to derive the foundation stiffness β.

The first limitation may be easily overcome because, as shown in [64, 65], the buckling mode is associated with the minimum value of the ratio $(\beta/N)_i$ at section i along the deck. The critical section at the location where (β/N) is at a minimum, defines the buckling mode of an equivalent BEF column with constant axial force and for a constant stiffness foundation β_i. Thus, the critical load parameter $\lambda_i = (N_{cr}/N)_i$ can still be obtained by using Eq. (6.200), with the foundation stiffness β_i of this i deck section.

The second limitation may be considered by dividing β by a coefficient $\mu > 1.0$ (tends to 1.0 for rigid masts/lateral spans and approaches 2.0 for a very flexible mast/lateral spans [65]). If $\mu = 2$, then the N_{cr} obtained by Eq. (6.200) is reduced by a factor of $\sqrt{2}$ with respect to the case of very rigid masts and deck lateral spans. In any case, the associated buckling length L_e of the deck and the number of half waves n are given by:

$$L_e = \sqrt{\pi^2 EI/N_{cr}} \quad \text{and} \quad n = \frac{L}{\pi}\sqrt[4]{\beta_i/EI} \tag{6.201}$$

The buckling length is usually much longer than the distance between two adjacent anchor points at the deck level.

Torsion buckling of axial suspended decks may also be investigated [42] but due to the shape of the deck cross section, usually a box girder deck adopted for axial suspension, it turns out torsional buckling is not a critical issue.

In general, several studies have shown that static stability is not usually a critical issue for most cable-stayed bridges in service, even for very slender decks [59]. However, aerodynamic stability may be critical, particularly for axial suspended decks, namely during construction by the balanced cantilever method. The basic aspects of deck aerodynamic instabilities have been dealt with in Chapter 3, Section 3.11.3. Gallop and flutter in a single torsion mode or classical flutter with interaction of bending and torsion deformations should be verified. For most cable-stayed bridges, at least for medium to large spans, aerodynamic appraisal is usually done on the basis of wind-tunnel tests and in most cases sectional tests, as referred to in Chapter 3. A simplified assessment, at least for pre-design, is usually done on the basis of simplified formulas discussed in Section 3.11.3. To do this in a cable-stayed bridge, first the fundamental bending and torsion frequencies should be determined. The bending fundamental frequency, obtained by the Rayleigh Method and adjusted to numerous cable-stayed bridges, yields

$$f_b = \frac{1.1}{2\pi}\sqrt{p/\delta_{\max}} \tag{6.202}$$

where p is the permanent load of the deck and δ_{max} is the maximum vertical deck deflection under permanent loads; the factor 1.1 results from the adjustment above referred. For the torsion fundamental frequencies, for lateral suspended slender bridges, the torsion stiffness mainly results from the transverse distance b_s between stays, while for axial suspended decks it results from section uniform torsion stiffness assuming free warping over supports

$$f_t = \frac{b_s}{2i_{0m}} f_b \left(\text{lateral suspension}\right) \tag{6.203}$$

$$f_t = \frac{1}{2L}\sqrt{\frac{GJ}{I_{0m}}} \left(\text{axial suspension}\right) \tag{6.204}$$

where I_{0m} is the mass polar moment of inertia of the deck cross section ($I_{0m} = \rho\ (I_y + I_z)$ where ρ is the unit mass) and $i_{0m} = \sqrt{I_{0m}/A}$ the correspondent polar mass radius of gyration.

The alternative to these approximate formulas determining vibration frequency is to adopt a numerical FEM. An example is shown in Figure 6.146 for the vibration modes of the cable-stayed bridge of Figure 6.145b with the deck cross section represented in Figure 3.26.

The numerical and analytical approximate values are very close for bending frequencies (0.90 Hz analytical and 0.83 Hz numerical) and have a difference of approximately

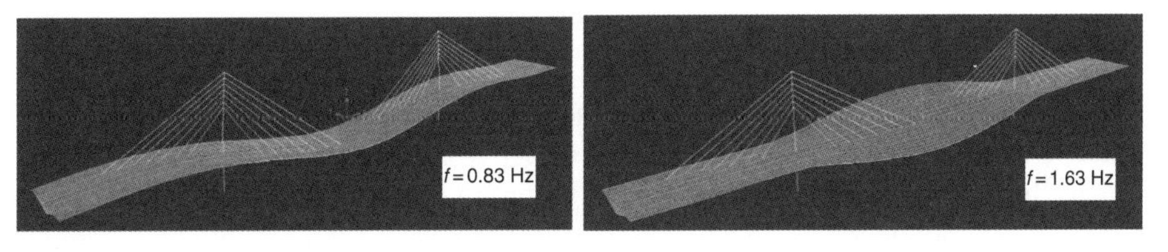

Figure 6.146 Fundamental bending and torsion frequencies and mode shapes obtained by with a FEM model, for Figure 6.145, a cable-stayed bridge.

20% for the torsion frequencies (1.90 Hz analytical and 1.63 Hz numerical). The reason is the effect of cross sectional deformations that are taken into account in the shell FEM model. By knowing the bending and torsion frequencies, the appraisal of critical wind speeds to induce aerodynamic instabilities may be done on the basis of the simplified formulas referred to in Chapter 3. A simple application to the design case here referred may be seen in [67]. Sectional wind-tunnel tests are required in any case to validate the approximate analytical results. The use of numerical models to appraise aerodynamic instabilities of cable-stayed bridge decks is nowadays more and more applicable, although wind-tunnel tests tend to be the most reliable approach.

6.10.3.4 Stays and Stay Cable Anchorages

As for bow string arch bridges, the stay cable system is nowadays made of Parallel Strand Cables (PSC) or Full Lock Coils (FLC). The former solution is the most adopted, allowing a double corrosion protection because the individual strands (0.5″ = 12.5 mm, or 0.6″ = 15 mm) are made of hot dip galvanized individual wires and a high density polyethylene (HDPE) sheath extruded around the strands. Also, the interstices between the wires and the strands and the sheath are filled with petroleum wax. The external HDPE sheet has specific surface helical fillets to avoid *rain induced vibrations* due to vortex shedding around the stay cable under wind flow. Dampers near the anchorages are usually needed to decrease vibrations of long stay cables.

A stay cable is designed for a maximum of 50% of the ultimate load under characteristic load combination and up to 45% for the frequent load combination. The ultimate stress of the individual strands is usually f_u = 1770–1860 MPa and its modulus of elasticity E = 195 GPa. The most adopted strands are usually 15 mm in diameter with an ultimate resistance in general F_u = 265 KN and a design maximum load, for frequent load combinations, of 120 kN (i.e. 45% of F_u). For the characteristic load combination, the design force may be up to 50% of the F_u.

The stay cables may have usually between 12 strands and approximately 200 strands reaching design forces in the order of 50 000 kN. For light bridges (e.g. pedestrian bridges) it is sometimes necessary to adopt stay cables with a reduced number of strands (usually 4–19 strands of 15 mm), with *fork anchorages* made of machined steel allowing tension adjustments.

Specific anchorages depending on the fabrication system are available allowing fixed and movable anchorages that may be installed at the deck level, but preferably at the mast. The last option simplifies the tension operation, nowadays made strand by strand with a mono strand jack. A specific system for controlling the forces at each strand, in order to ensure the same forces are obtained at each strand at the end, is available to the majority of stay suppliers. A final adjustment of the stay cables should be possible with a multi-strand jack. At the anchorages, there are *deviators* to keep the strands parallel and a *rubber damping device* to increase the fatigue life resistance of stays under vibration.

An efficient method to control stay forces during erection is the vibration method illustrated with a design example in Figure 6.147. It is based on the string vibration frequency formula relating the stay force F, its length L and mass per unit length μ, and number of mode shape n = 1, 2, 3, … [66]. Usually, the most difficult parameter to adjust is the equivalent free string length, mainly for short length stay cables. It is possible to correlate the frequencies (a linear correlation) measured from the power spectrum with

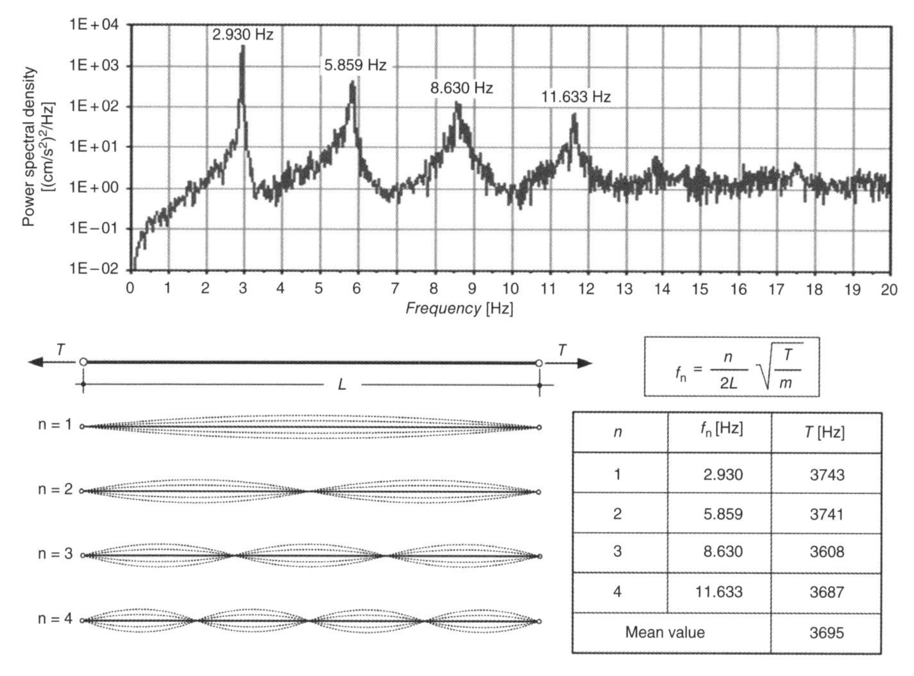

n	f_n [Hz]	T [Hz]
1	2.930	3743
2	5.859	3741
3	8.630	3608
4	11.633	3687
Mean value		3695

Figure 6.147 Control of stay forces during erection, by the vibration method. Design example for $m = 58.59\,\text{kg}\,\text{m}^{-1}$ and $L = 43.13\,\text{m}$.

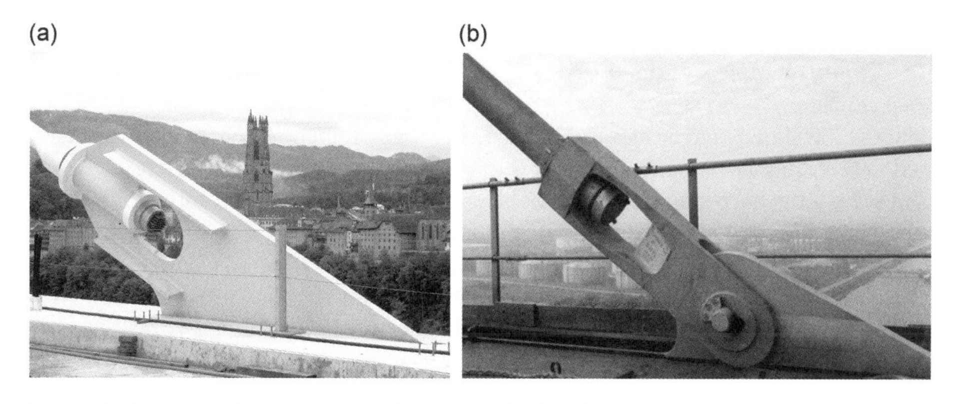

Figure 6.148 Stay cable anchorages at the deck level: (a) at the Poya bridge with a composite double plate girder deck, and (b) at the Normandy bridge with a steel deck at the main span (*Source:* Photograph by José O. Pedro).

the vibration mode number n to adjust the ideal length L to be adopted in the string Figure 6.147.

Several schemes are adopted for the stay cable anchorages at the deck level, which is usually different for concrete or steel/composite deck bridges. For latter, the anchorages are usually made on the top of the deck (Figure 6.148) and for prestressed concrete decks the anchorages are usually positioned underneath the deck.

Figure 6.149 Wandre bridge in Belgium – First bridge with a steel-concrete composite stay cable anchorage at the top of the mast (*Source:* Photograph by René Beideler / https://commons. http:// wikimedia.org / Public Domain).

The anchorages at the mast may be done directly by cable crossing (with deviation saddles, for small bridges only) or attached to the concrete walls of a tubular mast that should be prestressed to withstand large horizontal components of the stay forces. However, the most adopted system nowadays, introduced by R. Greisch and J-M. Cremer in the Wandre Bridge (Figure 6.149) in Belgium, consists of adopting a steel box at the top of the mast working in composite action with the concrete walls through headed studs.

Figure 6.150 shows two design examples of the type of anchorages at the mast mentioned before. In Figure 6.150a, in the cable-stayed bridge of Figure 4.19 (Bridge over the River Lis), the harp stay arrangement is made by a set of two adjacent parallel stays for the main span crossing at the top of the mast with a single plane of back stays; in Figure 6.150b the anchorage steel box at the top of the mast for the viaduct with a an axial semi-fan stay arrangement and a multi-cell box girder deck cross section from Figure 6.145b, is shown.

6.10.3.5 Analysis of the Superstructure

The analysis of the superstructure of a cable-stayed bridge is usually done on the basis of a FEM. For permanent load actions, the analysis of internal forces was discussed in Section 6.10.3.2. For live loads, in particular traffic loads, influence lines may be taken for positive and negative bending moments in the deck on the basis of a linear FEM (Figure 6.151). The deck works as a BEF, as previously mentioned. At least at a pre-design stage, a simplified procedure based on the BEF analogy may be adopted for defining the influence lines and internal forces due to traffic loading. Defining the foundation stiffness by β as in Section 6.10.3.3 (Eq. 6.200), influence lines for maximum positive bending moments M^+ may be estimated with quite good accuracy, as shown in Figure 6.151. From BEF, the 'elastic length' is defined as $L_e = \sqrt[4]{4EI/\beta}$.

(a)

(b)

Figure 6.150 (a) External anchorages at the masts in a small span cable-stayed bridge (bridge over River Lis – Figure 4.19) with an harp stay configuration and (b) internal anchorages with a steel box in composite action with a tubular concrete mast in a cable-stayed viaduct (deck cross section of Figure 3.26) (*Source:* Courtesy of GRID, SA).

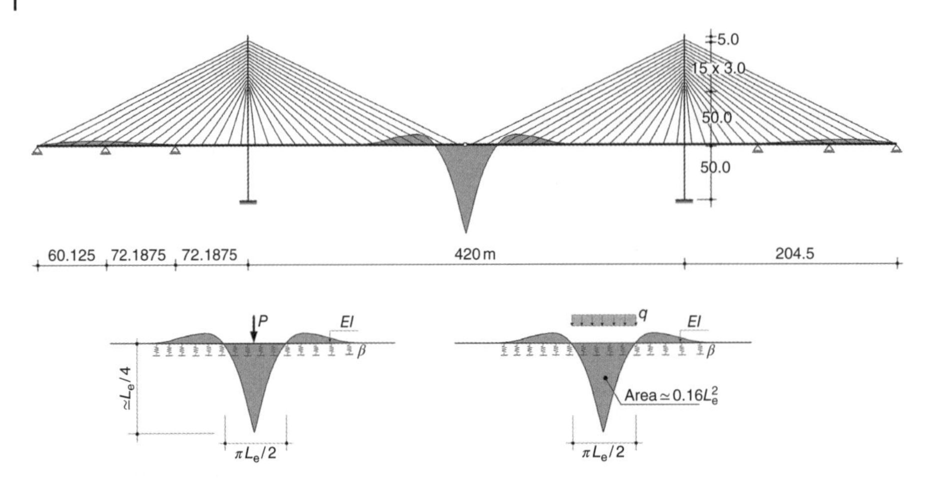

Figure 6.151 Typical influence lines for positive bending moments in a cable-stayed deck.

For a concentrated load P at the midpoint between two stays, at distance a, the positive bending moment $M^+ = P\,L_e\,/4$. For a uniformly distributed load live load q acting between two consecutive zero points of the influence line for M^+ (Figure 6.151), the area under the influence line is approximately $A = 0.16\ L_e^{\ 2}$, yielding $M^+ = 0.16\ qL_e^{\ 2}$. These two formulas generally allow good accuracy for estimating positive bending moments in the deck due to live loads. The accuracy increases for reduced bending stiffness of the deck. For negative bending moments in the deck, due to live loads, the accuracy of the BEF analogy is not so good.

In cable-stayed decks geometrical nonlinearities due to large displacements effects and, in particular, second order effects induced by large axial forces induced by the stays in the deck working as a beam-column, should be taken into consideration. A detailed analysis on these effects is presented in [68] for the specific case of composite cable-stayed decks. For medium to large spans and in particular at ULS the nonlinear effects in the deck may increase the bending moments in the deck significantly. Also, time dependent deformations (creep and shrinkage) induce redistribution effects, reducing axial forces in the deck and stay cable forces. The reader is referred to specific literature on this subject [62, 63] as well for the subject of effective slab widths for shear lag effects.

References

1 Reissner, E. (1996). *Selected Works in Applied Mechanics and Mathematics*, 624. Johns and Barlett Publishers.
2 Mason, J. (1977). *Concreto Armado e Protendido*, 203. Livros Técnicos e Científicos.
3 Beg, D., Kuhlmann, H., Davaine, L., and Braun, B. (2010). *Design of Plated Structures. ECCS Eurocode Design Manuals*. Ernst & Son.
4 EN1993-1-1 (2005). *Eurocode 3: Design of Steel Structures – Part 1–1: General Rules and Rules for Buildings*. Brussels: CEN.

5 EN 1993-1-5 (2006). *Eurocode 3 – Design of Steel Structures – Part 1–5: Plated Structural Elements: Steel Bridges*. Brussels: CEN.
6 EN 1991-2 (2005). *Eurocode 1: Actions on Structures – Part 2: Traffic Loads on Bridges*. Brussels: CEN.
7 EN 1992-2 (2005). *Eurocode 2 – Design of Concrete Structures – Part 2: Concrete Bridges*. Brussels: CEN.
8 EN 1994-2 (2005). *Eurocode 4 – Design of Composite Steel and Concrete Structures – Part 2: General Rules and Rules for Bridges*. Brussels: CEN.
9 Ghali, A., Favre, R., and Elbadry, M. (2002). *Concrete Structures: Stresses and Deformations*, 3e. London and New York: E&FN Spon.
10 Oucher, A. (1977). *Einflußfelder elastischer Platten/Influence Surfaces of Elastic Plates*. Springer-Verlag.
11 Homberg, H. (1976). Orthotropic Plates – Values Influencing Bending Moments and Shearing Forces of Discontinuous Systems (No. Rept No. 205).
12 Bakht, B. and Jaeger, L.G. (1985). *Bridge Analysis Simplified*, 294. McGraw-Hill.
13 Leonhardt, F. (1979). *Construções de Concreto – Vol. 6 – Princípios Básicos da Construção de Pontes de Concreto*, 241. Editora Interciência.
14 AISC Pelikan Esslinger Method Tables Wolchuk, R (1963). *Design Manual for Orthotropic Steel Plate Deck Bridges*. Chicago, IL: American Institute of Steel Construction.
15 Mason, J. (1977). *Pontes Metálicas e Mistas Em Viga Reta – Projeto e Cálculo*, 203. Livros Técnicos e Científicos.
16 Bares, R. and Massonnet, Ch. (1966). Le Calcul des grillages de poutres et dalles orthotropes: selon la méthode Guyon-Massonnet-Bares. Ed. SNTL-Maison d'edition technique. 431 pp. Paris.
17 Quiroga, A.S. (1983). *Calculo de estructuras de puentes de hormigon*, 463. Editorial Rueda.
18 Cusens, A.R. and Pama, R.P. (1975). *Bridge Deck Analysis*, 294. Wiley.
19 Homberg, H.a. and Trenks, K. (1962). *Drehsteife Kreuzwerke*. Springer Verlag.
20 Timoshenko, S. (2015). *Theory of Plates & Shells*, 580. McGraw-Hill Education (India) Pvt Limited.
21 Hambly, E.C. (1991). *Bridge Deck Behaviour*, 2e, 313. E & FN Spon.
22 Reddy, J.N. (1985). *An Introduction to the Finite Element Method*. International Student Edition. Mc Graw-Hill Book Company.
23 Desai, C. (1972). *Introduction to the Finite Element Method: A Numerical Method for Engineering Analysis*. Van Nostrand Reinhold.
24 Zienkiewicz, O., Taylor, R., and Zhu, J.Z. (2005). *The Finite Element Method: Its Basis and Fundamentals*, 6e, 752. Butterworth-Heinemann.
25 Crandall, S.H. (1956). *Engineering Analysis: A Survey of Numerical Procedures*. New York: McGraw-Hill.
26 Allen, H.G. and Bulson, P.S. (1980). *Background to Buckling*, 532. McGraw-Hill.
27 Reis, A. and Camotim, D. (2001). *Estabilidade Estrutural*, 470. McGraw-Hill.
28 Arantes e Oliveira, E. (1999). *Elementos da Teoria da Elasticidade*, 2e, 176. IST Press.
29 Clough, R.W. and Penzien, J. (1993). *Dynamics of Structures*. New York: McGraw-Hill.
30 C.R.C. of Japan (1971). *Handbook of Structural Stability*. Tokyo: Corona Publishing Company.
31 Bowles, J. (1997). *Foundation Analysis and Design*, 5e. The McGraw-Hill Companies Inc.

32 Kollbrunner, C.F. and Basler, K. (2013). *Torsion in Structures: An Engineering Approach*. Springer Science & Business Media.
33 Schlaich, J. and Scheef, H. (1982). *Structural Engineering Documents SED 1: Concrete box-girder bridges*. Zurich, Switzerland: IABSE, International Association for Bridge and Structural Engineering.
34 Lebet, J.-P. and Hirt, M. (2009). *Traité de Génie Civil volume 12: Ponts en acier – Conception et dimensionnement des ponts métalliques et mixtes acier-béton*. Lausanne: PPUR presses polytechniques.
35 Mathivat, J. (1983). *Construction Par Encorbellement des Ponts en Beton Precontraint*, 341. Wiley.
36 Podolny, W. Jr. and Muller, J.M. (1982). *Construction and Design of Prestressed Concrete Segmental Bridges*. New York, NY: Wiley.
37 Virtuoso, F. and Reis, A. J. (1988). Deslocamentos e redistribuições de esforços em pontes de betão pré-esforçado construídas por consolas sucessivas. 2ª Jornadas Portuguesas de Engenharia de Estruturas, Lisboa, 25–28 Novembro.
38 Menn, C. (1990). *Prestressed Concrete Bridges*, 536. Springer.
39 EN 1990 (2002). *Eurocode – Basis of Structural Design, and Amendment A1 (2005) Annex A2 – Application for Bridges*. Brussels: CEN.
40 SETRA (2006). *Footbridges – Assessment of | Vibrational Behaviour of Footbridges Under Pedestrian Loading – Practical Guidelines*, 127. Department for Transport, Roads and Bridges Engineering and Road Safety.
41 Fujino, Y. and Siringoringo, D. (2013). Vibration Mechanisms and Controls of Long-Span Bridges: A Review. SEI 3/2013, (August) pp. 248–268.
42 Reis, A.J. and Pedro, J.O. (2011). Composite truss bridges: trends, design and research. *Steel Construction – Design and Research* 4 (3): 176–182.
43 de Fomento, M. (1996). *RPX-95 – Recomendaciones para el proyecto de puentes mixtos para carreteras*, 257. Serie normativas.
44 Hirt, M., Bez, R., and Nussbaumer, A. (2006). *Traité de Génie Civil volume 10: Construction métallique. Notions Fondamentales et Méthodes de Dimensionnement*. Lausanne: PPUR Presses Polytechniques.
45 Nussbaumer, A., Borges, L., and Davaine, C. (2011). *Fatigue Strength Fatigue Design of Steel and Composite Structures. ECCS Eurocode Design Manual*, 1e. European Convention for Constructional Steelwork.
46 EN 1993-1-9 (2009). *Eurocode 3 – Design of Steel Structures – Part 1–9: Fatigue*. Brussels: CEN.
47 Reis, A. and Camotim, D. (2012). *Estabilidade e dimensionamento de estruturas*. Lisboa: Orion.
48 Hoglund, T. (1979). Shear buckling resistance of steel and aluminium plate girders. *Thin Walled Structures* 29 (1–4): 13–30.
49 Galéa, Y. and Martin, P.-O. (2007). Presentation Manual of EBPlate. COMBRI project, RFCS Contract RFS-CR-03018, Document COMBRI-Report-CTICM-005.
50 SIA 263:2013 *Construction en acier. Norme Suisse*, 108. Zürich: SIA.
51 Klöppel, K., Sheer, J., and Möller, K.H. (1960). *Beulwerte ausgesteifter Rechteckplatten: Kurventafeln zum direkten Nachweis der Beulsicherheit fuer verschiedene Steifenanordnungen und Belastungen*. 1, Ernst & Sohn, Berlin.
52 Koiter, W. and Pignataro, M. (1976). A general theory for the interaction between local and overall buckling of stiffened panels. Report no. 556, Delft University.

53 Tvergaard, V. (1972). Influence of post buckling behaviour on the optimum design of stiffened panels. Ed. Danish Center for Applied Mathematics and Mechanics, Technical University of Denmmark, Report no. 35.
54 Reis, A.J. and Roorda, J. (1979). Post-buckling behavior under mode interaction. *Journal of the Engineering Mechanics Division* 105 (4): 609–621.
55 Tveit, P. (2011). *About the Network Arch*, 2e. Norway: Agder University.
56 Reis, A.J., Cremer, J.-M., Lothaire, A., and Lopes, N. (2010). The steel design for the new railway bridge over the river Sado in Portugal. *Steel Construction – Design and Research* 3 (4): 201–211.
57 Petersen, C. (1982). *Statik und Stabilität der Baukonstruktionen*, 2e. Braunschweig: Vieweg.
58 Pflüger, A. (1951). Ausknicken des Parabelbogens mit. Versteifungsträger. *Stahlbau* 20: 117–120.
59 Reis, A. J. and Pedro, J. O. (2011). Axially suspended decks for road and railway bridges. 35th IABSE International Symposium on Bridge and Structural Engineering. London, 2011.
60 Gimsing, N.J. and Georgakis, C. (2012). *Cable Supported Bridges – Concept and Design*, 3e, 590. Wiley.
61 Reis, A. J. and Pereira, A. (1994). Socorridos bridge: a cable-panel stayed concept. IABSE/FIB Conference: Cable-Stayed and Suspension Bridges. Deauville, 1994, Deauville.
62 Walther, R., Houriet, B., Isler, W. et al. (1999). *Cable Stayed Bridges*, 2e, 320. Thomas Telford.
63 Svensson, H. (2012). *Cable-Stayed Bridges – 40 Years of Experience*, 430. Germany: Wilhelm Ernst & Sohn.
64 Klein, J-F. (1991). Ponts Haubanés à Tablier Mince. Comportement et Stabilité. Etude Théorique. Ecole Polytechnique Fédéral de Lausanne, Rapport N° 81–11.04. Lausanne.
65 Pedro, J.O. and Reis, A.J. (2016). Simplified assessment of cable-stayed bridges buckling stability. *Engineering Structures* 114: 93–103. doi: 10.1016/j.engstruct.2016.02.001.
66 de Sá Caetano, E. (2007). *Cable Vibrations in Cable-Stayed Bridges (SED – Vol. 9)*, 188. IABSE.
67 Reis, A. J., Pedro, J. O. and Feijóo, R. (2014). Tabuleiros mistos com suspensão axial para pontes atirantadas e "bowstring arches" – concepção e análise estrutural. Revista da Estrutura de Aço do CBCA – Centro Brasileiro de Construção em Aço, Vol. 3 – no. 1, pp. 89–108, ISSN 2238–9377.
68 Pedro, J.O. and Reis, A.J. (2010). Nonlinear analysis of composite steel-concrete cable-stayed bridges. *Engineering Structures* 32 (9): 2702–2716.
69 EN 1993-2 (2006). *Eurocode 3 – Design of steel structures – Part 2: Steel Bridges*. Brussels: CEN.
70 Landau, L. and Lifchitz, F. (1967). *Physique théorique – Tome VII: Théorie de l'élasticité*. Moscou: MIR.

7

Substructure

Analysis and Design

7.1 Introduction

The substructure of a bridge includes the piers, abutments and foundations as described in Chapters 1 and 4 of this book. The connections and force transfer of the superstructure to the piers and abutments is generally made by bridge bearings. The foundations of piers and abutments may be made by footings, piles or other special types of foundations, like caissons for very large bridges. The design of the various elements of the substructure requires the knowledge of the distribution between piers and abutments, whether due to applied forces or due to imposed displacements by the superstructure.

This chapter will address the basic aspects for the analysis and design of piers, foundations and abutments. Basic design concepts for structural bearings and seismic dampers are also discussed.

7.2 Distribution of Forces Between Piers and Abutments

7.2.1 Distribution of a Longitudinal Force

A longitudinal bridge model is shown in Figure 7.1. The deck is connected to an abutment E1 with a longitudinal stiffness K_{E} and it is simply supported at the other abutment E2. The connection to the piers is made by fixed bearings (P1 and P3), movable bearings (P4) or through a monolithic connection (P2). Assuming the deck to be rigid, the distribution of an horizontal force H is made according to the longitudinal stiffness of each substructure element, that is

$$K_{\mathrm{Pi}}=\frac{3EI_{\mathrm{i}}}{l_{\mathrm{i}}^{3}},\ i=1,3\quad K_{\mathrm{P2}}=\frac{12EI_{2}}{l_{2}^{3}} \tag{7.1}$$

Here, EI_{i} is the flexural rigidity of the pier i with length l_{i} admitted to be built in at the foundation level and at the deck level in the case of pier P2. The horizontal force is distributed by the abutments and piers as follows:

$$H_{\mathrm{i}}=\frac{K_{\mathrm{Pi}}}{K_{\mathrm{E}}+\sum_{\mathrm{i}=1}^{3}K_{\mathrm{Pi}}}H \tag{7.2}$$

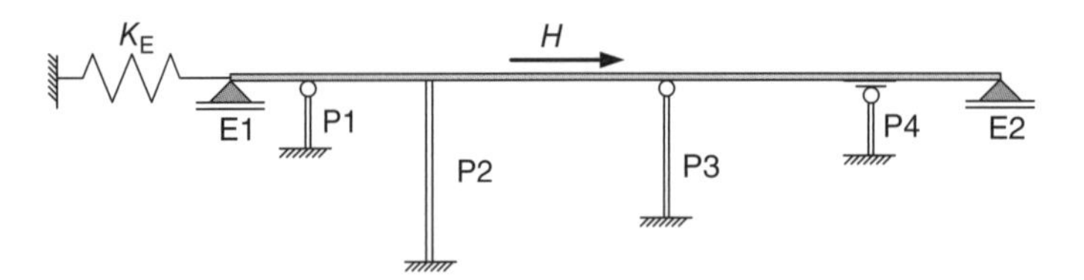

Figure 7.1 Distribution of a longitudinal horizontal force between pier abutments.

where i = 1, 2 or 3 for piers P1, P2 and P3. In general $K_E \gg K_1$, K_2 and K_3, and most of the force H will be installed at the abutment. For a rigid abutment $K_E = \infty$, hence $R_{E1} = H$. It should be noted that when the force H is the longitudinal seismic action, the force H depends on the stiffness K_E of the abutment and K_i of the piers. The *spectral acceleration or the seismic coefficient,* defined in the seismic code, increases with the *fundamental natural frequency, f,* of vibration. Assuming the deck to be rigid, one has for the equivalent one DOF system,

$$f = \frac{1}{2\pi}\sqrt{\frac{K}{M}} \tag{7.3}$$

where M is the mass of the deck and associated mass of the piers in the equivalent one DOF oscillator. Another important aspect is the axial compressive forces in the piers, reducing K_{Pi} stiffness due to *2nd order effects* (effects of instability/geometrically non-linear effects). This is particularly important for analysing the overall stability of the bridge or, in particular, to assess the buckling lengths to be adopted in piers design.

7.2.2 Action Due to Imposed Deformations

Under the action of imposed deformation of the deck on the piers – deformations due to thermal effects, shrinkage and creep in the case of a concrete deck, the end sections of the piers undergo longitudinal displacements u_i. Assuming the time dependant deformations – *shrinkage and creep,* induce uniform strains ε_{CS} and ε_{CC} along the deck and the thermal strain is uniform and defined by $\varepsilon\Delta_T$, the displacements u_i vary linearly with distance x_i from the *stiffness centre* CR. This point corresponds to the 'centre of gravity' of the stiffness's (K_E, K_{Pi}). Its X_{CR} coordinated with regard to any point, for example in relation to abutment E1, is

$$X_{CR} = \frac{\sum_i K_{Pi} X_i}{K_E + \sum K_{Pi}} \tag{7.4}$$

In the case of a bridge deck, fixed to a rigid abutment ($K_E = \infty$), CR coincides with the abutment. The displacements are given by

$$u_i = \varepsilon x_i \quad \text{with} \quad x_i = X_i - X_{CR} \quad \text{and} \quad \varepsilon = \varepsilon_{CS} + \varepsilon_{CC} + \varepsilon_{\Delta T} = \alpha \Delta T_{eq} \quad x_i = X_i - X_{CR} \tag{7.5}$$

where ΔT_{eq} is the *equivalent temperature variation* to induce a total strain ε and α the linear coefficient of thermal expansion. The forces generated at the abutment and at each pier, are given by

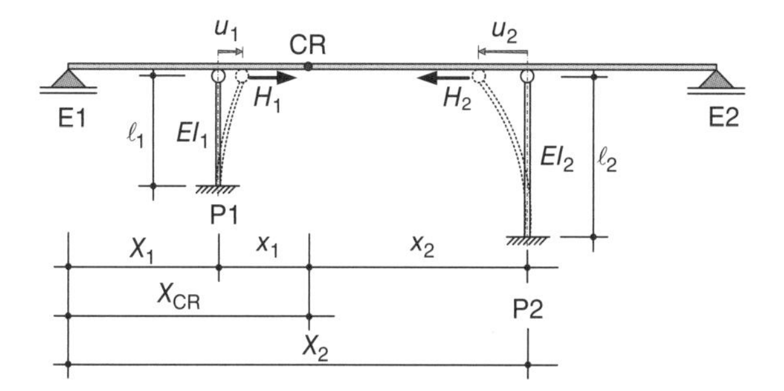

Figure 7.2 Centre of stiffness and forces induced in piers due to an imposed displacement by the deck.

$$H_{\mathrm{i}} = K_{\mathrm{i}}\, u_{\mathrm{i}} = K_{\mathrm{i}} \varepsilon x_{\mathrm{i}} = \alpha \Delta T_{\mathrm{eq}} \left(X_{\mathrm{i}} - X_{\mathrm{CR}} \right) \text{ with } i = 1,2,3 \tag{7.6}$$

In Figure 7.2 these forces are illustrated for the case of a bridge deck supported by two piers, for which the following results are obtained:

$$x_{\mathrm{CR}} = \frac{(I_1/\ell_1)3x_1 + (I_2/\ell_2)3x_2}{(I_1/\ell_1)^3 + (I_2/\ell_2)^3} \quad H_{\mathrm{i}} = 3E\left(\frac{I_{\mathrm{i}}}{\ell_{\mathrm{i}}}\right)^3 \Delta T_{\mathrm{eq}} \left(X_{\mathrm{i}} - X_{\mathrm{CR}} \right) \text{ with } i = 1,2 \tag{7.7}$$

7.2.3 Distribution of a Transverse Horizontal Force

Let the action of a transverse horizontal force (Figure 7.3) on the plane of the deck, be considered. The deck has a transverse displacement and rotation in its plane around the stiffness centre and transverse displacements at the top of piers and abutments (stiffness K_1 and K_5).

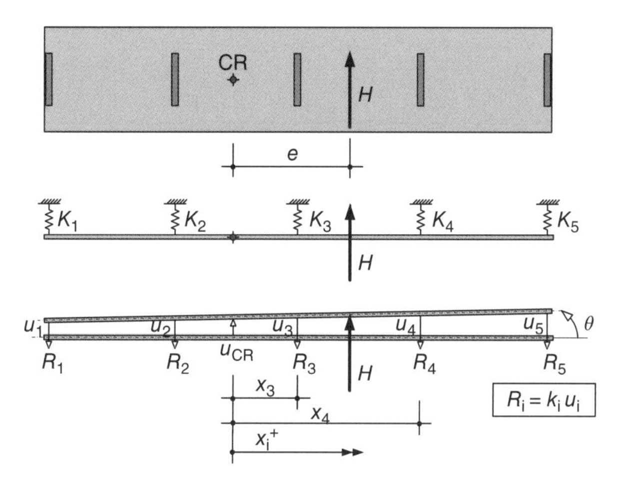

Figure 7.3 Distribution of an horizontal force between piers and abutments.

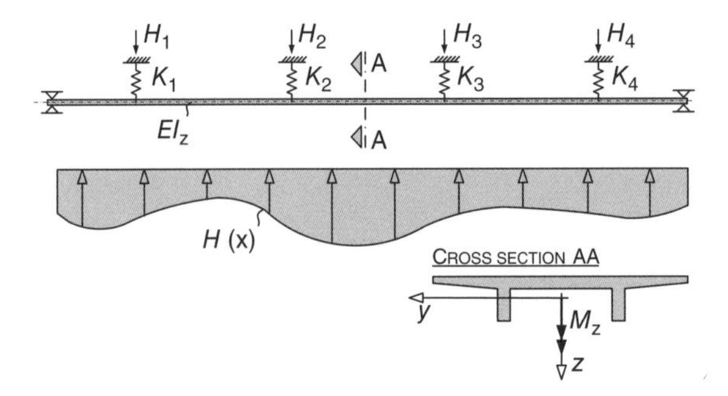

Figure 7.4 Effect of deformability of the deck between piers and abutments.

From Eq. (7.7) the location of CR is determined and the coordinates x_i are defined. For equilibrium one has

$$\Sigma K_i u_i = H \qquad \Sigma K_i u_i x_i = He \text{ with } u_i = u_{CR} + \theta x_i \tag{7.8}$$

Hence,

$$\Sigma K_i (u_{CR} + \theta x_i) = H \qquad \Sigma K_i (u_{CR} + \theta x_i) x_i = He \tag{7.9}$$

Taking into consideration $\Sigma K_i u_i = 0$ (by definition of CR) one obtains

$$u_{CR} = \frac{H}{\Sigma K_i} \tag{7.10}$$

$$\theta = \frac{He}{\Sigma K_i x_i^2} \tag{7.11}$$

The induced forces in piers and abutments are

$$H_i = \left(\frac{K_i}{\Sigma K_i} + \frac{K_i e x_i}{\Sigma K_i x_i^2} \right) H \tag{7.12}$$

This approach is valid under the simplified assumption of a rigid deck in its plane. In the case of a long deck, the distribution of forces is greatly affected by the transverse deformability of the deck. Hence, a beam model with bending stiffness EI_z supported by elastic supports (defined from the transverse stiffness of the piers) should be adopted. It is important to simulate the distribution of the horizontal force (seismic force, wind) along the deck (Figure 7.4). The abutments may generally be regarded as rigid in the transverse direction due to the large transverse stiffness introduced by the front walls. If that is not the case, the transverse deformability of the abutment is considered as well. The stiffness of the deck EI_z may take into account possible effects of *shear lag* (see Chapter 6) and cracking of the deck slab under the action of transverse bending moments M_z (Figure 7.4).

In the general case of a curved bridge deck (Figure 7.5), the rigid deck model may still be adopted to estimate the distribution of global horizontal actions H_x and H_y on the deck.

In this case, it is appropriate to begin by defining the rigidities K_{xi}, K_{yi} for each pier P_i, in the x and y directions, which may be done from the longitudinal and transverse stiffnesses on the *local axes* of piers cross sections. The effect of torsional stiffness of piers may be considered for tubular piers only. The resulting in plane displacements are

$$u_i = u + u^R{}_i,\; u^R{}_i = -r_i\ \theta \sin\alpha = -y_i\ \theta \tag{7.13}$$

$$v_i = v + v^R{}_i,\; v^R{}_i = r_i\ \theta \cos\alpha = x_i\ \theta \tag{7.14}$$

Establishing the overall equilibrium of forces, with respect to any point O, where the components of the overall forces and the moment are H_x, H_y and M, one obtains:

$$\begin{aligned} &\sum_i \left(K^i_{xx}\, u_i + K^i_{xy}\, v_i\right) = H_x \\ &\sum_i \left(K^i_{yx}\, u_i + K^i_{yy}\, v_i\right) = H_y \\ &\sum_i \left[-\left(K^i_{xx}\, u_i + K^i_{xy}\, v_i\right) y_i + \left(K^i_{yx}\, u_{i+} K^i_{yy}\, v_i\right) x_i\right] = M \end{aligned} \tag{7.15}$$

where $K^i_{xy} = K^i_{yx}$ are non-diagonal terms of the stiffness matrix of the top of pier i. As u_i and v_i can be expressed in terms of the three degrees of freedom u, v and θ of the deck, the previous system of equations allows to obtain the kinematic unknowns of the problem. Knowing u, v and θ, the displacements u_i and v_i of each pier are determined. This problem admits a very simple solution with some practical application, when:

1) it is admissible to neglect non diagonal terms $K^i{}_{xy} = K^i{}_{yx}$ of the stiffness matrix K of the top of the piers associated to displacements ($u_i = 1$; $v_i =$ o) and ($u_i = 0$, $v_i = 1$);
2) x_i and y_i are coordinates referenced to the centre of stiffness.

$$X_{CR} = \frac{\sum K_{xi}\, X_i}{\sum K_{xi}};\quad Y_{CR} = \frac{\sum K_{yi} Y_i}{\sum K_{yi}} \qquad \text{with} \quad K^i_{xx} \equiv K_{xi}\, e\, K^i_{yy} \equiv K_{yi} \tag{7.16}$$

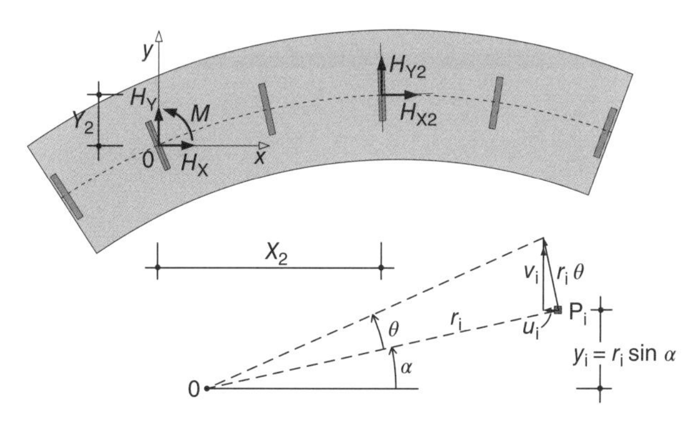

Figure 7.5 Distribution of a horizontal transverse force between piers and abutments in a curved bridge.

Under these assumptions, one obtains from Eq. (7.15), taking into consideration x_i and y_i are referred to CR ($\Sigma\ K_{xi}\ y_i = \Sigma\ K_{yi}\ x_i = 0$), the forces at each pier

$$H_{xi} = K_{xi}\left[\frac{H_x}{\Sigma K_{xi}} - y_i \frac{M}{\Sigma\left(K_{xi}\, y_i^2 + K_{yi} x_i^2\right)}\right]$$
$$H_{yi} = K_{yi}\left[\frac{H_y}{\Sigma K_{yi}} + x_i \frac{M}{\Sigma\left(K_{xi}\, y_i^2 + K_{yi} x_i^2\right)}\right] \tag{7.17}$$

The application of these expressions requires knowing the coordinates of CR and the stiffness coefficients K^i_{xx}, K^i_{yy} e $K^i_{xy} = K^i_{yx}$ obtained from the local axes coefficients (Figure 7.6) K_ξ and K_η. This is a simple problem, if a rotation of the local axes into the global ones, is considered. The rotation matrix is:

$$T = \begin{bmatrix} \cos\alpha & -\sin\alpha \\ \sin\alpha & \cos\alpha \end{bmatrix} \tag{7.18}$$

Resulting in the stiffness matrix

$$K' = T^T K\, T \qquad \text{where} \qquad K = \begin{bmatrix} K_\xi & O \\ O & K_\eta \end{bmatrix} \qquad K' = \begin{bmatrix} K_{xx} & K_{xy} \\ K_{yx} & K_{yy} \end{bmatrix} \tag{7.19}$$

This axis rotation may also be done through Mohr's circumference (Figure 7.7).

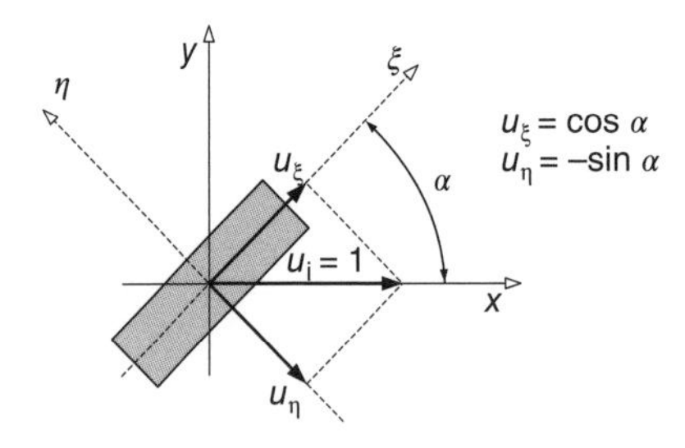

Figure 7.6 Definition of local pier axes for determining stiffness coefficients.

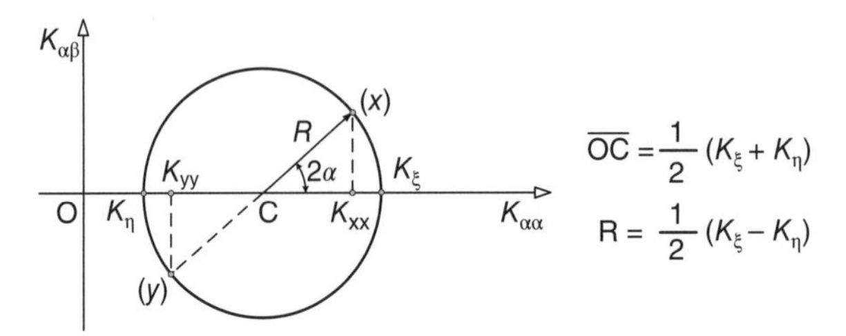

Figure 7.7 Representation of the axis rotation through Mohr's circumference

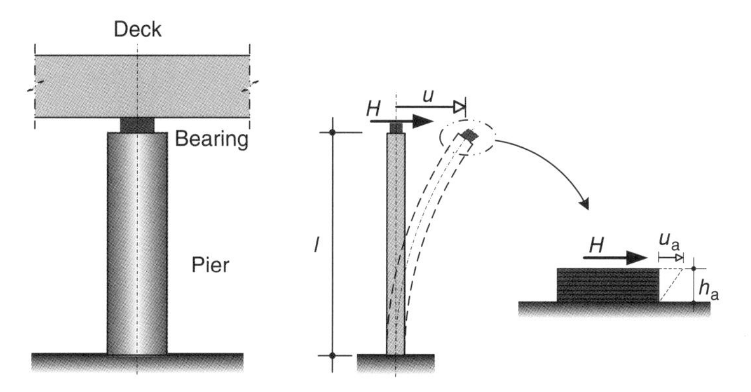

Figure 7.8 Effect of deformability of bridge bearings on the stiffness of the pier top sections.

7.2.4 Effect of Deformation of Bearings and Foundations

In the calculation of the stiffness coefficients K_i of the piers and abutments, due account should be taken of the influence of the bearings. An elastomeric bearing, for example, is a simple 'neoprene' block with area A_a, shear modulus G_a and height h; it contributes (Figure 7.8) to the deformation of the top of a pier or abutment, by

$$u_a = \gamma h_a = \frac{\tau}{G_a} h_a = \frac{H h_a}{G_a A_a} \tag{7.20}$$

$$\ell + h_a \cong \ell$$

The total displacement is

$$u = u_{\text{bearing}} + u_{\text{pier}} = \frac{H h_a}{G_a A_a} + \frac{H \ell^3}{3EI} = \left(\frac{h_a}{G_a A_a} + \frac{\ell^3}{3EI} \right) H \tag{7.21}$$

Hence, the equivalent stiffness coefficient of the top of the pier, is defined by

$$K = \frac{H}{u} = \frac{1}{\left(h_a / G_a A_a\right) + \left(\ell^3 / 3EI\right)} \tag{7.22}$$

For the influence of the deformability of the foundations on the displacements at the top of the piers (Figure 7.9), only the horizontal displacement and rotations are considered. The result is

$$u = u_{\text{bearing}} + u_{\text{pier}} + u_{\text{fondation}} = \left(\frac{h_a}{G_a A_a} + \frac{\ell^3}{3EI} \right) H + \frac{H}{K_u^f} + \frac{H\ell}{K_\theta^f} \ell \tag{7.23}$$

Hence, the stiffness coefficient at the top of the pier is

$$K = \frac{1}{\left(h_a / G_a A_a\right) + \left(\ell^3 / 3EI\right) + \left(1 / K_u^f\right) + \left(l^2 / K_\theta^f\right)} \tag{7.24}$$

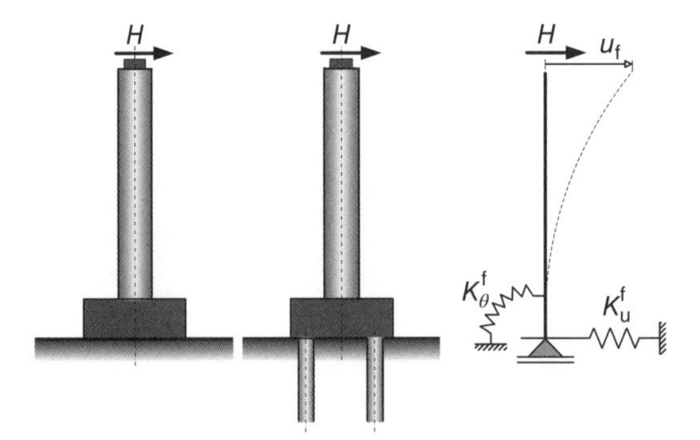

Figure 7.9 Effect of deformability of foundations on piers stiffness at the top section.

7.3 Design of Bridge Bearings

7.3.1 Bearing Types

Bearing devices between deck and piers and abutments, may be: (i) elastomeric bearings; (ii) elastomeric pot bearings; (iii) metal bearings or (iv) concrete hinges. Bridge bearings are discussed in detail in [1–3] and specific design recommendations are given in [4].

7.3.2 Elastomeric Bearings

The elastomeric bearings are made from various elastomer layers – natural or artificial rubber (synthetic); the latter being made of polychloroprene reinforced by thin steel sheets – 2–4 mm thick (Figure 7.10). The elastomeric layers, generally of 8–16 mm thickness, are vulcanized together with the reinforcement steel sheets. The unreinforced elastomeric bearings are only suitable for small loads and are not generally adopted for bridges. Artificial rubber, commercially known as *neoprene,* generally exhibits better resistance than natural rubber; in particular, resistance to ozone, ultraviolet radiation and ageing. However, *vulcanized natural rubber* has excellent characteristics and is preferred in some countries and for certain types of bearing devices, particularly for *seismic dampers.* The function of the steel sheets, in a reinforced neoprene bearing, is to improve the resistance to transverse deformations when subjected to vertical loads, as illustrated in Figure 7.11. The steel sheets increase the vertical stiffness of the neoprene bearing.

The shear stresses, τ, must be controlled, at the risk of causing cracks in the elastomer, yielding the penetration of ozone and accelerating the ageing process and deterioration of the bearing. The shear stress increases towards the edges, while the normal stresses tend to reduce towards the edges. The *average normal stress,* σ_m, is adopted for the designing the area required for the bearing.

The specification of elastomeric bearings for a bridge includes

- physical and mechanical characteristics for the elastomer.
- bearing capacity, referred to both vertical and horizontal forces (longitudinal and transverse).
- maximum allowable displacements and rotations at the bearings.
- specific requirements for the device connections to the bridge structure.
- requirements for metal corrosion protection – bearing plates and fasteners (anchors).

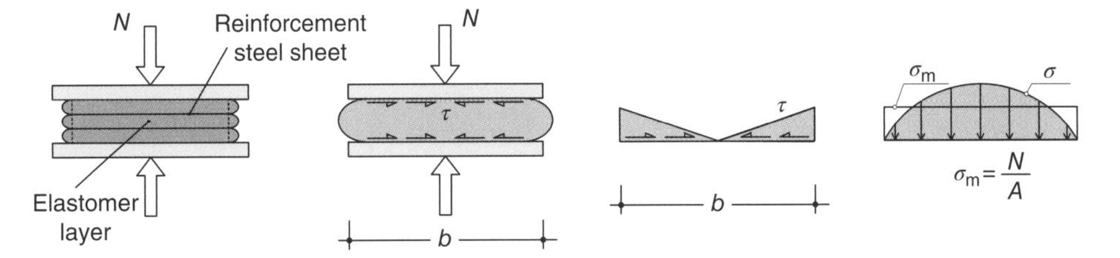

Figure 7.10 Elastomeric bearing: deformability and stresses.

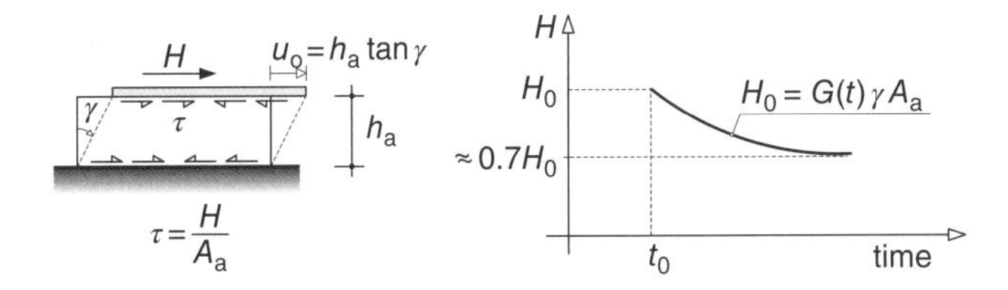

Figure 7.11 Relaxation effect on an elastomeric bearing.

These characteristics are generally defined in manufacturers' specifications. The European standard (EN 1337-3) [4] specifies the conditions for materials of elastomeric bearings, tolerances, design criteria and tests to be performed. The special conditions to be met by the elastomer are as follows [5]:

- Hardness 'shore' $A = 60° \pm 5°$
- Tensile strength ≥15 MPa
- Elongation at break ≈ 350%
- Permanent compressive strain ≤25%
- Shear modulus $G = 1$ MPa ± 0.2 MPa
- Adherence in elastomer-steel connection and resistance at the distortion test – up to distortions (γ) with maximum $\tan(\gamma) = 2$.

The transverse modulus of elasticity (shear modulus G) is variable with time; *creep deformations*, generally stabilizing after about six months, reach about 50% of their final value after a few days (around 10 days). The *relaxation effects* tend to reduce horizontal reaction forces, H, due to permanent imposed deformations. It is equivalent to reduce, in the long term, the initial shear modulus of elasticity, G_0, to a value in the order of $0.7G_o$ (Figure 7.11).

The shear modulus, G, at normal temperature varies between 0.7 and 1.15 MPa. For design purposes it may be assumed as 0.9 MPa for slow permanent actions and 1.8 MPa for instantaneous (high speed) applied actions. The stress variations do not practically cause any volume change[1] in the elastomer because its Poisson coefficient is 0.5. The modulus of elasticity, E, is $E \approx 2(1+\nu)\ G \approx 3G$. Another set of characteristics that should be specified in the design refers to the resistance to ageing as defined by the following limits:

1 The volumetric deformation $\varepsilon = \Delta V/V$ is related to the isotropic pressure p through the modulus of compressibility k, such that $\varepsilon = p/k$ with $k = E/[3(1-2\nu)]$.

- Changes by ageing (laboratory tests)
- Hardness 'shore' $A + 15°$
- Traction breaking strength 15%
- Elongation at break 40%
- Ageing in ozone, characterized by the absence of visible cracks at an amplification of seven times.

As mentioned before, the specification of bearings in bridge design is generally based on manufacturers' catalogues taking into consideration the loads, displacements and rotations corresponding to the permanent and variable actions. The design done by the manufacturers is based on criteria (set in catalogues) and tests at factory. Generally speaking, these criteria are as follows: average normal stress $-\sigma_m = N/A_a$ where $A_a = a\ \ b$ is the in plan area of the bearing; it is generally comprised between 10 and 15 MPa for the characteristic combination of serviceability limit states (ELS). This value depends on the so-called bearing 'shape factor', K_f, that is the ratio between the in-plan area A_a and the area of the lateral surface of the bearing:

$$K_f = \frac{a \times b}{2h_a\left(a+b\right)} \tag{7.25}$$

In general, σ_m values up to 15 MPa are acceptable for high K_f. For pre-design of the required in plan area A_a one takes $\sigma_m = 12$ MPa.

The stresses σ_m do not directly control the scaling of the area A_a of the bearing, but the need to limit shear stresses τ that develop transversely in the horizontal plans of the elastomer strip. These stresses, for a vertical load N, are related to σ_m by:

$$\tau_N = 1.5\frac{\sigma_m}{K_f} = 1.5\frac{N}{K_f\,A_a} \tag{7.26}$$

wherein the coefficient 1.5, experimentally determined, takes into account the variability of stresses having a detrimental effect on the bearing. Under an imposed rotation α by the deck (Figure 7.12) shear stresses τ_α are induced

$$\tau_\alpha = \frac{Ga^2}{2h_a^2}\frac{\alpha}{n} \tag{7.27}$$

where n is the number of elastomer layers. Under the action of an horizontal force H or an imposed distortion γ, the shear stresses (Figure 7.10) are given by

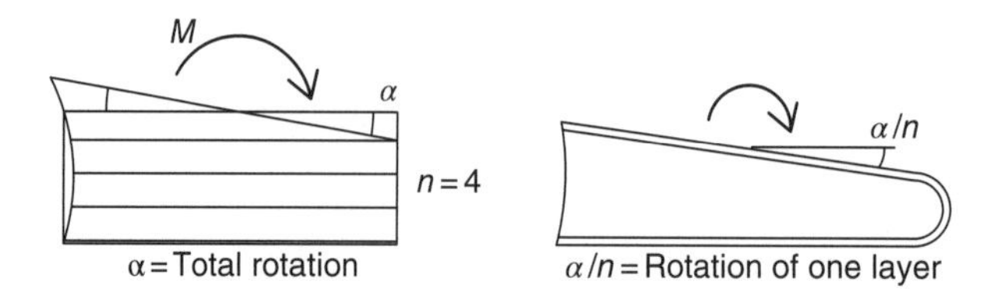

Figure 7.12 Effect of an imposed rotation on an elastomeric bearing.

$$\tau_H = \frac{H}{A_a} = G \tan\gamma \tag{7.28}$$

The total shear stress due to τ_N, τ_α and τ_H, should be limited by the resistance of the elastomer and by its connection to the steel layers as well. As a design criterion, for the characteristic load combination one may adopt, $\tau = \tau_N + \tau_\alpha + \tau_H < \tau_{lim} \approx 5\,G \approx 5\,\text{MPa}$.

On the other hand, the maximum horizontal displacement at the bearing is a function of the maximum allowable distortion, usually defined by $\gamma_{max} = 0.5$ or 0.7, for permanent and characteristic load combinations, respectively. Hence, maximum horizontal displacements for a bearing with n layers, each one of thickness, t, are limited to $u_{max} < 0.7\,nt$.

Under imposed rotations α the induced moment M at the bearing is usually quite small; for that it is convenient to adopt for the dimension, in the direction of the rotation, the smallest bearing in plan dimension, denoted by a. The moment M is

$$M = \frac{\alpha}{n}\frac{a^5 b}{ct^3}G \tag{7.29}$$

where the coefficient $c = 75$ or 60, depending on the codes.

The total height of the bearing is limited to $h_a < a/5$ for stability. The minimum permanent bearing (contact) pressure, to avoid sliding between the steel base plate and the concrete surface, should be $\sigma_{perm} > 3\,\text{MPa}$. If this last condition is not satisfied, some anchor devices, as shown in Figure 7.13, should be adopted.

Some steel devices ('steel guides') may be adopted to obtain unidirectional moving bearings (Figure 7.14) allowing displacements in one direction only. Fixed bearings may be obtained by blocking in-plan movements in two directions.

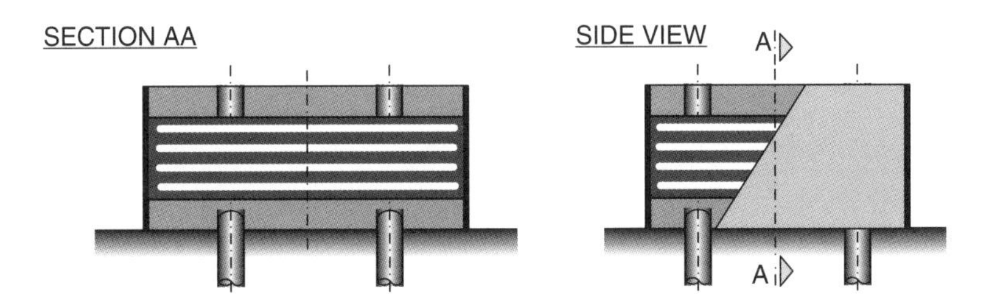

Figure 7.13 Elastomeric bridge bearings, anchored to the base.

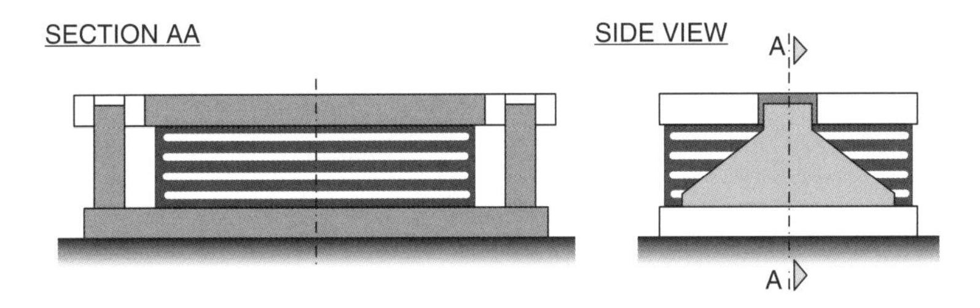

Figure 7.14 Elastomeric unidirectional bearing.

7.3.3 Neoprene-Teflon Bridge Bearings

The use of 'neoprene' bearings of the type here described is generally conditioned by two factors: the in-plan area to accommodate the device, and maximum distortion generated by imposed displacements – shrinkage, creep and temperature or horizontal forces – braking, wind, earthquakes. Regarding the first factor, it should be noted that the capacity for vertical forces of neoprene bearing devices, described so far, is generally limited by manufacturers to less than 1000 kN, which already corresponds to in-plan dimensions of approximately 900 mm.

Concerning the second factor, the maximum distortion capacity is limited by the height h of the bearing; for long bridges, the distance of the centre of stiffness to some of the bearing devices, is too large. Consequently, the imposed displacements cannot be absorbed by distortion of the neoprene layers. The forces generated by the bearing distortion are too high. In this case the solution is to adopt a 'movable support', unidirectional or multidirectional, obtained by introduction of a polytetrafluoroethylene (PTFE) layer commercially known as 'teflon'. This PTFE layer is placed between the neoprene and a stainless steel sheet, bonded to top support steel plate (Figure 7.15); alternatively, the top plate is a polished stainless steel plate.

Teflon is a polymer, thermally stable, not easy to deform and very resistant to chemicals and weathering agents. The life span of PTFE is usually predicted as being between 40 and 60 years. It has been adopted for bridge bearings from 1960 on, succeeding bronze as a sliding material for bearings. It should be noted that the difference between the friction generated in the former bronze plates, sliding against steel plates (even in well-polished sheets) and current teflon layers, is significant. In bronze, the static friction coefficients are in the order of 20%, while in teflon layers, these will be in the order of 2–4%. The *friction coefficient* of teflon, is reduced under increasing applied normal stresses (Figure 7.16).

That reduction, approximately linear for $\sigma > 2$MPa, yields friction coefficients of 2% for high normal stresses. Hence, in neoprene teflon bridge bearings it is interesting to

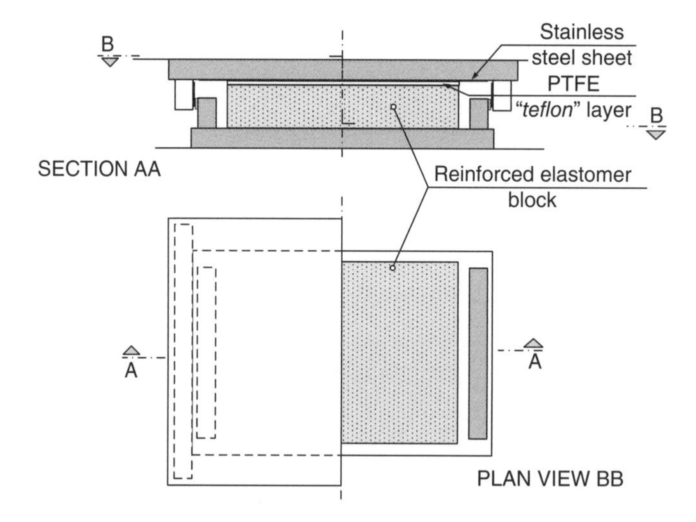

Figure 7.15 Unidirectional neoprene – teflon bearing.

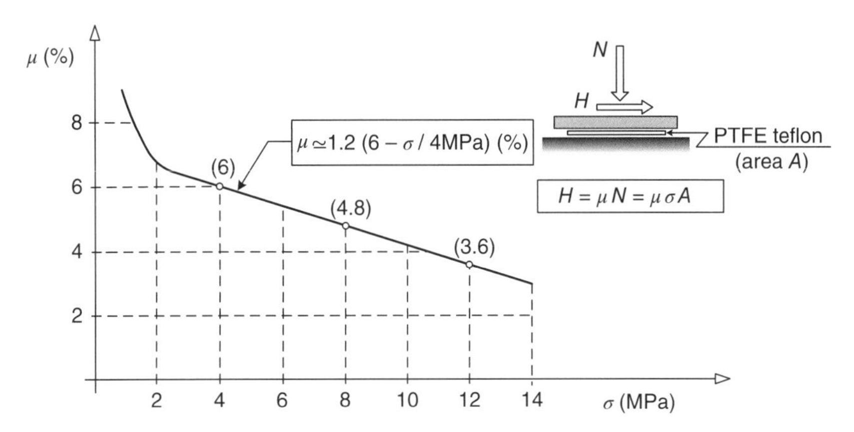

Figure 7.16 Variation of teflon friction coefficient with the applied normal stress.

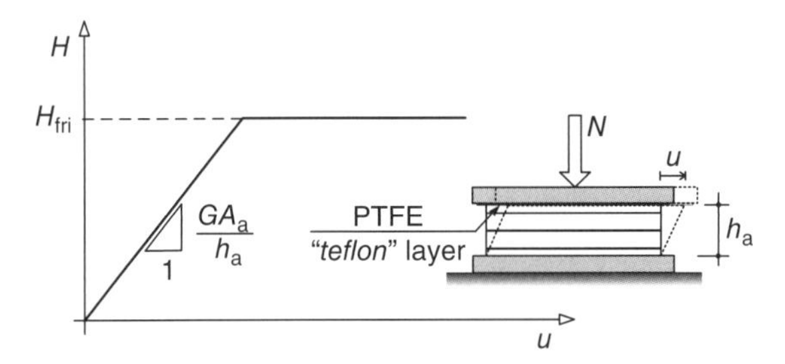

Figure 7.17 Horizontal force developed on a neoprene – teflon bridge bearing.

have teflon working under high permanent stresses; for $\sigma_{\mathrm{perm}} > 8\mathrm{MPa}$ one adopts $\mu \approx 5\%$. Design values between 3 and 5% are recommended by bridge bearing fabricators. The following conditions may be adopted for technical specifications of teflon:

- tensile resistance ≥ 17.5 MPa
- ultimate deformation ≥ 275%

In a neoprene-teflon bearing, the horizontal force due to an imposed displacement, u, is as shown in Figure 7.17,

$$H = G\,A_{\mathrm{a}}\frac{u}{h_{\mathrm{a}}} < H_{\mathrm{fri}} = \mu N \tag{7.30}$$

One may adopt fixed (Figure 7.18), multidirectional or unidirectional neoprene-teflon bearings. The last case introduces steel blocking devices, as shown in Figure 7.19.

7.3.4 Elastomeric 'Pot Bearings'

An elastomer (natural rubber or neoprene), when confined and subjected to high pressures, behaves as a viscous fluid allowing high compressive stresses and deformations by rotation with small induced moments. This is the principle adopted for bridge 'pot bearings'. A soft elastomer (shore hardness A of 50° and distortional modulus

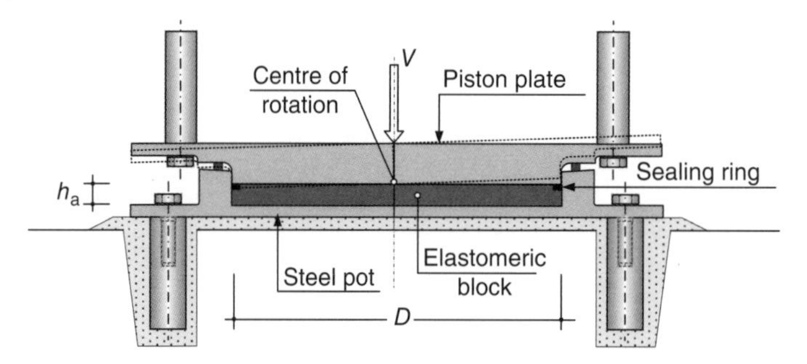

Figure 7.18 Elastomeric fixed 'pot bearing'.

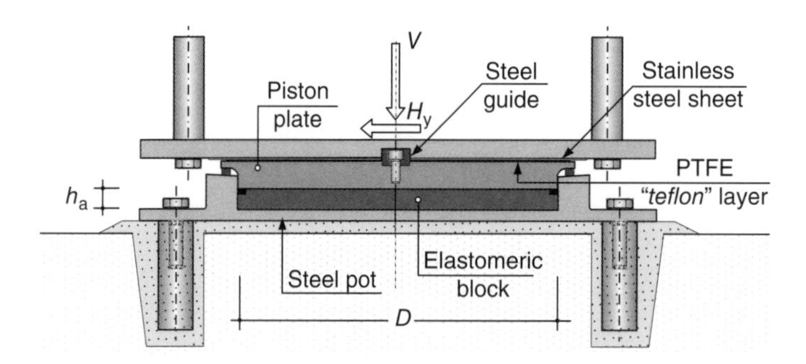

Figure 7.19 Neoprene – teflon unidirectional 'pot bearing'.

$G = 80$ MPa) is inserted in a steel cylindrical pot with an internal diameter D and with a lid (Figure 7.18) working as a piston. The lid is sealed against the internal surface of the pot by a bronze ring that is a key piece for good behaviour of the bearing device. Otherwise, the neoprene will not be contained. Inspection and quality control of pot bearings should take this into consideration.

Neoprene in pot bearings works at pressures between 30 and 50 MPa behaving as an incompressible viscous fluid [6]. Values reaching 90 MPa have already been obtained. The maximum pressure in a neoprene pot bearing is limited to the ring sealing capacity and concrete (if that is the case) bearing pressure at the steel base plate.

The elastomer thickness, h_a, depends on the required rotation capacity, α, and limitation of the induced moment, α. This moment decreases with the ratio h_a/D of the bearing and may be estimated by [7]

$$M_\alpha = 1.3\left[c \tan\alpha + 0.005\, p\right) D^3 \tag{7.31}$$

where c (MPa) is approximately 7.5 MPa for $D/h_a = 10$, is 17 MPa for $D/h_a = 15$ and 50 MPa for $D/h_a = 20$. The bearing capacity for centring the loads may be referred to the induced eccentricity $e = M_\alpha/N$.

Example 7.1 Let a bridge neoprene pot bearing, with values of $D = 400$ mm and h_a, working at a compressive stress of 20 MPa, be considered. For a rotation $\alpha = 1\%$, one has

$M_\alpha = 1.3\ [7.5 \tan(0.01) + 0.005 \times 20]\ 400^3 \text{Nmm} = 15 \text{kNm}$

$$N = \left(\pi \times \frac{400^2}{4}\right) \times 20\,\text{N} = 2512\,\text{kN}$$

$$e = 6\ \text{mm} = 0.015D$$

Hence, a very small eccentricity is obtained even under a large rotation of 1%; rotations as large as 2% (i.e. = 1/50) may be reached with this type of bearing. Due to time-dependent effects (relaxation) the moment, α, is reduced to approximately 50% of the instantaneous value at long term.

This type of bearings (Figure 7.18) is adopted for *fixed bearings*. The horizontal force capacity H_{max} is dependent on the steel pot thickness and on the shear connectors between the base plate and supports. Standard catalogue bridge bearings are fabricated for $H_{max} = 0.1\ V_{max}$ but higher values may be obtained on demand.

When unidirectional or multidirectional horizontal movements are needed, pot bearings with teflon (Figure 7.19) may be adopted. The PTFE layer, with a polished stainless steel sheet, is inserted between the upper surface of the lid and the top plate fixed to the deck. For unidirectional pot bearings, a steel central guide is inserted between the lid and the top plate, as shown in the figure.

Friction between sliding surfaces is reduced through a grease type product made on the basis of silicone. For design and bearing pressures in the order of p = 20MPa and p = 30MPa, one may take μ = 4% and μ = 3%, respectively.

It is convenient to fix the pot bearings to the concrete basis (pier and abutments seating bearing surfaces) through anchor bars in spite of large friction resistance (μ = 0.5) that could be developed, under permanent vertical forces, between the steel and concrete surfaces. If no anchors are adopted, a safety factor of 1.5 at least, should be adopted for design. Anchor devices should always be adopted for bridges in seismic zones.

7.3.5 Metal Bearings

Metal bearings have been adopted in bridges since long time ago. They allow rotations between the deck and the bridge piers and abutments, and in some cases relative horizontal movements as well. In the rehabilitation of bridges, very often 100 years old or even more (Figure 7.20), these bearings may be preserved by simple cleaning and painting rehabilitation works.

A metal bearing is made of a high strength metal cylinder or a steel cylinder with a hard metal layer at the contact surface rolling over a plate (Figure 7.21).

At the contact zone, a tri-axial stress state is developed, this being the maximum stress calculated through the classical Hertz formula

$$\sigma_{max} = 0.418\sqrt{\frac{NE}{b_o R}} \tag{7.32}$$

where b_o is the width of the bearing, N the normal force, E the modulus of elasticity and R the cylinder radius (Figure 7.22).

Maximum stress, σ_{max}, usually refers to the ultimate uniaxial steel stress (f_u) on the basis of an allowable stress $\sigma_{adm} = 5\ f_u/\gamma$ in which γ is the safety factor. Sometimes, σ_{max} has referred to approximately one half of the Brinell hard shore of the bearing metals, that is, σ_{max} = 950 MPa for steel S355, and σ_{max} = 650MPa for steel S235.

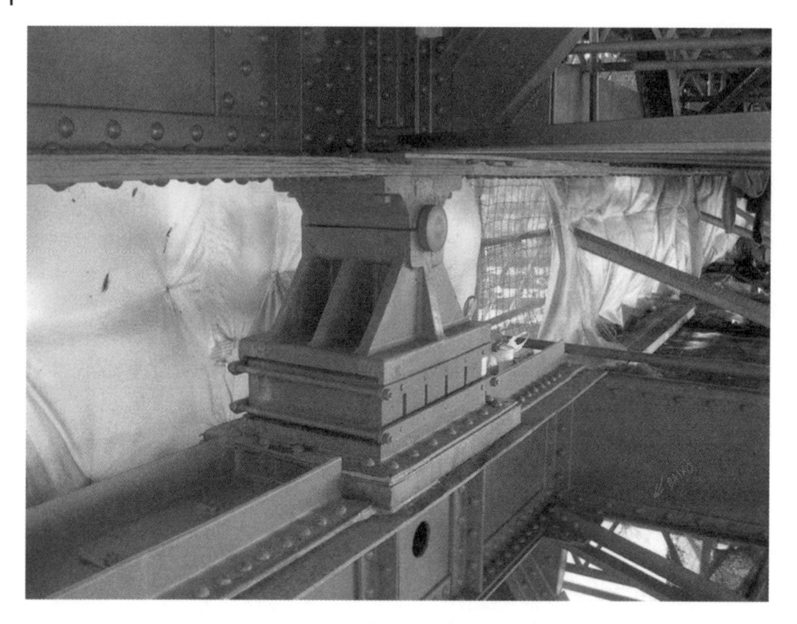

Figure 7.20 Original metal bearings in Luiz I Bridge, in Oporto.

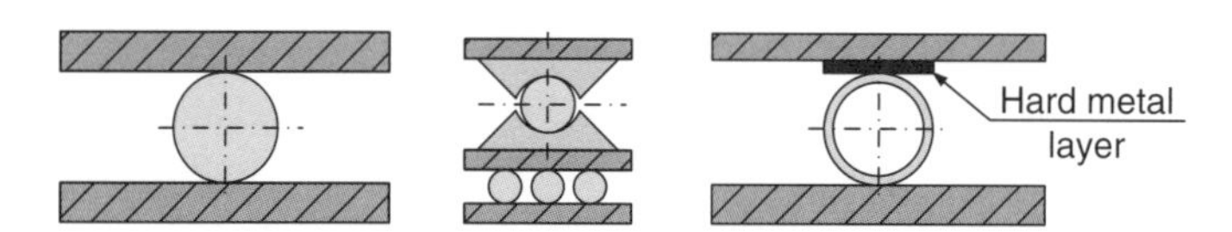

Figure 7.21 Different types of metal bearings.

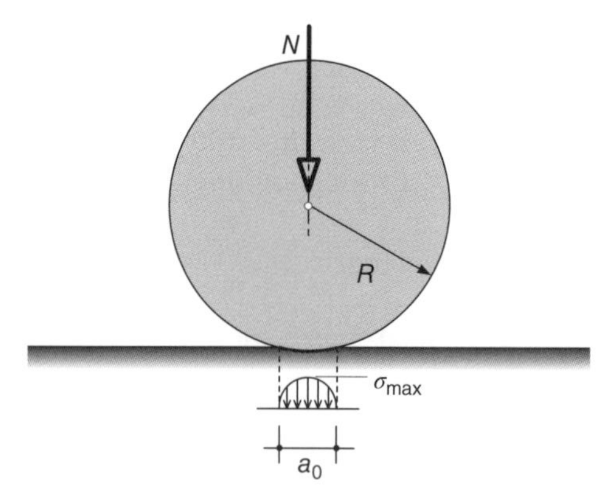

Figure 7.22 Contact area in a rolling metal bearing.

Nowadays, spherical or cylindrical steel bearings with teflon (Figure 7.23) are often adopted as an alternative to 'pot bearings'. They are made of special steel polished pieces sliding against *stainless steel* plates. Friction coefficients are very low and multidirectional, unidirectional or fixed bearings may be adopted. The in-plan dimensions, at least in one direction, are lower than equivalent pot bearings for the same load capacity.

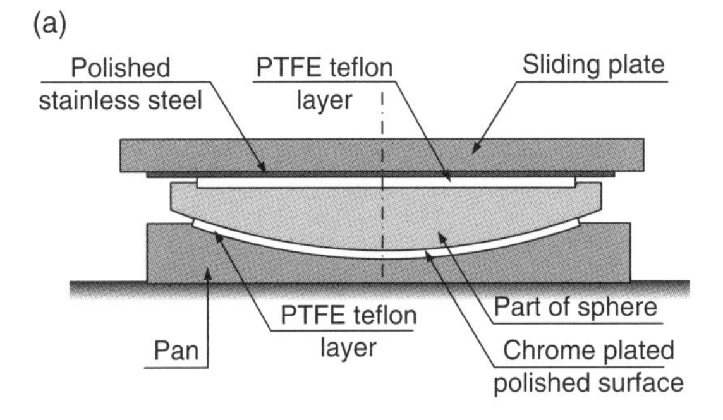

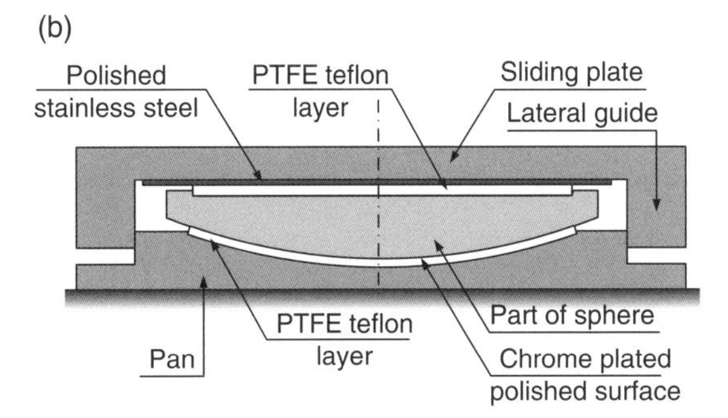

Figure 7.23 Spherical metal bearings with teflon (a) multidirectional and (b) unidirectional.

Teflon in these bearings may work at 30 MPa stresses. Standard metal bearings for loads up to 60 000 kN are available on the market, with in-plan dimensions reaching 2 m. High level rotations (order of ±3°) are allowed in spherical bearings. This type of bearing is recommended when the parallel state between the contact planes (underneath bridge deck and pier/abutment seat bearing) is difficult to reach in execution, namely in bridges with prefabricated girders for rail bridges.

7.3.6 Concrete Hinges

Concrete hinges, known as 'Freyssinet hinges' after the Canedlier Bridge, designed in 1923 by Eugéne Freyssinet, are the simplest hinges for concrete bridges. These are rotation bearings where no relative translational sliding movements are allowed.

In Figure 7.24 a scheme of a concrete hinge is presented. This is based on the principle of allowing rotations up to 1% when plasticity is reached in concrete subjected to large compressive stresses. A cylindrical hinge is reached in that case.

Under a compressive force N at the hinge, with in-plan dimensions at the reduced section $a_o \times b_o$, the average stress is $\sigma_m = N/(a_o \times b_o)$ and the maximum stress approximately $\sigma_{max} = 1.5\sigma_m$. A tri-axial compressive stress state is induced in the hinge, reducing the transverse Poisson deformations to zero due to plasticity. Small indents at the transverse direction should be adopted as shown in the figure. The load bearing capacity may reach average values in the order of five times the compressive resistance of concrete, that is, values of approximately 150 MPa or even more.

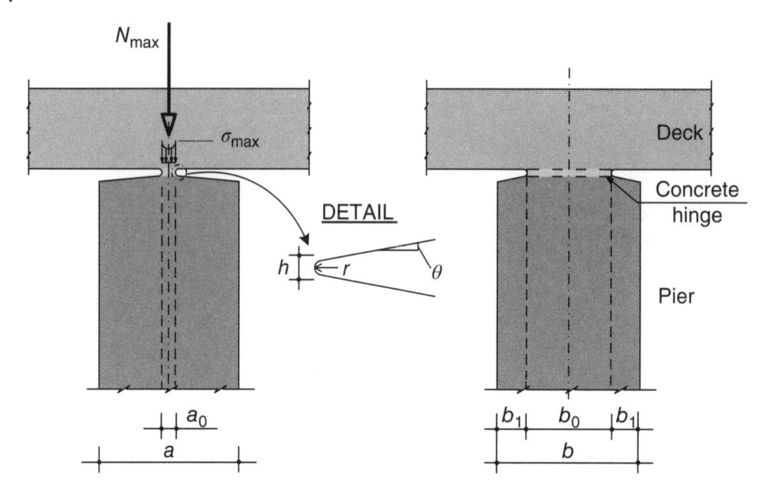

Figure 7.24 Concrete hinges – Freyssinet type.

The following design criteria are defined for concrete hinges [7]:

- $a_o \leq 0.3\,a$, with $a_o \geq 5\,\text{cm}$
- $h \leq 0.2\,a_o$, with $h < 2\,\text{cm}$
- $r = 0.5\,h$
- $b_1 \geq 0.7\,a_o$, but $b_1 < 5\,\text{cm}$

$$2\,f_{ck} < \sigma_{max} = \frac{N_{max}}{a_o\,b_o} < 100\,\text{MPa}$$

In this expression, f_{ck} is concrete resistance referred to cubic samples. For a_o, values in the order of 15 cm are currently adopted.

Shear resistance in a Freyssinet hinge may be assured by the concrete shear resistance only and the allowable value is given by $V_{adm} < 0.25\,N_{max}$.

However, this shear resistance may be increased by reinforcement steel bars crossing the hinge. Details are presented by Leonhardt [7] and the design of hinges for a specified maximum rotation α as well. Where α_g and α_q are the rotations under permanent and variable actions, then N_g and N_{max} the permanent and maximum values of N, the effective hinge area A_o should be between the following limits

$$A_{o\,min} = \frac{N_{max}}{0.85 f_{ck}\left[1+\lambda\left(1-1.47\eta\;\alpha_E\,/\,\sqrt{f_{cu}}\right)\right]} \tag{7.33}$$

and

$$A_{o\,max} = \frac{N_g}{1.25\alpha_e\sqrt{f_{cu}\times 10^3}}\;[\text{N, MPa, ‰, mm}^2] \tag{7.34}$$

with

$$\lambda = \left(1.2 - 4\frac{a_o}{a}\right) \le 0.8 \tag{7.35}$$

$$\alpha_e = \frac{1}{2}\alpha_g + \alpha_q \text{ in ‰} \tag{7.36}$$

$$\eta = \frac{N_{max}}{N_g} \tag{7.37}$$

In these expressions, N_g should be at a maximum $1.5N_{min}$ and the rotations α_g should include shrinkage and prestressing effects. Only 50% of the permanent rotation is accounted for in α_e to consider creep deformations of the structure.

The secondary moments induced at concrete hinges are quite small for most cases. Transverse movements at the hinge may induce tensile stresses that should be resisted by reinforcement steel bars, centred at the axis and crossing the concrete hinge.

7.4 Reference to Seismic Devices

7.4.1 Concept

In last three decades, there has been a great use of seismic devices in bridges. Bridges may be *seismically isolated* or *seismically protected* by dampers. Isolation is based on the principle of lowering the natural frequencies by reducing the stiffness of the structure. The seismic forces are reduced but displacements are increased. In the limit case of a superstructure that is connected to the substructure by a 'zero stiffness' bearings, like sliding bearings ideally with a zero friction coefficient, movements of the superstructure will not induce any forces on the substructure. However, in this case no resistance to horizontal forces exists for braking or wind action effects that would cause uncontrolled deck movements in practice.

Reducing the natural frequency, or equivalently increasing the natural period of vibration, reduces the acceleration in the seismic spectrum (see Chapter 3, Section 3.17). In a bridge superstructure this may be achieved by supporting the deck on elastomeric bearings allowing the horizontal stiffness to be reduced and consequently reducing the *effective accelerations* of the deck and seismic forces, as shown in Figure 7.25. The main difficulty of this approach is how to deal with long continuous deck structures because the displacements imposed to elastomeric bearings located at long distances from the *stiffness centre of the bearings*, will be too large.

7.4.2 Seismic Dampers

Protecting the structure by seismic dampers consists of increasing the damping for deck oscillations. The seismic dampers, introduced between the superstructure and the substructure, are nowadays usually *viscous fluid dampers* with 'silicone' in a piston as shown in Figure 7.26. After a certain load is reached, the deck can oscillate horizontally as in the previous case, but with a lot of damping, reducing the seismic forces in the spectrum. It should be noted that the response and natural frequency of a damped

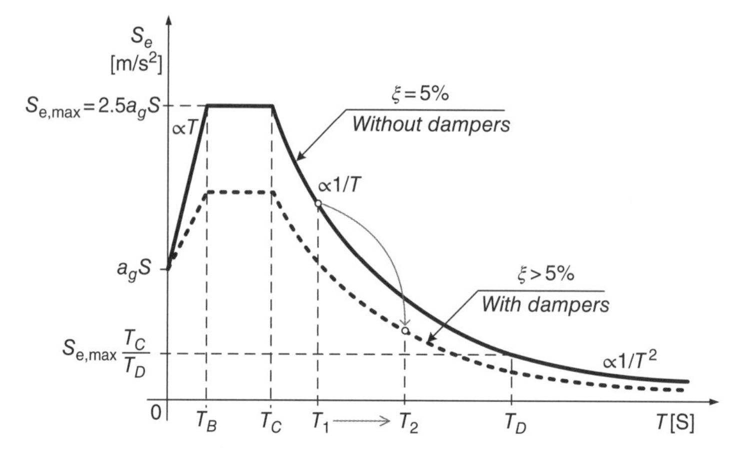

Figure 7.25 Elastic response spectrums for the original structure and for the seismically protected structure.

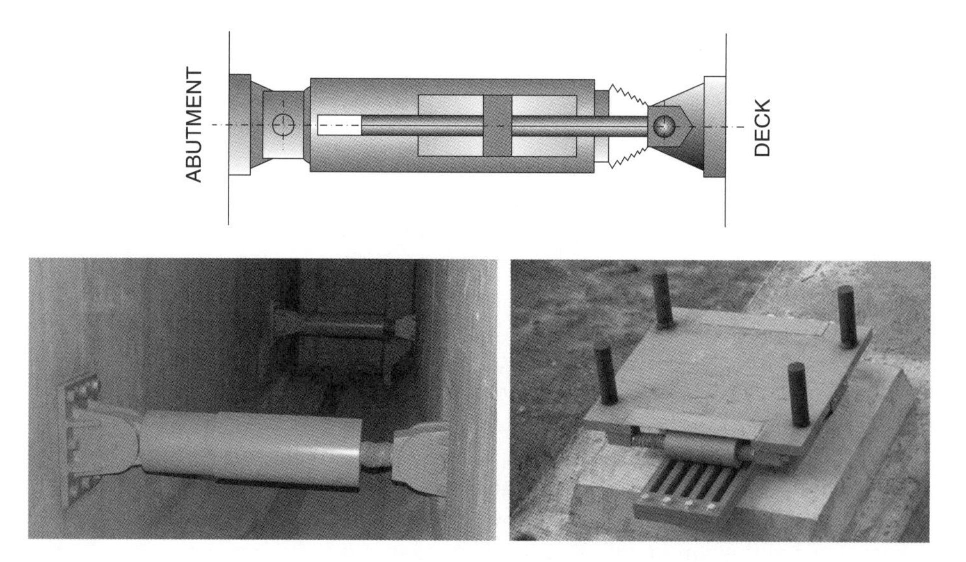

Figure 7.26 Viscous damper – schematic drawing and Sorraia Viaduct devices at the abutment and pier bearings.

system were defined in Chapter 3 by Eq. (3.16) and Eq. (3.19). Seismic energy may be thought as being transferred from one part to the structure energy and the other part as damper energy dissipation.

If seismic dampers are introduced, for example, at both abutments, and only a fixed bearing is retained at the one of the central piers, slow movements of displacements due to thermal actions, shrinkage and creep effects in the deck are allowed at the dampers without imposing any force to the abutments. If no seismic action occurs, the horizontal actions due to wind or braking forces are taken by the pier with fixed bearings.

However, if a seismic event occurs, this induces a high frequency movement in the deck that is restricted by the dampers to a certain force level. If this level is exceeded, the deck oscillates longitudinally with a large damping under a constant force transferred from the dampers to the abutments and to the pier with fixed bearings. Of course, the fixed bearings at the pier may also be avoided but in this case it will be the seismic devices and the friction forces at the bearings to resist the wind and braking forces.

Instead of applying the seismic dampers between the deck and the abutments, a series of dampers between the deck and piers may also be adopted.

Seismic dampers shall be designed to allow the maximum displacements occurring during a seismic event. The maximum force that can be developed in the dampers is a design choice taking into consideration the substructure resistance capacity. The load–displacement relationship in a damper may be considered to be an elastoplastic behaviour (Figure 7.27), where the 'yield' force is a function of the velocity V of the high frequency movements u in the damper, defined by a constitutive law

$$F = CV^{\alpha} \tag{7.38}$$

Here, C and α are parameters to be defined in design to control the maximum force in the structure and maximum displacements. The seismic device may be designed in such a way that it is blocked until the force reaches a certain value F_0. Besides, it may incorporate a spring system in order to keep a certain elastic stiffness even after the occurrence of the earthquake. If a *preloaded spring* is adopted (Figure 7.27) in conjunction with the viscous damper, the system returns to the original position after the earthquake. The constitutive law is defined for this case by

$$F = Ku + F_0 + CV^{\alpha} \tag{7.39}$$

Seismic devices may be applied to allow longitudinal or (and) transverse displacements of the deck at the pier tops and abutments. This means a deck may be totally protected by seismic devices. Seismic devices should be designed to control maximum movements of the deck compatible with links to substructure and expansion joints.

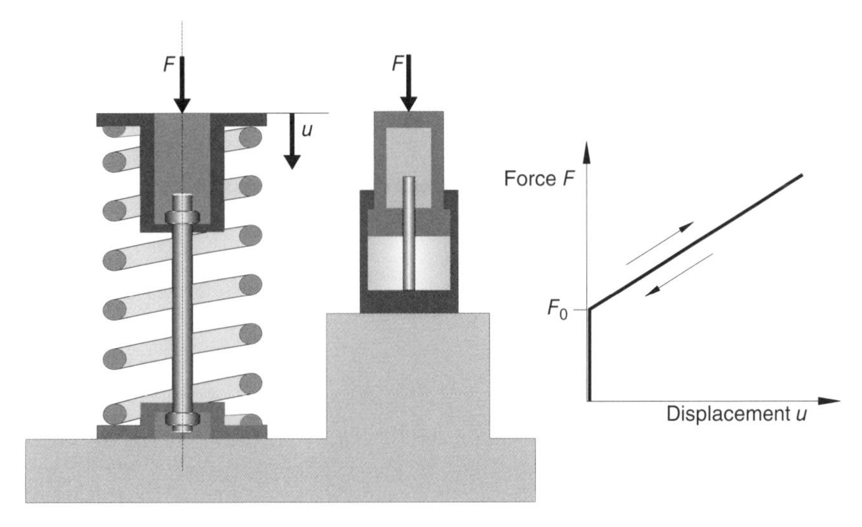

Figure 7.27 Scheme of a viscous damper with a pretensioned spring system for movement recovering after an earthquake.

To evaluate maximum deck movements when seismic devices are adopted, a nonlinear dynamic analysis introducing the constitutive laws of the dampers has to be carried out. This is usually done on the basis of a time history analysis for accelerograms of *artificial earthquakes* obtained from the *seismic power spectrum*. The reader is addressed to specific references [8] for additional information on this subject.

7.5 Abutments: Analysis and Design

7.5.1 Actions and Design Criteria

Different types of abutments have been discussed in Chapter 4. The actions to be considered for designing abutments are as follows (Figure 7.28):

- dead load of the abutments and of the soil enclosed in the abutment structure.
- actions due to the bridge deck transmitted through the bridge bearings (vertical and horizontal bearing reaction forces R_v and R_h).
- earth pressures against the abutment walls associated to at rest or active soil impulses, I_0 (at rest soil coefficient K_0) or I_a (active soil coefficient K_a) including live load effects S.
- horizontal forces H induced at the expansion joint zone of the abutment.

The reaction of the transition slab R_{TS} induces a favourable action effect for the abutment stability. However, it is difficult to quantify since it depends on the deformability of the slab supported by the soil. A model of a slab on an elastic foundation may be developed to estimate the slab forces induced at the abutment. The passive impulse at the front of the abutment is favourable and very often neglected.

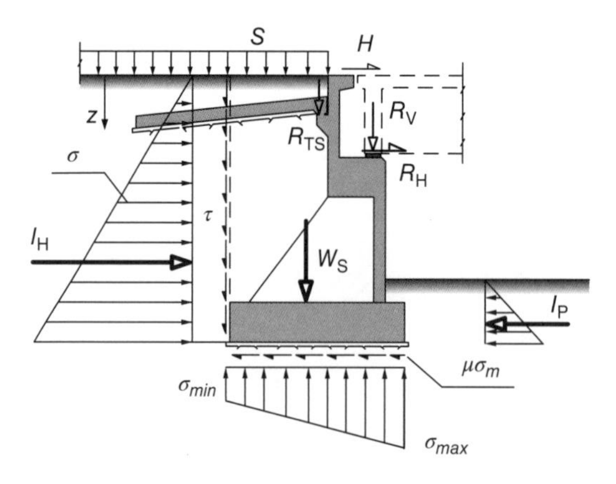

R_V,R_H – deck reactions; H – traffic actions on the abutment; S – traffic live loads
W_S – soil dead weight; R_{TL} – reaction of the transition slab; I_H – soil impulse on the abutment
τ – shear stresses; σ – normal stresses; I_P – passive impulse on the front

Figure 7.28 Actions on a bridge abutment.

To quantify the soil impulses on the abutment, the horizontal displacements, in particular, due to foundation rotation and structural deformations of the abutment front walls should be estimated. From soil mechanics, it is known active impulses may be considered for foundation rotations of the abutment reaching values of the order 10^{-3} to 10^{-2} rad, for granular to cohesive dense hard soils. The rotations are so small that at equilibrium ultimate limit state (ULS) the active soil pressure may be taken for overall (sliding and overturning) stability verifications of the abutment. For other cases, such as for making design verifications at the serviceability limit state (SLS), the at rest impulse should be considered. For the wing/lateral walls, the soil is being contained by the structure of the abutment. Hence, the at rest impulse should always be taken on a conservative design basis.

For abutments seated on a pile foundation, the horizontal movement of the pile cap may be sufficient to reduce at rest the earth pressures to active soil impulses. However, imposed horizontal deformations on the pile cap may induce large pile crack widths that should always be verified. The effect of a large soil compaction at the back of the abutment may even be considered, which may induce earth pressures higher than the at rest soil impulses. This has been measured so far and should be taken into account when dealing with abutments on pile foundations. For these cases, at rest soil impulses should at least be accounted for. The passive earth pressures at the back of the abutment, when a backwards movement is induced, for example due to a seismic force, are not considered for design since the movement of the abutment may not be sufficient to achieve a passive earth pressure.

The following soil mechanics formulae are adopted to evaluate the soil (unit weight γ) impulses, active, at rest and passive, on the back walls of abutment structures:

$$\sigma = K\left(\gamma Z + S\right) \quad \tau = \sigma \tan\delta \tag{7.40}$$

with

$$K = K_a = \tan^2\left(\frac{\pi}{4} - \frac{\phi}{2}\right) \tag{7.41}$$

$$K = K_o = 1 - \sin\phi \tag{7.42}$$

$$K = K_p = \tan^2\left(\frac{\pi}{4} + \frac{\phi}{2}\right) \tag{7.43}$$

where K_a, K_0 and K_p, are, respectively, the active, at rest and passive soil impulse coefficients and ϕ the soil friction angle and δ the friction angle between the soil and the internal surface of the front wall.

For the abutment type in Figure 7.29, soil pressures acting on the upper parts of the abutment and the abutment counterforts should be distinguished. The soil surrounding the counterforts may be mobilized with them and may be considered for pressure evaluation as corresponding to an exposed area with three times the width of the counterfort.

The overall stability of the abutment (overturning and sliding) is verified for the active earth pressure, since the abutment movements have to be sufficiently large to reach an ULS of Equilibrium. The friction effect between the soil filling the abutment and the front wall

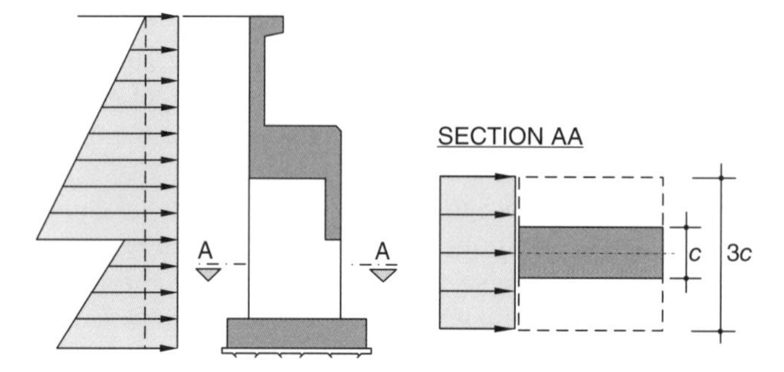

Figure 7.29 Soil impulse on an abutment with a partial front wall.

inducing friction stresses τ (Figure 7.28) has a stabilizing effect. It is not recommended to adopt a value larger than 10° for δ. Friction between the footing and the foundation soil or rock is defined through the friction coefficient, μ. Values between 0.4 and 0.6, depending on the foundation material, may be adopted.

7.5.2 Front and Wing Walls

Figure 7.30 shows the earth pressures against the walls of an abutment. These horizontal pressures are defined at each height.

Earth pressure p may be conservatively calculated on the basis of the at rest pressure soil coefficient K_o. For the geometry as shown in Figure 7.30, a continuous beam model may be adopted. To assume unidirectional bending of the wall (in the transverse direction), the height between the upper surface of the footing and the underneath of the seating beam should be at least $2b$.

Drainage backwards of the abutment walls should be made using specific drain materials as adopted in standard retaining walls; otherwise, water pressures should be account for. The submersed unit weight of the soil γ_{subm} is reduced with respect to dry unit weight γ, but the hydrostatic pressure should be added as shown in Figure 7.31.

Pressure against the front wall is transferred as the reaction of the wall panel on the counterforts (Figure 7.32). These are designed for shear and bending moment as the counterforts of a standard retaining wall.

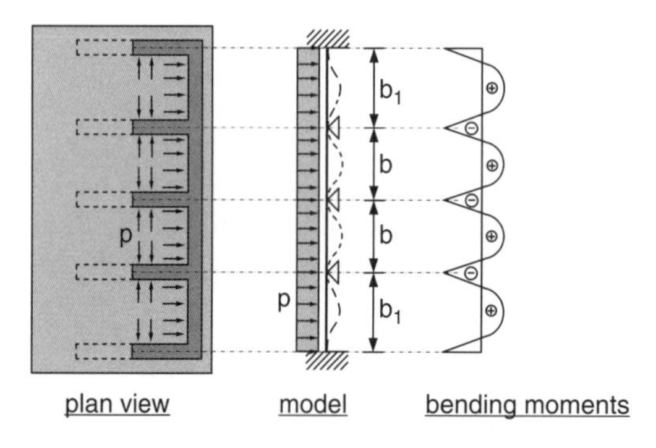

Figure 7.30 Bending moments in a front wall of a bridge abutment.

Figure 7.31 Effect of water pressures on the backwards of an abutment.

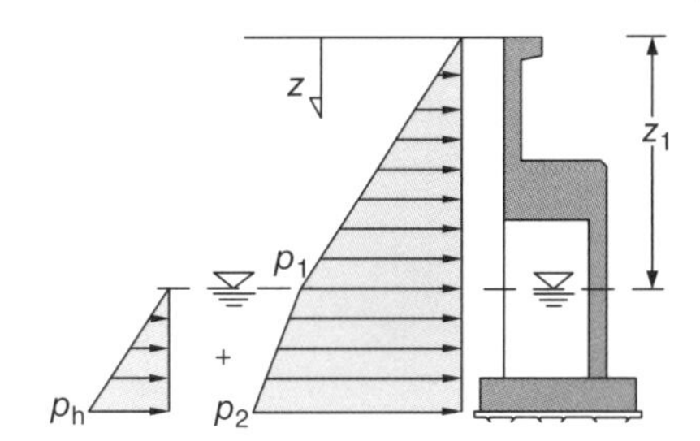

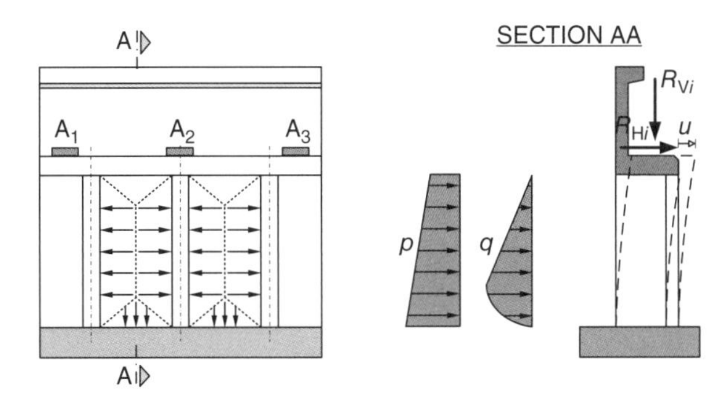

Figure 7.32 Soil pressure distribution p on the abutment front wall and reaction q on the counterfort. Reactions R_V and R_H of bridge bearings, respectively.

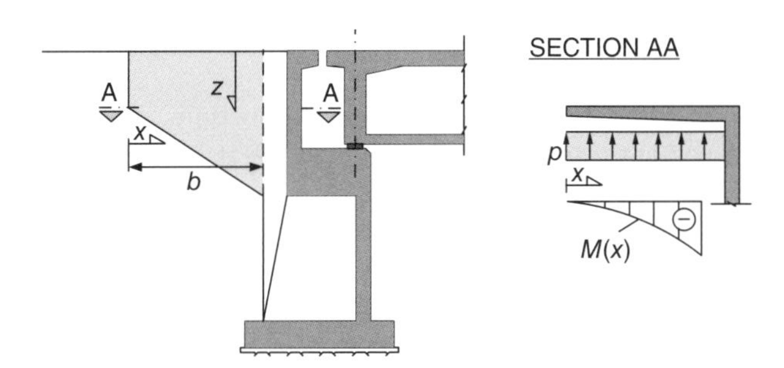

Figure 7.33 Soil pressures and bending moments $M(x)$ in abutment wing walls.

At the seating beam, at least two bearings exist. Even if several bearings are adopted, the stiffness of the seating should be such as to assume equal horizontal displacements u at all bearings. Equal forces are then induced in the counterforts, if assumed with the same geometry, due to the displacement, u, or the total horizontal force $H = \Sigma R_{\mathrm{Hi}}$ induced by the deck on the abutment.

Wing walls may be modelled by a horizontal cantilever beam (Figure 7.33) subjected to the earth pressure evaluated from at rest coefficient K_o.

The width b of the wing wall does not exceed 5.0 m in general (Figure 7.33). Its thickness may be variable across the length. The lateral counterfort may have a vertical stiffener, as shown in Figure 7.34.

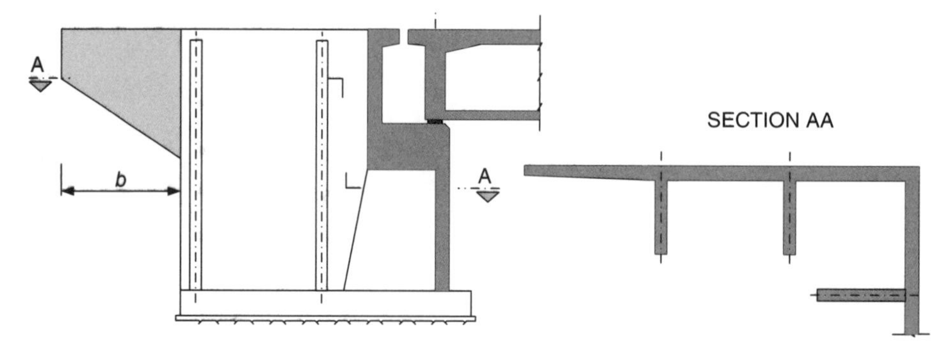

Figure 7.34 Counterforts in front and wing walls.

7.5.3 Anchored Abutments

Whenever the deck is fixed to one of the abutments, that is, rotations only between the deck and the abutment are allowed (by fixed pot bearings for example) the induced longitudinal forces due to braking traffic actions or seismic forces, are mainly taken by the abutment even if some piers with fixed bearings exist. The stiffness of the abutments is much larger than the one of the piers.

If the longitudinal force cannot be resisted by the abutment, one possible solution is to adopt an anchored abutment, as shown in Figure 7.35. This may be also convenient for an abutment with a pile foundation. As rack piles are not recommendable for seismic zones, vertical piles may be quite penalized by the longitudinal force at the abutment. Anchoring the abutment may be a convenient solution for this case. The anchorages are distributed transversally, being fixed at the level of the seating beam or at the footing (Figure 7.35).

One may adopt pretensioned anchorages in which, at the cable end sections, a sealing device is executed by injection of cement grout. That should be done at a sufficiently large distance from the abutment, as shown in Figure 7.35.

For the cases of very bad foundation conditions, the execution of the earthfill inside the abutment, namely the effects of compaction inducing horizontal forces on the piles, should be taken into account. If soft soil layers exist, the problem is aggravated. It is convenient to accelerate the soil layer vertical deformations under vertical loading, by using geodren, before piles are executed.

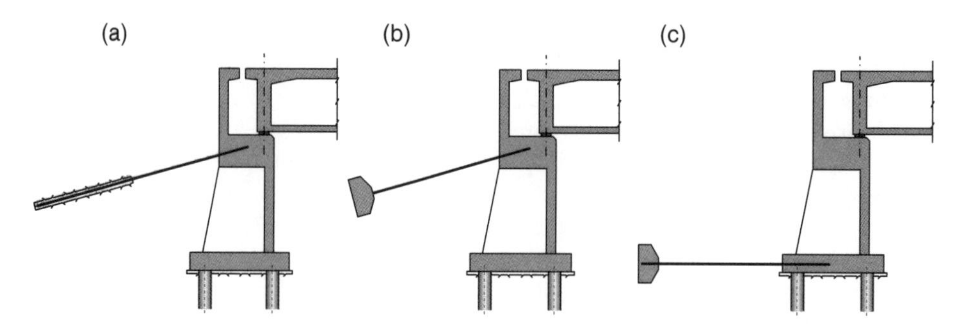

Figure 7.35 Anchored abutments.

The design and safety verifications of the pre-tensioned anchorages or anchorages by using the dead weight of large concrete blocks at the back of the abutments, as shown in Figure 7.35, should be made by adopting models from soil mechanics.

7.6 Bridge Piers: Analysis and Design

7.6.1 Basic Concepts

In Chapter 4, some basic design rules for bridge piers design were referred to. In the present chapter, the concept design is developed that is basically controlled by the following aspects,

- Formal and aesthetics concepts influencing pier geometry, namely external cross section shapes and their variation along the height;
- Stability, influencing pier slenderness limits at SLS and ULS including execution phases as well;
- Ultimate resistance as a result of interaction of geometrical and material nonlinearities, associated with overall instability effects and cross sectional behaviour as influenced by the nonlinear behaviour of concrete, for the case of concrete piers currently adopted;
- Behaviour at serviceability limit states influenced by cracking effects and time dependent deformations; namely creep, relaxation (under imposed deformations), shrinkage and thermal actions.

7.6.1.1 Pre-design

As already referred to in Chapter 4, piers up to 30 m height are currently made with constant cross section geometries. Piers above 50 m in height are preferred with variable cross sections, at least in one direction. Figure 5.17 shows how this can be made without inducing a warping in the formwork of a pier with a polygonal cross section variable in the transverse direction. High piers are made with tubular cross sections with a minimum wall thickness of 0.25–0.30 m to facilitate concreting operations. The longitudinal reinforcement is currently made with 20 or 25 mm diameter bars spaced at 10–20 cm, and horizontal stirrups of 8–16 mm spaced at 10–30 cm along the height. The amount of reinforcement in a tall pier may currently reach 200–300 kg m^{-3} of concrete. The transverse reinforcement should include closed shape stirrups. The reader is referred to references [7, 9] for reinforcement details in concrete bridge piers.

7.6.1.2 Slenderness and Elastic Critical Load

The slenderness of a bridge pier and the elastic critical loads, for buckling in longitudinal (x–z plane) or transverse (x–y plane) directions are, respectively, defined by

$$\lambda = \frac{\ell_e}{i} \rightarrow P_{cr} = \frac{\pi^2 EI}{\ell_e^2} \text{ with } i = \sqrt{\frac{I}{A}} \tag{7.44}$$

where i is the relevant radius of gyration of the cross section, I is the correspondent moment of inertia of the cross section, with area A and E the modulus of elasticity of the material. The buckling length l_e is defined in Figure 7.36 for different boundary conditions of the pier model.

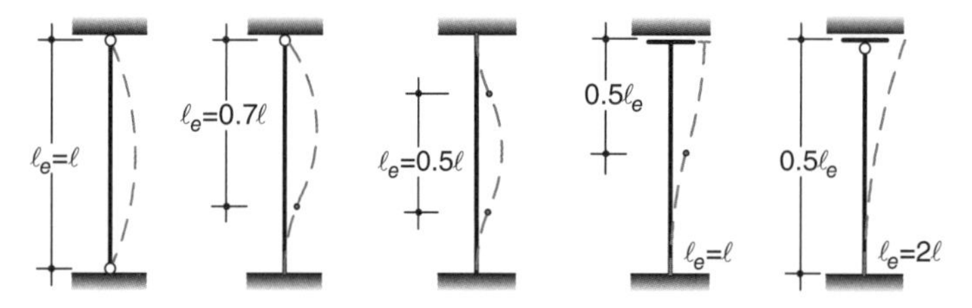

Figure 7.36 Buckling length of bridge piers for different boundary conditions.

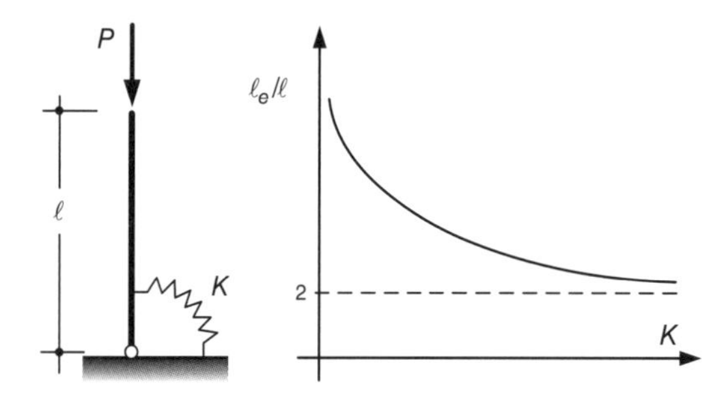

Figure 7.37 Influence of the rotation deformability of the foundation (stiffness K) on pier buckling lengths.

An important aspect to be taken into account in pier effective length is the effect of deformability of the foundation as illustrated in Figure 7.37.

7.6.1.3 The Effect of Geometrical Initial Imperfections

Due to initial geometrical imperfections, deviations of the ideal pier axis or load eccentricities, bridge piers under a load P are subjected to instability effects even in the elastic range. Primary and secondary bending moments are induced by a vertical load P (Figure 7.38) and the transverse displacements tend to increase nonlinearly under increasing load P approaching the ideal critical load P_{cr}.

7.6.1.4 The Effect of Cracking in Concrete Bridge Piers

In a concrete bridge pier, cracking and the nonlinear behaviour of concrete along the nonlinear equilibrium path ($P-w$) represented in Figure 7.39 come into play. The ULS is reached at one of the cross sections when the nonlinear N-M curve intersects the concrete N-M interaction curve for axial and bending moments, as shown in Figure 7.40. As previously referred to, $M = M_1 + M_2$, where M_1 and M_2 are, respectively, the primary and secondary bending moments. The resistant ULS of a bridge pier, with constant cross section and constant reinforcement, is reached at the base section.

At point C the ultimate resistance of the cross section is attained, which happens for maximum load P_{max}; after, the equilibrium state is unstable. That means under imposed load it is only possible to find an equilibrium state if the load is reduced, along the equilibrium path $P-w$ of Figure 7.40, with respect to P_{max}. An elastoplastic equilibrium path (Figure 7.40) exists resulting from the interaction of geometrical and material nonlinearities.

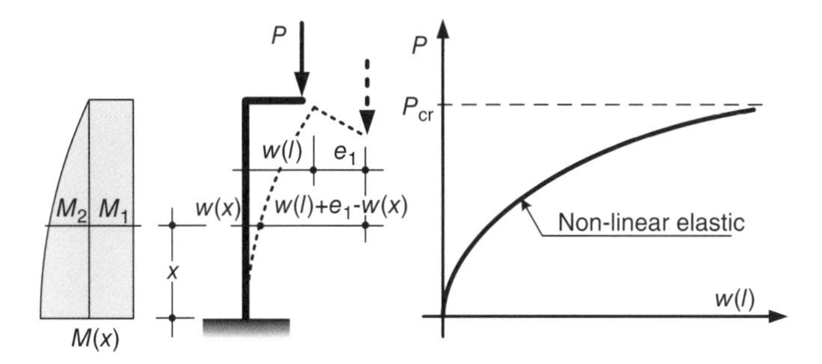

Figure 7.38 Elastic nonlinear behaviour (load P – transverse displacement w) of a bridge pier under increasing eccentric loading.

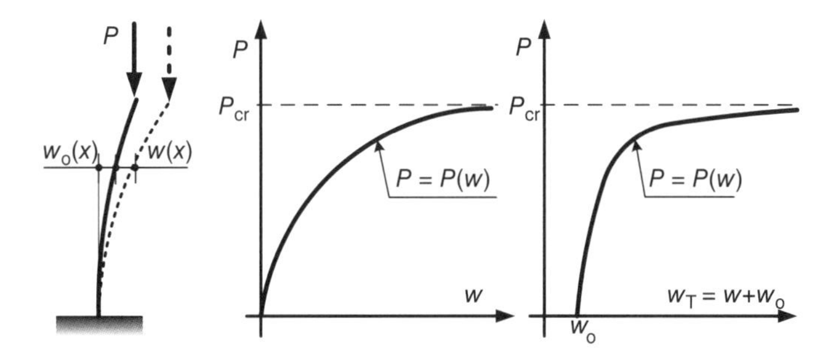

Figure 7.39 Behaviour and ULS of a bridge pier under eccentric loading P.

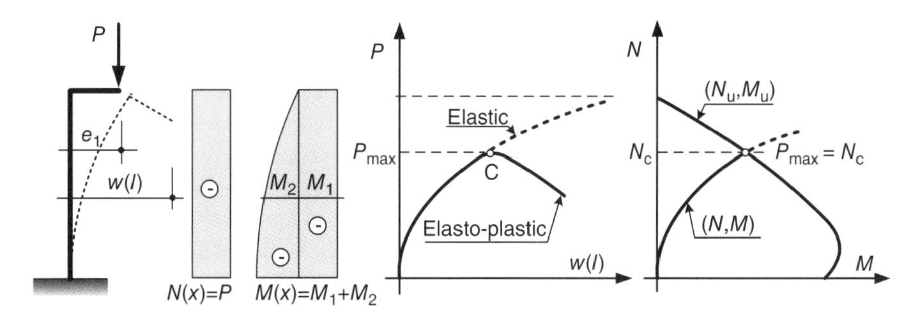

Figure 7.40 Elastic equilibrium path in piers with initial geometrical imperfections.

7.6.1.5 Bridge Piers as 'Beam Columns'

The behaviour of a pier under load P with eccentricity e (Figure 7.40) is equivalent to a pier under an axial load P and an applied moment at the top section ($M = Pe$).

Real piers have geometrical initial imperfections $w_o(x)$ inducing bending moments, even for $e = 0$, from the onset of loading (Figure 7.41) The additional displacements $w(x)$, even in the elastic range, tend to increase nonlinearly with P due to geometrically nonlinear effects. If the eccentricity e or initial imperfections $w_o(x)$ are not present but transverse loads are applied along or at the top of the pier, the behaviour is similar to previously discussed cases.

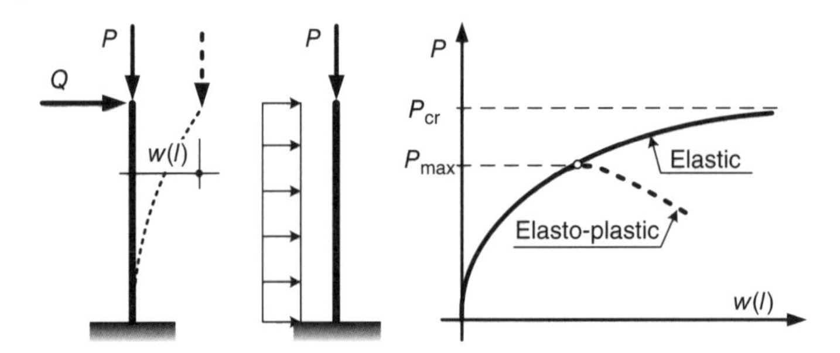

Figure 7.41 Elastic and elastoplastic equilibrium paths of piers under axial and transverse loading.

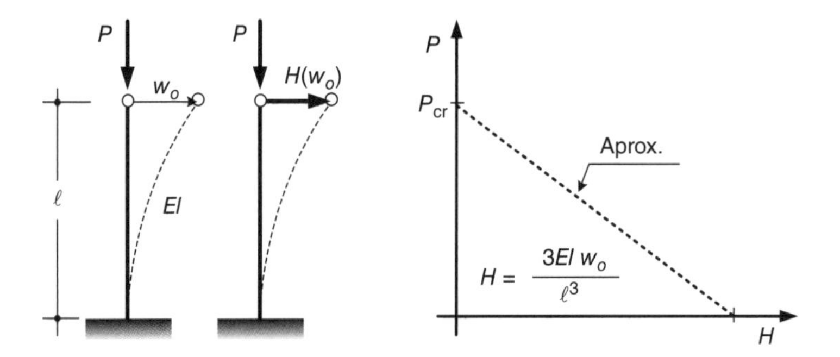

Figure 7.42 Elastic behaviour of a bridge pier under an imposed displacement at its top cross section.

A bridge pier under axial or eccentric loads at the top, or transverse loads along its height, behaves like a 'beam-column' according to theory of elastic stability. There are primary and secondary bending moments and the effects of geometric imperfections. Since we are currently talking about concrete bridge piers, the ending rigidity of the cross sections EI is affected by concrete cracking effects and should be reduced in conformity. The lower limit of the bending stiffness of the cross section is EI_y, associated with a cracked section when the steel reinforcement reaches the yield stress. The inelastic nonlinear behaviour of bridge piers is dealt with in one of the following sections after discussing elastic behaviour.

7.6.1.6 The Effect of Imposed Displacements

Up to now, only bridge piers under direct loading have been discussed. However, it has been shown that the bridge deck induces imposed displacements (due to thermal, creep and shrinkage effects) at the top sections of bridge piers. Under imposed displacements, deformations are amplified through instability effects induced by axial loads.

Let a bridge pier, of constant cross section and elastic bending stiffness EI, be considered under an imposed displacement w_o at its top cross section. This is equivalent to applying a horizontal load H inducing w_o. The transverse displacements along the bridge length are amplified by P (Figure 7.42).

For $P = 0$, one has $H = 3EI/\ell^3$. Under increasing P values approaching the elastic critical load P_{cr}, the force H at the top reduces linearly to keep the imposed displacement w_o (this is an approximate result shown in Section 7.4.2) with P. When P reaches P_{cr}, there is no stiffness at the top of the pier and $H = 0$.

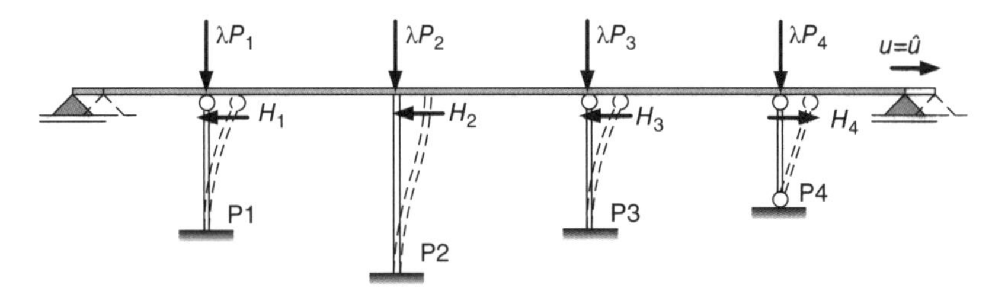

Figure 7.43 Overall stability of a bridge structure.

7.6.1.7 The Overall Stability of a Bridge Structure

Up to now, isolated bridge piers have been considered. When integrated in a bridge structure, the overall structure should be stable by itself independently of the ultimate resistance of the piers that has been verified concerning its elastoplastic stability considering the cracking effects of cross sections.

Let a bridge structural model be considered as shown in Figure 7.43. The piers have different boundary conditions due to the variety of connections to the deck. One may differentiate:

- monolithic piers, with rigid connections to the deck
- articulated piers to the deck, by fixed bearings
- piers supporting the deck with horizontally movable bearings
- bi-articulated piers, at the deck and the foundations – *pendular piers*

With different lengths, cross sections and boundary conditions, the overall stability of the bridge structure, if assumed to be simply supported at the abutments, may be defined according to its capacity to be stable under a horizontal load at the deck level. As previously discussed, the stiffness of each pier at the top cross section is reduced by its axial load P_i, more precisely through its axial internal force N_i. The limit condition of stability is reached when the total horizontal stiffness of the deck is decreased to zero under increasing axial load N_i in the piers (Figure 7.43).

Assuming the axial loads are increased with a single load parameter λ, the horizontal stiffness of the structure is reduced to zero when λ increases up to λ_{cr}. In the theory of elastic stability, this may be investigated on the basis of the so called 'adjacent equilibrium criterion' [10]. For that, a small (infinitesimal) horizontal displacement $u = \hat{u}$ at the deck level is imposed, the horizontal forces at the top of the piers are obtained at the deformed configuration and the horizontal equilibrium requirement is

$$\Sigma_i \; H_i \left(\lambda N_i\right) = 0 \tag{7.45}$$

where λN_i represents axial force at pier i. Equation (7.45) is highly nonlinear, since it involves trigonometric functions for defining the forces H_i as a function of the imposed displacement $\hat{u}$ by means of the so called *stability functions* [10, 11]. However, since $\hat{u}$ can be a very small virtual displacement, the stability functions may be linearized.

7.6.1.8 Design Bucking Length of Bridge Piers

After obtaining the critical value of the load parameter λ_{cr} by solving Eq. (7.45), the critical load is defined and the associated buckling length, respectively

$$\lambda_{cr}N_i = \frac{\pi^2 EI_i}{\ell_{ei}^2} \text{ and } \ell_{ei} = \pi\sqrt{\frac{EI_i}{\lambda_{cr}N_i}} \tag{7.46}$$

If one attends for example for the bridge model of (Figure 7.43), and assuming identical cross sections for the bridge piers under equal axial loads, pier P1 tends to stabilize piers P2 and P3. One shall be back to this topic at the last section of this chapter.

7.6.2 Elastic Analysis of Bridge Piers

Let the bridge pier of Figure 7.44 be considered with initial geometrical imperfections $w_o(x)$, a horizontal force H_o and an applied moment M_o at the top section, assumed to be induced by the deck; a horizontal distributed transverse load due to wind $q(x)$ is applied as well.

Primary bending moments are given by

$$M_1 = -H_o(\ell - x) - M_o - M_q \tag{7.47}$$

Where M_q is the bending moment due to q, determined from

$$M_q = \int_x^{\ell}(s - x)q\,ds \tag{7.48}$$

where s is a variable for integration between x and l. If q is uniform ($q = q_o$) one has

$$M_q = \frac{q_o}{2}(\ell - x)^2 \tag{7.49}$$

The total bending moment at one typical section is the sum of primary and secondary (M_2) bending moments and it is designated as moment of second order M_{II}

$$M = M_{II} = M_1 + M_2 = M_1 - P\{w(\ell) - [w_o(x) + w(x)]\} \tag{7.50}$$

As $M = -EI\dfrac{d^2w}{dx^2}$ one obtains from these equations

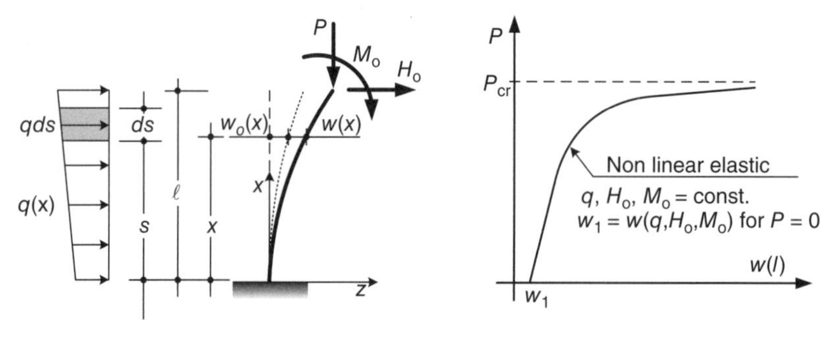

Figure 7.44 Bridge pier subjected to primary bending moments, due to H_0, M_0, $q(x)$ and vertical force P; w_0 are geometrical initial imperfections.

$$EI\frac{d^2w}{dx^2}-P\left\{w(\ell)-\left[w_o(x)+w(x)\right]\right\}=-M_1 \tag{7.51}$$

By differentiating twice and adopting the simplified notation for derivatives $dw/dx = w'$, one obtains the differential equation of transverse elastic displacements $w(x)$ of the pier

$$\left(EIw''\right)''+Pw''=-Pw_o''-M_1'' \tag{7.52}$$

Knowing w_0 and the boundary conditions, the integration of Eq. (7.52), valid for any type of boundary conditions and variable section piers, allows one to determine $w(x)$ and the total bending moments M from Eq. (7.50). The applied forces and moments H_o and M_o may be interpreted as the hyperstatic action forces from the deck on the pier obtained from the global structural analysis. That is currently obtained from an elastic linear frame model. For constant section piers, Eq. (7.52) yields for $w_o=0$

$$EIw^{IV}+Pw''=-M_1'' \Leftrightarrow EIw^{IV}+Pw''=q(x) \tag{7.53}$$

as $M_1''=-q(x)$

This is a constant coefficient differential equation with a general solution of the form

$$w(x)=A\sin Kx+B\cos Kx+Cx+D+f(x) \tag{7.54}$$

Here, A, B, C and D are constants, to be determined from the boundary conditions, $K=\sqrt{P/EI}$ and $f(x)$ is the particular solution dependent on $q(x)$. Knowing $w(x)$, the second order bending moment distribution in the bridge pier, $M_{II}\ (x) = M_1+M_2$ (total moments), is obtained from Eq. (7.50). The parameter K in these equations may be rewritten in terms of the ideal Euler buckling load P_E, as

$$K=\frac{\pi}{\ell}\sqrt{\frac{P}{P_E}} \text{ with } P_E=\frac{\pi^2 EI}{\ell^2} \tag{7.55}$$

The buckling modes, for the bridge pier, with assumed boundary conditions, are determined from Eq. (7.53) with $q(x)=0$ which has nonzero solutions for specific values of $P=P_{cr}$, the critical loads associated with the modes. This is a standard eigenvalue problem, well known from study of elastic stability and strength of materials [10–12].

Example 7.2 Let the bridge pier in Figure 7.45 be considered, subjected to P and H_0 only. In this case, one has $w_0=0$ and $q(x)=0$. The boundary conditions are as follows

$$w(0)=w'(0)=0 \tag{7.56}$$

$$w''(\ell)=0;\ EI\,w'''(\ell)+P\,w'(\ell)=-H_o \tag{7.57}$$

Where last condition corresponds to the shear force equilibrium $V=-EI\,w'''(x)$, being

$$V(\ell)=H_o\cos\alpha+P\sin\alpha\approx H_o+P\tan\alpha=H_o+P\,w'(\ell) \tag{7.58}$$

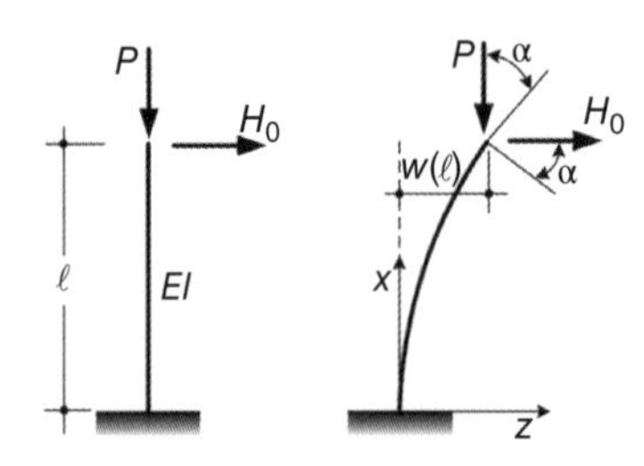

Figure 7.45 A bridge pier under vertical and horizontal loads at the top cross section.

At the end top section one has

$$V(\ell) = -EI\,w'''(\ell) = H_o + P\,w'(\ell) \tag{7.59}$$

From the general solution and boundary conditions, (7.54) and (7.55) one obtains

$$A = \frac{H_o}{K^3\,EI}; B = -\frac{H_o}{K^3\,EI}\tan K\ell; C = \frac{H_o}{K^2\,EI}; D = \frac{H_o}{K^3\,EI}\tan K\ell \tag{7.60}$$

Hence

$$w(x) = \frac{H_o}{K^3\,EI}\left(\sin Kx - \tan K\ell \cos Kx - Kx + \tan K\ell\right) \tag{7.61}$$

For $x = \ell$, the maximum transverse deflection is obtained

$$w_{max} = w_1\,\frac{3(\tan 2\beta - 2\beta)}{8\beta^3} \tag{7.62}$$

where $\beta = K\ell/2$ and w_1 is the primary displacement due to H_o, that is, $w_1 = H_0\ell^3/(3EI)$. Total bending moments are therefore given by,

$$M(x) = -EI\,w''(x) = -\frac{H_0}{K}\left(\sin Kx - \tan K\ell \cos Kx\right) \tag{7.63}$$

From which, the maximum bending moment is obtained for $x = 0$,

$$M_{max} = +\frac{H_o}{K}\tan K\ell = \frac{H_o}{2\beta/\ell}\tan 2\beta = M_1\,\frac{\tan 2\beta}{2\beta} \tag{7.64}$$

where $M_1 = H_o\ell$ is the primary bending moment.

The load parameter $\beta = K\ell/2$ in w_{max} and M_{max} expressions may be rewritten in terms of the ideal elastic critical load P_{cr}, as

$$\beta = \frac{K\ell}{2} = \frac{\pi}{2}\sqrt{\frac{P}{P_E}} = \frac{\pi}{4}\sqrt{\frac{P}{P_{cr}}} \tag{7.65}$$

with

$$P_{cr} = \frac{\pi^2\,EI}{\ell_e^2} = \frac{\pi^2\,EI}{(2\ell)^2} = \frac{P_E}{4} \tag{7.66}$$

APPROXIMATE FORMULAE

By developing in a power series the trigonometric functions in the previous expressions for w_{max} and M_{max}, the following approximate expressions are obtained,

$$w_{max} \approx \left[\frac{1}{1-P/P_{cr}}\right] w_1 \tag{7.67}$$

$$M_{max} \approx \left[\frac{1-0.18(P/P_{cr})}{1-(P/P_{cr})}\right] M_1 \tag{7.68}$$

The multiplying factors [10–12] for w_1 and M_1 play the role of amplification factors of primary deflections and bending moments due to second order effects. These expressions may be generalized for any types of boundary conditions and loading in beam-columns [10–12], with the result

$$w_{max} \approx \left[\frac{1}{1-P/P_{cr}}\right] w_1 \tag{7.69}$$

$$M_{max} \approx \left[\frac{1+\psi(P/P_{cr})}{1-(P/P_{cr})}\right] M_1 \tag{7.70}$$

Here, ψ is a negative value coefficient (=−0.18 in previous example) varying between −0.4 and 0 for most cases [10–12] and P_{cr} takes into consideration the boundary conditions. Hence, one may take, conservatively, $\psi = 0$. In Figure 7.46, these expressions, representing the variation of w and M (total bending moment, i.e. second bending moment M_{II}) under increasing load P are represented. The secondary bending moment is given by $M_2 = M - M_1$.

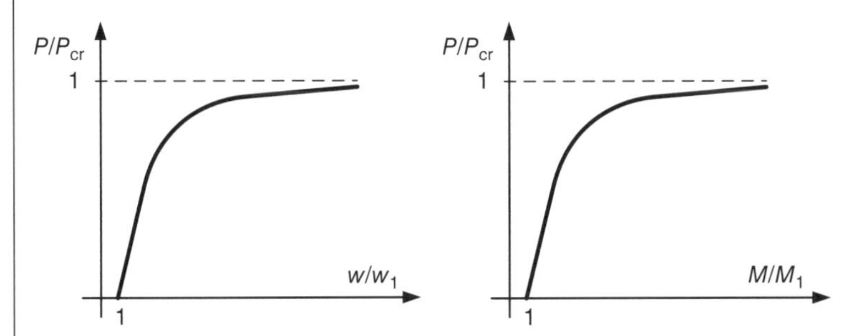

Figure 7.46 Variation of maximum deflection and bending moment with P for the example of Figure 7.42.

Example 7.3 A Tubular Bridge Pier
The bending moments distribution (secondary moments included), obtained from Eq. (7.70), is exemplified for the case of a constant section tubular pier represented in Figure 7.47. One assumes $P = 0.2P_{cr}$ and $H = 0.02P$. The geometrical characteristics of the pier and uncracked section properties are as follows:

$$\ell = 30\,\text{m} \quad A = 5.76\text{m}^2 \quad I = 3.25\text{m}^4 \quad W = 3.25\text{m}^3 \quad E = 32\times10^6\ \text{kNm}^{-2}$$

$$P = 0.2 \quad P_{cr} = 56967\text{kN} \quad H = 0.02P \quad M_{max} = \frac{1-0.18\times0.2}{1-0.2}M_1 = 1.21\,M_1 = 41358\,\text{kNm}$$

The cracking bending moment of the cross section, for a tensile concrete resistance $f_{ctm} = 3\text{MPa}$, is $M_{cr} = 3.25\left(3000 + \dfrac{56967}{5.76}\right) = 41893\text{kNm} > M_{max}$

Example 7.4 Elastic Effect of an Imposed Displacement
For the effect of an imposed displacement w_o, the example of Figure 7.42, is retaken. The force H induced at the top of the pie is H_o for $P = 0$ and H at a certain axial load level P. From the approximate expression (7.69) with $w_{max} = w_o$ one has

$$w_o \approx \frac{H\ell^3}{3EI}\left(\frac{1}{1-(P/P_{cr})}\right) \text{ with } w_o = \frac{H_0\ell^3}{3EI} \tag{7.71}$$

Hence, one has

$$\frac{H}{H_0} \approx 1 - \frac{P}{P_{cr}} \tag{7.72}$$

The relationship $H = H(P)$ is the linear interaction anticipated as an approximation in Figure 7.42. The imposed displacement at the top of the pier $w(\ell)$, taken from Figure 7.48, is the result of thermal and differed effects (creep and shrinkage) of the deck. The induced horizontal force H at the top of the pier is also influenced by the second order effects and may be evaluated from

$$w(\ell) = \int_0^\ell \frac{M\bar{M}}{EI}dx, \text{ with } \bar{M} = -1(\ell - x) \tag{7.73}$$

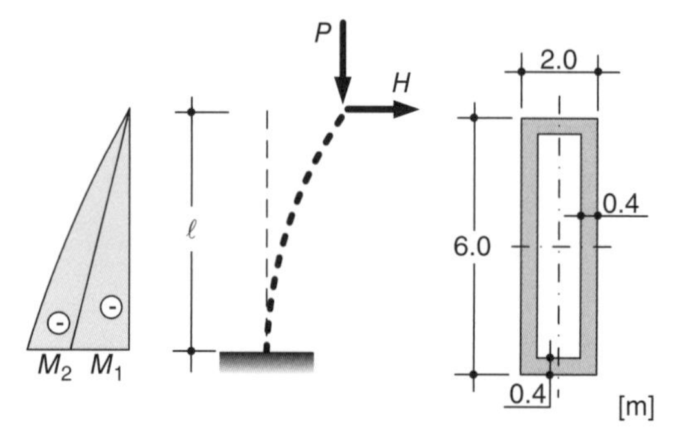

Figure 7.47 Example – A tubular bridge pier; primary and second order bending moments M_1 and M_2, for $P = 0.2\,P_{cr}$ and $H = 0.02\,P$.

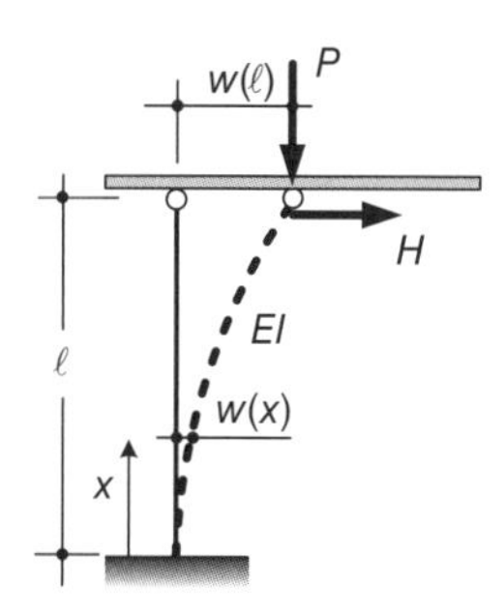

Figure 7.48 Imposed displacements $w(l)$ by the deck and second order effects.

where M and $\bar{M}$ are the bending moments with second order effects included, that is,

$$M(x) = -P\left[w(\ell) - w(x)\right] - H(\ell - x) = -P\left[w(\ell) - w_o(x)\frac{1}{1 - P/P_{cr}}\right] - H(\ell - x) \quad (7.74)$$

where $w_0(x)$ is the initial geometrical imperfections in the pier that are amplified by second order effects through the amplification factor. Since $w(\ell)$ is known from Eq. (7.73), by replacing $M(x)$ in Eq. (7.74), the horizontal force H acting at the top of the pier is evaluated.

In a bridge pier with a neoprene-teflon bearing, if the induced force H due the movement of the deck is not sufficient to reach the friction force resistance (coefficient of friction μ approximately 5%) under permanent loads G, that is, $H < \mu G$, the pier will work as if it was articulated to the deck and the buckling length is the $\ell_e << 2\ell$. However, that can only be taken into account by assuming a conservative low value for μ, say $\mu = 1\%$.

7.6.3 Elastoplastic Analysis of Bridge Piers: Ultimate Resistance

As discussed in Section 7.6.1, material nonlinear behaviour, cracking effects included, yields the following:

- the bending stiffness EI is reduced under increasing bending moment, M
- the ultimate load corresponds to the intersection of the curve $(M - N)$ (Figure 7.40) with the resistance interaction curve $(M_{Rd} - N_{Rd})$ at the most loaded section

The first aspect is illustrated on Figure 7.49, by moment M – curvature $(1/R)$ diagram for one specific section. The *cracking bending moment* is here designated by M_e. For $0 < M < M_e$, one has a linear behaviour; with an elastic stiffness EI. For $M > M_e$, one has a cracked section and yielding in the steel reinforcement at a stress f_y is reached at M_y. From this moment on, $M > M_e$, the curvatures increase quickly with M, until the ultimate resistance moment M_r is reached. In a bridge pier, if at one section $M_y < M < M_r$, large secondary bending moments are induced due to increased displacements w under increasing curvatures. Hence, one may consider the ultimate bending moment of a cross section, the value at which yielding is reached at the steel reinforcement; that is, M_y.

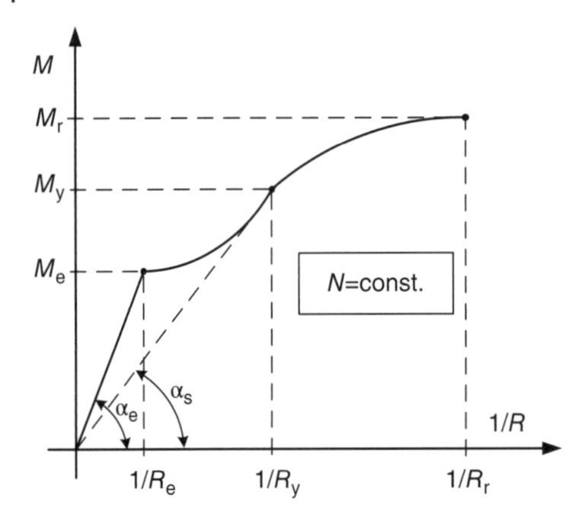

Figure 7.49 Moment–curvature curve for a reinforced concrete section under bending.

For $M_e < M < M_y$ one may refer to the 'tangent stiffness' $\tan\alpha_T = dM/d(1/R)$ or to the 'secant stiffness' EI_s. As a conservative approach, for the bridge pier one may specify its secant stiffness at yielding, that is, EI_y at the most loaded section (critical section). If this stiffness is spread throughout the pier, one overestimate the secondary bending moments due to instability effects. Hence, one takes conservatively,

$$EI = EI_y = \frac{M_y}{1/R_y} \approx \frac{M_r}{1/R_y} \tag{7.75}$$

Here, R_y represents the curvature at yielding at a constant axial load level N. Normalized (M,N) interaction resistance diagrams, for reinforced concrete cross-sections of bridge piers under bending and axial load, are currently given by diagrams where the values R_y are specified. Computer programs for generating the interaction curves for tubular reinforced concrete sections are also available. In Figure 7.50 an example is shown.

For the ULS of a bridge pier of constant stiffness EI, the second order displacements and bending moments may be approximately evaluated by formulae (7.69) and (7.70), being the elastic critical load P_{cr} evaluated with the reduced stiffness EI_y.

The procedure for the verification of the ULS resistance of a bridge pier may be summarized as follows:

- for a defined cross section, predesign the steel reinforcement;
- for the design action axial force and predesigned steel reinforcement, the curvature at yielding and the ultimate resistant moment are determined from the axial force-bending moment diagram and consequently the cross section bending stiffness $EI_y = M_r/(1/R_y)$.

Second order displacements w_{max} and bending moments M_{max} may then be determined by the approximate formulae (7.69) and (7.70). If $M_{max} < M_r$ the pre-designed

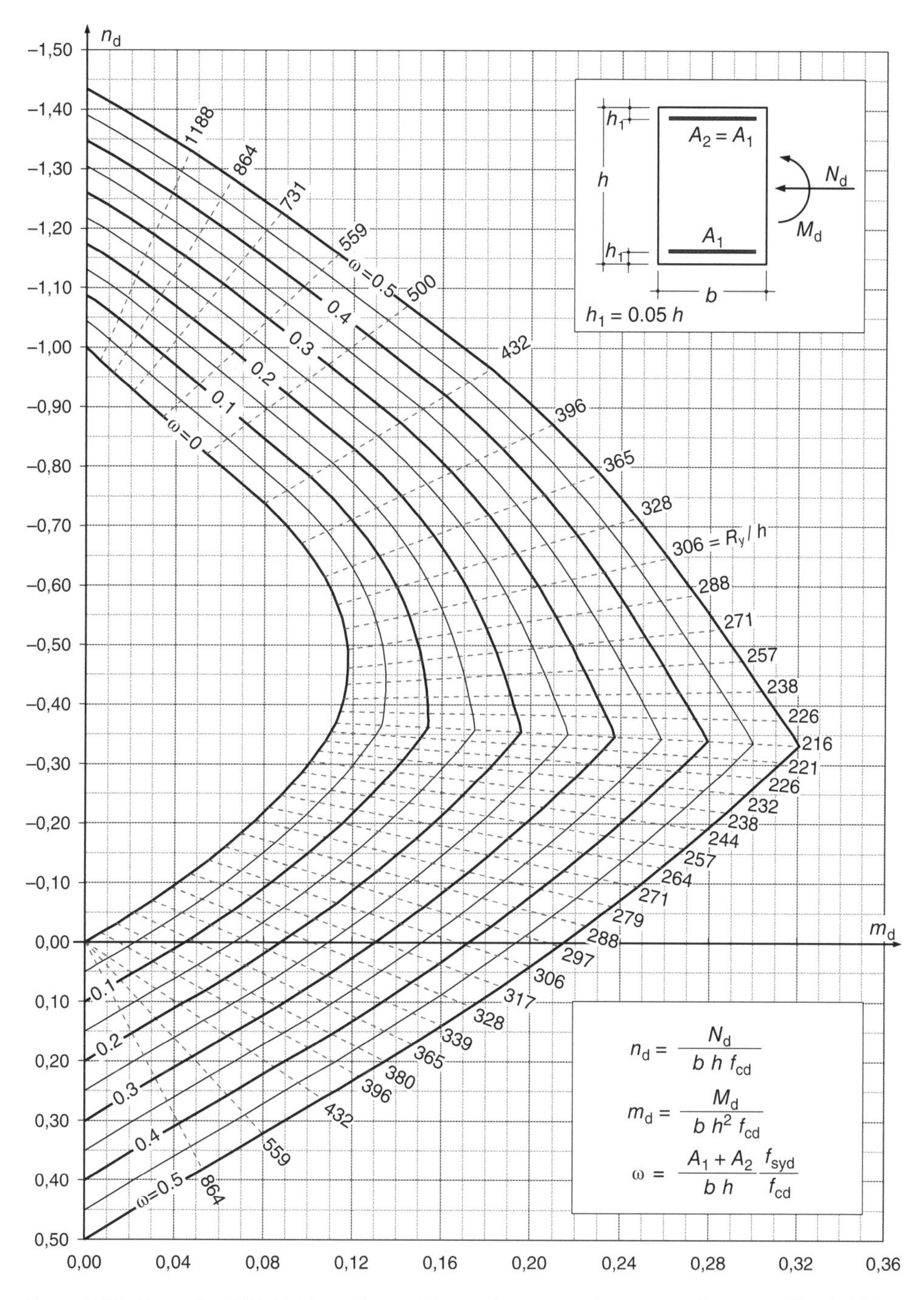

Figure 7.50 Normalized (M, N) interaction resistance diagram and corresponding normalized yielding radius of curvature for a rectangular reinforced concrete cross section (concrete parabola-rectangle stress–strain diagram with a $f_{cd} = \alpha_{cc}\, f_{ck} / \gamma_c = 1.0\, f_{ck}/1.5$, and reinforcing steel bi-linear diagram with $f_{syd} = f_{sy} / \gamma_s = f_{sy}/1.15$, $E_s = 200$ GPa and $\varepsilon_{ud} = 2.2‰$).

steel reinforcement is sufficient; otherwise ($M_{max} > M_r$) so the steel reinforcement should be increased and the previous procedure repeated; in some cases, a more straightforward iterative procedure may be applied as shown in Example 7.5.

As previously mentioned, the procedure as proposed is conservative since the second order bending moments are overestimated by assuming the yielded stiffness EI_y at the critical section as constant through the pier length. For uncracked zones a larger stiffness exists, $EI > EI_y$, because $M < M_e$. For the pier zones where $M_e < M < M_y$, the cross sections may be cracked but the steel reinforcement is not at yield stress. The cross section curvatures $1/R(x)$ are varying along the pier length and the stiffness is

$$EI(x) = M(x)\ R(x) \tag{7.76}$$

This effect may be relevant for tall piers, usually designed with a variable cross section. The procedure referred to previously can only be applied by a discrete segment model of the pier and adopting a numerical approach.

Base case: A bridge pier with initial geometrical imperfections $w_0(x)$, under compression – let the pier of Figure 7.44 be considered with $H_0 = 0$, $q(x) = 0$ and $w_0 \neq 0$, subject to the vertical load P. The second order bending moment at the critical section is,

$$M = Pw_{max} = Pw_{o,max} \frac{1}{1 - P/P_{cr,y}} \tag{7.77}$$

Where $$P_{cr,y} = \frac{\pi^2 EI_y}{\ell_e^2}, \quad \text{with } \ell_e = 2\ell \tag{7.78}$$

and $$EI_y \approx \frac{M_r}{I/R_y} = M_r\, R_y \tag{7.79}$$

By replacement in the M expression, one has

$$M = Pw_{o,max} \frac{1}{1 - \left(P\ell_e^2/\pi^2 M_r R_y\right)} \tag{7.80}$$

When the equilibrium state is reached for the design load P_d, one has $M_r = M_d$. By introducing this condition in Eq. (7.80) for M, one has after solving for M_d

$$M_d = P_d \left(W_{o,max} + \frac{\ell_e^2}{\pi^2 R_y} \right) \tag{7.81}$$

The second term in this equation represents the additional eccentricity associated with the secondary bending moment. However it should be noted in this expression R_y depends on $(M, N) = (M_d, P_d)$. An iterative procedure may be adopted as follows:

- A percentage of steel reinforcement ω may be initially adopted by defining the steel ratio parameter $\omega = \left((A_s + A_s')/bh\right)\left(f_{sy}/f_c\right)$ for a rectangular section.
- One enters into the interaction diagram with ω and the reduced axial force $n_d = (N_d/bhf_c)$; hence R_y and M_d are determined by Eq. (7.81).

- A new reinforcement ratio ω is assumed and for (ω, n_d) a new R_y and new M_d are determined. The process is repeated until M_d is stabilized.
- In general, one may start with ω = 0.20 and then increase the steel reinforcement in order to stabilize M_d.

Example 7.5 Let a bridge pier with a rectangular cross section ($3 \times 1\,m^2$) with a length L = 13 m, subject to a vertical load P = 10 000 kN, applied at the free top section, be considered. The pier is built in at the base section. The concrete is C35/45 (f_{cd} = 23.3 MPa) and steel reinforcement B500 B. The assumed initial geometrical imperfection, is a modal imperfection, that is, in the form of the buckling mode shape, being the maximum deviation at the top cross section

$$w_{0,\max} = \frac{\ell_e}{300} = \frac{2 \times 13}{300} = 0.09\,\text{m}$$

Assuming

$$\omega = 0.20 \rightarrow A_s + A_s' = 0.20 \times bh \frac{f_{cd}}{f_{syd}} = 321\,\text{cm}^2 \rightarrow n_d = \frac{\gamma P}{bh\, f_{cd}} = \frac{-1.5 \times 10000}{3.0 \times 1.0 \times 23300} = -0.2$$

From the (M, N) interaction diagram of Figure 7.50 for the rectangular cross-section, one has ω = 0.20, h'/h = 0.05, n_d = $-0.2 \rightarrow R_y/h$ = 264
and

$$M_r = 0.155\, bh^2 f_{cd} = 10850\,\text{kNm}$$

From Eq. (7.81) $M_d = 15000\left[0.09 + \frac{(2 \times 13)^2}{\pi^2 \times 264}\right] = 5242\,\text{kNm}$

As $M_d \ll M_r \rightarrow$ one takes for a second iteration

$$\omega = 0.15 \rightarrow R_y/h = 267, M_r = 0.135\, bh^2 f_{cd} = 9450\,\text{kNm}$$

From Eq. (7.81) M_d = 5198 kNm
As $M_d < M_r \rightarrow$ one assumes for a third iteration

$$\omega = 0.10 \rightarrow R_y/h = 271, M_r = 0.115\, bh^2 f_{cd} = 8050\,\text{kNm}$$

From Eq. (7.81) $M_d = 15000\left[0.09 + \frac{(2 \times 13)^2}{\pi^2 \times 271}\right] = 5141\,\text{kNm}$

As M_d/M_r = 0.64, one accepts the assumed steel reinforcement, with the result

$$2A_s = 0.10\, bh \frac{f_{cd}}{f_{syd}} = 160\,\text{cm}^2 \rightarrow A_s = \frac{\phi 20}{0.10}$$

Example 7.6 Let the same bridge pier of Example 7.5 be considered, but with an additional horizontal force H_0 at the top, as shown in Figure 7.45. At the base section, that is the critical section, one has

$$M_{max} = M_1 + M_2 = H_o \ell + P\left[w_{o,max} + \frac{H_o \ell^3}{3EI_y}\right]\frac{1}{1 - P/P_{cr,y}} \tag{7.82}$$

with EI_y e $P_{cr,y}$ as defined in the previous example. For H_0 one takes an equivalent bearing friction force $H_o = \mu P$ with $\mu = 5\%$. Hence, $H_o = 500\,\text{kN}$.

Assuming $\omega = 0.20 \rightarrow n_d = 0.2$, $R_y/h = 264$, $M_r = 10850\,\text{kNm} \rightarrow EI_y = M_r R_y = 2.864 \times 10^6\,\text{kNm}^2$

$$P_{cr,y} = \frac{\pi^2 EI_y}{\ell_e^2} = \frac{\pi^2 \times 2864400}{26^2} = 41820\,\text{kN}$$

$$\text{For } P = P_d = 15000\,\text{kN} \rightarrow P/P_{cr,y} = 0.36$$

$$w_{o,max} = \frac{\ell_e}{300} = 0.09\,\text{m}$$

$$M_{max} = 500 \times 13 + 15000\left[0.09 + \frac{500 \times 13^3}{3 \times 2\,864\,400}\right]\frac{1}{1 - 0.36}$$

$$= 6500 + 15000\left[0.09 + 0.13\right] \times 1.56$$

$$= 6500 + 5097 = 11597\text{ kNm} > M_r = 10850\text{ kNm}$$

As $M_{max} > M_r$ (slightly higher) the steel reinforcement should be a little bit increased. One should note the secondary moment $M_2 = 5097\,\text{kNm}$, that is 78% of the primary bending moment.

A final comment may be made for this example, referring the action values of the internal forces. The load P is the result of the permanent G and variable loads Q on the deck.

Hence

$$N_d = \gamma_g G + \gamma_q Q \tag{7.83}$$

$$M_d = \gamma_q H_o \ell + N_d\left[W_{o,max} + \frac{\gamma_q H_o \ell^3}{3EI_y}\right]\frac{1}{1 - (N_d / P_{cr,y})} \tag{7.84}$$

being the action safety coefficients taken generally,

$$\gamma_g = 1.35 \text{ e } \gamma_q = 1.35 \text{ or } 1.50 \text{ depending on the codes} \tag{7.85}$$

The design resistance of the cross section is taken in the general form

$$R_d = R(f_{cd}, f_{syd}) \text{with } f_{cd} = \frac{f_{ck}}{\gamma_{mc}} \text{ and } f_{syd} = \frac{f_{syk}}{\gamma_{ms}} \tag{7.86}$$

This means the resistance stress safety coefficients for concrete and steel reinforcement γ_{mc} and γ_{ms} affect the M_d values through EI_y and $P_{cr,y}$. However, tolerances and execution deviations, like cross section deviations, affecting the values of EI_y and $P_{cr,y}$ are not directly taken into account and should be covered through the γ_f safety coefficients on the actions. This may justify an additional safety coefficient directly on the values of EI_y, in particular when the permanent loads are factored by 1.35 and not 1.50. A reduced value for the cross section design bending stiffness $EI_{yd} = EI_y/1.2$ has been proposed for bridge piers to account for the nonlinear effects previously referred to. In any case, it should be kept in mind that EI_y is not the elastic bending stiffness, but the bending stiffness of the cracked cross section at yield of the steel reinforcement.

7.6.4 Creep Effects on Concrete Bridge Piers

Concrete creep effects increase deformations under constant stress, mathematically represented by the viscoelastic strain–stress as discussed in Chapter 3. Equations (3.63) and (3.64) define the creep function $\Phi(t, t_0)$ and the creep coefficient $\varphi(t, t_0)$.

In bridge piers, the second order displacements and bending moments will increase under creep, affecting the ultimate moment resistance of the critical cross section.

The simplest models to consider creep effects in the design of bridge piers, consist of replacing the elastic modulus of concrete E_c by the effective elastic modulus $E_{c,\,eff}$ defined by Eq. (3.74). However, as was discussed in previous sections, bending stiffness of concrete EI should be taken with its cracked value at yielding of steel reinforcement EI_y. Hence, adding creep effects on the pier stability problem may be done by reducing EI_y through a reduction creep coefficient K_φ, and defining the cracked stiffness

$$EI_y(\varphi) = K_\varphi EI_y \tag{7.87}$$

The reduction coefficient $K_\varphi < 1.0$ depends on the creep coefficient φ and on the level of axial force, as shown in Figure 7.51 in terms of the long term creep coefficient $\varphi \equiv \varphi_\infty$ and normalized axial force $n_r/(1+\omega)$ where ω is the percentage of steel reinforcement. The elastic critical load $P_{cr,\,y}$ is evaluated now with $EI_y(\varphi)$, yielding a critical load $P_{cr,\,\varphi}$.

It should be noted that only the permanent loads G induce creep effects and not the variable loads Q. The second order displacements may be evaluated from the first order displacements by the adopting the amplification factor

$$\frac{1}{1-\big(P(G)+P(Q)\big)/P_{cr,\varphi}} \tag{7.88}$$

An iterative procedure is then adopted as referred to in Section 7.6.3. The reader is addressed to Ref. [9] for additional details on the procedure.

7.6.5 Analysis of Bridge Piers by Numerical Methods

The structural analysis of bridge piers with variable cross section, as usually adopted for tall piers, requires the pier to be divided in constant bar section segments, and a numerical approach based on the procedures established on Section 7.6.3 to be adopted. This numerical approach allows geometrical and material nonlinearities as well as creep effects to be considered. A finite element model (FEM) may be adopted to deal with all kind of nonlinearities and the reader is addressed to specific literature on this subject.

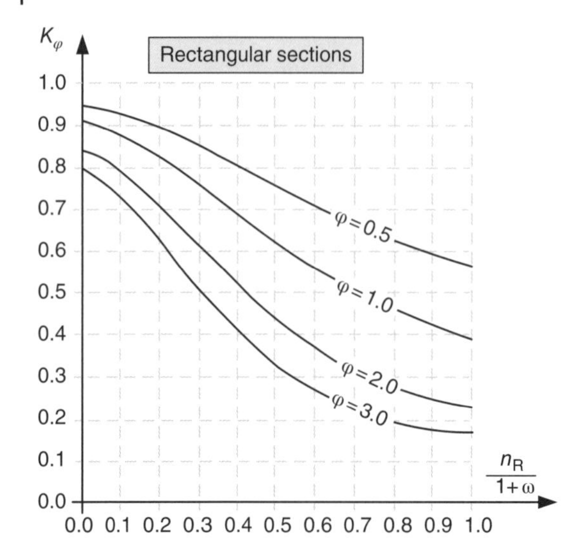

Figure 7.51 Reduction coefficient for creep effects in concrete piers of rectangular section subjected to its self-weight (adapted from [9]).

The FEM may be seen as a generalization of the Raleigh–Ritz Method (RRM), a powerful method to transform a stability problem in a continuous system in a discrete model. The fundaments of the RRM in stability problems are considered in many text books [9–13]. The main difference between the RRM and the FEM is the consideration of a single element in the RRM compared to multiple elements in FEM.

In both methods the displacement field is taken in a discrete form through a certain number of shape functions ϕ_i and generalized coordinates (degrees of freedom) q_i. For the particular case of a bridge pier, the transverse displacement field $w(x)$ is taken as

$$w(x) \approx \overline{w}(x) = q_1 \phi_1(x) + q_2 \phi_2(x) + \cdots + q_n \phi_n(x) \tag{7.89}$$

The shape functions ϕ_i shall satisfy all the *cinematic boundary conditions* of the problem but not necessarily the *static boundary conditions*. To explain the difference between the two types of boundary conditions, let the example of Figure 7.45 be retaken.

At the base section of the column the cinematic boundary conditions require a zero displacement and rotation, that is, $\phi_i = 0$ and $\phi_i' = d\phi_i/dx = 0$. At the top one has static boundary conditions requiring a zero bending moment $M = -EI(l)w''(l) = 0$, and a shear force and axial force that can be written in terms of P, H and w'.

In the RRM, the generalized coordinates q_i are determined under the Stationary Total Potential Energy Principle for equilibrium states,

$$\frac{\partial \overline{V}}{\partial q_i} = 0, \text{ with } \overline{V} = V\left[\overline{w}(x)\right] \tag{7.90}$$

Equation (7.90) represents a system of n linear equations with respect to n unknowns (q_i). In the following, the RRM is presented through an example for a geometrically nonlinear problem of a bridge pier.

Example 7.7 Bridge pier built in at the base section and with a vertical force P and horizontal force H at the top (Figure 7.45). The total potential energy is

$$V = U + V_e \tag{7.91}$$

in which U is the elastic strain energy and V_e the potential energy of the external forces. Hence, one has

$$U = \frac{1}{2}\int_o^\ell M\frac{1}{R}dx = \frac{1}{2}\int_o^\ell EI\left(\frac{1}{R}\right)^2 dx = \frac{1}{2}\int_o^\ell EI\, w''^2\, dx \tag{7.92}$$

$$V_e = -H\, w(\ell) - Pu(\ell) \tag{7.93}$$

Here, $(-V_e)$ represents the work done by the external forces under the horizontal and vertical displacements $w(\ell)$ and $u(\ell)$. The last one may be expressed in terms of the transverse displacement $w(x)$ assuming no axial deformations (Figure 7.52)

$$ds = \sqrt{(dx)^2 + (dw)^2} = \sqrt{1 + w'^2}\, dx \approx (1 - \frac{1}{2}w'^2)\, dx \tag{7.94}$$

Therefore, one has

$$\text{u}(\ell) = \ell - \int_o^\ell ds = \ell - \int_o^\ell \left(1 - \frac{1}{2}w'^2\right)dx = \frac{1}{2}\int_o^\ell w'^2\, dx \tag{7.95}$$

Replacing in the total potential energy expression yields

$$V\left[w(x)\right] = \int_o^\ell \left(\frac{EI}{2}w''^2 - \frac{P}{2}w'^2\right)dx - H\, w(\ell) \tag{7.96}$$

Assuming the approximate displacement field (7.89) with one degree of freedom only in the form

$$\bar{w}(x) = q_1\left(1 - \cos\frac{\pi \text{x}}{2\ell}\right) \tag{7.97}$$

where the shape function $\phi_1(x)$ satisfying all the kinematic boundary conditions, since

$$\phi_1(0) = 0 \text{ and } \phi'_1(0) = 0 \tag{7.98}$$

Replacing $\bar{w}(\text{x})$ in V expression yields,

$$V \approx \bar{V}(q_1) = \int_o^\ell \left\{\frac{EI}{2}\left[q_1\left(\frac{\pi}{2\ell}\right)^2 \cos\frac{\pi x}{2\ell}\right]^2 - \frac{P}{2}\left(q_1\frac{\pi}{2\ell}\sin\frac{\pi x}{2\ell}\right)^2\right\}dx - H\bar{w}(\ell) \tag{7.99}$$

Here $\bar{w}(\ell) = q_1$ is the transverse displacement at the top cross section. Taking into consideration

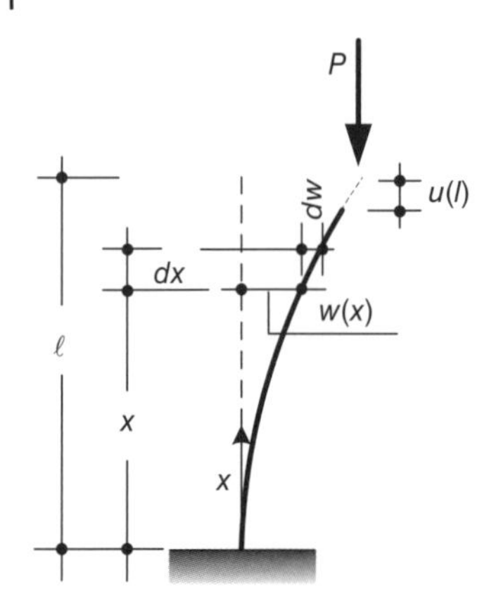

Figure 7.52 Buckling of a column – notation.

$$\int_0^\ell \cos^2 \frac{\pi x}{2\ell} dx = \int_0^\ell \frac{1+\cos \pi x/\ell}{2} dx = \frac{\ell}{2} \tag{7.100}$$

$$\int_0^\ell \sin^2 \frac{\pi x}{2\ell} dx = \int_0^\ell \frac{1-\cos \pi x/\ell}{2} dx = \frac{\ell}{2} \tag{7.101}$$

One has for $\bar{V}(q_1)$

$$\bar{V}(q_1) = \left(\frac{\pi^4 EI}{64\ell^3} - \frac{P\pi^2}{16\ell} \right) q_1^2 - Hq_1 \tag{7.102}$$

The stationary total potential energy condition requires for equilibrium states

$$\frac{d\bar{V}}{dq_1} = 0 \rightarrow 2\left(\frac{\pi^4 EI}{64\ell^3} - \frac{P\pi^2}{16\ell} \right) q_1 - H = 0 \tag{7.103}$$

The horizontal displacement at the top is therefore given by

$$q_1 = \frac{H}{\left(\pi^4 EI/32\ell^3\right) - \left(P\pi^2/8\ell\right)} \tag{7.104}$$

For $q_1 = \infty$ the elastic critical load is obtained, that is, P_{cr} for a column with $\ell_e = 2\ell$, is

$$\frac{\pi^4 EI}{32\ell^3} - \frac{P\pi^2}{8\ell} = 0 \rightarrow P = P_{cr} = \frac{\pi^2 EI}{4\ell^2} \tag{7.105}$$

that is, P_{cr} for a column with $\ell_e = 2\ell$. The displacement at the pier top cross section may be expressed in terms of the primary top cross section displacements, that is,

$$w(\ell) = \frac{w_1}{\left(\pi^4/96\right)\left[1 - \left(P/P_{cr}\right)\right]} \approx \frac{w_1}{1 - P/P_{cr}} \tag{7.106}$$

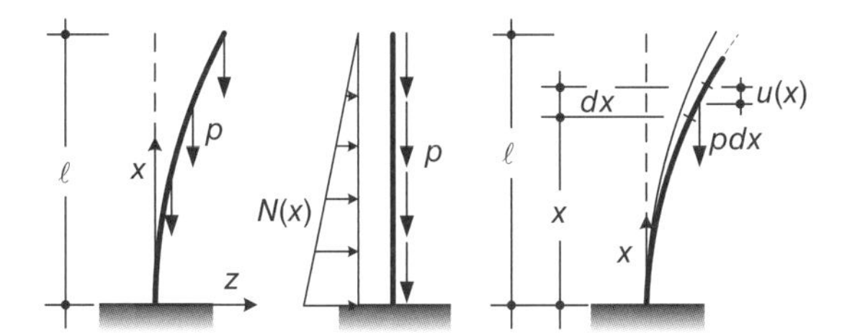

Figure 7.53 Bridge pier subjected to its self-weight.

Where

$$w_1 = \frac{H\ell^3}{3EI} \tag{7.107}$$

This expression for $w(\ell)$, obtained by the RRM, has been already determined in Section 7.6.2 as an approximation of the exact solution of the equilibrium differential equation.

Example 7.8 Stability of a Bridge Pier During Construction (Figure 7.53)
Let a bridge pier, built in at the base section and free at the top, be considered under its own weight p per unit length. The cross section is assumed to be constant and p as a consequence.

The total potential energy is

$$V = \int_o^\ell \frac{EI}{2} w''^2 dx - \int_o^\ell p u(x) dx \tag{7.108}$$

As determined for $u(l)$ in Example 7.7, one has

$$u(x) = \frac{1}{2}\int_o^x w'^2 dx \tag{7.109}$$

Applying the RRM with the approximate solution

$$w \approx \overline{w} = q_1\left(1 - \cos\frac{\pi x}{2\ell}\right) \tag{7.110}$$

One has

$$\overline{w}' = q_1 \frac{\pi}{2\ell} \sin\frac{\pi x}{2\ell}; \ \overline{w}'' = q_1 \frac{\pi}{4\ell^2} \cos\frac{\pi x}{2\ell} \tag{7.111}$$

$$\int_o^x w'^2 dx = q_1^2 \frac{\pi^2}{4\ell^2}\int_o^x \sin^2\frac{\pi x}{2\ell} dx = q_1^2 \frac{\pi^2}{4\ell^2}\int_o^x \frac{1-\cos \pi x/\ell}{2} dx = q_1^2 \frac{\pi^2}{8\ell^2}\left(x - \frac{\ell}{\pi}\sin\frac{\pi x}{\ell}\right) \tag{7.112}$$

For the total potential energy, one has

$$V = \int_o^{\ell} \left[\frac{EI}{2} q_1^2 \frac{\pi^4}{16\ell^4} \cos^2 \frac{\pi x}{2\ell} - \frac{p}{2} q_1^2 \frac{\pi^2}{8\ell^2} \left(x - \frac{\ell}{\pi} \sin \frac{\pi x}{\ell} \right) \right] dx \tag{7.113}$$

Since

$$\int_o^{\ell} \cos^2 \frac{\pi x}{2\ell} dx = \frac{\ell}{2} \text{ and } \int_o^{\ell} \sin \frac{\pi x}{2\ell} dx = \frac{2\ell}{\pi} \tag{7.114}$$

one has

$$\begin{aligned} V &= \frac{EI}{32} \frac{\pi^4}{\ell^4} \frac{\ell}{2} q_1^2 - p \frac{\pi^2}{16\ell^2} \left(\frac{\ell^2}{2} - \frac{\ell}{\pi} \frac{2\ell}{\pi} \right) q_1^2 = \left[\frac{EI}{64} \frac{\pi^4}{\ell^3} - p \frac{\pi^2}{16} \left(\frac{1}{2} - \frac{2}{\pi^2} \right) \right] q_1^2 \\ &= \left[\frac{EI}{64} \frac{\pi^4}{\ell^3} - p \frac{\pi^2}{16} \left(\frac{1}{2} - \frac{2}{\pi^2} \right) \right] q_1^2 \end{aligned} \tag{7.115}$$

The equilibrium condition is

$$\frac{dV}{dq_1} = 0 \rightarrow \left[\frac{EI}{64} \frac{\pi^4}{\ell^3} - p \frac{\pi^2}{16} \left(\frac{1}{2} - \frac{2}{\pi^2} \right) \right] q_1 = 0 \tag{7.116}$$

From this equation one has two solutions

$$\begin{aligned} &q_1 = 0 \\ &q_1 \neq 0 \text{ for } p = p_{cr} = \frac{EI}{2\ell^3} \frac{\pi^4}{\pi^2 - 4} \end{aligned} \tag{7.117}$$

The non-zero solution is indeterminate as is usual in a linear stability problem (eigenvalue problem) and this happens for $p = p_{cr.}$ One may define the correspondent total load $P = p_{cr}\ \ell$, equivalent to the dead weight inducing column buckling, when applied at the top cross section of the bridge pier:

$$P_{cr} = p_{cr}\ \ell = \frac{2\pi^2}{\pi^2 - 4} \frac{\pi^2 EI}{(2\ell)^2} \cong 3.4 \frac{\pi^2 EI}{\ell_e^2} \tag{7.118}$$

The conclusion is the critical value of the dead weight is 3.4 times the critical of an equivalent column loaded at the top cross section with its total dead weight. Hence, the dead weight of the column for stability analysis is equivalent to apply at the top cross section (1/3.4) $p\ell \approx p\ell/3$, that is, approximately one-third of its total self-weight, as Figure 7.54 shows.

One may define the critical height of a bridge pier to induce elastic instability under its self-weight. By solving Eq. (7.118) for P_{cr} with respect to ℓ, one has

$$\ell_{cr} = 2.025 \sqrt[3]{\frac{EI}{p}} \tag{7.119}$$

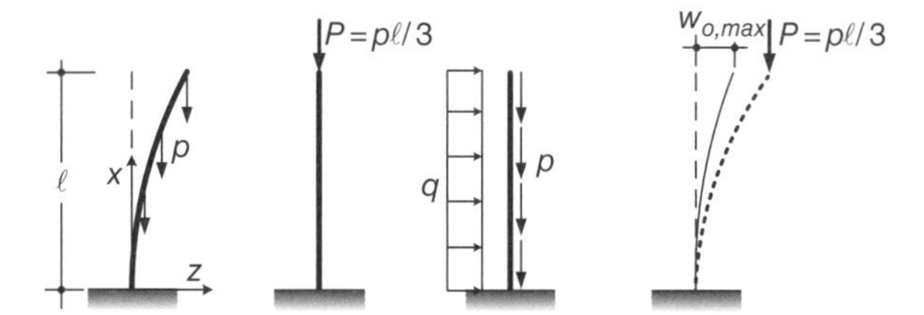

Figure 7.54 Equivalent load to the self-weight, for stability analysis, applied at the top cross section of a bridge pier and wind lateral load.

During the construction phase, bridge piers are subjected to lateral loads due to wind pressures q (Figure 7.54). Designating the maximum lateral deviation due to geometrical imperfections by $w_{o,\,max}$, one has for the maximum lateral displacement under self-weight and wind loads, with 2nd order effects included,

$$w_{max} = \left(w_{o,max} + \frac{q\ell^4}{8\,EI} \right) \frac{1}{1-\left(p\ell/3/P_{cr}\right)} \tag{7.120}$$

For design verifications cracking effects should be included by taking $EI = EI_y$ and load safety coefficients taken into account, that is, $q_{sd} = \gamma_q\ q$ and $p_{sd} = \gamma_g\ p$. The internal forces at the pier base section are

$$\begin{aligned} N_{sd} &= -\gamma_g\, p\ell \\ M_{sd} &= -\frac{1}{2}\gamma_q\, q\ell^2 - \frac{1}{3}\gamma_g\, p\ell\, w_{max} \end{aligned} \tag{7.121}$$

Here, w_{max} is evaluated from Eq. (7.120) with the factored loads q_{sd} and p_{sd}.

7.6.6 Overall Stability of a Bridge Structure

In Section 7.6.1 the need to consider a global stability analysis of a bridge structure to estimate piers buckling lengths was justified. That overall stability analysis allows determining the critical load factor by solving Eq. (7.45) if the loads H_i that are developed at the top of the piers, for a small horizontal displacement of the entire bridge deck $u = \hat{u}$, are known. For that it is necessary to know the stiffness coefficients at the top of the bridge piers taking into account the second order effects of the axial forces.

The values of the referred stiffness coefficients for bridge piers under different boundary conditions [9] are indicated in the Figure 7.55. Limit conditions of articulated or built in were considered at the connections of the piers with the foundation and with the bridge deck.

i) Built in pier at the top and base sections (Figure 7.55a):

$$\begin{aligned} H &= \frac{EI}{\ell^3}\, F_1\, u \\ K &= \frac{EI}{\ell^3}\, \frac{\varphi^3 \sin\varphi}{2-2\cos\varphi-\varphi\sin\varphi} \end{aligned} \tag{7.122}$$

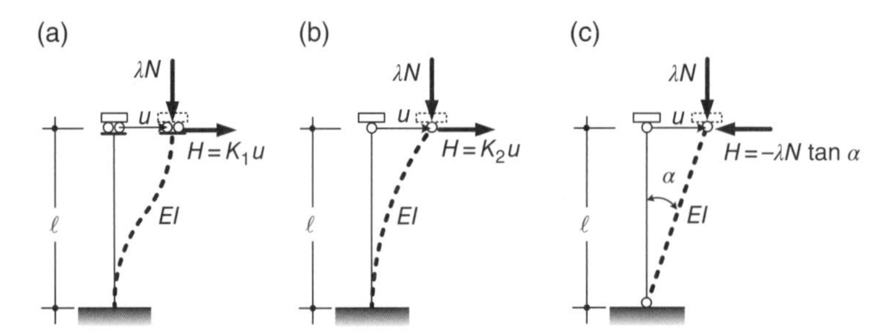

Figure 7.55 Stiffness coefficient K for: (a) a built in pier at the top and base cross sections; (b) a built in pier at the base and articulated at the top cross sections and (c) a pendulum pier.

Here and in the sequel, the load parameter is redefined as

$$\varphi = \pi\sqrt{\frac{\lambda N}{N_E}}, \text{ with } N_E = \frac{\pi^2 EI}{\ell^2} \tag{7.123}$$

ii) A pier built in at the base section and articulated at the top section (Figure 7.55b)

$$\begin{aligned} H &= \frac{EI}{\ell^3} F_2 u \\ K &= \frac{EI}{\ell^3} \frac{\varphi^3 \sin\varphi}{\sin\varphi - \varphi\cos\varphi} \end{aligned} \tag{7.124}$$

iii) A bi-articulated bridge pier (pendulum) (Figure 7.55c)

$$\begin{aligned} H &= -\lambda N \tan\alpha \\ \tan\alpha &= \frac{u}{\ell} \\ K &= -\frac{\lambda N}{\ell} \end{aligned} \tag{7.125}$$

In this last case, a pendulum pier, a force inducing instability is generated increasing with the inclination of the pier under the assumed horizontal displacement. This force should be stabilized by the positive reactions at the top of the other piers.

In general, for a bridge structure with different boundary conditions at the top and base of the bridge piers as shown in Figure 7.43, assuming n_1 built-in piers at the top and base sections, n_2 piers built in at the base and articulated at the top sections and n_3 pendulum type piers, the condition for critical equilibrium (Eq. (7.45)) is a homogeneous equation with nonzero solutions for the critical value of the load parameter $\lambda = \lambda_{cr}$

$$\sum_{i=1}^{n_1} \frac{EI_i}{\ell_i^3} F_{1i} + \sum_{i=1}^{n_2} \frac{EI_i}{\ell_i^3} F_{2i} - \sum_{i=1}^{n_3} \frac{\lambda N_i}{\ell_i} = 0 \tag{7.126}$$

In this equation, functions F_{1i} and F_{2i} are dependent on the axial forces in the piers λN_i through the load factor ϕ. Equation (7.126) involves trigonometrical functions and should be solved with respect to λ to determine the associated critical load value λ_{cr}. An iterative procedure is required by successive iterations under increasing values of the load parameter. The left-hand side of Eq. (7.126) passes from a positive value to a negative value under increasing λ values.

It should be noted that λ_{cr} represents the safety factor with respect to the elastic critical load only. The safety factor with respect to the design ultimate resistance of the bridge structure should be determined taking into account the elastoplastic effects on bridge piers, namely cracking effects, the effect of initial imperfections and the horizontal loads acting on the bridge.

References

1 Calgaro, J.-A. (2000). *Projet et Construction des Ponts – Tome 1: Généralités – Fondations – Appuis – Ouvrages courants*, 3e. Presses Ponts et Chaussées.
2 Parke, G. and Hewson, N. (2008). *ICE Manual of Bridge Engineering*, 2e. Thomas Telford Lim.
3 Ramberg, G. (2002). Structural Bearings and Expansion Joints for Bridges. *Structural Engineering Documents*, Ed. by IABSE, Switzerland. ISBN: 3-85748-105-6.
4 EN 1337 (2005). Structural bearings. Part 3: Elastomeric bearings; Part. 4: Roller Bearings; Part 5: Pot Bearings. CEN - European Committee for Standardization.
5 SETRA (2007). Guide technique: Appareils d'appui en élastomère fretté Utilisation sur les ponts, viaducs et structures similaires. ISBN: 978-2-11-095820-4.
6 SETRA (2007). Guide technique Appareils d'appui à pot utilisation sur les ponts, viaducs et structures similaires. ISBN: 978-2-11-094622-5.
7 Leonhardt, F. (1979). *Construções de Concreto – Vol. 6 – Princípios Básicos da Construção de Pontes de Concreto*, 241. Editora Interciência.
8 Clough, R.W. and Penzien, J. (1993). *Dynamics of Structures*. New York: McGraw Hill.
9 Menn, C. (1990). *Prestressed Concrete Bridges*, 536. Springer.
10 Reis, A. and Camotim, D. (2001). *Estabilidade Estrutural*. McGraw Hill.
11 Godfrey Allen, H. and Bulson, P.S. (1980). *Background to Buckling*, vol. 582. McGraw-Hill.
12 Reis, A. and Camotim, D. (2012). *Estabilidade e Dimensionamento de Estruturas*, 704 pp. Publ. by Orion, Lisboa.ag
13 Crandall, S.E. (1956). *Engineering Analysis: A Survey of Numerical Procedures*. New York: McGraw Hill.

8

Design Examples

Concrete and Composite Options

8.1 Introduction

Two design solutions are presented for a continuous three span bridge deck. A twin-girder deck is adopted for both solutions: a prestressed concrete and a steel-concrete composite deck. The aim is to present the design steps of a road bridge deck with straight alignment, initially discussing possible solutions and justifying different options at a conceptual design level. Then, key aspects of structural safety and serviceability for both options are checked.

The bridge, its function and the materials adopted are presented in Section 8.2; hazard scenarios and actions are referred to in Section 8.3.

Section 8.4 covers the pre-design of the concrete solution, the structural analysis and safety verifications for ultimate and serviceability limit states (ULS and SLS) of the deck. Section 8.4.3 highlights the distribution of internal moments and forces, explains how to obtain the hyperstatic effects for a longitudinal prestressing layout and discusses the influence of the construction stages.

Section 8.5 presents the pre-design of the steel-concrete composite solution. ULS and SLS design verifications are presented for concrete slab and steel girders. Section 8.5.3 highlights the distribution of internal moments and forces, and explains the effects of slab shrinkage and imposed deformations.

Iterative design procedures are not covered by this example, with checks being carried out only for the final proposed geometries. Moreover, structural analysis is limited to calculations for the superstructure, with traffic as the leading action of hazard scenarios. The emphasis of this example is on the overall process of pre-design/design of a bridge deck, based on simple structural models.

8.2 Basic Data and Bridge Options

8.2.1 Bridge Function and Layout

The bridge carries two road traffic lanes, travelling in opposite directions, as well as two footpaths for pedestrians and cyclists. The bridge design working life is 100 years.

Bridge Design: Concepts and Analysis, First Edition. António J. Reis and José J. Oliveira Pedro.

The elevation of the bridge, over an existing railway line, is shown in Figure 8.1. The deck total length is 83.6 m split into three spans of 22.8, 38 and 22.8 m. The deck is straight in plan with no skew and has a 2.5% longitudinal slope.

The need for a central span of 38 m is due to the requirement for crossing a double track railway with due account taken for horizontal clearance requirements. To comply with geometrical requirements, 22.8 m side spans are adopted. The side/internal span ratio is 22.8/38 = 0.60. This is a minimum desirable value to have a good balance between permanent bending moments at the side and internal span, avoiding any up-lift at the bearings located at the abutments.

As a design requirement, the construction of the deck should not interfere with railway operation. An incremental launching construction scheme may be adopted, preferably with a steel-concrete composite deck (SC). Alternatively, a prestressed concrete deck (PC) executed with a formwork supported from the ground and spanning over the railway may be adopted. These are the options considered next.

8.2.2 Typical Deck Cross Sections

According to Figure 8.2, the deck includes two traffic lanes of 3.5 m width, with 0.25 m drainage channels for transition to 1.35 m elevated pedestrian sidewalks. With fascia beams 0.50 m width at each sidewalk, the total deck width is 11.2 m. The BN4 safety barriers are assumed to be non-structural, since they are not continuously attached to the deck. The slab has a transverse slope of 2.5% and the sidewalks are inclined to the inside at 4% for drainage.

Structural slab – this is 10.88 m since the fascia beams contribute 0.15 m additional width at each side, and a 1 cm gap is left between the cantilever tips and the fascia beams for construction proposes (Figure 8.2). The overhangs are 2.69 m width and the main girders are at 5.50 m distance apart. A slab span ratio 2.69/5.50 = 0.49 is within the range of recommended values. For both deck solutions, the slab is 300 mm thick over the main girders (with an additional 100 mm over the top steel flanges for the composite option). The slab is 280 mm thick at its centre and 200 mm thick at the cantilever ends. The average slab thickness, h_c, for the composite deck section is 293 mm (common values often adopted for pre-design of road and railway decks are, respectively, 300 and 350 mm) and has 0.9% longitudinal reinforcement at span sections and 1.6% at intermediate support sections.

Main girders and cross beams – Typical cross sectional dimensions for the main girders are shown in Figure 8.2 for the PC and SC decks. The deck depth h is 2.20 m for both solutions, which corresponds to a deck slenderness of 38/2.2 = 17.3, within typical L/h values between 15 and 20.

The deck has a constant depth as recommended for the construction scheme envisaged. The steel girder depth $h_a = 1.8$ m is constant over the whole length of the structure. The width and thickness of the flanges, as well as the web thickness, are shown in Figure 8.2.

Table 8.1 summarizes the area, A, second moments of area, I_a (steel) and I_b (composite, or concrete), elastic modulus, W_{inf}, W_{sup}, and centre of gravity positions, z_{inf}, z_{sup}, for both support and span deck sections. For the composite solution, values are given for different stages: steel section only; steel-concrete composite mid-span section at the

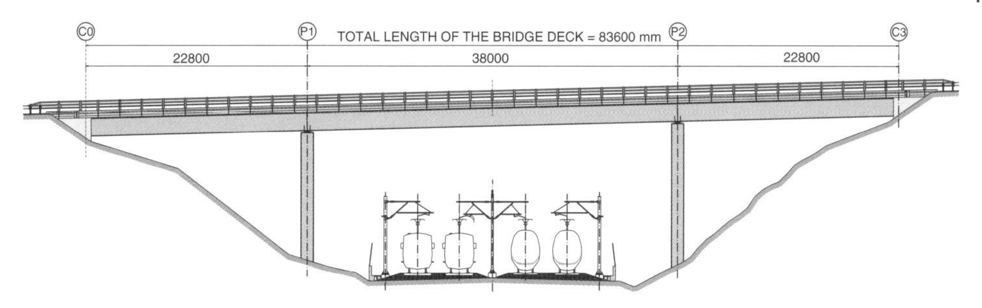

Figure 8.1 Elevation of the bridge for the pre-design example.

short and long term, shrinkage effects, imposed deformations and cracked composite support sections, including the steel girders and slab reinforcement.

The longitudinal PC girders are cross braced only at the abutments and intermediate supports.

The steel girders are braced at the abutments, at intermediate supports and span sections through welded section cross beams at 7.6 m centres, as shown in Figure 8.3. At the supports, cross beams are made of 1.0 m high welded sections and, at the span, rolled sections IPE 750 × 147 are adopted. These beams are located at the mid-depth of the main girders.

At the supports, the cross beams include a flat plate of $250 \times 25\,\mathrm{mm}^2$ as shown in the Figure 8.2b and a T section with a $250 \times 25\,\mathrm{mm}^2$ web and a $330 \times 25\,\mathrm{mm}^2$ flange from the inside the section, which are welded to each side of the webs of the main girders. The connection of the cross beams at span sections is also made through a T section, which is welded to the inside surface of the main girders web. Additional intermediate flat plate stiffeners $250 \times 25\,\mathrm{mm}^2$ are located only at the web panels near the supports, and welded to the inside of the girders at 2.535 m from the axes of the bearings. During launching, a temporary bracing is adopted (Figure 8.3). This is made of a truss composed by the main girders (chords), the cross beams (uprights) and diagonals (cables ϕ 30) located in the plane defined at mid-height of the cross beams. At the final state, the concrete slab fulfils the function of in-plan bracing.

8.2.3 Piers, Abutments and Foundations

Piers and foundations are made of reinforced concrete. Pier shafts have a $2.50 \times 1.25\,\mathrm{m}^2$ section envelope, with pier caps to accommodate the deck pot bearings. The piers cross section, identical for the two deck options, corresponding to a longitudinal slenderness of 52 for a maximum pier height of 21 m and assuming the deck is free to slide in the longitudinal direction at both abutments. Pier foundations are $4.0 \times 7.0\,\mathrm{m}^2$ with concrete footings 1.5 m high. Reinforced concrete abutments are made of seating beams 1.5 m high and 11.3 m long directly supported on the soil.

8.2.4 Materials Adopted

Piers, abutments and foundations are made of concrete grade C30/37 XC2 (EN 206 standard [1]) with steel reinforcement bars B500 B (EN 10080 standard [2]).

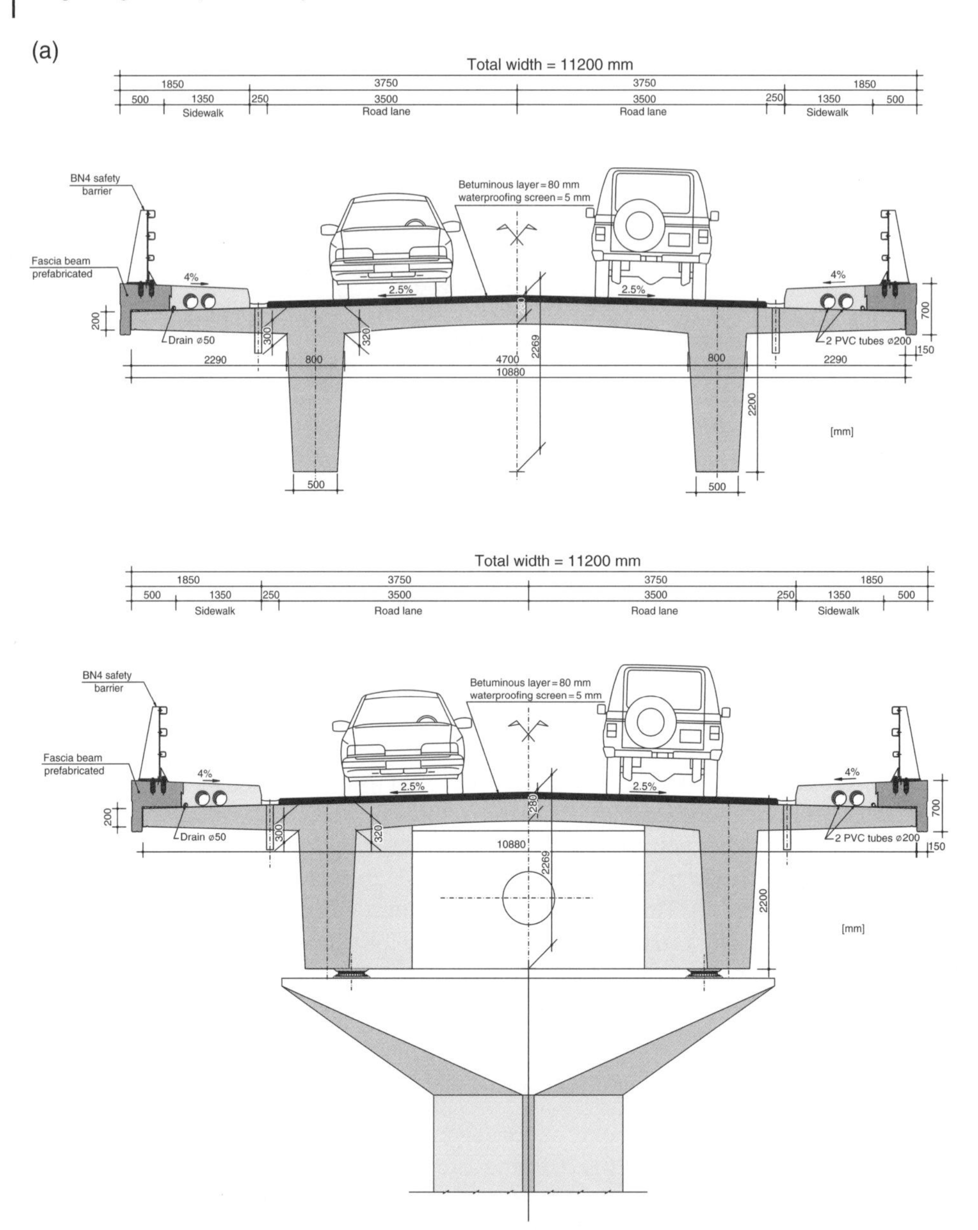

Figure 8.2 Typical deck cross sections at a support and in span regions for both: (a) the prestress concrete solution and (b) the steel-concrete composite solution.

8.2.4.1 Prestressed Concrete Deck

Concrete grade C35/45 XS1 is adopted for the prestressed concrete deck with steel reinforcement B500 B (EN 206, EN 10080). Concrete characteristic resistance in compression is $f_{ck} = 35$ MPa, the average elastic modulus $E_{cm} = 34$ GPa (Table 4.1). Steel characteristic yield strength is $f_{yk} = 500$ MPa and the modulus of elasticity is $E_s = 200$ GPa. Prestressing cables are made of seven-wire bonded strands with 15.7 mm nominal diameter (T15 super) with a nominal tensile strength of $f_{pk} = 1860$ MPa, a proof stress of

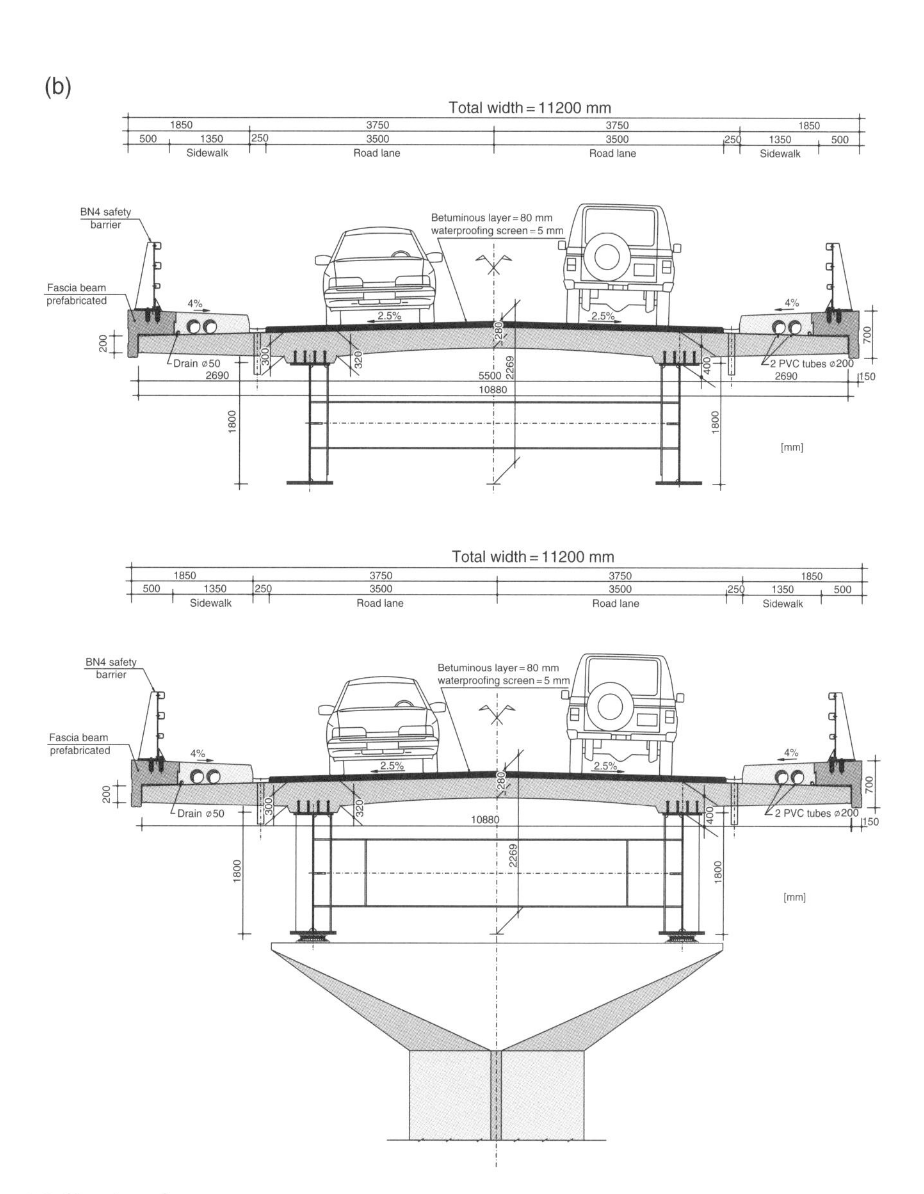

Figure 8.2 (Continued)

Table 8.1 Deck cross sections proprieties for the concrete and composite solutions.

Cross section	A (m^2)	z_{nf} (m)	z_{sup} (m)	W_{inf} (m^3)	W_{sup} (m^3)	I_y (m^4)
Prestress concrete	5.700	1.562	0.638	1.549	3.792	2.420
Steel	0.154	0.789	1.011	0.108	0.084	0.085
Composite short term	0.686	1.686	0.114	0.148	2.188	0.249
Composite shrinkage effect	0.407	1.508	0.292	0.143	0.738	0.215
Composite long term	0.320	1.389	0.411	0.139	0.470	0.193
Composite imposed deformations	0.287	1.325	0.475	0.137	0.383	0.182
Steel + slab reinforcement	0.208	1.089	0.711	0.128	0.196	0.139

For the composite solution, the elastic modules W_{inf}, W_{sup} and centre of gravity positions z_{inf}, z_{sup} are given for the steel girders lower and upper fibres, respectively.

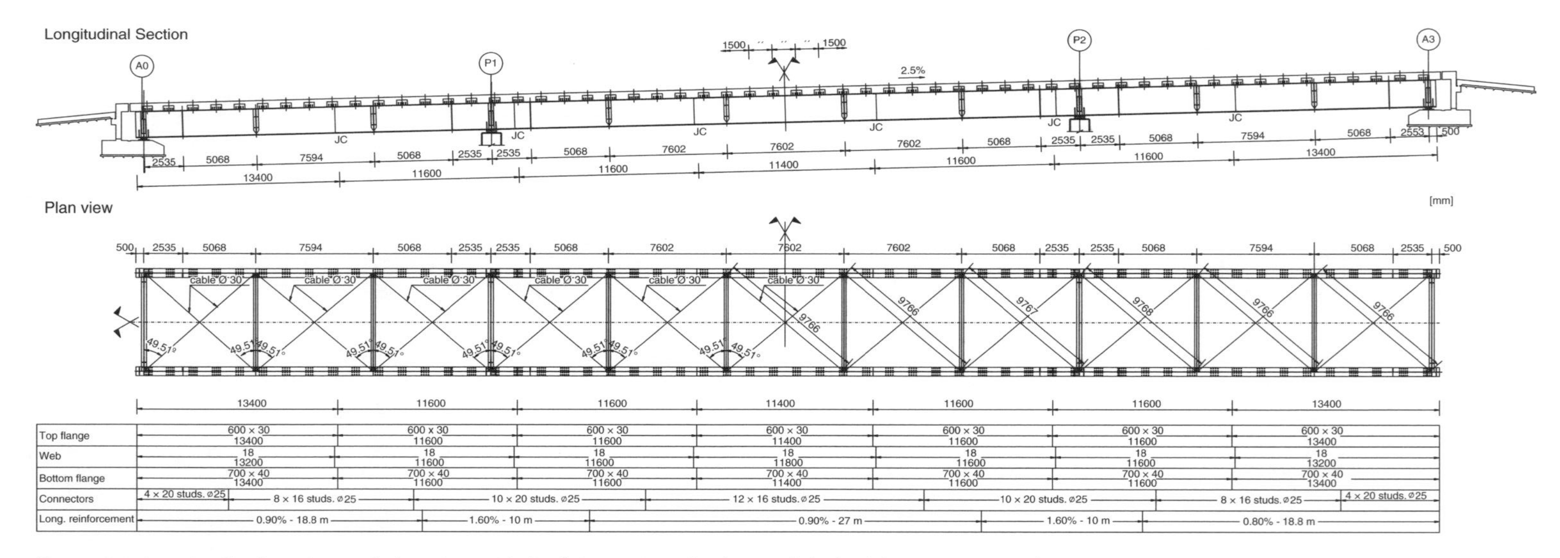

Figure 8.3 Longitudinal section and plan view with final dimensions for the steel deck of the composite solution.

$f_{p0.1k} = 1680$ MPa and a modulus of elasticity is $E_{sp} = 195$ GPa for prestressing steel Y1860 (in accordance with the EN 10138-3 [3]).

8.2.4.2 Steel-concrete Composite Deck

The main girders, stiffeners and cross beams are made of steel grade S355J2+N. The yield strength of the individual steel plates depends on the thickness (following EN 10025-3 [4]): $f_y = 355$ MPa, for $t \leq 16$ mm, $f_y = 345$ MPa for 16 mm $< t \leq 40$ mm and $f_y = 335$ MPa for 40 mm $< t \leq 63$ mm); the elastic modulus is $E_s = 210$ GPa.

The shear stud steel is cold drawn grade S235 J2G3 + C450, which has an ultimate strength of 450 MPa (ISO 13918 [5] and EN 1994-2 [6]). The deck slab is made of concrete grade C40/50 XS1 and B500 B reinforcement (EN 206-1, EN 10080). Concrete characteristic resistance in compression is $f_{ck} = 40$ MPa, its average elastic modulus is $E_{cm} = 35$ GPa (Table 4.1) and its average shear modulus is $G_c = E_{cm}/[2(1+\nu)] = 35/(2(1+0.2)) = 14.58$ GPa. Reinforced steel has a yield strength of 500 MPa and an elastic modulus $E_a = 200$ GPa, according to EN 1992-1 [7]. For simplification, EN 1994-2, §3.2(2) [6] allows the use of a single design value of 210 GPa for the elastic modulus of the structural steel and reinforcement.

Creep is taken into account by reducing the value of the elastic modulus of concrete and assuming a creep coefficient $\varphi = 2.0$ for the precast slabs. This means the short term elastic modular ratio $n_0 = E_s/E_{cm}$ varies depending on the nature of the loads acting on the composite section (following §5.4.2.3 of EN 1994-2 [6]):

for short term loading: $n_0 = E_s/E_{cm} = 210/35 = 6.0$
for shrinkage effects: $n_S = E_s/[E_{cm}/(1+\varphi\,\psi_S)] = 210/[35/(1+2\cdot 0.55)] = 12.6$
for long term loading: $n_L = E_s/[E_{cm}/(1+\varphi\,\psi_L)] = 210/[35/(1+2\cdot 1.10)] = 19.2$
for imposed deformations: $n_D = E_s/[E_{cm}/(1+\varphi\,\psi_D)] = 210/[35/(1+2\cdot 1.50)] = 24.0$

8.2.5 Deck Construction

Prestressed concrete deck – First, the lateral spans are executed over steelwork scaffoldings, extended to construction joint sections at the main span 7.6 m distant from the piers' axes. After stressing the prestressing cables, the remaining 22.8 m of the main span is concreted with a formwork suspended from temporary steel truss girders, working from above and supported near the tip of the cantilever's deck. A second set of cables are then stressed.

Steel-concrete composite deck – The structural steelwork members are fabricated in a workshop, transported to site and welded together at a provisional platform executed at the back of one abutment. The entire steel work is launched into place using a front nose to overpass the piers and a rear nose to make the push reaction. A precast segmental slab is adopted. Segments are placed on the top of the steel girders by a small crane placing 1.50 m long segments one by one. Cast in place joints between segments are executed before the finishing works.

8.3 Hazard Scenarios and Actions

The hazard scenarios and limit states for checking structural safety and serviceability of the deck are defined, with their related load cases; then design actions are quantified.

8.3.1 Limit States and Structural Safety

Ultimate limit states – Table 8.2 present the hazard scenario for checking structural safety of the bridge superstructure for persistent and transient design situations. The ultimate limit states (ULS) associated with the design of the superstructure is of the STR type (cross sectional resistance, collapse mechanism of the load carrying structure). Deck fatigue resistance is excluded at the present design level.

Serviceability Limit States – The SLS considered are summarized in Table 8.3. The frequent combination is adopted for prestressing cables design and slab cracking control, while the characteristic combination is adopted for checking the stress limits of the composite deck.

8.3.2 Actions

Actions are quantified for the case of a bridge near Lisbon and using the relevant parts of EN 1991-1 and EN 1991-2 [8] for the specific bridge loads.

8.3.2.1 Permanent Actions and Imposed Deformations

Self-weight of the deck – The PC deck has a constant cross sectional area, $A_c = 5.70\,m^2$. Adopting the weight per unit volume of reinforced concrete as $25\,kN\,m^{-3}$, one has $g_c = 142.5\,kN\,m^{-1}$ or $71.25\,kN\,m^{-1}$ per beam.

The SC deck has steel girders with constant cross sections $A_s = 0.154\,m^2$; for the steel unit weight $78.5\,kN\,m^{-3}$ the dead weight per unit length is $g_{s1} = 12.09\,kN\,m^{-1}$. Cross beam dead weight is $g_{s2} = 1.06\,kN\,m^{-1}$ and stiffener weight is estimated as being 7% of the weight of the main girder. Hence, total steelwork dead weight is $g_s = 12.09 \times 1.07 + 1.06 = 14\,kN\,m^{-1}$.

Both solutions adopt the concrete slab shown in Figure 8.2. Slab thicknesses are constant along the bridge length and result in a slab cross sectional area $A_{cs} = 3.188\,m^2$. The self-weight (SW) of the slab is $g_{cs} = 79.7\,kN\,m^{-1}$. Deck total self-weight is $g_s + g_{cs} = 93.7\,kN\,m^{-1}$ or $46.85\,kN\,m^{-1}$ per beam (the SC deck section is 34% lighter than the PC deck section).

Superimposed dead loads – Non-structural elements comprise the surfacing, safety barriers, sidewalks/pavements, waterproofing screen and fascia beams. Characteristic values are given in Section 3.2. The surfacing has a nominal thickness of 80 mm although 40 mm additional thickness is considered for possible recharges. Density of the bituminous layer as well as of the sidewalks made with normal density concrete is $24\,kN\,m^{-3}$. The total superimposed dead load (SDL) is:

Bituminous layer (total 8 + 4 cm)	$21.60\,kN\,m^{-1}$
Sidewalks (normal concrete)	$2 \times 7.25\,kN\,m^{-1}$
BN4 Safety barriers	$2 \times 1.40\,kN\,m^{-1}$
Fascia beams	$2 \times 7.35\,kN\,m^{-1}$
Waterproofing screen	$1.40\,kN\,m^{-1}$
SDL/beam = $27.5\,kN\,m^{-1}$	Total SDL = $55\,kN\,m^{-1}$

Imposed deformations – For the SC option, an imposed deformation of 100 mm at intermediate supports is introduced at the end of construction when the temporary supports are replaced by the final pot bearings. This action induces a positive curvature of the deck and an internal redistribution of forces due to dead loads, increasing

Table 8.2 Hazard scenarios for structural ULS safety checks in the final state.

Hazard scenario	Permanent actions		Leading variable action		Accompanying variable action	
	G_k	γ_G	Q_{k1}	γ_Q	Q_{ki}	$\psi_{0i}\gamma_{Qi}$
ULS structural resistance	Dead loads	1.35/1.00	LM 1 – Road UDL system	1.35	Temperature gradient – concrete deck	0.00
	Prestress	1.20/0.90	LM 1 – Tandem system	1.35	Temperature gradient – composite deck	0.6 × 1.5
	Shrinkage Imp. Defor.	1.00/0.00	LM 1 – Pedestrian uniform load	1.35		

the sagging bending moments and reducing the hogging bending moments at piers sections.

Shrinkage and creep – For the PC option, shrinkage induces an imposed axial deformation on the deck (ε_{cs}) and, consequently, longitudinal prestressing losses. For the SC option, concrete deck slab shrinkage induces tension on the slab that shall be anchored through shear connectors at the ends of the composite girders. Shrinkage induces a negative curvature and, due to support restraints, hyperstatic effects. Shrinkage deformation (ε_{cs}) is quantified only for the composite deck, being the sum of autogenous shrinkage (ε_{ca}) and drying shrinkage (ε_{cd}), determined in accordance with the EN 1992-1 [7]. The drying shrinkage is given by:

$$\varepsilon_{cd}(t)=\beta_{ds}(t,t_s)k_h\,\varepsilon_{cd,0}\ \text{and}\ \beta_{ds}(t,t_s)=\frac{(t-t_s)}{(t-t_s)+0.04\sqrt{h_0^3}}$$

where:

A_c is the cross sectional area of the concrete slab or deck ($A_{cs}=3.188\,\text{m}^2$);

u is the slab perimeter exposed to drying effects, obtained by subtracting from the actual perimeter of the cross section by the lengths that are not permanently in contact with the atmosphere (i.e. the top surface of the slab and the width of the upper steel flanges); after pre-casting the slabs u_1 = 22.38 m; after executing the finishing works: u_2 = 9.10 m;

h_0 is the equivalent thickness given by $2A_c/u$ (h_{01} = 285 mm; h_{02} = 701 mm);

k_h is a value which depends on h_0 and is obtained from Table 3.3 of EN 1992-1 $k_{h1}=0.765$; $k_{h2}=0.70$;

$\varepsilon_{cd,0}$ is determined according to appendix B.2 of the EN 1992-1 by:

$$\varepsilon_{cd,0}=0.85\left[\left(220+110\alpha_{ds1}\right)\times\exp\left(-\alpha_{ds2}\frac{f_{cm}}{f_{cm0}}\right)\right]10^{-6}\beta_{RH};$$

$$\beta_{RH}=1.55\left[1-\left(\frac{RH}{RH_0}\right)^3\right]=0.756\ ;\ f_{cm}=48\,MPa;\ f_{cm0}=10\,MPa$$

with ambient relative humidity: $RH=80\%$; $RH_0=100\%$

Table 8.3 Hazard scenarios for structural SLS safety checks in the final state.

	Permanent actions		Leading variable action		Accompanying variable action	
Hazard scenario	G_k	γ_G	Q_{k1}	ψ_1	Q_{ki}	ψ_{2i}
SLS – Frequent comb.	Dead loads	1.00	LM 1 – Road UDL system	0.40	Temperature gradient – concrete deck	0.50
	Prestress	1.00	LM 1 – Tandem system	0.75	Temperature gradient – composite deck	0.50
	Shrinkage Imp.Defor.	1.00/0.00	LM 1 – Pedestrian uniform load	0.40		
	G_k	γ_G	Q_{k1}	γ_Q	Q_{ki}	ψ_{0i}
SLS – Characteristic comb.	Dead loads	1.00	LM 1 – Road UDL system	1.00	Temperature gradient – concrete deck	0.60
	Prestress	1.00	LM 1 – Tandem system	1.00	Temperature gradient – composite deck	0.60
	Shrinkage Imp.Defor.	1.00/0.00	LM 1 – Pedestrian uniform load	1.00		

Cement Class R (high resistance at short term): $\alpha_{ds1} = 6$; $\alpha_{ds2} = 0.11$

Table 8.4 gives the shrinkage deformation at time of casting the joints between the slab segments and the steel girders (4 months after casting the slab segments $t_{ini} = 120$ days) and for the long term ($t_{inf} = 100$ years).

The autogenous shrinkage is given also in Table 8.4 using:

$$\varepsilon_{ca}(t) = \beta_{as}(t)\varepsilon_{ca}(\infty)$$

$$\beta_{as}(t) = 1 - \exp\left(-0.2t^{0.5}\right) \; ; \; \varepsilon_{ca}(\infty) = 2.5\left(f_{ck} - 10\right)10^{-6}$$

The total shrinkage deformation, transmitted from the slab to the steel girders, is therefore given by $\varepsilon_{cc} + \varepsilon_{ca} = [(228.9 - 98.1) + (75.0 - 66.6)] \times 10^{-6} = 139.2 \times 10^{-6}$, that is, an equivalent uniform temperature variation in the slab of -13.9°C.

8.3.2.2 Variable Actions

Traffic Loading – Actions due to road traffic are determined in accordance with paragraph 3.3, that is, total width if platform 7.5 m, divided in two notional lanes of 3 m (number of notional lanes $n = \text{int}(7.5/3) = 2$) and the remaining area is 1.5 m. No exceptional traffic is considered. Horizontal forces due to acceleration and braking are not relevant for checking the superstructure. Only load model 1 (LM1) needs to be considered. Individual loads are taken from Table 3.1. Additionally, to distribute uniformly the road live load, the sidewalks are loaded with pedestrian and cycle-track live loads as per Table 8.5.

Table 8.4 Drying and autogenous shrinkage for the pre-casted slabs of the composite deck.

Time	β_{RH}	$\varepsilon_{cd,0}[10^{-6}]$	(t, t_s)	h_0 (mm)	$\beta_{ds}(t, t_s)$	ε_{cd} $[10^{-6}]$
t_{ini} = 120 days	0.756	333.7	120	285	0.384	98.1
t_{inf} = 100 years	0.756	333.7	36 500	701	0.980	228.9
Time	β_{as}		$\varepsilon_{ca}(\infty)$ (10^{-6})		f_{ck}(MPa)	$\varepsilon_{ca}(10^{-6})$
t_{ini} = 120 days	0.888		75.0		40	66.6
t_{inf} = 100 years	1.000		75.0		40	75.0

Table 8.5 Vertical traffic loads for the deck according to EN1991-2 [8].

Location	Tandem system TS Loads per axe Q_{ik} (kN)	Uniform distributed load UDL q_{ik} (or q_{rk}) (kN m^{-2})
Notional lane 1	2 axes of 300	9.0
Notional lane 2	2 axes of 200	2.5
Remaining area (q_{rk})	0	2.5
Sidewalks	0	3.0

Table 8.6 Temperature gradients for the concrete and the composite deck according to EN1991-1-5 and Portuguese NA [9].

	Concrete Deck (°C)		Composite Deck (°C)	
Location = Lisbon	ΔT_{heat}	ΔT_{cool}	ΔT_{heat}	ΔT_{cool}
Uniform component	+25	−15	+28	−18
Vertical linear gradient	+15	−5	+15	−15

The values of the vertical forces of LM1 are multiplied by the adjustment factors α_{Qi}, α_{qi} and α_{qr}. The recommended values $\alpha_{Qi} = \alpha_{qi} = \alpha_{qr} = 1.0$ of the EN 1991-2, corresponding to a heavy industrial international traffic and representing a large part of the total traffic of heavy vehicles, are considered. For each lane, the TS is composed by two axes 1.2 m distance apart in the longitudinal direction.

Wind actions – the horizontal drag and vertical forces are determined according to paragraph 3.11 and figure 3.27. The vertical forces can be neglected compared to traffic loads. The horizontal wind forces are generally adopted to design the substructure and the steel in-plan bracing of the SC deck solution, during the launching phase (not covered by the example).

Snow – In this example, snow loads at the bridge site can be neglected compared to traffic loads.

Temperature gradients – Temperature effects are taken into account according to paragraph 3.13. Table 8.6 values are obtained for the bridge site without correction due to altitude and a mean air temperature during the year of 15°C. Additional gradient of ±10°C is considered at the slab thickness, for local verifications.

Earthquake – Seismic forces are relevant for the substructure only, and therefore not considered in this example.

8.4 Prestressed Concrete Solution

8.4.1 Preliminary Design of the Deck

At a preliminary design stage, dimensions of the most highly stressed cross sections of the concrete girder and slab should be defined, before progressing to the overall structural verifications. Traffic is the leading action to be combined with permanent loads, for obtaining the internal forces used for preliminary verifications. But design is an iterative process and deck dimensions chosen at preliminary design stages should be re-checked at each stage, to ensure requirements relative to ultimate safety and serviceability, to arrive to the most convenient ones. For this example, the results of such a procedure correspond to the distribution of materials, as shown in Figure 8.2 for the typical deck cross section. For the preliminary design, one has:

1) Total deck depth $h = 2.20$ m (deck slenderness of 38/2.2 = 17.3);
2) Overhangs span length $a = 2.69$ m, with spacing between the main girders of $b = 5.5$ m; slab span ratio of $a/b = 2.69/5.5 = 0.49$ (in the recommended range);
3) Thickness at tip of the overhang = 0.20 m – recommended minimum for steel reinforcement covering – thickness at the cantilever support = 0.30 m, within the pre-design rule $a/(8 \text{ to } 10) = 0.27$ to 0.34;
4) Thickness at the mid-span central slab = 0.28 m, within the pre-design rule $b/(18 \text{ to } 25) = 0.22$ to 0.31;
5) Bottom thickness of the main girders = 0.60 m, to be able to group to prestressing cables with 19–22 strands with maximum eccentricity;
6) Top thickness of the main girders = 0.80 m, yielding a small inclination of the girder lateral faces for aesthetics and easiness of removing formwork.

8.4.2 Structural Analysis and Slab Checks

An elastic structural analysis is made on the basis of analytical models. Bending moments resulting from dead loads, temperature gradients and uniform distributed live loads assume a cylindrical bending curvature and a beam sectional model of the deck 1 m width is considered. Influence surfaces are adopted for concentrated traffic loads, located on the bridge slab at the most unfavourable positions for each of the slab panels both transversally and longitudinally.

Overhangs – The bending moments at the cantilever support are evaluated at the axes of the main girder, resulting in a 2690 mm span. Applying the SW, superimposed dead loads (SDL), uniformly distributed road traffic and pedestrian live loads (UDL), the transverse bending moments of Table 8.7 (Figure 8.4) are obtained. The temperature gradients do not induce bending moments at the cantilever.

Table 8.7 Transverse bending moments and safety checks at the cantilever support.

	Actions			
Cantilever section	**SW**	**SDL**	**UDL**	**TS**
m (kNm m)	−21.1	−33.7	−9.3	−60.3
ULS of resistance	$m_{Ed} = -168\,\text{kNm}\,\text{m}^{-1}$		$m_{Rd} = -208\,\text{kNm}\,\text{m}^{-1a}$	
SLS of crack widths	$m_{freq} = -104\,\text{kNm}\,\text{m}^{-1}$		$m_{cr} = -57\,\text{kNm}\,\text{m}^{-1} \rightarrow$ $w_k = 0.20\,\text{mm}^a < 0.3\,\text{mm}$	

[a] Values obtained for the reinforcement {ϕ16//0.15 + ϕ12//0.15} and C35/45, B500B, cover = 45 mm.

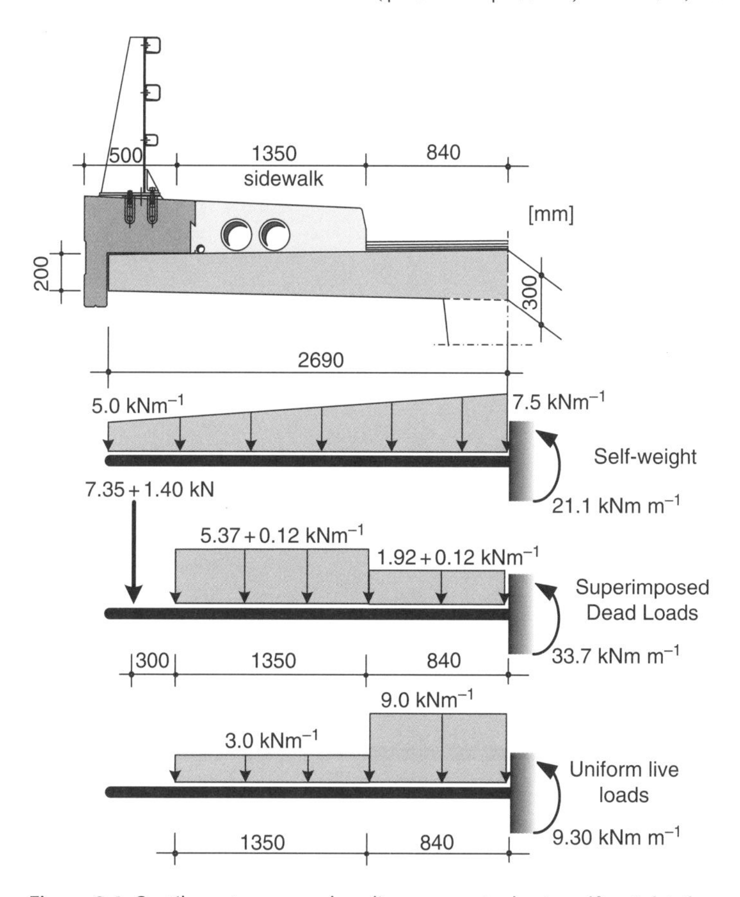

Figure 8.4 Cantilever transverse bending moments due to self-weight, the superimposed dead loads and the road and pedestrian uniform live loads.

For obtaining the bending moments due to the heaviest vehicle (with concentrated loads of $Q_1 = 150\,\text{kN}$), two positions are considered for loading the influence surfaces of Figure 8.5. It is interesting to note that using an influence surface of a slab with variable thickness increases about 7.5% the bending moment with respect to the same loading applied on a constant thickness slab (Table 8.8). Table 8.7 also summarizes the ULS

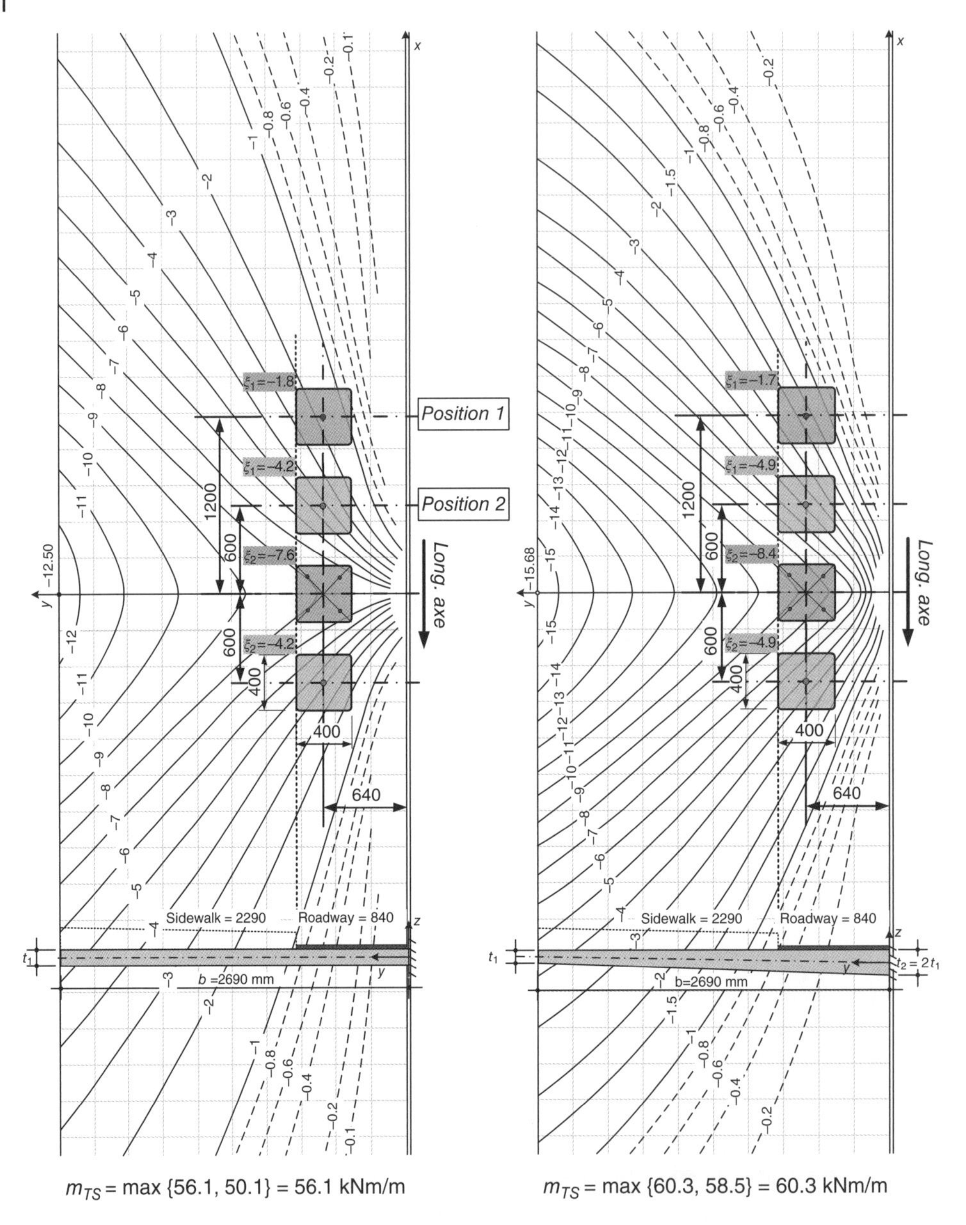

Figure 8.5 Transverse bending moments at the cantilever support for concentrated traffic loads, using influence surfaces with: (a) uniform thickness and (b) variable thickness $t_2 = 2t_1$.

bending moment and resistance for the reinforcement of Figure 8.6 with a concrete cover of 45 mm. For the frequent bending moment, crack widths in the long term are estimated to be $w_k = 0.20$ mm, lower than the 0.30 mm limit, for a XS1 exposure class of concrete.

Slab between the webs – The bending moments at the slabs between girders are evaluated at mid-span and support sections. An equivalent transverse beam with a span of $b = 5.5$ m between the axes of the main girders is adopted (Figure 8.7). Applying the SW,

Table 8.8 Transverse bending moments at the cantilever support for concentrated traffic loads.

	Action – Concentrated traffic load (TS)			
	Slab with uniform thickness		Slab with variable thickness	
Cantilever section	**Position 1**	**Position 2**	**Position 1**	**Position 2**
Wheel 1 – ξ_1	−1.8	−4.2	−1.7	−4.9
Wheel 2 – ξ_2	−7.6[a]	−4.2	−8.4[a]	−4.9
$m_{TS} = Q_1/8\pi\Sigma\xi_j$ (kNm m^{-1})	−56.1	−50.1	−60.3	−58.5

[a] Value obtained using four Gauss integration points and the 0.4 m square wheel.

Support cross-section

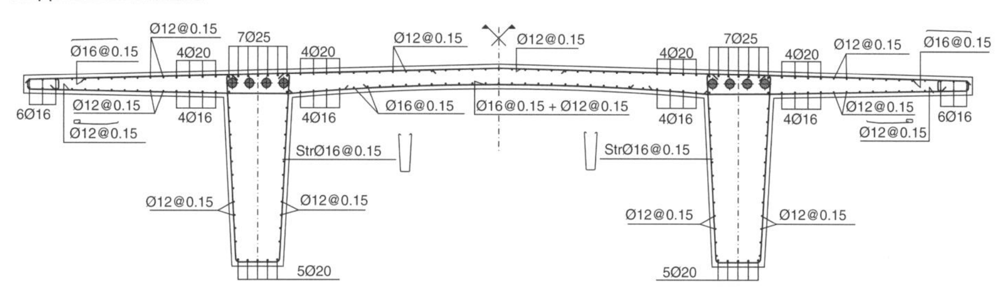

Span cross-section

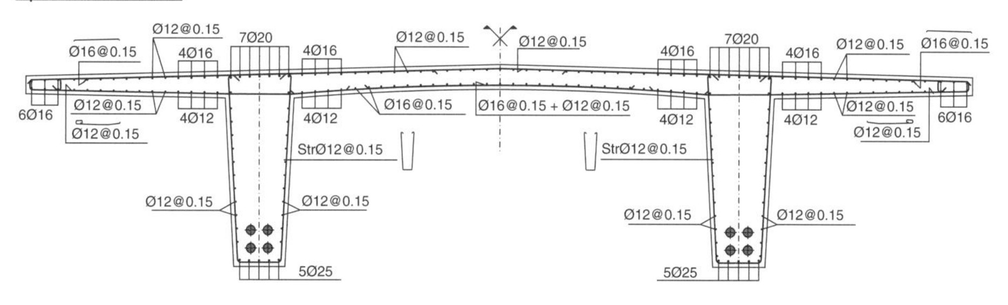

Figure 8.6 Reinforcement layout for the span and support PC deck cross sections.

SDL and uniformly distributed live loads (UDL) one obtains the transverse bending moments of Table 8.9, assuming a simply supported or perfectly restrained beam (Figure 8.6). The temperature gradients only induce bending moments for restrained slab edges. For a transverse beam model, with an equivalent thickness $h_c = 0.30\,\text{m}$, one has:

$$m_{\Delta T} = \frac{\alpha_T\,\Delta T}{h_c}\frac{E_c h_c^3}{12} = \frac{10^{-5}/°\text{C}\cdot\pm 10°\text{C}}{0.30\,\text{m}}\cdot\frac{34\cdot 10^6\,\text{m}^{-2}\cdot 0.30^3\,\text{m}^3}{12} = \pm 25.5\,\text{kNm m}^{-1}$$

For bending moments due to concentrated loading (TS), two vehicles with four wheels of $Q_1 = 150\,\text{kN}$ and $Q_2 = 100\,\text{kN}$ are applied at 3 m wide lanes drawn over the influence surfaces of Figure 8.8. Table 8.10 presents the contribution of each vehicle for the overall transversal bending moment at the slab mid-span section. The vehicle positions usually

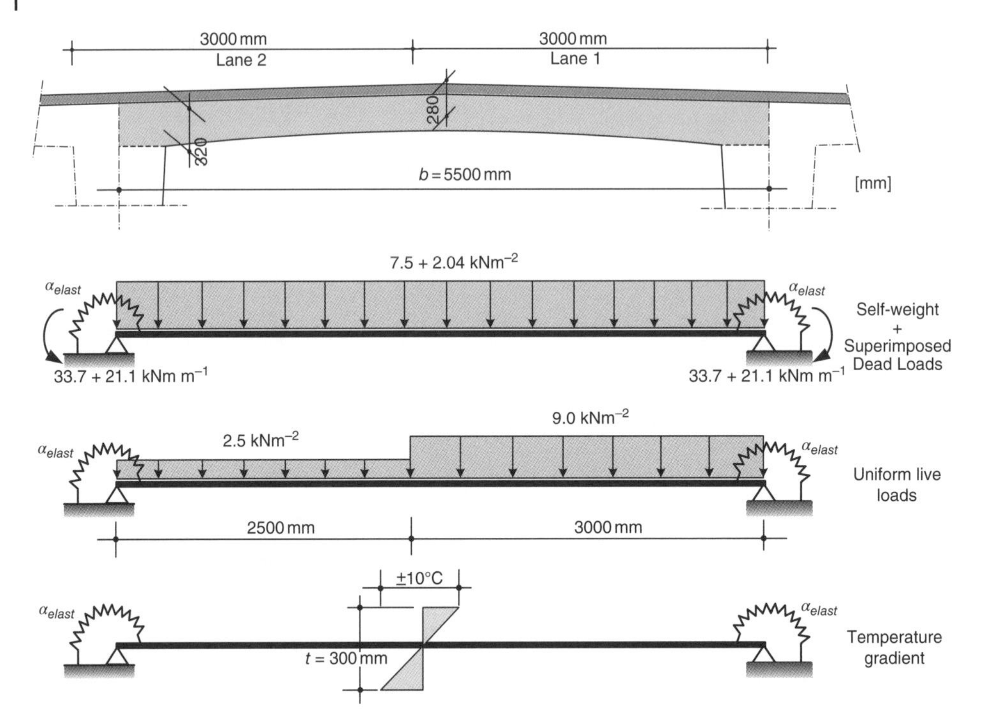

Figure 8.7 Slab between girders Transverse model for self-weight, the superimposed dead loads, the road uniform live loads and thermal gradient.

Table 8.9 Transverse bending moments and safety checks at the central slab mid-span section.

		Actions			
Slab mid-span section		**SW + SDL**	**UDL**	**TS**	**ΔT**
m (kNm m^{-1})	*Two supported edges*	+36.1	+24.5	+123.4	0
	Two restrained edges	+12.0	+8.6	+80.0	+25.5
	Two elastic restrains	−11.6[a]	+20.8	+113.4	+5.9
ULS of resistance		$m_{Ed} = 170\,\text{kNm m}^{-1}$		$m_{Rd} = 190\,\text{kNm m}^{-1}$[b]	
SLS of crack widths		$m_{freq} = 85\,\text{kNm m}^{-1}$		$m_{cr} = 49\,\text{kNm m}^{-1} \rightarrow$ $w_k = 0.17\,\text{mm}$[b] $< 0.3\,\text{mm}$	

[a] With the contribution of the cantilever bending moments given by $-54.8 \times (1 - \alpha_{elast}) = -42.2\,\text{kNm}^{-1}$.
[b] Values obtained for the reinforcement {ϕ16//0.15 + ϕ12//0.15}, and C35/45, B500B, cover = 45 mm.

yielding the highest moments have one of the heaviest wheels (from the lane 1 vehicle) at the maximum value of the influence lane, although this cannot be assumed as a rule.

As explained in Chapter 6, bending moments at the slab mid-span are taken by interpolation between the two extreme support conditions. Evaluating the elastic restrains given by the main girders to the central slab by the Eq. (6.19) one has:

$$\alpha_{elast} = \frac{1}{1 + 0.60 I_s / J_g \cdot L^2 / b} = 0.23$$

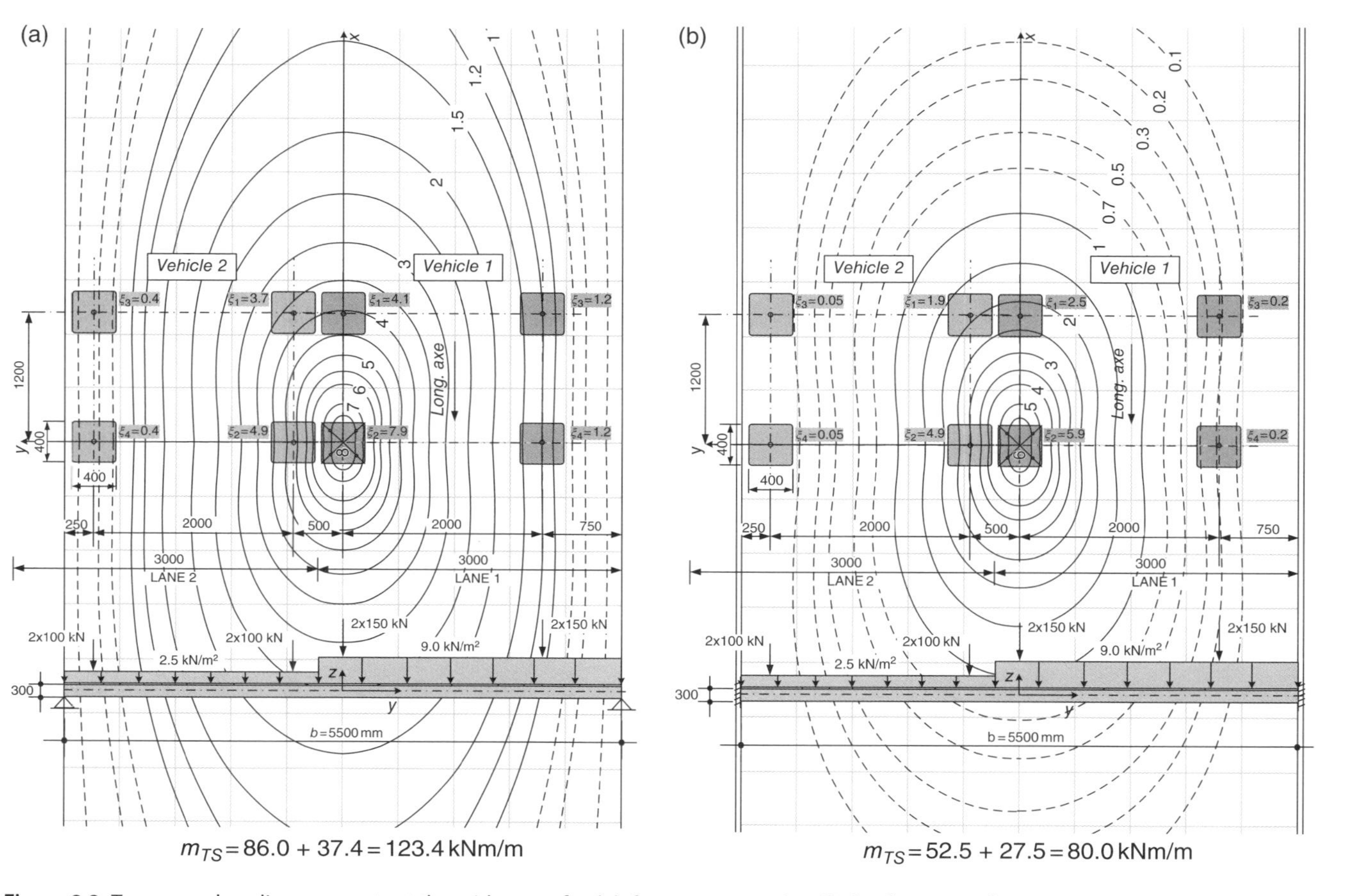

Figure 8.8 Transverse bending moments at the mid-span of a slab for concentrated traffic loads, using influence surfaces with: (a) two supported edges and (b) two restrained edges.

Table 8.10 Transverse bending moments at the central slab mid-span section for concentrated traffic loads.

	Action – Concentrated traffic load (TS)			
	Slab with supported edges		Slab with restrained edges	
Slab mid-span section	**Vehicle 1**	**Vehicle 2**	**Vehicle 1**	**Vehicle 2**
Wheel 1 – ξ_1	4.1	3.7	2.5	1.9
Wheel 2 – ξ_2	7.9[a]	4.9	5.9[a]	4.9
Wheel 3 – ξ_3	1.2	0.4	0.2	0.05
Wheel 4 – ξ_4	1.2	0.4	0.2	0.05
$m_{TS,i} = Q_i/8\pi\Sigma\xi_j$ (kNm m^{-1})	86.0	37.4	52.5	27.5

[a] Value obtained using four Gauss integration points and the 0.4 m square wheel.

with

Girders span between cross beams L = 38 m
Span of the central slab b = 5.5 m
Moment of inertia of the slab for t_{eq} = 0.3 m I_s = 0.00225 m^4
Torsional constant for the cracked girder $J_g = 50\% \cdot 0.21 = 0.11$ m^4

As per Chapter 6, one has

$$m_{elast} = \alpha_{elast} \cdot m_{rest} + (1 - \alpha_{elast}) \cdot m_{supp}$$

Table 8.9 summarizes ULS action and resistant bending moments for the reinforcement of Figure 8.6. For the frequent value of the bending moment, crack widths in the long term are estimated to be $w_k = 0.17\,\text{mm} < 0.30\,\text{mm}$.

For the slab support section, Table 8.11 and Figure 8.9 summarize the results for highest bending moment. Similar safety verifications to the previous case are presented in Table 8.11.

At the mid-span section the slab can have large bending moments in the longitudinal direction, mainly due to the concentrated traffic loading. Bending moments from the vehicles adopting the influence surfaces are shown in Figure 8.10 and assume the same elastic restraint as in the transverse direction. All other moments in Table 8.12 are estimated as 20% of the transverse bending moments at this section due to the cylindrical curvature of the slab. Temperature gradient produces the same bending moment in the two directions.

8.4.3 Structural Analysis of the Main Girders

For a prestressed concrete girder bridge deck, the internal moments and shear forces are determined by a first order elastic analysis. The traffic loads are located on the bridge deck at most severe positions, both transversally and longitudinally, as explained next.

Table 8.11 Transverse bending moments and safety checks at the central slab support section.

Slab support section		Actions			
		SW + SDL	UDL	*TS*	ΔT
m (kNm m^{-1})	*Two supported edges*	0	0	0	0
	Two restrained edges	−24.1	−18.5	−119.0	−25.5
	Two elastic restrains	-47.7^{a}	−4.3	−27.4	−5.9
ULS of resistance		$m_{\mathrm{Ed}} = 107$ kNm m^{-1}		$m_{\mathrm{Rd}} = 226$ kNm m^{-1b}	
SLS of crack widths		$m_{\mathrm{freq}} = 73$ kNm m^{-1}		$m_{\mathrm{cr}} = 64$ kNm m^{-1} → $w_k = 0.11$ mm^{b} < 0.3 mm	

[a] With the contribution of the cantilever bending moments given by $-54.8 \times (1 - \alpha_{elast}) = -42.2$ kNm m^{-1}.
[b] Values obtained for the reinforcement {ϕ16//0.15 + ϕ12//0.15}, and C35/45, B500B, cover = 45 mm.

Since the live loads may not be applied in a symmetrical form, the longitudinal model considers the geometrical proprieties of one beam with one half of the slab. Following paragraph 5.3.2.1(4) of EN1992-1 [7], for structural analysis, a constant effective width from the span section, which corresponds to the width of the concrete slab, is assumed over the whole span.

8.4.3.1 Traffic Loads: Transverse and Longitudinal Locations

Transversally, for the analysis of a twin-girder bridge, one uses the transverse influence line to identify the most unfavourable position of the live loads. This transverse influence line for the loads is determined according to Section 6.6.5.3 and particularly Eqs. (6.104) and (6.115), allowing to obtain the coefficients η_1, η_2 as a function of the parameter α:

$$\alpha^2 = \frac{G_c J_c}{2EI_y}\left(\frac{l}{b}\right)^2 = 0.88 \quad \eta_{1,2} = \frac{1}{2} \pm \frac{1}{\alpha^2}\left(1 - \frac{1}{\cosh\alpha}\right) = 0.50 \pm 0.37 \quad \eta_1 = 0.87 \quad \eta_2 = 0.13$$

The following material and geometrical data are taken

G_c Distortional modulus of concrete = $E_c/2.4$
E Average elastic modulus of concrete, $E_{cm} = 34$ GPa
I_y Second moments of area of one girder + slab = 1.21 m^4
J_c Uniform torsion constant of the deck (assumed cracked on the surface due to torsion) = 50% × 0.44 = 0.22 m^4
l Equivalent simply supported span≈0.70·38 = 26.6 m
b Spacing of the main girders = 5.5 m

Strictly, one should determine a different transverse influence line for each span and each cross section. Applying the same equations to the lateral span for l = 19.4 m, $\eta_1 = 0.92$ and $\eta_2 = 0.08$ are obtained. However, in practice, this distinction is rarely made

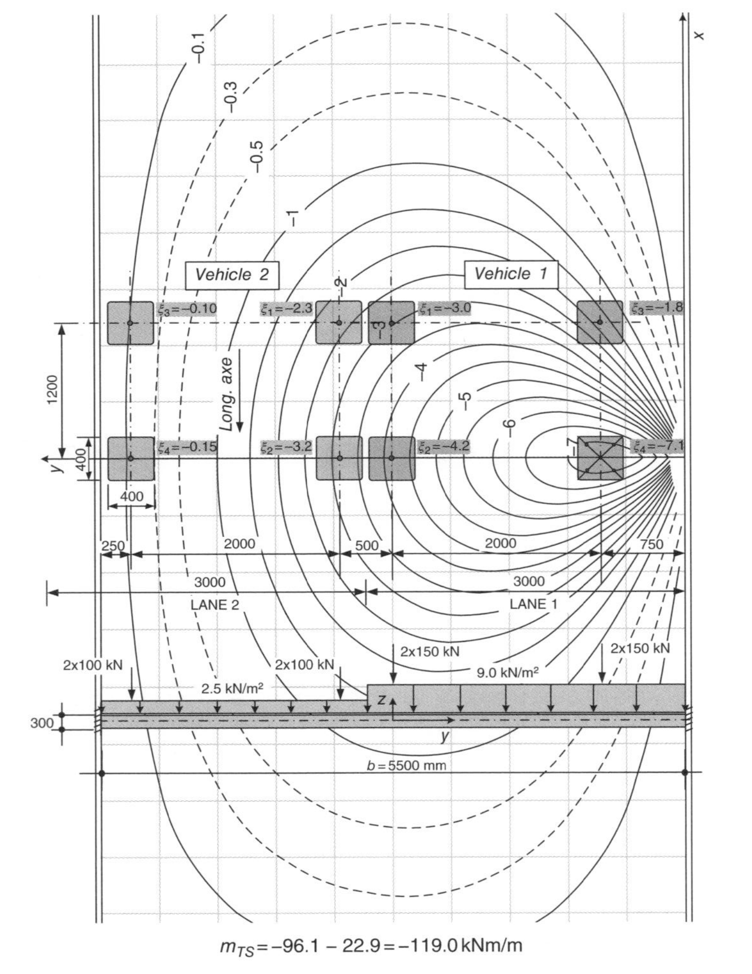

$m_{TS} = -96.1 - 22.9 = -119.0\,\text{kNm/m}$

Figure 8.9 Transverse bending moment at the support of a slab with two restrained edges for concentrated traffic loads.

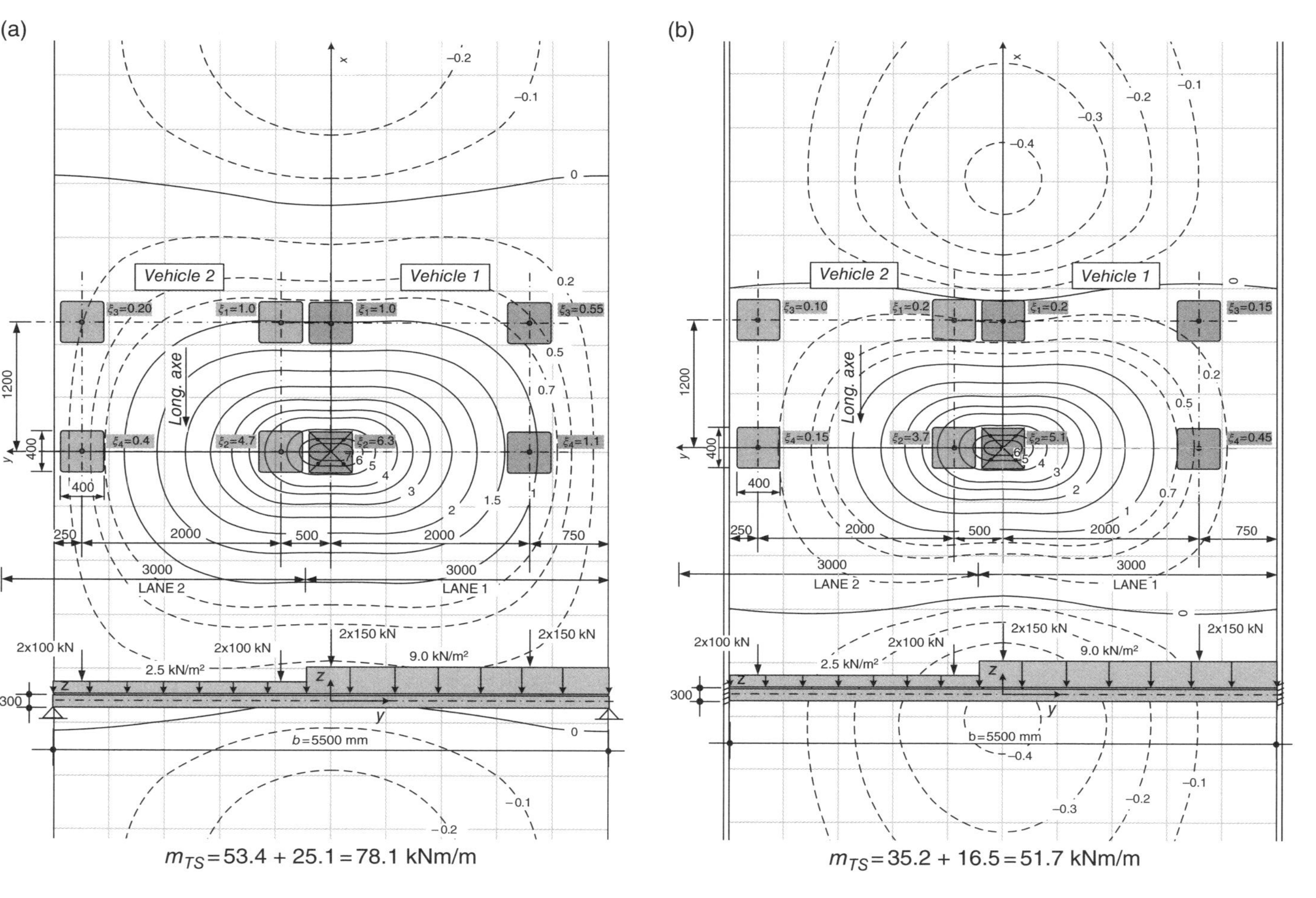

Figure 8.10 Longitudinal bending moment at the mid-span of a slab for concentrated traffic loads, using influence surfaces with: (a) two supported edges and (b) two restrained edges.

Table 8.12 Longitudinal bending moments and safety checks at the central slab span section.

Slab mid-span section		Actions			
		SW+SDL	UDL	TS	ΔT
m (kNm m^{-1})	*Two supported edges*	+7.2	+4.9	+78.5	0
	Two restrained edges	+2.4	+1.7	+51.7	+25.5
	Two elastic restrains	−2.3[a]	+4.2	+72.3	+5.9
ULS of resistance		$m_{\text{Ed}} = 101\,\text{kNm}\,\text{m}^{-1}$		$m_{\text{Rd}} = 116\,\text{kNm}\,\text{m}^{-1}$[b]	
SLS of crack widths		$m_{\text{freq}} = 57\,\text{kNm}\,\text{m}^{-1}$		$m_{\text{cr}} = 49\,\text{kNm}\,\text{m}^{-1} \rightarrow$ $w_{\text{k}} = 0.25\,\text{mm}$[b] $< 0.3\,\text{mm}$	

[a] With the contribution of the cantilever bending moments given by $-11 \times (1 - \alpha_{elast}) = -8.4\,\text{kNm}\,\text{m}^{-1}$.
[b] Values obtained for the reinforcement ϕ16//0.15, and C35/45, B500B, cover = 45 mm.

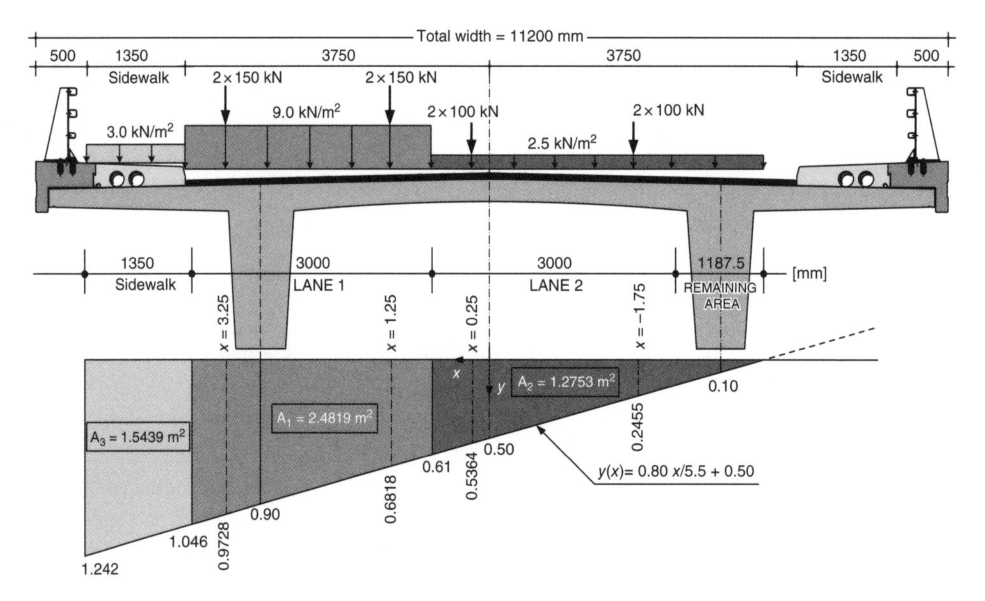

Figure 8.11 Transverse influence line of the left girder loaded with the uniform and concentrated live loads.

and the same values are assumed to be applied throughout the bridge length. Thus, approximate values $\eta_1 = 0.90$, $\eta_2 = 0.10$ are adopted in pre-design.

Loading the transverse influence line of Figure 8.11, one can evaluate the load per metre q_k due to distributed traffic loads and Q_k due to the concentrated loads (placed centred at each lane):

$$q_k = 9.0A_1 + 2.5A_2 + 3.0A_3 = 30.2\,\text{kN}\,\text{m}^{-1}$$

$$Q_k = 2\times150(0.9728 + 0.6818) + 2\times100(0.5364 + 0.2455) = 2\times326.4\,\text{kN}$$

This concentrated load produces an effect on the girder, with two groups of loads at a longitudinal distance of 1.20 m, as shown in Figure 3.6.

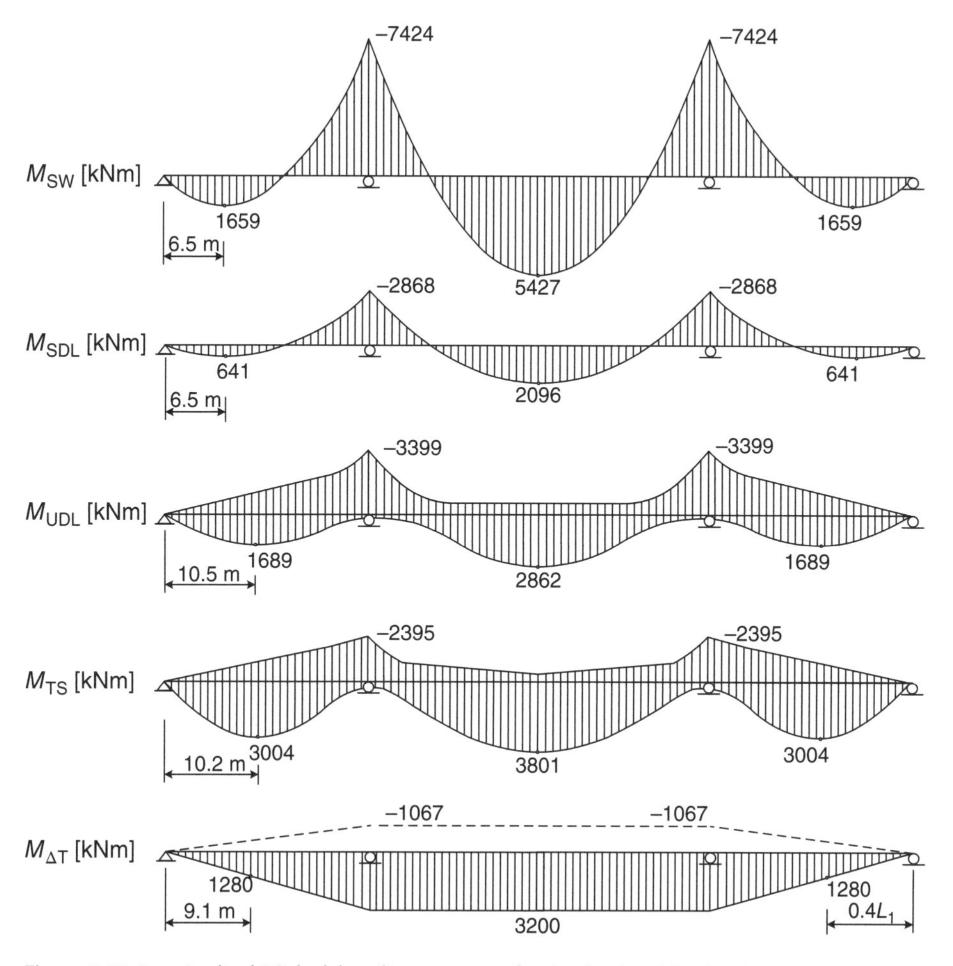

Figure 8.12 Longitudinal PC deck bending moments for the dead and live loads.

Longitudinally, the live loads are placed unfavourably according to the influence lines corresponding to the different internal moments and forces being considered. An envelope of the internal forces is therefore obtained.

8.4.3.2 Internal Forces

Figure 8.12 summarizes the bending moment distributions for one girder of the deck, obtained using a simple three span beam model. The maximal values at the lateral spans do not occur all at the same deck section. As usual, maximum values are adopted for design, assuming all were determined at 0.4 L_1 = 9.1 m from the lateral support.

8.4.3.3 Prestressing Layout and Hyperstatic Effects

The layout of the prestressing cables is defined taking into consideration bending moment distributions. For this design example, one can quickly verify that the bending moments at the lateral span are much lower than over intermediate supports and the mid-central span. Thus two groups of cables are adopted: Tendon layout 1 composed of two cables will run from one end to the other of the bridge deck, and tendon layout 2 also composed of two cables will start at the lateral span, approximately at $1/5\,L_1$ of the intermediate support, and run until the opposite lateral span section.

Adopting the maximum eccentricities at the design sections (S1 – lateral span section, S2 – intermediate support sections and S3 – central mid-span) and inflexion points at distances from supports between 10 and 20% of the span length, the prestressing layout of Figure 8.13 is obtained. An adjustment is made at the lateral span, moving up the two tendons because it was found there was no need to have the maximum eccentricity at the span section of the lateral spans. The equivalent action of the prestressing cables is obtained by applying its equivalent loads on a beam model of Figure 8.14, considering initially P_∞ = 1000 kN for each tendon:

$$\text{Layout 1} - V_1 = 10.2\,p_1 = -205.88\,\text{kN} \qquad M_1 = V_1 e_o = -62\,\text{kNm}$$

$$p_1 = 8P_\infty f_1 / l_1^2 = +20.185\ \text{kNm}^{-1} \qquad p_2 = 8P_\infty f_2 / l_2^2 = +30.234\,\text{kNm}^{-1}$$

$$p_3 = 8P_\infty f_3 / l_3^2 = -60.468\ \text{kN/m} \qquad p_4 = 8P_\infty f_4 / l_4^2 = -52.632\,\text{kN/m}$$

$$p_5 = 8P_\infty f_5 / l_5^2 = +13.158\,\text{kNm}^{-1}$$

$$\text{Layout 2} - V_3 = 4.2\,p_3 = +204.76\,\text{kN} \qquad M_3 = V_1 e_o = +58\ \text{kNm}$$

$$p_3 = 8P_\infty f_3 / l_3^2 = -48.753\,\text{kNm}^{-1} \quad p_4 = 8P_\infty f_4 / l_4^2 = -23.546\,\text{kNm}^{-1}$$

$$p_5 = 8P_\infty f_5 / l_5^2 = +15.697\,\text{kNm}^{-1}$$

Adopting these loads, one should first verify that the isostatic effect of the cable is reproduced in an isostatic beam. This is easily done by suppressing the intermediate supports from the model. The moment's distribution should be equal to $P_\infty e$ at each section of the deck, as presented in Figure 8.15a. After checking the input data and reintroducing the intermediate supports on the beam model, one obtains the global prestressing effect of the cables (Figure 8.15b). The difference between the two diagrams is the hyperstatic effect of the prestressing cables, due to the intermediate support's vertical restraint. Hence, this linear diagram is simply obtained by multiplying the vertical reaction of these supports by L_1 (Figure 8.15c).

8.4.3.4 Influence of the Construction Stages

The internal moments resulting from the erection of the deck are shown in Figure 8.16. The SW of the deck is calculated according to the construction joints at 7.6 m distant from the piers' axes. The hyperstatic effects from the prestressing cables are obtained using the tension sequence of Figure 8.13. Only the second construction stage cables produce hyperstatic effects. Comparing the bending moments from the end of construction with the ones obtained assuming the SW and prestressing applied at the final structure (designated as the 'elastic distribution'), one can conclude that only small differences occur. Therefore, the construction stage bending moments are not considered at this design stage.

8.4.4 Structural Safety Checks: Longitudinal Direction

8.4.4.1 Decompression Limit State – Prestressing Design

Using a frequent combination of actions, Table 8.13 presents the bending moments for sections S1–S3 of one girder, as well as the minimum prestressing value to verify the decompression limit state of these three deck cross-sections. A net value of P_∞ = 2900 kN at the

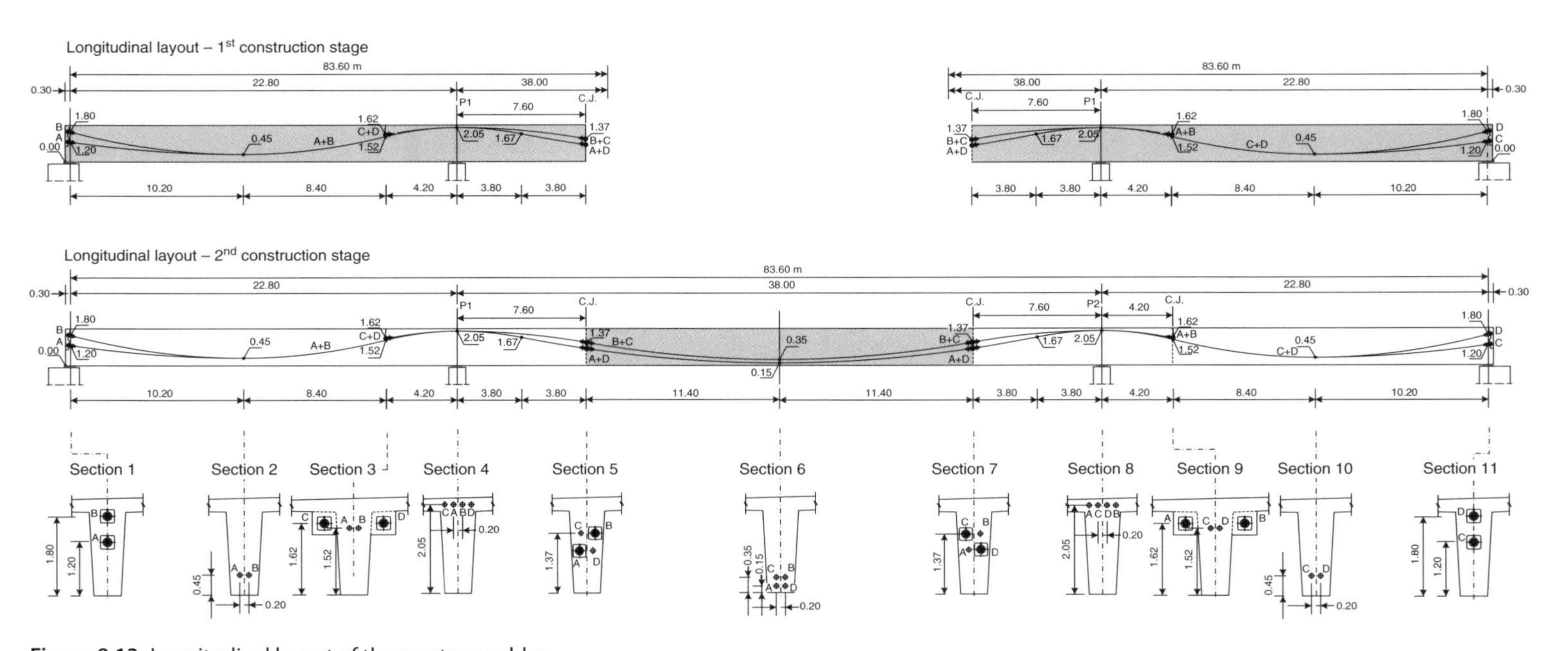

Figure 8.13 Longitudinal layout of the prestress cables.

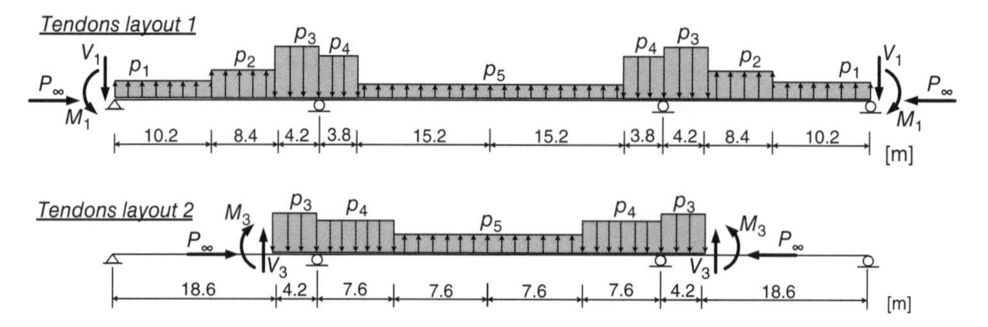

Figure 8.14 Longitudinal prestress action – equivalent loads for tendon layouts 1 and 2.

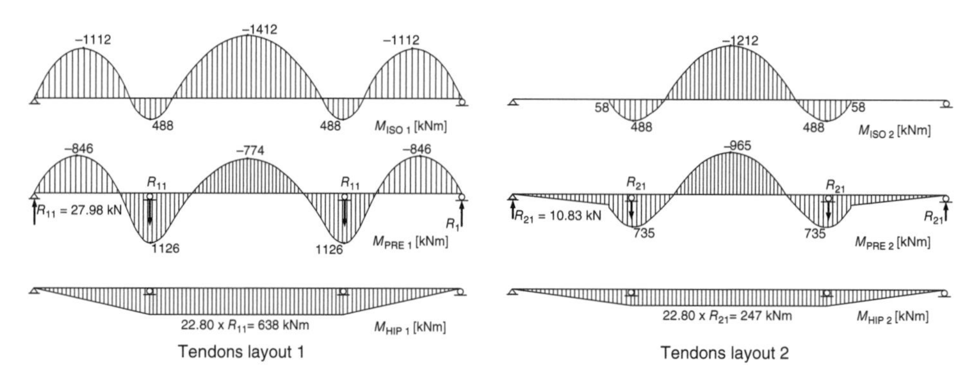

Figure 8.15 Longitudinal prestress action – bending moments for tendon layouts 1 and 2 and P = 1000 kN.

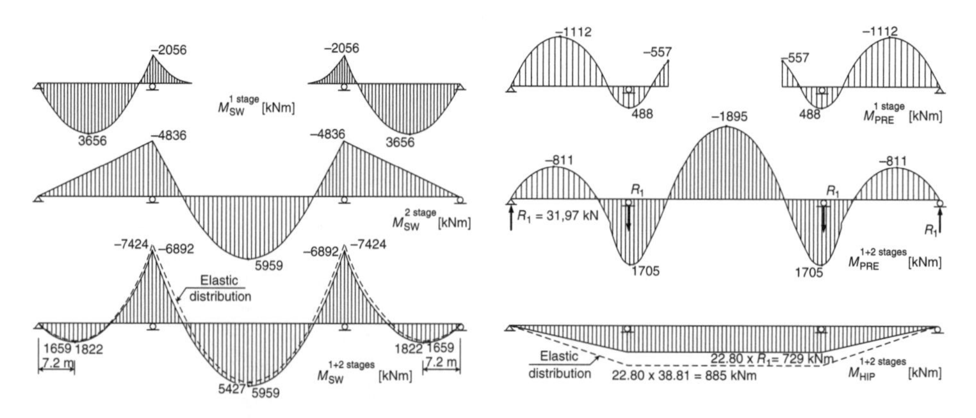

Figure 8.16 Bending diagrams of self-weight and prestress layouts (1+2) with $P = 1000\,\text{kN}$, during construction stages.

Table 8.13 Longitudinal bending moments and prestress adopted for verifying the decompression limit state at the design sections of one girder of the deck.

Action		Design section		
		S1 – Lateral span	*S2 – Interm. support*	*S3 – Central mid-span*
M (kNm)	SW + SDL	2300	−10 292	7523
	UDL	1689	−3399	2862
	TS	3004	−2395	3801
	ΔT	1280	−1067	3200
M_{freq} (kNm)		5869	−13 981	13 119
M_{HIP} (kNm)		0.1770 P_∞	0.4424 P_∞	0.4424 P_∞
A (m^2)		2.850	2.850	2.850
W_{sup} (m^3)		−1.896	−1.896	−1.896
W_{inf} (m^3)		0.775	0.775	0.775
Exc (m)		1.112	−0.488	1.312
$P_{\infty,\,min}$(kN)		4862	8762	11 493
$P_{\infty,\,adp}$(kN)		2 × 2900 = 5800	4 × 2900 = 11 600	4 × 2900 = 11 600
σ_{freq-}(MPa)	Top	−0.209	**−2.389**	−2.159
	Bottom	−6.502	−8.184	−8.746
σ_{perm}(MPa)	Top	−0.929	−4.335	−2.718
	Bottom	−4.741	−3.423	−7.379
σ_{freq+}(MPa)	Top	−2.811	−5.375	−5.669
	Bottom	**−0.136**	−0.878	**−0.158**

long term and after all loses occur is adopted for each cable, which can be obtained by 19 strands of 0.6″. The same table presents the stresses at the top and bottom fibres of the deck at the design section, confirming no tensions occur for the frequent combinations.

8.4.4.2 Ultimate Limit States – Bending and Shear Resistance

Usually, well-designed prestressed concrete girders verify by a good margin the ULS of resistance. Typically, these verifications are performed for the more stressed mid-span and support sections and allow defining the main longitudinal and shear reinforcement of the girders.

For the reinforcement layout of Figure 8.6, bending and shear resistances of Table 8.14 are evaluated as per EN 1992-1 [7]. These resistances compare with the design values of the applied bending moments and shear forces, confirming the safety margins given by S_{Rd}/S_{Ed}.

The ULS bending resistances given in Table 8.14 adopt the effective slab widths of b_{eff} = 5.44 m, for each girder at span regions, and b_{eff} = 3.55 m at internal supports as per Figures 6.7 and 6.8. The shear resistance uses the angle "θ" = 26° between the concrete compression strut and the beam axis perpendicular to the shear force.

Table 8.14 Longitudinal ULS bending and shear verifications for the design sections of the deck.

	Design section S2 – Intermediate support		*Design sections S1/S3 – Mid-spans*
ULS of	*Shear* (kN)	*Bending* (kNm)	*Bending* (kNm)
Design value S_{Ed}	4217	−17 096	11 904/25 310
Resistance value S_{Rd}	4625	−23 414	14 244/29 949
$FS = S_{Rd}/S_{Ed}$	1.10	1.37	1.20/1.18

8.5 Steel–Concrete Composite Solution

8.5.1 Preliminary Design of the Deck

As for the concrete deck solution, at the preliminary design of the steel-concrete composite deck one defines the dimensions of the most highly stressed cross sections of the steel girder and concrete slab, following an iterative design process. The internal moments and forces are determined for an assumed geometry. The dimensions must subsequently be checked to ensure they also satisfy all requirements relative to ultimate safety and serviceability.

The steel plate girder cross sections should be pre-designed. Starting with the girder depth, the design values of the bending moments at the supports and at span may be adopted to obtain a first assess of the flange dimensions of the main girders. Bending moments due to SW and traffic is divided by the depth of the beam to obtain an approximation of the resulting flange axial forces. Dividing this force by the design strength of the steel, an initial estimation of flange cross sectional areas is obtained.

To evaluate the web thickness at span, a first assess can be based on the flange induced buckling slenderness limits (Table 4.4 – paragraph 4.5.4, and EN1993-2 [10]), respecting a minimum web thickness of 12 mm. For a first estimate of web area at the supports, the maximum shear force may be divided by shear strength of around 100 MPa. Following these procedures, and using steel grade S355, one has (Figure 8.17):

1) Girders depth $h = 1.80$ m, that is, deck slenderness of $38/1.8 = 21.1$, an acceptable and usual value for twin steel girder decks with no height restraints; this depth is constant over the total length of the bridge deck;
2) Bottom flange width and thickness = 700×40 mm^2, at the limit of class 2 section and using the 40 mm plate thickness to avoid further reduction of the yielding stress according with EN 10025-3 [4];
3) Top flange width and thickness = 600×30 mm^2, yielding a class 3 section to avoid plate buckling during construction, while the top flange is not connected to the slab;
4) Web thickness = 18 mm, both at span and support regions, which corresponds to a plate slenderness of 96, at the limit of class 3 in pure bending (at the span region, thinner webs could be considered, but the option of having a constant web thickness was taken based on constructability reasons and to ensure patch loading resistance during incremental launching).

The bottom compressed flange design, over the supports, should also take into consideration lateral torsional buckling. A reduced factor of the yield stress is adopted

Support cross-section

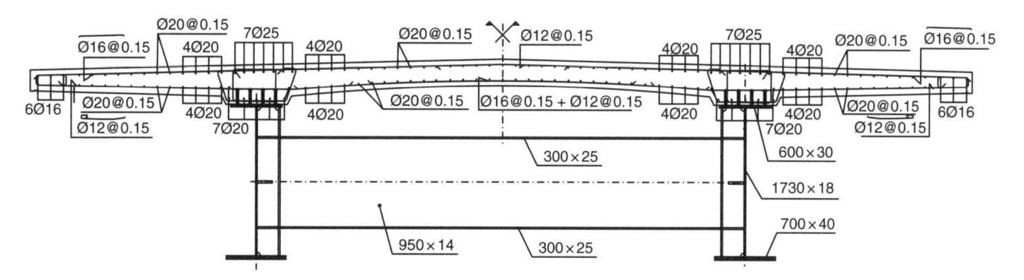

Span cross-section

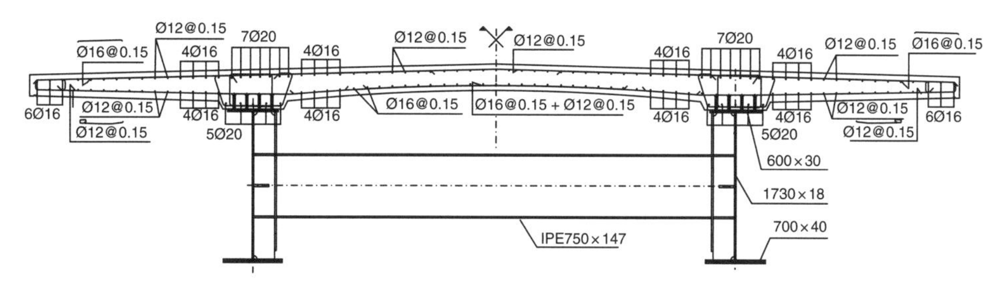

Figure 8.17 Reinforcement and plate girder layout for the span and support composite deck cross sections.

for preliminary design. At span regions, to evaluate the bottom flange cross sectional area, some contribution from the web to the bending resistance may also be taken into account and plastic resistance can be used.

8.5.2 Structural Analysis and Slab Design Checks

The structural analysis is similar to that done for the concrete deck option. The cantilever support is verified using a free span up to the axe of the main girder, resulting in a 2690 mm span. Transverse bending moments from Table 8.7 are still valid, as well as the correspondent verifications and reinforcements (Figure 8.17).

The bending moments at the central slab between the girders are evaluated at the mid-span. An equivalent transverse beam with a span of b = 5.5 m between the axes of the main girders is adopted. Applying the SW, SDL and the road traffic and pedestrian uniform live loads (UDL), one obtains the bending moments of Table 8.15, adopting the conservative assumption that main girders do not restrain the slab rotation. For the same reason, it is assumed temperature gradients do not introduce slab bending moments. Table 8.15 also summarizes the ULS and SLS bending verifications and reinforcements adopted (Figure 8.17).

8.5.3 Structural Analysis of the Main Girders

For a composite deck, the internal moments and forces in the main steel girders are determined using a first order elastic analysis and considering the construction stages and the deck proprieties at each stage. Due to shear lag of the slab, effective widths are

Table 8.15 Bending moments and safety checks at the central slab mid-span composite section.

	Actions			
Slab mid-span section	**SW + SDL**	**UDL**	**TS**	**ΔT**
$m_{transverse}$ (kNm m^{-1})	−18.8[a]	+24.5	+123.4	0
ULS of transverse resistance	m_{Ed} = 181 kNm m^{-1}		m_{Rd} = 190 kNm m^{-1}[b]	
SLS of crack widths	m_{freq} = 84 kNm m^{-1}		m_{cr} = 49 kNm m^{-1} → w_k = 0.17 mm[b] < 0.3 mm	
$m_{longitudinal}$ (kNm m^{-1})	−3.8	+4.9	+78.5	0
ULS of longitudinal resistance	m_{Ed} = 109 kNm m^{-1}		m_{Rd} = 116 kNm m^{-1}[c]	
SLS of crack widths	m_{freq} = 57 kNm m^{-1}		m_{cr} = 49 kNm m^{-1} → w_k = 0.25 mm[c] < 0.3 mm	

[a] With the contribution of the cantilever bending moments given by −54.8 kNm m^{-1}.
[b] Values obtained for the reinforcement {ϕ16//0.15 + ϕ12//0.15}, and C40/50, B500B, cover = 45 mm.
[c] Values obtained for the reinforcement ϕ16//0.15, and C40/50, B500B, cover = 45 mm.

adopted to evaluate deck mechanical proprieties at each stage (Table 8.1). In Figures 6.7 and 6.8, b_{eff} values for one girder are 5.30/5.44 m at lateral/central span regions and 4.25 m at the internal support region.

The analysis assumes the slab is fully cracked in a zone of 15% of the span lengths on both sides at each internal support. According to the §5.4.2.3 of EN 1994-2 [6], this assumption should be verified, since imposed deformations are used at internal supports to initially reduce the deck's negative bending moments over these supports. The normal procedure is to execute first a global 'uncracked analysis' (with the concrete strength of the slab in all deck cross sections) to evaluate the deck cross sections where the upper fibre of the slab has a stress above $2f_{ctm}$ for a characteristic SLS combination of actions. Then the concrete slab of these deck cross sections should be considered as cracked (reducing the stiffness to the reinforcing steel) in the second global 'cracked analysis.'

At service, the traffic loads are placed on the bridge deck most unfavourable locations, as previously done for the concrete option. Since the live loads are not applied in a transversally symmetrical form, the longitudinal model considers the geometrical proprieties of a cross section formed by one beam connected to its effective slab.

8.5.3.1 Traffic Loads Transverse and Longitudinal Positioning

As for the concrete deck, for the analysis of the twin-girder composite bridge deck one uses the transverse influence line to identify the most unfavourable position of the live loads. The correspondent η_1, η_2 coefficients are a function of the parameter α. Using $b = 5.5$ m, $l = 26.6$ m; $E = 210$ GPa; $I_y = 0.1245$ m^4; $G_c = 35/2.4 = 14.58$ GPa and $J_c = 0.092$ m^4, one obtains $\alpha = 0.7747$ and, as a result, $\eta_1 = 0.90$, $\eta_2 = 0.10$. Thus, the distributed and concentrated traffic loads, $q_k = 30.2$ kN m^{-1} and $Q_k = 2 \times 326.4$ kN m^{-1} of one girder, can still be obtained by loading the transverse influence line of Figure 8.11

Longitudinally, as for the concrete deck, live loads are placed unfavourably using the influence lines to obtain envelopes of internal moments and forces.

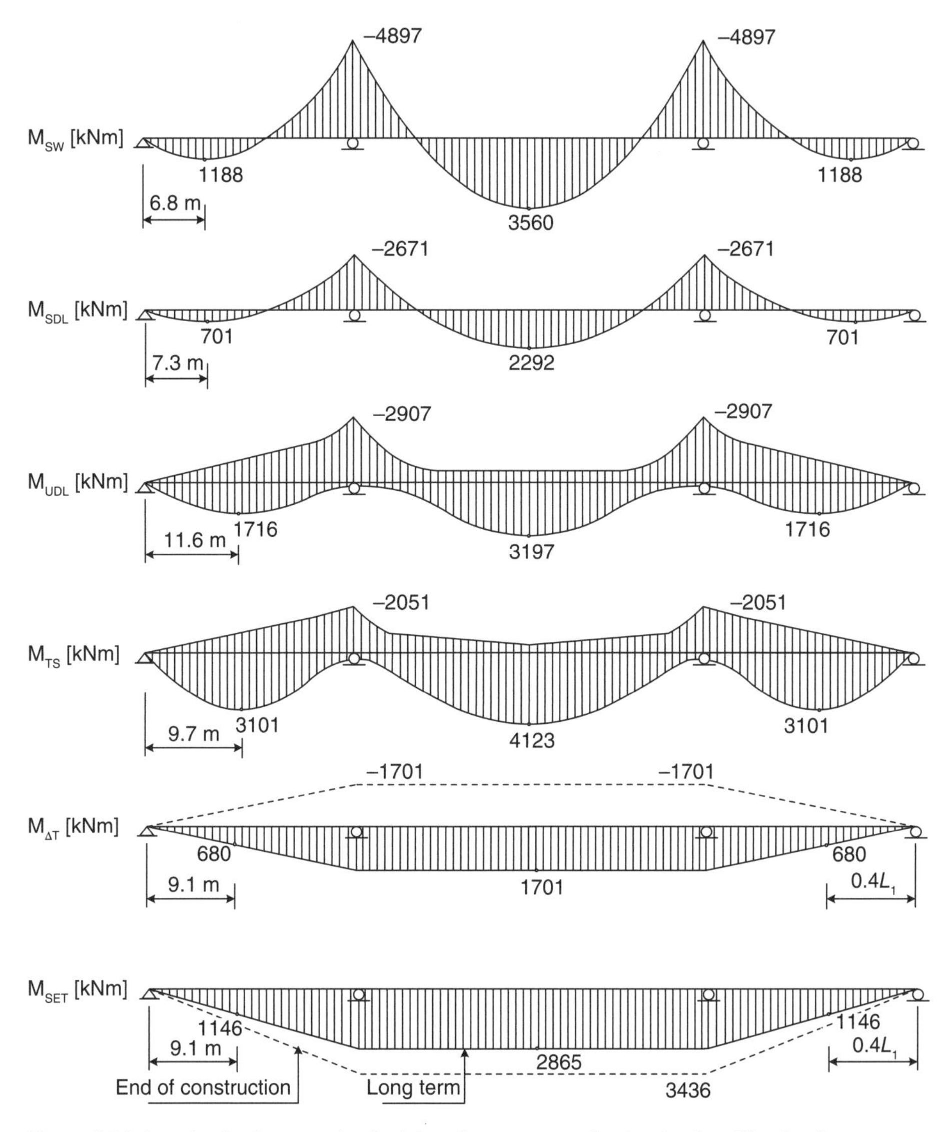

Figure 8.18 Longitudinal composite deck bending moments for the dead and live loads.

8.5.3.2 Internal Forces

Figure 8.18 presents the bending moment distributions for one girder of the deck obtained using a simple three-span beam model. The maximum values at the lateral spans do not all occur at the same deck cross section. As before for the concrete option, the maximum values are used for verifying section $0.4\,L_1 = 9.1$ m from the lateral support.

8.5.3.3 Shrinkage Effects

The restraint to free shrinkage deformation of the slab implies primary stresses in the composite section and secondary, or hyperstatic, internal forces and moments in the structure. The hyperstatic shrinkage effects in the structure are determined by applying

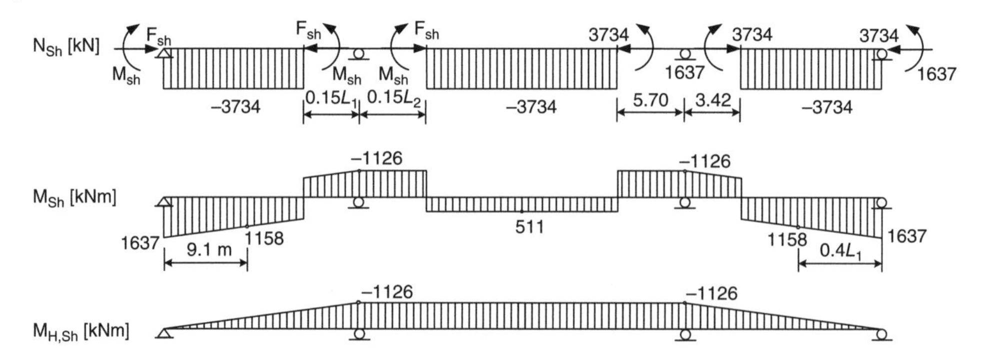

Figure 8.19 Composite deck shrinkage action and primary and secondary hyperstatic bending moments.

in the model the equivalent forces inducing the free shrinkage deformation in the slab. These forces are determined taking the effect of creep into account on the equivalent concrete modulus of elasticity as per paragraph 6.7.4:

$$E_{eq} = \frac{E_{cm}}{1+\psi_s \varphi} = 16.67\,\text{GPa}$$

with the creep multiplier $\psi_s = 0.55$ and the creep coefficient $\varphi = 2.0$. Hence, the force equivalent to fully restraining deformation of the concrete slab of area A_c associated to one girder is: $F_{sh} = \varepsilon_{sh} \times E_{eq} \times A_c = -13.9 \times 10^{-5} \times 16.67 \times 10^6 \times 1.612 = 3734\,\text{kN}$.

The equivalent effect at the centroid of the composite section is given by the same force and the bending moment $M_{sh} = F_{sh} \times exc = 3734 \times 0.4383 = 1637\,\text{kNm}$. These forces and moments are applied in the longitudinal model of one girder to determine the global internal bending moments at each section due to shrinkage (Figure 8.19). The primary effects of shrinkage may be neglected in regions where the concrete is assumed to be cracked. So, the forces referred here are not applied at support regions, but at the sections located at $0.15L$ from the support. Therefore, in section analysis of supports, where the concrete slab is cracked, only the hyperstatic effects of shrinkage are taken into consideration. According to EN 1994-2 [6], at span sections, two different situations may be considered. If an elastic section analysis is made, primary stresses and the hyperstatic effects of shrinkage should be considered. However, if a plastic section analysis is made, only the hyperstatic effects of shrinkage should be used.

8.5.3.4 Imposed Deformation Effect

At the end of construction, an imposed deformation (controlled settlement) of 100 mm is introduced at intermediate supports when the temporary supports are replaced by the final bearings. This action generates a positive curvature of the deck at the end of construction and therefore an internal redistribution of forces due to dead loads. Due to creep effect, these bending moments are reduced during time; Figure 8.18 presents M_{SET} distribution in the long term.

8.5.3.5 Influence of the Construction Stages

The steel structure is constructed by incremental launching. The bending moments of one girder, during launching, are presented on Figure 8.20. As can be seen, the values are not high, even without the effect of the lighter launching nose to reduce the highest

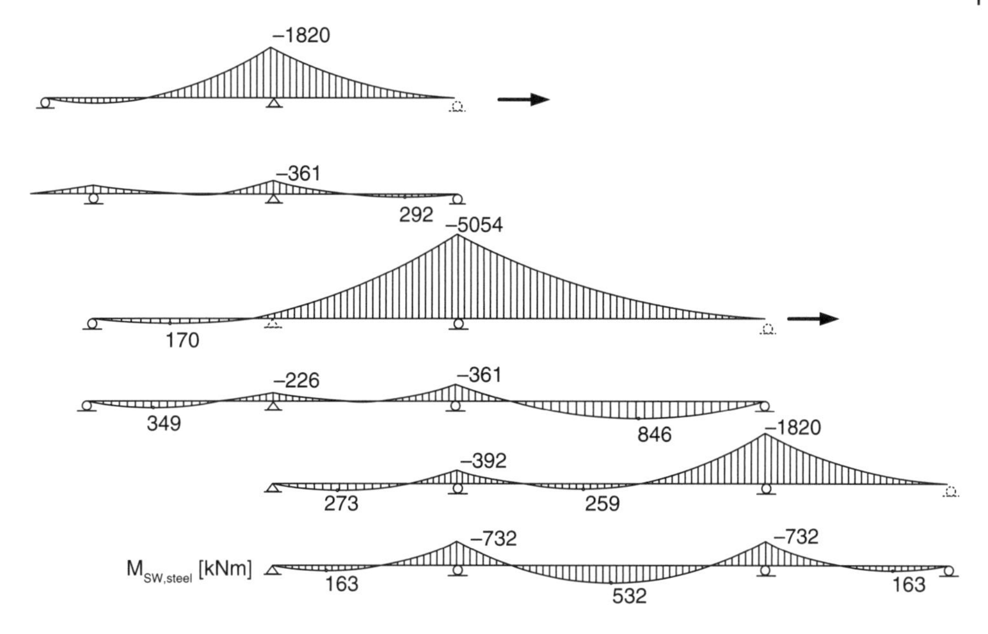

Figure 8.20 Longitudinal steel self-weight bending moments during incremental launching.

hogging bending moment when the cantilever deck arrives at pier P2. At this stage, stresses at the extreme fibres of one girder are: $\sigma_{top} = +60$ MPa, $\sigma_{bot} = -46.6$ MPa, very far from the yielding stress multiplied by the reduction factor to account for lateral torsion buckling.

8.5.4 Safety Checks: Longitudinal Direction

8.5.4.1 Ultimate Limit States – Bending and Shear Resistance

Usually, composite deck girders verify by a good margin the ULS of sagging bending resistance. Typically these verifications are performed for the more stressed mid-span and support sections and allow checking of the steel girders geometry and longitudinal reinforcement of the slab over the internal supports, using the stress-strain diagrams of Figure 8.21.

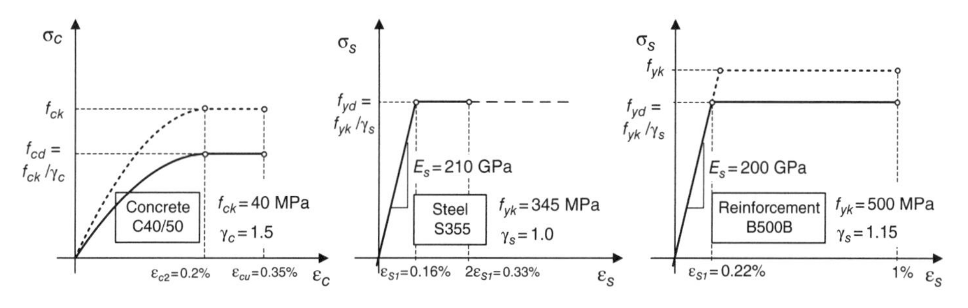

Figure 8.21 Stress–strain diagrams for concrete, structural steel and reinforcement of the composite deck.

Table 8.16 Longitudinal bending moments and ULS limit state of bending resistance for the design sections of the deck.

		Design section		
	Action	*S1 – Lateral span*	*S2 – Interm. support*	*S3 – Central mid-span*
M (kNm)	SW + SDL	1889	−7568	5852
	UDL	1716	−2907	3197
	TS	3101	−2051	4123
	ΔT	680	−1701	1701
	Shrinkage	0[a]	−1126[b]	0[a]
	Imp. Def.	1374[a]	2865[b]	3436[a]
M_{Ed} (kNm)		11 040	−16 702	22 750
M_{Rd} (kNm)		27 990	−19 690	27 990
$FS = M_{Rd}/M_{Ed}$		2.54	1.18	1.23

[a] At the end of construction, assuming no shrinkage has occurred.
[b] In the long term.

For the mid-span sections, since the neutral axe is near the top flange of the girder, the class 3 web is almost fully in tension contributing to the bending resistance. Using the limited plastic criterion explained in Section 6.9.2 at span sections (limiting the maximum strain at the bottom flange to two times the yield strain ε_y), and an elastic criterion at the supports, bending resistances are given in Table 8.16 and Figure 8.22. For internal supports, the bottom flange stress is limited to $0.9 f_{sy} = 310.5$ MPa, to account for lateral torsional buckling.[1] These values compare with the design value of the applied bending moments, confirming the safety margins given by M_{Rd}/M_{Ed}. There is no need to reduce the span resistance by a factor 0.9 because the ratio of span lengths is 0.6.

Following Section 6.9.5.1, the shear resistance of one girder's web over the internal support is $V_{w,Rd} = 4152$ kN (with $t_w = 18$ mm, $a = 2535$ mm, $h_w = 1730$ mm, $k_\tau = 7.2$, $\chi_w = 0.736$, $f_{yw} = 345$ MPa and $\gamma_{M1} = 1.10$). The highest shear design force at the most stressed panel is $V_{Ed} = 3583$ kN; the safety factor is therefore $FS = V_{Rd}/V_{Ed} = 1.16$.

The interaction between bending and shear at the support section also verifies the condition given in paragraph 6.9.5.4, as per §7.1 of EN 1993-1-5 [11]:

$$\bar{\eta}_1 + \left(1 - \frac{M_{f,Rd}}{M_{Rd}}\right) \cdot \left(2\bar{\eta}_3 - 1\right)^2 = 0.93 \leq 1.0 \text{ with } \bar{\eta}_1 = \frac{M_{Ed}}{M_{Rd}} 0.848,\ \bar{\eta}_3 = \frac{V_{Ed}}{V_{w,Rd}} = 0.863$$

1 From Figure 8.22, the support web section has: $\psi = -0.678$, $k_\sigma = 16.57$, $\bar{\lambda}_p = (h_w/t_w)/(28.4\,\varepsilon\,\sqrt{k_\sigma}) = 1.01$, $\rho = 0.88 < 1$; thus, a small reduction of the width under compression of the web should be considered; this effect was not taken into consideration at this design stage.

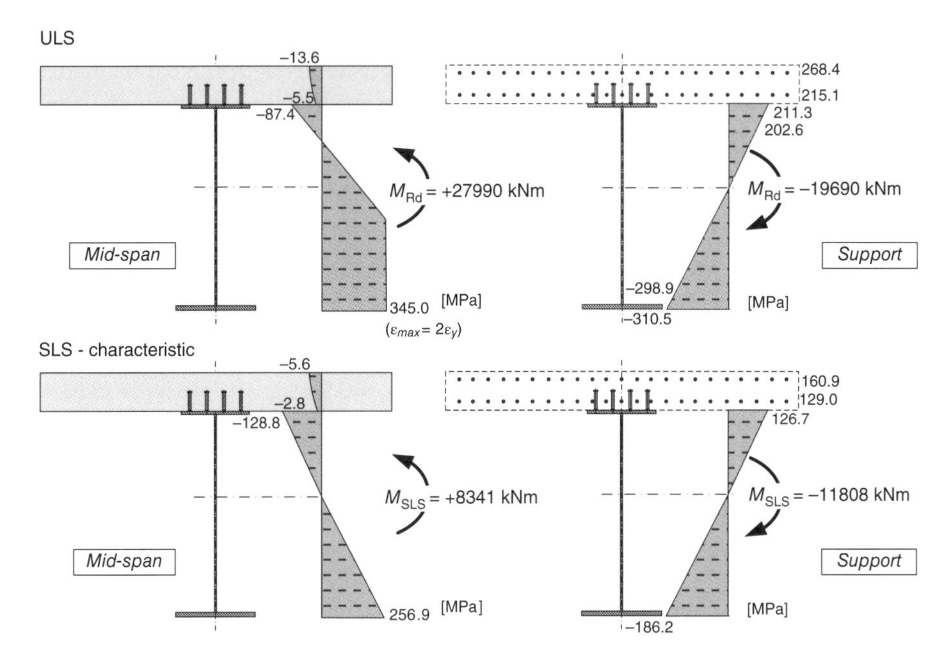

Figure 8.22 Stresses at ULS and SLS-characteristic for mid-span and internal support cross sections.

It should be finally pointed out that: (i) the shear and interaction (M, V) verifications do not need be checked for the cross sections located less than $h_w/2$ from a support with a vertical stiffener, as was applied to the most stressed section over the support and (ii) the maximum bending moment and shear force adopted to check interaction should be concomitant, and not resulting from the live load envelope.

Further to this, deck fatigue resistance should be assessed according to paragraph 6.9.4, but this will not be covered at the design stage considered in this example.

8.5.4.2 Serviceability Limit States – Stresses and Crack Widths Control

Figure 8.22 presents the normal stress distribution at the main span and internal support deck sections, taking the superposition of the construction stages into consideration. Stresses with elastic assumptions are limited in the structural steel at characteristic SLS, as in the concrete slab and in the reinforcing steel bars. Given the ULS verifications, these stress limitations do not normally govern the design. Span section is verified at the end of construction while support section is usually governed by the long term check. The stress limits at characteristic SLS are:

- $\sigma_s \leq 1.0 f_{yd}$ = 345 MPa – for structural steel[2]
- $\sigma_c \leq 0.6 f_{ck}$ = 24 MPa – for concrete under compression
- $\sigma_s \leq 0.8 f_{sk}$ = 400 MPa – for reinforcement

2 Stress σ_s should be evaluated using the Von Mises criterion $\sqrt{\sigma^2 + 3\tau^2}$; Conservatively, σ_s can be assessed by considering the non concomitant maximum normal stress in the flange with the average shear stress in the web; for the support cross section: $\sqrt{186.2^2 + 3 \times 85.2^2} = 237.6 < 345\,MPa$.

A simplified and conservative limitation of crack widths can be achieved by ensuring a minimum reinforcement area for a certain concrete tensile strength and bar diameter. This corresponds to verify that the crack widths remain less than the design limit, using the indirect method in the tensile sections of the slab for characteristic SLS combination of actions. In this section, subjected to significant tension due to restraint of imposed deformations (e.g. primary and secondary effects of shrinkage), in combination or not with the effects of direct loading, §7.4.2 of EN1994-2 [6] defines a minimum reinforcement area as: $A_s = k_s\, k_c\, k\, f_{ct,\,eff}\, A_c/\sigma_s$.

For span sections, taking into consideration the use of the equivalent bar diameter $\phi^* = 12 \times 2.9/3.5 = 10\,\text{mm}$ and for a design crack width $w_k = 0.3\,\text{mm}$, the maximum allowed reinforcement stress is obtained from Table 7.1 of EN1994-2: $\sigma_s = 320\,\text{MPa}$. The required minimum reinforced area at span section $A_s = 0.9 \times 1.0 \times 0.8 \times 3.5 \times A_c/320 = 0.8\%\, A_c$ is therefore lower than the $0.9\%\, A_c$ adopted.

For the internal support section, using $\phi^* = 20 \times 2.9/3.5 = 16.7\,\text{mm}$ and $w_k = 0.3\,\text{mm}$, the maximum allowed reinforcement stress is $\sigma_s = 237\,\text{MPa}$. Hence, the required minimum reinforced area at support section, $A_s = 0.9 \times 1.0 \times 0.8 \times 3.5 \times A_c/237 = 1.06\%\, A_c$, is also much lower than the $1.6\%\, A_c$ adopted.

References

1 EN 206 (2013). *Concrete – Specification, Performance, Production and Conformity*. Brussels: CEN.
2 EN 10080 (2005). *Steel for the Reinforcement of Concrete*. Brussels: CEN.
3 EN 10138-3 (2005). *Prestressing Steels – Part 3: Strand*. Brussels: CEN.
4 EN 10025-3 (2004). *Hot Rolled Products of Structural Steels – Part 3: Technical Delivery Conditions for Normalized/Normalized Rolled Weldable Fine Grain Structural Steels*. Brussels: CEN.
5 ISO 13918 (2008). *Welding – Studs and Ceramic Ferrules for Arc Stud Welding*. Switzerland: ISO.
6 EN 1994-2 (2005). *Eurocode 4 – Design of Composite Steel and Concrete Structures – Part 2: General Rules and Rules for Bridges*. Brussels: CEN.
7 EN 1992-1 (2004). *Eurocode 2 – Design of Concrete Structures – Part 1: Part 1-1: General Rules and Rules for Buildings*. Brussels: CEN.
8 EN 1991-2 (2005). *Eurocode 1: Actions on Structures – Part 2: Traffic Loads on Bridges*. Brussels: CEN.
9 NP EN 1991-1-5 (2009). *Eurocode 1: Actions on Structures – Part 1–5: General Actions – Thermal Actions*. Portugal: ISQ.
10 EN 1993-2 (2006). *Eurocode 3 – Design of Steel Structures – Part 2: Steel Bridges*. Brussels: CEN.
11 EN 1993-1-5 (2006). *Eurocode 3 – Design of Steel Structures – Part 1–5: Plated Structural Elements: Steel Bridges*. Brussels: CEN.

Annex A

Buckling and Ultimate Strength of Flat Plates

A.1 Critical Stresses and Buckling Modes of Flat Plates

A flat plate of length a and width b, with a thickness t is assumed to be supported along the four edges and loaded at the transverse edges under a non-uniform normal stress field (σ_1 and σ_2) and a uniform shear stress distribution τ as shown in Figure A.1. This is a quite usual stress field in webs of plate girders, as referred to in Section 6.9.5.1, with normal stresses being due to bending moments and axial load and shear stresses due to shear force.

Let the case $\tau = 0$ be considered first. The elastic critical buckling stress σ_{cr} is given by [1, 2]

$$\sigma_{cr} = k\frac{\pi^2 E}{12\left(1-\nu^2\right)}\left(\frac{t}{b}\right)^2 \tag{A.1}$$

where the buckling coefficient $k = k(\alpha,\psi)$, with $\alpha = a/b$ the aspect ratio of the plate and $\psi = \sigma_2/\sigma_1$ the stress ratio, as per Figure A.1.

A.1.1 Plate Simply Supported along the four Edges and under a Uniform Compression ($\psi = 1$)

The plate buckles in a single transverse half wave and with multiple longitudinal m half waves. The coefficient k, taking values defined by the buckling mode $m = 1,2,3, \ldots$ depending on the aspect ratio α, as shown in Figure A.2, is given by

$$k = k_m = \left(m\frac{b}{a} + \frac{1}{m}\frac{a}{b}\right)^2 \tag{A.2}$$

From Figure A.2 it is concluded that $k \geq 4.0$. Besides, for long plates (say $\alpha \geq 4.0$) as in the case of plates integrated in welded or rolled structural sections (I shapes, tubular sections, sub-panels of stiffened plates, see Figure A.3), one may take conservatively $k = 4.0$ to evaluate the elastic critical stress. For long plates, the boundary conditions along the transverse edges ($x = 0$ and $x = a$) do not play any role in the values of the buckling coefficients due to the large number of half waves of the buckling mode. However, these conditions are quite relevant for the post-buckling behaviour of flat plates, as will be seen in the next section.

Bridge Design: Concepts and Analysis, First Edition. António J. Reis and José J. Oliveira Pedro.

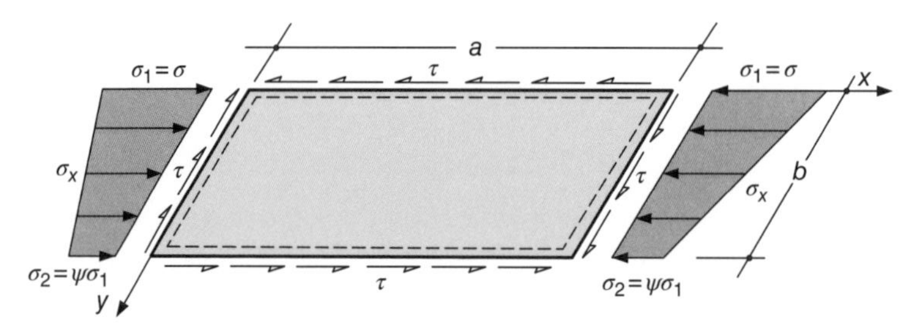

Figure A.1 Flat plate under a normal and shear stress field.

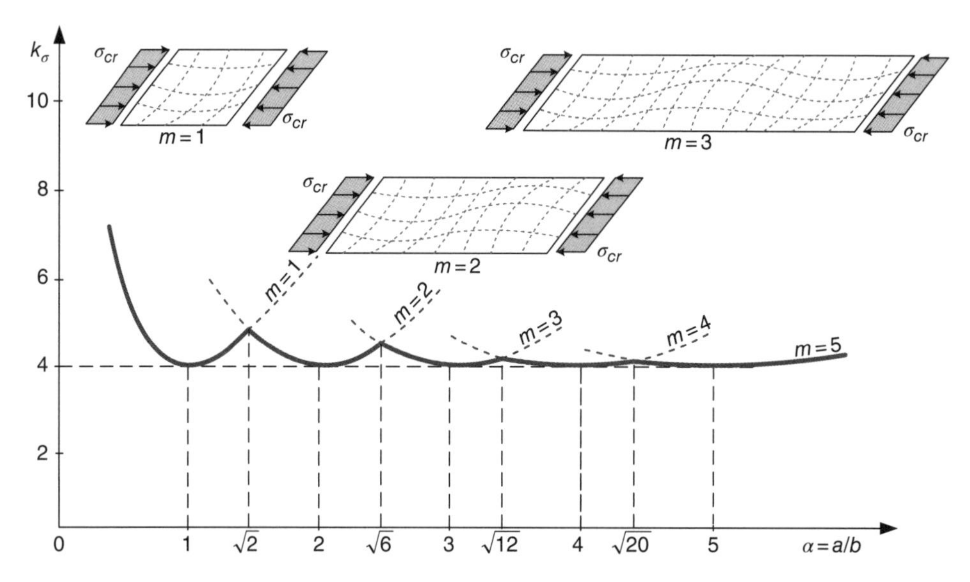

Figure A.2 Buckling modes and buckling mode coefficient for rectangular plates simply supported along the four edges under uniform compression.

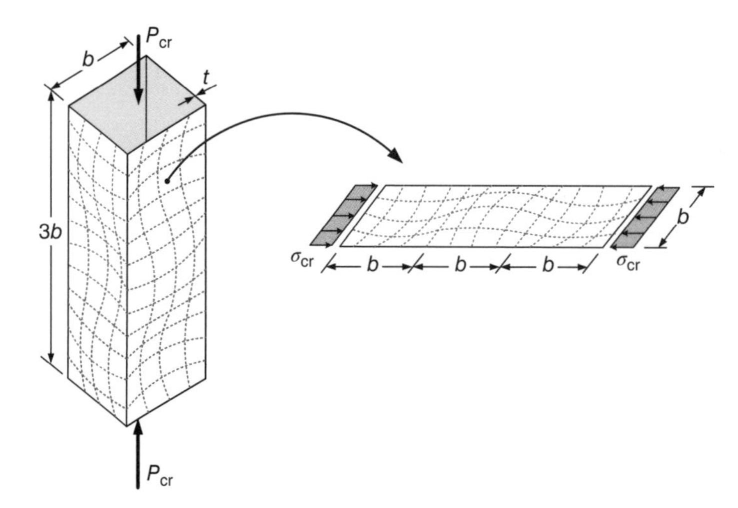

Figure A.3 First buckling mode of long rectangular plate under uniform compression.

A.1.2 Bending of Long Rectangular Plates Supported at both Longitudinal Edges or with a Free Edge

These cases represent the example of web and flanges of steel plate girders (Figure 4.58). For long plates $\alpha > 4$, supported on both longitudinal edges, as in the case of a web of a plate girder, for pure bending ($\psi = -1$) one has a buckling coefficient $k = 23.9$ if the web is considered to be simply supported in the flanges due to its low torsional stiffness (Table A.1). If the web has been considered fully restrained by the flanges, one has for pure bending $k = 39.52$. Only the first boundary conditions are considered in practice for design purposes. For this case, if the web is under variable compression, the buckling coefficient k may be evaluated from the values given in Table A.1 [3].

The flange of a plate girder is a plate with a free edge being the other longitudinal edge elastically restrained by the web, to a degree varying between 0 (simply supported flange in the web) and ∞ (flange fully restrained by the web) (Figure 4.58). Only the former case is usually considered in practice and for uniform compression ($\psi = 1$), the buckling coefficient is taken as $k = 0.43$, compared to $k = 1.28$ for the case of a fully restrained flange in the web.

If the flange is under a variable compressive stress distribution, with σ_1 being the stress at the free edge and σ_2 the stress at the simply supported longitudinal edge, the buckling coefficient may be evaluated from Table A.2 [3].

If σ_2 is the stress at the free edge and σ_1 the stress at the simply supported longitudinal edge, the buckling coefficient may be evaluated from Table A.3 [3].

A.1.3 Buckling of Rectangular Plates under Shear

Let the case of the plate of Figure A.1, with $\sigma_1 = \sigma_2 = 0$ and $\tau \neq 0$, be considered. The longitudinal edges are taken to be simply supported. This is the case of a web panel with a depth $b \equiv d$ and with transverse stiffeners (spaced at a) assumed with zero torsion stiffness (open section stiffeners). The web panels tend to buckle due to the diagonal

Table A.1 Buckling coefficient k for a rectangular plate supported at both longitudinal edges.

$\psi = \sigma_2/\sigma_1$	1	$1 > \psi > 0$	0	$0 > \psi > -1$	-1	$-1 > \psi > -3$
k	4.0	$8.2/(1.05 + \psi)$	7.81	$7.81 - 6.29\psi + 9.78\psi^2$	23.9	$5.98\,(1 - \psi)^2$

Table A.2 Buckling coefficient k for a rectangular plate supported at one longitudinal edge with σ_1 the stress at the free edge and σ_2 the stress at the simply supported longitudinal edge.

$\psi = \sigma_2/\sigma_1$	1	0	-1	$0 > \psi > -3$
k	0.43	0.57	0.85	$0.57 - 0.21\psi + 0.07\psi^2$

Table A.3 Buckling coefficient k for a rectangular plate supported at one longitudinal edge with σ_1 the stress at the simply supported longitudinal edge and σ_2 the stress at the free edge.

$\psi = \sigma_2/\sigma_1$	1	$1 > \psi > 0$	0	$0 > \psi > -1$	-1
k	0.43	$0.578/(0.34 + \psi)$	1.70	$1.7 - 5\psi + 17.1\psi^2$	23.8

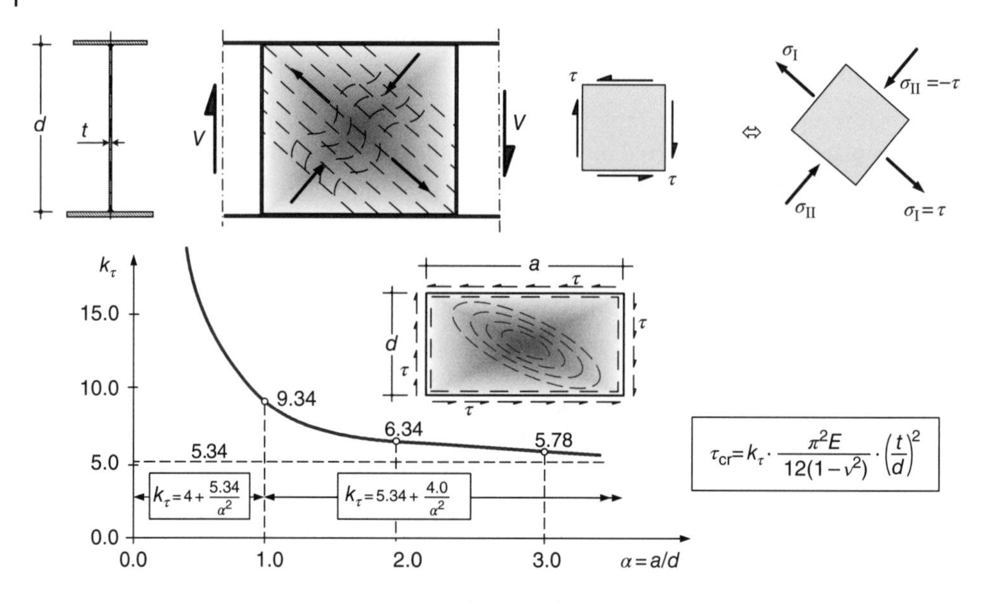

Figure A.4 Buckling coefficient k_τ for plate panels under shear stresses.

compressive stresses induced by the shear stress field. The buckling coefficient k_τ is given in terms of the aspect ratio $\alpha = d/a$ in Figure A.4. From this figure, and for common stiffener spacing $\alpha \geq 0.5$, k_τ varies between 25.36 and 5.34.

A.2 Buckling of Stiffened Plates

In Chapters 4 and 6, designs of stiffened plates for webs and flanges of plate and box girders have been considered, as shown in Figures 4.78 and 6.119. The basic case of a longitudinal stiffened plate under uniform compressive stresses (Figure 4.78) is considered here. The local buckling mode between longitudinal stiffeners is associated with the critical stress:

$$\sigma_{cr,loc} = k_{loc} \frac{\pi^2 E}{12\left(1-\nu^2\right)}\left(\frac{t}{b}\right)^2 \tag{A.3}$$

where k_{loc} depends on the torsion stiffness GJ_s of the stiffeners. If GJ_s tends to zero, k_{loc} tends to 4.0 for wide and long panels ($a/b > 4$) as per Section A.1. If so, the panel buckles in this local mode in a series of longitudinal waves (Figure 4.79) with a half wave length μb with $\mu = 1$ for $GJ_s = 0$. This local mode occurs if the stiffeners are sufficiently rigid in bending in order to avoid a plate buckling mode involving stiffeners and plating (Figure 4.78). The problem is usually treated in terms of the parameters

$$\gamma = {EI_{sl}}\big/{bD} \approx {I_{sl}}\big/{I_p} \quad \text{and} \quad \delta = {EA_{sl}}\big/{EA_p} = {A_{sl}}\big/{A_p} \tag{A.4), (A.5}$$

already introduced in Section 6.9.7 defining the relative bending and axial stiffness of the cross section of the stiffener (EI_{sl} and EA_{sl}) and of the plate (bD and EA_p). To take the torsional stiffness of the stringers GJ_{sl} into account, a third parameter is introduced

$$\theta = {GJ_{sl}}/{bD} \tag{A.6}$$

The analysis of stiffened plates under compression is usually done for the following cases: (i) plates with one stiffener, (ii) plates with two stiffeners and (iii) plates with multiple stiffeners (three or more). The reason is that for the first two cases specific methods should be adopted to solve the buckling equations (involving compatibility between the plate and the stiffeners). For the case of multiple stiffeners, these may be smeared off through the plating and orthotropic plate theory can be adopted [1, 4]. The case of one or two stiffeners is quite common in webs of plate girders.

A.2.1 Plates with One Longitudinal Stiffener at the Centreline under Uniform Compression

The stiffener is considered symmetric with respect to the middle surface of the plate. All the edges are assumed to be simply supported. Depending on the bending stiffness EI_{sl}, the buckling modes are as shown in Figure A.5. The critical stress of the stiffened plate is given by:

$$\sigma_{cr,p} = k_p \frac{\pi^2 E}{12\left(1-\nu^2\right)}\left(\frac{t}{b}\right)^2 \tag{A.7}$$

where b is the total plate width. The buckling coefficient k_p shown in Figure A.6, depends on γ and $\alpha = a/b$. Note that for large α and γ, k_p tends to 16 associated with the local buckling coefficient of a plate of width ($b/2$).

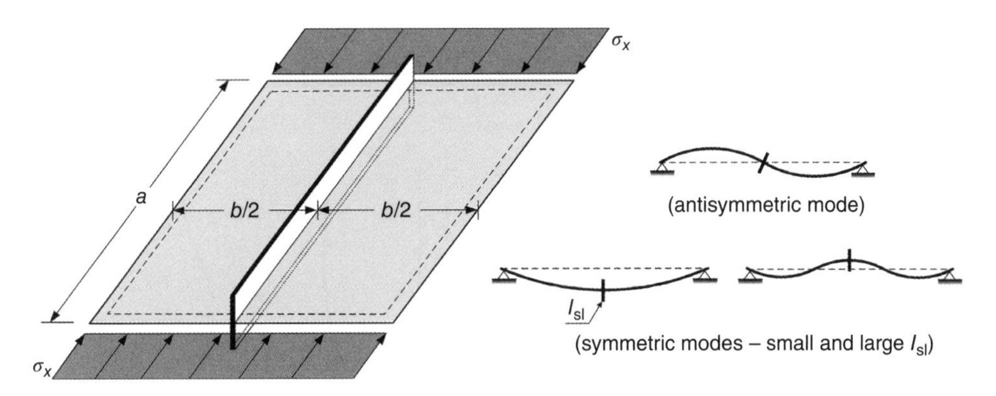

Figure A.5 Buckling modes for a plate with one longitudinal stiffener at the centreline under uniform compression.

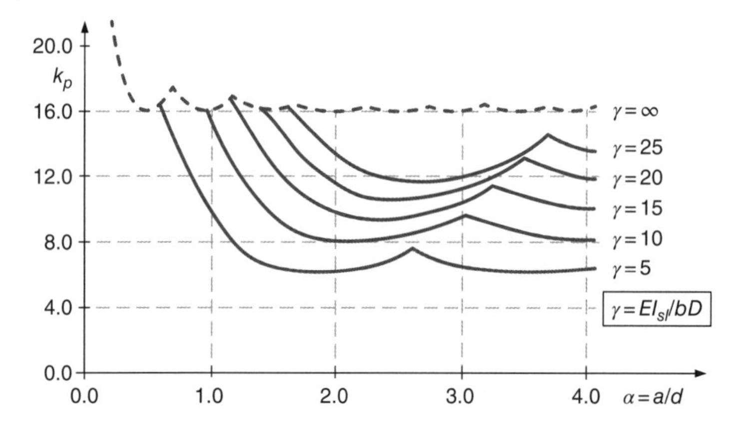

Figure A.6 Buckling coefficient k_p for a plate with one longitudinal stiffener at the centreline under uniform compression.

A.2.2 Plate with Two Stiffeners under Uniform Compression

The stiffeners are considered symmetric with respect to the middle surface of the plate. All the edges are assumed to be simply supported, see Figure A.7.

A solution for a stiffened plate under compression with simply supported edges has been obtained by Barbré [5] solving the differential equation for the buckling modes taking into consideration the compatibility conditions at the interface plate/stiffeners. Based on an energy method, a solution has been obtained for the same problem by Timoshenko [4] with the buckling coefficient k_p (Figure A.8) given by

$$k_p = \frac{\left(1+\alpha^2\right)^2 + 3\gamma}{\alpha^2\left(1+3\delta\right)} \tag{A.8}$$

It should be noted that Timoshenko's approximate solution [4] considers only the symmetric mode, which is the fundamental critical buckling mode for long panels. For this mode, when δ and γ tend to zero, k_p tends towards the coefficient k_m (Eq. (A.2)) for an unstiffened plate.

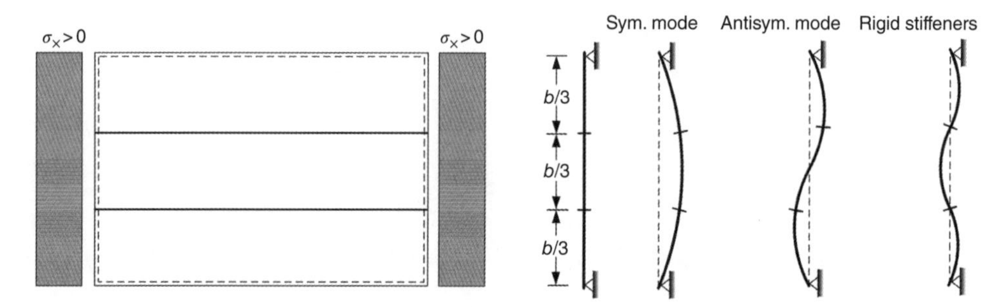

Figure A.7 Buckling modes for a plate with two longitudinal stiffeners at the centreline under uniform compression.

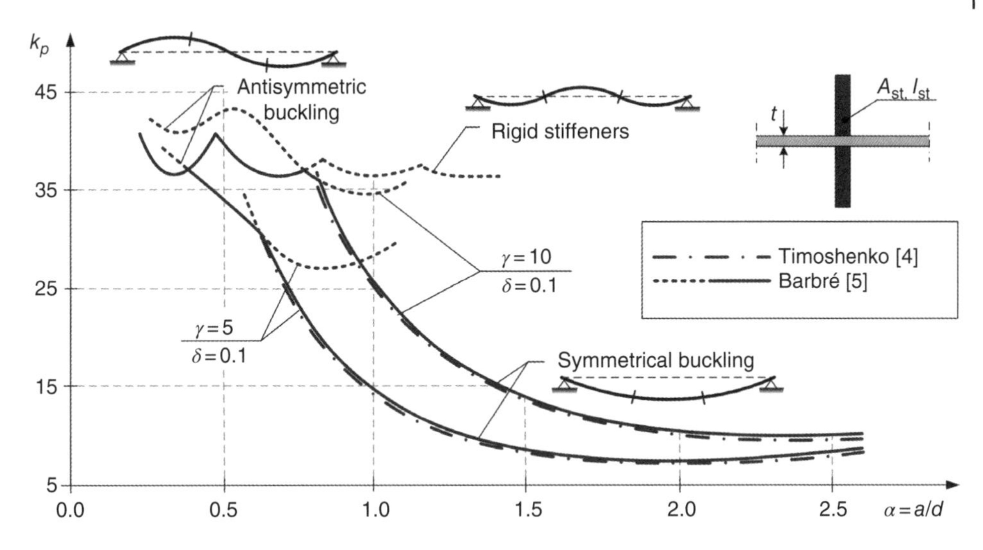

Figure A.8 Buckling coefficient k_p for a plate with two longitudinal stiffeners at the centreline under uniform compression.

A.2.3 Plates with Three or More Longitudinal Stiffeners

As previously mentioned, an orthotropic plate model is considered for these cases with longitudinal and transverse bending rigidities D_x and D_y and an average torsion rigidity H. The plate model is assumed to be under a membrane stress field $\{N_x^f, N_y^f, N_{xy}^f\}$ along the fundamental (unbuckled) equilibrium path. For the case of a compressed panel along the longitudinal direction, one has $N_x^f = -\sigma t$. A Fourier series solution yields the buckling modes [1, 2], and associated critical stresses for the orthotropic plate $\sigma_{cr,p}$. Minimum values of these critical stresses occurs for a plate ratio $a/b = (D_x/D_y)^{1/4}$ yielding

$$\sigma_{cr,p}^{min} = \frac{2\pi^2}{b^2 t}\left(\sqrt{D_x D_y} + H\right) \tag{A.9}$$

For a panel with equally spaced stiffeners at a transverse distance c and with bending and torsion rigidities EI_{sl} and GJ_{sl}, respectively, one has

$$D_x = \frac{EI_{sl}}{c} + \frac{Et^3}{12\left(1-\nu^2\right)};\ D_y = \frac{Et^3}{12\left(1-\nu^2\right)} \qquad \text{(A.10), (A.11)}$$

and

$$H = D_y + \frac{1}{2}\cdot\frac{GJ_{sl}}{c} \tag{A.12}$$

The final result for the critical stress is given by Eq. (A.7) with

$$k_p = 4\left[\frac{1}{2} + \frac{1}{2}\sqrt{1 + 12\left(1-\nu^2\right)\frac{I_{sl}}{ct^3}}\right] \tag{A.13}$$

The term within the brackets represents the increase of the critical plate stress due to the effect of the stiffeners. This critical plate stress occurs for non-rigid stiffeners. Otherwise, one has a local buckling mode, that is, plate buckling between stiffeners, with $\sigma_{cr} = \sigma_{cr,loc}$.

A.2.4 Stiffened Plates under Variable Compression. Approximate Formulas

For multiple equally spaced stiffeners, when the membrane stresses N_x are not constant through the panel, a stress ratio $\psi = \sigma_2/\sigma_1$ is defined as referred to in Chapter 6 (Section 6.9.7) and approximate formulas for the bucking coefficient k_p are given in [3, 7] as referred to in Section 6.9.7, Eq. (6.187).

Approximate formulas for the critical plate stress may be obtained by considering the stiffener as a column on elastic foundation with a foundation modulus of reaction obtained from the plate stiffness in the transverse direction [3, 7].

A.3 Post-Buckling Behaviour and Ultimate Strength of Flat Plates

A long flat plate, simply supported along the longitudinal edges and subjected to compressive stresses σ is considered as shown in Figure A.2. After reaching the critical stress σ_{cr} ($k_p = 4$) the plate goes into the post-buckling range and the equilibrium path is as shown in Figure A.9, for plates with and without geometrical imperfections (i.e. an ideal and real plate). One of the characteristics of plates is the *stable post-buckling path*; that is, the applied load tends to increase with buckling deflections. An approximate solution for the post-buckling path is given by [2]

$$\frac{\sigma}{\sigma_{cr}} = 1 + \frac{3}{8}\left(1-\nu^2\right)\left(\frac{w}{t}\right)^2 \tag{A.14}$$

where w denotes the amplitude of the buckling deflection as shown in Figure A.9. For a deflection of the same magnitude of the plate thickness, and taking $\nu = 0.3$ for steel, the load increases by 34%. This post-buckling strength of plates in the elastic range is taken

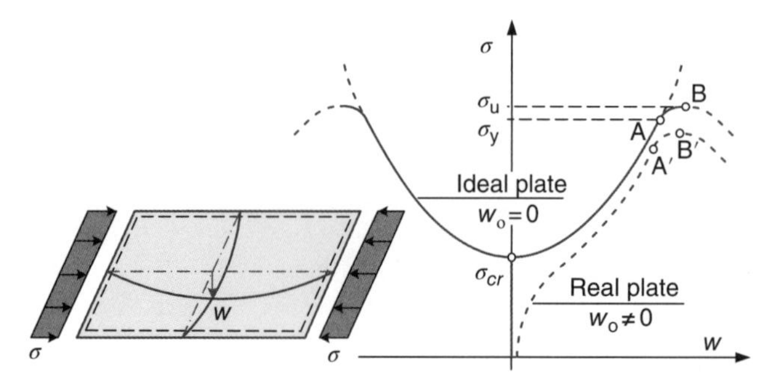

Figure A.9 Load carrying capacity path up to failure of a simply supported slender plate.

into consideration in the design with due account for the effect of plasticity and imperfections. It should be noted only slender plates have a post-buckling strength. To define the slenderness a non-dimensional parameter

$$\bar{\lambda}_p = \sqrt{\frac{f_y}{\sigma_{cr}}} = \frac{1}{\pi}\sqrt{\frac{12\left(1-\nu^2\right)}{k}\frac{f_y}{E}}\left(\frac{b}{t}\right) \tag{A.15}$$

is defined, where f_y denotes the material yield stress. The non-dimensional slenderness parameter $\bar{\lambda}_p$ of a plate is proportional to the geometrical slenderness (b/t) of the plate. A slender plate has $\sigma_{cr} << f_y$ (large $\bar{\lambda}_p$ values), while a stocky plate $\sigma_{cr} >> f_y$ (low $\bar{\lambda}_p$ values).

Real plates have material and geometrical imperfections. The former are, for example, residual stresses due to welding, and the last are taken in general as deviations from the flat configuration of the ideal shape. To combine these two types of imperfections in plated structures, the concept of *equivalent geometrical imperfection* is introduced. This is a fictitious geometrical imperfection to combine material and geometrical deviations, and to be introduced in analytical or numerical models. The aim is to achieve, with a single type of imperfection, the same load carrying capacity of the real plate. The equivalent geometrical imperfections are taken in the form of the buckling mode of the perfect plate, that is a *modal imperfection*. For the example under discussion (Figure A.9), the buckling mode has is a multiple longitudinal sinusoidal shape, with a half wave length equal to the plate width. This allows study of a single half wave length equivalent to a square plate of length b and with a modal imperfection $w_0 = \varepsilon\ t \sin\ (\pi x/b)$ being ε a non-dimensional amplitude parameter. Plates with geometrical imperfection have transverse deflections from the beginning of the loading process as shown in Figure A.9.

A slender perfect plate ($\sigma_{cr} << f_y$) reaches yielding at point A of the elastic post-buckling path above the critical stress σ_{cr}, reaching the ultimate stress at point B after the spread of plasticity through the plate thickness (Figure A.9). The effect of geometrical imperfections is to lower the stress by changing point A to point A′ and point B to point B′.

A stocky plate has ($\sigma_{cr} >> f_y$) and the ultimate load will be reached at a stress value $\sigma_m < f_y$. In the intermediate slenderness, when σ_{cr} and f_y are of the same order of magnitude, the plate is very sensitive to imperfections. This behaviour is presented in Figure A.10 for three ranges of slenderness.

A.3.1 Effective Width Concept

The stress field in the post-buckling range is as shown in Figure A.11, assuming the longitudinal edges remains straight. The average stress in the longitudinal direction is denoted by σ_x. These stresses σ_x tend to concentrate at the longitudinal edges reducing at the centreline. Along the longitudinal edges, tensile and compressive stresses are balanced with a zero stress resultant as it should be to equilibrate the external loads.

The longitudinal stress distribution may be replaced by a statically equivalent uniform stress distribution (Figure A.11a) equal to the edge stress σ_e ($=\sigma_{x,max}$), allowing one to introduce the *concept of effective width*, b_e, from the condition

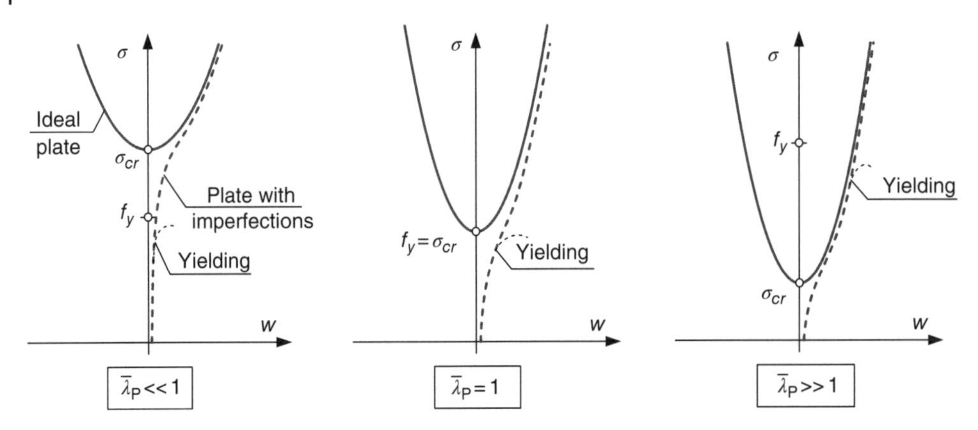

Figure A.10 Load carrying capacity path of three plates as a function of the plate slenderness.

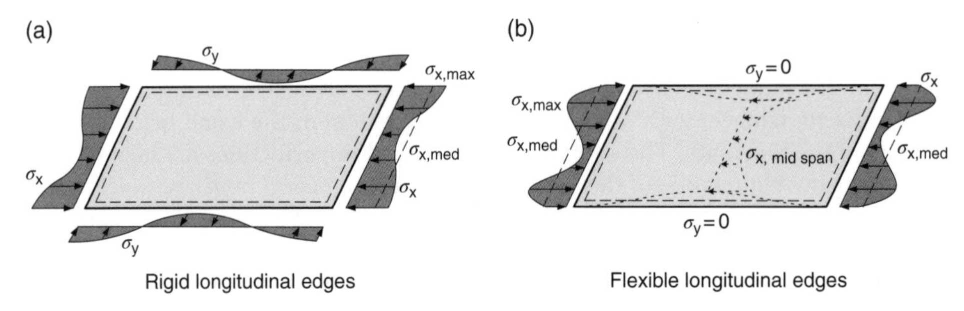

Figure A.11 Stress distribution in the post-buckling range of a compressed flat plate with: (a) rigid longitudinal edges and (b) flexible longitudinal edges.

$$\sigma_e b_e = \int_{-b/2}^{+b/2} |\sigma_x| d_y \equiv \sigma_m b \tag{A.16}$$

To evaluate the effective width b_e the stress distribution σ_x in the post-buckling range shall be known. This stress distribution depends on material and geometrical imperfections. The effective width concept may be generalized for plates with longitudinal flexible edges (Figure A.11b).

A.3.2 Effective Width Formulas

Based on experimental results, von Karman proposes a semi-analytical criterion to determine the effective width b_e (= b_{eff}) of a flat plate with supported edges (Figure A.12). It consists of assuming the edge stress σ_e (= $\sigma_{x,max}$) is equal to the critical stress of a fictitious plate with a width equal to b_{eff}. From this condition one obtains *von Karman's formula*

$$\frac{b_{eff}}{b} = \sqrt{\frac{\sigma_{cr}}{\sigma_e}} \leq 1 \tag{A.17}$$

valid for $\sigma > \sigma_{cr}$. For $\sigma \leq \sigma_{cr}$ one has of course $b_{eff} = b$. According to this formula, the plate reaches the ultimate strength when $\sigma_e = f_y$, allowing one to calculate directly the value of b_{eff} at collapse, as

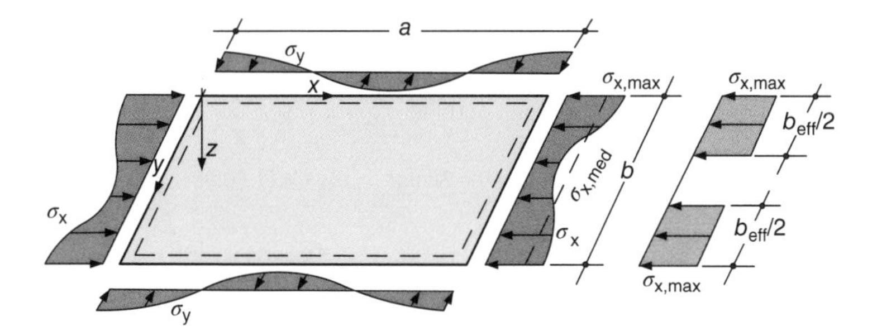

Figure A.12 Effective width concept of a compressed flat plate.

$$\rho \equiv \left(\frac{b_{eff}}{b}\right)_{colapse} = \sqrt{\frac{\sigma_{cr}}{f_y}} = \frac{1}{\bar{\lambda}_p} \leq 1 \tag{A.18}$$

For a long plate simply supported along the longitudinal edges and under uniform compression at the transverse edges, one has $k = 4$ in Eq. (A.18) for σ_{cr}, with the result

$$\left(\frac{b_{eff}}{t}\right)_{collapse} = \pi\sqrt{\frac{E}{3\left(1-\nu^2\right)f_y}} \tag{A.19}$$

This expression yields for steel S235, an effective width value $b_{eff} = 57\ t$, and for steel S355 $b_{eff} = 46\ t$. In conclusion, the effective at collapse decreases with the increase of the yield stress (that is the steel grade).

The von Karman's formula was improved by the Winter's formula, based on experimental results:

$$\frac{b_e}{b} = \sqrt{\frac{\sigma_{cr}}{\sigma_e}}\left(1-0.22\sqrt{\frac{\sigma_{cr}}{\sigma_e}}\right) \leq 1 \tag{A.20}$$

This formula, widely used in design codes [7], can be rewritten at collapse using $\sigma_e = f_y$, as

$$\rho \equiv \left(\frac{b_{eff}}{b}\right)_{collapse} = \frac{1}{\bar{\lambda}_p}\left(1-\frac{0.22}{\bar{\lambda}_p}\right) = \frac{1}{\bar{\lambda}_p^2}\left(\bar{\lambda}_p - 0.22\right) \leq 1 \tag{A.21}$$

The effective width ratio (b_{eff}/b) when the edge stress reaches the yield stress ($\sigma_e = f_y$), is equivalent to the stress ratio $\rho = (\sigma_m / f_y)$ at yielding (Figure A.13).

From Figure A.13, and taking Winter's formula, it may be concluded that a plate is fully effective if $\bar{\lambda}_p \leq 0.673$ and displays a post-buckling strength if $\bar{\lambda}_p > 1.22$. The plate limit geometric slenderness $(b/t)_{\lim}$, to be fully effective under uniform compression ($b_{eff} = b$),is obtained under the condition $\bar{\lambda}_p = 0.673$ with σ_{cr} defined for $k = 4.0$. The result is:

$$\left(\frac{b}{t}\right) \leq 1.278\sqrt{\frac{E}{f_y}} \tag{A.22}$$

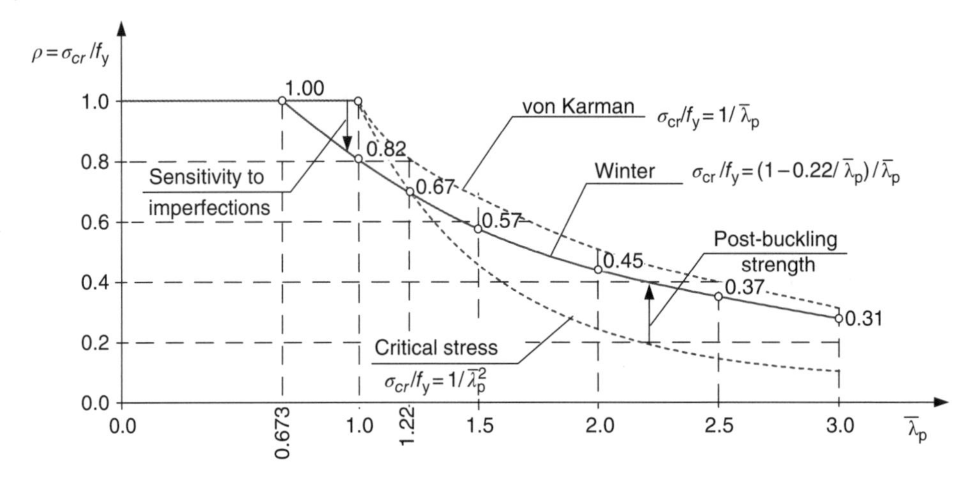

Figure A.13 Effective width $b_{eff} = \rho\, b$ given by von Karman's and Winter's formulas as a function of the slenderness $\bar{\lambda}_p$.

that can be rewritten as $(b/t) \leq 38.2\varepsilon$ with $\varepsilon = \sqrt{235/f_y}$ and f_y in N mm^{-2}. In EN 1993.1.1 [6] the limit between class 3 and 4, for uniformly compressed elements, is slightly higher $(b/t) \leq 42\varepsilon$.

Winter's formula (A.21) for flat plates under uniform compression was generalized for the case of plates with linearly variable compressive stresses (Figure A.1 with $\tau = 0$) [3, 7], with the following result

$$\rho \equiv \left(\frac{b_{eff}}{b}\right) = \frac{1}{\bar{\lambda}_p} - \frac{0.055(3+\psi)}{\bar{\lambda}_p^2} \leq 1 \tag{A.23}$$

valid for

$$\bar{\lambda}_p \geq 0.5 + \sqrt{0.085 - 0.055\psi} \tag{A.24}$$

If $\bar{\lambda}_p$ is less than this value $\rho = 1$, that is, the plate is fully effective. Under uniform compression, $\psi = 1.0$, Eq. (A.23) yields Winter's formula (A.21). For the case of plate elements with a free edge, like flanges of plate girders, Winter's formula (A.21) is replaced by

$$\rho \equiv \left(\frac{b_{eff}}{b}\right) = \frac{1}{\bar{\lambda}_p} - \frac{0.188}{\bar{\lambda}_p^2} < 1 \quad \text{for} \quad \bar{\lambda}_p > 0.748 \tag{A.25}$$

If $\bar{\lambda}_p \leq 0.748$, one has $\rho = 1.0$. Hence, a flange of width c is fully effective if it has a non-dimensional slenderness $\bar{\lambda}_p \leq 0.748$. This condition yields, after introducing σ_{cr} with $k = 0.43$ in Eq. (A.25), the result in Section 4.5.4.2, Eq. (4.10): $c/t \leq 0.466\sqrt{E/f_y}$.

References

1 Allen, H.G. and Bulson, P.S. (1980). *Background to Buckling*, 532 pp. McGraw-Hill.
2 Reis, A. and Camotim, D. (2001). *Estabilidade Estrutural*, 470 pp. Mc Graw Hill.
3 Beg, D., Kuhlmann, H., Davaine, L., and Braun, B. (2010). *Design of Plated Structures. ECCS Eurocode Design Manuals*. Ernst & Son.
4 Timoshenko, S. and Gere, J. (1961). *Theory of Elastic Stability*, 557 pp. New York: McGraw-Hill.
5 Barbré, R (1939). "Stability of rectangular plates with longitudinal or transverse stiffeners under uniform compression. Technical Memorandum no. 904 from NACA – National Advisory Committee for Aeronautics. Washington, DC.
6 EN1993-1-1 (2005). *Eurocode 3: Design of Steel Structures – Part 1–1: General Rules and Rules for Buildings*. Brussels: CEN.
7 EN 1993-1-5 (2006). *Eurocode 3 – Design of Steel Structures – Part 1–5: Plated Structural Elements: Steel Bridges*. Brussels: CEN.

Index

Bridge Design: Concepts and Analysis, First Edition. António J. Reis and José J. Oliveira Pedro.